Methods in Cell Biology

VOLUME 64
Cytometry
Third Edition, Part B

Series Editors

Leslie Wilson
Department of Biological Sciences
University of California, Santa Barbara
Santa Barbara, California

Paul Matsudaira
Whitehead Institute for Biomedical Research and
Department of Biology
Massachusetts Institute of Technology
Cambridge, Massachusetts

Methods in Cell Biology

Prepared under the Auspices of the American Society for Cell Biology

VOLUME 64

Cytometry
Third Edition, Part B

Edited by

Zbigniew Darzynkiewicz

Brander Cancer Research Institute
New York Medical College
Hawthorne, New York

Harry A. Crissman

Cell and Molecular Biology Group
Los Alamos National Laboratory
Los Alamos, New Mexico

J. Paul Robinson

Purdue Cytometry Laboratories
Purdue University
West Lafayette, Indiana

ACADEMIC PRESS

A Harcourt Science and Technology Company

San Diego   San Francisco   New York   Boston   London   Sydney   Tokyo

Academic Press
A Harcourt Science and Technology Company
525 B Street, Suite 1900, San Diego, California 92101-4495, USA
http://www.academicpress.com

Academic Press
Harcourt Place, 32 Jamestown Road, London NW1 7BY, UK
http://www.academicpress.com

International Standard Book Number: 0-12-544167-3 (case)
International Standard Book Number: 0-12-203054-0 (combbound)

PRINTED IN THE UNITED STATES OF AMERICA
00 01 02 03 04 05 EB 9 8 7 6 5 4 3 2 1

CONTENTS

Contents of Volume 63   xii

Contributors   xv

Preface to the Third Edition   xix

Preface to the Second Edition   xxiii

Preface to the First Edition   xxvii

PART VII   Cytogenetics and Molecular Genetics

30. Sorting of Plant Chromosomes

Jaroslav Doležel, Martin A. Lysák, Marie Kubaláková, Hana Šimková, Jiří Macas, and Sergio Lucretti

I. Introduction   3

II. Application   5

III. Materials   6

IV. Procedures   12

V. Critical Aspects of the Procedures   18

VI. Instruments   21

VII. Results   21

VIII. Conclusions and Perspectives   26

References   28

31. Quantitative DNA Fiber Mapping

Heinz-Ulli G. Weier

I. Introduction   33

II. Materials   35

III. Protocols   38

IV. Critical Aspects of the Procedure   46

V. Results and Discussion   47

References   52

32. Primed *in Situ* Labeling

Johnny Hindkjaer, Lars Bolund, and Steen Kølvraa

I. Introduction   55

II. Applications   56

III. Materials   56

IV. Protocols   58

V. Critical Aspects of the Procedures   66
References   67

33. Measurements of Telomere Length on Individual Chromosomes by Image Cytometry

Steven S. S. Poon and Peter M. Lansdorp

I. Introduction   70
II. Background   72
III. Methods   77
IV. Results   88
V. Discussion   92
References   94

34. Detection of Chromosome Translocation Products in Single Interphase Cell Nuclei

Jingly Fung, Santiago Munné, and Heinz-Ulli G. Weier

I. Introduction   98
II. Materials   99
III. Protocols   102
IV. Results and Discussion   108
V. Critical Aspects of the Procedure   111
VI. Applications   112
References   113

PART VIII   Cell Function and Differentiation

35. Analysis of Mitochondria by Flow Cytometry

Martin Poot and Robert H. Pierce

I. Introduction   118
II. Materials and Methods   121
III. Critical Aspects   125
References   126

36. Analysis of RNA Synthesis by Cytometry

Peter Østrup Jensen, Jacob Larsen, and Jørgen K. Larsen

I. Introduction   129
II. Background   130
III. Methods for Analysis of 5′-Bromouridine Incorporation and DNA Content   131
IV. Results of Labeling RNA with 5′-Bromouridine   134
V. Applications   136
References   137

37. Flow Cytometry of Erythropoiesis in Culture: Bivariate Profiles of Fetal and Adult Hemoglobin

Ralph M. Böhmer

I. Introduction	139
II. Details of the Method	141
III. Sample Experiments	143
IV. Problems and Limitations	149
References	151

38. Flow Cytometric Analysis of Human Hemopoietic Progenitor Differentiation by Assessing Cell Division Rate and Phenotypic Profile

Luca Pierelli, Giovanni Scambia, and Andrea Fattorossi

I. Introduction	153
II. Background	155
III. Critical Aspects of Methodology	156
IV. Functionally Distinct Circulating Hemopoietic Progenitor Subsets Can Be Assessed during Cytokine-Driven Differentiation	157
V. Concluding Remarks	167
References	169

PART IX   Experimental Oncology

39. Cytometry of Antitumor Drug–Intracellular Target Interactions

Paul J. Smith and Marie Wiltshire

I. Introduction	173
II. General Classification of Cytotoxic Anticancer Agents	174
III. Establishment of Quality Control Parameters	178
IV. Drug–DNA Interactions	183
V. Conclusions	188
References	190

40. Monitoring of Cellular Resistance to Cancer Chemotherapy: Drug Retention and Efflux

Awtar Krishan

I. Introduction and Background	193
II. Applications	196
III. Cell Lines, Efflux, Multiple Drug Resistance Drugs, and Blockers	196
IV. Staining Protocols	197
V. Critical Aspects	198
VI. Controls, Standards, and Instruments	199
VII. Results and Conclusions	201
References	206

41. Resistance of Tumor Cells to Chemo- and Radiotherapy Modulated by the Three-Dimensional Architecture of Solid Tumors and Spheroids

Ralph E. Durand and Peggy L. Olive

I. Introduction and Historical Perspective	211
II. Background	213
III. Methods	215
IV. Results and Discussion	217
V. Conclusions and Future Directions	227
References	228

42. Analysis of DNA Damage in Individual Cells

Peggy L. Olive, Ralph E. Durand, Judit P. Banáth, and Peter J. Johnston

I. Introduction	235
II. The Development of the Single Cell Gel Electrophoresis/Comet Assay	236
III. Comet Preparation and Analysis	238
IV. Types of Damage Detected by the Comet Assay	242
V. Using DNA Damage to Predict Cell Survival in Complex Systems	244
VI. Future Directions	246
References	247

43. Cytometric Methods to Analyze Ionizing-Radiation Effects

William D. Wright, Isabelle Lagroye, Peng Zhang, Robert S. Malyapa, and Joseph L. Roti Roti

I. Introduction	251
II. Applications	252
III. Methods for Measuring DNA Damage on a Cell-by-Cell Basis	253
IV. Summary	267
References	267

44. Cytometric Methods to Analyze Thermal Effects

Robert P. VanderWaal, Ryuji Higashikubo, Mai Xu, Douglas R. Spitz, William D. Wright, and Joseph L. Roti Roti

I. Introduction	269
II. Application	270
III. Methods to Measure Nuclear and Nuclear Matrix Protein Content	271
IV. Identification of Altered DNA Replication Patterns Following Heat Shock	276
V. Nuclear Localization of hsp70	278
VI. Prooxidant Measurement	280
VII. Results	281
References	285

PART X   Clinical Oncology

45. Multiparameter Data Acquisition and Analysis of Leukocytes by
Flow Cytometry

Carleton C. Stewart and Sigrid J. Stewart

I. Introduction	289
II. Correlated List Mode Data	290
III. Verification of Instrument Performance	293
IV. Optical Filtration and Spectral Compensation	295
V. Multiparameter Data Analysis	308
VI. Summary	311
References	311

46. Immunophenotyping of Hematological Malignancies by Laser
Scanning Cytometry

Richard J. Clatch

I. Immunophenotyping of Leukemia and Lymphoma: General Considerations	313
II. Description of Laser Scanning Cytometry	316
III. Immunophenotyping by Laser Scanning Cytometry	316
IV. Extensions of the Method	327
V. Conclusions	340
References	340

47. Immunophenotyping of Acute Leukemia: Utility of CD45 for Blast
Cell Identification

J-P. Vial and F. Lacombe

I. Introduction	344
II. Background	344
III. Methods	345
IV. Results and Comparison with Other Methods	348
V. Critical Aspects of the Methodology	352
VI. Pitfalls and Misinterpretation of the Data	353
VII. Future Directions	355
References	357

48. Cell Proliferation Markers in Human Solid Tumors: Assessing Their
Impact in Clinical Oncology

Maria Grazia Daidone, Aurora Costa, and Rosella Silvestrini

I. Introduction	359
II. Proliferation Markers	361
III. Biological Studies	365
IV. Clinical Studies	366
V. Conclusions on First Generation Translational Studies with Proliferation Markers	377
References	378

49. Detection of Minimal Residual Disease

Andrzej Deptala and Sharon P. Mayer

I. Introduction	385
II. Tissue Sources to Detect Minimal Residual Disease	387
III. Methods to Detect Minimal Residual Disease	388
IV. Technical Problems	414
V. Concluding Remarks	416
References	416

50. Analysis of Human Tumors by Laser Scanning Cytometry

Wojciech Gorczyca, Andrzej Deptala, Elżbieta Bedner, Xun Li, Myron R. Melamed, and Zbigniew Darzynkiewicz

I. Introduction	422
II. Analysis of Cellular DNA Content by Laser Scanning Cytometry	423
III. Analysis of Apoptosis by Laser Scanning Cytometry	425
IV. Analysis of Proliferation Associated Antigens by Laser Scanning Cytometry	430
V. Analysis of Estrogen Receptors by Laser Scanning Cytometry	435
VI. Analysis of Transcription Factors by Laser Scanning Cytometry	437
VII. Measurement of Nucleoli Using Laser Scanning Cytometry Fluorescence *in Situ* Hybridization Protocol	438
References	441

51. Laser Cytometry of Human Tissues and Tumors: Proliferation and Therapeutic Applications

David A. Rew

I. Introduction	446
II. Ploidy and Proliferation in Surgical Oncology	447
III. Cytometric Studies of Proliferation	448
IV. The Halogenated Pyrimidines in Cell Proliferation Research	449
V. Clinical Studies of Cell Production Rates with Thymidine Analogs	452
VI. Further Applications of Cytometry in Clinical Oncology	474
VII. Conclusions	478
References	478

52. Prediction and Precise Diagnosis of Diseases by Data Pattern Analysis in Multiparameter Flow Cytometry: Melanoma, Juvenile Asthma, and Human Immunodeficiency Virus Infection

Günter Valet, Hanna Kahle, Friedrich Otto, Edeltraut Bräutigam, and Luc Kestens

I. Introduction	488
II. Material and Methods	489
III. Results	493
IV. Discussion	506
References	507

PART XI Microorganisms and Infectious Diseases

53. Flow Cytometric Analysis of Microorganisms

S. A. Sincock and J. Paul Robinson

I. Introduction	511
II. Experimental Approaches	514
III. Applications in Medical and Food Microbiology	526
IV. Conclusion	531
References	532

54. Staining and Measurement of DNA in Bacteria

Harald B. Steen

I. Introduction	539
II. Basic Considerations	540
III. Experimental Methods	543
IV. Standards and Controls	548
References	550

55. Flow Cytometric Monitoring of Bacterial Susceptibility to Antibiotics

Mette Walberg and Harald B. Steen

I. Introduction	553
II. Fluorescent Dyes	554
III. Uptake of Fluorescent Dyes	555
IV. Effects of Antibiotics	558
V. Assessment of Drug Effects	558
VI. Applications of Flow Cytometry to Medical Microbiology	563
References	563

56. Flow Cytometry for Evaluation and Investigation of Human Immunodeficiency Virus Infection

Thomas W. Mc Closkey

I. Introduction	567
II. Application of Flow Cytometry to Monitor HIV Infection	568
III. Application of Flow Cytometry to Investigate HIV Disease	582
IV. Conclusions	584
References	584

Index	593
Volumes in Series	609

Contents of Volume 63
Cytometry, Third Edition, Part A

PART I   Principles of Cytometry and General Methods

1. A Brief History of Flow Cytometry and Sorting
Myron R. Melamed

2. Principles of Flow Cytometry: An Overview
Alice L. Givan

3. Laser Scanning Cytometry
Louis A. Kamentsky

4. Principles of Confocal Microscopy
J. Paul Robinson

5. Optical Measurements in Cytometry: Light Scattering, Extinction, Absorption, and Fluorescence
Howard M. Shapiro

6. Flow Cytometric Fluorescence Lifetime Measurements
Harry A. Crissman and John A. Steinkamp

7. Principles of Data Acquisition and Display
Howard M. Shapiro

8. Time as a Flow Cytometric Parameter
Larry Seamer and Larry A. Sklar

9. Protein Labeling with Fluorescent Probes
Kevin L. Holmes and Larry M. Lantz

PART II   Cell Preparation

10. Preparation of Cells from Blood
J. Philip McCoy, Jr.

11. Cell Preparation for the Identification of Leukocytes

Carleton C. Stewart and Sigrid J. Stewart

12. Strategies for Cell Permeabilization and Fixation in Detecting Surface and Intracellular Antigens

Steven K. Koester and Wade E. Bolton

PART III   Standardization, Quality Assurance

13. Stoichiometry of Immunocytochemical Staining Reactions

James W. Jacobberger

14. Standardization and Quantitation in Flow Cytometry

Robert A. Hoffman

PART IV   Cell Proliferation

15. Methods to Identify Mitotic Cells by Flow Cytometry

Gloria Juan, Frank Traganos, and Zbigniew Darzynkiewicz

16. Cell Cycle Kinetics Estimated by Analysis of Bromodeoxyuridine Incorporation

Nicholas H. A. Terry and R. Allen White

17. Flow Cytometric Analysis of Cell Division History Using Dilution of Carboxyfluorescein Diacetate Succinimidyl Ester, a Stably Integrated Fluorescent Probe

A. Bruce Lyons, Jhagvaral Hasbold, and Philip D. Hodgkin

18. Antibodies against the Ki-67 Protein: Assessment of the Growth Fraction and Tools for Cell Cycle Analysis

Elmar Endl, Christiane Hollmann, and Johannes Gerdes

19. Detection of Proliferating Cell Nuclear Antigen

Jørgen K. Larsen, Göran Landberg, and Göran Roos

20. Lymphocyte Activation Associated Antigens

Andrea Fattorossi, Alessandra Battaglia, and Cristiano Ferlini

PART V   Cell Death/Apoptosis

21. Analysis of Mitochondria during Cell Death
Andrea Cossarizza and Stefano Salvioli

22. Cytometry of Caspases
Steven K. Koester and Wade E. Bolton

23. Analysis of Apoptosis in Plant Cells
Iona E. Weir

24. Difficulties and Pitfalls in Analysis of Apoptosis
Zbigniew Darzynkiewicz, Elżbieta Bedner, and Frank Traganos

PART VI   Cell–Cell, Cell–Environment Interactions

25. Analysis of Cell Migration
Nicole Dodge Zantek and Michael S. Kinch

26. Three-Dimensional Extracellular Matrix Substrates for Cell Culture
Sherry L. Voytik-Harbin

27. Three-Dimensional Imaging of Extracellular Matrix and Extracellular Matrix–Cell Interactions
Sherry L. Voytik-Harbin, Bartlomiej Rajwa, and J. Paul Robinson

28. Cytometric Analysis of Cell Contact and Adhesion
Michael S. Kinch

29. Invadopodia: Unique Methods for Measurement of Extracellular Matrix Degradation *in Vitro*
Emma T. Bowden, Peter J. Coopman, and Susette C. Mueller

CONTRIBUTORS

Numbers in parentheses indicate the pages on which the authors' contributions begin.

Judit P. Banáth (235), Medical Biophysics Department, British Columbia Cancer Research Centre, Vancouver, British Columbia, Canada V5Z 1L3

Elżbieta Bedner (421), Department of Pathology, Pomeranian School of Medicine, Szczecin, Poland

Ralph M. Böhmer (139), Department of Pediatrics, Division of Genetics, New England Medical Center, Boston, Massachusetts 02111

Lars Bolund (55), Institute of Human Genetics, University of Aarhus, 8000 Århus C, Denmark

Edeltraut Bräutigam (487), Pathologisches Institut, Klinikum Görlitz, D-02828 Görlitz, Germany

Richard J. Clatch (313), Department of Pathology, Highland Park Hospital, Highland Park, Illinois 60035

Aurora Costa (359), Istituto Nazionale per lo Studio e la Cura dei Tumori, 20133 Milano, Italy

Maria Grazia Daidone (359), Istituto Nazionale per lo Studio e la Cura dei Tumori, 20133 Milano, Italy

Zbigniew Darzynkiewicz (421), Brander Cancer Research Institute, New York Medical College, Hawthorne, New York 10532

Andrzej Deptala (385, 421), Department of Hematology, Oncology and Internal Medicine, Warsaw Medical University, 02-097 Warsaw, Poland

Jaroslav Doležel (3), Institute of Experimental Botany, CZ-772 00 Olomouc, Czech Republic

Ralph E. Durand (211, 235), Medical Biophysics Department, British Columbia Cancer Research Centre, Vancouver, British Columbia, Canada V5Z 1L3

Andrea Fattorossi (153), Institute of Obstetrics and Gynecology, Universitá Cattolica del Sacro Cuore, 00136 Rome, Italy

Jingly Fung (97), Department of Obstetrics, Gynecology and Reproductive Sciences, University of California, San Francisco, San Francisco, California 94143; and Life Sciences Division, University of California, Berkeley, E. O. Lawrence Berkeley National Laboratory, Berkeley, California 94720

Wojciech Gorczyca (421), Department of Pathology, New York Medical College, Valhalla, New York 10595

Ryuji Higashikubo (269), Mallinckrodt Institute of Radiology, Radiation Oncology Center, Section of Cancer Biology, Washington University School of Medicine, St. Louis, Missouri 63108

Johnny Hindkjaer (55), Centre of Preimplantation Genetic Diagnosis, Fertility Clinic, Århus University Hospital, 8200 Århus N, Denmark

Peter Østrup Jensen (129), Finsen Laboratory, Finsen Center, Rigshospitalet, DK-2100 Copenhagen, Denmark

Peter J. Johnston (235), Medical Biophysics Department, British Columbia Cancer Research Centre, Vancouver, British Columbia, Canada V5Z 1L3

Hanna Kahle (487), Cell Biochemistry Group, Max-Planck-Institute für Biochemie, D-82152 Martinsried, Germany

Luc Kestens (487), Prince Leopold Institute for Tropical Medicine, Pathology & Immunology, B-2000 Antwerp, Belgium

Steen Kolvraa (55), Institute of Human Genetics, University of Aarhus, 8000 Århus C, Denmark

Awtar Krishan (193), Division of Experimental Therapeutics, Radiation Oncology Department, University of Miami School of Medicine, Miami, Florida 33136

Marie Kubaláková (3), Institute of Experimental Botany, CZ-772 00 Olomouc, Czech Republic

Francis Lacombe (343), Laboratory of Hematology, University Hospital Haut-Lévêque, 33604 Pessac, France

Isabelle Lagroye (251), Mallinckrodt Institute of Radiology, Section of Cancer Biology, Washington University School of Medicine, St. Louis, Missouri 63108

Peter M. Lansdorp (69), Terry Fox Laboratory, British Columbia Cancer Agency, Vancouver, British Columbia, Canada V5Z 1L3; and Department of Medicine, University of British Columbia, Vancouver, British Columbia, Canada V6T 2B5

Jacob Larsen (129), Finsen Laboratory, Finsen Center, Rigshospitalet, DK-2100 Copenhagen, Denmark

Jørgen K. Larsen (129), Finsen Laboratory, Finsen Center, Rigshospitalet, DK-2100 Copenhagen, Denmark

Xun Li (421), Department of Pathology, New York Medical College, Valhalla, New York 10595; and Brander Cancer Research Institute, New York Medical College, Hawthorne, New York 10532

Sergio Lucretti (3), ENEA, Casaccia Research Center, Plant Biotechnology Division, Rome, Italy

Martin A. Lysák (3), Institute of Experimental Botany, CZ-772 00 Olomouc, Czech Republic

Jiří Macas (3), Institute of Plant Molecular Biology, CZ-370 05 České Budějovice, Czech Republic

Robert S. Malyapa (251), Mallinckrodt Institute of Radiology, Section of Cancer Biology, Washington University School of Medicine, St. Louis, Missouri 63108

Sharon P. Mayer (385), Departments of Pediatrics and Pathology, New York Medical College, Valhalla, New York 10595

Thomas W. Mc Closkey (567), Department of Pediatrics, Division of Allergy and Immunology, North Shore University Hospital, New York University School of Medicine, Manhasset, New York 11030

Myron R. Melamed (421), Department of Pathology, New York Medical College, Valhalla, New York 10595

Santiago Munné (97), The Institute for Reproductive Medicine and Science, Saint Barnabas Medical Center, West Orange, New Jersey 07052

Peggy L. Olive (211, 235), Medical Biophysics Department, British Columbia Cancer Research Centre, Vancouver, British Columbia, Canada V5Z 1L3

Friedrich Otto (487), Fachklinik Hornheide, Abteilung für Tumorforschung, D-48157 Münster, Germany

Robert H. Pierce (117), Department of Pathology, Wright–Patterson Medical Center, Wright–Patterson Air Force Base, Dayton, Ohio 45433

Luca Pierelli (153), Institute of Hematology, Universitá Cattolica del Sacro Cuore, 00136 Rome, Italy

Steven S. S. Poon (69), Terry Fox Laboratory, British Columbia Cancer Agency, Vancouver, British Columbia, Canada V5Z 1L3

Martin Poot (117), Department of Pathology, University of Washington, Seattle, Washington 98195

David A. Rew (445), Royal South Hants Cancer Centre, Southampton University Hospitals, Southampton SO14 0YG, England

J. Paul Robinson (511), Purdue Cytometry Laboratories, Department of Basic Medical Sciences, School of Veterinary Medicine, and Department of Biomedical Engineering, Purdue University, West Lafayette, Indiana 47907

Joseph L. Roti Roti (251, 269), Mallinckrodt Institute of Radiology, Radiation Oncology Center, Section of Cancer Biology, Washington University School of Medicine, St. Louis, Missouri 63108

Giovanni Scambia (153), Institute of Obstetrics and Gynecology, Universitá Cattolica del Sacro Cuore, 00136 Rome, Italy

Rosella Silvestrini (359), Istituto Nazionale per lo Studio e la Cura dei Tumori, 20133 Milano, Italy

Hana Šimková (3), Institute of Experimental Botany, CZ-772 00 Olomouc, Czech Republic

S. A. Sincock (511), Purdue Cytometry Laboratories, Department of Basic Medical Sciences, School of Veterinary Medicine, Purdue University, West Lafayette, Indiana 47907

Paul J. Smith (173), Department of Pathology, University of Wales College of Medicine, Heath Park, Cardiff CF4 4XN, United Kingdom

Douglas R. Spitz (269), Mallinckrodt Institute of Radiology, Radiation Oncology Center, Section of Cancer Biology, Washington University School of Medicine, St. Louis, Missouri 63108

Harald B. Steen (539, 553), Department of Biophysics, Institute for Cancer Research, 0310 Oslo, Norway

Carleton C. Stewart (289), Laboratory of Flow Cytometry, Roswell Park Cancer Institute, Buffalo, New York 14263

Sigrid J. Stewart (289), Laboratory of Flow Cytometry, Roswell Park Cancer Institute, Buffalo, New York 14263

Günter Valet (487), Cell Biochemistry Group, Max-Planck-Institute für Biochemie, D-82152 Martinsried, Germany

Robert P. VanderWaal (269), Mallinckrodt Institute of Radiology, Radiation Oncology Center, Section of Cancer Biology, Washington University School of Medicine, St. Louis, Missouri 63108

J-P. Vial (343), Laboratory of Hematology, University Hospital Haut-Lévêque, 33604 Pessac, France

Mette Walberg (553), Institute of Medical Microbiology, National Hospital, University of Oslo, 0027 Oslo, Norway

Heinz-Ulli G. Weier (33, 97), Department of Subcellular Structures, Life Sciences Division, E. O. Lawrence Berkeley National Laboratory, University of California, Berkeley, Berkeley, California 94720

Marie Wiltshire (173), Department of Pathology, University of Wales College of Medicine, Heath Park, Cardiff CF4 4XN, United Kingdom

William D. Wright (251, 269), Mallinckrodt Institute of Radiology, Radiation Oncology Center, Section of Cancer Biology, Washington University School of Medicine, St. Louis, Missouri 63108

Mai Xu (269), Mallinckrodt Institute of Radiology, Radiation Oncology Center, Section of Cancer Biology, Washington University School of Medicine, St. Louis, Missouri 63108

Peng Zhang (251), Mallinckrodt Institute of Radiology, Section of Cancer Biology, Washington University School of Medicine, St. Louis, Missouri 63108

PREFACE TO THE THIRD EDITION

This is the third edition of cytometry volumes in the *Methods in Cell Biology* series. The first, single-volume edition (*Flow Cytometry, Methods in Cell Biology,* Volume 33, 1990) appeared a decade ago. The continuing rapid growth of this methodology prompted us to prepare the second, two-volume edition (*Flow Cytometry, Methods in Cell Biology,* Volumes 41 and 42, 1994), which introduced a variety of new methods developed since the publication of the first edition. The growth and applications of this methodology have continued at an accelerating pace. This progress and the demand for the first two editions, which have become the "bible" for researchers who utilize the presented methods in a variety of fields of biology and medicine, prompted us to prepare the third edition.

This two-volume set differs from the earlier editions in several respects. The title is changed to *Cytometry* to indicate its wider scope. Several chapters describe methods and instrumentation that are not particular to cell analysis in "flow." Also changed are the scope and specifics of many chapters. Specifically, with the appearance of similar series of books on methods by other publishers (e.g., *Current Protocols in Cytometry* by Wiley-Liss or the "Practical Approach" books by Oxford Press), there was no point in duplicating them by focusing on presentation of individual methods in a cookbook form only. The authors, therefore, were requested to prepare their chapters in a form that presented not only technical protocols but also different aspects of the methodology that cannot be included in the protocols format. Thus, theoretical foundations of the described methods, their applicability in experimental laboratory and clinical settings, traps and pitfalls common to particular methods, problems with data interpretation, comparison with alternative assays, etc., are all presented in greater detail in many chapters. Furthermore, some chapters review applications of cytometry and complementary methodologies to particular biological problems or clinical tasks.

With few exceptions, nearly all 56 chapters in the present edition are novel, describing methods that were not included in the earlier editions. The present edition thus complements rather than merely updates the earlier edition. Because most of the methods described in Volumes 41 and 42 have not changed much since publication and are still in wide use, the combination of the earlier two volumes and these two new volumes becomes the most comprehensive collection of all methods in cytometry ever published.

The chapters presented in *Cytometry* cover a wide range of topics. The first several chapters are introductory. They describe principles of flow cytometry, laser scanning cytometry (LSC), confocal microscopy, and general approaches in cell measurement and data acquisition. Newcomers to the field of cytometry may find these chapters particularly useful, as they provide the foundation needed to understand specific methods and more complex data analysis. The next chapters address the issue of cell preparation for analysis by cytometry, quality assurance, and standardization. Of special interest may be the chapters focused on strategies for cell permeabilization and fixation to detect intracellular components, quantitation of the immunocytochemical staining reactions, and standardization in cytometry in general. Unfortunately, these important issues are neglected in many studies utilizing cytometric methods.

Analysis of cell proliferation is the subject of several other chapters. All the methods presented in these chapters are used extensively in experimental and clinical research. The methods include measurements of mitotic activity by flow cytometry, assays of cell kinetics by analysis of BrdU incorporation, analysis of the history of cell proliferation of the progeny cells from geometric dilution of the probe integrated into parent cells, applications of Ki-67 and PCNA antibodies as proliferation markers, and analysis of the lymphocyte activation antigens. Further chapters are devoted to methods of analysis of cell death, primarily by apoptosis. They include probing of mitochondria, activation of caspases, and analysis of apoptosis in the plant kingdom. A review of common problems, difficulties, and pitfalls encountered in analysis of apoptosis, with the key information of how to avoid them, also is provided.

Another group of methods is focused on analysis of cell-to-cell interactions and interactions of cells with the extracellular matrix. This is an exciting cross-disciplinary area that is revealing new directions for cytometry. This section includes cell migration assays, cytometric analysis of cell contact and adhesion, and measurement of extracellular matrix degradation. Also included in this group is a review of extracellular matrix substrates for culturing cells and analysis of cell interactions in three dimensions.

The field of cytogenetics and molecular genetics is represented by several chapters that present methods for sorting plant chromosomes, quantitative DNA fiber mapping, primed *in situ* (PRINS) methodology, individual chromosome telomere length analysis, and approaches to detecting products of chromosomal translocation in individual interphase cells. Functional cell assays, such as probing mitochondria with new markers of the electrochemical transmembrane potential, and measurement of RNA synthesis by immunocytochemical detection of the incorporated BrU, as well as new approaches to monitoring erythropoiesis or proliferation and differentiation of human progenitor cells, are all the subjects of additional chapters.

A large group of chapters is devoted to applications of cytometry in experimental oncology. Presented here are the methods to study interactions between antitumor drugs and intracellular targets, monitoring cellular resistance to chemotherapy in solid tumors and in spheroids related to their three-dimensional architecture, and analyzing DNA damage in individual cells caused by ionizing radiation. The methods specifically designed for studying effects of hyperthermia on tumor cells are also presented.

The most numerous chapters are those on applications of cytometry in the clinic. Indisputably, immunophenotyping is the most common application of cytometry in the clinical setting, and three chapters are devoted to this subject. A very exhaustive chapter on multiparametric analysis of human leukocytes describes the approaches to identifying the cells in different hematological malignancies. Adaptation of laser scanning cytometry to achieve similar tasks is the topic of another chapter on this subject. The third chapter addresses the specific issue of utility of a CD45 gate for identification of malignant cells in acute leukemias. Two insightful and exhaustive reviews, one that critically assesses clinical impact of analysis of different proliferation markers in human solid tumors and another that covers applications of flow cytometry and complementary methodologies in detection of minimal residual disease in leukemias, will be of great value for oncologists. There are also two reviews on applications of laser scanning cytometry to analysis of human tumors, one presented from the perspective of the pathologist and another from the surgeon's perspective. Applications of flow cytometry to monitoring HIV-infected patients is also a subject of thorough review. The last chapter in this group (Chapter 53) may be of particular interest to all researchers, regardless of the discipline. This chapter presents a unique approach to multiparameter data analysis in the clinic that often reveals unexpected correlations with high impact on disease prognosis.

The last group of chapters describes the methods and applications of cytometry in studies of microorganisms. They include flow cytometric analysis of microorganisms and monitoring of bacterial susceptibility to antibiotics. Applications of these assays are expected to rapidly expand and become routine tools in microbiology with wide application in the field of infectious diseases as well as in monitoring environmental contaminations.

As in the earlier editions, the chapters were prepared by colleagues who developed the described methods, contributed to their modification, or found new applications and have extensive experience in their use. The list of authors, as before, represents a "Who's Who" directory in the field of cytometry. On behalf of the readers, we express our gratitude to all contributing authors for the time they devoted to sharing their knowledge and experience.

Zbigniew Darzynkiewicz
Harry A. Crissman
J. Paul Robinson

PREFACE TO THE SECOND EDITION

The first edition of this book appeared four years ago (*Methods in Cell Biology,* Vol. 33, *Flow Cytometry,* Z. Darzynkiewicz and H. A. Crissman, Eds., Academic Press, 1990). This was the first attempt to compile a wide variety of flow cytometric methods in the form of a manual designed to describe both the practical aspects and the theoretical foundations of the most widely used methods, as well as to introduce the reader to their basic applications. The book was an instant publishing success. It received laudatory reviews and has become widely used by researchers from various disciplines of biology and medicine. Judging by this success, there was a strong need for this type of publication. Indeed, flow cytometry has now become an indispensable tool for researchers working in the fields of virology, bacteriology, pharmacology, plant biology, biotechnology, toxicology, and environmental sciences. Most applications, however, are in the medical sciences, in particular immunology and oncology. It is now difficult to find a single issue of any biomedical journal without an article in which flow cytometry has been used as a principal methodology. This book on methods in flow cytometry is therefore addressed to a wide, multidisciplinary audience.

Flow cytometry continues to rapidly expand. Extensive progress in the development of new probes and methods, as well as new applications, has occurred during the past few years. Many of the old techniques have been modified, improved, and often adapted to new applications. Numerous new methods have been introduced and applied in a variety of fields. This dramatic progress in the methodology, which occurred recently, and the positive reception of the first edition, which became outdated so rapidly, were the stimuli that led us to undertake the task of preparing a second edition.

The second edition is double the size of the first one, consisting of two volumes. It has a combined total of 71 chapters, well over half of them new, describing techniques that had not been presented previously. Several different methods and strategies for analysis of the same cell component or function are often presented and compared in a single chapter. Also included in these volumes are selected chapters from the first edition. Their choice was based on the continuing popularity of the methods; chapters describing less frequently used techniques were removed. All these chapters are updated, many are extensively modified, and new applications are presented.

From the wide spectrum of chapters presented in these volumes it is difficult to choose those methods that should be highlighted because of their novelty, possible high demand, or wide applicabilities. Certainly those methods that offer new tools for molecular biology belong in this category; they are presented in chapters on fluorescence *in situ* hybridization (FISH), primed *in situ* labeling

(PRINS), mRNA species detection, and molecular phenotyping. Detection of intracellular viruses and viral proteins and analysis of bacteria, yeasts, and plant cells are broadly described in greater detail than before in separate chapters. The chapter on cell viability presents and compares ten different methods for identifying dead cells and discriminating between apoptosis and necrosis, including a new method of DNA gel electrophoresis designed for the detection of degraded DNA in apoptotic cells. The chapter describing analysis of enzyme kinetics by flow cytometry is very complete. The subject of magnetic cell sorting is also described in great detail.

Numerous chapters that focus on the analysis of cell proliferation also should be underscored. The subjects of these chapters include univariate DNA content analysis (using a variety of techniques and fluorochromes applicable to cell cultures, fresh clinical samples, or paraffin blocks), the deconvolution of DNA content frequency histograms, multivariate (DNA vs protein or DNA vs RNA content) analysis, simple and complex assays of cell cycle kinetics utilizing BrdUrd and IdUdr incorporation, and studies of the cell cycle based on the expression of several proliferation-associated antigens, including the G_1- and G_2-cyclin proteins. Approaches to discriminating between cells having the same DNA content but at different positions in the cell cycle (e.g., noncycling G_0 vs cycling G_1, G_2 vs M, and G_2 of lower DNA ploidy vs G_1 of higher ploidy) are also presented.

Many of the methods described in these volumes will be used extensively in the fields of toxicology and pharmacology. Among these are the techniques designed for analysis of somatic mutants, formation of micronuclei, DNA repair replication, and cumulative DNA damage in sperm cells (DNA *in situ* denaturability). The latter is applicable as a biological dosimetry assay. A plethora of methods for analysis of different cell functions (functional assays) will also find application in toxicology and pharmacology.

The largest number of chapters is devoted to methods having clinical applications, either in medical research or in routine practice. Chapters dealing with lymphocyte phenotyping, reticulocyte and platelet analysis, analysis and sorting of hemopoietic stem cells, various aspects of drug resistance, DNA ploidy, and cell cycle measurements in tumors are very exhaustive. Diagnosis and disease progression assays in HIV-infected patients, as well as sorting of biohazardous specimens, new topics of current importance in the clinic, are also represented in this book.

Individual chapters are written by the researchers who developed the described methods, contributed to their modification, or found new applications and have extensive experience in their use. Thus, the authors represent a "Who's Who" directory in the field of flow cytometry. This ensures that the essential details of each methodology are included and that readers may easily learn these techniques by following the authors' protocols. We express our gratitude to all contributing authors for sharing their knowledge and experience.

The chapters are designed to be of practical value for anyone who intends to use them as a methods handbook. Yet, the theoretical bases of most of the techniques are presented in detail sufficient for teaching the principle underlying the described methodology. This may be of help to those researchers who want to modify the techniques, or to extend their applicability to other cell systems. Understanding the principles of the method is also essential for data evaluation and for recognition of artifacts. A separate section of most chapters is devoted to the applicability of the described method to different biological systems. Another section of most chapters covers the critical points of the procedure, possible pitfalls, and experience of the author(s) with different instruments. Appropriate controls, standards, instrument adjustments, and calibrations are the subjects of still another section of each chapter. Typical results, frequently illustrating different cell types, are presented and discussed in yet another section. The Materials and Methods section of each chapter is exhaustive, providing a detailed, step-by-step description of the procedure in a protocol or cookbooklike format. Such exhaustive treatment of the methodology is unique; there is no other publication on the subject of similar scope.

We hope that the second edition of *Flow Cytometry* will be even more successful than the first. The explosive growth of this methodology guarantees that soon there will be the need to compile new procedures for a third edition.

Zbigniew Darzynkiewicz
Harry A. Crissman
J. Paul Robinson

PREFACE TO THE FIRST EDITION

Progress in cell biology has been closely associated with the development of quantitative analytical methods applicable to individual cells or cell organelles. Three distinctive phases characterize this development. The first started with the introduction of microspectrophotometry, microfluorometry, and microinterferometry. These methods provided a means to quantitate various cell constituents such as DNA, RNA, or protein. Their application initiated the modern era in cell biology, based on quantitative—rather than qualitative, visual—cell analysis. The second phase began with the birth of autoradiography. Applications of autoradiography were widespread and this technology greatly contributed to better understanding of many functions of the cell. Especially rewarding were studies on cell reproduction; data obtained with the use of autoradiography were essential in establishing the concept of the cell cycle and generated a plethora of information about the proliferation of both normal and tumor cells.

The introduction of flow cytometry initiated the third phase of progress in methods development. The history of flow cytometry is short, with most advances occurring over the past 15 years. Flow cytometry (and associated with it electronic cell sorting) offers several advantages over the two earlier methodologies. The first is the rapidity of the measurements. Several hundred, or even thousands, of cells can be measured per second, with high accuracy and reproducibility. Thus, large numbers of cells from a given population can be analyzed and rare cells or subpopulations detected. A multitude of probes have been developed that make it possible to measure a variety of cell constituents. Because different constituents can be measured simultaneously and the data are recorded by the computer in list mode fashion, subsequent bi- or multivariate analysis can provide information about quantitative relationships among constituents either in particular cells or between cell subpopulations. Still another advantage of flow cytometry stems from the capability for selective physical sorting of individual cells, cell nuclei, or chromosomes, based on differences in the variables measured. Because some of the staining methods preserve cell viability and/or cell membrane integrity, the reproductive and immunogenic capacity of the sorted cells can be investigated. Sorting of individual chromosomes has already provided the basis for development of chromosomal DNA libraries, which are now indispensable in molecular biology and cytogenetics.

Flow cytometry is a new methodology and is still under intense development, improvement, and continuing change. Most flow cytometers are quite complex and not yet user friendly. Some instruments fit particular applications better than others, and many proposed analytical applications have not been extensively tested on different cell types. Several methods are not yet routine and a certain degree of artistry and creativity is often required in adapting them to new biological material, to new applications, or even to different instrument designs. The methods published earlier often undergo modifications or improvements. New probes are frequently introduced.

This volume represents the first attempt to compile and present selected flow cytometric methods in the form of a manual designed to be of help to anyone interested in their practical applications. Methods having a wide immediate or potential application were selected, and the chapters are written by the authors who pioneered their development or who modified earlier techniques and have extensive experience in their application. This ensures that the essential details are included and that readers may easily master these techniques in their laboratories by following the described procedures.

The selection of chapters also reflects the peculiarity of the early phase of method development referred to previously. The most popular applications of flow cytometry are in the fields of immunology and DNA content–cell cycle analysis. While the immunological applications are now quite routine, many laboratories still face problems with the DNA measurements, as is evident from the poor quality of the raw data (DNA frequency histograms) presented in many publications. We hope that the descriptions of several DNA methods in this volume, some of them individually tailored to specific dyes, flow cytometers, and material (e.g., fixed or unfixed cells or isolated cell nuclei from solid tumors), may help readers to select those methods that would be optimal for their laboratory setting and material. Of great importance is the standardization of the data, which is stressed in all chapters and is a subject of a separate chapter.

Some applications of flow cytometry included in this volume are not yet widely recognized but are of potential importance and are expected to become widespread in the near future. Among these are methods that deal with fluorescent labeling of plasma membrane for cell tracking, flow microsphere immunoassay, the cell cycle of bacteria, the analysis and sorting of plant cells, and flow cytometric exploration of organisms living in oceans, rivers, and lakes.

Individual chapters are designed to provide the maximum practical information needed to reproduce the methods described. The theoretical bases of the methods are briefly presented in the introduction of most chapters. A separate section of each chapter is devoted to applicability of the described method to different biological systems, and when possible, references are provided to articles that review the applications. Also discussed under separate subheads are the critical points of procedure, including the experience of the authors with different instruments, and the appropriate controls and standards. Typical results, often illustrating different cell types, are presented and discussed in the "Results" section. The "Materials and Methods" section of each chapter is the most extensive, giving a detailed description of the method in a cookbook format.

Flow cytometry and electronic sorting have already made a significant impact on research in various fields of cell and molecular biology and medicine. We hope that this volume will be of help to the many researchers who need flow cytometry in their studies, stimulate applications of this methodology to new areas, and promote progress in many disciplines of science.

Zbigniew Darzynkiewicz
Harry A. Crissman

Cytogenetics and Molecular Genetics

Sorting of Plant Chromosomes

Jaroslav Doležel,[*] Martin A. Lysák,[*] Marie Kubaláková,[*] Hana Šimková,[*] Jiří Macas,[†] and Sergio Lucretti[‡]

[*]Institute of Experimental Botany
CZ-772 00 Olomouc, Czech Republic

[†]Institute of Plant Molecular Biology
CZ-370 05 České Budějovice, Czech Republic

[‡]ENEA
Casaccia Research Center
Plant Biotechnology Division
Rome, Italy

I. Introduction
II. Application
III. Materials
    A. Plant Material
    B. Reagents and Stock Solutions
    C. Equipment and Other Materials
IV. Procedures
    A. Sample Preparation
    B. Flow Cytometric Analysis and Sorting
V. Critical Aspects of the Procedures
    A. Sample Preparation
    B. Flow Karyotyping and Chromosome Sorting
VI. Instruments
VII. Results
    A. Sample Preparation
    B. Flow Cytometric Analysis and Sorting
VIII. Conclusions and Perspectives
    References

I. Introduction

Techniques for flow cytometric analysis and sorting of mitotic chromosomes were originally developed for the Chinese hamster and subsequently modified for humans and several other animal species (Gray *et al.*, 1975; Carrano *et al.*, 1979; Dixon *et al.*, 1992). Since then, flow cytometric classification of chromosomes (flow karyotyping) has proved to be a useful tool for detection of numerical and structural chromosome aberrations (Otto, 1988; Cooke *et al.*, 1988; Boschman *et al.*, 1992). In addition to chromosome analysis, flow cytometry permits isolation of single chromosome types in large quantities. DNA of sorted chromosomes has been shown to be unique material for gene mapping (Lebo, 1982), construction of chromosome-specific DNA libraries (Van Dilla and Deaven, 1990), and chromosome painting probes (Carter, 1994).

Most economically important plant species have large nuclear genomes (Bennett and Smith, 1976). It is clear that gene mapping and isolation would be greatly simplified by fractionation of genomes to well-defined parts (e.g., chromosomes). Until recently, the progress in flow cytometric analysis and sorting of plant chromosomes (plant flow cytogenetics) has been slow. The initial delay was mainly due to problems with preparation of suspensions of intact chromosomes (cf. Doležel *et al.*, 1994, 1995a). Although other experimental systems have been employed, a procedure based on mechanical release of chromosomes from hydroxyurea-synchronized root tips of seedlings (Doležel *et al.*, 1992) has been most frequently used (Table I). The use of root tips for chromosome isolation offers several advantages over other systems: seedlings are easy to handle, root meristems are karyologically stable, and root tips can be synchronized to obtain a high proportion of metaphase cells (Lucretti and Doležel, 1995; Doležel *et al.*, 1999). In the original high-yield procedure, chromosomes were isolated from formaldehyde-fixed roots. Kaeppler *et al.* (1997) introduced a modified version of the procedure in which the fixation step is omitted.

Other problem with the development of plant flow cytogenetics concerned discrimination of individual chromosomes. Most plant species have similarly sized chromosomes that cannot be resolved. To overcome this problem, Lucretti *et al.* (1993) suggested using chromosomal stocks in which chromosome morphology is changed due to translocations and/or deletions. This approach has been found very useful in *Vicia faba* (Doležel and Lucretti, 1995), *Pisum sativum* (Neumann *et al.*, 1998), and *Hordeum vulgare* (Číhalíková *et al.*, 1998; Lysák *et al.*, 1999). Other approaches have been tested, including bivariate analysis after simultaneous staining with AT- and GC-specific dyes. Although this type of analysis permits the resolution of almost all chromosomes in humans and in other mammals (Langlois *et al.*, 1982), the number of resolvable chromosomes is not considerably increased in plants (Lucretti and Doležel, 1997; Schwarzacher *et al.*, 1997). A novel method to improve chromosome discrimination is based on fluorescent labeling of repetitive sequences that are not homogeneously distributed within the genome. This may be achieved using a method called primed *in situ* DNA labeling *en* suspension (PRINSES)

Table I
List of Plant Species for Which Flow Cytometric Analysis of Mitotic Chromosomes Has Been Reported

Species	Material	n^a	N^b	Sorting	Reference
Haplopappus gracilis	Suspension cells	2	2	Yes	De Laat and Blaas (1984), De Laat and Schel (1986)
Hordeum vulgare	Root meristems	7	1 (4)	Yes	Číhalíková *et al.* (1998), Lysák *et al.* (1999)
Lycopersicon esculentum	Suspension cells	12	2	Yes	Arumuganathan *et al.* (1991)
Lycopersicon pennellii	Suspension cells	12	2	Yes	Arumuganathan *et al.* (1994)
Melandrium album	Hairy root meristems	12	2	Yes	Veuskens *et al.* (1992, 1995)
Nicotiana plumbaginifolia	Mesophyll protoplasts	10	?	No	Conia *et al.* (1989)
Petunia hybrida	Mesophyll protoplasts	7	1	Yes	Conia *et al.* (1987, 1988)
Pisum sativum	Root meristems	7	2	Yes	Gualberti *et al.* (1996)
	Hairy root meristems	7	2 (4)	No	Neumann *et al.* (1998)
Secale cereale	Root meristems	7	1	No	Doležel *et al.* (1995b)
Triticum aestivum	Suspension cells	21	?	Yes	Wang *et al.* (1992), Schwarzacher *et al.* (1997)
	Root meristems	21	?	Yes	Lee *et al.* (1997)
Vicia faba	Root meristems	6	1 (6)	Yes	Lucretti *et al.* (1993), Doležel and Lucretti (1995)
	Hairy root meristems	6	1 (6)	No	Doleželová *et al.* (1996)
Zea mays	Root meristems	10	2	Yes	Lee *et al.* (1996)

^{a} Number of chromosomes in a haploid set.

^{b} Number of single chromosome types unambiguously discriminated. The numbers in parentheses give the maximum number of chromosomes discriminated in a chromosome translocation line.

(Macas *et al.,* 1995). Although it has great potential, until now the procedure has not been applied to species other than *V. faba* (Pich *et al.,* 1995).

II. Application

A common procedure for chromosome isolation and flow sorting consists of three basic steps: (a) cell cycle synchronization and metaphase accumulation; (b) preparation of chromosome suspensions; and (c) flow cytometric analysis and sorting. This chapter outlines high-yield procedures for preparation of chromosome suspensions from synchronized root tips in two legume and two cereal species. Procedures for univariate and bivariate flow cytometric analysis of isolated chromosomes are given, including those for chromosome sorting. Furthermore, two different methods are provided for testing the identity and purity of sorted chromosome fractions, namely, fluorescence microscopy and polymerase chain reaction (PCR).

III. Materials

A. Plant Material

1. Seeds of Field Bean (*Vicia faba* L. ssp. *faba* var. *equina* Pers., $2n = 12$)

Seeds of *V. faba* cv. 'Inovec' with standard karyotype were obtained from Dr. M. Vavák (PBS, Horná Streda, Slovakia). Seeds of *V. faba* line EF with reconstructed karyotype were kindly provided by Dr. I. Schubert (IPK, Gatersleben, Germany). The line carries translocations between chromosomes 2 and 3, and between chromosomes 4 and 5 (Schubert and Rieger, 1991).

2. Seeds of Pea (*Pisum sativum* L., $2n = 14$)

Seeds of *P. sativum* cv. 'Ctirad' with standard karyotype were obtained from Dr. I. Hasalová (Semo, Smržice, Czech Republic). Seeds of *P. sativum* line JI 2332 with reconstructed karyotype were obtained from Dr. M. J. Ambrose (JIC, Norwich, England). The line carries a translocation between chromosomes 4 and 7 (Simpson *et al.*, 1990).

3. Seeds of Rye (*Secale cereale* L., $2n = 14$)

Seeds of *S. cereale* cv. 'Daňkovské' were provided by F. Macháň (ARI, Kroměříž, Czech Republic). Seeds of rye with B chromosomes were provided by Prof. R. N. Jones (University of Wales, Aberystwyth, U.K.).

4. Seeds of Barley (*Hordeum vulgare* L., $2n = 14$)

Seeds of *H. vulgare* cv. 'Akcent' with standard karyotype were obtained from Dr. P. Svačina (Plant Select, Hrubčice, Czech Republic). Seeds of *H. vulgare* line T2-6y with reconstructed karyotype was kindly provided by Prof. G. Künzel (IPK). The line carries a translocation between chromosomes 2 and 6 (Marthe and Künzel, 1994).

B. Reagents and Stock Solutions

Amiprophos-methyl (APM) was a gift from the Agricultural Chemical Division, Mobay Corp. (Kansas City, MO); 4',6-diamidino-2-phenylindole (DAPI) (Catalog No. D-1306) was purchased from Molecular Probes (Eugene, OR); formaldehyde solution (37%, Catalog No. 1.04003) was purchased from Merck (Darmstadt, Germany); mithramycin A (Catalog No. 6891) was purchased from Sigma (Saint Louis, MO); all nucleotides were purchased from Boehringer Mannheim (Mannheim, Germany); *Taq* DNA polymerase was obtained from Promega (Madison, WI); Vectashield antifade solution was purchased from Vector Laboratories (Burlingame, CA). Hydroxyurea (Catalog

No. H-8627) and remaining reagents used but not specifically mentioned were purchased from Sigma.

1. Hoagland's Nutrient Solution

a. Solution A

280 mg	H_3BO_3
340 mg	$MnSO_4 \cdot H_2O$
10 mg	$CuSO_4 \cdot 5H_2O$
22 mg	$ZnSO_4 \cdot 7H_2O$
10 mg	$(NH_4)_6Mo_7O_{24} \cdot 4H_2$

Adjust volume to 100 ml with deionized H_2O and store at 4°C.

b. Solution B

0.5 ml	concentrated H_2SO_4

Adjust volume to 100 ml with deionized H_2O and store at 4°C.

c. Solution C

3.36 g	Na_2EDTA
2.79 g	$FeSO_4$

Adjust volume to approximately 400 ml. Heat the solution to 70°C while stirring until the color turns yellow-brown. Cool, adjust the volume to 500 ml, and store at 4°C.

d. Hoagland's Stock Solution (10×)

4.7 g	$Ca(NO_3)_2 \cdot 4H_2O$
2.6 g	$MgSO_4 \cdot 7H_2O$
3.3 g	KNO_3
0.6 g	$NH_4H_2PO_4$
5 ml	solution A
0.5 ml	solution B

Adjust volume to 500 ml with deionized H_2O and store at 4°C.

e. Hoagland's Nutrient Solution (1×)

100 ml	stock solution
5 ml	solution C

Adjust volume to 1000 ml with deionized H_2O. Prepare the solution just before use.

f. Hoagland's Nutrient Solution (0.1×)

10 ml	stock solution
0.5 ml	solution C

Adjust volume to 1000 ml with deionized H_2O. Prepare the solution just before use.

2. Amiprophos-Methyl Solutions

a. Stock Solution (20 mM)

Dissolve 60.86 mg amiprophos-methyl in 10 ml of ice-cold acetone. Store at $-20°C$ in 1-ml aliquots.

b. Treatment Solution

Add amiprophos-methyl stock solution to the Hoagland's nutrient solution with continuous stirring (the volume added depends on the final concentration needed; see Table II). Prepare the solution just before use.

3. Tris Buffer

Final concentrations are given in parentheses.

0.606 g	Tris	(10 mM)
1.861 g	Na_2EDTA	(10 mM)
2.922 g	NaCl	(100 mM)

Adjust volume to 500 ml with deionized H_2O. Adjust the final pH to 7.5 using 1 N NaOH.

4. Formaldehyde Fixative

Final concentrations are given in parentheses.

0.303 g	Tris	(10 mM)
0.931 g	Na_2EDTA	(10 mM)
1.461 g	NaCl	(100 mM)
250 μl	Triton X-100	(0.1%, v/v)

Adjust volume to 200 ml with deionized H_2O. Adjust the final pH to 7.5 using 1 N NaOH. Add formaldehyde (the volume added depends on the final concentration needed; see Table III). Adjust volume to 250 ml with deionized H_2O. Prepare the fixative just before use.

Table II
Treatment of Seedlings for Synchronization and Accumulation of Dividing Root Tip Cells in Metaphase

Species	Root length (cm)	Hoagland's nutrient solution	Hydroxyurea treatment		Recovery after hydroxyurea treatment (hr)	APM treatment	
			Concentration (mM)	Incubation time (hr)		Concentration (μM)	Incubation time (hr)
Vicia faba	4–5	1×	1.25	18.5	4.5	2.5	
Pisum sativum	3–4	1×	1.25	18	3	10.0	2
Hordeum vulgare	2–3	0.1×	2.00	18	6.5	2.5	2
Secale cereale	2–3	1×	2.50	18	5.5	0.5	2

Table III
Parameters for Formaldehyde Fixation and Mechanical Homogenization of Root Tips

| Species | Formaldehyde fixation | | Homogenization (9500 rpm) time (sec) |
	Formaldehyde concentration (%, v/v)	Fixation time (min)	
Vicia faba	4	30	15
Pisum sativum	3	20	15
Hordeum vulgare	2	20	10
Secale cereale	2	30	10

5. LB01 Lysis Buffer

Final concentrations are given in parentheses.

0.363 g	Tris	(15 mM)
0.149 g	Na$_2$EDTA	(2 mM)
0.348 g	spermine · 4HCl	(0.5 mM)
1.193 g	KCl	(80 mM)
0.234 g	NaCl	(20 mM)
200 μl	Triton X-100	(0.1%, v/v)

Adjust volume to 200 ml with deionized H$_2$O. Adjust final pH to 7.5 using 1 N HCl. Filter through a 0.22-μm filter to remove small particles. Add 220 μl β-mercaptoethanol and mix well. Store at -20°C in 10-ml aliquots.

6. DAPI Stock Solution (0.1 mg/ml)

Dissolve 5 mg DAPI in 50 ml deionized H$_2$O by stirring for 60 min. Filter through a 0.22-μm filter to remove small particles. Store at -20°C in 0.5-ml aliquots.

7. Mithramycin Stock Solution (1 mg/ml)

Dissolve 50 mg mithramycin A in 50 ml deionized H$_2$O by stirring for 60 min. Filter through a 0.22-μm filter to remove small particles. Store at -20°C in 0.5-ml aliquots.

8. Sheath Fluid SF 50

Final concentrations are given in parentheses.

5.96 g KCl	(40 mM)
1.17 g NaCl	(10 mM)

Adjust volume to 2000 ml with deionized H_2O.

9. Magnesium Sulfate Stock Solution (100 mM)

Dissolve 1.23 g of $MgSO_4 \cdot 7H_2O$ in 50 ml of deionized H_2O, filter through a 0.22-μm filter to remove small particles, and store at 4°C.

10. PRINS Reaction Mix

Final concentrations in parentheses are calculated for the volume of 50 μl. Note that this 10× *Taq* DNA polymerase buffer contains 15 mM $MgCl_2$.

5 μl	10× *Taq* DNA polymerase buffer	(1×)
5 μl	25 mM $MgCl_2$	(4 mM)
2.5 μl	2 mM dCTP, dGTP	(0.1 mM)
1 μl	0.1 mM fluorescein-12-dUTP	(2 μM)
1 μl	0.1 mM fluorescein-15-dATP	(2 μM)
1.7 μl	1 mM dTTP	(34 μM)
1.7 μl	1 mM dATP	(34 μM)
1.5 μl	66 μM primer A	(2 μM)
1.5 μl	66 μM primer B	(2 μM)
0.6 μl	5 U/μl *Taq* DNA polymerase	(3 U/50 μl)
33.5 μl	H_2O (5 μl for evaporation)	

11. Stop Buffer for PRINS Reaction

Final concentrations are given in parentheses.

2.923 g	NaCl	(0.5 M)
1.861 g	Na_2EDTA	(0.05 M)

Adjust volume to 100 ml with deionized H_2O. Adjust final pH to 8.0 using 1 N NaOH before autoclaving. Store at 4°C.

12. Wash Buffer for PRINS Reaction

Final concentrations are given in parentheses.

1.161 g	maleic acid	(0.1 M)
0.876 g	NaCl	(0.15 M)
0.5 ml	Tween 20	(0.05%, v/v)

Adjust volume to 100 ml with deionized H_2O. Adjust final pH to 7.5 using 1 N NaOH before autoclaving. Store at 4°C.

13. PCR Premix

Final concentrations in parentheses are given for the volume of 50 μl. Note that this 10× *Taq* DNA polymerase buffer does not contain $MgCl_2$.

5 μl	10× *Taq* DNA polymerase buffer	(1×)
	(the buffer does not contain mgCl$_2$	
3 μl	25 mM MgCl$_2$	(1.5 mM)
1 μl	10 mM dNTPs	(0.2 mM)
1 μl	50 μM primer A	(1 μM)
1 μl	50 μM primer B	(1 μM)
0.5 μl	5 U/μl *Taq* DNA polymerase	(2.5 U/50 μl)
18.5 μl	sterile deionized H$_2$O	

Vortex and centrifuge the ingredients briefly. Prepare the premix shortly before use.

C. Equipment and Other Materials

1. Flow cytometer and sorter [equipped with an ultraviolet (UV) laser and pulse processing]
2. Fluorescence microscope
3. PCR cycler
4. Biological incubator (heating/cooling), internal temperature adjusted to 25° ± 0.5°C
5. Cooled water bath, temperature adjusted to 5° ± 0.5°C
6. Mechanical homogenizer (e.g., Polytron PT1200 with a PT-DA 1205/5 probe)
7. Aquarium bubbler with tubing and aeration stones
8. Calibration beads (e.g., Polysciences BB beads 4.5 μm, Catalog No. 18340) selected for low coefficient of variation
9. Calibration beads (e.g., Polysciences YG beads 2 μm, Catalog No. 18338) selected for low coefficient of variation
10. Perlite (inert substrate for seed germination)
11. A plastic tray, 4000 ml (e.g., 25 cm length, 15 cm width, 11 cm height), for seed germination in perlite
12. Glass petri dishes (18 cm diameter) for seed germination on filter paper
13. A plastic tray, 750 ml (e.g., 14 cm length, 8 cm width, 10 cm height), including an open-mesh basket to hold germinated seeds
14. Frame-Seal incubation chambers, volume 25 μl (MJ Research, Watertown, MA)
15. Nylon mesh (pore size 50 μm and 20 μm), squares 4 × 4 cm

IV. Procedures

The procedures described have been developed for analysis and sorting of plant mitotic chromosomes. Suspensions of intact chromosomes are prepared from root tips of young seedlings. Prior to chromosome isolation, meristem cells are accumulated at metaphase by a sequential treatment with a DNA synthesis inhibitor hydroxyurea and a mitotic spindle inhibitor amiprophos-methyl. Root tips accumulated at metaphase are mildly fixed with formaldehyde, and intact chromosomes are released into a lysis buffer by mechanical homogenization of root tip meristems. Suspensions of intact chromosomes are stained by DNA fluorochromes, and their fluorescence is analyzed by flow cytometer. Chromosomes can be sorted on a slide for microscopic observation and/or collected into a tube for PCR and other applications.

A. Sample Preparation

1. Seed Germination

Two different procedures are used to germinate seeds of legumes and cereals. In both cases, the germination is performed in the dark at $25° \pm 0.5°C$.

a. Germination in Perlite (Vicia faba, Pisum sativum)

Imbibe the seeds for 24 hr in deionized H_2O with aeration (~30 seedlings are needed to prepare one sample).

Wet the perlite with the Hoagland's nutrient solution and put it into a plastic tray.

Wash the seeds in deionized H_2O, layer them onto the surface of perlite, cover them with a layer of 1–2 cm of wet perlite.

Cover the tray with aluminum foil and leave the seeds to germinate for 2–3 days to achieve proper length of primary roots (Table II).

Remove the seedlings from perlite and wash them in deionized H_2O.

b. Germination on a Filter Paper (Hordeum vulgare, Secale cereale)

Place several layers of paper towels into a glass petri dish (18 cm diameter) and top them with a single sheet of filter paper.

Moisten the paper layers with deionized H_2O until runoff.

Spread the seeds on the paper surface (~50 seedlings are needed to prepare one sample).

Cover the petri dish and leave the seeds to germinate for 2–3 days to achieve proper length of primary roots (Table II).

2. Accumulation of Root Tip Cells in Metaphase

Adjust the temperature of all solutions to $25° \pm 0.5°C$ prior to use. Perform all incubations in the dark in a biological incubator at $25° \pm 0.5°C$. Aerate all solutions.

Select seedlings with the proper length of primary roots (Table II).

Thread seedling roots through the holes of the open-mesh basket positioned on a plastic tray filled with deionized H_2O.

Transfer the basket with seedlings to a plastic tray containing the hydroxyurea solution and incubate for periods listed in Table II.

Wash the roots vigorously in several changes of deionized H_2O.

Incubate in hydroxyurea-free Hoagland's working solution for recovery periods listed in Table II.

Transfer the basket with seedlings to a tray filled with amiprophos-methyl treatment solution and incubate for periods indicated in Table II.

In case of cereal species (*S. cereale, H. vulgare*), transfer the basket with seedlings to a container filled with a mixture of ice cubes and deionized H_2O (1°–2°C). Place the container in a refrigerator, and treat the roots overnight.

3. Preparation of Chromosome Suspensions

Harvest root tips (1 cm) and transfer them into deionized H_2O.

Fix the roots by transferring them into a formaldehyde fixative, and incubate them at 5°C for periods given in Table III.

Wash the roots in Tris buffer three times for 5 min at 5°C.

Excise root meristems (1–2 mm, depending on a species) and transfer them into a 5-ml polystyrene tube containing 1 ml of LB01 lysis buffer.

Isolate chromosomes by homogenizing at 9500 rpm for periods given in Table III.

Filter the suspension through a 50-μm nylon mesh into a polystyrene tube.

Store the suspension on ice.

Chromosomes can also be isolated by chopping the root tips with a sharp scalpel in a glass petri dish containing 1 ml of LB01 lysis buffer. In this case, the suspension must be syringed once through a 22-gauge needle to disperse chromosome clumps. This method is more laborious and inconvenient in species with small root tips.

4. Chromosome Staining

a. DAPI Staining for Univariate Analysis

Stain the chromosomes in suspension by adding DAPI stock to a final concentration of 2 μg/ml.

Filter through a 20-μm nylon mesh.

Examine the quality of the suspension under a fluorescence microscope.

b. Mithramycin/DAPI Staining for Bivariate Analysis

Add $MgSO_4$ stock to final concentration of 10 mM.

Stain the chromosomes in suspension by adding DAPI stock to final concentration of 1.5 μg/ml and by adding mithramycin stock to final concentration of 20 μg/ml.

Leave to equilibrate for 30 min on ice.

Filter through a 20-μm nylon mesh.

Examine the quality of suspension under a fluorescence microscope.

B. Flow Cytometric Analysis and Sorting

1. Modeling of Theoretical Flow Karyotypes

Prepare theoretical flow karyotypes using either a spreadsheet or dedicated computer software (Conia *et al.*, 1989; Doležel, 1991).

Determine the resolution (coefficient of variation of chromosome peaks) needed to discriminate individual chromosome types.

2. Flow Karyotyping

Switch on the laser(s) [for univariate analysis, the argon ion laser is operated at multi UV mode (351.1–363.8 nm) with 300 mW output power; for bivariate analysis, the first laser (argon ion) is operated at multi UV mode (351.1–363.8 nm) with 300 mW output power, and the second laser (argon ion) is operated at 457.9 nm with 300 mW output power].

Let the laser(s) stabilize for 30 min, then peak the laser optics for maximum light output.

Empty the waste container and fill the sheath container with sterile sheath fluid SF50.

Adjust sheath fluid pressure to 11 psi, and leave the fluid running to fill all plastic lines and filters of the instrument.

Install a nozzle (70 μm orifice) and check for air bubbles.

a. Instrument Alignment

Install appropriate optical filters [for univariate analysis, band-pass 424/44 nm in front of the fluorescence 1 (FL1) detector; for bivariate analysis, band-pass 530/30 nm in front of the FL1 detector and band-pass 585/42 nm in front of fluorescence 4 (FL4) detector].

Set up the trigger signal to forward scatter height (FSC-H), and select a threshold level.

Run calibration beads (for univariate analysis, Polysciences BB beads; for bivariate analysis, Polysciences YG beads) at flow rate of 200 particles/sec.

Display the data on a dot plot of FSC-H and FL1 pulse height (FL1-H) and on histograms of FSC-H and FL1-H.

Align the instrument to achieve maximum signal intensity and minimum coefficient of variation of FSC-H and FL1-H signals.

For bivariate analysis, use a histogram of FL4 pulse height (FL4-H), and align the second laser to achieve maximum signal and the lowest coefficient of variation. Change only settings specific for the second laser—do not adjust other controls.

Adjust dual-laser delay and dead time parameters as needed.

b. Univariate Analysis

Make sure that the 424/44 nm band-pass filter is placed in front of the FL1 detector.

Run a dummy sample (LB01 buffer containing DAPI) to equilibrate the sample line.

Introduce the sample and let it stabilize at appropriate flow rate (e.g., 200 particles/sec).

Set a gating region on a dot plot of FSC-H and FL1-H to exclude debris, nuclei, and large clumps.

Adjust photomultiplier voltage and amplification gains so that chromosome peaks are evenly distributed on a histogram of FL1 pulse area (FL1-A).

Analyze 20,000–50,000 chromosomes and save the result on a disk.

c. Bivariate Analysis

Use a half-mirror to split the fluorescence of DAPI to the FL1 detector through the 424/44 nm band-pass filter and mithramycin fluorescence to the FL4 detector through a 490 nm long-pass filter.

Run a dummy sample (LB01 buffer containing DAPI and mithramycin) to equilibrate the sample line.

Run the sample and let it stabilize at an appropriate flow rate (e.g., 200 particles/sec).

Set a gating region on a dot plot of FSC-H and FL1-H to exclude debris, nuclei, and large clumps.

Adjust photomultiplier voltages and amplification gains so that chromosome peaks are evenly distributed on histograms of the FL1-A and FL4 pulse area (FL4-A).

Display the data on a dot-plot of FL1-A versus FL4-A.

Analyze 20,000–50,000 chromosomes and save the result on a disk.

3. Chromosome Sorting (after Univariate Analysis)

Switch on the sorting module and adjust the drop drive frequency and drop drive amplitude to break the stream at a suitable distance from the laser intercept point (check for satellite drops).

Adjust the drop drive phase to obtain single side streams (test mode and test sort must be turned on).

Calculate drop delay and perform its optimization.

Select the sort mode and sort envelope according to the required purity, number of chromosomes to be sorted, and desired volume for the sorted fraction.

Run the sample, and display the signals on a dot plot of FL1 pulse width (FL1-W) versus FL1-A.

Adjust the FL1-W amplifier gain and width offset as needed to achieve optimal resolution of the width signal.

Define sorting regions on the FL1-W versus FL1-A dot plot.

Check for stability of the break-off point and of the side streams.

Sort the required number of chromosomes onto a microscope slide or into a polystyrene tube as needed.

If reanalysis of sorted chromosome fractions and/or two-step sorting is required, sort at least 5×10^4 chromosomes into 250 μl of LB01 in a polystyrene tube. The sorted fraction can be analyzed after adding DAPI to a final concentration of 2 μg/ml.

4. Chromosome Identification and Purity of Sorted Fraction

The content of chromosome peaks on flow karyotypes and the purity of the sorted chromosome fraction can be estimated by fluorescence microscopy or after PCR with chromosome-specific primers.

a. Fluorescence Microscopy

Pipette 15 μl of LB01 buffer containing 10% (v/v) sucrose onto a clean microscope slide, sort 1000 chromosomes to the drop, and air dry.

Drain excess fluid and mount in Vectashield antifade solution. Blot and seal with rubber cement.

Examine the slide with an epifluorescence microscope equipped with a filter set for DAPI.

b. Primed in Situ DNA Labeling

Pipette 15 μl of LB01 buffer containing 10% (v/v) sucrose onto a clean microscope slide, sort 1000 chromosomes to the drop, and air dry in an aseptic box.

Stick Frame-Seal chamber (25 μl) to the slide over the specimen area.

Pipette 25 μl of PRINS reaction mix into the frame, and place the polyester cover over the frame.

Run the PRINS reaction: the first cycle consists of 5 min at 94°C, 5 min at 55°C, 10 min at 70°C; the following eight cycles consist of 1 min at 94°C, 1 min at 55°C, and 3 min at 70°C. In the final cycle, prolong the annealing to 5 min and extension to 10 min at 70°C.

Remove the cover, add 100 μl of the PRINS stop buffer, and incubate for 2 min at 70°C.

Remove the stop buffer and transfer the slide to a petri dish. Add 100 μl of the PRINS wash buffer and incubate at room temperature for 5 min. Repeat the washing step twice.

Counterstain slide with DAPI (0.2 μg/ml) made in the PRINS wash buffer.

Drain excess fluid, and mount in Vectashield antifade solution. Blot and seal with rubber cement.

Examine the slide with epifluorescence microscope equipped with filter sets for DAPI, fluorescein isothiocyanate (FITC), and a dual filter set for simultaneous observation of DAPI and FITC fluorescence.

c. Polymerase Chain Reaction

Prepare PCR tubes containing 19 μl of sterile deionized H_2O (final volume after sorting will be approximately 20 μl).

Sort 500 chromosomes into each tube.

Store at -20°C, or freeze and melt the chromosomes to use them immediately for PCR.

Add 30 μl of PCR premix, vortex, and spin briefly.

Perform PCR: initial denaturation for 2 min at 94°C followed by 35 cycles consisting of 1 min denaturation at 94°C, 1 min annealing at the temperature appropriate for the chosen primer pair, 2 min extension at 72°C, and the final extension for 10 min.

Take equal amounts of PCR products (4–8 μl) from each tube and run electrophoresis on a 1.5% agarose gel.

Stain with ethidium bromide and photograph.

V. Critical Aspects of the Procedures

A. Sample Preparation

The quality of chromosome suspension is of prime importance. It depends both on the proportion of metaphase cells in the root meristem and on the conditions under which chromosomes are released from root tip cells.

1. Cell Cycle Synchronization and Metaphase Accumulation

The procedures described here are simple, and usually no problems occur if they are performed carefully. Cell synchronization is based on the inhibition of cell cycle progression at the G_1/S interface by the action of hydroxyurea. The effect of the treatment depends critically on the hydroxyurea concentration, which must be strictly followed (Table II). Although higher frequencies of metaphases may be obtained after longer treatments with amiprophos-methyl, a short (2-hr) treatment is preferable to decrease the proportion of single chromatids and to avoid chromosome decondensation. The main advantage of amiprophos-methyl over other mitotic spindle poisons is that it is very effective and thus can be used in micromolar concentrations. If needed, the compound may be replaced by other spindle inhibitors, including oryzalin, triflurain, and colchicine. This will require determination of optimal concentration for given species. In cereal species (*S. cereale, H. vulgare*) overnight incubation of synchronized seedlings in ice water improves the spreading of chromosomes within cells. Cell cycle kinetics is sensitive to external factors, including temperature. It is thus important to carefully adjust the temperature of all solutions prior use. Special care must be paid to the washing away of remnants of treatment solutions prior to the transfer of seedlings to other solutions. The length of all treatments must be followed exactly. It is recommended that one check the frequency of metaphases in root tips after the treatment, on Feulgen-stained squash preparations.

2. Preparation of Chromosome Suspensions

The procedure for chromosome isolation described here is based on the observation that intact chromosomes can be released mechanically from root meristems that have been mildly fixed with formaldehyde (Doležel *et al.,* 1992). In this procedure, chromosome morphology and yield depend critically on the extent of the fixation. The concentration of formaldehyde, the temperature of fixative, and the length of the fixation must be strictly followed (Table III). If the chromosome suspension is prepared by chopping root tips with a scalpel, the suspension should be syringed once through a 22-gauge needle to disperse chromosome clumps. The suspension should not be syringed if prepared by mechanical homogenization. Storage of fixed root tips in Tris buffer prior to chromosome isolation may decrease chromosome yield, and it is advisable to perform the isolation immediately after the fixation. Although suspensions of isolated chromosomes may be stored at 4°C, the best resolution of flow karyotypes is usually obtained with freshly prepared samples.

B. Flow Karyotyping and Chromosome Sorting

1. Flow Karyotyping

In addition to the quality of the sample, the resolution of flow karyotypes depends critically on the setup of the instrument. The flow cytometer must be well aligned with suitable fluorescent microspheres to achieve the lowest coefficient of variation of fluorescent peaks (1.5% or less; note that the coefficient of variation will depend also on the quality of the beads). A dirty nozzle is the most frequent reason for poor resolution. Nozzles can be conveniently cleaned in a small sonication water bath (do not sonicate the nozzle holder). However, the remaining parts of the fluidics system also critically influence the resolution and must be clean and free of air bubbles. To achieve the best discrimination of chromosome peaks, the sample should be run at low speed (200–500 chromosomes/sec). Because the isolated chromosomes vary greatly in the degree of condensation, and hence also in length, better resolution is obtained on histograms of fluorescence pulse area rather than fluorescence pulse height.

2. Chromosome Sorting

When sorting, the instrument should be triggered on the fluorescence (FL1-H) signal and the threshold set to a minimum value so that all DNA-containing particles are detected. It is important to recognize particles that should not be sorted, and to avoid contamination of the sorted fraction. Precise focusing of the sorted sample is of prime importance. The drop drive phase should be adjusted to obtain single side streams without fanning. Correct position for the sorting stream must be assessed by examining the tube. During the sorting, charged droplets of the same polarity may repel each other. If this happens, sorting tubes should be grounded to avoid a loss in recovery of the sorted fraction. The stability of the break-off point must be followed, and sorting must be stopped if any change is observed.

3. Chromosome Identification and Purity of Sorted Fraction

The usefulness of the sorted chromosome fraction depends on its purity that is primarily determined by the degree of discrimination of chromosome peaks. However, the purity is also affected by the presence of debris particles, chromatids, and chromosome clumps in the chromosome suspension. The purity of the sorted fraction is best monitored by a combination of two approaches, fluorescence microscopy and PCR with chromosome-specific primers.

Microscopic observation of chromosomes stained with DNA fluorochrome can be used to check the recovery after sorting. However, as the degree of condensation varies greatly among sorted chromosomes, simple fluorescent staining is not suitable to identify the chromosome content of peaks or to determine the purity of the sorted fraction. This is best done after fluorescent labeling of sorted chromosomes using either chromosome-specific DNA sequences or sequences that show chromosome-specific labeling pattern. Compared to fluorescence *in situ* hybridization, PRINS is more rapid and thus suitable for fast analysis. However, the method may require careful optimization of several parameters to achieve specific labeling (Kubaláková and Doležel, 1998).

Polymerase chain reaction with primers derived from chromosome-specific DNA sequences is a sensitive tool to identify chromosome content of peaks on the flow karyotype and to detect contamination of sorted fraction by other chromosomes. When using PCR, negative and positive controls should be always included to check whether the PCR ran properly with the given primer pair.

VI. Instruments

The high-yield procedure for preparation of chromosome suspensions described here involves formaldehyde fixation. Our earlier results demonstrated that the fixation resulted in lower resolution of flow karyotypes after staining with DNA intercalators such as propidium iodide. On the other hand, high resolution flow karyotypes could be obtained after staining with DNA fluorochromes such as DAPI that bind preferentially to AT-rich regions (Doležel and Lucretti, 1995). These fluorochromes are excited by UV, and thus the procedures were developed for use with a Becton Dickinson FACSV antage instrument equipped with an UV laser at the first position. For bivariate (DAPI/mithramycin) analyses, the second laser was tuned to 457.9 nm. However, the procedures described here are applicable to all commercial flow cytometers and sorters provided they are equipped with an UV laser at the first position.

VII. Results

A. Sample Preparation

1. Cell Cycle Synchronization and Metaphase Accumulation

The procedure for cell cycle synchronization using hydroxyurea described here is very effective, and mitotic indices ranging from 45 to 55% are routinely obtained. The time when the maximum number of root tip cells undergo mitosis depends on the species and hydroxyurea concentration (Table II). Treatment of synchronized cells with amiprophos-methyl at optimal concentrations (Table II) leads to the accumulation of at least 50% of the cells in metaphase (Fig. 1). Although the treatment alone is effective in achieving high metaphase indices, overnight treatment in ice water improves chromosome spreading and thus reduces the occurrence of chromosome clumps in *H. vulgare* and *S. cereale*.

2. Preparation of Chromosome Suspensions

The extent of formaldehyde fixation critically determines chromosome yield and the quality of suspensions, in general. Weak fixation results in poor preservation of chromosome morphology: Most isolated chromosomes are damaged, and the suspensions contain a large amount of chromosome debris. If, on the other

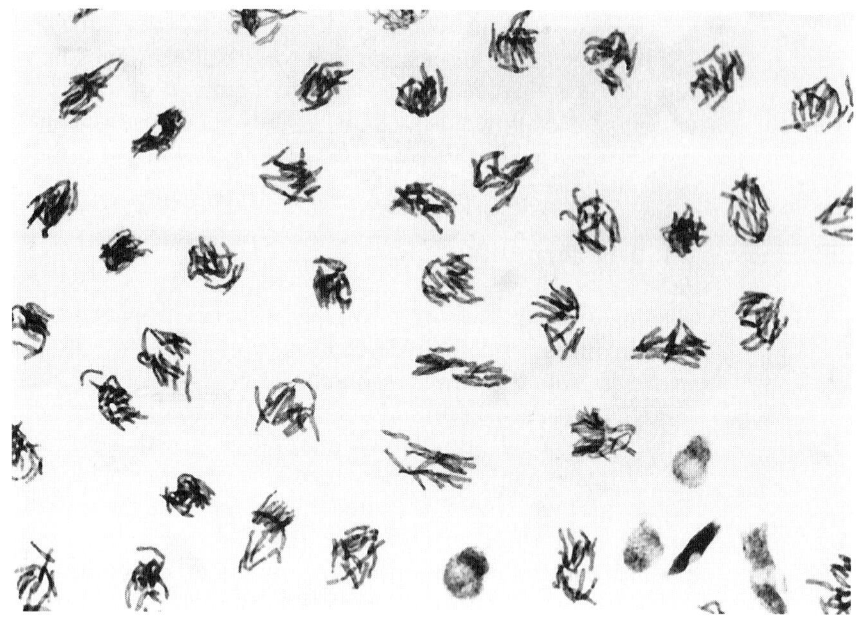

Fig. 1 Metaphase cells in *Vicia faba* root tip accumulated after the treatment of hydroxyurea-synchronized cells with amiprophos-methyl (2.5 $\mu M/2$ hr).

hand, the fixation is too strong, the suspensions obtained after homogenization contain large numbers of chromosome clumps and intact cells. The number of isolated chromosomes depends on the species. More than 1×10^6 chromosomes can be isolated from 30 root tips of *V. faba* and *P. sativum*. Approximately 5×10^5 chromosomes can be isolated from 50 root tips of *S. cereale* and *H. vulgare*. Isolated chromosomes have well-preserved morphology and are suitable for flow cytometric analysis and sorting.

B. Flow Cytometric Analysis and Sorting

1. Theoretical Flow Karyotypes

Theoretical flow karyotypes may be calculated using DNA content or relative chromosome lengths. The models are very useful in planning flow karyotyping experiments and for identification of chromosomes that may be discriminated and sorted. The effect of the resolution (coefficient of variation of chromosome peaks) on chromosome discrimination can be easily analyzed (Fig. 2).

2. Chromosome Analysis (Flow Karyotyping)

a. Univariate Analysis

The analysis of chromosome suspension results in distribution of relative fluorescence intensity or flow karyotype. Flow karyotypes obtained according to the

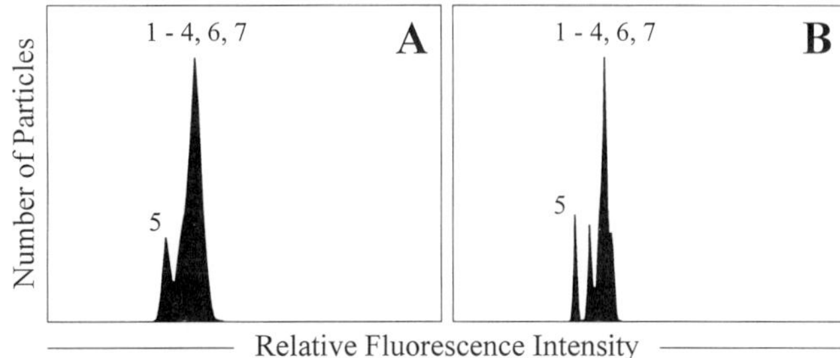

Fig. 2   Theoretical flow karyotypes modeled for *Hordeum vulgare* ($2n = 14$) with two different coefficients of variation of chromosome peaks. The coefficient of variation of 3% is not sufficiently low to discriminate chromosome 5 (A); this can be achieved at a coefficient of variation of 1.5% (B). The karyotypes were modeled using the KARYOSTAR software (Doležel, 1991).

present procedure contain low amounts of chromatids and debris background. The coefficient of variation of chromosome peaks ranges from 1.5 to 3%. Figures 3, 4, and 5 show the results of univariate analysis of DAPI-stained chromosomes in *V. faba, P. sativum,* and *H. vulgare.* Although only one or a few chromosomes can be resolved in lines with standard karyotype, the number of resolvable chromosomes increases in lines where karyotypes have been changed due to one or more chromosome translocations.

b. Bivariate Analysis

Figure 6 shows DAPI versus mithramycin bivariate flow karyotypes obtained in two chromosome translocation lines of *V. faba.* The distribution of chromosome peaks over the two dimensions is very limited, and most of them lie on a straight diagonal line, suggesting only minor differences in the AT:GC ratio among the chromosomes. Bivariate analysis did not help to resolve chromosomes

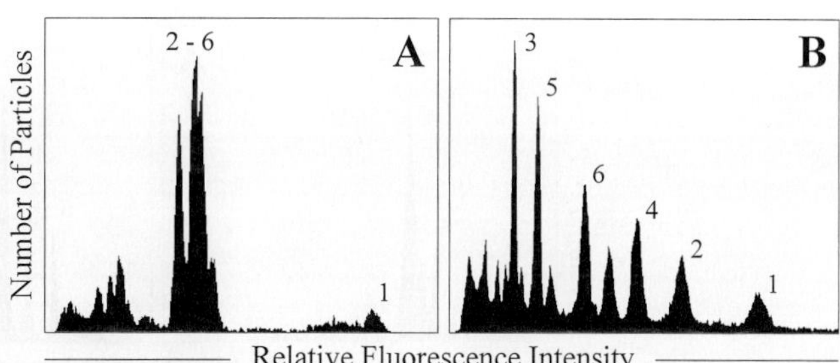

Fig. 3   Univariate flow karyotypes obtained after the analysis of chromosomes isolated from two different lines of *Vicia faba* ($2n = 12$). (A) Line with a standard (wild-type) karyotype—only chromosome 1 can be discriminated, remaining chromosomes (2–7) form a composite peak. (B) Line EF with a reconstructed karyotype—all six chromosomes can be discriminated.

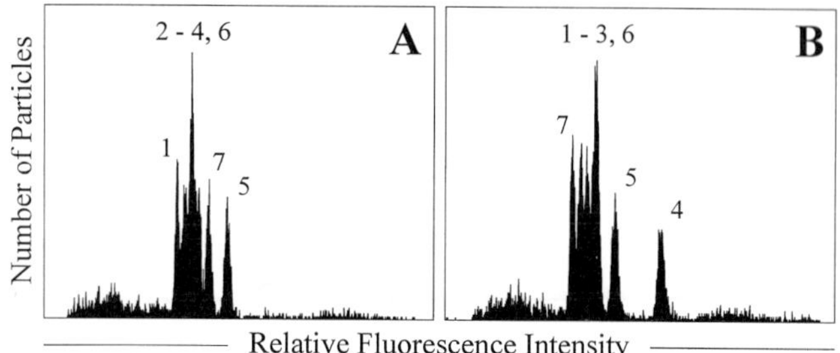

Fig. 4 Univariate flow karyotypes obtained after the analysis of chromosomes isolated from two different lines of *Pisum sativum* (2*n* = 14). (A) Line with a standard (wild-type) karyotype—only chromosome 5 can be fully discriminated. (B) Translocation line JI 2332 with a reconstructed karyotype—in addition to chromosome 5, chromosome 4 can be discriminated.

4 and 6 in the line JF that were not discriminated using univariate analysis. Similar results are obtained with other translocation lines and indicate the limited potential of bivariate flow karyotyping in this species.

3. Chromosome Sorting

a. Chromosome Yield

When the sample is run at a flow rate allowing for highest resolution, sorting rates ranging from 5 to 25 per second can be achieved. The rate of sorting and the chromosome yield (recovery) depend on the frequency of the chromosome in suspension. Longer chromosomes are more sensitive to breakage during the isolation, and therefore lower yield is expected. The large metacentric chromosome 1 in the standard karyotype of *V. faba* represents an extreme example. A population of the chromosome selected using a sort window set on a dot plot

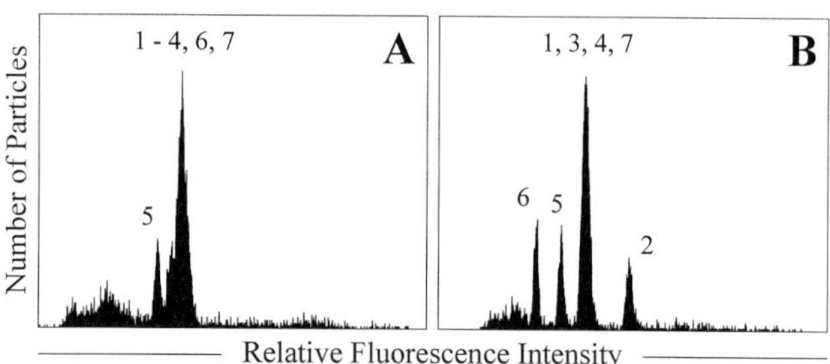

Fig. 5 Univariate flow karyotypes obtained after the analysis of chromosomes isolated from two different lines of *Hordeum vulgare* (2*n* = 14). (A) Line with a standard (wild-type) karyotype—only chromosome 5 can be discriminated. (B) Three chromosomes (2, 5, 6) can be discriminated in line T2-6y with a reconstructed karyotype.

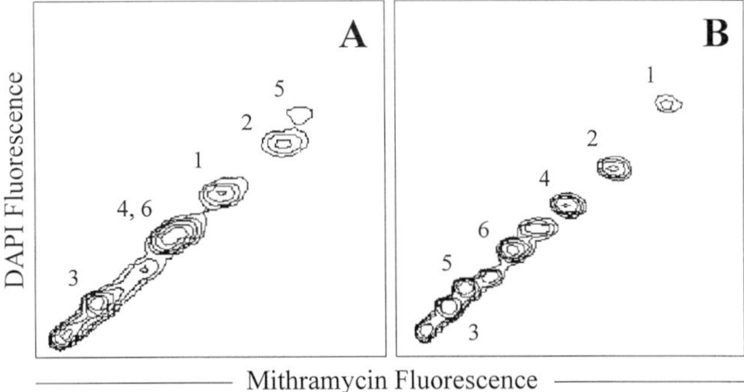

Fig. 6 Bivariate flow karyotypes obtained after the analysis of chromosomes isolated from two translocation lines of *Vicia faba* ($2n = 12$): line JF (A) and line EF (B). Note that with the exception of JF chromosome 5, all chromosomes lie on a straight line indicating small differences in AT/GC ratio.

FL1-W versus FL1-A represents only about 2.5% of the total chromosome population instead of the theoretically expected 16.7% (Fig. 7A).

b. Purity

The purity of the sorted fraction depends on the degree of resolution of chromosome peaks. It may be compromised by the occurrence of chromatids, chromosome fragments, and chromosome doublets that have the total fluorescence similar to that of chromosomes. It is important to select appropriate sorting windows to avoid these particles. Figure 7A shows a strategy used to discriminate

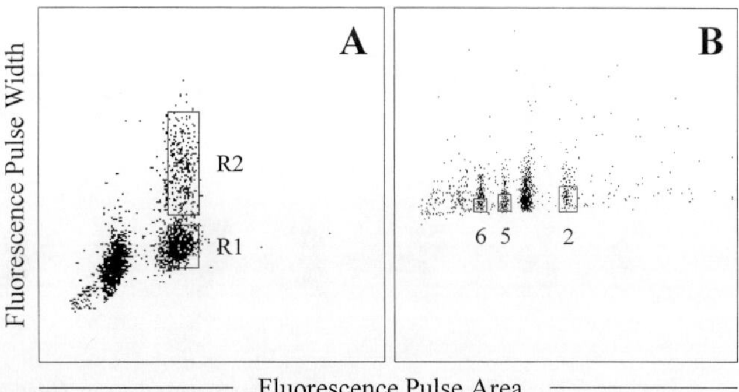

Fig. 7 Dot plots of fluorescence pulse area versus fluorescence pulse width. (A) Analysis of chromosomes isolated from *Vicia faba* ($2n = 12$) line with a standard (wild-type) karyotype. Region R1 contains doublets of small acrocentric chromosomes and metacentric chromosome 1 with folded arms, region R2 contains long metacentric chromosome 1. (B) Analysis of chromosomes isolated from *Hordeum vulgare* ($2n = 14$) translocation line T2-6y. Sorting windows used to sort chromosomes 2, 5, and 6 are shown.

between doublets of small acrocentric chromosomes and a large metacentric chromosome 1 in *V. faba*. Under these conditions, and with a two-step sorting procedure employed, chromosome 1 can be sorted with purity higher than 93%. Figure 7B shows sorting windows used to sort chromosomes 2, 5, and 6 from *H. vulgare* translocation line T2-6y. All three chromosomes were clearly discriminated and could be sorted with a high purity. Analysis of sorted chromosomes after labeling of GAA microsatellites (Fig. 8) showed that the contamination of sorted fractions was lower than 3%. A very high purity of sorted *H. vulgare* chromosome fractions was confirmed by PCR with chromosome-specific primers (Fig. 9).

VIII. Conclusions and Perspectives

Although developing slowly, plant flow cytogenetics has become a reality. The number of species where chromosome analysis and sorting has been reported increased dramatically, and the list now includes major cereals and legumes (Table I). The high-yield procedure for preparation of chromosome suspensions based on the use of root meristems is relatively simple and can be modified to different plant species. The use of so-called hairy root cultures (Tempé and Casse-Delbart, 1989) represents an elegant solution for those genotypes that are difficult to propagate via seeds. Doleželová *et al.* (1996) and Neumann *et al.* (1998) demonstrated that the current protocol could also be used with hairy root cultures.

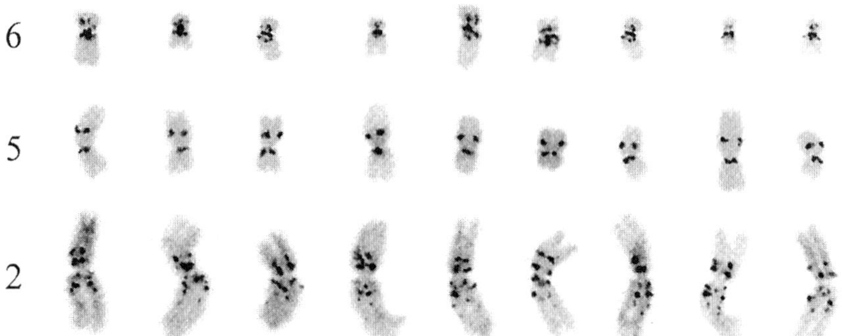

Fig. 8 Three different chromosome types sorted from *Hordeum vulgare* ($2n = 14$) translocation line T2-6y. Sorted chromosomes were labeled with PRINS using a pair of primers for GAA microsatellites and fluorescein-labeled nucleotides, and counterstained with DAPI. The images of fluorescein and DAPI fluorescence were acquired separately with a black and white CCD camera. Adobe Photoshop software was used to superimpose the images and to convert them to gray levels. Due to a specific labeling pattern, which corresponds to N banding, each barley chromosome can be identified. Images of nine chromosomes of each type, showing the range of sizes and labeling pattern, are shown.

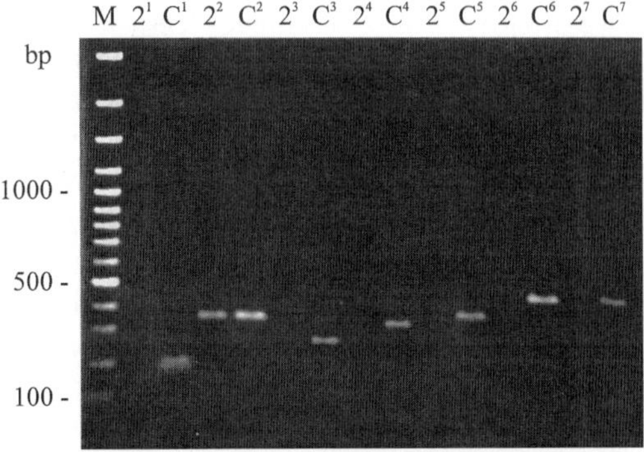

Fig. 9 Identification and analysis of purity of chromosome 2 fraction sorted from *Hordeum vulgare* ($2n = 14$) translocation line T2-6y. Agarose gel electrophoresis was performed after PCR with DNA of sorted chromosomes and primers specific for each of the seven *H. vulgare* chromosomes. The reactions were performed with 500 (number 2) chromosomes (lanes marked 2), and 3500 chromosomes (~500 chromosomes of each type) were used as positive control (lanes marked C). The specificity of primers for individual chromosomes are shown as superscripts. Note that besides the positive controls, amplification product was obtained only with primers specific for chromosome 2. DNA ladder (100 bp) was used as the size marker (M).

Chromosomes isolated according to the protocol outlined here are mechanically stable and withstand two-step sorting, which may be required in some cases to achieve a high purity of the sorted fraction (Lucretti *et al.*, 1993). The extracted DNA is of high molecular weight and suitable for gene mapping and construction of chromosome-specific DNA libraries. In *V. faba,* sorted chromosomes were used for PCR with sequence-specific primers to localize seed-specific protein genes (Macas *et al.*, 1993) and, for the first time in plants, to construct a complete set of chromosome-specific DNA libraries (Macas *et al.*, 1996). The use of reconstructed karyotypes permitted sorting of all chromosomes within karyotype and enabled subchromosomal gene localization. Chromosome-specific DNA libraries could be effectively enriched for microsatellites (Kobližková *et al.*, 1998). This opened a way for effective production of molecular markers needed to increase the density of genetic maps. The procedure for isolation of large quantities of intact chromosomes also stimulated progress in other areas of plant chromosome research. Isolated chromosomes were shown suitable for scanning electron microscopy (Schubert *et al.*, 1993), for high resolution mapping of DNA sequences using *in situ* hybridization (Fuchs *et al.*, 1994) and PRINS (Kubaláková *et al.*, 1997), and for analysis of chromosomal proteins (Binarová *et al.*, 1998).

In addition to the ability to modify the procedure to other crops, the future of plant flow cytogenetics will depend on the capability to discriminate and sort specific chromosomes. At present, the use of various chromosomal stocks (e.g., translocation, deletion) appears to be the most powerful approach. The procedure for specific labeling of chromosomes in suspension (PRINSES) developed in *V.*

faba (Macas *et al.*, 1995; Pich *et al.*, 1995), which permits chromosome sorting independent of their size, needs to be modified for other crops. Furthermore, the method has potential to be used for accurate quantification of the number of repeats of repetitive sequences on individual chromosomes within a karyotype.

Provided the purity of the sorted chromosome fraction will approach 100% (as shown here for *H. vulgare*), flow cytogenetics will be more convenient for physical mapping using PCR when compared to chromosome microdissection. However, the real power of the method lies in a possibility to sort large quantities of chromosomes needed for direct cloning and construction of large insert DNA libraries (e.g., YAC or BAC). Although the attempts to use sorted chromosomes to prepare chromosome painting probes have been unsuccessful (Fuchs *et al.*, 1996), other approaches to isolate chromosome-specific probes and paints should be tested. Construction of an artificial plant chromosome (cf. De Jong *et al.*, 1998) and a limited gene transfer via chromosome transplantation (Houben and Schlegel, 1991) represent two additional areas of potential application of flow cytogenetics.

Acknowledgments

We are grateful to Prof. G. Künzel for the supply of seeds of *H. vulgare* translocation line T2-6y and Dr. L. Korzun for chromosome-specific PCR primers. We thank Dr. I Schubert and Dr. M. Vavák for the supply of *V. faba* seeds, Dr. M. J. Ambrose and Dr. I. Hasalová for the *P. sativum* seeds, Prof. N. Jones and Dr. F. Machaň for *S. cereale* seeds, and P. Svačina for *H. vulgare* seeds. We also thank Dr. J. Číhalíková, Dr. P. Neumann, and Dr. J. Vrána for chromosome isolation and analysis, and O. Blahoušek, M.Sc. for photography. The excellent technical assistance of J. Weiserová, B.Sc., and Mrs. R. Tušková is gratefully acknowledged. A gift sample of amiprophos-methyl from the Mobay Corporation is acknowledged. This work was supported by a research grant (No. 521/96/K117) from the Grant Agency of the Czech Republic, and a grant (No. 6046) of the National Agency for Agricultural Research, Czech Republic.

References

Arumuganathan, K., Slattery, J. P., Tanksley, S. D., and Earle, E. D. (1991). Preparation and flow cytometric analysis of metaphase chromosomes of tomato. *Theor. Appl. Genet.* **82,** 101–111.

Arumuganathan, K., Martin, G. B., Telenius, H., Tanksley, S. D., and Earle, E. D. (1994). Chromosome 2-specific DNA clones from flow-sorted chromosomes of tomato. *Mol. Gen. Genet.* **242,** 551–558.

Bennett, M. D., and Smith, J. B. (1976). Nuclear DNA amounts in angiosperms. *Philos. Trans. R. Soc. London B* **274,** 227–274.

Binarová, P., Hause, B., Doležel, J., and Dráber, P. (1998). Association of γ-tubulin with kinetochore/centromeric region of plant chromosomes. *Plant J.* **14,** 751–757.

Boschman, G. A., Manders, E. M. M., Rens, W., Slater, R., and Aten, J. A. (1992). Semiautomated detection of aberrant chromosomes in bivariate flow karyotypes. *Cytometry* **13,** 469–477.

Carrano, A. V., Gray, J. W., Langlois, R. G., Burkhart-Schultz, K. J., and Van Dilla, M. A. (1979). Measurement and purification of human chromosomes by flow cytometry and sorting. *Proc. Natl. Acad. Sci. U.S.A.* **76,** 1382–1384.

Carter, N. P. (1994). Cytogenetic analysis by chromosome painting. *Cytometry* **18,** 2–10.

Conia, J., Bergounioux, C., Perennes, C., Muller, P., Brown, S., and Gadal, P. (1987). Flow cytometric analysis and sorting of plant chromosomes from *Petunia hybrida* protoplasts. *Cytometry* **8,** 500–508.

Conia, J., Bergounioux, C., Brown, S., Perennes, C., and Gadal, P. (1988). Caryotype en flux biparametrique de *Petunia hybrida.* Tri du chromosome numero I. *C. R. Acad. Sci. Paris* **307,** 609–615.

Conia, J., Muller, P., Brown, S., Bergounioux, C., and Gadal, P. (1989). Monoparametric models of flow cytometric karyotypes with spreadsheet software. *Theor. Appl. Genet.* **77,** 295–303.

Cooke, A., Gillard, E. F., Yates, J. R. W., Mitchell, M. J., Aitken, D. A., Weir, D. M., Affara, N. A., and Ferguson-Smith, M. A. (1988). X Chromosome deletions detectable by flow cytometry in some patients with steroid sulphatase deficiency (X-linked ichtyosis). *Hum. Genet.* **79,** 49–52.

Číhalíková, J., Lysák, M., and Doležel, J. (1998). Cell cycle synchronisation and chromosome isolation in barley (*Hordeum vulgare*). *In* "Plant Cytogenetics" (J. Maluszynska, ed.). Proceedings of the Spring Symposium, pp. 153–158. Wydawnictwo Uniwersytetu Šlaskiego, Katowice, Poland.

De Jong, G., Telenius, A., Telenius, H., Perez, C., and Hadlaczky, G. (1998). Large-scale isolation and purification of mammalian artificial chromosomes. *Cytometry* **9**(Suppl), 88.

De Laat, A. M. M., and Blass, J. (1984). Flow-cytometric characterization and sorting of plant chromosomes. *Theor. Appl. Genet.* **67,** 463–467.

De Laat, A. M. M., and Schel, J. H. N. (1986). The integrity of metaphase chromosomes of *Haplopappus gracilis* (Nutt.) Gray isolated by flow-cytometry. *Plant Sci.* **47,** 145–151.

Dixon, S. C., Miller, N. G. A., Carter, N. P., and Tucker, E. M. (1992). Bivariate flow cytometry of farm animal chromosomes: A potential tool for gene mapping. *Anim. Genet.* **23,** 203–210.

Doležel, J. (1991). KARYOSTAR: Microcomputer program for modelling of monoparametric flow karyotypes. *Biologia* **46,** 1059–1064.

Doležel, J., and Lucretti, S. (1995). High-resolution flow karyotyping and chromosome sorting in *Vicia faba* lines with standard and reconstructed karyotypes. *Theor. Appl. Genet.* **90,** 797–802.

Doležel, J., Číhalíková, J., and Lucretti, S. (1992). A high-yield procedure for isolation of metaphase chromosomes from root tips of *Vicia faba* L. *Planta* **188,** 93–98.

Doležel, J., Lucretti, S., and Schubert, I. (1994). Plant chromosome analysis and sorting by flow cytometry. *Crit. Rev. Plant Sci.* **13,** 275–309.

Doležel, J., Lucretti, S., and Macas, J. (1995a). Flow cytometric analysis and sorting of plant chromosomes. *In* "Kew Chromosome Conference IV" (P. Brandham and M. D. Bennett, eds.), pp. 185–200. Royal Botanic Gardens, Kew, England.

Doležel, J., Číhalíková, J., Macas, J., and Jones, R. N. (1995b). Isolation and flow cytometric analysis of mitotic chromosomes in rye. *Chrom. Res.* **3**(Suppl. 1), 99.

Doležel, J., Číhalíková, J., Weiserová, J., and Lucretti, S. (1999). Cell cycle synchronization in plant root meristems. *Methods Cell Sci.* **21,** 95–107.

Doleželová, M., Macas, J., Lucretti, S., and Doležel, J. (1996). The use of *Vicia faba* hairy root cultures for preparation of chromosome suspensions. *Biologia* **51**(Suppl. 4), 159.

Fuchs, J., Joos, S., Lichter, P., and Schubert, I. (1994). Localization of vicilin genes on field bean chromosome II by fluorescent *in situ* hybridization. *J. Hered.* **85,** 487–488.

Fuchs, J., Houben, A., Brandes, A., and Schubert, I. (1996). Chromosome 'painting' in plants—a feasible technique? *Chromosoma* **104,** 315–320.

Gray, J. W., Carrano, A. V., Steinmetz, L. L., Van Dilla M. A., Moore II, D. H., Mayall, B. H., and Mendelsohn, M. L. (1975). Chromosome measurement and sorting by flow systems. *Proc. Natl. Acad. Sci. U.S.A.* **72,** 1231–1234.

Gualberti, G., Doležel, J., Macas, J., and Lucretti, S. (1996). Preparation of pea (*Pisum sativum* L.) chromosome and nucleus suspensions from single root tips. *Theor. Appl. Genet.* **92,** 744–751.

Houben, A., and Schlegel, R. (1991). Chromosomen-Transfer bei Pflanzen. *Wissenschaft Fortschrift* **41,** 358–360.

Kaeppler, H. F., Kaeppler, S. M., Lee, J.-H., and Arumuganathan, K. (1997). Synchronization of cell division in root tips of seven major cereal species for high yields of metaphase chromosomes for flow-cytometric analysis and sorting. *Plant Mol. Biol. Rep.* **15,** 141–147.

Koblížková, A., Doležel, J., and Macas, J. (1998). Subtraction with 3′ modified oligonucleotides eliminates amplification artifacts in DNA libraries enriched for microsatellites. *BioTechniques* **25,** 32–38.

Kubaláková, M., and Doležel, J. (1998). Optimisation of detection telomeric sequences in broad bean (*Vicia faba* L.) chromosomes by cycling-PRINS. *Biol. Plant.* **41,** 177–184.

Kubaláková, M., Macas, J., and Doležel, J. (1997). Mapping of repeated DNA sequences in plant chromosomes by PRINS and C-PRINS. *Theor. Appl. Genet.* **94,** 758–763.

Langlois, R. G., Yu, L. C., Gray, J. W., and Carrano, A. V. (1982). Quantitative karyotyping of human chromosomes by dual beam flow cytometry. *Proc. Natl. Acad. Sci. U.S.A.* **79,** 7876–7880.

Lebo, R. V. (1982). Chromosome sorting and DNA sequence localization: A review. *Cytometry* **3,** 145–154.

Lee, J.-H., Arumuganathan, K., Kaeppler, S. M., Kaeppler, H. F., and Papa, C. M. (1996). Cell synchronization and izolation of metaphase chromosomes from maize (*Zea mays* L.) root tips for flow cytometric analysis and sorting. *Genome* **39,** 697–703.

Lee, J.-H., Arumuganathan, K., Yen, Y., Kaeppler, S., Kaeppler, H., and Baezinger, P. S. (1997). Root tip cell cycle synchronization and metaphase-chromosome isolation suitable for flow sorting in common wheat (*Triticum aestivum* L.). *Genome* **40,** 633–638.

Lucretti, S., and Doležel, J. (1995). Cell cycle synchronization, chromosome isolation, and flow-sorting in plants. *In* "Methods In Cell Biology" (D. W. Galbraith, D. P. Bourque, and H. J. Bohnert, eds.), Vol. 50, pp. 61–83. Academic Press, New York.

Lucretti, S., and Doležel, J. (1997). Bivariate flow karyotyping in broad bean (*Vicia faba*). *Cytometry* **28,** 236–242.

Lucretti, S., Doležel, J., Schubert, I., and Fuchs, J. (1993). Flow karyotyping and sorting of *Vicia faba* chromosomes. *Theor. Appl. Genet.* **85,** 665–672.

Lysák, M. A., Číhalíková, J., Kubaláková, M., Šimková, H., Künzel, G., and Doležel, J. (1999). Flow karyotyping and sorting of mitotic chromosomes of barley (*Hordeum vulgare* L.). *Chrom. Res.* **7,** 431–444.

Macas, J., Doležel, J., Lucretti, S., Pich, U., Meister, A., Fuchs, J., and Schubert, I. (1993). Localization of seed storage protein genes on flow-sorted field bean chromosomes. *Chrom. Res.* **1,** 107–115.

Macas, J., Doležel, J., Gualberti, G., Pich, U., Schubert, I., and Lucretti, S. (1995). Primer-induced labeling of pea and field bean chromosomes *in situ* and in suspension. *Bio Techniques* **19,** 402–408.

Macas, J., Gualberti, G., Nouzová, M., Samec, P., Lucretti, S., and Doležel, J. (1996). Construction of chromosome-specific DNA libraries covering the whole genome of field bean (*Vicia faba* L.). *Chrom. Res.* **4,** 531–539.

Marthe, F., and Künzel, G. (1994). Localization of translocation breakpoints in somatic metaphase chromosomes of barley. *Theor. Appl. Genet.* **89,** 240–248.

Neumann, P., Lysák, M., Doležel, J., and Macas, J. (1998). Isolation of chromosomes from *Pisum sativum* L. hairy root cultures and their analysis by flow cytometry. *Plant Sci.* **137,** 205–215.

Otto, F. J. (1988). Assessment of persisting chromosome aberrations by flow karyotyping of cloned Chinese hamster cells. *Z. Naturforsch. C: Biosci* **43,** 948–954.

Pich, U., Meister, A., Macas, J., Doležel, J., Lucretti, S., and Schubert, I. (1995). Primed *in situ* labelling facilitates flow sorting of similar sized chromosomes. *Plant J.* **7,** 1039–1044.

Schubert, I., and Rieger, R. (1991). Catalogue of chromosomal and morphological mutants of faba bean in the Gatersleben Collection. *FABIS Newslett.* **28/29,** 14–22.

Schubert, I., Doležel, J., Houben, A., Schertan, H., and Wanner, G. (1993). Refined examination of plant metaphase chromosome structure at different levels made feasible by new isolation methods. *Chromosoma* **102,** 96–101.

Schwarzacher, T., Wang, M. L., Leitch, A. R., Miller, N., Moore, G., and Heslop-Harrison, J. S. (1997). Flow cytometric analysis of the chromosomes and stability of a wheat cell-culture line. *Theor. Appl. Genet.* **94,** 91–97.

Simpson, P. R., Newman, A.-M., Davies, D. R., Ellis, T. H. N., Matthews, P. M., and Lee, D. (1990). Identification of translocations in pea by *in situ* hybridization with chromosome-specific DNA probes. *Genome* **33,** 745–749.

Tempé, J., and Casse-Delbart, F. (1989). Plant gene vectors and genetic transformation: Agrobacterium Ri plasmids. *In* "Cell Culture and Somatic Cell Genetics of Plants" (F. Constabel and L. K. Vasil, eds.), Vol. 6, pp. 25–49. Academic Press, New York.

Van Dilla, M. A., and Deaven, L. L. (1990). Construction of gene libraries for each human chromosome. *Cytometry* **11,** 208–218.

Veuskens, J., Marie, D., Hinnisdaels, S., and Brown, S. C. (1992). Flow cytometry and sorting of plant chromosomes. *In* "Flow Cytometry and Cell Sorting" (A. Radbruch, ed.), pp. 177–188. Springer-Verlag, Berlin and Heidelberg.

Veuskens, J., Marie, D., Brown, S. C., Jacobs, M., and Negrutiu, I. (1995). Flow sorting of the Y sex-chromosome in the dioecious plant *Melandrium album. Cytometry* **21,** 363–373.

Wang, M. L., Leitch, A. R., Schwarzacher, T., Heslop-Harrison, J. S., and Moore, G. (1992). Construction of a chromosome-enriched HpaII library from flow-sorted wheat chromosomes. *Nucleic Acids Res.* **20,** 1897–1901.

CHAPTER 31

Quantitative DNA Fiber Mapping

Heinz-Ulli G. Weier

Department of Subcellular Structures
Life Sciences Division
E. O. Lawrence Berkeley National Laboratory
University of California, Berkeley
Berkeley, California 94720

I. Introduction
II. Materials
    A. Chemicals and Buffers
    B. Instruments
III. Protocols
    A. Preparation of Aminopropyltriethoxysilane Slides
    B. Purification of High Molecular Weight DNA
    C. Preparation of DNA Fibers
    D. Generation of Probes from Cloned DNA Fragments
    E. Generation of Probes by *in Vitro* DNA Amplification
    F. Probe Labeling via Random Priming
    G. Fluorescence *in Situ* Hybridization
    H. Digital Image Analysis
IV. Critical Aspects of the Procedure
V. Results and Discussion
    A. Gene Mapping by Hybridization onto Genomic DNA Fibers
    B. Mapping Cloned Probes onto Linear, Randomly Broken, or Circular DNA Molecules
    C. cDNA Mapping
References

I. Introduction

High resolution physical maps are indispensable for large-scale, cost-effective gene discovery. The construction of such maps of the human genome and model organisms therefore is one of the major goals of the human genome project

METHODS IN CELL BIOLOGY, VOL. 64

(Collins and Galas, 1993). The precise localization of cloned DNA fragments within much larger genomic fragments and knowledge about the extent of overlap between two clones are needed to assemble high resolution physical maps. As demonstrated in this chapter, fluorescence *in situ* hybridization (FISH) provides this critical information.

Isolation of DNA from cell nuclei and preparation of some sort of DNA "fibers," that is, chromatin or 30-nm fibers representing bundles of DNA (Heng *et al.*, 1992; Parra and Windl, 1993; Haaf and Ward, 1994), improves accessibility of the DNA target for probes as well as detection reagents and thus increases the hybridization efficiency. Furthermore, if the DNA molecules can be stretched in some way, they may provide linear templates for visual mapping (Heiskanen *et al.*, 1994). FISH applied to most preparations of decondensed nuclear or isolated cloned DNA allows visualization of probe overlap and provides some information about the existence and size of gaps between clones (Heiskanen *et al.*, 1994; Florijn *et al.*, 1995). However, none of these techniques provides sufficiently accurate information about the extent of clone overlap or the separation between elements in the map because the chromatin onto which clones are mapped is condensed to varying degrees from site to site in these preparations.

We demonstrated previously that cloned DNA fragments can readily be mapped by FISH onto DNA molecules prepared by the hydrodynamic action of a receding meniscus, and, referring to its quantitative nature, we termed our technique quantitative DNA fiber mapping (QDFM) (Weier *et al.*, 1995a).

In QDFM, a solution of purified DNA molecules is placed on a flat surface prepared so that the DNA molecules slowly attach at one or both ends. The DNA solution is then spread over a larger area by placing a coverslip on top. DNA molecules are allowed to bind to the surface. During drying, the molecules are straightened and uniformly stretched by the hydrodynamic action of the receding meniscus. Molecules prepared in this manner are stretched with remarkable homogeneity. We estimate a properly stretched molecule should extend about 2.3 kb/μm, that is, approximately 30% over the length predicted for a double-stranded DNA molecule of the same size (Bensimon *et al.*, 1994; Weier *et al.*, 1995a; Hu *et al.*, 1996). We previously showed that QDFM can be applied to DNA molecules ranging in size from a few kilobases to more than 1 Mbp, which allowed us to map small probes with near kilobase resolution onto entire yeast chromosomes and large (mega) YAC (yeast artificial chromosome) clones from the CEPH/Genethon library (Chumakov *et al.*, 1992; Wang *et al.*, 1996; Duell *et al.*, 1997).

Applications of QDFM extend beyond map assembly and can provide valuable information for quality control, clone validation, definition of a minimal tiling path, as well as for the sequence assembly process. Furthermore, due to the high hybridization efficiency obtained with DNA fibers, QDFM is also the method of choice to optically map expressed sequences in limited genomic intervals.

II. Materials

A. Chemicals and Buffers

3-Aminopropyltriethoxy silane (APS, Sigma, St. Louis, MO).

β-Mercaptoethanol (Sigma).

β-Agarase (New England Biolabs, Beverly, MA).

Agarose (GIBCO/BRL, Life Technologies, Rockville, MD).

Antibodies against digoxigenin, rhodamine-conjugated, made in sheep (Boehringer Mannheim, Indianapolis, IN) stock solution is 1 mg/ml in PNM, dilute 1:50 with PNM prior to use. Store at 4°C.

Antibodies against fluorescein isothiocyanate (FITC), made in mouse (Dako, Carpinteria, CA) stock solution is 1 mg/ml in PNM (see p. 37), dilute 1:50 prior to use. Store at 4°C.

Anti-mouse antibodies, FITC conjugated, made in horse (Vector Laboratories, Burlingame, CA), stock solution is 1 mg/ml in PNM, dilute 1:50 prior to use. Store at 4°C.

Anti-avidin antibodies, biotinylated, made in goat (Vector Laboratories), stock solution is 1 mg/ml in PNM, dilute 1:50 prior to use. Store at 4°C.

Avidin conjugated to 7-amino-4-methyl coumarin-3-acetic acid (AMCA) (Vector Laboratories), stock solution is 2 mg/ml in PNM, dilute 1:500 prior to use. Store at 4°C.

Chloroform/isoamyl alcohol: 24/1 (v/v) (GIBCO/BRL).

4′,6-Diamino-2-phenylindole (DAPI) (Calbiochem, La Jolla, CA), 0.05 μg/ml in antifade solution. Store at −20°C.

dATP, dCTP, dGTP, dTTP: 100 mM each (Boehringer Mannheim or Pharmacia Biotech, Piscataway, NJ). Store at −20°C.

Digoxigenin-11-dUTP (dig-dUTP), 1 mM (Boehringer Mannheim). Store at −20°C.

10X dNTP mix: dATP, dCTP, dGTP, and dTTP, 10 mM each.

EDTA (ethylenediaminetetraacetic acid), 0.5 M (pH 8.0) (GIBCO/BRL).

Ethidium bromide (EB), 10 mg/ml (GIBCO/BRL).

Fluorescein avidin DCS (avidin-FITC, Vector Laboratories), dilute stock solution (2 mg/ml in PNM) 1:100 with PNM. Store at 4°C.

Fluorescein-12-dUTP, 1mM (Boehringer Mannheim). Store at −20°C.

Formamide (FA, GIBCO/BRL or Boehringer Mannheim). Store according to the manufacturer's recommendation, that is, at −20°C or 4°C.

Glycogen, 20 mg/ml (Boehringer Mannheim). Store at −20°C.

Human COT1 DNA, 1 mg/ml (GIBCO/BRL). Store at −20°C.

Lambda phage DNA (Boehringer Mannheim or GIBCO/LTI). Prepare 2.5 ng/μl in 2X SSC (see p. 37). Store at 4°C.

Low melting point (LMP) agarose (Bio-Rad, Hercules, CA).

Phenol/chloroform/isoamyl alcohol: 25/24/1 (v/v/v)(GIBCO/LTI). Store at 4°C.

Random-priming kit: BioPrime kit (GIBCO/BRL). Store at −-20°C.

Proteinase K, 20 mg/ml in 10 mM Tris-HCl, pH 7.5 (Boehringer Mannheim). Store at −20°C.

RNase (Boehringer Mannheim), DNase-free: boil at 100°C for 10 min, aliquot, and store at −20°C.

Salmon sperm DNA, 20 mg/ml, (3′–5′). Store at −20°C.

Sodium dodecyl sulfate (SDS) (Na salt, Sigma): 10% in water.

Thermus aquaticus (*Taq*) DNA polymerase, 5 U/μl (Perkin-Elmer, Foster City, CA). Store at −20°C.

Ultrapure water (Mallinckrodt, Phillipsburg, NJ, Catalog No. H453).

Yeast artificial chromosome (YAC) library (Research Genetics, Huntsville, AL). Store at −80°C.

YOYO-1, stock is 1 mM in DMSO (Molecular Probes, Eugene, OR). Dilute 1:1000 with water prior to use. Store at −20°C and discard diluted dye after 1 week.

Zymolase, 70,000 U/g, 10 mg/ml in 50 mM KH$_2$PO$_4$ (pH 7.8), 50% glycerol (Sigma). Store at −20°C.

Buffers and Other Solutions:

Acid hydrolized casein (AHC) medium (BIO 101): Add 36.7 g of AHC powder per liter of purified water. Autoclave at 121°C for 15 min.

AHC agar medium (BIO 101): Add 53.7 g of AHC powder per liter of purified water. Autoclave at 121°C for 15 min. Cool to 50°C, mix well, and pour plates. Store plates at 4°C.

Alkaline lysis (AL) solutions (Birnboim and Doly, 1979) sufficient for 12 preparations using 20 ml cell culture.

  AL Solution I: 50 mM glucose, 10 mM EDTA, 25 mM Tris-HCl, pH 8.0. Add 4 ml of 0.5 M glucose, 0.8 ml of 0.5 M EDTA, and 1 ml of 1 M Tris to 34.2 ml ultrapure water. Store at 4°C.
  AL Solution II: 0.2 N NaOH, 1% SDS. Add 1.4 ml of 10 N NaOH, 7 ml of 10% SDS to 61.6 ml ultrapure water.
  AL Solution III: 3 M sodium acetate, pH 4.8.

Antifade solution: 1% *p*-phenylenediamine, 15 mM NaCl, 1 mM H$_2$PO$_4$, pH 8.0, 90% glycerol. Store at −80°C.

Cell fixative: acetic acid/methanol, 1/3 (v/v). Make fresh before use.

DB 0.5 solution: 0.5 M EDTA (pH 8.0), 1.0% *N*-lauroyl sarcosine (Sigma), 0.5 mg/ml proteinase K (Boehringer Mannheim).

Denaturing Solution: 70% FA, 2× SSC, pH 7.0. Prepare fresh at least every 2 weeks.

ES Buffer: 0.5 M EDTA (pH 8.0), 1% sarcosyl.

Gel loading dye: 1% bromophenol blue in 30% glycerol.

Hybridization master mix (MM 2.1): 14.3% (w/v) dextran sulfate, 78.6% FA, 2.9× SSC, pH 7.0. For 10 ml MM 2.1, mix 1.45 ml of 20× SSC with 0.7 ml ultrapure water, dissolve 1.43 g dextran sulfate (Calbiochem), incubate overnight, then add 7.86 ml formamide. Divide into aliquots in 1.5-ml microcentrifuge tubes and store at −20°C.

Lysis buffer: 1% Triton X-100, 20 mM Tris-HCl, 2 mM EDTA, pH 8.5.

Modified nucleotide mix (10X) for labeling in combination with 1 mM dig-dUTP or FITC-dUTP: Combine 5 μl each of 100 mM dATP, 100 mM dGTP, and 100 mM dCTP with 2.5 μl of 1 M Tris-HCl, pH 7.5, 0.5 μl of 0.5 M EDTA, pH 8.0 (GIBCO/LTI), and 232 μl ultrapure water for a total of 250 μl. Store at −20°C. The concentration of nucleoside triphosphates is 2 mM each.

PNM: Dissolve 5 g of nonfat dry milk (Carnation, Glendale, CA) in 100 ml PN buffer (PN buffer is 0.1 M sodium phosphate, pH 8.0, 0.1% Nonidet P-40), incubate at 50°C overnight, add 1/50 volume sodium azide, spin at 1000 g for 30 min, aliquot clear supernatant into 1.5-ml tubes, and store at 4°C. Spin at 2000 g for 30 sec prior to use.

SCE: 1 M sorbitol, 0.1 M sodium citrate, 10 mM EDTA, pH 7.8.

Slide blocking solution (5× SSC containing 2% Blocking Reagent, 0.1% N-lauroyl sarcosine): Combine 0.05 g N-lauroyl sarcosine (sodium salt, Sigma) and 1 g Blocking Reagent (Boehringer Mannheim, Catalog No. 1096-176) with 12.5 ml of 20× SSC (pH 7.0), add 30 ml ultrapure water, heat to 60°C while stirring, and bring the final volume to 50 ml with ultrapure water, when the Blocking Reagent is dissolved. Divide into aliquots in 1.5-ml tubes, spin at 2000 rpm for 10 min, and store at 4°C.

SSC: 20× SSC is 3 M NaCl, 300 mM sodium citrate, pH 7.0.

10× *Taq* amplification buffer: 500 mM KCl, 100 mM Tris-HCl, pH 8.3, 10 mM MgCl$_2$. Store at −20°C.

TBE (Tris/borate/EDTA) buffer: 10X is 890 mM Tris base, 890 mM boric acid, 20 mM EDTA.

TE (Tris/EDTA) buffer: 1X is 10 mM Tris-HCl, 1 mM EDTA, pH 7.4, 7.5, or 8.0.

TE 50 buffer: 10 mM Tris-HCl, 50 mM EDTA, pH 7.8.

Tris-HCl [tris(hydroxymethyl)aminomethane]: 1 M, pH 7.5 or 8.0.

B. Instruments

Centrifuge (MP4R, IEC, Needham Heights, MA)

Dry bath (heat block) (Model 2001, Labline Instruments, Melrose Park, IL)

Fluorescence microscope (Axioskop, Carl Zeiss) equipped with 40× and 63× oil immersion lenses

Incubator oven (set to 37°C)

Pulsed field gel electrophoresis (PFGE) system (Bio-Rad)

Shaking incubators (New Brunswick Scientific, Edison, NJ): 30°C for yeast cell culture, 37°C for culture of *Escherichia coli*

Thermal cycler for *in vitro* DNA amplification

Water bath (Model 188, Precision Scientific, Winchester, VA)

III. Protocols

A. Preparation of Aminopropyltriethoxysilane Slides

1. Slide Preparation

1. Preclean glass slides mechanically by repeated rubbing with wet cheesecloth.
2. Rinse several times with ultrapure water.
3. Immerse slides in boiling ultrapure water for 10 min.
4. Air dry.
5. Immerse slides in 18 *M* sulfuric acid (T.J. Baker, Phillipsburg, NJ) for at least 30 min to remove organic residues.
6. Immerse in boiling water for 1–2 min.
7. Air dry and store until further use.

2. Silane Modification

1. Immerse precleaned dry slides in a solution of 0.1% APS in 95% ethanol for 10 min.
2. Remove slides from the silane solution.
3. Rinse several times with water, and immerse in ultrapure water for 2 min.
4. Dehydrate by immersing in absolute ethanol.
5. Dry in air at 65°C on a heat plate standing upright for 10 min.
6. Store slides for 2–6 weeks at 4°C in a sealed box under nitrogen prior to use.

Coverslip preparation is performed as described above for slides. Briefly, coverslips are rinsed with distilled water and dehydrated in 100% ethanol. Coverslips are derivatized with a 0.1% solution of APS in 95% ethanol for 2 min.

B. Purification of High Molecular Weight DNA

All YAC clones are from the CEPH/Genethon library (Weissenbach *et al.*, 1992; Cohen *et al.*, 1993; Munné *et al.*, 1998). Information about the exact size of nondeleted clones is available for most clones from either the CEPH/Genethon www server at URL http://www.genethon.fr/genethon_en.html/ or the Massachusetts Institute of Technology (MIT) server (URL http://www.genome.wi.mit.edu/) (Hudson *et al.*, 1995). Agarose plug preparation and pulsed field gel electrophoresis using a CHEF electrophoresis system (Bio-Rad) followed standard protocols. Typically, five to fifteen individual YAC colonies are tested to account for deletions. In most cases, the largest clone carries the least deletion(s).

The P1/PAC/BAC (Ioannou *et al.*, 1994; Shizuya *et al.*, 1992) clones show fewer deletions, and it is often sufficient to pick two to three colonies from a plate, grow the cells overnight in Luria broth (LB) medium (Maniatis *et al.*, 1986), and extract the DNA using the alkaline lysis protocol. The DNA can then be loaded directly into the PFGE well using a common gel loading dye.

The DNA is recovered from the gel by excising the appropriate band with a razor blade. High molecular weight DNA is then prepared by β-agarase digestion (New England Biolabs) of the gel slices.

1. Pulsed Field Gel Electrophoresis

a. Preparation of Plugs Containing YACs

1. Spin down cells from 5 ml AHC medium at 400 rpm for 6 min. Resuspend cells in 0.5 ml of 0.125 M EDTA, pH 7.8. Spin again and remove supernatant.
2. Resuspend the cell pellet (~70 μl) in 500 μl of SCE. Mix with an equal volume of 1.5% LMP agarose preheated to 43°C. Quickly pipette up and down, then vortex for 1–2 sec to mix. Pipette into plug molds (Bio-Rad) and allow to solidify at room temperature or on ice.
3. Remove plugs from molds, incubate samples in 2 ml SCE containing 100 μl of zymolase, and shake at 150 rpm at 30°C for 2.5 hr to overnight.
4. Remove SCE and add 2 ml of ES containing 100 μl of proteinase K (20 mg/ml). Shake 5 hr to overnight at 50°C.
5. Remove ES and rinse five times with 6 ml of TE 50 for 30 min each rinse. Store the plugs at 4°C until use.

b. PFGE Running Conditions

1. YACs: voltage gradient, 6 V/cm; switching interval, 79 sec forward, 94 sec reverse; running time, 38 hr; agarose concentration, 1.0% LMP agarose; running temperature, 14°C; running buffer, 0.5× TBE.
2. P1/PAC/BAC clones: voltage gradient, 6 V/cm; switching interval, 2 sec forward, 12 sec reverse; running time, 18 hr; agarose concentration, 1.0% LMP agarose; running temperature, 14°C; running buffer, 0.5× TBE.

c. For Probe Production and Determination of Optimal PFGE Conditions

Stain the gel with EB (0.5 μg/ml in water) and cut out a gel slice containing the target band. Transfer gel slice to a 14-ml polystyrene tube (Applied Scientific,

South San Francisco, CA, Catalog No. AS-2264). Wash slice with ultrapure water for 30 min, and then wash with $1\times$ agarase buffer for 30 min.

d. For High Molecular Weight DNA Isolation

Run duplicate samples on the right and left side of the gel, respectively. After the predetermined run time, cut gel in half, and stain one half with EB. Measure the migrated distance, cut a gel slice at the approximate position from the unstained half, and proceed as described in Section III,B,1,c.

2. Recovery of High Molecular Weight DNA from Gel Slices

The procedure for recovering high molecular weight DNA from gel slices is as follows.

1. Melt the gel completely by incubating 10 min at 85°C.
2. Transfer the molten agarose to a 43°C water bath.
3. Add 1 μl β-agarase for every 25 μl of molten agarose.
4. Incubate at 43°C for 2 hr.
5. Add an equal volume of 200 mM NaCl.
6. Store the sample at 4°C until used.

3. Genomic DNA

Genomic high molecular weight (HMW) DNA is isolated from exponentially growing human cells such as the C32 melanoma cell line (American Type Culture Collection, Rockville, MD) or diploid fibroblast cells using standard procedures. Briefly, about 5×10^5 cells are washed in phosphate-buffered saline (PBS). The cells are then resuspended in 0.5 ml of PBS and mixed with 1.2% low melting point agarose previously melted in PBS and allowed to cool down to 43°C. Aliquots of 100 μl are dispensed into plug molds and allowed to set for 30 min at 4°C. Agarose plugs are then placed into DB 0.5 solution and incubated overnight at 50°C. Next, plugs are washed four to six times for 30 min each in 50 mM Tris-HCl, 1 mM EDTA and stored at 4°C. The HMW DNA is released by digestion of the plugs with β-agarase according to the manufacturer's instructions (New England Biolabs).

C. Preparation of DNA Fibers

We have used different methods for the stretching of DNA on APS-pretreated glass and mica surfaces. In our experience, the quality of the resulting DNA fibers is influenced more by the quality of DNA preparation and by the properties of the modified glass surface than by the method of DNA stretching. In a typical experiment, 1–2 μl of clonal or genomic DNA are mixed with an equal amount of YOYO-1 (1 or 0.1 μM) and 8 μl water. One or two microliters of this diluted DNA is then applied to an untreated coverslip, which is placed (DNA side down) on the APS-treated slide or coverslip. The DNA concentration can now be estimated in the fluorescence microscope using a filter set for FITC and adjusted as needed. As early as after 2 min of incubation at room temperature, the untreated coverslip can be removed slowly from one end, allowing the meniscus between the coverslips to stretch the bound DNA molecules (fibers) in one direction (Hu *et al.*, 1996). Alternatively, the slide or coverslip sandwich can be allowed to dry overnight at room temperature, after which the untreated coverslip is removed. Slides or coverslips carrying DNA fibers are rinsed briefly with water, drained, allowed to dry at room temperature, and "aged" in ambient air at 20°C for 1 week before hybridization.

D. Generation of Probes from Cloned DNA Fragments

1. Alkaline Lysis Protocol and Purification of DNA from P1, PAC, or BAC Clones

This protocol describes the isolation of DNA from ~20-ml overnight cultures using 40-ml Oakridge centrifugation tubes. The protocol can be scaled down to accommodate smaller volumes.

1. Grow culture overnight in ~30 ml LB or Terrific broth (TB) medium containing the recommended amount of antibiotic.
2. Prepare Oakridge tubes. Write the clone identification (ID) on a small piece of tape and stick it to the cap. Spin 18.5 ml of culture at 2000 *g* for 10 min at 4°C and discard the supernatant.
3. Resuspend the pellet in 2340 μl of AL Solution I, then add 100 μl of lysozyme stock (50 mg/ml in 10 m*M* Tris, pH 7.5) to each tube. Lysozyme needs to be stored at −20°C in 100 μl aliquots. Do not refreeze, instead discard the remainder. Incubate tubes for 5 min at room temperature; then, place the tubes on ice.
4. Add 5.2 ml of AL Solution II. The mixture should now become clear. Mix gently by inverting the tubes several times. Incubate for 5 min on ice.
5. Add 3.8 ml of AL Solution III and mix gently by inverting the tubes several times. Incubate for 10 min on ice.
6. Spin for 15 min at high speed (11,500 rpm = 14,000 *g*).
7. Transfer 10.4 ml of supernatant into a new Oakridge tube, add 5.8 ml of isopropanol, and mix gently by inverting tubes several times. Use the old cap (with the ID sticker) on the new tube.
8. Spin for 5 min at ~10,000 *g* and discard the supernatant. Watch the pellet.
9. Wash the pellet in cold 70% ethanol. Let the pellets dry briefly, that is, at ~20–40 min at room temperature or at 37°C.
10. Resuspend the pellet in 0.8 ml of TE buffer and split the volume into two 1.5-ml microcentrifuge tubes.

11. Add 400 μl phenol/chloroform/isoamyl alcohol to each tube. All centrifugations during the following phenol/chloroform extraction are done at 12,000 g.

12. Vortex for 15 sec and spin down for 3 min.

13. Remove most of the bottom layer and spin again for 3 min.

14. Transfer the top layer to new microcentrifuge tubes and add 400 μl chloroform/isoamyl alcohol.

15. Vortex well for 15 sec, spin down for 3 min, and remove most of the bottom layer. Do a second centrifugation for 3 min.

16. Transfer top layer to a new microcentrifuge tube, add 2.5 volumes, that is, 1 ml of 100% ethanol, and let the DNA precipitate for 30 min at $-20°C$.

17. Spin down for 15 min, discard the supernatant, and wash the pellet in ice-cold 70% ethanol, spin again briefly, remove supernatant, and air dry the pellet.

18. Resuspend the pellet in 20–40 μl TE, pH 7.4, containing 10 μg/ml RNase.

19. Incubate 30 min at 37°C (in water bath); then, store at $-20°C$ until used.

2. Preparation of DNA from Yeast Artificial Chromosome Clones

Grow the selected yeast clone (containing the YAC) on AHC agar for 2–3 days at 30°C. Pick colonies from the plates and culture the clones in up to 35 ml AHC medium at 30°C for 2–3 days.

DNA extraction, phenol purification, and alcohol precipitation:

1. Centrifuge cells (in ~35 ml AHC medium) at 2000 g at 4°C for 5 min.

2. Decant the supernatant and resuspend cells in 3 ml total of 0.9 M sorbitol, 0.1 M EDTA, pH 7.5, containing 4 μl β-mercaptoethanol, followed by addition of 100 μl of zymolase (2.5 mg/ml), and then incubate at 37°C for 60 min.

3. Pellet the cells at 2000 g and 4°C for 5 min and decant supernatant.

4. Resuspend pellet in 5 ml of 50 mM Tris, pH 7.4, 20 mM EDTA. Add 0.5 ml of 10% SDS and mix gently. Incubate at 65°C for 30 min.

5. Add 1.5 ml of 5 M potassium acetate and place on ice for 60 min.

6. Spin at 12,000 g for 15 min at 4°C, and transfer the supernatant to a new tube.

7. Mix the supernatant gently with 2 volumes of 100% ethanol by inverting tube a few times. Spin at 12,000 g for 15 min at room temperature.

8. Prepare 12 sets of 1.5-ml microcentrifuge tubes.

9. Decant supernatant and air dry the pellet. Resuspend pellet in 3 ml of 1$\times$ TE, pH 7.5.

10. Transfer the DNA solution to four 1.5-ml microcentrifuge tubes.

11. Add an equal volume of phenol/chloroform/isoamyl alcohol (25/24/1, pH 8.0), vortex well, and spin at high speed (10,000 *g*) for 3 min.

12. Transfer the top layer to new 1.5-ml microcentrifuge tubes and add an equal volume of chloroform/isoamyl alcohol (24/1). Vortex well and centrifuge at high speed (10,000 *g*) for 3 min.

13. Transfer the top layer to new 1.5-ml microcentrifuge tubes. Add 40 μl of RNase (1 mg/ml, DNase free) to each of the four tubes and incubate at 37°C for 30 min.

14. Add 1 volume of isopropanol and gently mix by inversion. Centrifuge at high speed (10,000 *g*) for 20 min.

15. Decant supernatant and wash pellet with 1 volume of cold 70% ethanol, and centrifuge at high speed (10,000 *g*) for 3 min.

16. Decant the 70% ethanol and air dry the pellet.

17. Resuspend pellet in 20–30 μl 1X TE, and measure DNA concentration after the pellet is completely dissolved.

E. Generation of Probes by *in Vitro* DNA Amplification

1. Cloning Vector-Specific Probes

The generation of P1/PAC-, BAC-, and YAC-vector probe DNA takes advantage of the access to published vector sequences. Polymerase chain reaction (PCR) primers are typically designed to amplify fragments of 1100–1400 bp of vector sequence. Several such oligonucleotide pairs have been designed in several laboratories including ours and are used either in single pairs or in combination. The PCR usually follows standard conditions, that is, the buffer contains 1.5 m*M* MgCl$_2$, annealing temperatures range from 50° to 60°C, and 1 unit *Taq* DNA polymerase is used per 50 μl reaction.

On the other hand, the YAC cloning vectors pJs97 and pJs98, cloned in plasmid vectors (GIBCO/BRL), can be used to prepare probes useful to determine the orientation of the YAC insert (Duell *et al.*, 1997). For this purpose, plasmid DNA is extracted using a standard kit (Qiagen, Valencia, CA; FMC, Rockland, ME; or Schleicher & Schuell, Keene, NH) and labeled by random priming as described later.

2. Mixed Base Oligonucleotide-Primed PCR

The DNA probes for counterstaining of the P1/PAC/BAC or YAC DNA fibers are generated by mixed base oligonucleotide primed PCR (sometimes also referred to as DOP-PCR) (Telenius *et al.*, 1992; Weier *et al.*, 1993; Cassel *et al.*, 1997). An aliquot of the appropriate HMW DNA obtained by PFGE for fiber preparation is PCR amplified for a total of 42 cycles with oligonucleotide primers that anneal about every 200–800 nucleotides. In our scheme, we use two different DNA amplification programs. Initially we perform a few manual PCR cycles using a thermolabile DNA polymerase to extend the oligonucleotide primers at a relatively low temperature. Then, the DNA copies prepared in those first cycles are amplified using the thermostable *Taq* DNA polymerase and rapid thermal cycling.

In the first amplification stage, T7 DNA polymerase (Sequenase II, Amersham, Arlington Heights, IL is used in five to seven cycles to extend the mixed base primer JUN1 (5′-CCAAGCTTGCATGCGAATTCNNNNCAGG-3′, N = A, C, G, or T) (Weier *et al.*, 1993) that was annealed at low temperature (Kroisel *et al.*, 1994). Briefly, 2–3 μl of HMW DNA solution is removed from the bottom of each tube and PCR amplified using the following conditions: denaturation at 92°C for 3 min, primer annealing at 20°C for 2 min, and extension at 37°C for 6 min. Sequenase II must be added after each denaturation. (Please see Chapter 34 of this volume, for details.)

In the second amplification stage, 20 μl of the reaction product are then resuspended in a 200 μl *Taq* amplification reaction buffer and amplified with primer JUN15 (5′-CCCAAGCTTGCATGCGAATTC-3′) with the following PCR conditions: denaturation at 94°C for 1 min, primer annealing at 50°C for 1 min, and extension at 72°C for 2 min, repeated for 35 cycles. After precipitation in isopropanol, the product is resuspended in 30 μl of TE buffer. Subsequently, 1.5 μl of this solution is labeled in a 25-μl random priming reaction incorporating FITC-dUTP.

F. Probe Labeling via Random Priming

1. Measurement of DNA Concentration

The concentration of PCR products can be estimated from the agarose gels run to confirm target amplification. If a sufficient amount of clonal or genomic DNA is available, one or two microliters can be used to accurately determine the concentration using Hoechst 33258 fluorometry using a TK100 fluorometer (Hoefer/Pharmacia, San Francisco, CA).

2. Random Priming

1. Add 250 ng of DNA to ultrapure water to a final volume of 7 μl in a 0.5-ml microcentrifuge tube.
2. Boil DNA at 100°C for 5 min, then quickly chill on ice.
3. For labeling with either dig-dUTP or FITC-dUTP, add:

   2.5 μl 10× Modified Nucleotide Mixture
   3.25 μl 1 m*M* dTTP
   1.75 μl Dig-11-dUTP or FITC-12-dUTP (1 m*M*, Boehringer Mannheim, Catalog No. 1093-088)
   10 μl 2.5× Random primers (BioPrime kit, GIBCO, part YO1393)

For labeling the DNA with biotin, add 2.5 μl of 10× dNTP mix provided with the BioPrime kit (containing biotin-14-dCTP), 5 μl ultrapure water, and 10 μl of 2.5× random primers.

4. Mix well and add 0.5 μl DNA polymerase I Klenow fragment (40 units/μl, GIBCO/BRL, Part YO1396).

5. Incubate tube at 37°C for 60–120 min.

6. Add 2.5 μl of 10× stop buffer (GIBCO/BRL, Part YO1107, part of the BioPrime kit).

7. Store probe at −20°C until used.

G. Fluorescence *in Situ* Hybridization

All hybridizations are carried out overnight at 37°C in a moist chamber. Fiber hybridizations include a biotin- or FITC-labeled DNA probe at a comparatively low concentration to counterstain the fibers. This highlights the otherwise invisible DNA fiber and allows competitive displacement by the probes to be mapped along the DNA fiber (Weier *et al.*, 1995a; Duell *et al.*, 1997). Additionally, one or two cloning vector-specific probes are included to allow the assessment of the orientation of the insert.

The hybridization procedure is very similar to protocols used with metaphase spreads:

1. Hybridization mix: combine 1 μl of each probe, 1 μl of human COT1 DNA (optional), 1 μl of salmon or herring sperm DNA, and 7 μl of hybridization master mix (MM 2.1).

2. Apply the hybridization mixture to the slide and coverslip.

3. Denature the slide at 88°–92°C for 90 sec on a hot plate.

4. Transfer the slide to a moist chamber and incubate overnight at 37°C.

Wash and detection steps are not very different from the protocols used for FISH to interphase and metaphase cells and have been described in sufficient detail before (Duell *et al.*, 1997, 1998; Smith *et al.*, 1997; Wang *et al.*, 1996; Weier *et al.*, 1993, 1994, 1995a,b):

1. After hybridization, wash the slide three times in 2X SSC at 20°C for 10 min each.

2. Incubate the slide with 100 μl PNM buffer under a plastic coverslip at 20°C for 5 min.

3. The slide is then incubated at room temperature for 30 min with 100 μl PNM buffer containing AMCA-avidin (Pharmacia), anti-digoxigenin-rhodamine (Boehringer Mannheim), and a mouse antibody against FITC (Dako). [If only two labels are used, that is, biotin and digoxigenin, bound probes are detected with avidin-FITC DCS (Vector Laboratories) and anti-digoxigenin-rhodamine, respectively (Weier *et al.*, 1995a).]

4. The slide is washed two to three times in $2\times$ SSC for 15 min each at 20°C with constant motion on a shaking platform.

5. If necessary, signals are amplified using a biotinylated antibody against avidin raised in goat (Vector Laboratories) followed by another layer of AMCA-avidin, a Texas Red-labeled antibody against sheep raised in rabbit (Sigma), and a horse–anti-mouse antibody conjugated to FITC (Vector Laboratories) (Wang *et al.,* 1996).

6. The slide is mounted in 8 μl of DAPI (0.05 μg/ml in antifade solution) and covered by a 22 mm $\times$ 22 mm coverslip.

H. Digital Image Analysis

Images are acquired using a standard fluorescence microscope (Zeiss Axioskop) equipped with a $63\times$, 1.25 numerical aperature (NA) and $40\times$, 1.2 NA objectives, and filter sets for excitation and observation of DAPI, Texas Red/ rhodamine, FITC, and CY5 fluorescence, respectively (ChromaTechnology, Brattleboro, VT). The registration shifts between the red and green images are less than 0.1 μm (referred to the object) at all points in the digital image. The current filters are capable of excitation in single bands centered around 360, 405, 490, 555, and 637 nm, and visualization in multiple bands in the vicinities of 460 (blue), 520 (green), 600 (red), and 680 nm (infrared) (Jossart *et al.,* 1996). Images are collected using a charge coupled device (CCD) camera (Xilix, Hamamatsu, or Photometrics) connected to a computer workstation (Weier *et al.,* 1995a,b).

For determination of map positions, interactive software is available for various computers that allows the user to trace DNA fibers by drawing a segmented line and then calculates the length of the line in pixels (Wang *et al.,* 1996; Duell *et al.,* 1997). The pixel spacing is known from the microscope objective used in the experiment (we use $63\times$ magnification for molecules up to 100 kb, $40\times$ magnification for all larger molecules) and is converted into micrometers (or kilobases using the factor of 2.3 kb/μm). We typically measured all relevant distances along the hybridized fibers in triplicate. The results are provided in form of lists, which are then imported into Microsoft Excel (Microsoft Corp., Redmond, WA) and used to calculate average values for each fiber. The average values of several fibers are then used to calculate mean values and standard deviations for individual experiments. Relative standard deviations (CVs) are typically in the order of 5%. Higher CVs provide a simple mean to control the procedure. They prompt us to check the data analysis results for operator errors and undesirable images such as broken molecules or insufficiently stretched fibers.

IV. Critical Aspects of the Procedure

1. Slide pretreatment

Slides from different manufacturers and even of the same brand may produce very different qualities of fibers. Use a large batch of slides from one manufacturer; avoid slides that are painted on one end, since the paint may come off during pretreatment. Slides that have a sandblasted area are preferable.

2. Homogeneous stretching of DNA molecules

Different procedures have been described to stretch DNA molecules. In our hands, stretching involving a hydrodynamical force (meniscus) at 20° or 4°C has proved most reproducible. There is, however, no need to wait until the preparation has dried to completion. Once the DNA molecules have bound to the substrate, the coverslip can be lifted to exert the hydrodynamic stretching force (Hu *et al.*, 1996).

3. Immunocytochemical signal amplification

Never let the slides or part of them dry out during the immunocytochemical signal amplification. It is important to just drain the liquids from the slides, and then rapidly apply the next solution such as a blocking solution or the antibodies. If the slides are allowed to dry out, the level of background staining will increase to unacceptable levels.

4. Image acquisition

Most fluorochromes fade very quickly. Thus, expose the slides to excitation light for a minimum amount of time. Do not waste time studying the beauty of the molecules.

5. Image analysis

Always measure additional segments of the molecule such as the vector segment since these might provide additional information about the extent and homogeneity of DNA stretching.

V. Results and Discussion

Solid substrates for QDFM are prepared in batches of 20–50 by derivatization of standard microscope slides, coverslips, or sheets of mica with APS, which results in primary amino groups on the surface (Weier *et al.*, 1995a; Hu *et al.*, 1996). For DNA fiber stretching, a solution of target DNA molecules onto which probes are to be mapped is placed on an untreated coverslip and spread by placing the coverslip upside down on the APS-derivatized glass or mica surface. Binding of DNA to the substrate and the stretching effect can be monitored by staining the DNA with YOYO-1 prior to deposition. This also allows rejecting batches of slides that bind DNA too tightly. Following DNA binding and stretching, the coverslips are removed, the slides are rinsed briefly with double-distilled water, air dried, and stored at 4°C.

The DNAs from plasmid, cosmid, P1/PAC, and BAC clones are isolated using an alkaline lysis protocol and, in most instances, inserts are sized by PFGE. Linear high molecular weight DNA molecules can be prepared by digestion of DNA with a rare cutting restriction enzyme, but the alkaline lysis procedure provided sufficient amounts of nicked circular or randomly broken DNA suitable for QDFM (Wang *et al.*, 1996). In general, the DNA is loaded on a 1.0% low melting point agarose gel and electrophoresed for about 15 hr. The band containing the desired linear or circular DNA is then excised from the gel, and the gel slice is digested with agarase. Similarly, YAC DNA from various clones is purified by PFGE. The integrity of DNA molecules can be assessed by microscopic inspection of aliquots of DNA stained with 0.5 μM YOYO-1, before high molecular weight DNAs are used for DNA fiber or FISH probe preparation or stored at 4°C in 100 mM NaCl.

The density of DNA molecules after DNA fiber stretching can be adjusted by altering the concentration of the DNA molecules prior to binding. The goal is to find at least one intact YAC DNA molecule for every two to three fields of view and an even higher density when using smaller DNA molecules. Figure 1A shows the typical density of hybridized λ DNA molecules. In experiments where we deposited circular P1 and BAC DNA molecules, the fraction of intact DNA molecules reached ~80%. Although binding of DNA molecules in their circular form helps to maintain their integrity, it interferes with DNA fiber stretching, and the molecules are found to be stretched to varying degrees (Fig. 1B). We therefore applied mapping onto circular molecules for rough estimation of overlap, and mapping on linear fibers for high precision measurements. This can be done in a single experiment, because some circular DNA molecules are sheared during deposition, thus providing randomly broken linear DNA molecules.

A. Gene Mapping by Hybridization onto Genomic DNA Fibers

Preparations of genomic DNA fibers are well suited for mapping of clones in regions of the human or model genomes for which no low resolution YAC framework map is available. Good examples of this application are provided by regions that are frequently involved in the onset and progression of human tumors such as the immunoglobulin λ variable gene cluster (IGLV) on chromosome 22q (Duell *et al.*, 1997). This region is highly unstable, and YACs from this region are found deleted and/or rearranged.

Fig. 1 Quantitative DNA fiber mapping. (A) λ DNA molecules bound to an APS-derivatized glass slide were hybridized with red or green labeled restriction fragments. (B) Circular DNA molecules were excised from a PFGE gel and purified by agarase digestion. Staining with YOYO-1 reveals mostly closed circular DNA molecules in the presence of linear molecules of different length. (C) Mapping of cosmid clones onto genomic DNA fibers reveals the distance between linked clones. The arrow and idiogram to the right indicated the approximate map position of the clones on the long arm of human chromosome 22. (D) Determination of overlap between two P1 clones (clone A and B). Circular DNA molecules from clone A were hybridized with a P1 DNA probed prepared from the partially overlapping clone B (red). The hybridization mixture also contained probes to counterstain the P1 vector part in green and an ~1300 bp PCR probe that binds close to the T7 promoter within the vector part (red). The stretched P1 molecules are counterstained by

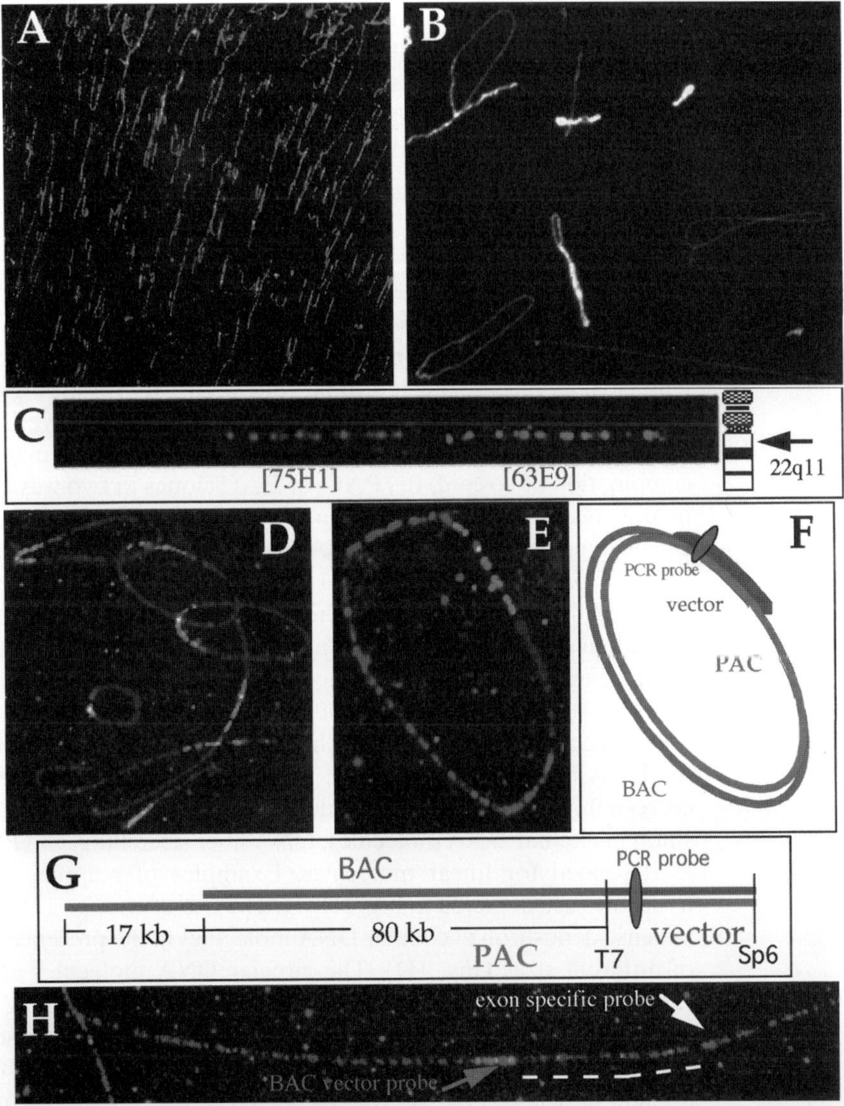

hybridization of a P1 probe (clone A) detected in blue. Overlap between the clones appears as a small red domain adjacent to the P1 vector part. (E–G) Measurement of overlap between a PAC and a BAC clone. The BAC DNA was labeled with digoxigenin and detected in red after hybridization to circular PAC molecules (counterstained in blue). Probes for the vector part contain P1 vector DNA (green) and a PCR-made probe (red) shown schematically in (F). The overlap between the clones extends for about 80 kb adjacent to the T7 promoter end of the vector part. (H) Cloned cDNA fragments can be mapped into larger genomic intervals by QDFM. Here, DNA fibers prepared from a BAC clone were hybridized with an ~2-kb insert of a plasmid specific for exon 2 of the human band 4.1 gene. The dashed line indicates the distance of the exon from the T7 end of the PAC vector part. (See color plates.)

Cosmid DNA probes labeled with either biotin or digoxigenin and hybridized to human genomic DNA fibers in the presence of 0.1 $\mu g/\mu l$ human COT1 blocking DNA are visualized after incubation with avidin-FITC and anti-dig-rhodamine. For example, the gap between the chromosome 22-specific clones 75H1 and 63E9 measured by hybridization onto genomic fibers is ~5 ± 2 kb (Fig. 1C) (Duell *et al.*, 1997). This result is in good agreement with DNA fingerprinting data suggesting little to no overlap between these clones (Frippiat *et al.*, 1995).

B. Mapping Cloned Probes onto Linear, Randomly Broken, or Circular DNA Molecules

The ability to localize and measure gaps in physical maps will greatly facilitate map closure and is expected to become highly useful during the finishing stages of genome mapping and sequencing projects. Quantitative DNA fiber mapping can facilitate the construction of high resolution physical maps comprised of any combination of cosmid, P1, PAC, or BAC clones in two ways: if a low resolution map is available, for example, in the form of a YAC contig, individual clones can be mapped directly onto DNA fibers prepared from the larger clones (Weier *et al.*, 1995a; Cheng and Weier, 1997). Alternatively, a high resolution map can be constructed by measuring the extent and orientation of overlap between individual clones (Fig. 1D–G). In most experiments, the applied scheme will be determined by the sources of the clones and might combine both schemes.

Preparation and purification of the circular DNA molecules is simple and fast. Following alkaline lysis, phenol/chloroform extraction, and ethanol precipitation, the DNA is loaded directly, that is, without digestion, on the PFGE gel. Bands corresponding to circular molecules are excised and digested with agarase. Deposition of circular DNA molecules, DNA fiber stretching, and FISH are performed as described for linear molecules. Examples of mapping onto circular PAC molecules are depicted in Fig. 1D–G. The use of circular DNA molecules results in dense deposition of circular DNA molecules in the presence of linear fragments of different sizes (Fig. 1D). The circular DNA molecules are found in a wide range of diameters representing different degrees of stretching. The largest circles are stretched to about ~2.3 kb/μm, but smaller, more condensed molecules could also be analyzed using the extent of the vector-specific green ~7 kb or ~17 kb domain on BACs or P1/PACs, respectively, as standard for normalization. The linear fragments found on the slides, on the other hand, are stretched more homogeneously, thus providing fibers without the need for normalization.

While the exact mechanism of coupling circular DNA molecules to APS-derivatized slides remains unknown, a large number of DNA molecules on the slides are found in the circular (complete) form. In some experiments, this fraction exceeded 80% of the DNA molecules, the remainder being randomly broken molecules. We attributed the observed high binding efficiency to single strand nicks in the circular DNA molecule induced during the alkaline lysis and DNA handling.

Figure 1 illustrates the application of QDFM for measurement of clone overlap by two examples: overlapping P1 clones (Fig. 1D) and the overlap between a PAC clone and a partially overlapping BAC clone (Fig. 1E). In both experiments, the DNA fiber is counterstained by a probe detected in blue. A P1/PAC vector-specific probe (green) and a PCR-generated probe of ~1300 bp (red) that binds close the T7 end of the vector are included to highlight the vector part of the fibers (Fig. 1D,E).

The two P1 clones overlap by approximately 2 kb, and the overlapping region is close to the SP6 promoter in the P1 vector (Fig. 1D).

The extent of overlap between a BAC clone (Fig. 1E, red) and the insert of a PAC molecule (Fig. 1E, blue) is ~80 kb located next to the T7 promoter in the PAC vector. About 17 kb of the PAC insert did not overlap with the BAC, and appears blue in Fig. 1E. Figure 1F shows a schematic representation of the probes and their relative map positions along the ~114 kb circular PAC molecule. These results are summarized in Fig. 1G.

C. cDNA Mapping

Expressed sequences can be mapped easily by QDFM, if each individual target extends for 500 bp or more. The common approach uses small genomic DNA fragments of 1–2 kb that contain known exons that are hybridized onto larger genomic DNA molecules. If the cDNA sequence and some information about intron–exon boundaries are available, such small DNA fragments can rapidly be generated from genomic DNA using PCR. We applied this strategy to map exon 2 of the human protein 4.1 gene (Chasis *et al.*, 1996; Peters *et al.*, 1998) onto a homologous BAC molecule (Fig. 1H). While this allowed the localization of the ~2-kb exon with near kilobase precision, hybridization of additional probes was needed and necessary to determine the orientation of the gene within the BAC insert.

A second approach relies on direct mapping of expressed sequences. In practice, DNA is isolated from cDNA clones, labeled, and hybridized onto a genomic DNA fiber. In the presence of blocking DNA, the cDNA probes will bind specifically to their complementary DNA targets, that is, expressed regions and 5'- or 3' untranslated regions (UTRs) along the fiber, leaving noncoding regions (introns, 5'- and 3'-flanking DNA) unstained (not shown). Simultaneous hybridization of fiber-specific probes in different colors makes these regions highly visible. Using FISH conditions usually applied to hybridization and probe detection on metaphase chromosomes, this system works well with cDNA probes of several kilobases.

Acknowledgments

This work was supported by a grant from the Director, Office of Energy Research, Office of Health and Environmental Research, U.S. Department of Energy, under Contract DE-AC-03-

76SF00098 and the LBNL/UCSF Training Program in Genome Research sponsored by the University of California Biotechnology Research and Education Program.

References

Bensimon, A., Simon, A., Chiffaudel, A., Croquette, V., Heslot, F., and Bensimon, D. (1994). High resolution mapping using fluorescence in situ hybridization to extended DNA fibers prepared from agarose-embedded cells. *Science* **265,** 2096–2098.

Birnboim, H. C., and Doly J. (1979). A rapid alkaline lysis procedure for screening recombinant plasmid DNA. *Nucleic Acids Res.* **7,** 1513–1523.

Cassel, M. J., Munné, S., Fung, J., and Weier, H.-U. G. (1997). Carrier-specific breakpoint-spanning DNA probes for pre-implantation genetic diagnosis in interphase cells. *Hum. Reprod.* **12,** 101–109.

Chasis, J. A., Coulombel, L., McGee, S., Lee, G., Tchernia, G., Conboy, J., and Mohandas, N. (1996). Differential use of protein 4.1 translation initiation sites during erythropoiesis: Implications for a mutation-induced stage-specific deficiency of protein 4.1 during erythroid development. *Blood* **87,** 5324–5331.

Cheng, J.-F., and Weier, H.-U. G. (1997). Approaches to high resolution physical mapping of the human genome. *In* "Biotechnology International" (C. F. Fox and T. H. Connor, eds.), pp. 149–157. Universal Medical Press, San Francisco.

Chumakov, I., Rigault, P., Guillou, S., Ougen, P., Billault, G., Guasconi, S., Gervy, P., Le Gall, I., Soularue, P., Grinas, P., Bougueleret, L., Bellanné-Chantelot, C., Lacroix, B., Barillot, E., Gesnouin, P., Pook, S., Vaysseix, G., Frelat, G., Schmitz, A., Sambucy, J. L., Bosch, A., Estivill, X., Weissenbach, J., Vignal, A., Riethman, H., Cox, D., Patterson, D., Gardiner, K., Masahira, H., Sahaki, Y., Ichikawa, H., Ohsi, M., Le Paslier, D., Heilig, R., Antonorakis, S., and Cohen, D. (1992). *Nature* **359,** 380–387.

Cohen, D., Chumakov, I., and Weissenbach , J. (1993). A first-generation physical map of the human genome. *Nature* **36,** 698–701.

Collins, F., and Galas, D. (1993). A new five-year plan for the U.S. human genome project, *Science* **262,** 43–46.

Duell, T., Wang, M., Wu, J., Kim, U.-J., and Weier, H.-U. G. (1997). High resolution physical map of the immunoglobulin lambda variant gene cluster assembled by quantitative DNA fiber mapping. *Genomics* **45,** 479–486.

Duell, T., Nielsen, L. B., Jones, A., Wang, M., Young, S. G., and Weier, H.-U. G. (1998). Construction of two near-klobase resolution maps of the 5′ regulatory region of the human apolipoprotein B gene by Quantitative DNA Fiber Mapping (QDFM). *Cytogenet. Cell. Genet.* **79,** 64–70.

Florijn, R. J., Bonden, L. A. J., Vrolijk, H., Wiegant, J., Vaandrager, J.-W., Baas, F., den Dunnen, J. T., Tanke, H. J., van Ommen, G.-J. B., and Raap, A. K. (1995). High-resolution DNA Fiber-FISH for genomic DNA mapping and colour bar-coding of large genes. *Hum. Mol. Genet.* **4,** 831–836.

Frippiat, J. P., Williams, S. C., Tomlinson, I. M., Cook, G. P., Cherif, D., Le Paslier, D., Collins, J. E., Dunham, I., Winter, G., and Lefranc, M. P. (1995). Organization of the human immunoglobulin lambda light-chain locus on chromosome 22q11.2. *Hum. Mol. Genet.* **4,** 983–991.

Haaf, T., and Ward, D. C. (1994). A new bacteriophage P1-derived vector for the propagation of large human DNA fragments. *Hum. Mol. Genet.* **3,** 29–633.

Heiskanen, M., Karhu, R., Hellsten, E., Peltonen, L., Kallioniemi, O. P., and Palotie, A. (1994). A new bacteriophage P1-derived vector for the propagation of large human DNA fragments. *BioTechniques* **17,** 928–933.

Heng, H. H., Squire, J., and Tsui, L. C. (1992). High-resolution mapping of mammalian genes by *in situ* hybridization to free chromatin. *Proc. Natl. Acad. Sci. U.S.A.* **89,** 9509–9513.

Hu, J., Wang, M., Weier, H.-U. G., Frantz, P., Kolbe, W., Olgletree, D. F., and Salmeron, M. (1996). Imaging of sinle extended DNA molecules on flat (aminopropyl) triethoxysilane-mica by atomic force microscopy. *Langmuir* **12,** 1697–1700.

Hudson, T. J., Stein, L. D., Gerety, S. S., Ma, J., Castle, A. B., Silva, J., Slonim, D. K., Baptista, R., Kruglyak, L., Xu, S. H., Hu, X., Colbert, A. M. E., Rosenberg, C., Reeve-Daly, M. P., Rozen, S., Hui, L., Wu, X., Vestergaard, C., Wilson, K. M., Bae, J. S., Maitra, S., Ganiatsas, S., Evans, C. A., DeAngelis, M. M., Ingalls, K. A., Nahf, R. W., Horton, L. T., Oskin-Anderson, M., Collymore, A. J., Ye, W., Kouyoumijian, V., Zemsteva, I. S., Tam, J., Devine, R., Courtney, D. F., Turner-Renaud, M., Nguyen, H., O'Connor, T. J., Fizames, C., Fauré, S., Gyapay, G., Dib, C., Morisette, J., Orlin, J. B., Birren, B. W., Goodman, N., Weissenbach, J., Hawkins, T. L., Foote, S., Page, D. C., and Lander, E. S. (1995). An STS-based map of the human genome. *Science* **270,** 1945–1954.

Ioannou, P. A., Amemiya, C. T., Garnes, J., Kroisel, P. M., Shizuya, H., Chen, C., Batzer, M., and de Jong, P. J. (1994). A new bacteriophage P1-derived vector for the propagation of large human DNA fragments. *Nat. Genet.* **6,** 84–89.

Jossart, G. H., O'Brien, B., Cheng, J.-F., Tong, Q., Jhiang, S. M., Duh, Q., Clark, O. H., and Weier, H.-U. G. (1996). A novel multicolor hybridization scheme applied to localization of a transcribed sequence (D10S170/H4) and deletion mapping in the thyroid cancer cell line TPC-1. *Cytogenet. Cell. Genet.* **75,** 254–257.

Kroisel, P. M., Ioannou, P. A., and de Jong, P. J. (1994). PCR probes for chromosome in situ hybridization of large-insert bacterial recombinants. *Cell Genet.* **65,** 97–100.

Maniatis, T., Fritsch, E. F., and Sambrook, J. (1986). "Molecular Cloning: A Laboratory Handbook." Cold Spring Harbor Laboratory, Cold Spring Harbor, New York.

Munné, S., Fung, J., Cassel, M. J., Márquez, C., and Weier, H.-U. G. (1998). Preimplantation genetic analysis of translocations: Case-specific probes for interphase cell analysis. *Hum. Genet.* **102,** 663–674.

Parra, I., and Windle, B. (1993). High resolution visual mapping of stretched DNA by fluorescent hybridization. *Nat. Genet.* **5,** 17–21.

Peters, L., Weier, H.-U. G., Walensky, L. D., Snyder, S., Parra, M., Mohandas, N., and Conboy, J.G. (1998). Four paralogous protein 4.1 genes map to distinct chromosomes in mouse and man. *Genomics* **54,** 348–350.

Shizuya, H., Birren, B., Kim, U. J., Mancino, V., Slepak, T., Tachiiri, Y., and Simon, M. (1992). Cloning and stable maintenance of 300-kilobase-pair fragments of human DNA in *Escherichia coli* using an F-factor-based vector. *Proc. Natl. Acad. Sci. U.S.A.* **89,** 8794–8797.

Smith, D. J, Stevens, M. E., Sudanagunta, S. P., Bronson, R. T., Makhinson, M., Watabe, A. M., O'Dell, T. J., Fung, J., Weier, H.-U. G., Cheng, J.-F., and Rubin, E. (1997). Functional screening of 2 Mb of human 21q22.2 in YAC transgenic mice implicates minibrain in learning defects. *Nat. Genet.* **16,** 28–36.

Telenius, H., Pelmear, A. H., Tunnacliffe, A., Carter, N. P., Behmel, A., Fergueson-Smith, M. A., Nordenskjold, M., Pfragner, R., and Ponder, B. A. (1992). Cytogenetic analysis by chromosome painting using DOP-PCR amplified flow-sorted chromosome. *Genes Chrom. Cancer* **4,** 257–263.

Wang, M., Duell, T., Gray, J. W., and Weier, H-U. G. (1996). High sensitivity, high resolution physical mapping by fluorescence in situ hybridization [FISH] on to individual straightened DNA molecules. *Bioimaging* **4,** 1–11.

Weier, H.-U. G., Miller, B. M., Yu, L. C., and Fuscoe, J. C. (1993). PCR cloning of a repeated DNA fragment from chinese hamster ovary (CHO) cell X chromosomes and mapping by fluorescence in situ hybridization. *DNA Sequence* **4,** 47–51.

Weier, H.-U. G., Matsuta, M., Zitzelsberger, H., Matsuta, M., and Gray, J. (1994). Generation of highly specific DNA hybridization probes for chromosome enumeration in human interphase cell nuclei: Isolation and enzymatic synthesis of alpha satellite DNA probes for chromosome 10 by primer directed DNA amplification. *Methods Mol. Cell. Biol.* **4,** 231–248.

Weier, H-U. G., Wang, M., Mullikin, J. C., Zhu, Y., Cheng, J-F., Greulich, K. M., Bensimon, A., and Gray, J. W. (1995a). Quantitative DNA fiber mapping. *Hum. Mol. Genet.* **4,** 1903–1910.

Weier, H-U. G., George, C. X., Greulich, K. M., and Samuel, C. E. (1995b). The interferon-inducible, double-stranded RNA-specific adenosine deaminage gene (DSRAD) maps to human chromosome 1q21.1–21.2. *Genomics* **30,** 372–375.

Weissenbach, J., Gyapay, G., Dib, C., Vignal, A., Morissette, J., Millasseau, P., Vaysseix, G., and Lathrop, M. (1992). Cloning and stable maintenance of 300-kilobase-pair fragments of human DNA in *Escherichia coli* using an F-factor-based vector. *Nature* **359,** 794–801.

CHAPTER 32

Primed *in Situ* Labeling

Johnny Hindkjaer,[*] **Lars Bolund,**[†] **and Steen Kølvraa**[†]

[*]Centre of Preimplantation Genetic Diagnosis
Fertility Clinic, Århus University Hospital
8200 Århus N, Denmark

[†]Institute of Human Genetics
University of Aarhus
8000 Århus C, Denmark

I. Introduction
II. Applications
III. Materials
    A. Chemicals
    B. Instruments
IV. Protocols
    A. Preparation of Cell Material for Primed *in Situ* Labeling
    B. Primed *in Situ* Labeling
    C. Visualization of Labeled DNA
V. Critical Aspects of the Procedures
    A. Preparation of Cell Material for Primed *in Situ* Labeling
    B. Primed *in Situ* Labeling
    C. Vizualization of Labeled DNA
References

I. Introduction

Primed *in situ* labeling (PRINS) is a fast and sensitive technique for sequence specific *in situ* detection of DNA. An unlabeled oligonucleotide probe is hybridized and used as primer for chain elongation *in situ* catalyzed by a DNA polymerase. Thus, the oligonucleotide primer binds by sequence-specific base pairing to its target sequence, which is subsequently labeled when labeled nucleotides are incorporated by the DNA polymerase, using the oligonucleotide as primer and the cellular DNA as template.

The fact that unlabeled probes (primers) are used in the PRINS reaction means that high concentrations can be used, since probe bound to cell-structures cannot function as a primer and will therefore not give rise to background signals—only probe hybridized correctly to DNA can function as primer for chain elongation. Thus, using high probe concentrations the hybridization is very fast. A PRINS reaction normally runs for only 5–30 min. Owing to the speed of the reaction the morphology of chromosomes is very well preserved, which makes it possible to obtain chromosome banding of good quality after a PRINS reaction.

PRINS was originally developed to demonstrate interchromosomal differences in the α-satellite DNA of human chromosomes (Koch *et al.*, 1989). It was observed that the monomers of different chromosomes had a conserved region common to all chromosomes, and a variable region with chromosome specific motifs. Studying the motifs of the chromosome specific regions of the monomers of α-satellite DNA and other satellite DNA repeats, chromosome specific PRINS primers have been constructed for all human chromosomes, except for chromosomes 14 and 22, which are detected simultaneously with a 14/22 PRINS primer (Koch *et al.*, 1995; Pellestor *et al.*, 1995; Hindkjaer *et al.*, 1996a). Also specific PRINS primers have been developed for other repearted DNA regions, for example, telomers, and primers for specific genes have also been published (Cinti *et al.*, 1993; Hindkjaer *et al.*, 1996b).

This chapter presents protocols for detection of a single primer (PRINS) or several primers by *in situ* labeling with different colors (multicolor-PRINS) (Hindkjaer *et al.*, 1994), a combination of PRINS and chromosome painting (PRINS–painting) (Hindkjaer *et al.*, 1995a), and a cyclic PRINS reaction (Repeated-PRINS) (Terkelsen *et al.*, 1993; Hindkjaer *et al.*, 1996b).

II. Applications

PRINS is used in cancer research (Hindkjaer *et al.*, 1995b, c; Pedersen *et al.*, 1997) and clinical genetics (Brandt *et al.*, 1993, 1994) as a supplement to traditional chromosome banding analysis for identification and quantification of aberrant methaphase chromosomes and interphase nuclei. Furthermore, PRINS can be used as a rapid screening procedure for chromosome aberrations in interphase nuclei for preimplantation genetic diagnosis (Pellestor *et al.*, 1996).

III. Materials

A. Chemicals

1. *Taq* DNA polymerase (Boehringer Mannheim, Indianapolis, IN, or Perkin Elmer Cetus, Foster, CA).

2. 10× *Taq* polymerase buffer: 500 mM KCl, 100 mM Tris-HCl (pH 8.3), 15 mM MgCl$_2$, 0.1% (w/v) bovine serum albumin (BSA) (or gelatin).

3. Labeled dUTP: digoxigenin-11-dUTP (Boehringer Mannheim), fluorescein-12-dUTP (Boehringer Mannheim), and biotin-16-dUTP (Boehringer Mannheim).

4. dVTP mixture: dATP, dCTP, dGTP, 3.3 mM each (Boehringer Mannheim).

5. ddNTP mixture: ddATP, ddCTP, ddGTP, ddTTP, 2.5 mM each (Boehringer Mannheim).

6. Glycerol, 87% (Merck, Darmstadt, Germany).

7. Blocking solution: 5% nonfat dry milk dissolved in washing buffer. Centrifuge for 2 min in an Eppendorf centrifuge and use supernatant.

8. Washing buffer: 4× SSC, pH 7.0, 0.05% Tween 20 (1× SSC is 150 mM NaCl, 15 mM sodium citrate).

9. Stop buffer: 50 mM NaCl, 50 mM EDTA, pH 8.0.

10. Phosphate-buffered saline (PBS), pH 7.2.

11. NaOH, 3 M.

12. Acified Thyrode's solution, pH 2.2 (Medi-Cult, Jyllinge, Denmark).

13. Ethanol, 99%.

14. Fixative (absolute methanol/glacial acetic acid, 3:1 v/v).

15. Paraformaldehyde, 3% in PBS.

16. Anti-digoxigenin-fluorescein, and anti-digoxigenin-rhodamin, Fab fragments (Boehringer Mannheim).

17. Fluorescein avidin DCS (Vector Laboratories, Burlingame, CA).

18. Biotinylated anti-avidin D (Vector Laboratories).

19. Antifade solution: 10 mg/ml *p*-phenylenediamine dihydrochloride in 80% (v/v) glycerol, 0.1 M Tris-HCl, pH 9.0.

20. Propidium iodide (Sigma, St. Louis, MO).

21. 4′, 6-Diamidino-2-phenylindole (DAPI) (Sigma).

22. Poly-L-lysine (Sigma, Catalog No. p8920).

23. Blastomere lysis buffer: 0.1 N HCl, 0.1% Tween 20.

B. Instruments

Staining jars.
Thermo-block.
Water bath with humidified chamber.
Thermal cycler with flat bed.
Phase contrast microscope.

Fluorescence microscope.

England Finder (Electron Microscopy Sciences, Catalog No. 68048-07)

<hr>

IV. Protocols

A. Preparation of Cell Material for Primed *in Situ* Labeling

1. Metaphase Chromosome Preparation

Standard methanol/acetic acid (3:1 v/v) fixed cells (Verma and Babu, 1989) are spread on a microscope slide which has been cooled in redistilled water at 4°C and drained of excess water (not dried). The best spreads are seen when two to four drops of cell suspension are dripped from a Pasteur pipette onto the slide from a distance of ~30–50 cm. The slide is cautiously drained of surplus water and cell suspension and air dried. The slide is now ready for a PRINS reaction. The slide can be stored for about a week; longer storage can give rise to background signals.

2. Preparation of Sperm Sample

1. A fresh ejaculate is allowed to liquefy for 30 min at room temperature, before it is diluted 1:10 with PBS and centrifuged for 8 min at 600 *g*.

2. The pellet is resuspended in 1 ml of methanol/acetic acid (3:1, v/v) and placed at −20°C for 1 hr.

3. Centrifuge at 600 *g* for 5 min and resuspend pellet in fresh methanol/acetic acid (3:1, v/v) in an appropriate volume.

4. The cell suspension is dripped on a microscope slide from a distance of 1 cm. The slide is drained and allowed to air dry. The slide is now ready for a PRINS reaction. The slide can be stored for about a week before use.

5. Immediately before the PRINS reaction, the slide is denatured in 3 *M* NaOH for 4 min at room temperature, followed by dehydration in an ice-cold (−20°C) ethanol series (70, 85, 99%, 2 min each), and then air dried. The denaturation step in the PRINS protocols is omitted when sperm cells are denatured in NaOH. Place the slide at annealing temperature, add PRINS mix, and cover with a 25 × 50 mm coverslip. The rest of the PRINS procedure is unchanged.

3. Preparation of Blastomeres or Whole Embryos

1a. A blastomere from an embryo biopsy is washed once in PBS in a 4-well dish (Nunc, Roskilde, Denmark).

1b. A whole embryo is placed in acidified Thyrode's solution until the zona perlucida is dissolved. The procedure is followed with a microscope.

2. The blastomere or embryo is applied to a marked position on a poly-L-lysine coated slide (according to manufacturer) with a drawn Pasteur pipette (H-11130, Swemed, Billdal, Sweden) and lysed in a small drop (10 μl) of blastomere lysis buffer that is placed on the marked position of the slide. The lysis is followed under a phase contrast microscope, and the nucleus is monitored as the lysis buffer dries out to be able to relocate the nucleus in the fluorescence microscope. An England Finder can be used to relocate the nucleus.

3. The slide is air dried, washed with PBS for 5 min, and dehydrated in an ethanol series (70, 85, 99%, 2 min each). The slide is now ready for a PRINS reaction.

B. Primed *in Situ* Labeling

1. Detection of Repeated Sequences Using Oligonucleotide Primers

Oligonucleotide primers synthezised from the consensus sequence of a repeat family can be used to achieve PRINS detection of the target sequences in the chromatin (Koch *et al.*, 1995; Pellestor *et al.*, 1995; Hindkjaer *et al.*, 1996a). Good PRINS signals have been obtained of telomeres, satellites I, II, III, and IV, Alu repeat, α-satellite DNA, and β-satellite DNA on human chromosomes (Koch *et al.*, 1995; Therkelsen *et al.*, 1995). Furthermore, specific labeling of characteristic repeat families of plants and animals has been reported (Macas *et al.*, 1995; Gu and Hindkjaer, 1996).

1. Reaction mixture is prepared in an Eppendorf tube: 0.5 μg of oligonucleotide DNA representing the target sequence, 1.5 μl dVTP nucleotide mixture (3.3 mM each), 0.5 μl a labeled dUTP (e.g., digoxigenin-11-dUTP, biotin-16-dUTP, fluorescein-12-dUTP) (1 mM), 2.5 μl of glycerol (87%), and 5 μl of 10x *Taq* polymerase buffer. Add redistilled water to a final volume of 50 μl.

2. Add 1 U of *Taq* DNA polymerase and mix gently.

3. Preheat the slide with the chromosome spread on a thermo-bloc at 94°C for about 15 sec.

4. For denaturation of chromosomal DNA, the reaction mixture is applied to the preheated slide, and spread with a 25 × 50 mm coverslip. Denature the chromosomal DNA for 4 min at 94°C.

5. For probe annealing and chain elongation, the slide is quickly transferred to a humidified chamber in a water bath at the stringent hybridization temperature according to the probe used. Incubate for 5–30 min according to the probe used. Steps 3–5 can be performed on a thermal cycler with a flat bed.

6. The PRINS reaction is terminated by washing the slide in 100 ml of stop buffer in a staining jar for 1 min at annealing temperature.

7. The slide is transfered to 50 ml of washing buffer in a staining jar with screw-cap and washed for 3 min at room temperature under gentle agitation.

A typical result of a PRINS reaction is shown in Fig. 1, using a primer detecting satellite III on chromosome 9 with fluorescein-12-dUTP as the labeled nucleotide.

2. Multicolor–PRINS

Multicolor detection of different target sequences in DNA offers the advantage that several probes can be analyzed simultaneously, which can make evaluation of the results easier and faster. Multicolor detection is especially valuable when only a few cells are avaliable for analysis, as, for example, in preimplantation genetic diagnosis, where often only one to two cells can be analyzed.

Multicolor-PRINS involves successive PRINS reactions on the same slide, using differently hapten-labeled dUTPs. After the multicolor-PRINS reaction, a staining procedure with different fluochromes is performed. Also fluorochrome-labeled dUTPs can be used directly in the reaction.

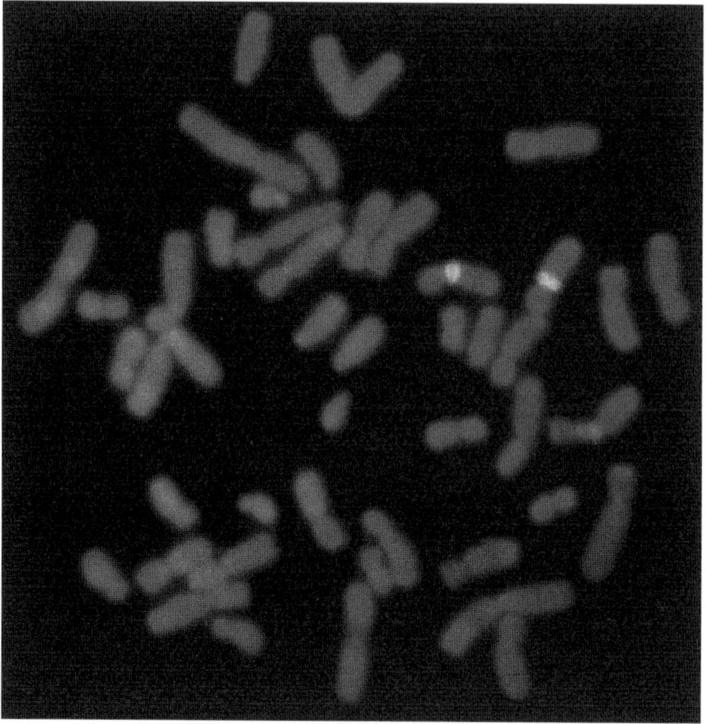

Fig. 1 A PRINS reaction performed with a primer (5′-CCATTCCATTCCAYTCGGGTT-3′) detecting satellite III DNA on chromosome 9. Fluorescein-12-dUTP was used as the labeled nucleotide. The annealing and chain elongation were performed at 60°C for 5 min. (See color plates.)

a. First PRINS Reaction

1. A PRINS reaction is performed according to steps 1–5 in Section IV,B,1.
2. The slide is washed for 1 min at annealing temperature in 100 ml of pre-heated stop buffer.
3. The slide is dehydrated in an ice-cold (−20°C) ethanol series (70, 85, and 99%, 2 min each in 100 ml) and air dried.

b. Blocking with Dideoxynucleotides (Optional)

1. Prepare the dideoxy reaction mixture by mixing 2 μl of ddNTP mixture, 2.5 μl of glycerol (87%), and 5 μl of 10× *Taq* polymerase buffer. Add redistilled water to a final volume of 50 μl.
2. Add 1 U of *Taq* DNA polymerase to the reaction mixture and mix gently.
3. For dideoxy blocking of 3′-ends, the dideoxy reaction mixture is applied to the preheated slide, which is incubated for 15 min in a humidified chamber at 55°C. The free 3′-ends are blocked by this reaction, preventing chain elongation to restart from these ends in the second PRINS reaction.
4. The slide is washed two times for 30 sec at 55°C in 100 ml of preheated stop buffer.
5. The slide is dehydrated in an ice-cold (−20°C) ethanol series (70, 85, and 99%, 2 min each).

c. Second PRINS Reaction

1. A PRINS reaction is performed according to steps 1–5 in Section IV,B,1, using a labeled dUTP that is different from the one used in the first PRINS reaction.
2. The slide is washed for 1 min at annealing temperature in 100 ml of pre-heated stop buffer. If PRINS with a third primer is performed the slide is dehydrated and air dried. The third PRINS reaction can now be performed with a labeled dUTP different from the ones used in the first and second PRINS reactions. After termination of the last PRINS reaction the result can be visualized with different fluorochromes (see Section V,C).

A typical multicolor-PRINS reaction is shown in Fig. 2. Multicolor PRINS is performed on sperm cells using a chromosome X specific α-satellite DNA primer in the first PRINS reaction with digoxigenin-11-dUTP as the labeled nucleotide and a Y chromosome (Yq) specific primer in the second PRINS reaction with biotin-16-dUTP as the labeled nucleotide.

3. PRINS–Painting

PRINS and chromosome painting can be combined in a PRINS–painting reaction (Hindkjaer *et al.*, 1995a). The method is especially useful in the detection

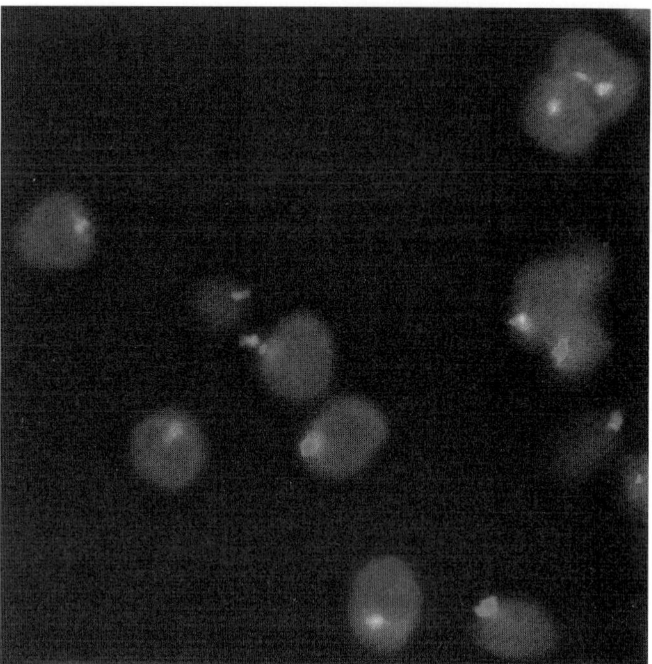

Fig. 2 Multicolor-PRINS on sperm cells. Sperm cells were denatured in 3 *M* NaOH for 4 min at room temperature. The first PRINS reaction was performed at 60°C for 30 min using an X chromosome specific primer (5′-CCTGAAAGCCTTTTCGCTTTATCTTCACAGAAAGA-3′) with digoxigenin-11-dUTP as the labeled nucleotide. The second PRINS reaction was performed at 60°C for 30 min using a Y chromosome specific primer (5′-TTTTTTTTTCATTTAAAAAATATGGATCT-TGGCTCAGGCC-3′) with biotin-16-dUTP as the labeled nucleotide. The dideoxy reaction was omitted. The X signal was visualized in red with anti-digoxigenin-rhodamine, and the Y signal was visualized in green with fluorescein-conjugated avidin. Magnification 1000×. (See color plates.)

of translocations of small DNA fragments. Evaluation of a reciprocal transloca-tion using two chromosome paints might be difficult, since a small fragment of foreign DNA in a chromosome might be hidden by the neighboring DNA mate-rial of the translocation chromosome. When only the centromere is labeled for chromosome identification in combination with one chromosome painting probe at a time, a small translocated DNA fragment is easier to recognize, since the neighboring DNA is unlabeled.

1. A standard PRINS reaction is performed according to Section IV,B,1. After termination of the PRINS reaction in stop buffer the slide is dehydrated in an ice-cold (−20°C) ethanol series (70, 85, and 99%, 2 min each).

2. The slide is air dried and preheated at the stringent hybridization tempera-ture of the chromosome painting probe for 30 sec before 10 μl of painting probe

is added and covered with a 18×18 mm coverslip, which is sealed with rubber solution. Prepare the painting probe according to the manufacturer's instructions while the PRINS reaction is performed, and apply the probe to the slide immediately when it is ready.

3. The hybridization is performed overnight at stringent hybridization temperature. After hybridization remove the rubber solution gently and wash the slide in $0.4\times$ SSC at $65°–72°C$ for exactly 2 min followed by a wash in $2\times$ SSC, 0.1% Tween 20 for 2 min at room temperature. Alternatively the slide can be washed in 50% formamide, 2X SSC for 10 min at a temperature $0°–5°C$ over the hybridization temperature followed by washing twice for 5 min in $2\times$ SSC at the same temperature. The slide is now ready for visualization (see Section IV,C).

A PRINS–painting of a translocation is shown in Fig. 3. A PRINS reaction is performed with a primer specific for the centromere (α-satellite DNA) of

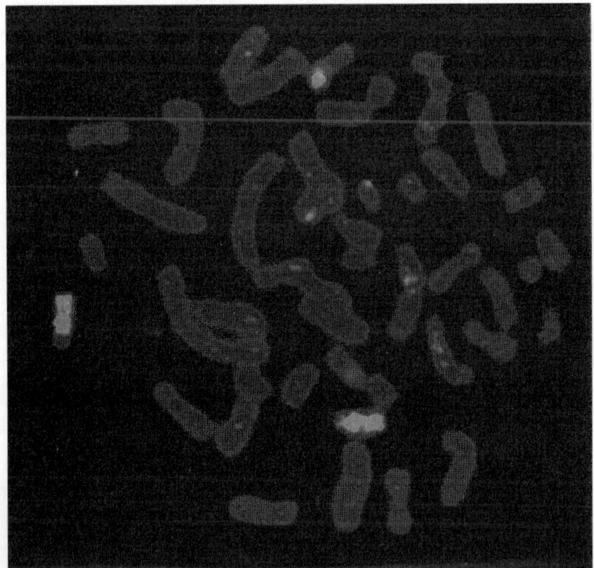

Fig. 3 PRINS–painting was used for clarification of complex structural rearrangements in bone marrow cells from a patient with acute lymphoblastic leukemia (ALL). PRINS was performed with an oligonucleotide representing the α-satellite DNA of chromosome 17 (5′-ACGTGAACTTT-GAAAGGAAAGTTCAACTCGGGGAT-3′) at 60°C for 30 min with digoxigenin-11-dUTP as the labeled nucleotide. After washing and dehydration a biotin-labeled chromosome 13 library was hybridized to the slide overnight at 37°C. The painting library was visualized with fluorescein-conjugated avidin, and the PRINS reaction was visualized with anti-digoxigenin-rhodamine, Fab fragments. Chromosomes were counterstained with DAPI. The result shows a derivative chromosome with a chromosome 17 centromere stained red and a part of the chromosome stained green by the chromosome 13 library, showing that the chromosome is a der(17)t(13;17). (See color plates.)

chromosome 17 and the chromosome painting with a chromosome 13 specific DNA library.

4. Repeated–PRINS

Repeated-PRINS is a cyclic reaction that is performed to obtain stronger PRINS signals than seen after a single PRINS reaction (Terkelsen *et al.*, 1993). It can be used to obtain signals from low copy number repeat sequences (Gosden and Hanratty, 1993) and even single copy genes (Hindkjaer *et al.*, 1996b).

Repeated-PRINS should always be optimized for the primer used, since reaction conditions such as $MgCl_2$ concentration, duration, and temperature of annealing are critical parameters for achieving good results. Results achieved on purified DNA is not necessarily working in cyclic-PRINS.

1. The reaction mixture is prepared in an Eppendorf tube by mixing 0.5 μg of oligonucleotide DNA representing the target sequence, 1.5 μl dVTP nucleotide mixture (3.3 mM each), 0.4 μl dTTP (3.3 mM), 0.5 μl of a labeled dUTP (e.g., digoxigenin-11-dUTP, biotin-16-dUTP, fluorescein-12-dUTP) (1 mM), 2.5 μl of glycerol (87%), and 5 μl of 10× *Taq* polymerase buffer. Add redistilled water to a final volume of 50 μl.

2. Add 2 U of *Taq* DNA polymerase and mix gently.

3. The slide is placed on the hot plate of the thermocycler heated to the annealing temperature. Add the PRINS mix and cover with a 24× 50 mm coverslip and seal with rubber solution. Start the program appropiate for the primer used.

Example: Program for the detection of the gene apo (a) located on chromosome 6 (6q26-27) with primers detecting the kringle IV motif of the gene (Mclean *et al.*, 1987; Hindkjaer *et al.*, 1996b). Two primers (0.25 μg each) (5′-CCCAGGCCT-TTGCTCAGTCGGTGC-3′ and 5′-GGGTGCAGGAGTGCTACCATGG-TAATGGAC-3′) separated by 8 kb were used in this repeated-PRINS reaction. Distinct signals were observed in about 70% of the metaphases analyzed (data not shown).

First cycle	Denaturation: 4 min at 94°C
	Annealing: 3 min at 63°C
	Elongation: 6 min at 70°C
Second to twentieth cycles	Denaturation: 1 min at 94°C
	Anncaling: 3 min at 63°C
	Elongation: 6 min at 70°C
Final cycle	Elongation 7 min at 70°C

4. The coverslip is removed and the reaction terminated in stop buffer heated to annealing temperature. The slide is now ready for visualization (see Section IV,C).

C. Visualization of Labeled DNA

1. Visualization of Digoxigenin–Labeled DNA

1. The slide is incubated with 100 μl of blocking solution for for 5 min under a 25 $\times$ 50 mm coverslip in a humidified box. All incubations and washings in this procedure are performed at room temperature.
2. Wash the slide briefly (15 sec) in 50 ml of washing buffer, drain, and add 50 μl of anti-digoxigenin-fluorescein (or rhodamine or AMCA), Fab fragments, in blocking solution (20 ng/μl) to the slide. Incubate for 30 min under a coverslip.
3. Wash 3 $\times$ 5 min in 50 ml of washing buffer under gentle agitation.
4. The slide is mounted with 20 μl of antifade solution containing propidium iodide (0.5 μg/ml) or DAPI (0.15 μg/ml).
5. The result is evaluated under a fluorescence microscope.

2. Visualization of Biotin–Labeled DNA

1. Incubate the slide in 100 μl of blocking solution for 5 min under a 25 $\times$ 50 mm coverslip. All incubations and washings in this procedure are performed at room temperature.
2. Wash the slide briefly (15 sec) in 50 ml of washing buffer, drain and add 100 μl of fluorescein avidin (or rhodamine- or AMCA avidin) in blocking solution (2 ng/μl), and incubate for 10–30 min in a humidified box.
3. Wash for 3 $\times$ 5 min in 50 ml washing buffer under gentle agitation.
4. The slide is mounted with 20 μl of antifade solution containing propidium iodide (0.5 μg/ml) or DAPI (0.15 μg/ml).
5. The result is evaluated under a fluorescence microscope. If staining is too weak, apply additional layers of fluorescein as described below.
6. Wash the slide for 2 $\times$ 5 min in washing buffer. The coverslip is removed after the first wash.
7. Incubate in 100 μl of anti-avidin antibodies in blocking solution (2 ng/μl) for 10–30 min.
8. Wash for 3 $\times$ 5 min in washing buffer.
9. Incubate in 100 μl of fluorescein avidin in blocking solution (2 ng/μl) for 10–30 min.
10. Wash for 3 $\times$ 5 min in washing buffer.
11. The slide is mounted in antifade solution containing propidium iodide or DAPI and evaluated under the microscope. If staining is still too weak, steps 6–11 can be repeated.

Digoxigenin- and biotin-labeled DNA from a multicolor-PRINS can be visualized at the same time mixing fluorochrome conjugated avidin and anti-digoxigenin (different fluorochromes) in one reaction using the concentrations mentioned. Normal staining time is 30 min.

V. Critical Aspects of the Procedures

A. Preparation of Cell Material for Primed *in Situ* Labeling

1. For metaphase chromosome preparations, the quality of the chromosome preparation is very important for good PRINS labeling. If too much cytoplasm is seen around the chromosomes then refix in methanol/acetic acid (3:2, v/v). More acetic acid tends to give a better chromosome spread. With too much acetic acid, however, the chromosomes can be spread too much so that some are missing in the metaphase. The slide with the chromosome spread should not be stored for more than 1 week. If nicks are introduced into the DNA they can act as primers for unspecific labeling.

2. Preparation of sperm cells: the time of denaturation in NaOH can vary with the age of the slide. Older slides normally require longer time of denaturation.

3. Preparation of blastomeres or whole embryos: it is important to monitor the nuclei in the blastomeres. When a blastomere lyses the nuclei is usually easy to see. Follow the nuclei in the microscope until the lysis buffer has dried. Write carefully the coordinates of the microscope to be able to find the nuclei again after the PRINS reaction. An England Finder can be used to relocate the nuclei.

B. Primed *in Situ* Labeling

1. When a new probe is introduced different hybridization temperatures, and different primer contrations should be tested to optimize the result. Normally, 0.5 μg of primer DNA works well.

2. Drying out of reaction mixture during incubation can be due to an ineffective humidified chamber or lack of glycerol in the reaction mixture.

3. An overall background on chromosomes can be seen in slides older than 1 week. This is most likely due to introduction of nicks in the DNA, which give rise to free 3'-ends that act as primers for chain elongation.

4. If the blocking with dideoxynucleotides is omitted in multicolor-PRINS, a double labeling of the primer used in the first PRINS reaction might be observed, since it might be further labeled with the hapten used in the second PRINS reaction.

5. In repeated-PRINS optimization of $MgCl_2$ and annealing temperature is very important, since the optimal reaction conditions can vary from what is seen with purified DNA.

6. If diffusion of the signals from repeated-PRINS is a problem, a postfixation in 3% paraformaldehyde for 10 min after washing in stop buffer can be used. After fixation the slide is washed for 5 min in washing buffer. Diffusion in normally not a problem with repeated-PRINS if the slides are evaluated immediately after the reaction, but can be a problem if the slides are stored before evaluation.

C. Visualization of Labeled DNA

We have used digoxigenin-, biotin-, and fluorescein-labeled dUTP for PRINS. Digoxigenin-labeled dUTP gives rise to the strongest signal closely followed by biotin-labeled dUTP. Fluorescein-labeled dUTP gives a somewhat weaker signal. However, with a good probe, it is advantageous to use fluorescein-12-dUTP, as results can be obtained faster in a smaller number of steps.

References

Brandt, C. A., Kierkegaard, O., Hindkjaer, J., Jensen, P. K. A., Pedersen, S., and Therkelsen, A. J. (1993). Ring chromosome 20 with loss of telomeric sequences detected by multicolour PRINS. *Clin. Genet.* **44,** 26–31.

Brandt, C. A., Djernes, B., Strømkjaer H., Petersen, M. B., Pedersen, S., Hindkjaer, J., Brinch-Iversen, J., and Bruun-Petersen, G. (1994). Pseudodicentric chromosome 18 diagnosed by chromosome painting and PRimed In Situ labelling (PRINS). *J. Med. Genet.* **31,** 99–102.

Cinti, C., Santi, S., and Maraldi, N. M. (1993). Localization of a single copy gene by PRINS technique. *Nucleic Acids Res.* **21,** 5799–5800.

Gosden, J., Hanratty, D. (1993). PCR *in situ* (PCR-IS), a rapid alternative to *in situ* hybridization for mapping short, low copy number sequences without isotopes. *Bio Technique* **15,** 78–80.

Gu, F., and Hindkjaer, J. (1996). Primed in Situ labeling (PRINS) detection of telomeric (CCCTAA)$_n$ sequences in chromosomes of domestic animals. *Mammal. Genome* **7,** 231–232.

Hindkjaer, J., Koch, J., Terkelsen, C., Brandt, C.A., Kølvraa, S., and Bolund, L. (1994). Fast and sensitive multicolor detection of nucleic acids in Situ by PRimed *In Situ* labeling (PRINS). *Cytogenet. Cell. Genet.* **66,** 152–154.

Hindkjaer, J., Brandt, CA., Koch, J., Lund, T. B., Kølvraa, S., and Bolund, L. (1995a). Simultaneous detection of centromere specific probes and chromosome painting libraries by a combination of PRimed IN situ labeling (PRINS) and chromosome painting (PRINS-painting). *Chromosome Res.* **3,** 41–44.

Hindkjaer, J., Hammoudah, S., Clausen, N., Koch, J., and Pedersen, B. (1995b). Partial triplication (q7) in a child with acute lymphoblastic leukemia demonstrated with conventional cytogenetics, PRINS and chromosome painting. *Cancer Genet .Cytogenet.* **84,** 19–23.

Hindkjaer, J., Hammoudah, A. F. M., Hansen, K. B., Jensen, P. D., Koch, J., Pedersen, B. (1995c). Translocation(1;16) identified by chromosome painting, and PRimed IN Situ labeling (PRINS). Report of two cases and review of the cytogenetic literature. *Cancer Genet. Cytogenet.* **79,** 15–20.

Hindkjaer, J., Brandt, C.A., Strømkjaer, H., Koch, J., Kølvraa, S., and Bolund, L. (1996a). PRimed IN Situ labelling (PRINS) as a rational strategy for identification of marker chromosomes using a panel of primers differentially tagging the human chromosomes. *Clin. Genet.* **50,** 437–441.

Hindkjaer, J., Terkelsen, C., Kølvraa, S., Koch, J., and Bolund, L. (1996b). Detection of nucleic acids (DNA and RNA) in situ by single and cyclic PRimed IN Situ labelling (PRINS): Two alternatives to traditional in situ hybridization methods. In *"In Situ* Hybridization" (M. Clarke, ed.), pp. 45–66. Chapman & Hall, London.

Koch, J. E., Kølvraa, S., Petersen, K. B., Gregersen, N., and Bolund, L. (1989). Oligonucleotide priming methods for the chromosome-specific labelling of α satellite DNA in situ. *Chromosoma* **98**, 259–265.

Koch, J., Hindkjaer, J., Kølvraa, S., and Bolund, L. (1995). Construction of a panel of chromosome specific oligonucleotide probes (PRINS-primers) useful for the identification of individual human chromosomes in situ. *Cytogenet. Cell Genet.* **71**, 142–147.

Macas, J., Dolezel, J., Gualberti, G., Pich, U., Shubert, I., and Lucretti, S. (1995). Primer-induced labeling of pea and field bean chromosomes in situ and in suspension. *BioTechniques* **3**, 402–408.

McLean, J. V., Tomlinson, J. E., Kuang, W-J., Eaton, D. L., Chen, E. Y., Fless, G. M., Scanu, A. M., and Lawn, R. M. (1987). CDNA sequence of human apolipoprotein(a) is homologous to plasminogen. *Nature* **330**, 132–137.

Pedersen, B., Koch, J., Bendix Hansen K., Hindkjaer, J., and Lindbjerg Andersen, C. (1997). The monosomy 7 clone in interphase and metaphase cell populations: A combined chromosome and primed in situ labeling (PRINS) study. *Acta Haematol.* **97**, 216–221.

Pellestor, F., Girardèt, A., Lefort, G., Andréo B., and Charlieu, J.P. (1995). Use of primed in situ labelling (PRINS) technique for a rapid detection of chromosomes 13, 16, 18, 21, X and Y. *Hum. Genet.* **95**, 12–17.

Pellestor, F., Girardèt, A., Lefort, G., Andréo, B., and Charlieu, J.P. (1996). Rapid chromosome detection in human gametes, zygotes, and preimplantation embryos using the PRINS technique. *J. Assist. Reprod. Genet.* **13**, 675–680.

Terkelsen, C., Koch, J., Kølvraa, S., Hindkjaer, J., Pedersen, S., and Bolund, L. (1993). Repeated primed *in situ* labeling: Formation and labeling of specific DNA sequences in chromosomes and nuclei. *Cytogenet. Cell. Genet.* **63**, 235–237.

Therkelsen, A. J., Nielsen, A., Koch, J., Hindkjaer, J., and Kølvraa, S. (1995). Staining of human telomers with PRimed IN Situ labeling (PRINS). *Cytogenet. Cell. Genet.* **68**, 115–118.

Verma, R. S. and Babu, A. (1989). "Husman Chromosomes: Manual of Basic Techniques." Pergamon, Oxford.

CHAPTER 33

Measurements of Telomere Length on Individual Chromosomes by Image Cytometry

Steven S. S. Poon[*] **and Peter M. Lansdorp**[*,†]

[*]Terry Fox Laboratory
British Columbia Cancer Agency
Vancouver, British Columbia, Canada V5Z 1L3

[†]Department of Medicine
University of British Columbia
Vancouver, British Columbia, Canada V6T 2B5

I. Introduction
    A. Structure and Function of Telomeres
    B. Conventional Method: Southern Analysis
    C. New Method: Quantitative Fluorescence *in Situ* Hybridization and Image Analysis
    D. Chapter Outline
II. Background
    A. Fluorescence Labeling of Sample
    B. Fluorescence Microscopy Imaging System
    C. Problems with Quantitating Fluorescence Images
III. Methods
    A. System Description
    B. Sample Preparation
    C. Telomere Segmentation and Integrated Fluorescence Intensity Measurements
    D. Chromosome Segmentation
    E. Calibration of Results
    F. Presentation of Results to User
IV. Results
    A. Algorithm Verification
    B. Comparison of Human Telomere Lengths with Southern Analysis

METHODS IN CELL BIOLOGY, VOL. 64

V. Discussion
    A. Validation of Telomere Integrated Fluorescence Intensity Method
    B. Chromosome Segmentation
    C. Biological Studies
    D. Future Suggestions
  References

I. Introduction

A. Structure and Function of Telomeres

The physical ends of chromosomes or telomeres are in most species composed of G-rich repeat sequences and associated proteins (Blackburn, 1995). Since 1990 much information has been obtained regarding the structure and function of telomeres in different species. The DNA of telomeres consists of specific repetitive sequences (TTAGGG in all vertebrates (Moyzis *et al.*, 1988) that are synthesized from an RNA template by a specialized reverse transcriptase called telomerase (see Greider, 1996, for review). Telomerase is essential to maintain the length of telomere repeats in the germ line (Blasco *et al.*, 1997). Most somatic cells do not express (sufficient levels of) telomerase and, as a result, telomeres in most somatic tissues shorten with each cell division (Harley *et al.*, 1990; Hastie *et al.*, 1990). The resulting loss of telomere repeats has been implicated in replicative senescence (Harley *et al.*, 1994). Recent data, showing that the life span of a normal diploid human cell can be increased by overexpression of the telomerase reverse transcriptase gene (Bodnar *et al.*, 1998; Vaziri and Benchimol, 1998) are in support of this notion. Additional interest in telomerase/telomere biology is based on the observation that tumor cells typically have increased levels of telomerase (Kim *et al.*, 1994) that could be used as a diagnostic marker and as a target for tumor therapy (Shay and Wright, 1996; Norton *et al.*, 1996).

Telomeres have a number of important functions in the cell. They provide a stable "protective" cap to the ends of chromosomes that prevents the fusions of chromosomes, and they shield chromosomes from attack by nucleases and components of the DNA repair machinery. Telomeres also provide a buffer against the "end replication" problem. It is understood that in order for conventional DNA polyerases to replicate the very 3' end of telomeric DNA, an RNA primer has to anneal to terminal repeats (Levy *et al.*, 1992) to initiate the synthesis of the novel 5' strand. At the end of the replication process, the RNA primer is removed. As a result, part of the telomeric DNA in the daughter cells is not duplicated, and the overall length of telomeric DNA is decreased. This end replication problem (Olovnikov, 1971; Watson, 1972), is resolved in most species using telomerase-mediated extension of the 3' telomere DNA strand. In higher organisms, telomerase extension of telomeres appears to be limited to cells of the germ line (Blasco *et al.*, 1997) and, possibly, germinal center B lymphocytes (Weng *et al.*, 1997). In most somatic human cells, however, several kilobases of telomeric repeats provide a buffer against the end replication problem and allow extensive but limited proliferation (Lansdorp, 1997a). Most cells appear to lose between 50 to 200 base pairs of telomeric DNA per cell division (Allsopp *et al.*, 1992; Harley *et al.*, 1994), allowing for up to a hundred cell divisions before telomere shortening is expected to compromise telomere function and signal cell senescence by currently poorly understood mechanisms. In this way, telomere-mediated replicative senescence appears to be a tumor suppressor mechanism that could potentially be very important to prevent the growth of preneoplastic cells.

The appreciation of the critical functions of telomeres/telomerase in the growth of normal and malignant cells has greatly increased the interest in this topic. In order to understand and test new ideas and hypothesis in the telomere field, tools to measure the length of telomeres in various cell types are of obvious critical importance.

B. Conventional Method: Southern Analysis

Conventionally, Southern analysis is used to estimate the average length of telomere repeat sequences. For this purpose, DNA extracted from a population of cells is digested with restriction enzymes and analyzed by gel electrophoresis (Allshire *et al.*, 1988; de Lange *et al.*, 1990). With this method DNA fragments are separated on the basis of size. The telomere repeat sequences present in a minor fraction of the DNA fragments are visualized after hybridization with a radiolabeled probe specific for telomeric DNA. After exposure of the gel to X-ray film, the final destination of the probes in the gel is recorded. The position of the hybridized telomere probe relative to molecular weight markers is then used to estimate the average size of terminal restriction fragments (TRF) containing telomeric DNA within the sample.

There are a number of drawbacks of the Southern analysis method for the analysis of telomere (TRF) length (de Lange, 1995). First, this method gives only a rough indication of the average telomere length in a population and not the telomere length of individual chromosomes. Second, a good representation of the average telomere length within a sample requires a large number of cells to yield enough telomere DNA (around 100,000 cells or more). Third, the restriction enzymes typically used in TRF analysis cleave chromosomal DNA at subtelomeric sites which are at a variable distance from the start of actual telomeric repeats. As a result, nontelomeric DNA in "telomeric fragments" which will vary from chromosome to chromosome, contributes to the telomere length estimate. Last, the Southern analysis method typically underestimates the size and number of short telomeres (as a result of the logarithmic relationship between fragment size and the migration distance in agarose gels). The higher mobility of short fragments results in a decreased local concentration (and decreased detection) of probe relative to large terminal restriction fragments. This problem is exaggerated by the use of telomeric probes that directly hybridize to T_2AG_3

repeats that are obviously present in lower numbers in short fragments. Hence, larger number of short telomeres are required to generate comparable signal intensities as long telomeres on the autoradiograph.

C. New Method: Quantitative Fluorescence *in Situ* Hybridization and Image Analysis

Information on the length of telomere repeats in individual chromosomes can also be extracted from digital images of metaphase chromosomes that were subjected to quantitative fluorescence *in situ* hybridization (Q-FISH) with directly labeled peptide nucleic acid (PNA) telomere probes (Lansdorp *et al.*, 1996). Instead of around 100,000 cells, less than 30 cells are needed to obtain an estimate of the telomere length distribution. Given that sufficient measurements are made, the length of telomere repeats is not under- or overestimated as telomere fluorescence signals are directly correlated with the length of repeats. In addition to greater sensitivity, the technique described in detail in this chapter enables studies on telomeres of specific chromosomes and on telomere length distributions of chromosomes within limited number of cells. With Q-FISH, the variation in telomere length between individual chromosomes and between different individuals has been, for the first time, determined (Zijlmans *et al.*, 1997; Martens *et al.*, 1998). Furthermore, modifications of the methods described here have allowed measurement of telomeric DNA in interphase cells by image cytometry (De Pauw *et al.*, 1998) and flow cytometry (Rufer *et al.*, 1998).

D. Chapter Outline

In this chapter, we first provide a description of the type of images that are used to estimate the length of telomeres in individual chromosomes within a cell. We then discuss the critical components of the fluorescence microscope based image acquisition system and the problems associated with quantitating the fluorescence intensities of microscopic objects with such a system. We next outline the method developed in our laboratory for measuring the integrated fluorescence intensity of individual telomeres and for linking the telomere information to specific chromosomes. We finally present the results and validation of our image analysis method and discuss some of the applications of this system.

II. Background

A. Fluorescence Labeling of Sample

Because we are interested in extracting information on the telomere length of individual chromosomes in a cell, the most practical current method is to observe metaphase chromosomes from a cell under a microscope following preparation of the sample using the Q-FISH technique (Lansdorp *et al.*, 1996). In this way, the chromosomes in the cell are spread out on the slide in a manner that allows each chromosome to be identified and karyotyped from one fluorescence image while the telomeres can be analyzed from another image captured at a different emission wavelength.

Although FISH techniques have been around for at least 15 years, it is not until relatively recently that synthetic PNA oglionucleotide probes have been developed (Nielsen *et al.*, 1991; Egholm *et al.*, 1993). Directly labeled $(CCCTAA)_3$ PNA probe was found to hybridize to telomeres with sufficient efficiency to consider quantitative analysis of the resulting fluorescence images (Lansdorp *et al.*, 1996). Relative to conventional FISH techniques with RNA or DNA probes, it was found that PNA probes can hybridize under (low-ionic strength) conditions that do not favor DNA to DNA and DNA to RNA hybridization. As a result, PNA probes can hybridize to denatured DNA target sequences without competition from identical complementary DNA sequences with the same sequence as the probe (Lansdorp *et al.*, 1997). By avoiding competition between target sequences hybridizing to probe or complementary DNA, the length of the target telomere repeat DNA becomes directly related to the number of hybridized probes. Fluorescence labeling of the probe will then result in fluorescence at individual telomeres that, in theory, should correlate to the number of probes hybridized to the telomere and hence to the length of the telomere.

The integrated fluorescence intensity (IFI) value is a measure of the total amount of fluorescence emitted from an object. In the majority of current FISH applications only a qualitative description of the hybridization result is needed, and hence the actual values of the fluorescence intensities are not important. In these applications, the resolution in the detection of specific fluorescence can be relatively coarse to enable the detection of the location and the number of fluorescence spots in a cell or object. However, a high degree of accuracy is required in Q-FISH analysis of telomeres IFI since the amount of fluorescence detected is used to estimate the actual length of telomere repeats.

B. Fluorescence Microscopy Imaging System

To capture images of metaphase chromosomes and telomeres for quantitative analysis, an integrated fluorescence microscope imaging system is used. Fluorescence images of a metaphase chromosome spread (e.g., stained with 4',6-diamidino-2-phenylindole, DAPI) and of telomeres (e.g., after Q-FISH with Cy3 labeled telomere probe) are captured by a digital camera and stored in a computer. The images are then preprocessed to remove the spatial and temporal distortions introduced by the system. The image is then segmented into the objects of interest (chromosomes and telomeres). Features and parameters are then extracted from the segmented objects. These features, then, form the basis for characterizing and classifying the objects.

The quality and accuracy in the initial step of the analysis have a large impact on the quality of the final results as errors introduced in the image acquisition steps are likely to propagate and be amplified in later steps of the analysis. Thus, efforts spent on optimizing the acquisition system to generate consistent and reproducible images for analysis are of utmost benefit and importance.

There are many commercial FISH image acquisition systems available today, all of which contain the following basic components: (i) the microscope, (ii) the camera, and (iii) the computing system. The fluorescence microscope is the key component in these imaging systems. It transforms and magnifies the fluorescence images of telomeres and chromosomes for visualization. The major components of the fluorescence microscope that affect the quality of the image are (i) the illumination source, (ii) the excitation and emission filters, and (iii) the objective lenses. Most brand name microscopes have similar performance. Once a microscope is chosen, however, the selection of components for optimization purposes are limited to those that are compatible with the microscope such that minimal modifications are required to integrate them into the system.

The choice of the illumination source is dependent on the fluorescence spectral characteristics of the fluorescent probes used. An ideal illumination source is one which (i) gives even (uniformly distributed) illumination, (ii) has sufficient intensity in the desired excitation wavelength for the probes used, and (iii) does not fluctuate over time. Most commercially available light sources for fluorescence microscopes (other than lasers) are less than ideal, and some sort of compensation and calibration is generally required.

The choice of the excitation and emission filters and the dichroic mirror also play an important role in the quality of the resulting image. To overcome the pixel shift problem of acquiring images from multiple filter blocks, one can use an emission filter that allows multiple bands of wavelengths of light to pass (Pinkel *et al.,* 1986). Hence, objects labeled with multiple fluorescent probes can be independently imaged by changing only the excitation wavelength. The excitation and emission filters are selected in conjunction with the probes used in order to minimize spectral cross talk in the resulting images. Since no optical components are moved in the imaging path, the shift between multispectrum images is small and is typically less than 2 pixels even at high magnifications. There are two drawbacks in using the ''Pinkel'' type of filter system that can be compensated for with adequate intensity illumination and appropriate selection of filters and probes. First, the amount of light that is allowed to pass through, at a selected wavelength, is diminished by approximately 10–50% compared to that of a single band-pass emission filter. Second, more noise is present as undesired light from other wavelengths, although minimal in most cases, is allowed to pass through.

Another critical component in the system that would affect the image quality is the objective lens. It plays an important role in determining (i) the spatial resolution, (ii) the chromatic response, (iii) the chromatic and spherical aberrations, and (iv) the intensity of light in the system. Spatial resolution in the objective lens is dependent on the wavelength as well as the numerical aperture of the objective lens. Spherical aberration relates to consistency in the size of the object at different points in the field of view. Chromatic aberration relates to the size of an object when seen under different wavelengths. The object size varies because the focal point is different for different wavelengths. For high quality objective lenses, such as apochromatic lenses, spherical and chromatic aberrations are relatively small in the central field of view where the objects are typically positioned and thus can be ignored. Last, the amount of light loss through the objective lens is of particular importance in FISH application, as the fluorescence signal intensities are typically low. In general, one should experiment with the lenses that are compatible with the microscope and select the best one for a particular application for optimum performance.

In addition to the microscope system, the quality of the image acquired is highly dependent on the camera used. The following considerations should be looked for in selecting a camera for quantitative microscopy: (i) high spatial resolution and large field of view, (ii) sufficient photometric resolution, high sensitivity, and large dynamic range, and (iii) multispectral image acquisition capability and relatively fast readout rates (Poon and Hunter, 1994). This could be realized using digital cameras with 12 bits of resolution and over 1000×1000 square pixels of approximately 7 μm in width each.

Finally, the computer system and acquisition software should be selected based not only on its features but also on its convenience regarding integration with the rest of the imaging system.

C. Problems with Quantitating Fluorescence Images

There are many problems associated with accurately quantitating the fluorescence intensity of objects. The majority of the problems are associated with obtaining consistent and reproducible images. Image distortions in captured images can propagate and transform into analysis errors. In addition, limitations in the object segmentation and subsequent feature extraction algorithms can also lead to inaccurate results.

1. Photobleaching

Photobleaching (fading) is an important aspect to consider in quantitative fluorescence microscopy. Antibleaching agents are often used to minimize the fading effects of the probes. If a sample was significantly photobleached (i.e., the sample fluorescence emitted was reduced by the duration of light exposure), a compensation method (such as Rigaut and Vassy, 1991) will need to be implemented to determine the amount or the stage of photobleaching in the sample so that the results can be correlated or compared with another sample captured at different times.

2. Stability of Illumination

Another problem in quantitative fluorescence microscopy is in obtaining a stable source of illumination. Temporal fluctuations are the main source of illumination inconsistency. Generally, the intensity of light output varies as the lightbulb ages. Hence, measurements taken over one day could give a different level of excitation illumination and consequently a different level of fluorescence emission from a similar sample analyzed on the next day. Some light sources simply vary too much in light output to be useful for quantitative purposes. Thus, one should perform experiments to determine the extent of the temporal fluctuation in illumination in the sample and determine if the variations of intensity measurements are acceptable (Poulin *et al.*, 1994). If the illumination is sufficiently stable over the course of an experiment session, then only one reference fluorescence sample would suffice for use in calibrating the data for that session.

Even with careful adjustment and alignment of the light source for Koehler illumination, spatial distortions in illumination of 5% or more can be observed over the field of view (hot spots from the bulb). This spatial variation in the acquired image $[I(x,y)]$ can be reduced using a flat-field compensation method similar to the following to generate the calibrated image $[C(x,y)]$:

$$C(x,y) = k\,\frac{I(x,y) - D}{B(x,y) - D} + D,$$

where $B(x,y)$ is the spatial distribution of intensities of a homogeneous bright fluorescence image, D is the spatial distribution of a dark image (which can generally be represented as an averaged value), and k is a constant scaling factor to scale the image such that the values are spread over the gray-scale intensity range. We found that this gives a flat illumination response over the field of view (standard deviation of less than 1 gray level) for different levels of intensity of illumination (Poon *et al.*, 1998a,b).

3. Variation in Preparation of Sample

Variation in the sample preparation will also contribute to variations in the fluorescence intensity of the objects in the captured image. Both variation in Q-FISH protocol (e.g., the duration of hybridization and wash steps) and variation between different samples will affect the amount of probe that is hybridized to telomeres. Even if the Q-FISH process is consistent, there may be variations in the efficiency by which the probe binds to telomere repeats depending on how exposed or tightly wrapped telomere sequences are for binding (accessibility) at a given site and how much telomeric DNA was lost in the hybridization process (Raap *et al.*, 1986).

4. Noise (Low Level Fluorescence Signals)

Because fluorescence signals are weak, the captured images are often very noisy (largely caused by photonic noise). As a result, a temporal variability in the intensities of the captured images can be observed. This noise can be reduced by averaging the intensities from a number of captured images. However, there are trade-offs to be considered, which include the time it takes to capture the images, the amount of photobleaching that occurs during the averaging process, and the amount of noise generated in other parts of the system.

5. Limited System Resolution and Sampling Error

The fluorescence objects that are imaged are generally very small in size. A typical telomere of 5 kilobases of TTAGGG repeats in metaphase chromosomes is believed to be less than 0.1 μm in size. Because telomere images are captured at the limit of resolution of the fluorescence microscope and the camera system, a slight movement of the object (e.g., displaced by one-half pixel) will result in differences in intensity values at each pixel and a different number of edge and object pixels from image to image. This problem is exacerbated because a significant portion of the telomere object pixels lie along the edge of the chromosome object, resulting in a further decrease in the precision of the segmentation and feature extraction.

III. Methods

A. System Description

The algorithm that was developed in the course of this study runs in a program (TFL-TELO) under the Microsoft Windows (version 3.1 and higher) operating system. The program can operate independently of the acquisition system and requires an image of metaphase chromosomes (stained with DAPI) and another image of the associated telomeres (labeled with Cy3). The program then generates telomere fluorescence intensity values of the chromosomes in the image. The TFL-TELO algorithm takes less than 1 min on a 100 MHz Pentium-based microcomputer to calculate the telomere fluorescence intensities of all chromosomes in a metaphase chromosome spread.

We chose a wide-field microscope over a confocal microscope for this application to avoid photobleaching (fading of the fluorescent probe caused by exposure to intense illumination) and because of cost considerations. Using the system available in our laboratory (Fig. 1) on typical Q-FISH samples, we observed that the fluorescence decay of the telomere probe is approximately 1% over 1 min of light exposure (the approximate time to set up and focus the telomere image). The fluorescence decay of beads is higher at approximately 3% per minute. Note that much higher values of bleaching are observed with standard mercury lamps that give a much more (> eight fold) intense illumination. As the variation due to photobleaching is much less than the variation in acquiring consecutive images (which can be up to 10%), we decided not to compensate for the photobleaching effects.

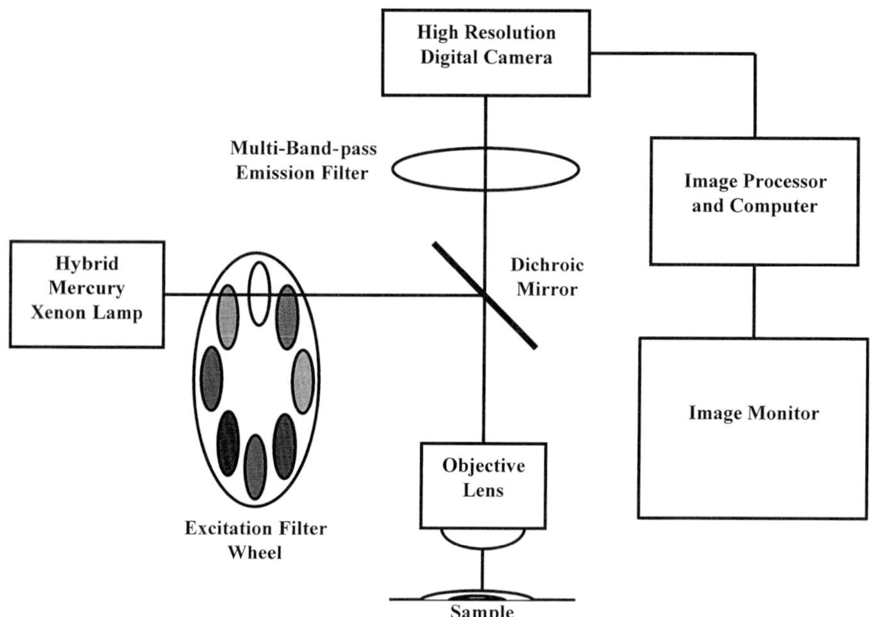

Fig. 1 Image acquisition system. Our system is based on a Zeiss fluorescence microscope with a 63×/1.4 Plan apochromatic objective lens. A 200 W hybrid mercury/xenon lamp is used as the illumination source. A filter wheel is employed to select the excitation wavelength. The image of the object is then captured via a multi-band-pass emission filter onto the Xillix high resolution digital camera. The digital image is then analyzed by a computer.

A Zeiss Axioplan fluorescence microscope was chosen because of its availability in the laboratory. For illumination stability, we equipped the microscope with a hybrid mercury/xenon lamp (200 W, OptiQuip distributed by Zeiss, Thornwood, NY). Its spectral characteristics is very similar to that of the (typical) mercury lamp but has a much longer bulb lifetime ($>$500 hr). This lamp has a number of intensity peaks including one at wavelengths of 546 nm that is used to excite the Cy3 dye coupled to the telomere PNA probes. We found that this hybrid lamp fluctuates less in time than the standard Zeiss mercury lamp and is more intense than the xenon lamp at the wavelengths of interest. Importantly, we observed that the illumination from this hybrid lamp fluctuates less than 1 gray level (out of 256) over the course of a typical 3-hr image acquisition session.

A filter wheel (Pacific Scientific, Rockford, IL) is used to select two of the eight possible excitation filters while a single multi-band-pass emission filter allows only the sample fluorescence light to pass. These excitation filters are used in combination with a DAPI/Cy3 dichroic mirror/emission filter set (Chroma Technology, Brattleboro, VT). We selected the fluorescence 63× magnification objective lens with a numerical aperture of 1.4 (Plan Apochromat 63×/1.4, Zeiss) because it has the best all around performance over other (Zeiss) 63× and 100×

magnification objectives for our purposes. We chose the MicroImager MI1400-12 digital camera (Xillix Technologies, Vancouver, Canada) for this project as it meets the camera requirements (described earlier in Section II,B) (Jaggi *et al.*, 1993; Poon and Hunter, 1994) and was available in the laboratory.

B. Sample Preparation

Metaphase chromosomes are prepared from dividing cells using standard cytogenetic techniques. Typically cells are cultured for several days in tissue culture medium supplemented with growth factors. Colcemid is added to cultures at, for example, 0.05 mg/ml prior to harvest of mitotic cells. The duration of colcemid treatment varies with the cell type (lymphocytes 1/2–2 hr/fibroblasts 6–24 hr). The cells are washed following colcemid incubation and treated with hypotonic KCl buffer (75 mM, 37°C) for 30 min. Cells are fixed in methanol/acetic acid according to standard procedures and stored at -20°C in fixative. Fixed cells can be stored for at least 3 months. However, prior to hybridization, cells should be washed twice in fresh fixative and concentrated at around 5×10^6/ml. Typically two drops are spotted on a clean (wiped with a cloth containing a 1 : 1 ether/ethanol solution) slide with a Pasteur pipette. Slides are air dried overnight. Prior to hybridization the cells are rehydrated in phosphate-buffered saline (PBS), followed by fixation with formaldehyde (4%) in PBS (2 min), wash steps with PBS (three times, 5 min), and treatment with pepsin (P-7000, Sigma, St. Louis, MO) at 1 mg/ml at pH 2.0 (acidified water, 10 min, 37°C). Formaldehyde fixation and wash steps are repeated prior to dehydration in ethanol (70, 90, and 100% for 5 min each). Slides are air dried and the hybridization mixture containing 70% formamide and 0.3 μg/ml Cy3 conjugated $(C_3TA_2)_3$ PNA probe (a custom order from Perseptive Biosystems, Framingham MA, see http://www.pbio.com/cat/synth/pna/custmpna.htm) is prepared. The hybridization mixture contains the following:

Formamide (ultra pure, pH 7.0–7.5)	70% (v/v)
blocking reagent (Dupont, Boston, MA)	0.25% (w/v)
0.2 M Tris	10 mM
PNA Tel Cy3	0.3 μg/ml
25 mM MgCl$_2$, 9 mM citric acid, 82 mM Na$_2$HPO$_4$ with double distilled water as diluent	5% (v/v)

Hybridization mixture (2×10 μl) is deposited on a coverslip (22×60 mm) and, avoiding air bubbles, lifted by positioning a slide (upside down) on top of it. Cells are denatured on a thick metal (e.g., 1 cm aluminum) plate in a preheated oven (80°C) for 3 min. Slides are then placed in a plastic slide box, and the box is positioned in a plastic beaker containing a damp paper towel. The beaker is covered with Parafilm, and hybridization is allowed to proceed for 1–2 hr at room temperature. The coverslip is carefully removed in wash solution I [70% formamide, 10 mM Tris, 0.1% bovine serum albumin (BSA), pH 7.0–7.5], and the slides are washed two times for 15 min in wash solution I, followed by three times for 5 min wash steps in wash solution II (0.1 M Tris, 0.15 M NaCl, 0.08%

Tween 20, pH 7.0–7.5). Cells are then dehydrated in ethanol (5 min 70% ethanol, 5 min 90% ethanol, 5 min 100% ethanol) and dried in air. Drops ($2 \times 10\ \mu\mathrm{l}$) of antifade solution (Vectashield, Vector Laboratories, Burlingame, CA) containing $0.2\ \mu\mathrm{g/ml}$ of DAPI are deposited on a (22×60 mm) coverslip, and the slide (upside down) is used to lift the coverslip. Slides can be stored for up to several weeks at room temperature prior to analysis in a light-protected storage box and are suitable for karyotyping studies on the telomere length of individual chromosomes in a cell.

DAPI was chosen to label chromosomes for two reasons. First, this DNA dye gives bright staining of double stranded DNA that facilitates karyotyping. Second, the spectral characteristics of DAPI are very different from that of Cy3. As a result, there is minimal interference of fluorescence signals between the two dyes, and intensity compensation for spectral overlap (Castleman, 1993) is not required.

C. Telomere Segmentation and Integrated Fluorescence Intensity Measurements

The major problem in accurately quantifying the IFI of telomeres from the captured images lies in their segmentation, that is, determining the exact boundaries of each telomere. Most telomeres are relatively easy to detect because they appear as bright spots. Approximate locations of these spots can be found by thresholding or edge detection methods (Poon *et al.*, 1993a; Russ, 1990). However, a problem arises in locating the exact location of the borders. If the estimated telomere borders are closer to the peak intensity, the IFI value will be underestimated. Conversely, if too much background intensity is included in the estimated border, the IFI value will be overestimated. There is also the problem of segmenting telomeres that are close to each other and determining which pixels belong to which telomere.

The first step of our segmentation algorithm is to find the location of each telomere in the image. For this purpose, we first use our average difference filter (Poon *et al.*, 1998a,b), which is similar to that of the Laplacian filter (Russ, 1990). The average intensity value of its surrounding pixels is subtracted from its intensity, $I(x,y)$ to generate an edge image, $E(x,y)$ as follows

$$E(x,y) = I(x,y) - \frac{1}{9} \sum_{i=-1}^{1} \sum_{j=-1}^{1} I(x-i, y-j).$$

We next apply a threshold to the average difference image, $E(x,y)$, to select the telomere objects from the background regions. Unlike other thresholding techniques, we do not look for a valley in the histogram like most other methods because there are instances where the histogram is not bimodal and hence does not have a valley. Instead, we assume that the intensity distribution of the background has a Gaussian distribution, and then select the threshold level that corresponds to the point in the object intensity histogram where 95% of the Gaussian background is removed.

This threshold level seems to be optimal for removing background noise pixels and also for preserving the relevant telomere peaks. At the center portion of a telomere, the average value of the surrounding pixels is generally less intense and hence the value $E(x,y)$ is positive and large. At the edges of the telomere, the average value of the surrounding pixels is generally the same as the pixel value since on average, half of the surrounding pixels have lower intensities than the central pixel and the other half have higher intensities. As a result, the value $E(x,y)$ is small and near zero. Similarly, at the background region, the average value of the surrounding pixels is similar to that of the central pixel. Hence, by using thresholding, the noise pixels in the background and the edge pixels of the telomeres are removed. Telomeres that are close to each other can also be separated using this technique. The reason is that the valley in between two nearby telomeres is lower in intensity than the average surrounding and thus can be removed by thresholding. An example of the use of our segmentation method on a typical telomere image is shown in Fig. 2.

The above algorithm removes the edges of telomeres. To recover these edges, we first need to dilate. However, this will combine or fuse multiple telomeres into one. To overcome this problem, we first perform labeling. In the labeling process, each continuous connected object is given a unique number and then

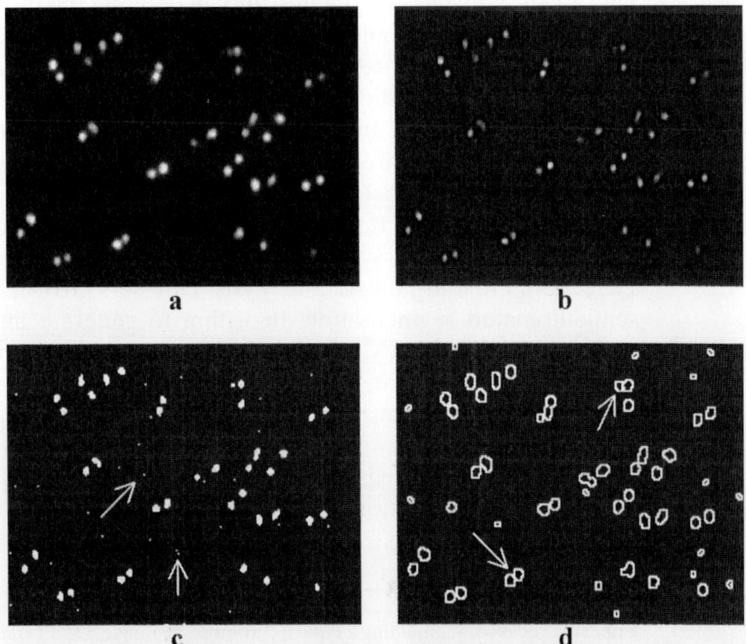

Fig. 2 Process of segmenting a typical telomere image. A typical fluorescence image of the telomeres (a) is processed with the average difference filter to generate image (b). A threshold level is then selected from the histogram of the processed image, and this threshold is applied to (b) to generate a binary image (c) of the telomeres. The resulting bright spots in (c) are first labeled and then dilated to generate the final mask image for the telomeres. The boundaries of the segmentation results (d) are then generated.

the size of each telomere mask is dilated by one to recover the lost edges of that telomere. Note that if dilation was performed before the labeling, objects that are closed together may be connected and considered as one object by the labeling process (those objects pointed out by the arrows in Fig. 2d). Finally, objects that are too small and whose intensities are similar to the background level (those objects pointed out by the arrows in Fig. 2c) are classified as artifacts and rejected from further analysis.

D. Chromosome Segmentation

We next describe how we segment fluorescence microscopy images of metaphase chromosomes. By segmenting and identifying each chromosome, the telomere lengths of individual chromosomes can be determined. Chromosome segmentation has always been a difficult task and no fully automated and accurate method is available despite intensive research efforts (Preston, 1976). The variability in the chromosome texture (intensity) within individual chromosomes and among different chromosomes makes it difficult to find the exact border for each chromosome in different metaphase preparations. Furthermore, correct segmentation is hindered by the high noise levels associated with low light level fluorescence images and the difficulty in defining the boundaries of touching and overlapping chromosomes. Although one can select metaphases where all the chromosomes are isolated from one another, such images are rare to find. Hence, one typically scans a slide to find a metaphase where most of the chromosomes are isolated from one another and some of which are touching or overlapping. Images are then acquired and used for analysis.

Due to the difficulties in chromosome segmentation, no single segmentation technique is available that can correctly segment all chromosomes. Commercial chromosome analysis systems (Applied Imaging, Santa Clara, CA; Biological Detection, Pittsburgh, PA; and Vysis, Downers Grove, IL) tend to use a simple semiautomated segmentation algorithm to generate an initial estimate of the chromosome borders. Most of the chromosomes are correctly segmented by the method described here. Hence, less user interaction is required in the manual verification process, resulting in a less tedious and a more economical overall interactive analysis.

Our chromosome segmentation algorithm (Fig. 3) consists of a combination of different segmentation methods because a single technique does not produce

---

Fig. 3 Process of segmenting a typical chromosome image. DAPI stained metaphase chromosomes (a) are first thresholded to generate a first approximation mask image (b). The masked chromosome region is then processed by an average difference filter to give (c) and subsequently a non-negative constraint is applied to give the results in (d). The resulting image is dilated (e) and the edges of the results are found (f). The borders found in (f) are removed from mask (e) to generate (g) such that objects that are still touching can be further split. The mask in (g) is then dilated to result in the borders of the segmented objects in (h).

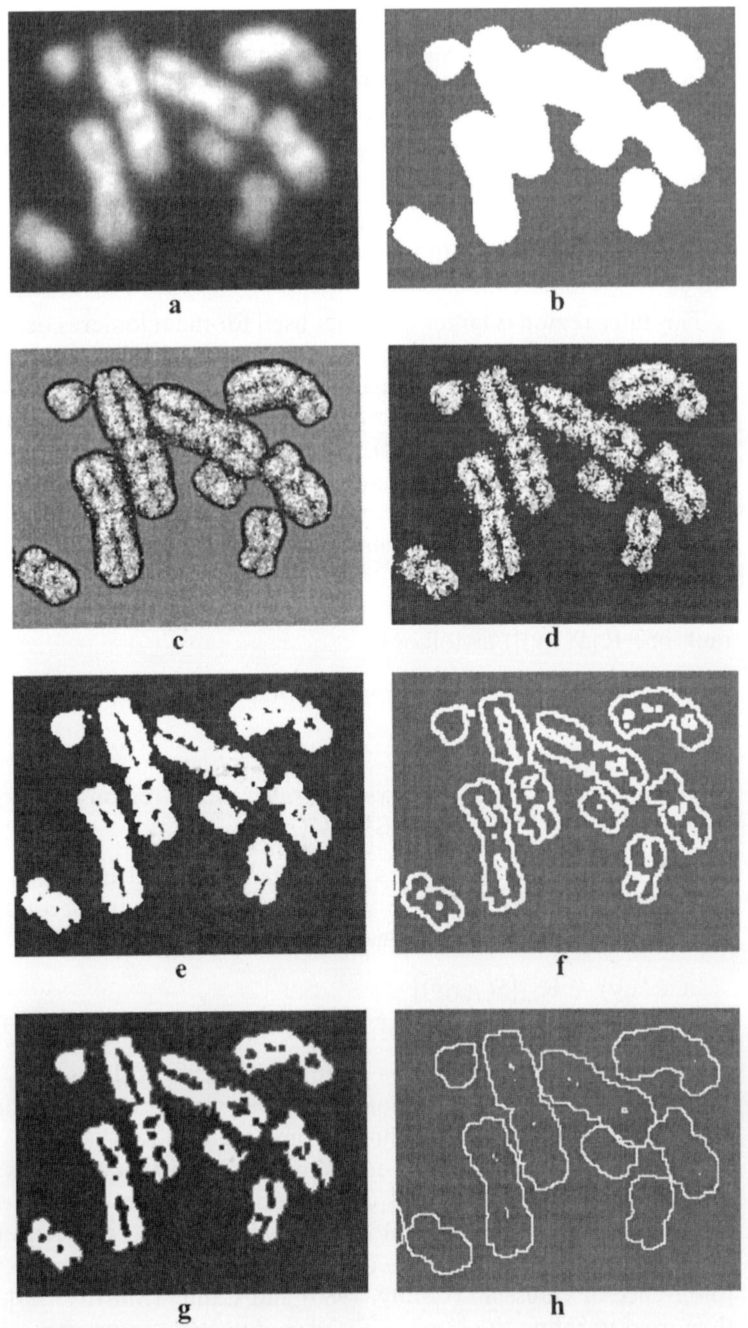

good results. Each method or step in the sequence improves on the results obtained by the previous step. Thresholding is first used to define the first approximation of the regions occupied by chromosomes (Fig. 3b). Texture information in the segmented region is then used to generate the second approximation of the chromosome region. In this step (Fig. 3c), we first detect the local high intensity pixels using the following average difference filter:

$$j(x,y) = i_1(x,y) - \frac{1}{25} \sum_{m=-2}^{2} \sum_{n=-2}^{2} i_1(x - m, y - n).$$

This filter region is larger than that used for the telomeres because it is needed to smooth out more noise and texture that is present in the chromosome images. We then impose a non-negative constraint on the difference image that sets the background regions that are close to the chromosomes, and most of the chromosome edges to 0 (Fig. 3d). Most of the points in-between touching chromosomes are also eliminated, because they have negative difference values.

We then use our rank difference filter as a morphological operator to merge detected pixels into different chromosome regions and at the same time further separate touching chromosomes (Poon *et al.*, 1998a,b). This filter is defined as the difference of two rank filters (i.e., an upper rank $R_u[S(i(x,y))]$ and a lower rank one $R_l[S(x,y)]$) as follows:

$$R_{u,l}[S(i(x,y))] = R_u[S(i(x,y))] - R_l[S(i(x,y))],$$

where

minimum rank $\leq l < u \leq$ *maximum rank* over a region $S(i(x,y))$ of the image $i(x,y)$.

The rank difference image is then binarized by setting all negative values to 0 and all others to 255. The resulting image $i_2(x,y)$ is the second approximation to the chromosome region (Fig. 3e) and is given by:

   a. $i_2(x,y) = R_{7,1}[S(j(x,y))]$
   b. if $i_2(x,y) > 0$, replace $i_2(x,y)$ with 255 otherwise replace $i_2(x,y)$ with 0.

Standard dilation and erosion morphological filters (Russ, 1990) were not used in this step because they do not perform as well as our filter for both merging and separating the different chromosome regions.

The final step first finds the edges of the chromosome regions using our rank difference filter again, but this time it functions as an edge detector. Our rank difference filter is simple, employing only integer operations, to implement and generate results comparable to or better than the more complex edge detectors [difference of Gaussian (Canny, 1986) and Canny (Marr, 1982)]. The edges are then used to refine the borders for each detected chromosome region.

The final step in determining the chromosome region is to refine the results of the previous approximation to obtain a better estimate of the chromosome border and to label or distinguish the region of one chromosome from another. At this stage, we can use either the difference of Gaussian, Canny, or rank difference filter because for binary pictures, they all produce good results. We chose to use our rank difference filter because of the following reasons. First, the algorithm is already available within the program. Second, the algorithm uses only integer operations, is less complex, and hence it is faster to compute. Last and most importantly, the rank difference filter gives thick edges at the appropriate locations such that some of the remaining touching chromosomes that are not segmented in previous approximations can be separated. In this instance, the rank difference filter is used as an edge detector instead of a selective dilation filter. The purpose of this filter operation is to determine the borders of the chromosome regions. This filter operates over a 3×3 neighborhood using 9 and 1 as the upper and lower rank numbers, respectively. This filter generates the boundary image $b(x,y)$ which is defined as follows (Fig. 3f):

$$b(x,y) = R_{9,1}[S(i_2(x,y))].$$

Since the input image is binary, the resulting image is also binary (values of 0 and 255). The filter generates thick edges (approximately 3 pixels wide) around the boundaries of the chromosome. Pixels in between touching chromosomes that have 4 or less continuous pixels now become an edge pixel and are set to 255. A logical arithmetic operation is then employed to separate the touching chromosomes. In this operation, a new chromosome region, $m(x,y)$ is generated based on the logical AND ($\cdot$) of the previous chromosome approximation region, $i_2(x,y)$, with the logical NOT (——) of the newly calculated boundary image, $b(x,y)$ as follows (Fig. 3g):

$$m(x,y) = i_2(x,y) \cdot \overline{b(x,y)}.$$

The resulting image, $m(x,y)$ then contains regions defined by the second approximation image $i_2(x,y)$ less those boundary pixels that lie in both $i_2(x,y)$ and $b(x,y)$. Although the regions found are smaller than the actual regions of the chromosomes, they are mostly distinct and isolated from one another. Each object is next labeled such that each isolated region is given a distinct number. The size of the region of each labeled object is then increased such that it is representative of the size of the chromosomes (Fig. 3h). This increasing process is accomplished by dilating each labeled region twice using a 3×3 dilation filter. In this dilation process, the center pixel in the 3×3 region is set to 255 if any of the pixels in the region has a value of 255. Otherwise, the center pixel is set to 0. As different label numbers are used in the dilation process, regions that touch one another after the dilation are kept distinct with different label numbers.

E. Calibration of Results

In order that data from one experiment can be meaningfully compared to a different experiment or to one performed at a different time, the data collected must be calibrated as the system may vary over time (e.g., day to day variation due to aging of the lamp, alignment of the optics). An initial experiment using different size plasmids and 0.1-μm beads is used to extract the calibration parameters for the system. From the plasmid experiment, one can deduce the parameters (slope M and intercept B of the linear relationship) to convert the calculated IFI values to telomere length estimates. We next measure the IFI value of beads and obtain the average IFI value. This value then serves as the normalization constant (K) for other experiments. In subsequent experiments, a normalization slide containing 0.1-μm beads is analyzed to generate IFI(bead) in addition to the chromosomes. The results of these experiments are then calibrated using the parameters determined from the initial calibration experiment to convert the measured IFI valued to telomere length estimate values as follows:

$$\text{Telomere length} = [\text{IFI(bead)}/K] \times [M \times \text{IFI(telomere)} + B].$$

F. Presentation of Results to User

The results are then presented to the user for interactive verification and editing. The key user-interaction features that were required as defined in consultation with the users are implemented into the analysis program. These factors are: (i) displaying an enhanced view of the banding structure in the chromosome for ease in karyotyping (chromosome classification), (ii) displaying the relative telomere positions on the chromosome image, (iii) marking and labeling each telomere in a chromosome, (iv) displaying the segmentation results, (v) marking and labeling each chromosome, (vi) displaying the results of the quantification (chromosome number, number assigned, telomere IFI for each chromosome, chromosome IFI, and area feature values), (vii) sorting the chromosome list based on the selected feature, and (viii) editing capabilities. The resulting implementation is shown in Fig. 4.

The resulting image displayed shows the borders of the telomeres overlaid onto the processed chromosome image. To accomplish this, the image of the chromosome is first thresholded such that the background pixels are set to a mid-intensity gray level of 128. This gray level allows the chromosome borders and both dark and bright intensity bands in the chromosomes to be seen on a gray background. The image is then inverted (linear map of 0–255 to 255–0). Contrast stretching is then performed on the chromosome to enhance the details of its banding structures.

The image of the telomere objects are next processed. To ensure that the telomeres lie at the ends of the chromosomes in the superimposed telomere–chromosome image, a pattern matching algorithm is then used to determine the placement of the detected telomere borders onto the chromosome image. In this matching algorithm, the telomere image is first shifted, pixel by pixel, from the chromosome image. For each pixel shift location, the number of detected telomeres that are within the borders of the detected chromosomes is determined.

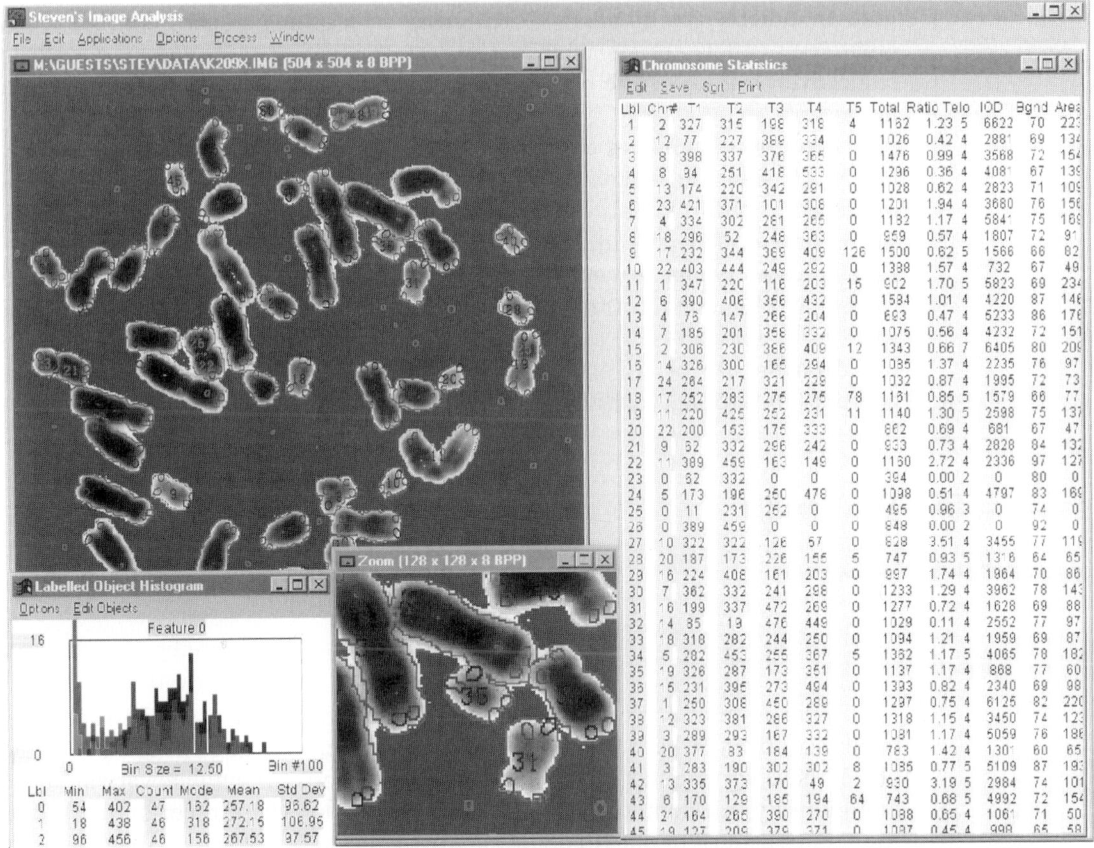

Fig. 4 Sample output of TFL-TELO program. The segmented telomere boundaries are superimposed on the segmented chromosome image. The IFI values of each chromosome are listed in the table. A 2×-magnified window and the different boundary colors help with the verification and editing of the results generated by the program. (See color plates.)

The pixel shift value (x,y) that generates the highest number of telomere/chromosome object matches becomes the pixel shift value for aligning the two images. The border of the telomeres are then superimposed onto the enhanced chromosome image. A different color is assigned to each of the telomeres in the chromosome for ease of user recognition (purple, red, blue, cyan, orange, and green are assigned to telomere number 1, 2, 3, 4, 5, and 6 or more). Although there are only four telomeres in each chromosome, two additional colors are used to facilitate the editing of the detected telomere results and to identify other non-telomeric probes (e.g., centromere of chromosome 17).

Finally, the border of the chromosomes are overlaid on the enhanced chromosome image with the highlighted telomere borders. Again, a different color is used to indicate the number of "telomere" objects in the chromosome (cyan, green, and yellow are assigned to 3 or less, 4 and 5 or more telomere objects, respectively). Each of the chromosome objects are then labeled with a unique number.

A corresponding list of chromosome and telomere IFI can be displayed in a window beside the image. An enlarged subregion of the image can also be displayed to aid in visualizing the details of the image. The user can then edit the information displayed. The editing capabilities include (i) the joining of chromosomes (which are improperly segmented), (ii) splitting telomere objects (touching telomeres), (iii) reassigning the telomere number in the chromosome (pair sister telomeres: 1 with 2 and 3 with 4), (iv) ranking the chromosomes based on its size or IFI value, (v) assigning a chromosome number to each chromosome (for karyotyping purposes), and (vi) adding a comment to the data.

IV. Results

A. Algorithm Verification

1. Simulated Objects

To validate our telomere IFI quantification algorithm, we used different simulated test objects where the relative IFI value of each is known. The simulated objects have varying shapes and intensity distributions (to simulate varying shape and intensity of telomeres) but the same IFI value. These objects are used to test the robustness of our IFI algorithm to see if similar IFI values are generated. The simulated objects consist of (i) a single point source whose dimensions are 1 pixel (object 1 in Table I), and (ii) point sources whose dimensions are greater than 1 pixel in the x directions (objects 2, 3, and 4 in Table I), greater than 1 pixel in the x and y directions (object 5 in Table I) and greater than 1 pixel in the z direction (objects 6, 7, 8, and 9 in Table I). Simulated images were generated by convolving test objects with the three-dimensional theoretical point spread function (PSF) of the system (Poon *et al.*, 1993b). Whereas the spatial distribution of each simulated object is different, the IFI value is the same for all objects (i.e., the sum of pixel intensity values for each object is the same as the sum of pixel intensity values for other objects). The values and shapes of the simulated test objects (in the x–z plane) are shown in Table I. The results of the IFI calculations (normalized to the single point object) are shown in Table I.

2. Fluorescent Beads and Plasmids

Images of different size fluorescence beads (0.1, 0.2, 0.5, and 1.0 μm) are acquired, and the IFI value at each focus plane are generated for each bead. If

Table I
Simulated Test Object Values, Shapes, and Their Calculated Integrated Fluorescence Intensity Values[a]

Object number	x = pixel 0	x = pixel 1	x = pixel 2	x = pixel 3	Calculated IFI
1		1.00			100
2		0.50	0.50		98
3	0.3333	0.3334	0.3333		98
4	0.25	0.25	0.25	0.25	95
5	0.13 0.22	0.25 0.15	0.06 0.19		95
6		0.50 0.50			98
7		0.3333 0.3334 0.3333			96
8		0.25 0.25 0.25 0.25			93
9		0.20 0.20 0.20 0.20 0.20			91

[a] For all objects, the horizontal direction represents the extent of luminance in the x direction. For object 5, the vertical direction represents the extent of luminance in the y direction. For objects 6, 7, 8, and 9, the vertical direction represents the extent of luminance in the z direction.

we assume that the fluorescence intensity of the bead is related to the three-dimensional size of the bead, then the IFI of the bead would be proportional to the cube of its one-dimensional size (diameter). The results of this experiment and the normalized expected theoretical IFI values are summarized in Table II and plotted in Fig. 5.

Table II
Comparison of Calculated and Theoretical Integrated Fluorescence Intensity Values of Different Size Beads

Diameter of bead (μm)	Best focus IFI (mean $\pm$ standard deviation)	Best focus normalized IFI	Theor. IFI
0.1	9.24 $\pm$ 2.64	1.00 $\pm$ 29%	1
0.2	99.1 $\pm$ 13.3	10.7 $\pm$ 13%	8
0.5	1162 $\pm$ 37	126 $\pm$ 3%	125
1.0	9344 $\pm$ 203	1011 $\pm$ 2%	1000

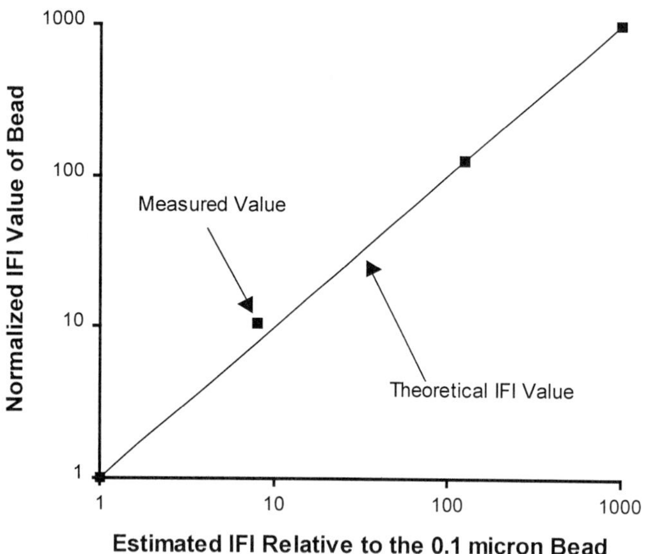

Fig. 5 Comparison of calculated and theoretical IFI distribution of different size beads. Solid squares represent measured values; the line represents theoretical values.

Similar to the analysis for the beads, images of different size telomere repeat sequences within plasmids (Hanish *et al.*, 1994; Martens *et al.*, 1998) (150, 400, 800, and 1600 base pairs) are acquired, and the total IFI value is generated for each plasmid. If we assume that the fluorescence intensity of the plasmid is related to the number of telomere base pairs present (such as in our assumption for telomeres in chromosomes), then the IFI of the plasmid would be proportional to the number of base pairs in the plasmid. The results of this experiment and the normalized expected theoretical IFI values are summarized in Table III and plotted in Fig. 6.

Table III
Comparison of Calculated and Theoretical Integrated Fluorescence Intensity Values of Different Size Plasmids

Size of plasmid (base pairs)	Normalized to 150 base pairs	Best focus IFI (mean + standard deviation)	Best focus normalized IFI
150	1.0	4.17 ± 1.31	1.00 ± 31%
400	2.7	11.2 ± 3.4	2.69 ± 29%
800	5.3	24.3 ± 4.6	5.83 ± 19%
1600	10.7	44.7 ± 5.7	10.7 ± 13%

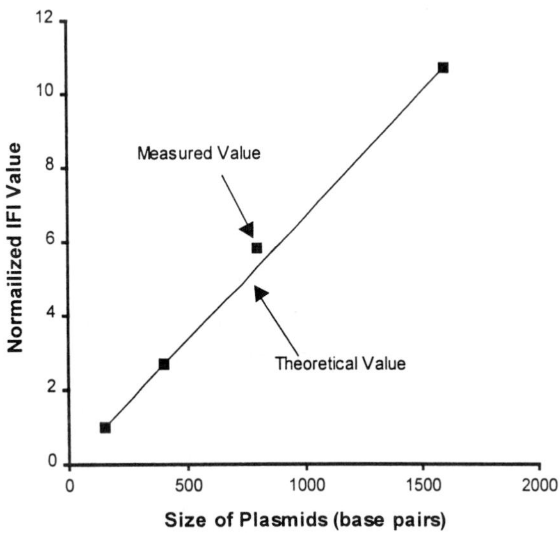

Fig. 6 Comparison of calculated and theoretical IFI distribution of different size plasmids. Solid squares represent measured values; the line represents theoretical values.

B. Comparison of Human Telomere Lengths with Southern Analysis

We next applied our algorithm to human cells. The measured distribution of the telomere IFI values is shown in Fig. 7. The validity of Q-FISH for the generation of telomere length estimates was confirmed by correlating the results of Southern analysis of cells (Lansdorp *et al.*, 1996) and sorted chromosomes (Martens *et al.*, 1998) with results generated by Q-FISH analysis.

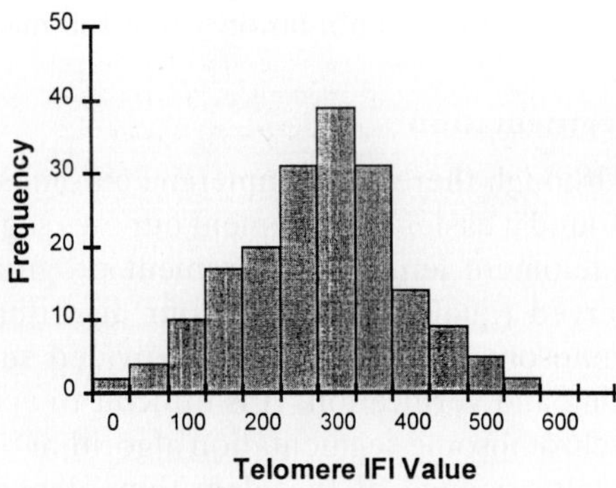

Fig. 7 Telomere IFI distribution in a cell.

V. Discussion

A. Validation of Telomere Integrated Fluorescence Intensity Method

No other method is currently available to determine the length of individual telomeres. Hence, no direct method for verifying the accuracy of our algorithms for telomere length measurements is available. For this reason, we resort to indirect methods (i.e., telomere-like objects of known fluorescence intensities) to validate our fluorescence measurements. These objects include (i) simulated objects of different shapes and sizes, (ii) fluorescence beads of known size and relative fluorescence intensities, and (iii) plasmids with known telomere insert lengths (which are typically an order of magnitude less in length than the telomeres in the cells).

Our algorithm estimated the IFI of simulated objects of varying shapes and sizes to within $\pm 3\%$. The estimated mean IFI values correlated well (correlation coefficient of 0.99) with the size of the fluorescence beads and with the length of the telomere insert in plasmids. The standard deviation in the estimation ranged from 2% for the 1-μm beads to 13% for the 0.2-μm beads to 29% for the 0.1-μm beads. The standard deviation was larger for the smaller beads primarily because of actual variations in size. The standard deviation for telomere inserts in plasmids was around 20% of the mean estimated IFI value. This variance is most likely due to the variable efficiency of the hybridization procedure (binding of the probe). Although the variation appears to be large, we observed that by averaging the results of 10 or more cells, a good indication (i.e., differences between groups with a significance level less than 0.05 using the Wilcoxon rank sum test) of the telomere length on a particular chromosome arm in a population of cells can be obtained. It is important to note that there are no other methods that can produce similar, let alone better results.

The telomere length distribution in adult cells is asymmetric, appears log normal (Oexle, 1998), and resembles the results from Southern analysis (Allshire *et al.*, 1988). The asymmetry in the distribution may be attributed to telomere breakage and recombination when telomeres reach a critical length.

B. Chromosome Segmentation

Although there are commercial chromosome segmentation systems available, we found it easier to implement our own segmentation algorithm than to integrate our telomere length measurement program with the commercial systems. We observed (qualitatively) that our algorithms correctly segmented most of the chromosomes and as a result provided substantial savings in time for manual editing and verification. It is difficult to accurately quantify the performance of our chromosome segmentation algorithm because of the following reasons. First, the performance of the algorithms depends largely on the samples that are analyzed, and these are typically manually selected. Second, we would need to purchase or have access to a commercial chromosome segmentation package in order to compare the performance of the systems. Last, there are cases where it is difficult to visually distinguish which portion of a touching or an overlapping chromosome belongs to which chromosome.

C. Biological Studies

Using our telomere analysis system, other experiments and investigations can now be performed to study and determine the role telomeres play in the aging process and in patients with cancer or genetic disorders. The advantage of our system is that significantly fewer cells (less than 30 cells) are required to obtain comparable results to the conventional Southern analysis that requires analysis of approximately 100,000 cells. This makes it possible to carry biological studies when only a limited number of cells are available for analysis. In addition, telomere length studies can now be carried out on individual cells as well as individual chromosomes in every cell.

For example, we found in one study (Martens *et al.*, 1998) that for an individual, the telomere lengths in a specific chromosome from a certain tissue are very similar to those of other tissues. However, the telomere lengths do vary from one individual to another. We also noted that the telomeres on the arms of chromosome 17p are consistently among the shorter telomeres in the cell from an experiment of 11 unrelated individuals.

In another study, we have found that there is a large variation in the telomere lengths in mouse cells (Zijlmans *et al.*, 1997). Previous studies in this area have shown that mice have long telomeres that do not appear to shorten as they age. This phenomena contradicts the concept that telomeres shorten with age. We also observed that there are specific chromosomes in bone marrow and skin fibroblast cells in individual mice that have similar telomere lengths. We also observed the presence of very short telomeres which may be the critical link in limiting the cell replication process (aging process) in mice. Other examples of studies in which Q-FISH was used can be found in the literature (Blasco *et al.*, 1997; Lansdorp *et al.*, 1997; Zhu *et al.*, 1998; Slijepcevic *et al.*, 1997; Lansdorp, 1997b; Hande *et al.*, 1998; Wan *et al.*, 1998; Katz *et al.*, 1998).

D. Future Suggestions

Although the system is currently used on a daily basis in a number of research centers, there are a number of improvements that can be made to the system. One improvement would be to provide an objective (automated) method of focusing rather than manually selecting the best focus plane (Poon *et al.*, 1992). This would result in a more consistent estimate of the telomere length because the IFI value can vary by 20% if the focus is 0.2 μm from its optimal position.

Another improvement would be to provide a more convenient means of analyzing images acquired by other image acquisition systems. Currently, the program supports the Microsoft's Window BMP file format. If the more common TIFF format is supported, then images from other acquisition systems can be more readily transported to our system for analysis.

Acknowledgments

Dr. Uwe M. Martens, Dr. J. Mark J.M. Zijlmans, and Elizabeth Chavez (Terry Fox Laboratory for Hematology/Oncology) are thanked for offering constructive criticism and feedback on the use of the telomere image analysis software during its development. Dr. Rabab Ward is thanked for discussions and guidance during various steps of the work.

Research Support: This work is supported from Grants AI29524 and GM56162 from the National Institutes of Health and by a grant from the National Cancer Institute of Canada with funds from the Terry Fox Run.

References

Allshire, R. C., Gosden, J. R., Cross, S. H., Cranston, G., Rout, D., Sugawara, N., Szostak, J. W., Fantes, P. A., and Hastie, N. D. (1988). Telomeric repeat from *T. thermophila* cross-hybridizes with human telomeres. *Nature* **332,** 656–659.

Allsopp, R. C., Vaziri, H., Patterson, C., Goldstein, S., Younglai, E. V., Futcher, A. B., Greider, C. W., and Harley, C. B. (1992). Telomere length predicts replicative capacity of human fibroblasts. *Proc. Natl. Acad. Sci. U.S.A.* **89,** 10114–10118.

Blackburn, E. H. (1995). Telomeres. *In* "Developmentally Programmed Healing of Chromosomes" (E. H. Blackburn and C. W. Greider, eds.), pp. 193–218. Cold Spring Harbor Laboratory, Plainview, New York.

Blasco, M. A., Lee, H.-W., Hande, M. P., Samper, E., Lansdorp, P. M., DePinho, R. A., and Greider, C. W. (1997). Telomere shortening and tumor formation by mouse cell lacking telomerase RNA. *Cell* **91,** 25–34.

Bodnar, A. G., Ouellette, M., Frolkis, M., Holt, S. E., Chiu, C.-P., Morin, G. B., Harley, C. B., Shay, J. W., Lichtsteiner, S., and Wright, W. E. (1998). Extension of life-span by introduction of telomerase into normal human cells. *Science* **279,** 349–353.

Canny, J. F. (1986). A computational approach to edge detection. *IEEE Trans. Pattern Analysis Maritime Intelligence* **6,** 679–698.

Castleman, K. R. (1993). Color compensation for digitized FISH images. *Bioimaging* **1,** 159–165.

de Lange, T. (1995). Telomere dynamics and genome instability in human cancer. *In* "Telomeres" (E. H. Blackburn and C. W. Greider, eds.), pp. 265–293. Cold Spring Harbor Laboratory, New York, New York.

de Lange, T., Shiue, L., Myers, R., Cox, D. R., Naylor, S. L., Killery, A. M., and Varmus, H. E. (1990). Structure and variability of human chromosome ends. *Mol. Cell Biol.* **10,** 518–527.

De Pauw, E. S. D., Verwoerd, N. P., Duinkerken, N., Willemze, R., Raap, A. K., Fibbe, W. E., and Tanke, H. J. (1998). Assessment of telomere length in hematopoietic interphase cells using in situ hybridization and digital fluorescence microscopy. *Cytometry* **32,** 163–169.

Egholm, M., Buchardt, O., Christensen, L., Behrens, C., Freier, S., Driver, D. A., Berg, R. H., Kim, S. K., Norden, B., and Nielsen, P. E. (1993). PNA hybridizes to complementary oligonucleotides obeying the Watson-Crick hydrogen bonding rules. *Nature* **365,** 566–568.

Greider, C. W. (1996). Telomere length regulation. *Annu. Rev. Biochem.* **65,** 337–365.

Hande, M. P., Samper, E., Lansdorp, P., and Blasco, M. A. (1999). Telomere length dynamics and chromosomal instability in cells derived from telomerase null mice. *J. Cell Biol.* **144,** 589–601.

Hanish, J. P., Yanowitz, J. L., and de Lange, T. (1994). Stringent sequence requirements for the formation of human telomeres. *Proc. Natl. Acad. Sci. U.S.A.* **91,** 8861–8865.

Harley, C. B., Futcher, A. B., and Greider, C. W. (1990). Telomeres shorten during ageing of human fibroblasts. *Nature* **345,** 458–460.

Harley, C. B., Kim, N. W., Prowse, K. R., Weinrich, S. L., Hirsch, K. S., West, M. D., Bacchetti, S., Hirte, H. W., Counter, C. M., Greider, C. W., Wright, W. E., and Shay, J. W. (1994). Cell immortality, telomerase, and cancer. *Cold Spring Harbor Symp. Quant. Biol.* **59,** 307–315.

Hastie, N. D., Dempster, M., Dunlop, M. G., Thompson, A. M., Green, D. K., and Allshire, R. C. (1990). Telomere reduction in human colorectal carcinoma and with ageing. *Nature* **346,** 866–868.

Jaggi, B., Pontifex, B., Swanson, J., and Poon, S. S. S. (1993). Performance evaluation of a 12-Bit, 8Mpel/s digital camera. *SPIE Proc. Cameras Scanners, Image Acquistion Syst.* **1901,** 99–108.

Katz, S. G., Schneider, S. S., Bartuski, A., Trask, B. J., Massa, H., Overhauser, J., Lalande, M., Lansdorp, P. M., and Silverman, G. A. (1999). An 18q⁻ syndrome breakpoint resides between the duplicated serpins *SCCA1* and *SCCA2* and arises via a cryptic rearrangment with satellite III DNA. *Hum. Mol. Genet.* **8,** 87–92.

Kim, N. W., Piatyszek, M. A., Prowse, K. R., Harley, C. B., West, M. D., Ho, P. L. C., Coviello, G. M., Wright, W. E., Weinrich, S. L., and Shay, J. W. (1994). Specific association of human telomerease activity with immortal cells and cancer. *Science* **266,** 2011–2015.

Lansdorp, P. M. (1997a). Self-renewal of stem cells. *Biol. Blood Marrow Transplant.* **3,** 171–178.

Lansdorp, P. M. (1997b). Lessons from mice without telomerase. *J. Cell Biol.* **139,** 309–312.

Lansdorp, P. M., Verwoerd, N. P., van de Rijke, F. M., Dragowska, V., Little, M.-T., Dirks, R. W., Raap, A. K., and Tanke, H. J. (1996). Heterogeneity in telomere length of human chromosomes. *Hum. Mol. Genet.* **5,** 685–691.

Lansdorp, P. M., Poon, S., Chavez, E., Dragowska, V., Zijlmans, M., Bryan, T., Reddel, R., Egholm, M., Bacchetti, S., and Martens, U. (1997). Telomeres in the haematopoietic system. *In* "Telomeres and Telomerease" (D. J. Chadwick and G. Cardew, eds.), pp. 209–222. Wiley, West Sussex, England.

Levy, M. Z., Allsopp, R. C., Futcher, A. B., Greider, C. W., and Harley, C. B. (1992). Telomere end-replication problem and cell aging. *J. Mol. Biol.* **225,** 951–960.

Marr, D. (1982). "Vision." Freeman, New York.

Martens, U. M., Zijlmans, J. M. J. M., Poon, S. S. S., Dragowska, W., Yui, J., Chavez, E. A., Ward, R. K., and Lansdorp, P. M. (1998). Short telomeres on human chromosome 17p. *Nat. Genet.* **18,** 76–80.

Moyzis, R. K., Buckingham, J. M., Cram, L. S., Dani, M., Deaven, L. L., Jones, M. D., Meyne, J., Ratliff, R. L., and Wu, J.-R. (1988). A highly conserved repetitive DNA sequence, (TTAGGG)ₙ, present at the telomeres of human chromosomes. *Proc. Natl. Acad. Sci. U.S.A.* **85,** 6622–6626.

Nielsen, P. E., Egholm, M., Berg, R. H., and Buchardt, O. (1991). Sequence-selective recognition of DNA by strand displacement with a thymine-substituted polyamide. *Science* **254,** 1497–1500.

Norton, J. C., Piatyszek, M. A., Wright, W. E., Shay, J. W., and Corey, D. R. (1996). Inhibition of human telomerase activity by peptide nucleic acids. *Nat. Biotechnol.* **14,** 615–619.

Oexle, K. (1998). Telomere length distribution and Southern blot analysis. *J. Theor. Biol.* **190,** 369–377.

Olovnikov, A. M. (1971). Principles of marginotomy in template synthesis of polynucleotides. *Dokl. Akad. Nauk SSSR* **201,** 1496–1499.

Pinkel, D., Straume, T., and Gray, J. W. (1986). Cytogenetic analysis using quantitative, high-sensitivity, fluorescence hybridization. *Proc. Natl. Acad. Sci. U.S.A.* **83,** 2934–2938.

Poon, S. S. S., and Hunter, D. B. (1994). Electronic cameras to meet the needs of microscopy specialists. *Adv. Imaging* **9,** 64–67.

Poon, S. S. S., Ward, R. K., and Palcic, B. (1992). Analysis of three-dimensional images in quantitative microscopy. *SPIE Proc. Biomed. Image Process Three-Dimensional Microscopy* **1660,** 178–185.

Poon, S. S. S., Ward, R. K., and Palcic, B. (1993a). Automated image detection and segmentation in blood smears. *Yearbook Med. Informatics* **93,** 271–279.

Poon, S. S. S., Lockett, S. J., and Ward, R. K. (1993b). Characterization of a 3D microscope imaging system. *SPIE Proc. Biomed. Image Process. Biomed. Visualization* **1905,** 121–128.

Poon, S. S. S., Martens, U. M., Ward, R. K., and Lansdorp, P. M. (1999a). Telomere length measurements using digital fluorescence microscopy. *Cytometry* **36,** 267–278.

Poon, S. S. S., Ward, R. K., and Lansdorp, P. M. (1999b). Segmenting telomeres and chromosomes in cells. IEEE International Conference on Acoustics, Speech, and Signal Processing, Phoenix, AZ, pp. 3413–3416.

Poulin, N., Harrison, A., and Palcic, B. (1994). Quantitative precision of an automated image cytometric system for the measurement of DNA content and distribution in cells labeled with fluorescent nucleic acid stains. *Cytometry* **16,** 227–235.

Preston, K., Jr. (1976). Digital picture analysis in cytology. *In* "Digital Picture Analysis" (A. Rosenfeld, ed.), pp. 209–293. Springer-Verlag, New York.

Raap, A. K., Marijnen, J. G. H., Vrolijk, J., and van der Ploeg, M. (1986). Denaturation, renaturation, and loss of DNA during *in situ* hybridization procedures. *Cytometry* **7,** 235–242.

Rigaut, J. P., and Vassy, J. (1991). High-resolution three-dimensional images from confocal scanning laser microscopy. Quantitative study and mathematical correction of the effects from bleaching and fluorescence attenuation in depth. *Anal. Quant. Cytol. Histol.* **13,** 223–232.

Rufer, N., Dragowska, W., Thornbury, G., Roosnek, E., and Lansdorp, P. M. (1998). Telomere length dynamics in human lymphocyte subpopulations measured by flow cytometry. *Nat. Biotechnol.* **16,** 743–747.

Russ, J. C. (1990). "Computer Assisted Microscopy: The Measurements and Analysis of Images." Plenum, New York.

Shay, J. W., and Wright, W. E. (1996). Telomerase activity in human cancer. *Curr. Opin. Oncol.* **8,** 66–71.

Slijepcevic, P., Hande, M. P., Bouffler, S. D., Lansdorp, P., and Bryant, P. E. (1997). Telomere length, chromatin structure and chromosome fusigenic potential. *Chromosoma* **106,** 413–418.

Vaziri, H., and Benchimol, S. (1998). Reconstitution of telomerase activity in normal human cells leads to elongation of telomeres and extended replicative life span. *Curr. Biol.* **8,** 279–282.

Wan, T. S. K., Martens, U. M., Poon, S. S. S., Tsao, S.-W., Chan, L. C., and Lansdorp, P. M. (1999). Absence or low number of telomere repeats at junctions of dicentric chromosomes. *Genes Chromosomes Cancer* **24,** 83–86.

Watson, J. D. (1972). Origin of concatameric T4 DNA. *Nature New Biol.* **239,** 197–201.

Weng, N.-P., Palmer, L. D., Levine, B. L., Lane, H. C., June, C. H., and Hodes, R. J. (1997). Tales of tails: Regulation of telomere length and telomerase activity during lymphocyte development, differentiation, activation, and aging. *Immunol. Rev.* **160,** 43–54.

Zhu, L., Hathcock, K. S., Hande, P., Lansdorp, P. M., Seldin, M. F., and Hodes, R. J. (1998). Telomere length regulation in mice is linked to a novel chromosome locus. *Proc. Natl. Acad. Sci. U.S.A.* **95,** 8648–8653.

Zijlmans, J. M. J. M., Martens, U. M., Poon, S. S. S., Raap, A. K., Tanke, H. J., Ward, R. K., and Lansdorp, P. M. (1997). Telomeres in the mouse have large inter-chromosomal variations in the number of T_2AG_3 repeats. *Proc. Natl. Acad. Sci. U.S.A.* **94,** 7423–7428.

CHAPTER 34

Detection of Chromosome Translocation Products in Single Interphase Cell Nuclei

Jingly Fung,[*,†] **Santiago Munné,**[‡] **and Heinz-Ulli G. Weier**[†]

[*]Department of Obstetrics, Gynecology and Reproductive Sciences
University of California, San Francisco
San Francisco, California 94143

[†]Life Sciences Division
E. O. Lawrence Berkeley National Laboratory
University of California, Berkeley
Berkeley, California 94720

[‡]The Institute for Reproductive Medicine and Science
Saint Barnabas Medical Center
West Orange, New Jersey 07052

I. Introduction
II. Materials
    A. Chemicals
    B. Buffers and Other Solutions
    C. Instruments
III. Protocols
    A. Cell Preparation
    B. Clone Selection
    C. Preparation of DNA from Yeast Artificial Chromosome Clones
    D. Probe Labeling via Random Priming
    E. Fluorescence *in Situ* Hybridization
IV. Results and Discussion
V. Critical Aspects of the Procedure
    A. Cell Biopsy and Fixation
    B. Clone Selection and Probe Preparation
    C. Slide Pretreatment and Fluorescence *in Situ* Hybridization
VI. Applications
    References

METHODS IN CELL BIOLOGY, VOL. 64

I. Introduction

Chromosome translocations occur at a frequency of about 0.1% (Jacobs *et al.*, 1974; Thompson *et al.*, 1991) in the general population. The frequency may reach as much as 1% in infertile men (Guichaoua *et al.*, 1990), and up to 3% in oligospermic men and couples requiring intracytoplasmic sperm injection (ICSI) (Abyholm and Stray-Pedersen, 1981; Testart *et al.*, 1996). Translocation carriers experience a higher incidence of infertility, and reproductive failures, as well as a higher risk of conceiving chromosomally abnormal offspring. The cause of misconception or fetal loss is believed to be impaired homolog pairing and/or disturbed karyokinesis during germ-cell development creating a chromosomal imbalance in germ cells (Srb *et al.*, 1965). Patients who are carriers of Robertsonian or reciprocal translocations may benefit from *in vitro* fertilization (IVF) followed by preconception and preimplantation genetic diagnosis (PGD).

Sophisticated PGD techniques have been developed so that cytogenetic investigations can be performed prior to the transfer of embryos to the woman, thus increasing the chances of successful nidation of unaffected embryos and ongoing pregnancies. PGD was first used to aid pregnancy in 1990 (Handyside *et al.*, 1990). Since then two critical techniques, the polymerase chain reaction (PCR) and fluorescence *in situ* hybridization (FISH), have been significantly refined (Griffin *et al.*, 1992; Grifo *et al.*, 1992; Handyside *et al.*, 1992; Harper *et al.*, 1995; Munné *et al.*, 1993a,b; Strom *et al.*, 1995; Verlinsky *et al.*, 1995). PGD usually involves the removal of one or two blastomeres from preimplantation embryos about 3 days after IVF, blastomere analysis by PCR or FISH, and the transfer of those embryos determined to be genetically normal. Similarly, preconception genetic diagnosis assays the first polar bodies of oocytes prior to fertilization to select normal eggs for IVF. Because few blastomere nuclei are found in metaphase, the chromosomes cannot be analyzed by conventional karyotyping, that is, banding, for PGD of structural abnormality. However, preimplantation embryo interphase nuclei can be examined by FISH.

The detection of translocation chromosomes in interphase nuclei from chromosomally balanced or aneuploid embryos prior to transfer to the womb has been rarely performed because case-specific probes for the patients were not easily obtainable. Three approaches for PGD of structural abnormalities based on FISH have been proposed: chromosome painting of polar bodies, probes mapping distal to the break points, and probes spanning the break points. When the carrier is female, the chromosomally normal versus abnormal oocytes can be determined by using whole chromosome painting probes hybridized to the first polar body chromosomes (Munné *et al.*, 1995). This technique can be improved by combining probes that bind distal to the respective break points with whole chromosome painting probes.

When the carrier is male, the only method available for PGD is the previously mentioned blastomere biopsy after IVF and *in vitro* culture of the embryo. Under these circumstances, probes mapping distal to or spanning the break points are required for interphase analysis. Probes localizing distal to the break points allow one to score chromosome segments and to identify unbalanced embryos. However, they do not allow one to differentiate between normal embryos and embryos carrying the balanced translocation. Furthermore, as demonstrated in this chapter and elsewhere (Cassel *et al.*, 1997; Munné *et al.*, 1998a,b), the analysis of embryos from patients carrying chromosome inversions or deletions requires probes that span the break points or map within the deleted region.

Our laboratory has prepared case-specific probes for patients carrying chromosome translocations and published their application to PGD of blastomeres (Cassel *et al.*, 1997; Fung *et al.*, 1998; Munné *et al.*, 1998a,b; Weier *et al.*, 1999). In this chapter, we describe rapid and inexpensive procedures to prepare case-specific probes for FISH-based PGD in germ cells or embryos by using yeast artificial chromosomes (YACs) as probes spanning or flanking translocation break points.

The process of break point mapping uses lymphocyte metaphase spreads from the translocation carrier. With an identical karyotype as the germ line, lymphocytes always contain the normal homologs in addition to the derivative chromosome. Classifying probes as mapping either proximal or distal to the break point is straightforward. After initial selection of YACs as DNA probes over a larger area, our probe search narrows the break point location until we isolate probe contigs that span individual break points. After labeling DNA probes with distinct reporter molecules and visualizing them in two different fluorescent colors specific for either break point, we can score normal versus derivative chromosomes in individual nuclei. Furthermore, the inclusion in the hybridization reaction of chromosome-specific centromeric probes labeled in a third color allows us to identify the origin of derivative chromosomes in interphase cell nuclei.

II. Materials

A. Chemicals

β-Agarase and buffer (New England Biolabs, Beverly, MA). Store at $-20°C$

Agarose [GIBCO/LTI (Life Technologies), Rockville, MD].

Antibodies against digoxigenin, rhodamine-conjugated, made in sheep (Boehringer Mannheim, Indianapolis, IN). Stock solution is 2 mg/ml, dilute 1:100 with PNM (see Section II,B) prior to use. Store at 4°C.

Bovine serum albumin (Sigma, St. Louis, MO). Store at 4°C.

Chloroform/isoamyl alcohol, 24/1 (v/v).

Colcemid (GIBCO/LTI), 10 mg/ml. Store at 4°C.

4′, 6-diamino-2-phenylindole (DAPI), 0.05 μg/ml in antifade solution (Calbiochem, La Jolla, CA). Store at $-20°C$.

10X dNTP mix: dATP, dCTP, dGTP, and dTTP (Boehringer Mannheim or Pharmacia, Piscataway, NJ) 10 mM each. Store at $-20°C$.

Digoxigenin-11-dUTP (dig-dUTP) (Boehringer Mannheim), 1 mM. Store at $-20°C$.

EDTA (ethylenediaminetetraacetic acid) (GIBCO/LTI), 0.5 M, pH 8.0.

Ethidium bromide (GIBCO/LTI), 10 mg/ml.

Fluorescein avidin DCS [avidin-fluorescein isothiocyanate(FITC), Vector Laboratories, Burlingame, CA]. Stock solution is 2 mg/ml, dilute 1:100 with PNM prior to use. Store at 4°C.

Fluorescein-12-dUTP (Boehringer Mannheim), 1 mM. Store at $-20°C$.

Formamide (FA) (GIBCO/BRL or Boehringer Mannheim). Store at 4°C.

Glycogen (Boehringer Mannheim), 20 mg/ml. Store at $-20°C$.

Human COT1 DNA (GIBCO/LTI), 1 mg/ml. Store at $-20°C$.

Low melting point (LMP) agarose (Bio-Rad, Hercules, CA).

β-Mercaptoethanol (Sigma).

Methotrexate (Sigma), 10^{-5} M.

Oligonucleotides used in *in vitro* DNA amplifications:

JUN1	5′-CCCAAGCTTGCATGCGAATTCNNNNCAGG-3′
JUN15	5′-CCCAAGCTTGCATGCGAATTC-3′
MSAlu1	5′-GGATTACAGGYRTGAGCCA-3′
MSAlu2	5′-RCCAYTGCACTCCAGCCTG-3′
PDJ33	5′-GCCTCCCAAAGTGCTGGGATTACAGGCGTGAGCCA-3′
PDJ34	5′-TGAGCCGAGATCGCGCCACTGCACTCCAGCCTGGG-3′

Phenol/chloroform/isoamyl alcohol (GIBCO/LTI): 25/24/1 (v/v/v), pH 8.0. Store at 4°C.

Phytohemagglutinin (GIBCO/LTI), 10 mg/ml.

Proteinase K (Boehringer Mannheim): 20 mg/ml in 10 mM Tris-HCl, pH 7.5. Store at $-20°C$.

Random-priming kit: BioPrime kit (GIBCO/BRL). Store at $-20°C$.

RNase (Boehringer Mannheim), DNase-free: boil at 100°C for 10 min. Divide into aliquots and store at $-20°C$

Salmon sperm DNA (3′-5′, Boulder, CO), 20 mg/ml. Store at $-20°C$.

Sodium dodecyl sulfate (SDS) (Sigma), sodium salt, 10% in water.

Sodium azide (Fisher Scientific, Pittsburg, PA).

T$_7$ Sequenase Version 2.0 DNA Polymerase (Amersham Pharmacia Biotech, Piscataway, NJ). Store at $-20°$ C.

Thermus aquaticus (*Taq*) DNA polymerase (Perkin-Elmer, Foster City, CA), 5 U/μl. Store at $-20°C$.

Thymidine (10^{-5} M, LTI).

Yeast artificial chromosome (YAC) library (Research Genetics, Huntsville, AL). Store at $-80°C$.

Zymolase, 70,000 U/g, 10 mg/ml in 50 mM KH$_2$PO$_4$ (pH 7.8), 50% glycerol. Store at $-20°$C.

B. Buffers and Other Solutions

Acid hydrolyzed casein (AHC) medium (BIO 101, Vista, CA): Add 36.7 g/liter of purified water. Autoclave at 121°C for 15 min.

AHC agar medium (BIO 101): Add 53.7 g/liter of purified water. Mix to dissolve dextrose. Autoclave at 121°C for 15 min. Cool to 50°C, mix well, and pour plates. Store plates at 4°C.

Antifade solution (1% p-phenylenediamine, 15 mM NaCl, 1 mM H$_2$PO$_4$, pH 8.0, 90% glycerol).

Denaturing solution: 70% formamide (FA), 2× SSC, pH 7.0. Prepare fresh at least every 2 weeks.

ES buffer: 0.5 M EDTA (pH 8.0), 1% Sarcosyl (Amresco, Solon, OH).

Fixative: acetic acid/methanol, 1/3 (v/v). Make fresh before use.

Hybridization Master Mix (MM 2.1): 14.3% (w/v) dextran sulfate, 78.6% FA, 2.9× SSC, pH 7.0. Prepare by combining 1.45 ml of 20× SSC, 0.7 ml ultrapure water, 1.43 g dextran sulfate (Calbiochem), and 7.86 ml formamide. Heat to 70°C to dissolve dextran sulfate. Dispense aliquots in 1.5-ml tubes and store at $-20°$C.

Lysis buffer: 1% Triton X-100, 20 mM Tris-HCl, 2 mM EDTA, pH 8.5. Store at 4°C.

Modified nucleotide mix (10× A4) for labeling in combination with 1 mM dig-dUTP or FITC-dUTP: Combine 5 μl each of 100 mM dATP, 100 mM dGTP, and 100 mM dCTP with 2.5 μl of 1 M Tris-HCl, pH 7.5, 0.5 μl of 0.5 M EDTA, pH 8.0, and 232 μl ultrapure water for a total of 250 μl. Store at $-20°$C. The concentration of nucleoside triphosphates is 2 mM each.

PNM: Dissolve 5 g of nonfat dry milk (Carnation, Glendale, CA) in 100 ml PN buffer (PN buffer is 0.1 M sodium phosphate, pH 8.0, 0.1% Nonidet P-40) and 0.1 ml of sodium azide, incubate at 50°C overnight, spin at 1000 g for 30 min, aliquot clear supernatant into 1.5-ml tubes, and store at 4°C. Spin at 2000 g for 30 sec prior to use.

SCE: 1 M sorbitol, 0.1 M Sodium citrate, 10 mM EDTA, pH 7.8.

SSC: 20× SSC is 3 M NaCl, 0.3 M trisodium citrate dihydrate, pH 7.0.

10× Taq buffer: 500 mM KCl, 100 mM Tris-HCl, pH 8.3, 15 mM MgCl$_2$. Store at $-20°$C.

TBE (Tris/borate/EDTA) buffer: 10× TBE is 890 mM Tris base, 890 mM boric acid, 20 mM EDTA.

TE (Tris/EDTA) buffer: 1× TE is 10 mM Tris-HCl, 1 mM EDTA, pH 7.4, 7.5, or 8.0.

TE 50 buffer: 10 mM Tris-HCl, 50 mM EDTA, pH 7.8.

Tris-HCl [tris(hydroxymethyl)aminomethane]: 1 M, pH 7.5 or 8.0.

Tyrode's solution, acidic (Sigma), pH 2.4. Store at 4°C. Make fresh before use.

C. Instruments

CDS-5 cytogenetic drying chamber (Thermotron Industries, Holland, MI).
Centrifuge [MP4R, International Equipment Co. (IEC), Needham Heights, MA].
Dry bath (heat block) (Model 2001, Labline Instruments, Melrose Park, IL).
Fluorescence microscope (Axioskop, Carl Zeiss, Thornwood, NY) equipped with
40× and 63× oil immersion lenses.
Incubator oven (set to 37°C).
Pulsed field gel electrophoresis (PFGE) system (Bio-Rad).
Shaking incubators (New Brunswick Scientific, Edison, NJ): 30°C for yeast cell
culture, 37°C for culture of *Escherichia coli.*
Thermal cycler for *in vitro* DNA amplification (Model 4800, Perkin-Elmer).
TKO 100 Mini-Fluorometer (Hoefer Scientific Instruments, San Francisco, CA).
Water bath (Model 188, Precision Scientific, Winchester, VA).

III. Protocols

A. Cell Preparation

1. Lymphocyte Metaphase Spreads

Metaphase spreads are made from short-term cultures of lymphocytes grown
for 72 hr in RPMI 1640 supplemented with 20% fetal calf serum, 2% penicillin,
and 4% phytohemagglutinin according to the procedure described by Harper
and Saunders (1981). Cultures are blocked for 17 hr with methotrexate, followed
by incubation in RPMI containing thymidine for 5 hr. Cells are blocked in mitosis
during a 10-min treatment with colcemid, harvested, and incubated in 75 mM
KCl for 15 min at 37°C. The cells are then pelleted and resuspended in freshly
made fixative. The fixative is changed twice, and cells are dropped on ethanol-
cleaned slides inside a CDS-5 cytogenetic drying chamber at 25°C and 47.5%
humidity. Slides are stored in slide boxes for at least 2 weeks at room temperature,
then sealed in plastic bags filled with nitrogen gas and stored at −20°C until used.

2. Blastomeres

Embryos for this study are donated for research by patients enrolled in the
IVF program of the Institute for Reproductive Medicine and Science of Saint
Barnabas Medical Center, West Orange, New Jersey. In accordance with guidelines set by the institutional review board of Saint Barnabas Medical Center,
written consent is obtained from the patients in each case.

Only nonarrested, monospermic embryos developing from bipronucleated zygotes are used for our translocation studies. A hole is drilled through the zona pellucida with acidified Tyrode's solution (Gordon and Talansky, 1986; Munné *et al.*, 1994) and one to two blastomeres are removed from each embryo by micromanipulation (Grifo *et al.*, 1992). Each blastomere is placed in a culture dish containing hypotonic solution (1% sodium citrate, 6 mg/ml bovine serum albumin in water) for 5 min at 25°C, and is then transferred into a small volume of hypotonic solution on a slide. All blastomeres are fixed individually as described (Tarkowski, 1966; Munné *et al.*, 1996, 1998b). One or two drops of fixative are applied to the blastomere until the cytoplasm breaks. At the right humidity (40–50%) and with gentle blowing following the cytoplasmic breakage, little debris remains on top of the nucleus. In this way, micronuclei do not get lost, as they might do if fixative is added after cytoplasmic rupture. Throughout the procedure, the nucleus is observed in a phase contrast microscope, and its position is marked using a carbide or diamond tip pen following fixation. The slide is then dehydrated in three consecutive baths of 70, 85, and 95% ethanol for 2 min each, before it is used for FISH or stored at −20°C.

B. Clone Selection

Our general steps for selection and optimization of probes for PGD are as follows:

1. Collect cytogenetic data.
2. Define or refine the approximate break point interval (initially +/− one chromosome band).
3. Select clones and make probes.
4. Map probes by FISH.
5. Determine if probes span the break point. If yes, continue with step 6. If not, note map position and repeat from step 2.
6. Optimize probes.
7. Perform PGD.

Typically, cytogenetic data localizing the approximate break points is available or can be generated rapidly by chromosome painting. We select YAC clones based on sequence tagged site (STS) markers that map approximately in these regions and from information provided in the Whitehead Institute for Biomedical Research/MIT Center for Genome Research data base (http://www.genome.wi.-mit.edu/) (Weissenbach *et al.*, 1992; Hudson *et al.*, 1995). We preferentially select YAC clones larger than 700–800 kb that appeared nonchimeric based on STS content mapping data. For the initial round of probe preparation, we choose clones in about 10–15 Mbp intervals to grossly estimate the location of the break points. In the second round of clone screening, we choose clones that map in approximately 2 Mbp intervals and classify them as proximal or distal to the break point. We continue to select and map YACs in smaller intervals until we identify clones that either span a break point or map very close to it. Besides the YAC information from the Whitehead Institute, we also use physical mapping data from the UCSF/LBNL Resource for Molecular Cytogenetics (RMC, http://rmc-www.lbl.gov/). This data base provides information about P1 and YAC clones physically mapped at RMC, as well as YAC mapping data from Dr. D. C. Ward's laboratory at Yale University School of Medicine. Both laboratories localize probes on the physical map based on fractional length measurements (Lichter *et al.*, 1993). In addition, most of the RMC P1 probes have STSs attached to them, so they can be integrated easily into the Whitehead Institute YAC map.

C. Preparation of DNA from Yeast Artificial Chromosome Clones

Grow the selected yeast clone (containing the YAC) on AHC agar for 2–3 days at 30°C. Pick colonies from the plates and culture the colonies in 5–35 ml AHC medium at 30°C for 2–3 days in a shaking incubator.

1. DNA Isolation, Phenol Extraction, and Isopropanol Precipitation of DNA

1. Pellet the cells from 35 ml AHC medium at 5000 rpm (2000 g) at 4°C for 5 min.
2. Decant the supernatant and resuspend cells in 3 ml total of 0.9 M sorbitol, 0.1 M EDTA, pH 7.5, containing 4 μl β-mercaptoethanol, followed by addition of 100 μl of zymolase (2.5 mg/ml), and then incubate at 37°C for 60 min.
3. Pellet the cells at 5000 rpm (2000 g) at 4°C for 5 min and decant supernatant.
4. Resuspend pellet in 5 ml of 50 mM Tris, pH 7.4, and 20 mM EDTA. Add 0.5 ml of 10% SDS and mix gently. Incubate at 65°C for 30 min.
5. Add 1.5 ml of 5 M potassium acetate and place on ice for 60 min.
6. Spin at 15,000 rpm (12,000 g) for 15 min at 4°C, and transfer the supernatant to a new tube.
7. Mix the supernatant gently with 2 volumes of 100% ethanol by inverting the tube a few times. Spin in 5000 rpm (2000 g) for 15 min at room temperature.
8. Prepare 12 sets of 1.5 ml microcentrifuge tubes.
9. Decant supernatant and air dry pellet. Resuspend in 3 ml of 1× TE to dissolve the pellet.
10. Transfer 700 μl of the solution to each of four 1.5-ml microcentrifuge tubes.
11. Add an equal volume of phenol/chloroform/isoamyl alcohol (25/24/1, pH 8.0), vortex well, and spin at high speed (10,000 g) for 3 min.
12. Transfer the top layer to new 1.5-ml microcentrifuge tubes and add an equal volume of chloroform/isoamyl alcohol (24/1). Vortex well and centrifuge at high speed (10,000 g) for 3 min.

13. Transfer the top layer to new 1.5-ml Eppendorf tubes. Add 40 μl of RNase (1 mg/ml, DNase-free) to each of the four tubes and incubate at 37°C for 30 min.

14. Add 1 volume of isopropanol and gently mix by inversion. Centrifuge at high speed (10,000 g) for 20 min.

15. Decant supernatant and wash pellet with 1 volume of cold 70% ethanol, and centrifuge at high speed (10,000 g) for 3 min.

16. Decant the 70% ethanol and air dry the pellet.

17. Resuspend pellet in 20–30 μl of 1× TE, and measure DNA concentration after the pellet is completely dissolved.

2. Pulsed Field Gel Electrophoresis

a. Plug Preparation

1. Pellet the yeast cells from 5 ml AHC medium at 400 rpm for 6 min. Resuspend cells in 0.5 ml EDTA (0.125 M), pH 7.8. Spin down again and take off supernatant.

2. Add 500 μl of SCE to a 70-μl pellet and resuspend. Mix with an equal volume of 1.5% LMP agarose at 43°C, quickly pipette up/down, and vortex for 1–2 sec to suspend. Pipette into molds (Bio-Rad) and allow to harden.

3. Remove plugs from molds. Incubate samples in 2 ml SCE with 100 μl of zymolase (10 mg/ml) and shake at 150 rpm at 30°C for 2.5 hr to overnight.

4. Take off SCE and add 2 ml of ES with 100 μl of proteinase K (20 mg/ml). Shake 5 hr to overnight at 50°C.

5. Take off ES and rinse five times with 6 ml of TE 50 for 30 min each rinse. Plugs are ready for running PFGE, or they can be stored at 4°C.

b. Electrophoresis

1. Perform PFGE under the following conditions: voltage gradient, 6 V/cm; switching interval, 79 sec forward, 94 sec reverse; running time, 38 hr; agarose concentration, 1.0% LMP agarose; running temperature, 14°C; running buffer, 0.5× TBE.

2. Stain the gel with ethidium bromide (0.5 μg/ml in water) and cut out the target band. Transfer gel slice to a clean tube.

3. Wash gel slices with water for 30 min, and then wash with 1× β-agarase buffer for 30 min.

4. Melt the gel completely by heating for 10 min at 85°C. Transfer the molten agarose to 43°C water bath. Add 1 μl β-agarase for every 25 μl molten agarose and continue incubation for 2 hr. The sample can now be used for the PCR reactions or stored at 4°C.

3. Inter–Alu PCR

1. Pick up colonies from AHC agar plate, and mix well in 20 μl of lysis buffer.
2. Place at 100°C for 10 min.
3. Spin in a microcentrifuge for 10 min at 2000 g, and transfer the supernatant to a fresh tube. The supernatant can be used for the PCR reactions or stored at -20°C.
4. In 0.5-ml tubes, mix in this order (total 50 μl)

33.2 μl water

5 μl 10× *Taq* buffer

5 μl 4× dNTPs

1 μl each MSAlu1/Alu2 or PDJ33/PDJ34

5 μl supernatant

0.8 μl *Taq* polymerase

5. Run the following program (Thermocycle file, Perkin-Elmer) for 35 cycles.

Temperature (°C)	Time (sec)
94	30
94	30
55	60
55	60
72	150 (plus 5-sec extension/cycle)
72	600

4. Degenerate Oligonucleotide–Primed PCR

There are two different degenerate oligonucleotide-primed polymerase chain reactions (DOP-PCR), namely, T7 DNA polymerase PCR and *Taq* PCR. The initial seven cycles of PCR use primer JUN1 (Weier *et al.*, 1993) with annealing at 37°C and extension by T7 DNA polymerase (Kroisel et al., 1994). The PCR products are then further amplified at higher annealing temperature (50°C) using primer JUN15 and *Taq* polymerase during 35 cycles.

a. T7 DNA Polymerase PCR (Seven Cycles)

1. Prepare the T7 mix (total 27 μl): 19.0 μl water, 5.2 μl of 5× Sequenase reaction buffer, 0.8 μl of 4× dNTP, 0.6 μl primer JUN1, and 0.4 μl Sequenase II (T7 DNA polymerase).
2. Denature 4 μl DNA (template) at 92°C for 6 min.
3. Add 8 μl T7 mix. Place at room temperature for 2 min, and then at 37°C for 6 min.
4. Denature at 92°C for 3 min.
5. Add 2 μl T7 mix. Place at room temperature for 2 min, and then at 37°C for 6 min.

6. Repeat steps 4 and 5 for five times total.

7. Denature DNA at 92°C for 3 min.

8. Add 8 μl T7 mix. Place at room temperature for 2 min, and then at 37°C for 10 min.

9. The Sequenase products can be used immediately for the *Taq* polymerase amplification reaction or stored at −20°C up to several years.

b. Taq Polymerase PCR (35 Cycles)

1. In 0.5 ml tubes mix in this order (total 200 μl)

152 μl water

20 μl 10× *Taq* buffer

4 μl 4× dNTPs

4 μl primer JUN15

20 μl Sequenase reaction product

0.67 μl *Taq*

2. Run the following PCR reaction (Thermocycle File, Perkin-Elmer) for 35 cycles.

Temperature (°C)	Time (sec)
94	30
94	60
52	60
52	60
72	120
72	120 (plus 5 sec extension/cycle)
72	600

3. Finally, the PCR products are precipitated in isopropanol and resuspended in 1× TE buffer.

D. Probe Labeling via Random Priming

1. Add 400 ng of DNA to water to a final volume of 14 μl in a 0.5-ml tube.

2. Boil tube at 100°C for 5 min, then quickly chill on ice. Add

5 μl	10× A4 mix
6.5 μl	1 mM dTTP
3.5 μl	Dig-11-dUTP or FITC-12-dUTP (1 mM)
20 μl	2.5× random primers.

(For biotin labeling add 5 μl of 10× dNTP (containing biotin-14-dCTP), 10 μl water, and 20 μl of 2.5× random primers.)

3. Mix well and add 1 μl DNA polymerase I Klenow fragment (40 units/μl).

4. Incubate tube at 37°C for 2 hr.

5. Add 5 μl of 10× stop buffer (EDTA).
6. Store at −20°C.

E. Fluorescence *in Situ* Hybridization

1. Hybridization

1. Prepare hybridization mix by combining 1 μl of each probe, 1 μl of human COT1 DNA, 1 μl of salmon sperm DNA, and 7 μl of hybridization master mix (MM 2.1).
2. Denature the hybridization mix at 76°C for 7 min, and allow it to preanneal at 37°C for 60 min.
3. Denature the slide at 76°C in denaturation solution (70% FA/2× SSC) for 3–5 min, then dehydrate in 70, 80, and 100% ethanol for 2 min each step, and allow the slide to air dry.
4. Apply the hybridization mix to the slide and cover with coverslip. Incubate overnight at 37°C in a moisture chamber.

2. Washing and Detection

1. After hybridization, the slide is washed three times in 50% FA/2× SSC for 10 min followed by two washes in 2× SSC for 10 min at 43°C.
2. The slide is incubated with 100 μl PNM buffer under a plastic coverslip at room temperature for 5 min.
3. The slide is then incubated with 100 μl PNM buffer containing fluorescein avidin DCS and/or anti-digoxigenin-rhodamine under a plastic coverslip at room temperature for 30 min.
4. The slide is washed three times in 2× SSC for 10 min each time at room temperature on a shaker.
5. The slide is mounted in 16 μl of DAPI (0.05 μg/ml in antifade solution) under a 24 mm × 50 mm coverslip to counterstain nuclei and chromosomes.
6. The slide is ready to be looked at in the fluorescence microscope. Fluorescence microscopy is performed with a filter set for simultaneous observation of Texas Red/rhodamine and FITC, and a separate filter for DAPI detection (ChromaTechnology). We acquire images using a cooled CCD camera (Photometrics, Tucson, AZ) and a Sun Sparc station (Weier *et al.*, 1995).

IV. Results and Discussion

As an example, we selected the case of a male patient carrying a balanced reciprocal translocation 46,XY,t(3;4)(p24;p15) (Fung *et al.*, 1998). Figure 2A shows the idiogram of one normal chromosome 3, one derivative chromosome 3, one normal chromosome 4, and one derivative chromosome 4, and break points on chromosome 3p24 and 4p15 are marked by thin horizontal lines.

The DNA probes to detect the translocations were prepared either by phenol extraction and isopropanol precipitation or by inter-Alu PCR. To find YAC clones spanning a break point, it normally takes two to five cycles of selection. For chromosome 3, 21 YAC clones were selected in five cycles, and 19 YAC clones could be localized by FISH. Eleven YACs clones (929g11, 786g9, 930h5, 864c11, 814h10, 827d2, 882h12, 810b7, 817a9, 810b7, and 762h1) mapped proximal to the break point; five YAC clones (932f3, 923c5, 921a3, 852b11, and 924g5) mapped distal to the break point; three YAC clones (955b5, 958b5, and 872h5) spanned the break point on chromosome 3p24. For chromosome 4, a total of four cycles and 14 YAC clones were needed, and 11 of these YAC clones could be localized by FISH. Six YAC clones (917c12, 964e7, 906b12, 747f3, 843d9, and 887e8) mapped distal to the break point. Three YAC clones (911e7, 931c2, and 817a3) mapped proximal to the break point, and two YAC clones (967c5 and 853c4) spanned the break point on chromosome 4p15.

When YAC clones spanning the break point are identified, we optimize the probe and hybridization parameters to increase hybridization signal intensity and contrast. In contrast to the probe selection step, we prepare YAC DNA from PFGE gels and DOP-PCR. Sometimes, contigs comprised of several YAC clones are necessary to improve the signal. Figure 1 shows the ideal probes for single interphase cell analysis. Optimal probes should be well centered, large, and contiguous. The probes are chosen such that the translocation divides the hybridization signal into approximately equal parts, allowing both derivative chromosomes to be identified. The ideal size of the probe or probe contig is

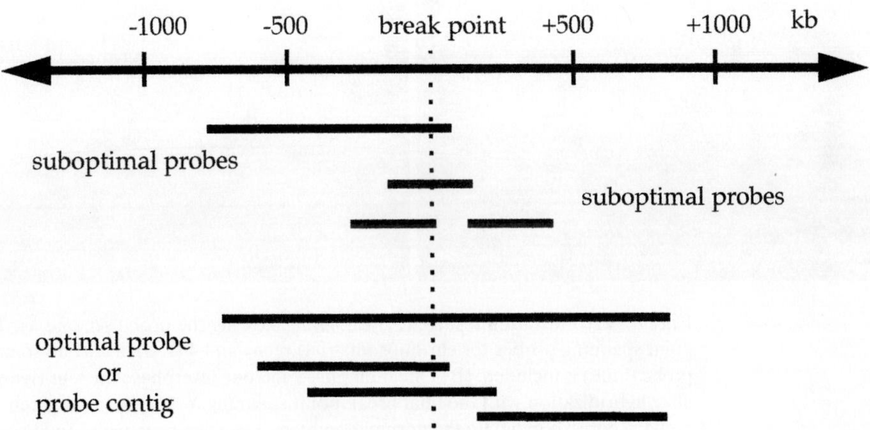

Fig. 1 The optimal and less desirable probes for detection of translocation chromosomes in interphase cell nuclei.

800–1000 kb of continuous DNA. In our t(3;4)(p24;p15) case, one YAC 958b5 for chromosome 3 and four YAC clones for chromosome 4 (2 flanking and 2 spanning the break point, 887e8, 931c2, 967c5, 853c4) provided optimal probes for the detection of translocation chromosomes in interphase. The probe for chromosome 3 was labeled with fluorescein-12-dUTP (green), and the probe mix for chromosome 4 was labeled with digoxigenin-11-dUTP detected by an anti-digoxigenin-rhodamine antibody (red). Hybridization results using lymphocytes carrying the t(3;4) are shown in Fig. 2B (one metaphase and one interphase). One chromosomal target appears green (normal chromosome 3) and one appears red (normal chromosome 4). Two chromosomes the derivative chromosome 3

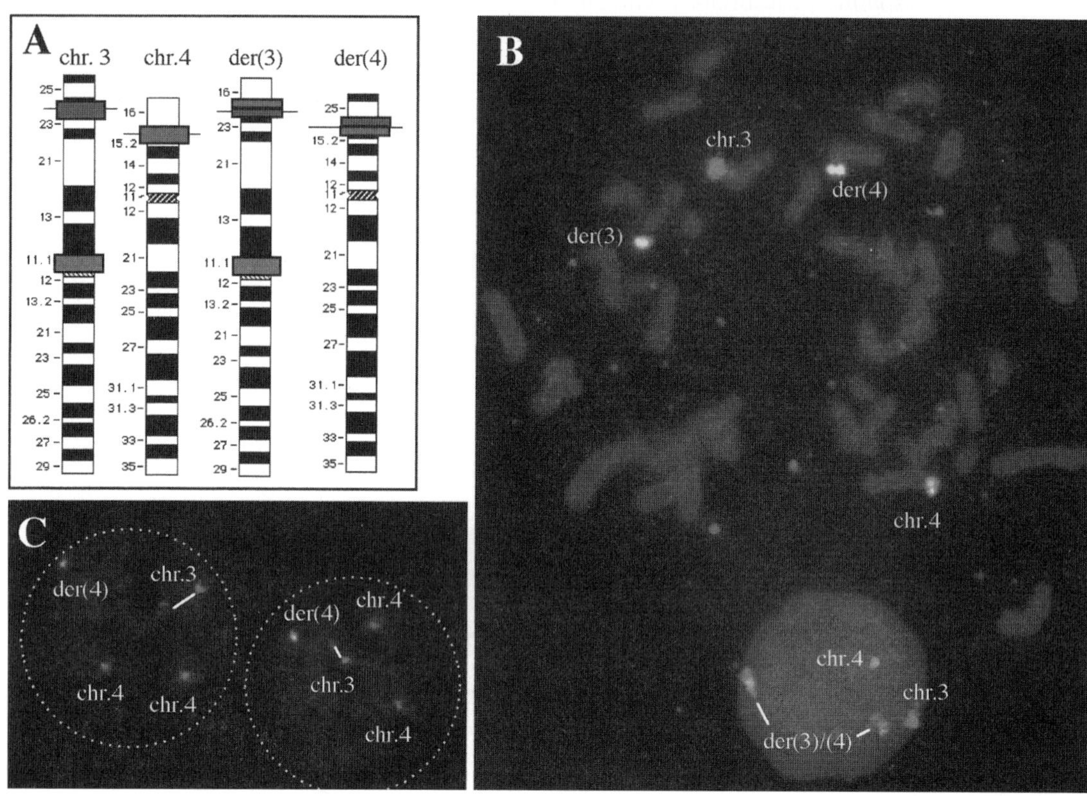

Fig. 2 Hybridization results. (A) Idiogram showing the probe scheme used; besides the two break point spanning probes for chromosomes 3 (green) and 4 (red), a chromosome 3 specific centromeric probe (blue) is included. (B) One metaphase and one interphase nucleus (lymphocytes) of the patient after hybridization with the final break point spanning YAC probes; a green and a red hybridization domain, corresponding to the normal homologs 3 and 4, respectively, and two domains with colocalizing red and green signals, corresponding to the derivative chromosomes. (C) Two abnormal blastomeric nuclei showing a derivative chromosome 4 in the presence of one copy of the normal chromosome 3 and two copies of the normal chromosome 4. (See color plates.)

[der(3)] and the derivative chromosome 4 [der(4)] are marked with red-green fusion signals.

Before analyzing patient blastomeres, the probes are tested on unrelated blastomeres obtained from a chromosomally normal embryo donated for research. This step is important to check and, if necessary, optimize the hybridization procedure. Hybridization efficiencies may vary between lymphocytes and blastomeres (Munné *et al.*, 1996). In the hybridization mixture, we include a commercially available centromeric probe (blue) as shown in Fig. 2A. This allows us to differentiate der(3) from der(4) in interphase nuclei. An extra blue signal with the red/green or yellow fusion domain indicates the presence of a der(3). Figure 2C presents hybridization results from an abnormal blastomeric cell derived from an embryo after fertilization with the translocation carrier's sperm. The binucleated abnormal blastomere carried an unbalanced translocation with one der(4) in addition to one normal homolog of chromosome 3 and two normal homologs of chromosomes 4 in each nucleus. This embryo was considered monosomic for most of chromosome 3 (except the part distal of the chromosome 3 break point) and trisomic for most of chromosome 4. Such extensive partial monosomy and trisomy is probably fatal, and this embryo was not transferred. In this case, a first IVF cycle produced a total of six embryos, five of which were found abnormal, and one balanced embryo was transferred (Munné *et al.*, 1998b).

V. Critical Aspects of the Procedure

A. Cell Biopsy and Fixation

The size of the hole in the zona pellucida should not be bigger than 30 μm to prevent loss of embryonic cell during embryo transfer to the womb. On the other hand, it should not be much smaller to avoid cell damage and DNA loss during biopsy. Cell transfer, hypotonic swelling, and fixation are critical steps for successful hybridization. Ideally, the blastomere is visible throughout the entire hypotonic treatment and fixation. One or two drops of fixative should be sufficient to fully disperse the cytoplasm. The position of the cell is marked with a diamond or carbide tip to allow easy retrieval.

B. Clone Selection and Probe Preparation

Clerical errors are common when working with clones from large libraries. Whenever possible, prepare labeled tubes and plates well in advance, make photocopies rather than transcribing lists, and double-check all clone identifiers.

The clones selected from a library for probe preparation must meet three requirements: correct STS content, no sign of chimerism, and large insert size. The selected probes hybridized to translocation carrier metaphase spreads allow one to rapidly classify clones as mapping either proximal or distal to the break point and thus define the genomic interval containing the break point. For each of the break points involved in a translocation, it is critical to obtain probes with highest hybridization efficiency for interphase analysis. Ideally, the break point should map to the midpoint of the respective probe.

C. Slide Pretreatment and Fluorescence *in Situ* Hybridization

The success of *in situ* hybridization and probe mapping depends on cell preparation and fixation. If cytoplasmic residues lowers hybridization efficiencies or produce unacceptable levels of background or unspecific hybridization, pretreatment with pepsin may eliminate the problem.

When dropping cells on the slides, temperature and humidity often affect the hybridization outcome. We drop cells on ethanol-cleaned slides at 25°C and 47.5% relative humidity. Slides are stored for at least 2 weeks in ambient air at room temperature, then placed under nitrogen in sealed plastic bags at −20°C until used.

VI. Applications

From 1995 to 1998, our laboratories studied 21 patients (J. Fung, S. Munné, and H.-U. G. Weier, unpublished data, 1998). It is interesting to note that the break point locations among our 21 patients appear randomly distributed across the entire genome. Only two of our 21 PGD patients (10%) carried intrachromosomal rearrangements: one patient had a pericentric inversion of chromosome 6 (karyotype: 46,XX, inv(6)(p23;q23.1) (Cassel *et al.*, 1997), while the second patient had an interstitial deletion on the short arm of chromosome 18. Using YACs as probes spanning or flanking translocation break points, we demonstrated the application in preconception (Munné *et al.*, 1998a) and preimplantation (Munné *et al.*, 1998b; Weier *et al.*, 1999) genetic diagnosis. This will greatly facilitate the selection of oocytes for IVF and embryos for transfer following blastomere analysis. On the other hand, the technology can also be utilized to delineate break points for position cloning of disease related genes (Lehmann *et al.*, 1997; Zitzelsberger *et al.*, 1999).

Acknowledgments

The authors gratefully acknowledge the support from Drs. J. D. Goldberg, R. A. Pedersen, R. T. Scott, and J. Cohen. We thank all patients who anonymously donated blood or embryos for this study. Furthermore, we thank the scientists at the Centre des Études du Polymorphisms Humain (CEPH), Paris, France, and the Whitehead Institute for Biomedical Research, Cambridge, Massachusetts, for sharing their resources and mapping results. J. F. was supported in part by a grant from the University of California Energy Institute.

All experiments were performed in compliance with the current laws in the United States.

References

Abyholm, T., and Stray-Pedersen, S. (1981). Hypospermiogenesis and chromosomal aberrations. A clinical study of azoospermic and oligospermic with normal and abnormal karyotype. *Int. J. Androl.* **4,** 546–558.

Cassel, M. J., Munné, S., Fung, J., and Weier, H.-U. G. (1997). Carrier-specific breakpoint-spanning DNA probes: An approach to preimplantation genetic diagnosis in interphase cells. *Hum. Reprod.* **12,** 2019–2027.

Fung, J., Munné, S., Duell, T., and Weier, H.-U. G. (1998). Rapid cloning of translocation breakpoints: From blood to YAC in 50 days. *J. Biochem. Mol. Biol. Biophys.* **1,** 181–192.

Gordon, J. W., and Talansky, B. E (1986). Assisted fertilization by zona drilling: A mouse model for correction of oligospermia. *J. Exp. Zool.* **239,** 347–354.

Griffin, D. K., Wilton, L. J., Handyside, A. H., Wiston, R. M. L., and Delhanty, J. D. A. (1992). Dual fluorescent in situ hybridization for simultaneous detection of X and Y chromosome-specific probes for the sexing of human preimplantation embryonic nuclei. *Hum. Genet.* **89,** 18–22.

Grifo, J. A., Tang, Y. X., Cohen, J., Gilbert, F., Sanyal, M. K., and Rosenwaks, Z. (1992). Ongoing pregnancy in a hemophilia carrier by embryo biopsy and simultaneous amplification of X and Y chromosome specific DNA from single blastomers. *JAMA* **6,** 727–729.

Guichaoua, M. R., Quack, B., Speed, R. M., Noel, B., Chandley, A. C., and Luciani, J. M. (1990). Infertility in human males with autosomal translocations: Meiotic study of a 14; 22 Robertsonian translocation. *Hum. Genet.* **86,** 162–166.

Handyside, A. H., Kontogianni, E. H., Hardy, K., and Winston, R. M. L. (1990). Pregnancies from biopsied human pre-implantation embryos sexed by Y-specific DNA amplification. *Nature* **344,** 768–770.

Handyside, A. H., Lesko, J. G., Tarin, J. J., Winston, R. M. L., and Hughes, M. R. (1992). Birth of a normal girl after in vitro fertilization and preimplantation diagnostic testing for cystic fibrosis. *N. Engl. J. Med.* **327,** 594–598.

Harper, J. C., Coonen, E., Handyside, A. H., Winston, R. M. L., Hopman, A. H. N., and Delhanty, J. D. A. (1995). Mosaicism of autosomes and sex chromosomes in morphologically normal, monospermic preimplantation human embryos. *Prenatal Diagn.* **15,** 41–49.

Harper, M., and Saunders, G. (1981). Localization of single copy DNA sequences on G-banded human chromsomes by in situ hybridization. *Chromosoma* **83,** 431–439.

Hudson, T. J., Stein, L. D., Gerety, S. S., Ma, J., Castle, A. B., Silva, J., Slonim, D. K., Baptista, R., Kruglyak, L., Xu, S. H., Hu, X., Colbert, A. M. E., Rosenberg, C., Reeve-Daly, M. P., Rozen, S., Hui, L., Wu, X., Vestergaard, C., Wilson, K. M., Bae, J. S., Maitra, S., Ganiatsas, S., Evans, C. A., DeAngelis, M. M., Ingalls, K. A., Nahf, R.W., Horton, L.T., Oskin-Anderson, M., Collymore, A. J., Ye, W., Kouyoumijian, V., Zemsteva, I. S., Tam, J., Devine, R., Courtney, D. F., Turner-Renaud, M., Nguyen, H., O'Connor, T. J., Fizames, C., Fauré, S., Gyapay, G., Dib, C., Morisette, J., Orlin, J. B., Birren, B. W., Goodman, N., Weissenbach, J., Hawkins, T. L., Foote, S., Page, D. C., and Lander, E. S. (1995). An STS-based map of the human genome. *Science* **270,** 945–1954.

Jacobs, P. A., Melville, M., Ratcliffe, S., Keay, A. J., and Syme, J. A. (1974). Cytogenetic survey of 11,680 newborn infants. *Ann. Hum. Genet.* **37,** 359–376.

Kroisel, P. M., Ioannou, P. A., and De Jong, P. J. (1994). PCR probes for chromosomal in situ hybridization of large-insert bacterial recombinants. *Cytogenet. Cell. Genet.* **65,** 97–100.

Lehmann, L., Greulich, K. M., Zitzelsberger, H., Spelsberg, F., Bauchinger, M., and Weier, H.-U. G. (1997). Cytogenetic and molecular genetic characterization of a chromosome 2 rearrangement in a case of human papillary thyroid carcinoma with radiation history. *Cancer Genet. Cytogenet.* **96,** 30–36.

Lichter, J. B., Difilippantonio, M. J., Pakstis, A. J., Goodfellow, P. J., Ward, D. C., and Kidd, K. K. (1993). Physical and genetic maps for chromosome 10. *Genomics* **16,** 320–324.

Munné, S., Weier, H.-U. G., Stein, J., Grifo, J., and Cohen, J. (1993a). A fast and efficient method for simultaneous X and Y in situ hybridization of human blastomeres. *J. Assisted Reprod. Genet.* **10,** 82–90.

Munné, S., Lee, A., Rosenwaks, Z., Grifo, J., and Cohen, J. (1993b). Diagnosis of major chromosome aneuploidies in human preimplantation embryos. *Hum. Reprod.* **8,** 2185–2191.

Munné, S., Weier, H.-U. G., Grifo, J., and Cohen, J (1994). Chromosome mosaicism in human embryos. *Biol. Reprod.* **51,** 373–379.

Munné, S., Dailey, T., Sultan, K. M., Grifo, J., and Cohen, J. (1995). The use of first polar bodies for preimplantation diagnosis of aneuploidy. *Hum. Reprod.* **10,** 014–20.

Munné, S., Dailey, T., Finkelstein, M., and Weier, H.-U. G. (1996). Reduction in signal overlap results in increased FISH efficiency: Implications for preimplantation genetic diagnosis. *J. Assisted Reprod. Genet.* **13,** 149–156.

Munné, S., Morrison, L., Fung, J., Márquez, C., Weier, U., Bahçe, M., Sable, D., Grundfelt, L., Schoolcraft, B., Scott, R., Cohen, J. (1998a). Reduction of spontaneous abortions after pre-conception genetic diagnosis of translocations. *J. Assisted Reprod. Genet.* **15,** 290–296.

Munné, S., Fung, J., Cassel, M. J., Márquez, C., Weier, H.-U. G. (1998b). Preimplantation genetic analysis of translocations: Case-specific probes for interphase cell analysis. *Hum. Genet.* **102,** 663–674.

Srb, A. M., Owen, R. D., and Edgar, R. S. (1965). "General Genetics," 2nd Ed. Freeman, San Francisco.

Strom, C. M., Verlinsky, Y., Milayeva-Rechitsky, S., Evsikov, S., Cieslak, J., Lifchez, A., Valle, J., Moise, J., Ginsberg, N., and Applebaum, M. (1995). Preconception genetic diagnosis for cystic fibrosis by polar body removal and DNA analysis. *Lancet* **336,** 306–307.

Tarkowski, A. K. (1966). An air drying method for chromosome preparation from mouse eggs. *Cytogenetics* **5,** 394–400.

Testart, J., Gautier, E., Brami, C., Rolet, F., Sedmon, E., and Thebault, A. (1996). Intracytoplasmic sperm injection in infertile patients with structural chromosome abnormalities. *Hum. Reprod.* **11,** 2609–2612.

Thompson, M. W., McInnes, R. R., and Willard, H. F. (1991). "Genetics in Medicine." Saunders, Philadelphia.

Verlinsky, Y., Cieslak, J., Frieidine, M., Ivakhnenko, V., Wolf, G., Kovalinskaya, L., White, M., Lifchez, A., Kaplan, B., Moise, J., Valle, J., Ginsberg, N., Strom, C., and Kuliev, A. (1995). Pregnancies following pre-conception diagnosis of common aneuploidies by fluorescence in-situ hybridization. *Hum. Reprod.* **10,** 1923–1927.

Weier, H.-U. G., Miller, B. M., Yu, L. C., and Fuscoe, J. C. (1993). PCR cloning of a repeated DNA fragment from chinese hamster ovary (CHO) cell X chromosomes and mapping by fluorescence in situ hybridization. *DNA Sequence* **4,** 47–51.

Weier, H.-U. G., Wang, M., Mullikin, J. C., Zhu, Y., Cheng, J. F., Greulich, K. M., Bensimon, A., and Gray, J. W. (1995). Quantitative DNA fiber mapping. *Hum. Mol. Genet.* **4,** 1903–1910.

Weier, H.-U. G., Munné, S., and Fung, J. (1999). Patient-specific probes for preimplantation genetic diagnosis (PGD) of structural and numerical aberrations in interphase cells. *J. Assisted. Reprod. Genet.* **16,** 182–191.

Weissenbach, J., Gyapay, G., Dib, C., Vignal, A., Morissette, J., Millasseau, P., Vaysseix, G., and Lathrop, M. (1992). A second-generation linkage map of the human genome. *Nature* **359,** 794–801.

Zitzelsberger, H., Lehmann, L., Hieber, L., Weier, H.-U. G., Janish, C., Fung, J., Negele, T., Spelsberg, F., Lengfelder, E., Demidchik, E. P., Salassidis, K., Kellerer, A. M., Werner, M., and Bauchinger, M. (1999). Cytogenetic changes in radiation-induced tumors of the thyroid. *Cancer Res.* **59,** 135–140.

PART VIII

Cell Function and Differentiation

Analysis of Mitochondria by Flow Cytometry

Martin Poot* and Robert H. Pierce[†]

Department of Pathology
University of Washington
Seattle, Washington 98195

[†]Department of Pathology
Wright–Patterson Medical Center
Wright–Patterson Air Force Base
Dayton, Ohio 45433

I. Introduction
    A. Principles of Fluorescent Detection of Mitochondria
    B. Xanthylium Dyes: Rhodamine 123 and CMXRosamine
    C. Symmetrical Carbocyanine Dyes: $DiOC_6(3)$ and JC-1
    D. Asymmetric Carbocyanine Dyes: MitoTracker Green and MitoFluor
    E. The Cardiolipin Dye: Nonyl Acridine Orange
    F. Reduced Dyes
    G. Subcellularly Targeted Green Fluorescent Protein
II. Materials and Methods
    A. Dyes
    B. Cell Preparation
    C. Reduced Nucleotides (NADH and FADH)
    D. Normalized Mitochondrial Membrane Potential
    E. Mitochondrial Oxidative Turnover
    F. Mitochondrial Protein
    G. Mitochondrial Membrane Phospholipid (Cardiolipin)
III. Critical Aspects
    A. Staining Parameters
    B. Cell Fixation
    C. Combinations of Dyes
    References

METHODS IN CELL BIOLOGY, VOL. 64

I. Introduction

A. Principles of Fluorescent Detection of Mitochondria

Most of the energy (ATP) needed for proper functioning of the cell is generated by the mitochondrion. This process involves oxidation of reduced nucleotides (e.g., NADH and FADH) and concomitant generation of a negative inside membrane potential. The complex biochemical reactions that ultimately lead to formation of ATP are not fully understood. However, the finding that mitochondrial dysfunction may lead to neuromuscular diseases, neurodegenerative disorders, and apoptosis (Wallace, 1995; Kroemer *et al.*, 1998; Green and Reed, 1998) spurred intense investigation of this organelle.

For study by flow cytometry, cells and their organelles have to be (made) fluorescent. Reduced nucleotides, such as NADH and FADH, emit blue and blue-green fluorescence after excitation with light in the ultraviolet (UV) range and 488 nm, respectively (Thorell, 1983). As a function of their negative inside membrane potential, mitochondria will take up fluorescent cations. Initially, fluorescent probes for mitochondria have been devised based on this reasoning. Meanwhile, two dyes have been described that stain mitochondria based on unique properties that do not depend on the mitochondrial membrane potential. Nonyl acridine orange (NAO) specifically stains the mitochondrial membrane lipid cardiolipin (Petit *et al.*, 1992). The MitoTracker Green FM (MTG) dye was found to specifically stain mitochondria of cells in which the mitochondrial membrane potential was dissipated after cell fixation (Hollinshead *et al.*, 1997). Table I lists the response of some of these dyes to perturbation of mitochondrial function.

Table I
Sensitivity to Mitochondrial Poisons[a]

Dye	Rotenone	Antimycin A	CCCP
Rhodamine 123	97 ± 15	65 ± 12	71 ± 5
CMXRos	100 ± 6	37 ± 3	82 ± 7
MTG	108 ± 8	94 ± 7	105 ± 3
NAO	107 ± 4	98 ± 8	92 ± 11
JC-1	103 ± 12	58 ± 8	66 ± 7

[a] Logarithmically growing cultures of lymphoblastoid cells were pretreated with rotenone, antimycin A, and carbonyl cyanide m-chlorophenyl-hydrazone (CCCP) and stained with 100 nM of rhodamine 123, CMXRos, or MTG or with 500 nM of nonyl acridine orange (NAO) or JC-1 according to methods described by Poot and co-workers (Poot *et al.*, 1996). With rhodamine 123, MTG, and NAO, green fluorescence was recorded (525 ± 20 nm); with CMXRos and JC-1, red fluorescence (above 640 nm) was recorded. All data are fluorescence intensities as percentages of untreated controls from three independent experiments each performed in triplicate.

B. Xanthylium Dyes: Rhodamine 123 and CMXRosamine

The first dye described as a specific probe for the mitochondrion was the xanthylium dye rhodamine 123 (Johnson *et al.*, 1981). The amount of fluorescence obtained with this dye responded to the physiological state of the mitochondrion, as was predicted from the assumption that the mitochondrial membrane potential was the driving force behind dye accumulation inside the cell. Its poor photostability notwithstanding, rhodamine 123 has found widespread application, and its use was described by Chen in an earlier volume of this series (Chen, 1989). To improve photostability of mitochondrial staining, the MitoTracker Red dye CMXRosamine (CMXRos) was developed. The bright and relatively photostable red fluorescence of CMXRos (excitation 594 nm, emission 608 nm) showed strong sensitivity toward manipulation of the mitochondrial membrane potential and colocalized with cytochrome *c* oxidase (Poot *et al.*, 1996). The enhanced dye retention after cell fixation has been linked to the presence of a chloromethyl moiety (Poot *et al.*, 1996), with which CMXRos can covalently modify reduced thiol groups of mitochondrial membrane proteins. In addition to improved photostability, CMXRos showed better retention in stained cells during washing with phosphate-buffered saline than rhodamine 123 (Poot *et al.*, 1996). The latter may be due to the increased lipophilicity of CMXRos. Because of this characteristic, CMXRos fluorescence is likely to be less sensitive to mitochondrial swelling than is rhodamine 123. Changes in fluorescence after rhodamine 123 staining may represent the sum of changes in the mitchondrial membrane potential and in the volume of the intramitochondrial sap (Vander Heiden *et al.*, 1997). Fluorescence after CMXRos staining, on the other hand, may be sensitive to the amount of mitochondrial protein. Rhodamine 123, CMXRos, and other cationic xanthylium dyes share a critical feature: they are fluorescent in any medium, which leads to a certain level of background fluorescence even in a cell that does not contain mitochondria.

C. Symmetrical Carbocyanine Dyes: $DiOC_6(3)$ and JC-1

To overcome the drawbacks of xanthylium dyes and to obtain stains that exhibit much less background fluorescence, symmetrical lipophilic carbocyanine dyes such as 3,3'-dihexyloxacarbocyanine iodide [$DiOC_6(3)$] and 5,5',6,6'-tetrachloro-1,1',3,3'-tetraethylbenzimidazoly iodide (JC-1) (Reers *et al.*, 1991) have been developed. The $DiOC_6(3)$ dye turned out to be sensitive to the membrane potential of both the plasma membrane, and the mitochondrial membrane, and it showed strong fluorescence enhancement in a hydrophobic versus a hydrophilic environment (Sims *et al.*, 1974). In an earlier volume of this series Shapiro described the use of $DiOC_6(3)$ to determine distributions of plasma membrane potential by flow cytometry (Shapiro, 1994). It is therefore no surprise that $DiOC_6(3)$ fluorescence does not accurately monitor changes in the mitochondrial membrane potential (Salvioli *et al.*, 1997).

The JC-1 dye exhibits green fluorescence (excitation 490 nm, emission 527 nm) if present in low concentrations, and red fluorescence (excitation

490 nm, emission 590 nm) if accumulating at higher concentrations. The intramitochondrial concentration of the dye, and thus its fluorescence maximum, depends on the mitochondrial membrane potential (Smiley *et al.*, 1991). In addition to accumulating inside the mitochondrion, carbocyanine dyes, such as $DiOC_6(3)$ and JC-1, were found to stain the endoplasmic reticulum (Terasaki *et al.*, 1984; Chen, 1989). In addition, JC-1 forms J-aggregates that can result in nonspecific speckled staining of the cytoplasm (Poot *et al.*, 1996). These findings cast doubt on the alleged specificity of these carbocyanine dyes for mitochondria. Thus, the use of these dyes for studies of mitchondrial physiology by flow cytometry does not appear to be warranted. In particular, the contention that during apoptosis the mitochondrial membrane potential decreases cannot be based on the observed changes in $DiOC_6(3)$ and JC-1 fluorescence (Cossarizza *et al.*, 1994; M'etivier *et al.*, 1998).

D. Asymmetric Carbocyanine Dyes: MitoTracker Green and MitoFluor

To overcome the numerous drawbacks associated with symmetric carbocyanine dyes, a novel family of asymmetric carbocyanine dyes, including MitoTracker Green FM (MTG) and MitoFluor, have been developed. Both MTG and MitoFluor dyes are well excited by the 488 nm line of the argon laser and emit in the green region of the spectrum with little fluorescence emission in the orange-red region. Since the fluorescence emission range of the MTG and MitoFluor dyes is much narrower than that of rhodamine 123 (which emits significantly at wavelengths up to 620 nm), these dyes are more suitable for multicolor applications. The MTG dye is equipped with a chloromethyl moiety, which causes it to be retained after fixative treatment, whereas MitoFluor fluorescence vanishes after cell fixation. Both dyes do not respond to alterations in the mitochondrial membrane potential (see Table I). Moreover, the MTG dye has been used to specifically stain mitochondria in fixed cells (Hollinshead *et al.*, 1997) and is believed to monitor possible changes in mitochondrial protein level (Poot and Pierce, 1999).

E. The Cardiolipin Dye: Nonyl Acridine Orange

Nonyl acridine orange (NAO) (Maftah *et al.*, 1989) accumulates in mitochondria due its specific, high affinity binding to cardiolipin (Petit *et al.*, 1992). On binding to monoacidic phospholipids NAO emits green fluorescence, whereas in the presence of the diacidic phospholipid cardiolipin additional red fluorescence arises (Gallet *et al.*, 1995). Since cardiolipin is situated at the inner mitochondrial membrane, the fluorescence of NAO depends on the amount of cardiolipin. In other words, the fluorescence intensity of NAO can be taken as a direct measure of the amount of mitochondrial membrane lipid in a cell (Petit *et al.*, 1992).

F. Reduced Dyes

Reduced forms of some dyes (e.g., dihydrorhodamine 123, H_2-CMXRos), which yield only a fluorescent response after they are oxidized in intact mitochondria, have become available (Whitaker *et al.*, 1991). These dyes allow monitoring the rate of oxidant formation in mitochondria (Poot and Pierce, 1999).

G. Subcellularly Targeted Green Fluorescent Protein

The fluorescence emission of the green fluorescent protein (GFP) from the jellyfish *Aequorea victoria* is sensitive to the calcium level (Rizzuto *et al.*, 1992) and the pH of its environment (Llopis *et al.*, 1998; Keen *et al.*, 1998). Via recombinant DNA techniques, fusion proteins containing GFP and subcellular targeting signals have been prepared. These fusion proteins have been shown to localize to the mitochondrion (Rizzuto *et al.*, 1992), the endoplasmic reticulum (Kendall *et al.*, 1992), and the nucleus (Rizzuto *et al.*, 1994). Thus, a set of reagents has been devised that can be used to detect rapid changes in subcellular calcium levels (Rizzuto *et al.*, 1994). Although this approach proved to be successful in image analysis, no flow cytometric protocol has been published yet.

By introducing amino acid substitutions, forms of the GFP with different pK_a values and fluorescence emission spectra have been created (Llopis *et al.*, 1998). By simultaneously detecting the fluorescence emission from the "cyan" and least pH-sensitive protein with the yellow and most pH-sensitive protein, changes in intracellular pH can be deduced from the ratio of fluorescence intensity in the two wavelength domains (Llopis *et al.*, 1998). Thus, intracellular pH can be measured by a fluorescence ratioing method. The genes for the cyan and yellow proteins were then fused with targeting sequences for the Golgi system and the mitochondrion. These fusion proteins were then expressed in HeLa cells. By confocal microscopy, it was shown that the modified proteins localized to the organelles of interest (Llopis *et al.*, 1998). This novel approach may be of great promise, but it has as yet not been implemented in flow cytometry.

In this chapter, protocols to determine changes in NADH level, mitochondrial membrane potential as normalized for possible differences in mitochondrial protein content, mitochondrial oxidative turnover, mitochondrial protein, and cardiolipin content will be described.

II. Materials and Methods

A. Dyes

Inside the mitochondrion NADH and FADH exist as natural fluorophores. Stock solutions of 2.5 or 5.0 mM of rhodamine 123 and NAO are prepared in phosphate-buffered saline and stored at 4°C in the dark. The MTG, MitoFluor, and CMXRos dyes are dissolved at 0.2 mM in dimethyl sulfoxide and stored at −20°C in the dark. Immediately before use these dyes are thawed at room temperature in the dark. In our experience these dyes do not suffer from repeated freeze–thaw cycles. Reduced dyes, such as dihydrorhodamine 123 and H$_2$-CMX-Ros, are dissolved in dimethyl sulfoxide and flushed with an inert gas (e.g., nitrogen) before storage at −20°C. At room temperature, reduced dyes will oxidize readily. It is therefore recommended to divide dye solutions into aliquots and not to refreeze them. Key features of the dyes used in the protocols below are displayed in Table II.

B. Cell Preparation

Cells are harvested by standard procedures. In case mitochondrial function during cell death by apoptosis is to be studied, care should be taken that all (attached and detached) cells are included. After harvesting, cells are resuspended in regular cell culture medium and warmed to 37°C.

C. Reduced Nucleotides (NADH and FADH)

In unstained cells, the UV-excited (360 nm) blue autofluorescence (around 450 nm) is proportional to the mitochondrial NADH content and can easily be detected by flow cytometry (Thorell, 1983). The NADH content of cells proved to be a very sensitive parameter for changes in mitochondrial metabolism during apoptosis (Poot and Pierce, 1999). Excitation with 488 nm light of isolated mitochondria gives green autofluorescence that is inversely proportional to their FADH content. In intact cells, FADH fluorescence is overshadowed by other sources of green autofluorescence (e.g., lipofuscin). Therefore, it is not possible to quantify FADH levels in intact cells based on green autofluorescence.

D. Normalized Mitochondrial Membrane Potential

Flow cytometry measures the total amount of fluorescence obtained from each individual cell. When cells are stained with dyes of which the fluorescence intensity is proportional to the mitochondrial membrane potential, the total

Table II
Features of Mitochondrial Dyes

Dye	Excitation maximum (nm)	Emission maximum (nm)	Fixability with aldehydes
Rhodamine 123	506	530	−
CMXRos	594	608	−
MTG	480	516	+
MitoFluor	480	516	−
NAO	497	519	−
JC-1	514	529, 590	+/−[a]

[a] The JC-1 dye is partially retained after formaldehyde fixation, though a covalent bond with cellular macromolecules appears unlikely (Poot et al., 1996).

cellular fluorescence will reflect the average of the mitochondrial membrane potential per mitochondrion multiplied by the amount of mitochondrial mass per cell. Thus, a cell with more mitochondrial mass will be more fluorescent than a cell with less mitochondria. To correct for possible differences in mitochondrial mass, it is necessary to have a measure for the amount of mitochondrial mass in each individual cell. This measure is provided by counterstaining the cells with a mitochondria-specific dye that is not sensitive to the mitochondrial membrane potential. This dye has to emit at wavelengths that are different from those of the mitochondrial membrane sensitive dye. That means that if the mitochondrial membrane sensitive dye emits in the green, the mitochondrial membrane insensitive dye has to emit in the red and vice versa. By dividing the fluorescence intensity of the membrane potential sensitive dye by the fluorescence intensity from the insensitive dye, the normalized mitochondrial membrane of each individual cell is obtained (Poot and Pierce, 1999). A dye pair that has worked successfully in this way is MTG (insensitive) and CMXRos (membrane potential sensitive) (Poot and Pierce, 1999).

To 1-ml aliquots of prewarmed cell suspensions, CMXRos and MTG are added to obtain a final concentration of maximally 100 nM. Cell suspensions are incubated for 30 min at 37°C in the dark. After incubation cell suspensions are placed in a melting ice bath and immediately assayed by flow cytometry. Both dyes are excitable with the 488 nm line of an argon laser; MTG emits fluorescence around 530 nm, whereas CMXRos is best detected at wavelengths above 610 nm. To minimize possible interference from MTG fluorescence a 640 long-pass filter is preferred.

The ratio of CMXRos to MTG fluorescence intensity for each individual cell represents its normalized mitochondrial membrane potential. Since the MTG and the CMXRos dyes emit green and red fluorescence after excitation with 488 nm laser light, it is possible to combine this assay with quantification of NADH by UV-excited blue fluorescence as described earlier. Thus, a combined assay for normalized mitochondrial membrane potential and normalized NADH content is obtained. Figure 1 shows a typical result of such an assay.

E. Mitochondrial Oxidative Turnover

To 1-ml aliquots of prewarmed cell suspensions, H_2-CMXros is added at a final concentration of 100 nM. Cell suspensions are incubated for 30 min at 37°C in the dark. After incubation cell suspensions are placed in a melting ice bath and immediately assayed by flow cytometry. After oxidation by mitochondrial metabolism H_2-CMXRos becomes CMXRos and can be excited with the 488 nm line of an argon laser; CMXRos is best detected at wavelengths above 610 nm. Since the amount of fluorescence generated by mitochondrial metabolism is usually low, it is not recommended to combine H_2-CMXros staining with MTG or any other dye that may generate some yellow and red fluorescence (e.g., NAO). Generation of fluorescence from H_2-CMXRos can be abolished by prein-

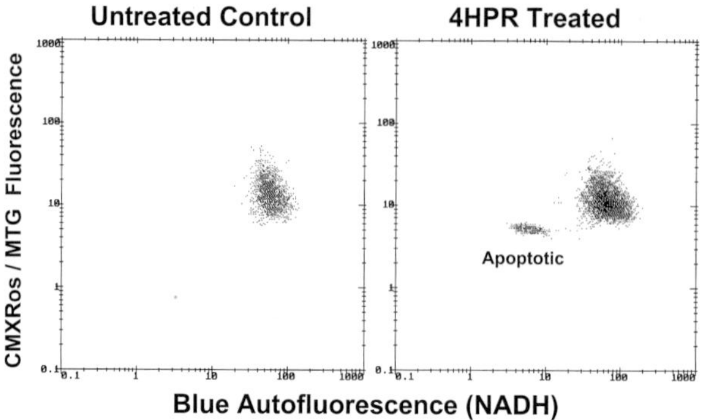

Fig. 1   Human breast cancer cells cultured for 4 days with 1 μM N-(4-hydroxyphenyl)retinamide (4HPR) and stained with CMXRos and MTG as described in the protocol for normalized mitochondrial membrane potential. Abscissa: UV-excited blue autofluorescence (which represents the NADH level); ordinate: the ratio of CMXRos to MTG fluorescence for each individual cell (which represents the normalized mitochondrial membrane potential). The signal dots labeled apoptotic represent cells that show a simultaneous decrease in NADH level and normalized mitochondrial membrane potential.

cubation with antimycin A, an inhibitor of mitochondrial electron flux (Poot and Pierce, 1999). Fluorescence generated in cells incubated with dihydrorhodamine 123 was found to be less sensitive toward inhibition of mitochondrial metabolism (M. Poot and R. H. Pierce, unpublished observation, 1997). Therefore, H_2-CMXros is now the preferred dye for measuring mitochondrial oxidative turnover.

F. Mitochondrial Protein

In fixed cells, the MTG dye stains mitochondria specifically (Hollinshead *et al.*, 1997). In flow cytometric assays the amount of fluorescence after MTG staining appears to correlate with cell volume (Poot and Pierce, 1999). In addition, mitochondrial proteins can be resolved as green bands by SDS–polyacrylamide gel electrophoresis (M. Poot, unpublished observation, 1997). Therefore, it is believed that the amount of MTG fluorescence represents the mitochondrial protein level. Incubation of cell suspensions with 100 nM of MTG for 30 min at 37°C followed by flow cytometric analysis is adequate to compare mitochondrial protein levels in different cell samples.

G. Mitochondrial Membrane Phospholipid (Cardiolipin)

Nonyl acridine orange (Maftah *et al.*, 1989) stains specifically the mitochondrial membrane lipid cardiolipin (Petit *et al.*, 1992). Cell suspensions are stained with 1 μM NAO during 30 min at 37°C, placed in a melting ice bath, and assayed immediately. NAO can be excited with the 488 nm line of an argon laser. The emission spectrum of NAO covers the green, yellow, and red domain. Because the red emission appears to most accurately reflect the mitochondrial cardiolipin level (Gallet *et al.*, 1995) a 640 nm long-pass filter is preferred (Pierce *et al.*, 2000).

<hr>

III. Critical Aspects

A. Staining Parameters

After harvesting and before staining, cells are resuspended in regular cell culture medium and warmed to 37°C. The latter is critical, because parameters of mitochondrial function, such as the mitochondrial membrane potential, and thus dye uptake, depend on the temperature. In order to not disturb mitochondrial function, regular cell culture medium should be used. If effects of certain environmental components are to be investigated, it is also possible to use special buffers or media. The staining protocols outlined are based on saturation of the target with dye. The recommended staining time is intended to be sufficient for this purpose. In case of doubt, it is recommended to test a series of staining times.

To avoid nonspecific staining it is important to keep the dye concentration as low as possible. With most cell types and with most flow cytometers, a sufficiently strong signal can be obtained with 100 nM of both CMXRos and MTG. If enough signal can be obtained, a lower dye concentration (<100 nM) may be preferable. A similar reasoning applies to the use of dihydrorhodamine 123, H$_2$-CMXRos, and NAO.

B. Cell Fixation

Some of the fluorescence of MTG and CMXRos is retained after cell fixation. However, the level of retained fluorescence may reflect the amount of protein thiols to which the dyes have bound covalently. This inference, which is derived from the behavior of chloromethyl moieties of other dyes (Poot *et al.*, 1991), has been supported experimentally (Gilmore and Wilson, 1999). One should not confuse the fluorescence intensities retained after cell fixation with the original fluorescence intensities, which are proportional with the mitochondrial membrane potential. It is clearly in error to assume that the retained fluorescence after cell fixation represents the mitochondrial membrane potential, as has been done in the literature (Macho *et al.*, 1996).

C. Combinations of Dyes

Since dyes to monitor mitochondrial parameters show different emission spectra, it is possible to use combinations of dyes in a single sample. This serves two purposes: first, concordant and discordant changes in metabolic parameters can be detected. Thus, a concordant change in mitochondrial membrane potential and cellular reduced glutathione level was elicited by exposing leukemia cells to

1-β-D-arabinofuranosylcytosine. This change preceded a high rate of reactive oxygen generation and an increase in intracellular calcium (Backway *et al.*, 1997). Second, it is possible to use the fluorescence emission of one dye to calibrate the fluorescence of other fluorochromes. Thus, ratios of fluorescence intensities on a per cell basis will represent "normalized" values for the parameters under investigation. Protocols for normalized mitochondrial membrane potential and normalized NADH levels were previously described. Such protocols give valid results only if the dyes used either do not interact with each other or if their interaction is constant. Owing to the small size of the mitochondrion, it is possible that dye molecules will bind in such a way that fluorescence resonance energy transfer may take place. Thus, it is conceivable that part of the fluorescence emission of MTG may be absorbed by CMXRos. It is therefore recommended to include samples stained with a single dye to verify that the differences between samples observed do not result from differences in efficiency of energy transfer between pairs of dyes.

Using a combination of NAO and CMXRos we have been able to document a change in the structure of the mitochondrial membrane during apoptosis (Poot and Pierce, 1999). Owing to the broad emission spectrum of NAO, combinations of NAO with other dyes require the use of control samples stained with a single dye or careful electronic compensation to eliminate cross-talk of NAO into the red fluorescence emission channel.

Acknowledgments

The authors are supported by a grant from the U.S. Army Medical Research and Material Command (DAMD: 17-94-J–4028) (M.P.), by the National Institute for Aging Research (Grant PO1 AG 01751) (M.P.), by the Nathan Shock Center of Excellence for the Basic Biology of Aging (Grant P30 AG 13240) (M.P.), by a Seed Money grant as part of an Institutional Research Grant of the American Cancer Society to the Fred Hutchinson Cancer Research Center (M.P.), and by grants from the National Institute of Environmental Health Sciences Environmental Pathology/Toxicology Training Program (Grant 5 T32 ES07032) and American Liver Foundation Irwin Arias Postdoctoral Research Fellowship (R.H.P.). We kindly acknowledge advice by W. T. Shen and Dr. P. S. Rabinovitch regarding the MPLUS software package. The views expressed in this article are those of the authors and do not reflect the official policy or position of the United States Air Force, Department of Defense, or the U.S. government.

References

Backway, K. L., McCulloch, E. A., Chow, S., and Hedley, D. W. (1997). Relationships between the mitochondrial permeability transition and oxidative stress during ara-C toxicity. *Cancer Res.* **57,** 2446–2451.

Chen, L. B. (1989). Fluorescent labeling of mitochondria. *Methods Cell Biol.* **29,** 103–123.

Cossarizza, A., Kalashnikova, G., Grassilli, E., Chiapelli, F., Salvioli, S., Capri, M., Barbieri, D., Troiano, L., Monti, D., and Franceschi, C. (1994). Mitochondrial modifications during rat thymocyte apoptosis: A study at the single cell level. *Exp. Cell Res.* **214,** 323–330.

Gallet, P. F., Maftah, A., Petit, J. M., Denis-Gay, M., and Julien, R. (1995). Direct cardiolipin assay in yeast using red fluorescence emission of 10-*N*-nonyl acridine orange. *Eur. J. Biochem.* **228,** 113–119.

Gilmore, K., and Wilson, M. (1999). The use of chloromethyl-X-rosamine (Mito Tracker Red) to measure loss of mitochondrial membrane potential in apoptotic cells is incompatible with cell fixation. *Cytometry* **36**, 355–358.

Green, D. R., and Reed, J. C. (1998). Mitochondria and apoptosis. *Science* **281**, 1309–1312.

Hollinshead, M., Sanderson, J., and Vaux, D. J. (1997). Anti-biotin antibodies offer superior organelle-specific labeling of mitochondria over avidin or streptavidin. *J. Histochem. Cytochem.* **45**, 1053–1057.

Johnson, L. V., Walsh, M. L., Bockus, B. J., and Chen, L. B. (1981). Monitoring of relative mitochondrial membrane potential in living cells by fluorescence microscopy. *J. Cell Biol.* **88**, 526–535.

Kreen, M., Farinas, J., Li, Y., and Verkman, A. S. (1998). Green fluorescent protein as a noninvasive intracellular pH indicator. *Biophys. J.* **74**, 1591–1599.

Kendall, J. M., Dormer, R. L., and Campbell, A. K. (1992). Targeting aequorin to the endoplasmic reticulum of living cells. *Biochem. Biophys. Res. Commun.* **189**, 1008–1016.

Kroemer, G., Dallaporta, B., and Resche-Rigon, M. (1998). The mitochondrial death/life regulator in apoptosis and necrosis. *Annu. Rev. Physiol.* **60**, 619–642.

Llopis, J., McCaffery, M., Miyawaki, A., Farquhar, M. G., and Tsien, R. Y. (1998). Measurement of cytosolic, mitochondrial, and Golgi pH in single living cells with green fluorescent proteins. *Proc. Natl. Acad. Sci. U.S.A.* **95**, 6803–6808.

Macho, A., Decaudin, D., Castedo, M., Hirsch, T., Susin, S. A., Zamzami, N., and Kroemer, G. (1996). Chloromethyl-X-Rosamine is an aldehyde-fixable potential-sensitive fluorochrome for the detection of early apoptosis. *Cytometry* **25**, 333–340.

Maftah, A., Petit, J. M., Ratinaud, M. H., and Julien, R. (1989). 10 *N*-Nonyl-acridine orange: A fluorescent probe which stains mitochondria independently of their energetic state. *Biochem. Biophys. Res. Commun.* **164**, 185–190.

M'etivier, D., Dallaporta, B., Zamzami, N., Larochette, N., Susin, S. A., Marzo, I., and Kroemer, G. (1998). Cytofluorometric detection of mitochondrial alterations in early CD95/Fas/APO-1-triggered apoptosis of Jurkat T lymphoma cells. Comparison of seven mitochondrion-specific fluorochromes. *Immunol. Lett.* **61**, 157–163.

Petit, J.-M., Maftah, A., Ratinaud, M.-H., and Julien, R. (1992). 10 *N*-Nonyl-acidine orange interacts with cardiolipin and allows the quantification of this phospholipid in isolated mitochondria. *Eur. J. Biochem.* **209**, 267–273.

Pierce, R. H., Campbell, J. S., Stephenson, A. B., Franklin, C. C., Chaisson, M., Poot, M., Kavanagh, T. J., and Fausto, N. (2000). Disruption of redox homeostasis in tumor necrosis factor induced apoptosis in murine hepatocytes. *Am. J. Pathol.* **157**, 221–236.

Poot, M., and Pierce, R. H. (1999). Detection of changes in mitochondrial function during apoptosis by simultaneous staining with multiple fluorescent dyes and correlated multiparameter flow cytometry. *Cytometry* **35**, 311–317

Poot, M., Kavanagh, T. J., Kang, H.-C., Haugland, R. P., Rabinovitch, P. S. (1991). Flow cytometric analysis of cell cycle-dependent changes in cell thiol level by combining a new laser dye with Hoechst 33342. *Cytometry* **12**, 184–187.

Poot, M., Zhang, Y.-Z., Krämer, J., Wells, K. S., Jones, L. J., Hanzel, D. K., Lugade, A. G., Singer, V. L., and Haugland, R. P. (1996). Analysis of mitochondrial morphology and function with novel fixable fluorescent stains. *J. Histochem. Cytochem.* **44**, 1363–1372.

Reers, M., Smith, T. W., and Chen, L. B. (1991). J-aggregate formation of a carbocyanine as a quantitative fluorescent indicator of membrane potential. *Biochemistry* **30**, 4480–4486.

Rizzuto, R., Simpson, A. W. M., Brini, M., and Pozzan, T. (1992). Rapid changes of mitochondrial Ca^{2+} revealed by specifically targeted recombinant aequorin. *Nature* **358**, 325–327.

Rizzuto, R., Brini, M., and Pozzan, T. (1994). Targeting recombinant aequorin to specific intracellular organelles. *Methods Cell Biol.* **40**, 339–358.

Salvioli, S., Ardizzoni, A., Franceschi, C., and Cossarizza, A. (1997). JC-1, but not DioC$_6$(3) or rhodamine 123, is a reliable fluorescent probe to assess $\Delta\Psi$ changes in intact cells: Implications for studies on mitochondrial functionality during apoptosis. *FEBS Lett.* **411**, 77–82.

Shapiro, H. M. (1994). Cell membrane potential analysis. *Methods Cell Biol.* **41**, 121–133.

Sims, P. J., Waggoner, A. S., Wang, C. H., and Hoffman, J. F. (1974). Studies on the mechanism by which cyanine dyes measure membrane potential in red blood cells and phosphatidylcholine vesicles. *Biochemistry* **13,** 3315–3330.

Smiley, S. T., Reers, M., Mottola-Hartshorn, C., Lin, M., Chen, A., Smith, T. W., Steele, G. D., Jr., and Chen, L. B. (1991). Intracellular heterogeneity in mitochondrial membrane potentials revealed by a J-aggregate-forming lipophilic cation JC-1. *Proc. Natl. Acad. Sci. U.S.A.* **88,** 3671–3675.

Terasaki, M., Song, J., Wong, J. R., Weiss, M. J., and Chen, L. B. (1984). Localization of endoplasmic reticulum in living and glutaraldehyde-fixed cells with fluorescent dyes. *Cell* **38,** 101–108.

Thorell, B. (1983). Flow-cytometric monitoring of intracellular flavins simultaneously with NAD(P)H levels. *Cytometry* **4,** 61–65.

Vander Heiden, M. G., Chandel, N. S., Williamson, E. K., Schumacker, P. T., and Thompson, C. B. (1997). Bcl-$_{xL}$ regulates the membrane potential and volume homeostasis of mitochondria. *Cell* **191,** 627–637.

Wallace, D. C. (1995). Mitochondrial DNA variation in human evolution, degenerative disease, and aging. *Am. J. Hum. Genet.* **57,** 201–223.

Whitaker, J. E., Moore, P. L., Haugland, R. L., and Haugland, R. P. (1991). Dihydrotetramethylrosamine: A long wavelength, fluorogenic peroxidase substrate evaluated in vitro and in a model phagocyte. *Biochem. Biophys. Res. Commun.* **175,** 387–393.

CHAPTER 36

Analysis of RNA Synthesis by Cytometry

Peter Østrup Jensen, Jacob Larsen, and Jørgen K. Larsen

Finsen Laboratory, Finsen Center
Rigshospitalet
DK-2100 Copenhagen, Denmark

I. Introduction
II. Background
III. Methods for Analysis of 5′-Bromouridine Incorporation and DNA Content
    A. 5′-Bromouridine Labeling
    B. Cell Preparation
    C. Staining
    D. Flow Cytometry
    E. Fluorescence Microscopy
    F. Critical Aspects of the Methodology
IV. Results of Labeling RNA with 5′-Bromouridine
    A. Specificity
    B. RNA Synthesis and the Cell Cycle
    C. Subcellular Distribution of Incorporated 5′-Bromouridine
V. Applications
   References

I. Introduction

The biological significance of RNA has been studied by different cytometric approaches, which can be divided into "static" approaches, describing the momentary distribution of the RNA content in the cell population, and "dynamic" approaches, describing the distribution of the content of RNA precursors that have been incorporated during a preset time period. In this chapter we will focus on the dynamic approach, with emphasis on the flow cytometric analysis limited to applications with mammalian cells.

METHODS IN CELL BIOLOGY, VOL. 64

II. Background

The RNA content of a cell can be estimated by direct staining with fluorochromes such as pyronin Y and acridine orange, the latter binding metachromatically to single-stranded RNA and DNA (Darzynkiewicz, 1994). The content of double-stranded RNA can be estimated by staining of DNase-pretreated cells with propidium iodide (Frankfurt, 1990). These static staining techniques only enable an indirect estimation of the net amount of synthesized RNA, as may be revealed by comparison of cell samples taken at different time points.

Dynamic measurements of RNA synthesis are based on a different approach. The object of interest, cell cultures or living organisms, must be labeled *in vitro* or *in vivo* with a detectable RNA precursor during a given time period. Thus, RNA synthesis has been traditionally investigated by incorporation of [^{3}H]uridine and counting the silver grains developed in an autoradiographic film over the individual cells (Fakan, 1986). As an alternative to this time-consuming and laborsome method, a faster and nonradioactive technology has emerged based on immunochemical detection. This method utilizes the incorporation of brominated RNA precursors, either 5′-bromouridine (BrUrd) (Jensen *et al.*, 1993a; Hozák *et al.*, 1994) or 5′-bromouridine triphosphate (BrUTP) (Jackson *et al.*, 1993), followed by immunofluorescent staining using anti-5′-bromodeoxyuridine (BrdUrd) antibodies. Alternatively, nascent RNA can be detected by fluorescein-UTP incorporation (LaMorte *et al.*, 1998).

For studies of RNA synthesis in experimental animals, *in vivo* labeling with BrUrd (or [^{3}H]uridine) is required. It is thought that the BrUrd enters the cell and that phosphate groups are attached by the cell's own anabolic machinery to produce BrUTP, which is suitable for incorporation into nascent RNA during transcription. For the incorporation of BrUTP, biotin-UTP, or fluorescein-UTP into nascent RNA by the RNA polymerases, access to the transcription sites in the cells has to be facilitated. Thus stripping of the plasma membrane (Jackson *et al.*, 1993), permeabilization (Dundr and Raška, 1993), microinjection (Wansink *et al.*, 1993; Carmo-Fonseca *et al.*, 1996; LaMorte *et al.*, 1998), or fusion with precursor-containing liposomes (Haukenes *et al.*, 1997) have been applied.

The BrUrd-substituted RNA may be detected by permeabilizing the cells and staining with certain anti-BrdUrd antibodies. Indeed, several of the antibodies marketed as anti-BrdUrd antibodies are produced by hybridomas that are derived from rodents that have been immunized with halogenated uridine, and not deoxyuridine. A FITC-conjugated anti-BrdUrd antibody may be used for visualization of the BrUrd-substituted RNA, or this may be indirectly stained with a fluorochrome-conjugated or gold-labeled secondary antibody. The secondary structure of RNA may influence the accessibility for the antibodies to the BrUrd-labeled RNA, in analogy with the well-known experience from staining of BrdUrd-labeled DNA, where staining with anti-BrdUrd antibodies requires that the BrdUrd is exposed in a denatured, single-stranded form (Dolbeare,

1995). Thus, it is to be expected that BrUrd-substituted single-stranded RNA is far more accessible to the anti-BrdUrd antibodies than is BrUrd-substituted double-stranded RNA.

Owing to their relatively short lifetime in comparison to other types of RNA molecules, specific RNA transcripts may provide estimates of the cellular expression of the corresponding genes at the RNA level. Specific mRNA molecules may be cytometrically recognized by hybridization with specific oligonucleotide probes. The bound probes may then be visualized by addition of fluorochrome-conjugated probes (fluorescence *in situ* hybridization, FISH) (Chapter 33) or by incorporation of fluorochrome-conjugated nucleotides where the specific oligonucleotide probes serve as primers (primed *in situ* labeling, PRINS) (Chapter 32).

III. Methods for Analysis of 5′-Bromouridine Incorporation and DNA Content

The described techniques have been developed for flow cytometric measurements of BrUrd incorporation, but they will also allow for investigations by fluorescence microscopy.

A. 5′-Bromouridine Labeling

Cell cultures of, for example, activated lymphocytes, leukemia cell lines, or adherent cell cultures are labeled by incubation for 1 hr with 1 mM BrUrd (03824HP, Aldrich, Steinheim, Germany). After harvest, samples of 10^6 cells are washed in cold PBS (Ca^{2+}- and Mg^{2+}-free Dulbecco's phosphate-buffered saline, pH 7.2).

For *in vivo* labeling of rats, 200 mg BrUrd dissolved in 2 ml PBS is injected intraperitoneally. After 1 hr the rats are anesthetized, and organ samples are quickly excised and placed on ice. A single cell suspension is prepared mechanically and stored at −20°C in a freezing buffer [250 mM sucrose, 40 mM trisodium citrate dihydrate, 5% (v/v) dimethyl sulfoxide, pH 7.6].

B. Cell Preparation

The cells must be fixed and permeabilized before staining. With methods A and B suspensions of nuclei are produced, and with methods C and D suspensions of entire cells.

For method A, according to Landberg and Roos (1991) and Jensen *et al.* (1993a), the cells are lysed for 15 min on ice with 0.5 ml of buffer [0.5% Triton X-100, 1% bovine serum albumin (BSA), and 0.2 μg/ml EDTA in PBS]. The nuclei are subsequently fixed by addition of 3 ml methanol at −20°C. Higher temperature increases the risk for clotting of the BSA. Store at −20°C.

For method B, modified after Otto (1990), the cells are lysed for 15 min on ice with 1 ml of buffer (2.1% citric acid and 0.5% Tween 20 in distilled water, pH 2). The nuclei are washed in PBS and fixed by addition of 1 ml methanol at $-20°C$. This method is suitable for release of nuclei from monolayer cultures. Store at $-20°C$.

For method C, modified after Carayon and Bord (1992) and according to Jensen *et al.* (1993b), the cells are fixed and permeabilized in 1 ml of buffer (1% formaldehyde and 0.05% Nonidet P-40 in PBS) at room temperature, slowly agitated for 15 min, and then stored at 4°C.

For method D, according to Li *et al.* (1994), the cells are fixed in 1 ml of 1% formaldehyde in PBS for 15 min. The fixed samples are washed in PBS and permeabilized by addition of 1 ml ethanol at $-20°C$. Store at $-20°C$.

C. Staining

Samples of approximately 5×10^5 fixed cells are washed in PBS, and 50 μl of monoclonal anti-bromodeoxyuridine antibody (ABDM, Partec, Münster, Germany), diluted 1:10 in PBS with 0.01% (v/v) Nonidet P-40, is added for 60 min. After washing once in PBS, a secondary, FITC-conjugated anti-mouse immunoglobulin antibody (e.g., F-313, DAKO) is added for 60 min. Finally, after washing once in PBS, 100 μl of 50 μg/ml propidium iodide (P-4170, Sigma, St. Louis, MO) in PBS is added for at least 15 min before flow cytometry. Samples are kept on an ice bath during and after staining.

In addition to the ABDM antibody, several antibodies exist that will bind to BrUrd in RNA: clone B-44 (7580, Becton Dickinson, San Jose, CA), and clone BMC9318 (1170376, Boehringer Mannheim, Mannheim, Germany), both from mouse, and clone BU1/75 (MAS 250c, Sera-Lab, Crawley Down, England), derived from rats. Interestingly, the clone BU20a anti-BrdUrd antibody (M-744, Dako, Copenhagen, Denmark), which is derived from mice immunized with BrdUrd, shows no cross-reactivity to BrUrd.

D. Flow Cytometry

Flow cytometric analysis of the BrUrd/DNA distribution is performed on a flow cytometer using laser excitation at 488 nm and collection of FITC fluorescence at 515–545 nm and of propidium iodide fluorescence at >615 nm. Recording of $1–2 \times 10^4$ cells is triggered by the propidium iodide signal. The instrument is adjusted to minimal coefficient of variation (CV) for the DNA measurement, using, for example, propidium iodide-stained chicken erythrocytes.

E. Fluorescence Microscopy

The subcellular distribution of the incorporated BrUrd is investigated by fluorescence microscopy of cells that are processed as described, but without the final staining with propidium iodide. Aliquots of the cell suspensions are either pipetted onto slides and air dried for 30 min, or spun onto microscope slides for 5 min at 800 g using 1-ml, 30-mm^2 cytocentrifuge containers (Zytokammer, Hettich Zentrifugen, Tuttingen, Germany). The slides are finally mounted with Fluoromount-G (Southern Biotechnology Associates, Birmingham, AL), containing 0.5 μg/ml 4′,6-diamidino-2-phenylindole (DAPI) (Serva Feinbiochemica, Heidelberg, Germany).

F. Critical Aspects of the Methodology

For control of the specificity of the immunofluorescence signal, a comparison with a BrUrd-unlabeled sample as well as a BrUrd-labeled and ribonuclease (RNase)-treated sample is recommended. For this purpose it is important to apply the RNase treatment before immunochemical staining of the BrUrd (Jensen *et al.*, 1993a). It has been reported that a BrUTP-specific cytoplasmic fluorescence can occur. This fluorescence could not be inhibited with actinomycin D and was only partially removed by treatment with RNase (Haukenes *et al.*, 1997). This problem might be omitted by using preparations of cell nuclei.

Because the BrUrd signal is quite stable against RNase treatment after immunostaining, addition of RNase to the propidium iodide staining solution (e.g., 5 mg/ml ribonuclease A, Sigma R-4875) is advantageous for increasing the specificity and resolution of the DNA measurement (Jensen *et al.*, 1993a).

In our experience, the flow cytometric detection level for BrUrd incorporation is at about 15 min incubation with 100 μM BrUrd. Application of BrUrd over extended periods of time ($\geq$50 μM for $\geq$24 hr) may induce cell cycle perturbation and apoptotic cell death, as shown in HL-60 and MOLT-4 cells by Li *et al.* (1994).

Limited access for the anti-BrdUrd antibody to bind to the incorporated BrUrd, which may be masked by proteins due to fixation with formaldehyde or not accessible due to incorporation into double-stranded RNA, may be an important factor in interpretation of the results. The problems due to masking of incorporated BrUrd may be omitted by alternative labeling with fluorescein isothrocyanate (FITC)-UTP (LaMorte *et al.*, 1998). Care should be taken to avoid contamination with RNases during handling of the cells.

The importance of the labeling efficiency of the artificial RNA precursors into the various types of RNA has not been completely investigated. It should be mentioned that BrUrd is incorporated unbiased in intact cells (Jackson *et al.*, 1998). Biotin-11-UTP and digoxigenin-11-UTP are less efficiently incorporated into RNA than BrUTP (Jackson *et al.*, 1993). For microscopic detection of BrUTP incorporation, labeling with 2 mM BrUTP for 2.5 min is sufficient (Jackson *et al.*, 1993).

Compared to [^{3}H]uridine incorporation, exposure to BrUrd has little effect initially on transcription rates (Jackson *et al.*, 1998), but the chemically modified uridines are known to inhibit the subsequent processing of RNA transcripts (Wansink *et al.*, 1994a). This implies that the very fast RNA processing is delayed.

However, in studies of the morphology of transcription this may turn out to be a practical advantage, as it provides the researcher the needed time to conduct the experiment.

IV. Results of Labeling RNA with 5′–Bromouridine

A. Specificity

In exponentially growing cultures of HL-60 cells, RNA synthesis can be demonstrated in the majority of the cells by flow cytometric analysis of nuclear BrUrd incorporation (Fig. 1). The specificity of the FITC fluorescence is indicated by the fact that RNase treatment before the immunochemical staining (Fig. 1C) induced a reduction of the signal from the labeled nuclei (Fig. 1A) to the level of the unlabeled nuclei (Fig. 1B). Attempts to measure the effect of ribonucleotide reductase activity, which might convert BrUrd to BrdUrd leading to labeling of DNA during replication, had no effect, as deliberately BrdUrd-labeled nuclei stained negatively (Fig. 1D), which is supported by Dundr and Raška (1993). Also, treatment of the cell culture with the RNA polymerase inhibitor actinomycin D prior to incubation with BrUrd decreased the labeling efficiency, in agreement with results from Dundr and Raška (1993) and Jackson *et al.* (1993).

B. RNA Synthesis and the Cell Cycle

Dual parameter analysis of FITC (BrUrd) and propidium iodide (DNA) fluorescence enables the correlation of transcriptional activity and cell cycle phase

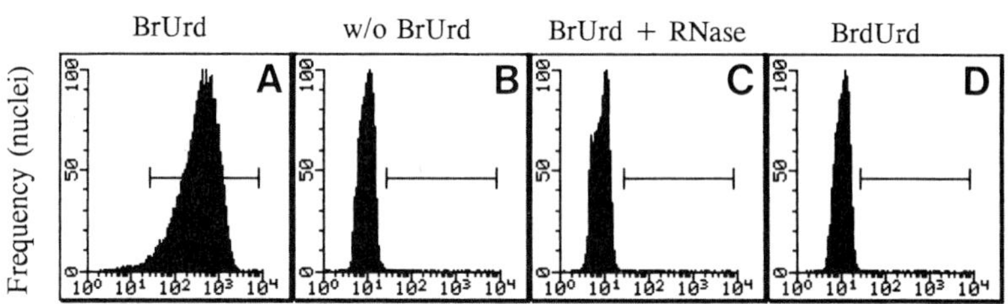

Fig. 1 Flow cytometric analysis of RNA synthesis according to BrUrd incorporation in exponentially growing HL-60 cells, demonstrating the specificity of the immunochemical staining. The cells were lysed and fixed according to method A before staining with ABDM anti-BrdUrd antibody (Partec) and secondary FITC-conjugated antibody. Flow cytometric analysis was performed on a FACS IV (Becton Dickinson), measuring the incorporated BrUrd by the log FITC-fluorescence of fixed nuclei. Markers indicate histogram regions of BrUrd positive nuclei. (A) Cells incubated with 1 mM BrUrd for 1 hr (98% positive nuclei). (B) Cells incubated without BrUrd (0.3% positive nuclei). (C) Cells incubated with 1 mM BrUrd for 1 hr, but the fixed nuclei were treated with RNase before staining with the antibody (0.3% positive nuclei). (D) Cells incubated with 10 μM BrdUrd instead of BrUrd (0.4% positive nuclei). Reprinted from Jensen *et al.* (1993b) with permission from the publisher.

distribution, as shown for HL-60 cells in Fig. 2A. Nuclei with S phase DNA content show a high level of BrUrd incorporation, whereas those with G_1 and G_2/M phase DNA content show a low or intermediate level. The horseshoe-shaped distribution, similar to the distribution found when labeling with the DNA precursor BrdUrd, indicates that the major part of RNA synthesis in this exponentially growing cell line is related to replication.

In vivo labeling of rat liver cells reveals a somewhat different distribution, shown in Fig. 2B. These largely noncycling and partially polyploid cells show a substantial labeling in the G_1 and G_2/M phases, thus indicating RNA synthesis in the noncycling state.

C. Subcellular Distribution of Incorporated 5′-Bromouridine

In Fig. 3, the FITC staining pattern of BrUrd incorporated after a labeling period of 1 hr may be compared with the localization of nuclei and nucleoli according to simultaneous staining of the DNA with DAPI. When cells were prepared according to method B, the nuclei showed intense staining of the nucleoli as well as several extranucleolar smaller foci in a darker nuclear matrix (Fig. 3a). This staining pattern is similar to labeling with [³H]uridine and FITC-UTP. In contrast, cells prepared with the milder extraction method A showed only extranucleolar staining. With methods C and D, the FITC staining in the nucleus was similarly confined to the extranucleolar portion, and in addition the cytoplasm contained FITC-stained foci.

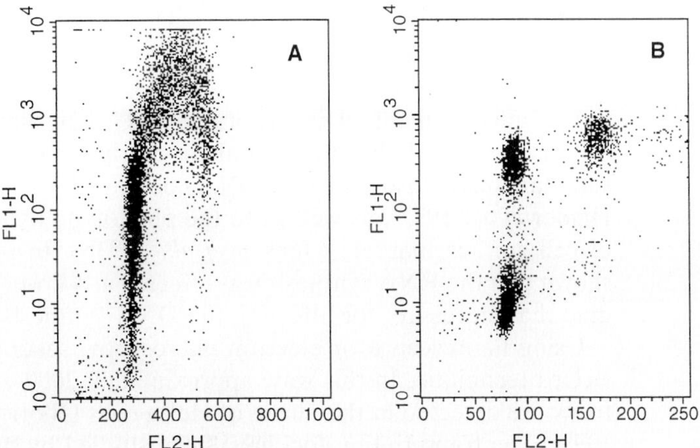

Fig. 2 Dual parameter flow cytometric analysis of log FITC (BrUrd) and linear propidium iodide (DNA) fluorescence. (A) Exponentially growing HL-60 cells were incubated with 1 m*M* BrUrd for 1 hr and prepared according to method B. (B) Rat liver cells harvested after *in vivo* labeling for 1 hr with 200 mg BrUrd and prepared according to method A.

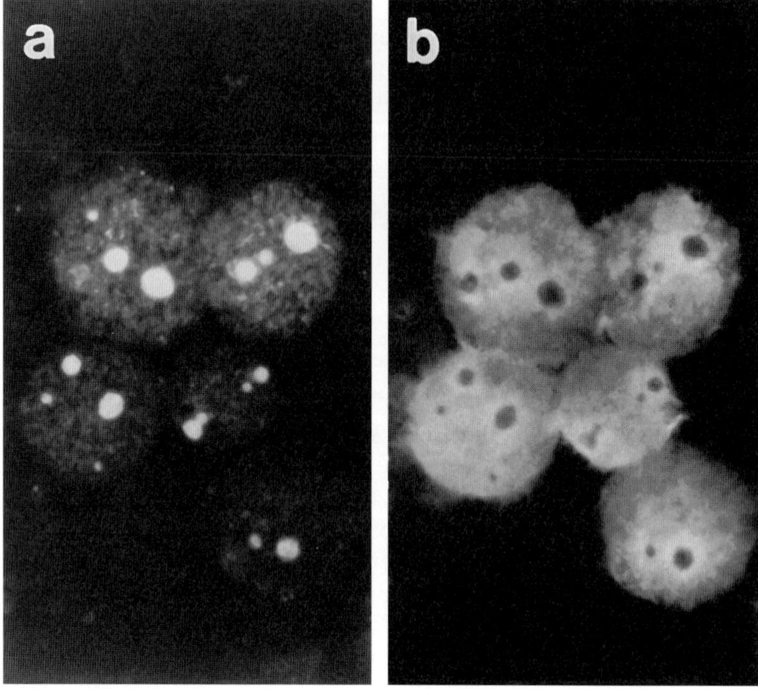

Fig. 3 Fluorescence microphotographs of exponentially growing HL-60 cells that were incubated with 1 m*M* BrUrd for 1 hr and prepared according to method B. (a) FITC-stained BrUrd. (b) DAPI-stained DNA.

V. Applications

The measurement of incorporated BrUrd by flow cytometry has provided a basis for correlating the overall transcriptional activity to the cell cycle by simultaneous measurement of DNA content (Jensen *et al.*, 1993a; Li *et al.*, 1994; Haider *et al.*, 1997), as well as to the phenotype by simultaneous measurement of cell surface markers (Jensen *et al.*, 1993b). In toxicological studies, it was shown that the RNA synthesis was affected at a lower concentration of 5-azacytidine than necessary for affecting the DNA synthesis (Murakami *et al.*, 1995).

Using fluorescence or electron microscopy, several studies have utilized the BrUrd technique. In this way, approximately 2000 extranucleolar transcription foci were detected in the nuclei of HeLa cells (Iborra *et al.*, 1996; Jackson *et al.*, 1998). The transcription foci have been studied by simultaneous staining of the splicing factors and nascent RNA (Jackson *et al.*, 1993; Pombo and Cook, 1996). When BrUTP incorporation into RNA was combined with biotin-dUTP or digitonin-dUTP incorporation into DNA, distributions of transcription sites could be compared to replication sites in the nuclei (Jackson *et al.*, 1993; Hassan *et al.,* 1994; Wansink *et al.,* 1994b). Implementing the BrUrd method in animal cells, nucleolar transcription sites were mapped to the dense fibrillar component (Dundr and Raška, 1993; Wansink *et al.,* 1994b) and additionally to the nucleolar fibrillary centers and vacuoles in plants (Melcak *et al.,* 1996).

Cytometric analysis in itself, only reporting the amount of recognized Br-substituted RNA, does not provide any information of the primary structure of the synthesized RNA. However, by exploiting the knowledge on BrUrd antigen–antibody reactivity in the field of methods that are applicable for cell lysates, the detection of RNA synthesis from specific genes is possible. Using immunoseparation of the RNA labeled with brominated precursors during a short labeling period, conditioned by specific external stimuli, and followed by reverse transcriptase polymerase chain reaction (RT-PCR) and Northern blotting or sequencing for relating the signal to specific genes, seems a promising path to follow (Haider *et al.,* 1997).

References

Carayon, P., and Bord, A. (1992). Identification of DNA-replicating lymphocyte subsets using a new method to label the bromodeoxyuridine incorporated into the DNA. *J. Immunol. Methods* **147,** 225–230.

Carmo-Fonseca, M., Cunha, C., Custódio, N., Carvalho, C., Jordan, P., Ferreira, J., and Parreira, L. (1996). The topography of chromosomes and genes in the nucleus. *Exp. Cell Res.* **229,** 247–252.

Darzynkiewicz, Z. (1994). Simultaneous analysis of cellular RNA and DNA content. *Methods Cell Biol.* **41A,** 401–420.

Dolbeare F. (1995). Bromodeoxyuridine: A diagnostic tool in biology and medicine, Part 1: Historical perspectives, histochemical methods and cell kinetics. *Histochem. J.* **27,** 339–369.

Dundr, M., and Raška, I. (1993). Nonisotopic ultrastructural mapping of transcription sites within the nucleolus. *Exp. Cell Res.* **208,** 275–281.

Fakan, S. (1986). Structural support for RNA synthesis in the cell nucleus. *Methods Achiev. Exp. Pathol.* **12,** 105–40.

Frankfurt, O. S. (1990). Flow cytometric analysis of double stranded RNA content distributions. *Methods Cell Biol.* **33,** 299–304.

Haider, S. R., Juan, G., Traganos, F., and Darzynkiewicz, Z. (1997). Immunoseparation and immunodetection of nucleic acids labeled with halogenated nucleotides. *Exp. Cell Res.* **234,** 498–506.

Hassan, A. B., Errington, R. J., White, N. S., Jackson, D. A., and Cook, P. R. (1994). Replication and transcription sites are colocalized in human cells. *J. Cell Sci.* **107,** 425–434.

Haukenes, G., Szilvay, A.-M., Brokstad, K. A., Kaestrøm, A., and Kalland, K.-H. (1997). Labeling of RNA transcripts of eukaryotic cells in culture with BrUTP using a liposome transfection reagent (DOTAP). *BioTechniques* **22,** 308–312.

Hozák, P., Cook, P. R., Schofer, C., Mosgoller, W., and Wachtler, F. (1994). Site of transcription of ribosomal RNA and intranucleolar structure in HeLa cells. *J. Cell Sci.* **107,** 639–648.

Iborra, F. J., Pombo, A., Jackson, D. A., and Cook, P. R. (1996). Active RNA polymerases are localized within transcription factories in human nuclei. *J. Cell Sci.* **109,** 1427–1436.

Jackson, D. A., Hassan, A. B., Errington, R. J., and Cook, P. R. (1993). Visualization of focal sites of transcription within human nuclei. *EMBO J.* **12,** 1059–1065.

Jackson, D. A., Iborra, F. J., Manders, E. M. M., and Cook, P. R. (1998). Numbers and organization of RNA polymerases, nascent transcripts, and transcription units in HeLa nuclei. *Mol. Biol. Cell* **9,** 1523–1536.

Jensen, P. Ø., Larsen, J., Christiansen, J., and Larsen, J. K. (1993a). Flow cytometric measurement of RNA synthesis using bromouridine labelling and bromodeoxyuridine antibodies. *Cytometry* **14,** 455–458.

Jensen, P. Ø., Larsen, J., and Larsen, J. K. (1993b). Flow cytometric measurement of RNA synthesis based on bromouridine labelling and combined with measurement of DNA content or cell surface antigen. *Acta Oncol.* **32,** 521–524.

LaMorte, V. J., Dyck, J. A., Ochs, R. L., and Evans, R. M. (1998). Localization of nascent RNA and CREB binding protein with the PML-containing nuclear body. *Proc. Natl. Acad. Sci. U.S.A.* **95,** 4991–4996.

Landberg, G., and Roos, G. (1991). Antibodies to proliferating cell nuclear antigen (PCNA) as S-phase specific probes in flow cytometric cell cycle analysis. *Cancer Res.* **51,** 4570–4575.

Li, X., Patel, R., Melamed, M. R., and Darzynkiewicz, Z. (1994). The cell cycle effects and induction of apoptosis by 5-bromouridine in cultures of human leukemic MOLT-4 and HL-60 cell lines and mitogen stimulated normal lymphocytes. *Cell Prolif.* **27,** 307–320.

Melcak, I., Risueno, M. C., and Raska, I. (1996). Ultrastructural nonisotopic mapping of nucleolar transcription sites in onion protoplasts. *J. Struct. Biol.* **116,** 253–263.

Murakami, T., Li, X., Gong, J., Bhatia, U., Traganos, F., and Darzynkiewicz, Z. (1995). Induction of apoptosis by 5-azacytidine: Drug concentration-dependent differences in cell cycle specificity. *Cancer Res.* **55,** 3093–3098.

Otto, F. (1990). DAPI staining of fixed cells for high-resolution flow cytometry of nuclear DNA. *Methods Cell Biol.* **33,** 105–110.

Pombo, A., and Cook, P. R. (1996). The localization of sites containing nascent RNA and splicing factors. *Exp. Cell Res.* **229,** 201–203.

Wansink, D. G., Schul, W., van der Kraan, I., van Steensel, B., van Driel, R., and de Jong, L. (1993). Fluorescent labeling of nascent RNA reveals transcription by RNA polymerase II domains scattered throughout the nucleus. *J. Cell Biol.* **122,** 283–293.

Wansink, D. G., Nelissen, R. L. H., and de Jong, L. (1994a). In vitro splicing of pre-mRNA containing bromouridine. *Mol. Biol. Rep.* **19,** 109–113.

Wansink, D. G., Manders, E. E., van der Kraan, I., Aten, J. A., van Driel, R., and de Jong, L. (1994b). RNA polymerase II transcription is concentrated outside replication domains throughout S-phase. *J. Cell Sci.* **107,** 1449–1456.

CHAPTER 37

Flow Cytometry of Erythropoiesis in Culture: Bivariate Profiles of Fetal and Adult Hemoglobin

Ralph M. Böhmer

Department of Pediatrics
Division of Genetics
New England Medical Center
Boston, Massachusetts 02111, USA

I. Introduction
II. Details of the Method
    A. Erythroid Cultures
    B. Cell Preparation and Labeling
    C. Flow Cytometer Setup
    D. Gating and DNA Histograms
    E. Hemoglobin Profiles
    F. Absolute Cell Counts
    G. Data Analysis
III. Sample Experiments
    A. Basic Flow Data
    B. Isolation of Fetal Nucleated Red Cells from Maternal Blood Cultures
    C. Effect of TGF-β on Proliferation and Hemoglobin Profiles
    D. Study of Sickle-Cell Erythropoiesis
IV. Problems and Limitations
    References

I. Introduction

The development of erythroid cells *in vivo* and *in vitro* comprises proliferation and maturation, from early committed progenitors to enucleated erythrocytes. An important marker of erythroid maturation is the expression and accumulation of hemoglobin in nucleated red cells. The pattern of hemoglobin accumulation is different in fetal and adult cells: fetal erythroid cells accumulate mostly the fetal form of hemoglobin (HbF, containing γ chains) (Stamatoyannopoulos *et al.*, 1979; Böhmer *et al.*, 1998). At about the time of birth, in parallel to and probably caused by the migration of hemopoiesis to the bone marrow, erythroid progenitors switch to β-chain expression to make adult hemoglobin (HbA, containing β chains). Thus, in cultures from adult stem cells, most nucleated red cells accumulate only HbA. However, some accumulate a combination of HbF and HbA (Papayannopoulou *et al.*, 1978, 1982; Migliaccio *et al.*, 1990; Böhmer *et al.*, 1998). Individual colonies, developing in semisolid media, usually contain both HbF$^+$ and HbF$^-$ cells (Migliaccio *et al.*, 1990; Constantoulakis *et al.*, 1990; Böhmer *et al.*, 1999), which means that both types are progeny from the same clonogenic cell. The proportions of adult F$^+$ cells developing in primary cultures from peripheral blood are not preprogrammed *in vivo* but depend on culture conditions. Therefore, a mechanism must exist whereby cells during the early stages of development in culture execute an option whether or not to express and accumulate HbF.

The possibility of manipulating adult erythroid stem and progenitor cell development toward fetal erythropoiesis is of great clinical interest because β-chain hemoglobin disorders such as sickle-cell anemia and the β-thalassemias are ameliorated by increased HbF production. Several agents are known to modify HbF *in vitro* and *in vivo*, acting via different mechanisms that are not yet fully understood (reviewed in Jane and Cunningham, 1998; Olivieri, 1996). A better understanding of these mechanisms may lead to better drugs and more effective treatments. Apart from the clinical applications, the hemoglobin switch is a good model system for basic studies on the regulation of gene expression.

Current methods to study hemoglobin synthesis in erythroid cultures include the measurement of relative hemoglobin contents [e.g., the ratio HbF/(HbF+HbA)], based on electrophoresis or high pressure liquid chromatography (HPLC), and the rates of γ and β-chain expression, based on mRNA quantitation or radioactive protein labeling (Stamatoyannopoulos and Nienhuis, 1985). These assays require substantial numbers of cells and are commonly done on bulk cell preparations from whole cultures or at least whole colonies. Individual HbF and HbA-containing cells can also be distinguished and enumerated by antibody labeling. Fluorescence microscopy (Horiuchi *et al.*, 1995) and single-parameter flow cytometry (Zheng *et al.*, 1995; Epstein *et al.*, 1996; Navenot *et al.*, 1998; Davis *et al.*, 1998; Campbell *et al.*, 1999) have been used to monitor fetal cells in maternal blood and F$^+$ cells in sickle-cell anemia patients. A problem with all those techniques is that they do not provide quantitative information about the correlation of different hemoglobin types and quantities in individual cells, and they do not allow the enumeration of cells with particular combinations of hemoglobin content. Such data are needed to help distinguish between various potential mechanisms by which erythropoiesis and hemoglobin accumulation is modified by drugs and other culture variables. For example, a modified proportion of HbF, measured after a week or more in culture, could be due to various mechanisms alone or in combination: a reversal of the hemoglobin switch (reactivation of γ-chain expression) at early or late stages of development, a selective change in the proliferative rates of F^+ or F^- cells, selective acceleration or delay of terminal differentiation (enucleation), or a selective deletion of early progenitors programmed to accumulate a particular combination of hemoglobin types or quantities.

To facilitate the study of erythropoiesis, we have used flow cytometry to measure two-parameter hemoglobin profiles (HbF versus HbA or the sickle-cell mutant HbS) of developing nucleated red cells, together with correlated DNA histograms and absolute cell counting by reference beads. Apart from studies on erythropoiesis in culture, the method appears also suitable for monitoring the erythropoiesis of patients with hemoglobinopathies in response to treatment. Although all this is standard multiparameter flow cytometry, it appears to have been overlooked in the past as a powerful tool to study erythropoiesis. In this chapter, I introduce the technique and show examples from several experimental projects that show the usefulness of the technique. Problems and limitations of the method are also discussed.

II. Details of the Method

A. Erythroid Cultures

In the examples given here, adult or fetal blood mononuclear cells (unenriched) were seeded in methylcellulose culture with cytokines and serum supplement that support erythropoiesis, as required for the particular experiment. For flow cytometry, whole cultures containing a large number of colonies (50–100) were harvested and mixed into a single-cell suspension. If the removal of apoptotic cells is desired, this cell suspension can be subjected to a density gradient (1.077) or enzymatic digestion with trypsin and DNase before the fixation step.

B. Cell Preparation and Labeling

Harvested cell samples were suspended in phosphate-buffered saline with bovine serum albumin (PBS/BSA), mixed 1:1 with 10% formaldehyde (FA) (methanol-free, in PBS), and incubated at 37°C for 1 hr, then stored in 5% FA solution at 4°C. Fixation with much lower FA concentrations (e.g., 1%), or without incubation at 37°C, resulted in lower signal levels, probably because hemoglobin was insufficiently cross-linked and leaked out of the cells. Stored in 5% FA, samples were stable for at least 3 months. For further processing, cells were sedimented in their Eppendorf tubes (4000 rpm, 1–2 min), resuspended in 100% methanol for 5 min at room temperature, washed twice in PBS/BSA, and then labeled with 1 μg of antibody per sample in 100 μl of the permeabilizing Solution B of the Caltag Fix & Perm kit (Caltag, Burlingame, CA). Phycoerythrin

(PE)-conjugated antibodies to the γ chain of hemoglobin (HbF) were from Cortex (San Leandro, CA). Fluorescein isothiocyanate (FITC)-conjugated antibodies specific for the normal β chain of hemoglobin (HbA) or the sickle-cell mutant of the β chain (HbS) were from Wallac (Acron, OH). After labeling at room temperature for 30 min, cells were washed twice in PBS/BSA and suspended in 1 ml of PBS with 1% formaldehyde and 0.2 μg/ml Hoechst 33342. The stained samples could be stored at 4°C for up to 2 weeks, but much longer storage resulted in some deterioration of the antibody signals. The coefficients of variation (CV) of DNA histograms tended to improve with time of storage.

C. Flow Cytometer Setup

Cells were processed in a Becton Dickinson Vantage flow cytometer/cell sorter with dual, displaced-beam laser excitation. Hoechst 33342 fluorescence was excited by ultraviolet light (UV) and measured at 430 nm. FITC and PE were excited with 488 nm and measured at 530 nm and 575 nm, respectively. List mode files of the three variables were recorded.

D. Gating and DNA Histograms

Hoechst fluorescence was recorded on a log scale when it was needed only to gate out debris and non-nucleated erythrocytes. When DNA histograms were desired for cell cycle analysis, the samples were rerecorded with Hoechst fluorescence on a linear scale. Cells with the full amount of DNA were selected for all display and numerical analysis. Most nonproliferative cells that had died (by apoptosis or necrosis) during the culture phase were not included in the DNA-gated profiles, because they tended to disintegrate in culture and/or during the multiple preparation steps from harvest to analysis. Therefore, apoptotic cell counts were considered of little use. However, due to the fixation with cross-linking formaldehyde that keeps DNA fragments from leaking out, cells within a certain stage of apoptotic death and decay are included in the analysis. By comparing cell counts in cell preparations with and without removing apoptotic cells by density gradient, this proportion of apoptotic cells was found to be small in most circumstances. Thus, profiles after gating for DNA contained mostly nucleated cells that were alive at the time of harvest.

E. Hemoglobin Profiles

FITC and PE fluorescence values were recorded on a four-cycle log scale, with color compensation set as appropriate. The accuracy of compensation over 4 logs of fluorescence values was limited.

F. Absolute Cell Counts

To determine absolute cell numbers in any area of the hemoglobin profiles, samples were transferred to stock "counting tubes" prepared with known amounts (usually 50,000 per tube) of fluorescent plastic beads (Böhmer, 1984). Coulter Immunobrite Level IV beads were used for this application, appearing in suitable positions relative to the cell clusters, and making very sharp peaks with both the UV and 488 excitation.

G. Data Analysis

Analysis regions were defined within the bivariate hemoglobin profiles, to provide the total number of events as well as the corresponding DNA histograms, if desired. The absolute cell numbers per sample in any region of interest could then be calculated from the ratio of cells to beads. Beads were quantitated from any dot plot where they did not overlap with cells, usually in a bivariate profile of DNA versus HbA or HbF. The shift of hemoglobin-specific fluorescence levels with culture conditions was assessed visually but not quantitated numerically, for reasons given in Section IV.

III. Sample Experiments

A. Basic Flow Data

Figure 1 shows a sample of the displays generated from a list mode file of PE, FITC, and Hoechst fluorescence. Mononuclear cells from adult blood were

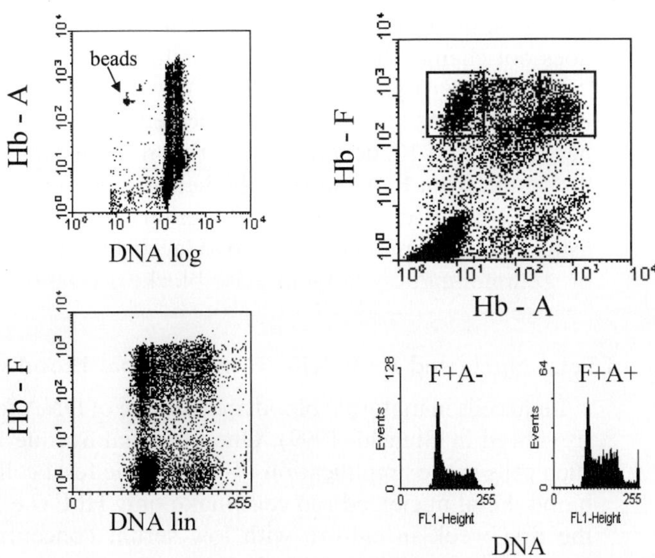

Fig. 1 Data displays from a list mode file of three parameters (HbA, HbF, and DNA). Data are from an 8-day culture grown under conditions that strongly stimulate HbF. See text for details.

cultured in methylcellulose medium with the cytokines erythropoietin (EPO), stem cell factor (SCF), interleukin 3 (IL-3), and 30% fetal calf serum (FCS). Analysis was after 8 days of culture.

In the left upper quarter (Fig. 1) is the bivariate profile of HbA versus DNA, with DNA on a logarithmic scale. The cluster of beads (indicated) is at a much lower 430 nm fluorescence than the cells. We use a very low concentration of Hoechst dye (0.2 μg/ml) to minimize the amount of background Hoechst-based fluorescence that tends to contribute to the "autofluorescence" at 530 and 575 nm in spite of the displaced-beam setup. At these low Hoechst concentrations, DNA-specific fluorescence at 430 nm decreases with increasing cell density in the sample. Overlap of the cell and bead clusters is avoided by choosing a sufficiently low cell density, either by taking only a fraction of the fixed samples for labeling or by diluting with extra dye solution after finding an overlap.

In the lower left quarter (Fig. 1) is the bivariate profile of HbF versus DNA on a linear scale, showing S-phase cells at all levels of HbF.

In the upper right quarter (Fig. 1) is the bivariate profile of HbF versus HbA. Cellular fluorescence levels span over about 3 logs in both directions, and are probably proportional to hemoglobin content at least over the first 2 logs. A calibration of fluorescence levels in terms of absolute hemoglobin content is not yet available and may be difficult to achieve (see Section IV). While there are cells at any combination of HbF and HbA, some clustering can be seen at F^-A^-, F^+A^-, F^-A^+, and F^+A^+. These clusters shift with culture time and conditions (see examples below). Owing to the diffuse cluster borders, region threshold settings for F^+ and A^+ are somewhat arbitrary. An effort was made to set the color compensation accurately, but it can be set according to taste, since this does not change the information contained in the profiles.

In the lower right quarter (Fig. 1), two DNA histograms are shown, from two regions at F^+A^- and F^+A^+, as indicated. Coefficients of variation (CV) ranged between 7 and 10, depending mostly on the machine alignment of the day. The histogram from F^+A^+ shows a higher proportion of S-phase cells than that from F^+A^-. Obviously, these DNA histograms are also suitable for cell kinetic studies using the single-parameter BUdR/Hoechst quenching technique (Böhmer, 1979) or stathmokinetics via metaphase blockers (Darzynkiewicz *et al.*, 1981).

B. Isolation of Fetal Nucleated Red Cells from Maternal Blood Cultures

Fetal cells in maternal blood are a source of DNA for prenatal genetic diagnosis (reviewed in Bianchi, 1999). One potential avenue to overcome the scarcity of such cells is the amplification of clonogenic fetal cells in cultures from maternal blood. Fetal nucleated red cells make only HbF (i.e., they are F^+A^-) for at least the first week in culture with low serum concentrations, while only a small proportion (typically <1%) of adult nucleated red cells become F^+A^-. Therefore, fetal cells can be enriched at least 100-fold by flow-sorting F^+A^- cells (Böhmer *et al.*, 1998, 1999). However, with only very few (if any) fetal clonogenic cells among typically up to 10,000 maternal clonogenic cells from 20 ml of blood, this enrichment is not satisfactory, and the selective suppression of maternal F⁺A⁻ cells becomes an essential pursuit.

Fetal calf serum and some cytokines are known to increase adult F⁺ cells in culture, but it is not known whether this stimulation leads to increased F⁺A⁻ cells in early cultures, nor whether it applies equally to fetal F⁺A⁻ cells, so that the proportions of fetal cells in cocultures may not be altered by tuning these variables. Therefore, we investigated whether there is a selective action of serum and cytokines in cocultures of fetal and adult cells. Figure 2 shows the effects of FCS and IL-3 on the proportions of fetal cells in the F⁺A⁻ window of a 7-day culture grown after spiking a very small amount of male fetal blood into female adult blood. The hemoglobin profile on the left (Fig. 2) demonstrates a cluster of F⁺A⁻ cells in a spiked culture (0.1% fetal blood, gestational age 21 weeks) with 1% charcoal-treated human cord serum (C-CHS 1) and supplemented with EPO, SCF, and IL-3. Cells from the indicated F⁺A⁻ window were sorted and the proportion of fetal cells determined by fluorescence *in situ* hybridization (FISH). The graph on the upper right (Fig. 2A) shows the effect of FCS in this system: FCS dramatically reduces the proportion of fetal F⁺A⁻ cells. This was found to be due to both a boost of adult F⁺A⁻ cells and a suppression of fetal F⁺A⁻ cell growth by serum (Böhmer *et al.*, 1999). The graph on the lower right (Fig. 2B) shows the effect of IL-3 in a spiked culture (0.01% fetal blood, gestational age 12 weeks) grown with C-CHS1, EPO, and SCF: the proportions of fetal F⁺A⁻ cells are dramatically reduced by the presence of IL-3, which was found to be due to a selective stimulation of adult F⁺A⁻ cells (Böhmer *et al.*, 2000b). Thus, to maximize the purity of flow-sorted fetal nucleated red cells from maternal blood cultures, the cultures should be grown in low concentrations of charcoal-treated serum and without IL-3.

C. Effect of TGF-β on Proliferation and Hemoglobin Profiles

Transforming growth factor β (TGF-β) has been shown to inhibit erythroid proliferation and accelerate hemoglobin accumulation (Krystal *et al.*, 1994). We

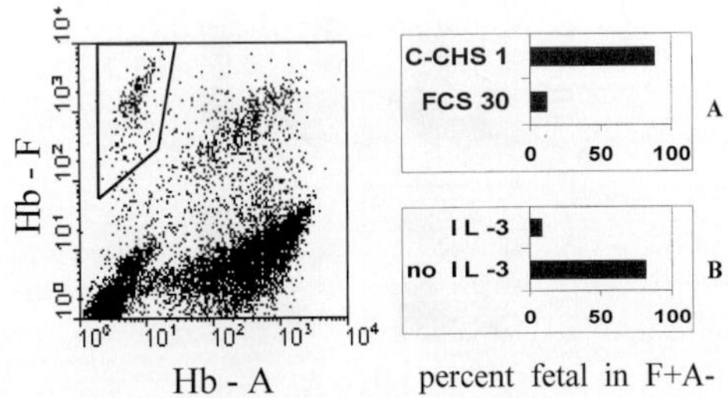

Fig. 2 Isolation of fetal nucleated red cells from cocultures with adult cells. The bivariate hemoglobin profile on the left shows a 7-day coculture from female adult blood spiked with a small amount (~0.1%) of 21-week male fetal blood. Cells in the indicated F⁺A⁻ window were sorted and the proportions of fetal cells determined by FISH, using a Y-chromosome-specific marker. (A) Effect of fetal calf serum (FCS). (B) Effect of IL-3. See text for further details.

tested if TGF-β would affect the proportions of F$^+$ cells, either suppress them and serve in maternal cell cultures, as described earlier, or boost them and potentially be of therapeutic use for the treatment of hemoglobinopathies (Böhmer *et al.*, 2000a).

The left part of Fig. 3 shows how TGF-β affects the development of hemoglobin profiles when added on day 5 of culture, at a time when the first hemoglobinized cells appear but have not yet developed into distinct clusters. TGF-β caused a dramatic (over fivefold) shift to higher HbA levels, equally so in the clusters of F$^+$A$^+$ and F$^-$A$^+$ cells (see Section IV for some caveats). The proportions of F$^+$ and F$^-$ cells appeared little affected. The right part of Fig. 3 shows absolute numbers of F$^+$ (open symbols) and F$^-$A$^+$ (closed symbols) cells as a function of time. One can see that the proliferation is dramatically reduced by TGF-β. The curves for F$^+$ and F$^-$A$^+$ run parallel and at the same distance in controls (squares) and TGF-β cultures (triangles), indicating that neither culture time nor TGF-β affected the ratio of F$^+$/F$^-$ cells. In some experiments (not shown), TGF-β induced a complete growth arrest, with secondary colonies (cultures were reseeded as single-cell suspension at the beginning of TGF-β treatment) at the 8–16 cell stage.

The cell cycle effect of TGF-β was further explored by DNA histograms and BUdR-induced fluorescence quenching. Figure 4 shows a set of histograms after 4 days of TGF-β treatment, with the normal DNA profiles (upper row) and the quenched profiles, with BUdR incorporation beginning 20 hr before the day 4 harvest (lower row). The TGF-β-treated cultures have a much lower S-phase fraction than controls, and only a small part of the population divided in the 20 hr before harvest.

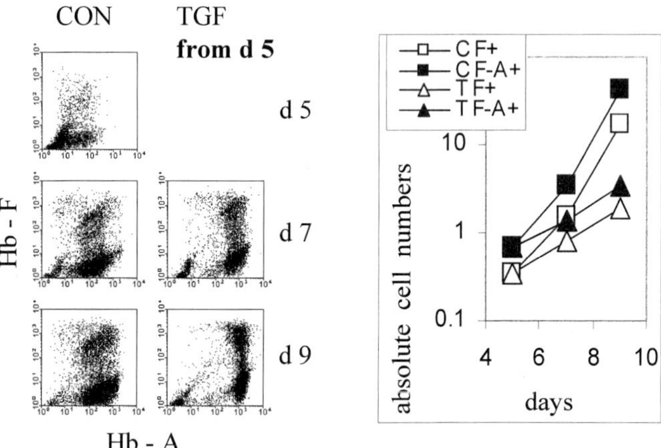

Fig. 3 Effect of TGF-β1 on erythropoiesis. TGF-β was added to cultures on day 5. The further development of hemoglobin profiles (days 7 and 9) is shown on the left, and the absolute numbers of F$^+$ cells (open symbols) and F$^-$A$^+$ cells (solid symbols) as a function of time is shown on the right. C = control (squares), T = TGF (triangles).

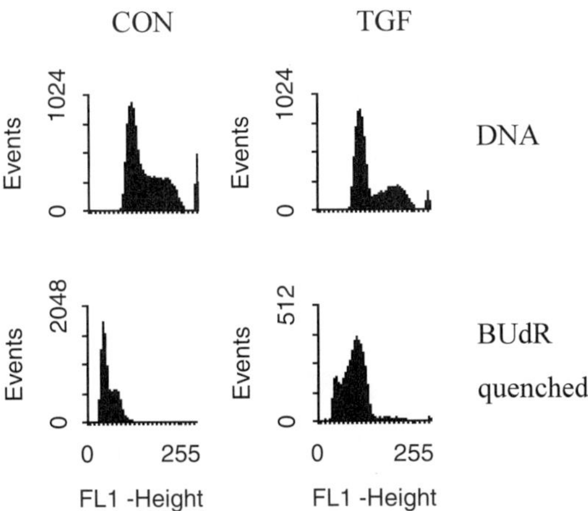

Fig. 4   DNA histograms (upper row) and BUdR-quenched Hoechst fluorescence histograms (lower row) on day 4 after the beginning of TGF-β treatment. BUdR was added 20 hr before harvest.

A very different effect was found when cultures were incubated with TGF-β during the first 4 days of culture, followed by reseeding without TGF-β and further subcultivation steps into fresh medium every 3 days (Fig. 5). The left part of Fig. 5 shows the development of hemoglobin profiles (days 7, 10, and 16) after a 4-day treatment. The proportions of F^+ cells are dramatically increased over controls, and up to day 10 a large proportion of F^+ cells are F^+A^- like fetal cells. The right part of Fig. 5 shows the production of F^+ and F^-A^+ cells over time (absolute numbers per culture, multiplied with the dilution factors from each subcultivation step), using the same symbols as in Fig. 3. Between days 7 and 10 of culture, F^+ cell numbers in controls and TGF-β cultures were the same, while the numbers of F^-A^+ cells were dramatically reduced in the TGF-β-treated cultures. This explains the increased proportions of F^+ cells (from about 25 to 75%) as a selective suppression of F^-A^+ progenitors during the phase of TGF-β treatment, rather than a switch in the programming of hemoglobin expression. Between days 7 and 10, the remaining cells (both F^+ and F^-A^+) proliferated equally in controls and TGF-β cultures. After day 10, the controls approached the end of their division potential, with unchanging proportions of F^+ cells (parallel curves). In contrast, the TGF-β-treated cultures kept proliferating (for more than 20 days in some cases), with some variability between blood donors. The proportions of F^+ cells remained high. Thus, by the end of the division potential of a culture, TGF-β treated cultures had produced 10–100 times as many F^+A^+ cells than controls. It will be interesting to see if this *in vitro* effect of TGF-β can somehow be translated into a therapeutic *in vivo* increase in the absolute and relative numbers of F^+ erythrocytes.

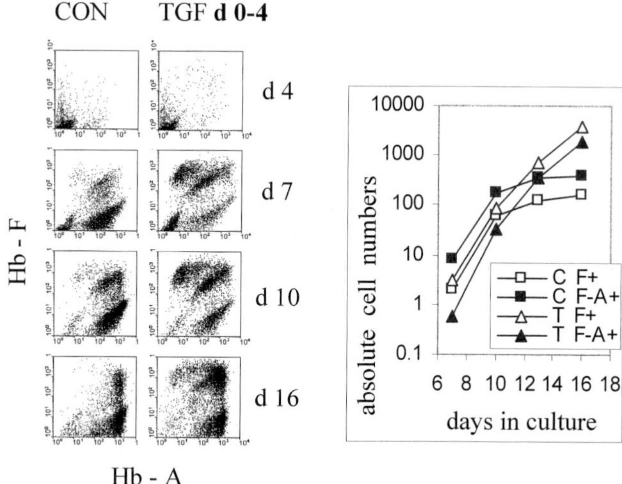

Fig. 5 Effect of TGF-β1 treatment during the early culture phase. Cultures were grown in TGF-β for the first 4 days, then washed and reseeded in control medium. Left: hemoglobin profiles on days 4, 7, 10 and 16. Right: absolute numbers of F$^+$ cells (open symbols) and F$^-$A$^+$ cells (full symbols) as a function of time. C = control (squares), T = TGF (triangles).

D. Study of Sickle-Cell Erythropoiesis

The β-chain defect that leads to the sickling of erythrocytes unfortunately does not affect the switch from γ-chain to β-chain expression around the time of birth, nor does it appear to affect erythropoiesis *in vitro*. Therefore, studies of drug effects on hemoglobin development in normal erythropoiesis are generally applicable to sickle-cell erythropoiesis. However, the study of sickle-cell erythropoiesis is necessary to monitor the condition and responses of sickle-cell patients to treatment. For this purpose, we used an antibody specific for sickle-cell hemoglobin (HbS), since the available HbA antibody binds only weakly to HbS. Figure 6 shows hemoglobin profiles from the blood of a 9-year-old patient with sickle disease. The upper left profile is from nucleated cells before culture. Because of the small proportion of nucleated red cells in peripheral blood, this dot plot was recorded with 100,000 cells instead of the usual 10,000–20,000. Nucleated red cells with all combinations of hemoglobin can be seen, with 20–30% F$^+$ cells (depending on region setting). An increase in F$^+$ erythrocytes due to patient treatment may be preceded by an increase in circulating F$^+$ nucleated red cells, so that the latter may represent an earlier marker for the effect of treatment. We propose to study whether the profile characteristics of circulating nucleated red cells can provide useful information about the conditions of sickle-cell patients. The other three profiles in Fig. 5 are from day 7 cultures and show the same effect of FCS and TGF-β that was found in normal erythropoiesis. The proportions of F$^+$ cells are indicated in each profile (as percentage of all

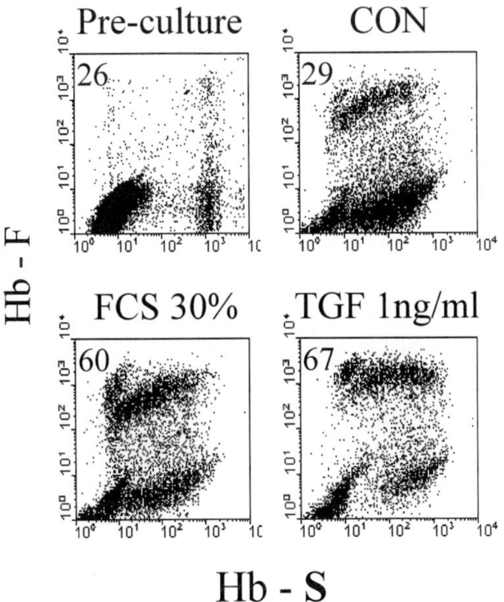

Fig. 6 Hemoglobin profiles from the blood of a 9-year-old sickle-cell patient. Upper left: nucleated cells before culture, with 100,000 cells displayed. Other: profiles from a 7-day culture with FCS, TGF-β, or control medium, as indicated. The proportions of F^+ cells (as percentage of all hemoglobin-containing cells) are indicated in the upper left corner of each profile.

hemoglobin-containing cells). Although this result was expected, it is not trivial because the stressed hemopoiesis of sickle-cell patients may result in circulating stem cells with altered maturity and different responses. Compared to the profile shapes seen in normal cultures, the two sickle-cell cultures that we had the opportunity to study showed subtle differences that cannot be expressed numerically by F^+ cell proportions. This needs further study.

IV. Problems and Limitations

We restricted our numerical evaluation of hemoglobin profiles to an enumeration of F^+ and F^-A^+ cells during a phase of culture development where F^+ and F^- cells were distinguishable and the border between them could be defined quite well due to some clustering. However, the bivariate hemoglobin profiles obviously contain a wealth of information beyond those numbers, and presently this is only evaluated at a qualitative level. Further numerical quantitation of hemoglobin levels and ratios would require a calibration of the fluorescence axes, which is problematic for several reasons:

1. At more advanced stages of erythroid maturation, the large amounts of hemoglobin per cell lead to saturation effects that prevent the resolution of quantitative differences in hemoglobin content by fluorescent antibody label. In other words, cells with any hemoglobin content above a certain level all appear at the same maximum fluorescence level, and further, potentially large differences in hemoglobin content are no longer resolved. We have not yet established the basis of this observed saturation. It could be caused by absorption of excitation or fluorescence light, or by limited antibody access to potential binding sites. A remedy is not yet known. On the other side, the flow cytometric detection of low hemoglobin levels was found to be much more sensitive than the detection by fluorescence microscopy.

2. A different type of fluorescence saturation occurred due to exhaustion of available antibody. We found that the large amounts of hemoglobin in mature red cells can easily cause antibody exhaustion at cell and antibody quantities that are commonly used for flow cytometry of other cell components such as surface receptors. Although subsaturation labeling might still reflect differences in hemoglobin contents of cells within an individual profile, the profiles from different samples can no longer be compared reliably. The data of Fig. 3 are a good example of the potential for artifacts: After 4 days of TGF-β treatment, cell numbers were approximately 10-fold less than in controls, while mean FITC fluorescence, supposed to reflect HbA content, was ca. fivefold higher and at the upper limit defined by the saturation effect described in the previous paragraph. This could have been due in part, or even completely, to a larger amount of antibody available per cell. We had to verify that the lower FITC fluorescence levels in the controls reflected lower HbA levels by repeating the labeling with an order of magnitude 10 times lower cell numbers and finding the same fluorescence levels. Thus, this antibody depletion effect can obviously be avoided by choosing a sufficiently small part of the sample for labeling and analysis. Increasing the antibody concentration is a very expensive option: considering the fluorescence scale over 3 logs, doubling the amount of antibody has more effect on the budget than on the profiles. As the average hemoglobin content per cell increases due to maturation, the cell numbers used for labeling have to be decreased accordingly.

3. The absolute fluorescence levels achieved by the antibody labeling was sensitively dependent on many variables of the procedure, such as the concentration, time, and temperature of formaldehyde in the fixation step, and the further incubation with methanol. Before any calibration between fluorescence and absolute Hb levels is meaningful, plateau values would have to be established for all variables of the procedure so that small sample-to-sample variations will not throw off the calibration.

4. In the lower ranges of cellular hemoglobin contents, which prevail up to day 8 of culture under conditions that maximally promote erythropoiesis, there is no saturation effect, so that the fluorescent antibody signal is likely to be approximately proportional to hemoglobin content. However, it must be noted

that the antibodies bind and measure β and γ chains, whether free or assembled in hemoglobin molecules. It appears worth investigating whether the relative quantities of γ and β chains reflect the relative quantities of assembled molecules throughout erythropoietic development.

In summary, the flow cytometric measurement of the correlated contents of different hemoglobin types is a powerful new technique for the study of early erythropoiesis. While the flow cytometric distinction between F^+ and F^- cells remains reliable throughout the erythroid development, the flow cytometric data do not inform us about the absolute or relative amounts of HbF and HbA per cell at the late stages of maturation toward enucleated erythrocytes. Of special interest in the context of sickle-cell disease is the relative amount of HbF in those cells that are F^+A^+. Unfortunately, flow-sorted F^+A^+ cells cannot easily be subjected to further protein analysis because of the strong cross-linking fixation that is required. Unless the problem of label saturation can be solved, flow cytometric data on hemoglobin contents in early erythropoiesis will have to be complemented by other assays measuring hemoglobin at the terminal stage of maturation.

Acknowledgments

Special thanks to Mr. Vincent Falco for skillfully maintaining our flow cytometry facility, to Dr. Diana Bianchi for critically reading the manuscript, and to Dr. Tomas Campbell for donations of special antibody samples and patient exchanges about labeling procedures and antibody specificities.

References

Bianchi, D. W. (1999). Fetal cells in the maternal circulation: Feasibility for prenatal diagnosis. *Br. J. Haematol.* **105,** 574–583.

Böhmer, R. M. (1979). Flow cytometric cell cycle analysis using the quenching of 33258 Hoechst fluorescence by bromodeoxyuridine incorporation. *Cell Tissue Kinet.* **12,** 101–110.

Böhmer, R. M. (1984). Two-step cell-death kinetics in vitro during *cis*-platinum, hydroxyurea and mitomycin incubation. *Cell Tissue Kinet.* **17,** 593–600.

Böhmer, R. M., Zhen, D., and Bianchi, D. W. (1998). Differential development of fetal and adult haemoglobin profiles in colony culture: Isolation of fetal nucleated red cells by two-colour fluorescence labeling. *Br. J. Haematol.* **103,** 351–360.

Böhmer, R. M., Zhen, D., and Bianchi, D. W. (1999). Identification of fetal nucleated red cells in co-cultures from fetal and adult peripheral blood: Differential effects of serum on fetal and adult erythrocytes. *Prenat. Diag.* **19,** 628–636.

Böhmer, R. M., Campbell T. A., and Bianchi, D. W. (2000a). Selectively increased growth of fetal hemoglobin-expressing adult erythroid progenitors after brief treatment of early progenitors with transforming growth factor β. *Blood* **95,** 2967–2974.

Böhmer, R. M., Johnson, K. L., and Bianchi, D. W. (2000b). Differential effects of interleukin-3 on fetal and adult erythroid cells in culture: Implications for the isolation of fetal cells from maternal blood. *Prenat. Diag.* in press.

Campbell, T. A., Ware, R. E., and Mason, M. (1999). Detection of hemoglobin variants in erythrocytes by flow cytometry. *Cytometry* **35,** 242–248.

Constantoulakis, P., Makamoto, B., Papayannopoulou, T., and Stamatoyannopoulos, G. (1990). Fetal calf serum contains activities that induce fetal hemoglobin in adult erythroid cell cultures. *Blood* **75,** 1862.

Darzynkiewicz, Z., Traganos, F., Xue, S. B., Staiano-Coico, L., and Melamed, M. R. (1981). Rapid analysis of drug effects on the cell cycle. *Cytometry* **1,** 279–286.

Davis, B. H., Olson, S., Bigelow, N. C., and Chen, J. C. (1998). Detection of fetal red cells in feto-maternal hemorrhage using a fetal hemoglobin monoclonal antibody by flow cytometry. *Transfusion* **38,** 749–756.

Epstein, N., Epstein, M., Boulet, A., Fibach, E., and Rodgers, G. P. (1996). Monoclonal antibody-based methods for quantitation of hemoglobins: Application to evaluating patients with sickle cell anemia treated with hydroxyurea. *Eur. J. Haematol.* **57,** 17–24.

Horiuchi, K., Osterhout, M. L., Kamma, H., Bekoe, N. A., and Hirokawa, K. J. (1995). Estimation of fetal hemoglobin levels in individual red cells via fluorescence image cytometry. *Cytometry* **20,** 261–267.

Jane, S. M., and Cunningham, J. M. (1998). Understanding fetal globin expression: A step towards effective HbF reactivation in haemoglobinopathies. *Br. J. Haematol.* **102,** 415–422.

Krystal, G., Lam, V., Dragowska, W., Takahashi, C., Appel, J., Gontier, A., Jenkins, A., Lam, H., Quon, L., and Lansdorp, P. (1994). Transforming growth factor β 1 is an inducer of erythroid differentiation. *J. Exp. Med.* **180,** 851–860.

Migliaccio, A. R., Migliaggio, G., Brice, M., Constantoulakis, P., Stamatoyannopoulos, G., and Papayannopulou, T. (1990). Influence of recombinant hemopoietin and of fetal bovine serum on the globin synthetic pattern of human BFUe. *Blood* **76,** 1150.

Navenot, J. M., Merghoub, R., Ducroq, R., Muller, J. Y., Krishnamoorthy, R., and Blanchard, D. (1998). New method for quantitative determination of fetal hemoglobin-containing red blood cells by flow cytometry: Application to sickle-cell disease. *Cytometry* **32,** 186–190.

Olivieri, N. F. (1996). Reactivation of fetal hemoglobin in patients with β-thalassemia. *Semin. Hematol.* **33,** 24–42.

Papayannopoulou, T., Nakamoto, B., Buckley, J., Kurachi, S., Nute, P. E., and Stamatoyannopoulos, G. (1978). Erythroid progenitors circulating in the blood of adult individuals produce fetal hemoglobin in culture. *Science* **199,** 1349.

Papayannopoulou, T., Kurachi, S., Nakamoto, B., Zanjani, E. D., Stamatoyannopoulos, G. (1982). Hemoglobin switching in culture: Evidence for a humoral factor that induces switching in adult and neonatal but not fetal erythroid cells. *Proc. Natl. Acad. Sci. U.S.A.* **79,** 6579.

Stamatoyannopoulos, B., and Rosenblum, B. B., Papayannopoulou, T., Brice, M., Nakamoto, B., and Shepard, T. H. (1979). HbF and HbA production in erythroid cultures from human fetuses and neonates. *Blood* **54,** 440.

Stamatoyannopoulos, G., Nienhuis, A. W. (eds.) (1985). "Experimental Approaches for the Study of Hemoglobin Switching." Liss, New York.

Zheng, Y. L., DeMaria, M., Zhen, D., Vadnais, T. J., and Bianchi, D. W. (1995). Flow sorting of fetal erythroblasts using intracytoplasmic anti-fetal haemoglobin: Preliminary observations on maternal samples. *Prenat. Diagn.* **15,** 897–905.

CHAPTER 38

Flow Cytometric Analysis of Human Hemopoietic Progenitor Differentiation by Assessing Cell Division Rate and Phenotypic Profile

Luca Pierelli, * **Giovanni Scambia,** [†] **and Andrea Fattorossi** [†]

*Institute of Hematology and
[†]Institute of Obstetrics and Gynecology
Universitá Cattolica del Sacro Cuore
00136 Rome, Italy

I. Introduction
II. Background
III. Critical Aspects of Methodology
IV. Functionally Distinct Circulating Hemopoietic Progenitor Subsets Can Be Assessed during Cytokine-Driven Differentiation
V. Concluding Remarks
References

I. Introduction

Hemopoiesis is a finely regulated and still not completely understood process in which multipotent undifferentiated hemopoietic progenitors (HP) possess both self-renewal ability and the capacity of differentiating into functionally mature cells following distinct maturative pathways (Metcalf, 1991). The fate of a single HP appears to be determined by both an intrinsic hemopoietic potential typical of each HP and the activity of various cytokines that regulate HP survival, proliferation, and differentiation (Morrison *et al.,* 1997). Whether these mechanisms are mutually exclusive or cooperative is still to be ascertained.

METHODS IN CELL BIOLOGY, VOL. 64

A large number of studies has supported either stochastic or deterministic models of hemopoiesis. According to the former model, HP would evolve into different lineages randomly, whereas according to the latter mode, HP would develop along a predictable sequence or hierarchy. Different roles have been proposed for the hemopoietic cytokines that would exhibit instructive or permissive activities (Morrison *et al.*, 1997). According to the former hypothesis, HP would be induced to choose one lineage at the expense of the others, whereas according to the latter hypothesis, HP would commit themselves to a given lineage independently of growth factors: these would act to promote the survival and/or proliferation of already committed HP.

In humans, most studies on hemopoiesis have been carried out using HP identified or purified by their expression of CD34, a 115-kDa glycoprotein containing nine potential sites of N-glycosylation and several sites of *O*-glycosylation in the extracellular domain. The relevance of this marker is enormous. CD34 is currently considered to be expressed by primitive HP (Civin *et al.*, 1989), albeit not exclusively (Watt *et al.*, 1987), and biological and clinical evidence indicates that human stem/progenitor cells are entirely retained in the CD34$^+$ fraction. Most important from a cytometric point of view, the degree of CD34 expression correlates directly with the hemopoietic potential (Krause *et al.*, 1996). Although its complete role in hemopoiesis is not fully understood, a putative function of the CD34 molecule might be in regulating the adhesion processes of HP to vascular endothelium and bone marrow stromal compartments through the interaction with L-selectin (Krause *et al.*, 1996). Intriguingly, the gene coding for CD34 protein is located on chromosome 1, where several genes coding for adhesion molecules are also present. The high expression of CD34 on more immature HP is very consistent with its downmodulation, eventually leading to complete disappearance, during HP amplification and committment along the various differentiating lineages. It should be noted, however, that although CD34 certainly represents a major indicator of the events taking place during the hemopoietic process and most likely plays a pivotal role in modulating the process, the exact mechanisms underlying the decision to self-renewal or differentiate on the part of HP are still completely unknown (Fackler *et al.*, 1995; Nakamura *et al.*, 1993). However, some experimental data suggest that alternative ways to hemopoiesis might also exist, since knock-out mice for CD34 have only minor hemopoietic defects (Krause *et al.*, 1996).

The plasma membrane of HP carries a series of molecules, usually referred to as secondary markers, capable of identifying distinct functional subsets. For example, the absence of CD33 identifies a more undifferentiated subset with an increased hemopoietic and multilineage potential, whereas its presence defines precursors committed to the myeloid lineage (Andrews *et al.*, 1989). The absence of class II major histocompatibility complex (MHC) molecules defines a more primitive subset (Rusten *et al.*, 1994), whereas an increased expression of Thy-1 is associated with a greater stem cell activity (Craig *et al.*, 1993). A particular relevance must be attributed to CD38. This marker reliably identifies committed

HP both in bone marrow and cord blood (Terstappen *et al.*, 1991; Hao *et al.*, 1995). Remarkably, the possibility of assigning selected functions to HP subsets identified on the basis of their phenotypic profile is clearly demonstrated for bone marrow and cord blood, whereas data on circulating HP are relatively scarce (Sakabe *et al.*, 1997). The vast majority of circulating HP are quite homogeneous in phenotype, sharing the same secondary markers, and thus suggesting that they are also homogeneous in function. More recent studies, however, suggest that the most primitive circulating HP are enriched in the $CD34^+/CD105^+$ fraction (Pierelli *et al.*, 1998b).

Further mention of the unresolved problems in hemopoiesis will not be made here, since the focus of this discussion is instrument-related issues. Because all studies described here started from HP identified/purified on the basis of CD34 expression, it will be operationally assumed that all HP are $CD34^+$, although there is evidence for the existence of $CD34^-$ HP (Osawa *et al.*, 1996).

II. Background

In recent years, studies on the biology of human hematopoietic progenitors have been greatly facilitated by the availability of recombinant human cytokines selectively leading these cells toward specific differentiating pathways in serum-free cultures (Lansdorp, 1993; Mayani *et al.*, 1993a,b). Most of these studies have been conducted by establishing single-cell cultures of purified HP which are then exposed to predefined cytokine combinations. By these approaches, important insights on the survival as well as proliferative and differentiative capability of HP have been obtained.

More recently, maturation of HP into functional hemopoietic cells *in vitro* has been monitored with success at the population level by the use of flow cytometry. This approach exploits the availability of monoclonal antibodies (MAb) to specific surface markers defining the maturative stage and the specific lineage a given precursor is committed to, as well as the capability of certain fluorescent probes to track cell division history. In these systems, cells are loaded with a probe that equally redistributes into daughter cells at each round of cell division, generating progenies with progressively halved fluorescence signal. Two probes have been the most widely described in the literature, PKH26 (Lansdorp and Dragowska, 1993; Young *et al.*, 1996; Yui *et al.*, 1998; Verfaillie and Miller, 1995) and carboxyfluorescein diacetate-succinimidyl ester (CFDA-SE) (formerly CFSE) (Lyons and Parish, 1994; Fattorossi *et al.*, 1996; Nordon *et al.*, 1997; Pierelli *et al.*, 1998a,b), which differ in terms of peak emission wavelength, 567 nm the former and 517 nm the latter. The fluorescein-like emission spectrum of CFDA-SE (an additional PKH dye, PKH2, has also been proposed by Ladd *et al.*, 1997, Traycoff *et al.*, 1998, as a fluorescein-like probe, but has received comparatively less attention) allows cells to be costained with MAb conjugated with a highly efficient fluorochrome, such as phycoerythrin (PE). This represents a distinct advantage, since developing HP typically do not express the various markers in an all-or-none fashion, and this produces a fairly large fluorescence distributions, from very dim to quite bright, making the distinction of weakly fluorescent cells from background difficult. Moreover, the spectral properties of CFDA-SE allow the use of far red emitting fluorochromes, such as tandem conjugates, typically PE-Cy5, and peridinium chlorophyll protein (PerCP). However, there is no conceptual difference between the usage of the various cell replication probes, and the choice rests on the personal preferences of the investigator. Examples presented in this chapter refer to data generated by CFDA-SE coupled with immunophenotyping.

The technical details regarding CFDA-SE usage in various models, including cell loading and troubleshooting, are reported in another chapter of this book. Here we will mainly focus on the problems the flow cytometrist faces when attempting to analyze variations in human HP phenotype and correlate them with CFDA-SE halving during *in vitro* cytokine-driven differentiation toward the major hemopoietic lineages, namely, granulocytic/monocytic, erythroid, and megakaryocytic. A close examination of the numerous membrane markers characterizing the development of the various hemopoietic lineages is beyond the scope of present discussion. Some basic reminding follows (useful for an easier understanding of data that will be presented later). The coexpression of CD15 and CD11b permits the monitoring of the maturative status of HP during differentiation toward the granulocytic/monocytic lineage, the immature precursors being negative for both markers, and promyelocytes and myelocytes expressing CD15 but not CD11b, band cells and granulocytes expressing both markers, and pro-monocytes and monocytes/macrophages expressing CD11b but not CD15; the concomitant presence of CD36 and glycophorin-A identifies erythroblasts and early normoblasts, while $CD36^-$/Glycophorin-A^+ cells represent more mature normoblasts and late normoblasts; the expression of CD41a, CD61, CD36, and thrombin receptor can be used for studying megakaryocytic differentiation.

III. Critical Aspects of Methodology

Monitoring HP development *in vitro* by combining CFDA-SE halving and MAb reactivity summarizes most of the difficulties that can be encountered in flow cytometry. This is a reflection of the fact that committment of HP to differentiate is a dynamic process and, consequently, the expression of differentiation markers not only occurs progressively but also with different kinetics depending on the differentiation step of each given subset. An exhaustive review of such difficulties is presented in another chapter of this volume dedicated to the immunophenotyping of activated/proliferating lymphocytes, and here we will only review a few of the major points.

1. There is heterogeneity in the background fluorescence signal owing to the simultaneous existence of various cell subsets as soon as HP have started to

differentiate. This heterogeneity reflects the progressive changes in size, with consequent modification in intrinsic components (autofluorescence), and surface area (number of nonspecific MAb binding sites). Remarkably, the major causes for cellular autofluorescence are pyridine and flavin nucleotides (Aubin, 1979), typically abundant in cells of the phagocytic lineage: HP developing along this pathway may therefore pose peculiar problems.

2. The unimodal and often dim expression of most antigens on the cell membrane generates histograms with poor resolution between positive and negative events.

3. There is variable but consistent cell mortality, particularly in the first days of culture.

4. There is a relative scarcity of cells of interest. Gating on a small cell subset, typically cells that have undergone a certain number of rounds of cell division, may reduce a list mode file of 10,000 events to a dozen or so.

All these points are complicated by the fact that CFDA-SE emission, while generally referred to as being fluorescein-like, actually spills quite well over the "orange" and "red" emission of PE and tandem conjugates or PerCP (>600 nm emitting fluorochromes). This has obvious adverse affects on color compensation and makes it necessary to first examine single-stained samples. In this regard, it is worth remembering that CFDA-SE fluorescence diminishes with time (it halves at each round of cell division), and compensation values notoriously remain valid only if the intensity of the stain remains relatively constant among samples.

It is advisable also to titrate any new batch of MAb to minimize nonspecific binding and establish the possible presence of free fluorochrome molecules (free fluorochrome molecules adsorb onto cells and contribute to an elevated background).

IV. Functionally Distinct Circulating Hemopoietic Progenitor Subsets Can Be Assessed during Cytokine-Driven Differentiation

It is clearly beyond the scope of the present discussion to extensively review the various cytokine combinations that can possibly be used to drive HP toward selective differentiation *in vitro* and of the large number of membrane markers used in the identification of the various progenies. Thus, rather than cover all information past and present in the field, we have chosen the point of view of a flow cytometrist sitting in front of the instrument and dealing with acquisition and analysis of an HP sample cultured in the presence of well-defined cytokine cocktails.

The sample is triple stained with CFDA-SE and MAb against lineage and nonlineage specific markers. Immediately after CFDA-SE loading, freshly isolated HP exhibit a narrow peak of fluorescence with a normal distribution of fluorescent events, typical of cells synchronyzed in G_{0-1} phase. In the first 3–4 days of culture, HP exposed to cytokines behave rather homogeneously in terms of proliferation.

Figure 1 depicts a typical situation in which CFDA-SE loaded HP purified by immunomagnetic beads from patients that underwent cytoreductive/mobilizing procedures were cultured in the presence of cytokines driving HP along granulo-monocytic differentiation and then costained with tandem conjugated (PE-Cy5)-CD34 and PE-CD33, or PE-CD38 or PE-HLA-DR. Most if not all CD34+ HP freshly collected from peripheral blood after the mobilizing procedure are CD33+,

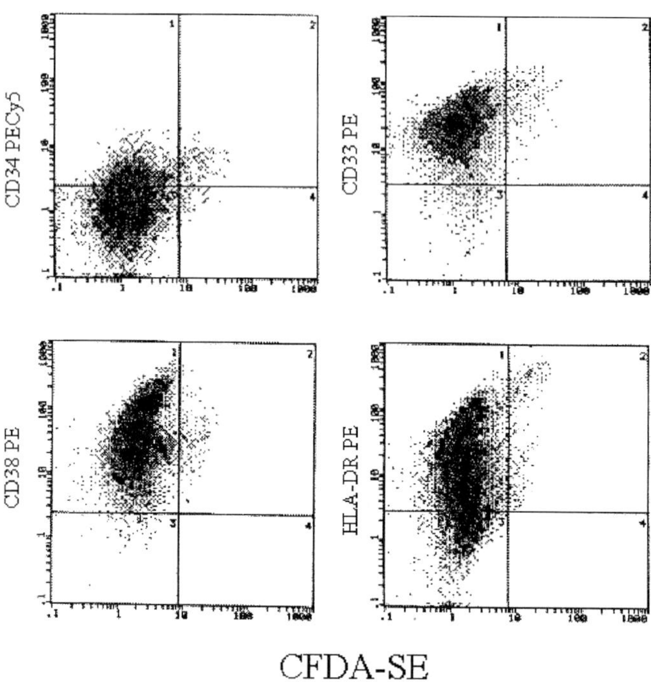

Fig. 1 Representative dual color dot plots generated from HP cultured for 6 days in the presence of recombinant cytokines [stem cell factor (SCF) 10 ng/ml, interleukin 3 (IL-3) 20 ng/ml, granulocyte-colony stimulating factor (G-CSF) 20 ng/ml, and granulocyte-monocyte colony stimulating factor (GM-CSF) 20 ng/ml] to induce granulocytic/macrophagic differentiation. Upper left plot, all non/low proliferating HP identified by their bright CFDA-SE fluorescence, express CD34; this marker is expressed by only a minor proportion of high proliferating cells, albeit there is no definite relationship with the number of cell replications as is the case for the erythroid condition (see Fig. 2). As shown at upper right, there is an inverse relationship between the expression of CD33 and the number of cell replications. At lower left, CD38 expression is enhanced in proliferating cells as compared to non/low proliferating cells. At lower right, non/low proliferating cells express more HLA-DR molecules than high proliferating cells. Percentages and absolute numbers referring to data presented here are detailed in Figs. 4 and 5.

CD38$^+$, and HLA-DR$^+$. In keeping with their quiescent status, CD34, which is typically lost by developing HP, was retained by the vast majority of non/low proliferating cells, and by a proportion of cells that underwent several rounds of cell replications. Similarly, the expression of CD33 and HLA-DR was higher in non/low proliferating cells. Analyzing dual color plots of CD34 versus the other antigens, formally confirmed that virtually all CD34$^+$ HP costained for the diverse antigens (not shown). In contrast, CD38 is maintained at the same level as freshly prepared HP by non/low proliferating cells, and upmodulated by actively proliferating cells.

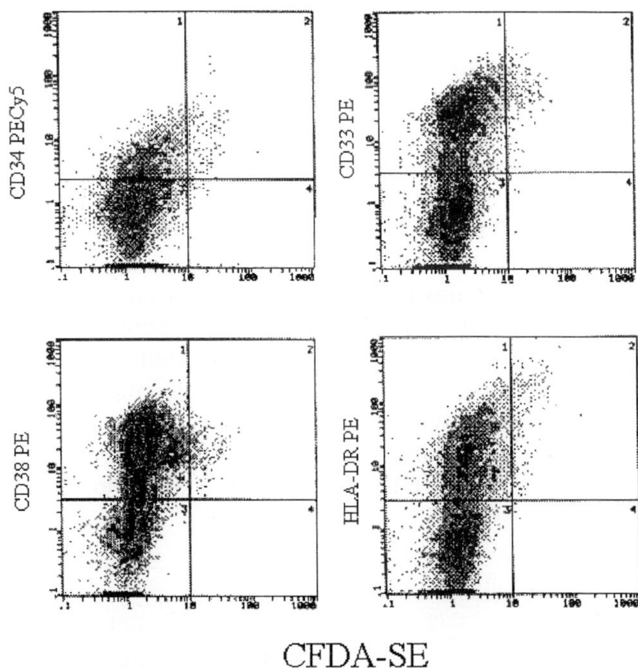

Fig. 2 Representative dual color dot plots generated from HP cultured for 6 days in the presence of recombinant cytokines (SCF 10 ng/ml and EPO 4 IU/ml) to induce erythroid, differentiation. In the upper left plot, all non/low proliferating HP express CD34; the intensity of expression of this marker shows a clear inverse relationship with the number of cell replications assessed by CFDA-SE intensity. In the upper right plot, the expression of CD33 is also inversely related to the number of cell replications, being higher in cells that have proliferated less. At lower left, CD38 expression is enhanced in proliferating cells as compared to non/low proliferating cells. At lower right, non/ low proliferating cells express more HLA-DR molecules than high proliferating cells. A comparison with data from Fig. 1 indicates that the pattern of expression of the various markers in non/low proliferating cells is essentially similar in the two differentiative pathways but profoundly differs in the high proliferating subset. This finding supports the view that irrespective of the differentiative lineage HP are driven into, some HP minimally responding to cytokines remain and maintain a progenitor phenotype. Percentages and absolute numbers referring to data presented here are detailed in Figs. 4 and 6.

Figure 2 refers to an HP population differentiating toward the erythroid lineage and demonstrates that the relationship between cell proliferation, that is, CFDA-SE halving, and the expression of CD34, CD33, CD38, and HLA-DR on low/non proliferating HP is similar to that observed in HP developing along the granulocytic/monocytic pathway, although the differentiated progeny presents a phenotypic pattern peculiar to the lineage (not shown). Figure 3 was generated by analyzing HP differentiating toward the megakaryocyte lineage and exemplifies the relationship betweeen cell proliferation and CD61, typically expressed by HP developing toward the megakaryocytic lineage. CD61 appears quite soon after exposing HP to cytokines, as indicated by the presence of CD61$^+$ cells with high CFDA-SE content, that is, non/low proliferating cells.

A possible drawback of the flow cytometric evaluation of CFDA-SE halving in these kind of cultures is the loss in the capacity to identify cells that still retain CFDA-SE after approximately 14 days of culture (the exact time period depends on the cytokine cocktail used). From this time onward, the overwhelming number of proliferating cells may reduce the proportion of non/low dividing cells below the discriminating capabilities of the instrument.

Data illustrated in Fig. 4 refer to two experiments performed in replicates to assess the behavior of HP when challenged with cytokine cocktails specifically driving HP toward granulocytic/monocytic, erythroid, and megakaryocytic differentiation. The non/low proliferating subset, that is, CFDA-SE^{bright} cells, minimally contributes to the total cell count, never exceeding 10% of total cells at any time point. Figures 5, 6, and 7 (referring to granulocytic/monocytic, erythroid, and megakaryocytic differentiation, respectively) report data obtained by combining the analysis of CFDA-SE dye intensity with the evaluation of CD34, CD33, CD38, HLA-DR, CD11b, glycophorin-A, and CD61 expression using

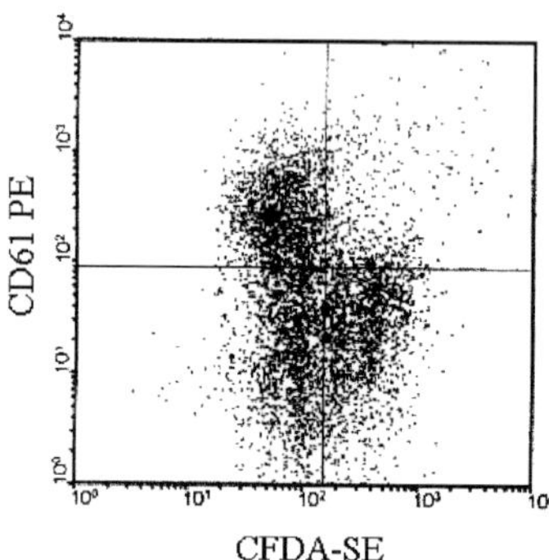

Fig. 3 Representative dual color dot plot generated from HP cultured for 6 days in the presence of recombinant cytokines (SCF 10 ng/ml, IL-6 1000 IU/ml, and megakaryocyte growth and development factor (MGDF) 10 ng/ml) to induce megakaryocytic differentiation. The typical megakaryocytic marker CD61 appeared early during differentiation, as indicated by its presence on non/low proliferating cells, but it was expressed on a much higher proportion of high proliferating cells. Percentages and absolute numbers referring to data presented here are detailed in Figs. 4 and 7.

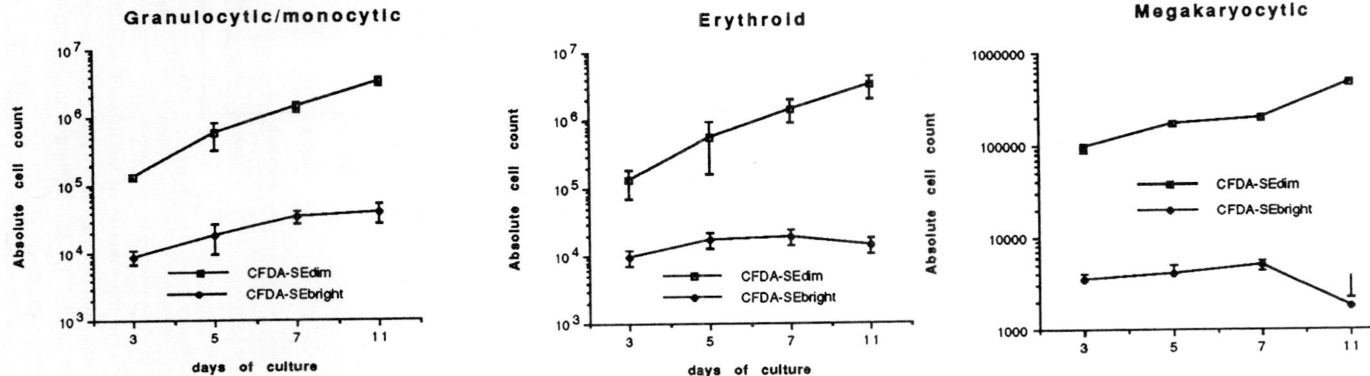

Fig. 4 Absolute cell counts of CFDA-SE^{dim} and CFDA-SE^{bright} (high and non/low proliferating HP, respectively) during differentiation toward granulocytic/monocytic, erythroid and megakaryocytic lineages. Data were collected at the indicated time points for 11 days. From this time point onward, the number of cells with no CFDA-SE fluorescence was so overwhelming as to prevent the identification of possible remaining CFDA-SE^{bright} cells. Thus, under the present experimental conditions, 11 days is the longest culture time still allowing non/low proliferating HP to be detected.

                                                                 Luca Pierelli *et al.*

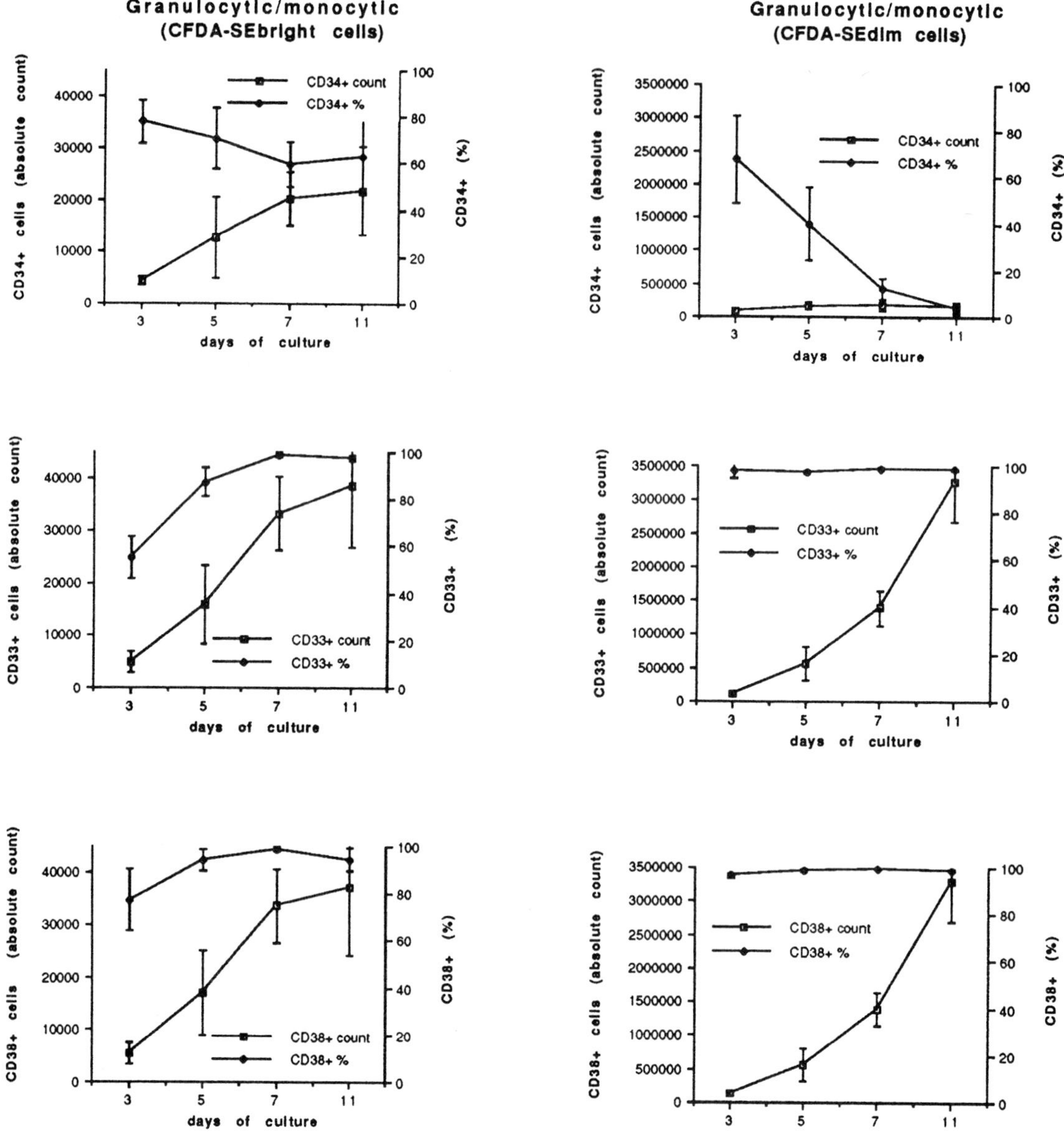

Fig. 5 Double *y* graphs showing the kinetics of modulation of CD34, CD33, CD38, HLA-DR, and CD11b expressed as absolute number (left axis) and percentages (right axis) of positive cells in CFDA-SE[bright] (left column) and CFDA-SE[dim] (right column) subsets during HP differentiation toward the granulocytic/monocytic lineage.

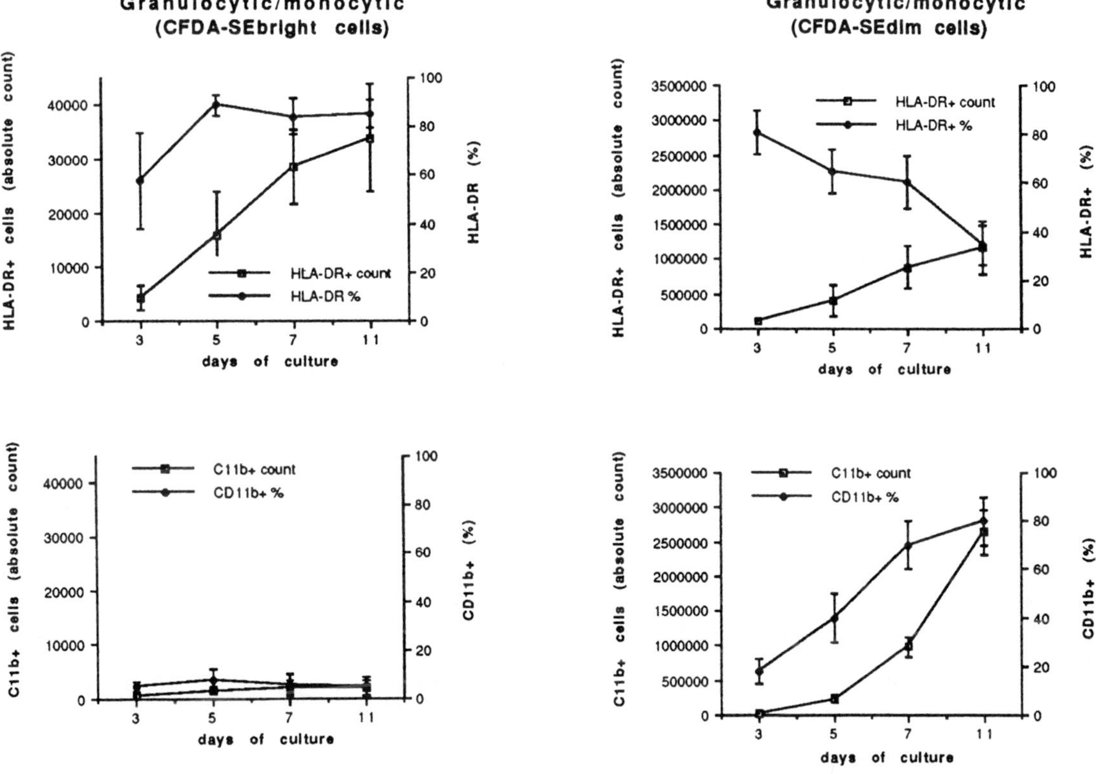

Fig. 5 (*continued*).

three color flow cytometric measurements. As depicted in the Figs. 5–7, the phenotypic profile of CFDA-SE^{bright} and CFDA-SE^{dim} cells profoundly diverges. Essentially, CFDA-SE^{dim} cells, representing the (more) proliferating fraction of the culture, behave as one would expect for proliferating/differentiating HP exposed to the specific cytokine cocktail. Particularly, the percentage of CD33$^+$ cells remains unchanged during the granulocytic/monocytic differentiation and decreases during erythroid and megakaryocytic differentiation. Similarly, the percentage of cells expressing CD38 remains unchanged during granulocytic/ monocytic differentiation and decreases during erythroid and megakaryocytic differentiation, the magnitude of the phenomenon being higher in the former. The percentage of HLA-DR$^+$ cells decreases irrespective of the differentiation pathway, albeit the decline is steeper in the erythroid and megakaryocytic than the granulocytic/monocytic pathway. The vast majority of cells acquired the specific differentiation markers CD11b, glycophorin-A, and CD61 during granu-

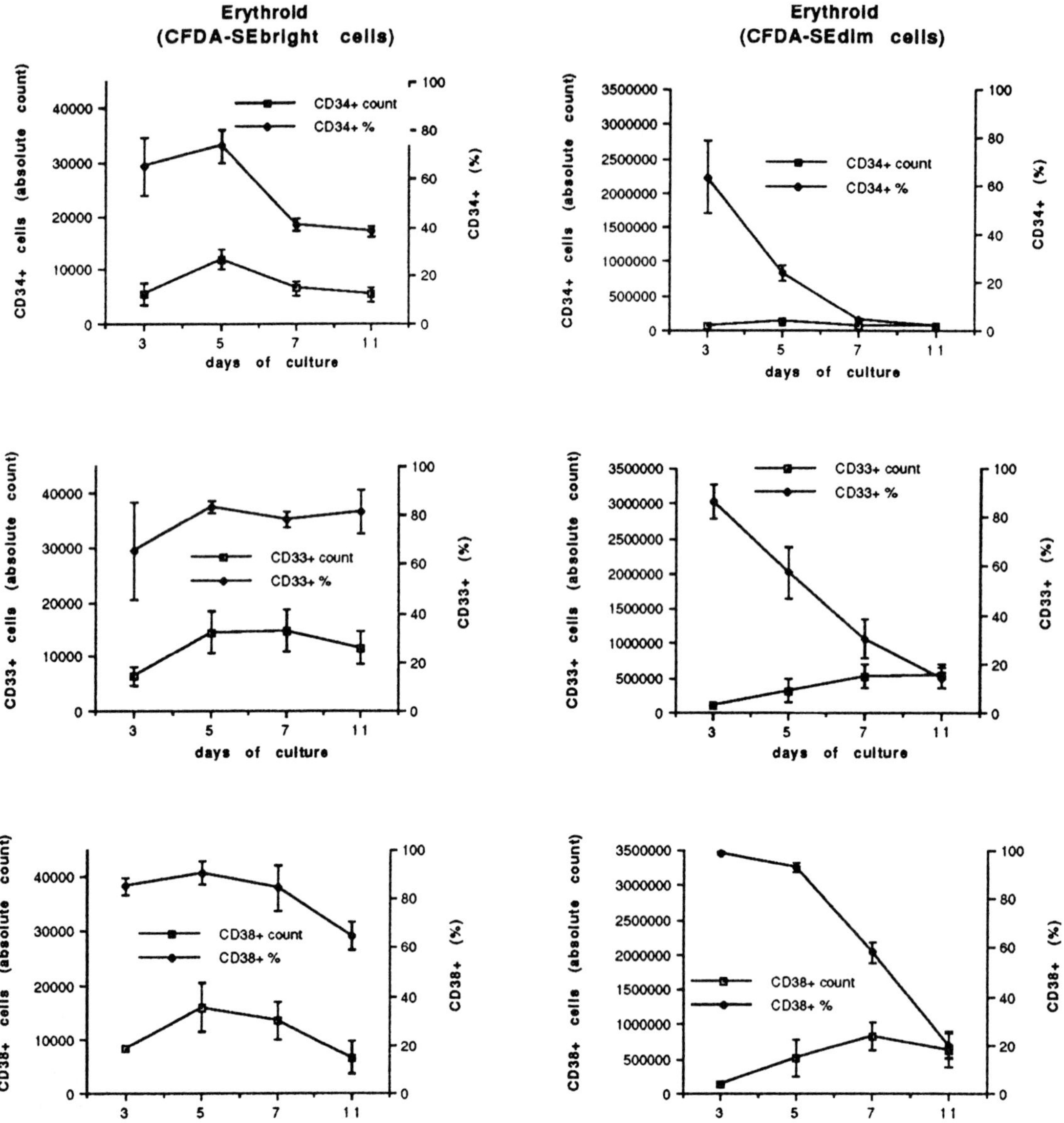

Fig. 6 Double *y* graphs showing the kinetics of modulation of CD34, CD33, CD38, HLA-DR, and glycophorin-A expressed as absolute number (left axis) and percentages (right axis) of positive cells in CFDA-SE[bright] (left column) and CFDA-SE[dim] (right column) subsets during HP differentiation toward the erythroid lineage.

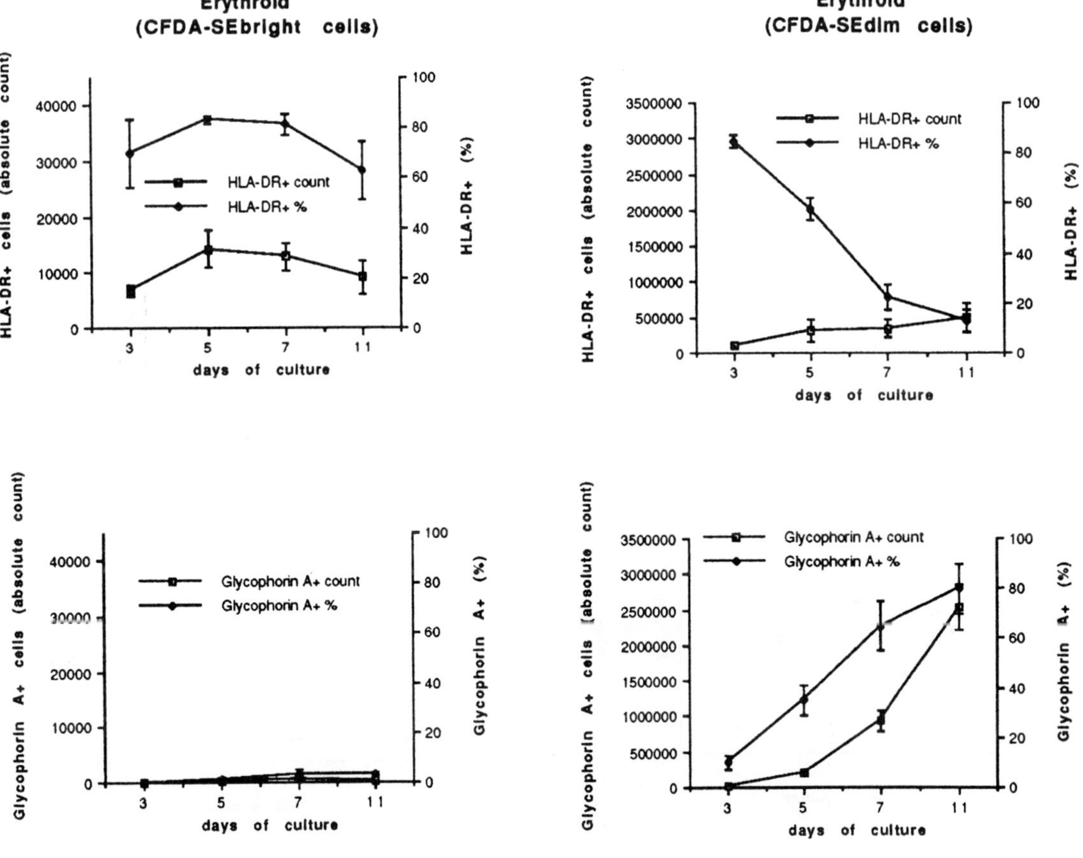

Fig. 6 (*continued*).

locytic/monocytic, erythroid, and megakaryocytic differentiating programs, respectively.

At variance with CFDA-SE^{dim} cells, CFDA-SE^{bright} cells show a resting HP profile irrespective of the cytokine cocktail they have been exposed to, and they may represent a functionally distinct class of precursors or a residual unstimulated population in a stochastic model of HP recruitment. To further investigate on the nature of these cells, the functional profile was assessed. CFDA-SE^{bright} and CFDA-SE^{dim} cells were sorted by a fluorescence activated cell sorter at day 8 of culture and were used in various assays. Compared to CFDA-SE^{dim} cells purified from the same culture, CFDA-SE^{bright} cells exhibited a considerably higher cloning efficiency, even superior to that of freshly isolated HP not exposed to cytokines, exhibited a lower percentage of cells in S and G_2–M, and, remark-

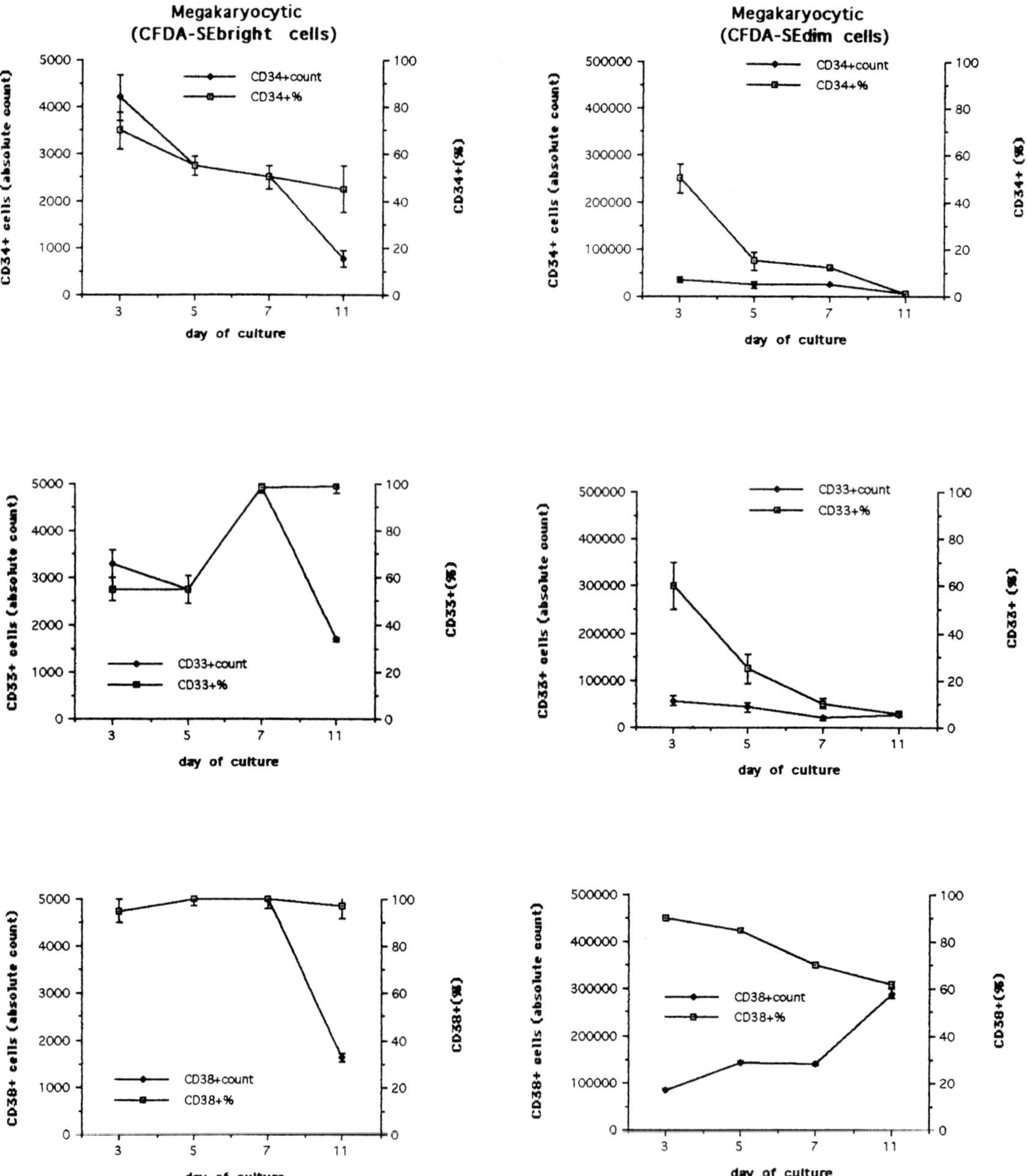

Fig. 7 Double *y* graphs showing the kinetics of modulation of CD34, CD33, CD38, HLA-DR, and CD61 expressed as absolute number (left axis) and percentages (right axis) of positive cells in CFDA-SE^bright (left column) and CFDA-SE^dim (right column) subsets during differentiation of circulating CD34^+ cells toward the megakaryocytic lineage.

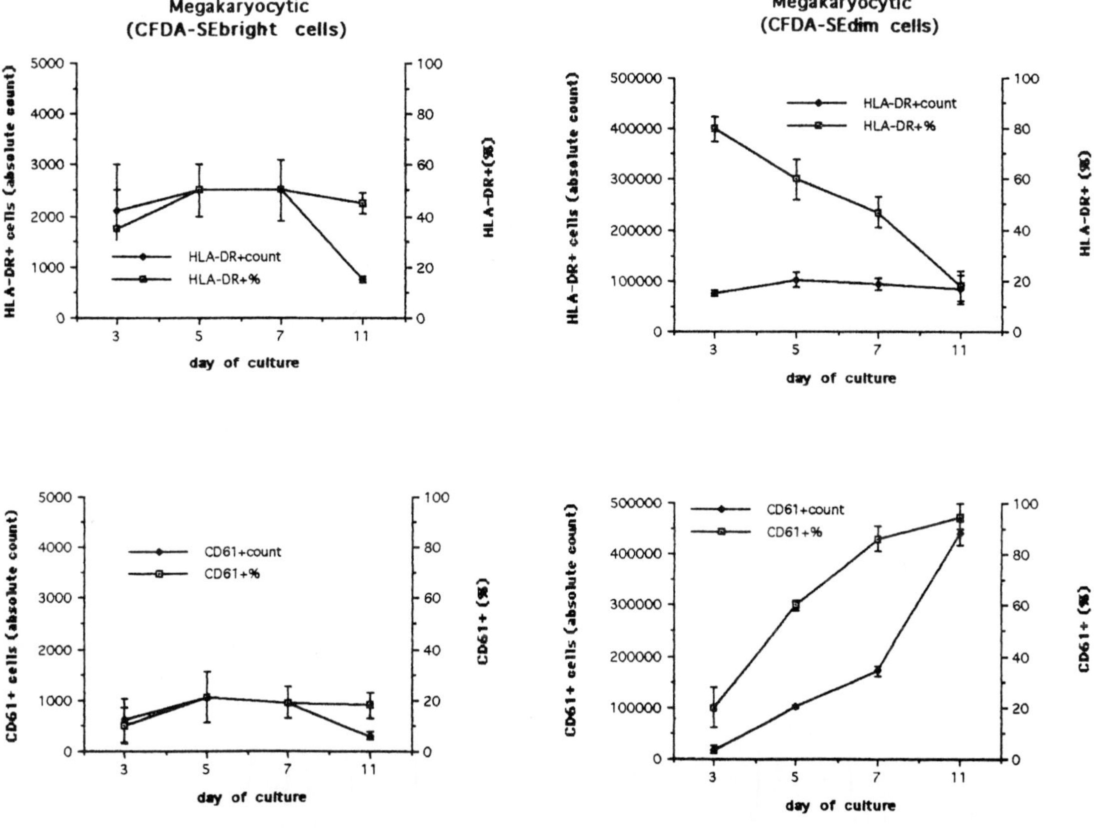

Fig. 7 (*continued*).

ably, maintained a high expression of the antiapoptotic gene *bcl-2* (Fig. 8). Together with the phenotypic profile, these data indicate that the non/low proliferating HP identified on the basis of their high CFDA-SE fluorescence, represent a subset highly enriched in the most primitive HP that resist to the differentiative stimulus of cytokines.

V. Concluding Remarks

The flow cytometric method described in this chapter requires some skills to be optimized, especially because of the problems of color compensation and the wide range of fluorescence intensities involved, but is certainly within the capacity

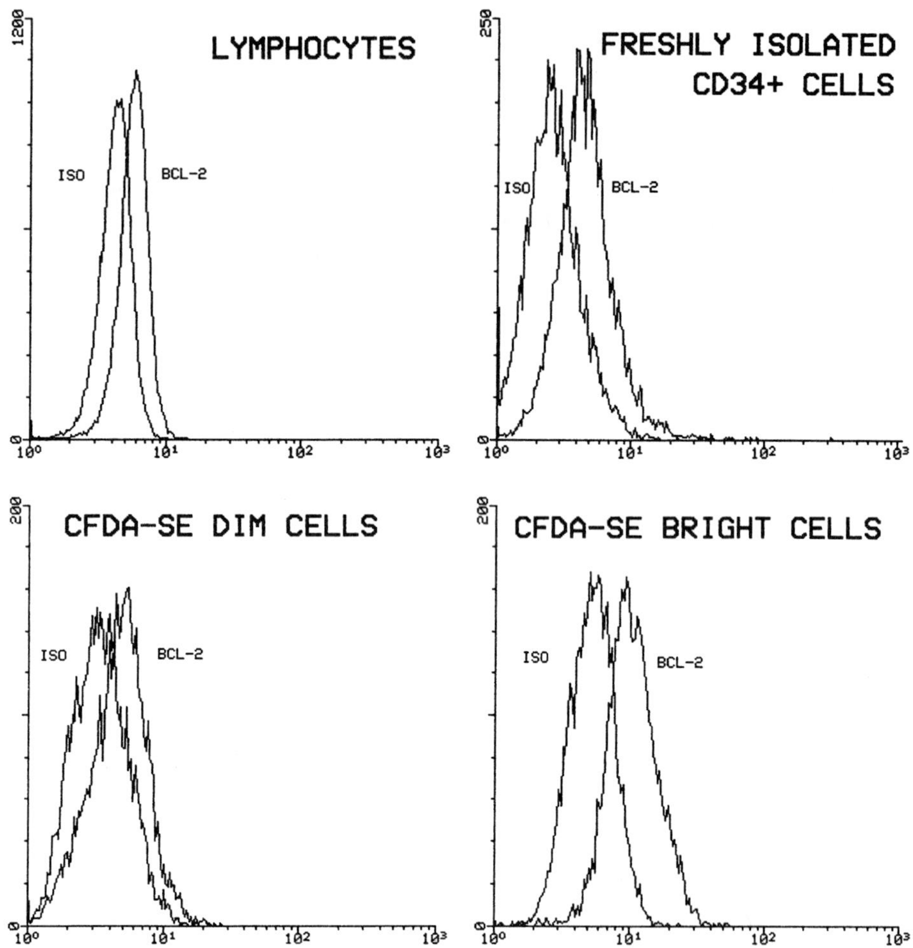

Fig. 8 CFDA-SE^{dim} and CFDA-SE^{bright} cells were sorted by a fluorescence activated cell sorter, permeabilized using methanol/acetone, and stained with fluorescein isothiocyanate (FITC) bcl-2. As comparison, freshly prepared lymphocytes from healthy subjects were also tested. Fluorescence histograms of isotypic control and FITC bcl-2 are depicted. The intensity of expression of the antigen is higher in CFDA-SE^{bright} than CFDA-SE^{dim} cells, indicating that this population contains cells with a more immature molecular profile.

of all cytometrists. It can be anticipated that determining the division history of human HP exposed to cytokines by the parallel evaluation of dye dilution and phenotype can provide valuable information on the behavior of HP and enlarge the possibility of investigating the complexities in the hematopoietic system.

References

Andrews, R. G., Singer, J. W., and Bernstein, I. D. (1989). Precursors of colony-forming cells in human can be distinguished from colony-forming cells by expression of the CD33 and CD34 antigens and light scatter properties. *J. Exp. Med.* **169,** 1721–1731.

Aubin, J. E. (1979). Autofluorescence of viable cultured mammalian cells. *J. Histochem. Cytochem.* **27,** 36–43.

Civin, C. I., Trischmann, T. M., Fadelo, M. J., Bernstein, I. D., Buhring, H. J., Campos, L., Greaves, M. F., Kamour, M., Katz, D. R., Lansdorp, P. M., Look, A. T., Seed, B., Sutherland, D. R., Tindle, R. W., and Uchanska-Ziegler, B. (1989). Report on the CD34 cluster workshop. *In* "Leukocyte Typing. IV." (W. Knapp, *et al.,* eds.), pp. 818–823. Oxford Univ. Press, New York.

Craig, W., Kay, R., Cutler, R. L., and Lansdorp, P. M. (1993). Expression of Thy-1 on human hematopoietic progenitor cells. *J. Exp. Med.* **177,** 1331–1342.

Fackler, M. J., Krause, D. S., Smith, O. M., Civin, C. I., and May, W. S. (1995). Full-length but not truncated CD34 inhibits hematopoietic cell differentiation of M1 cells. *Blood* **85,** 3040–3047.

Fattorossi, A., Pierelli, L., Scambia, G., Ciarli, M., Bonanno, G., Battaglia, A., Menichella, G., Benedetti-Panici, P., Bizzi, B., and Mancuso, S. (1996). A multiparametric flow cytometric approach to hematopoietic differentiation *in vitro. Cytometry* **S8,** 79.

Hao, Q.-L., Shah, A. J., Thiemann, F. T., Smogorzewska M. E., and Crooks, G. M. (1995). A functional comparison of CD34+CD38− cells in cord blood and bone marrow. *Blood* **86,** 3745–3753.

Krause, S. D., Fackler, M. J., Civin, C. I., and Stratford, M. W. (1996). CD34: Structure, bilogy and clinical utility. *Blood* **87,** 1–13.

Ladd, A. C., Pyatt, R., Gothot, A., Rice, S., McMahel, J., Traycoff, C. M., and Srour, E. F. (1997). Orderly process of sequential cytokine stimulation is required for activaction and maximal proliferation of primitive human bone marrow CD34+ hematopoietic progenitor cells residing in G0. *Blood* **90,** 658–668.

Lansdorp, P. M. (1993). In vitro properties of purified human stem cell candidates. *J. Hematother.* **2,** 329–332.

Lansdorp, P. M., and Dragowska, W. (1993). Maintence of hematopoiesis in serum-free bone marrow cultures involves sequential recruitment of quiescent progenitors. *Exp. Hematol.* **21,** 1321–1327.

Lyons, A. B., and Parish, C. R. (1994). Determination of lymphocyte division by flow cytometry. *J. Immunol. Methods* **171,** 131–137.

Mayani, H., Dragowska, W., and Lansdorp, P. M. (1993a). Cytokine-induced selective expansion and maturation of erythroid versus myeloid progenitors from purified cord blood precursor cells. *Blood* **81,** 3252–3258.

Mayani, H., Dragowska, W., and Lansdorp, P. M. (1993b). Characterization of functionally distinct subpopolations of CD34+ cord blood cells in serum-free long-term cultures supplemented with hematopoietic cytokines. *Blood* **82,** 2664–2672.

Metcalf, D. (1991). Control of granulocytes and macrophages: Molecular, cellular and clinical aspects. *Science* **254,** 529–533.

Morrison, S. J., Shah, N. M., and Anderson, D. J. (1997). Regulatory mechanisms in stem cell biology. *Cell* **88,** 287–298.

Nakamura, Y., Komano, H., and Nakauchi, H. (1993). Two alternative forms of cDNA encoding CD34. *Exp. Hematol.* **21,** 236–242.

Nordon, R. E., Ginsberg, S. S., and Eaves, C. J. (1997). High-resolution cell division tracking demonstrates the Flt3-ligand-dependence of human marrow CD34+CD38− cell production in vitro. *Br. J. Haematol.* **98,** 528–539.

Osawa, M., Hanada, K., Hamada, H., and Nakauchi, H. (1996). Long-term lymphohematopoietic reconstitution by a single CD34− low/negative hematopoietic stem cell. *Science* **273,** 242–245.

Pierelli, L., Scambia, G., Fattorossi, A., Bonanno, G., Battaglia, A., Perillo, A., Menichella, G., Benedetti-Panici, P., Leone, G., and Mancuso, S. (1998a). In vitro effect of amifostine on hematopoietic progenitors exposed to carboplatin and non-alkylating antineoplastic drugs: Hematoprotection acts as a drug-specific progenitor rescue. *Br. J. Cancer* **78,** 1024–1029.

Pierelli, L., Scambia, G., Fattorossi, A., Bonanno, G., Battaglia, A., Rumi, C., Marone, M., Mozzetti, S., Rutella, S., Menichella, G., Romeo, V., Mancuso, S., and Leone, G. (1998b). Functional, phenotypic and molecular characterization of cytokine-low responding circulating CD34$^+$ haemopoietic progenitors. *Br. J. Haematol.* **102,** 1139–1150.

Rusten, L. S., Jacobsen, S. E. W., Kaalhus, O., Veibi, O. P., Funderud, S., and Smeland, E. B. (1994). Functional differences between CD38$^-$ and DR$^-$ subfractions of CD34$^+$ bone marrow cells. *Blood* **84,** 1473–1481.

Sakabe, H., Ohmizono, Y., Tanimukai, S., Kimura, T., Mori, K. J., Abe, T., and Sonoda, Y. (1997). Functional differences between subpopulations of mobilized peripheral blood derived CD34$^+$ cells expressing different levels of HLA-DR, CD33, CD38 and c-kit antigens. *Stem Cells* **15,** 73–81.

Terstappen, L. W. M. M., Huang, S., Safford, D. M., Lansdorp, P. M., and Loken, M. R. (1991). Sequential generations of hematopoietic colonies derived from single nonlineage committed CD34$^+$CD38$^-$ progenitor cells. *Blood* **77,** 1218–1227.

Traycoff, C. M., Orazi, A., Ladd, A. C., Rice, S., McMahel, J., and Srour, E. F. (1998). Proliferation-induced decline of primitive hematopoietic progenitor cell activity is coupled with an increase in apoptosis of ex vivo expanded CD34$^+$ cells. *Exp. Hematol.* **26,** 53–62.

Verfaillie, C. M., and Miller, J. S. (1995). A novel single-cell proliferation assay shows that long-term culture-initiating cell (LTC-IC) maintence over time results from the extensive proliferation of small fraction of LTC-IC. *Blood* **86,** 2137–2145.

Watt, S. M., Karthi, K., Gatter, K., Furley, A. J. W., Katz, F. E., Healy, L. E., Atlass, L. J., Bradley, N. J., Sutherland, D. R., Levinsky, R. J., and Greaves, M. F. (1987). Distribution and epitope analysis of the cell membrane glycoprotein (HPCA-1) associated with human hemopoietic progenitor cells. *Leukemia* **1,** 417.

Young, J. C., Varma, A., DiGiusto, D., and Backer, M. P. (1996). Retention of quiescent hematopoietic cells with high proliferaive potential during ex vivo stem cell culture. *Blood* **87,** 545–556.

Yui, J., Chiu, C. P., and Lansdorp, P. M. (1998). Telomerase activity in candidate stem cells from fetal liver and adult bone marrow. *Blood* **91,** 3255–3262.

PART IX

Experimental Oncology

Cytometry of Antitumor Drug–Intracellular Target Interactions

Paul J. Smith and Marie Wiltshire

Department of Pathology
University of Wales College of Medicine
Heath Park, Cardiff CF4 4XN, United Kingdom

I. Introduction
II. General Classification of Cytotoxic Anticancer Agents
    A. Disruption of DNA Integrity
    B. Inhibitors of DNA Metabolism
    C. Inhibitors of Mitotic Spindle Function
    D. DNA Topoisomerase Inhibitors
III. Establishment of Quality Control Parameters
    A. Cell/Culture Integrity
    B. Target Enzyme Availability
IV. Drug–DNA Interactions
    A. Ligand Characteristics
    B. Spectrofluorimetry
    C. Analysis of Drug Uptake by Flow Cytometry
    D. Confocal Microscopy
    E. Flow Cytometric Analysis of Hoechst 33342–DNA Binding
V. Conclusions
    References

I. Introduction

The analysis of cellular systems is often aimed at gaining an understanding of the structure and function of subcompartments, the identity and behavior of particular molecular species, and the dynamic changes of such elements in response to perturbing variables. However, biological systems all depend critically on the coherence, control, complexity, and compartmentalization of

METHODS IN CELL BIOLOGY, VOL. 64

molecular interactions. Unfortunately, the opportunity to monitor and quantify interactions can be restricted by factors including: low event frequency, inadequate detection methodologies, low signal-to-noise ratios, and lack of information as to the nature and location of the interactions. Cytometric methods of monitoring the interaction of anticancer drugs with defined intracellular targets approach the difficult problem of the frequent heterogeneity of human tumors in terms of biological response and target presentation. Many cytotoxic anticancer agents have as their major targets nucleic acids, primarily genomic DNA, or pathways involved in DNA metabolism. A significant subset of anticancer cytotoxic agents participate in complex interactions with topoisomerase enzymes at their sites of action on DNA. This chapter will focus on selected DNA topoisomerase inhibitors to exemplify some of the issues involved for cytometric analysis.

Topoisomerase enzymes control conformational changes in DNA and aid the orderly progression of DNA replication, gene transcription, and the separation of daughter chromosomes at cell division. As noted, DNA topoisomerase II is a target for several classes of apparently unrelated anticancer drugs such as the epipodophyllotoxins and anthracyclines (Liu, 1989). Although drug–target interaction is an important aspect of cytostasis and cytotoxicity induced by topo-isomerase poisons (Smith, 1990), it appears that trapped topoisomerase complexes must combine with other cellular factors to effect either cell cycle arrest or cell death. It is important to assess target enzyme availability and sensitivity to drug-induced trapping (Smith and Makinson, 1989) to establish the potential for target interaction. However, in determining the behavior of a drug molecule within a targeted population, the effects of the agent on cellular integrity must also be taken into consideration. Methodological approaches are described for the monitoring of cell integrity, drug uptake and disposition, DNA targeting, enzyme availability, and complex trapping.

II. General Classification of Cytotoxic Anticancer Agents

To place these agents into context, a classification is given below for anticancer cytotoxic drugs according to their modes of action.

A. Disruption of DNA Integrity

This group contains alkylating agents, DNA cross-linking drugs, and various DNA damaging antibiotics. The alkylating agent group is extensive and includes nitrogen mustards (e.g., melphalan and cyclophosphamide), chloroethylni-trosoureas, triazenes (e.g., DTIC), dimethanesulfonates (busulfan), aziridines (e.g., diaziquone), and melamines (e.g., trimelamol). Classically alkylators act through their ability to damage DNA and interfere with cell replication. Detection of drug–DNA interactions can be made on the basis of antibody probes for alkylation damage or postlabeling procedures for defined lesions. Such approaches clearly require the some degree of cell disruption and are intrinsically retrospective.

In the case of some antitumor antibiotics, such as bleomycin (BLM), drug molecules can interact with DNA but require an intermediate reaction to occur for the generation of DNA damage. BLM is a cytotoxic drug, acting through the generation of DNA strand breaks via a ferrous oxidase cycle and radical formation. BLM has high intrinsic cytotoxicity once inside the cell. However, it is unable to diffuse through the plasma membrane, and very low amounts of BLM reach the cell interior (Mir *et al.,* 1996). Thus detection methods that rely on the identification of intracellular drug molecules are of limited use, while the presence of drug molecules at the target is no guarantee of DNA damaging activity.

B. Inhibitors of DNA Metabolism

Anticancer drugs such as folate antagonists and direct-acting inhibitors of thymidylate synthase are potent genotoxic antimetabolites. These agents are cytotoxic through interference with the control of DNA precursor metabolism, preventing efficient and faithful synthesis of DNA. The pharmacodynamics of methotrexate (MTX), one of the earliest cancer chemotherapy agents, demonstrates the pertinence of determining the intracellular form of the drug. MTX is taken into the cell via a carrier-mediated energy-dependent transport system and rapid polyglutamylation. Polyglutamates became the predominant form of intracellular drug, both free in the cytosol and bound to dihydrofolate reductase target enzyme (Schilsky, 1996).

C. Inhibitors of Mitotic Spindle Function

Microtubule organization is a target for a range of xenobiotic molecules (Chabner, 1992). Some anticancer agents can inhibit polymerization (e.g., Vinca alkaloids such as vinblastine), whereas others can prevent depolymerization (e.g., taxoids such as taxol).

D. DNA Topoisomerase Inhibitors

The DNA topoisomerase inhibitors comprise a group of disparate agents with the ability to inhibit one or both classes of DNA topoisomerases, generally through the disruption of nucleic acid–protein interactions (Table I). Identification of topoisomerases as critical targets for a range of anticancer drugs followed observations that such agents induced unique DNA lesions in cancer cells *in vitro* (for reviews, see Liu, 1989; Cummings and Smyth, 1993; Smith and Soues, 1994). The drugs that generate these DNA lesions interfere with the breakage–reunion reaction in the catalytic cycle of the enzyme, trapping the proteins as

Table I
Properties of Topoisomerase I and II Targeting Drugs[a]

				Mechanism of inhibition		
DNA binding mechanism	Drug class	Example	Target enzyme	Catalytic activity	Cleavable complexes	Enzyme–DNA complex
Intercalators	Acridines	Amsacrine (*m*-AMSA)	II	+	+	−
Intercalator/minor groove interaction	Actinomycins	Actinomycin D	I & II	+	+	+
Intercalators	Anthracyclines	Doxorubicin	II	+	+	−
Intercalators	Anthracenediones	Mitoxantrone	II	+	+	−
Intercalators	Benzisoquinolinediones	Amonafide, mitonafide	II	+	+	−
Intercalators	Ellipticines	2-Methyl-9-hydroxy-ellipticinium acetate	II	+	+	−
Weak intercalators	—	Saintopin	I & II	+	+	−
Minor groove binders	Lexitropsins	Netropsin, distamycin	I & II	+	−	+
Nonbinding	Camptothecins	9-Aminocamptothecin	I	+	+	−
Nonbinding weak	Epipodophyllotoxins	VP-16, VM26	II	+	+	−
Nonbinding	Isoflavanoids	Genistein	II	+	+	−
Nonbinding	Fostriecin analogs	Fostriecin	II	+	−	−
Nonbinding	Bis(2,6-dioxopiperazines)	ICRF-193, ICRF-154	II	+	−	−
Nonintercalators	Anthracenyl-peptides	NU-ICRF-500	I & II	+	−	−
Nonintercalators	Anthracenyl-peptides	Merbarone	II, weak I	+	−	−
Nonintercalators	Terpenoides	Terpentecin	II	+	+	−

[a] Information summarized from Anderson and Berger (1994), Chen and Liu (1994), Cummings and Smyth (1993), and Schneider *et al.* (1990).

reaction intermediates known as cleavable complexes. In general there are two types of topoisomerases with different functionalities and drug sensitivities. Type I DNA topoisomerases act to resolve topological problems through the transient introduction of a single strand break, strand passing, and religation, the process being energy-independent. Type II topoisomerases have the additional capacity to decatenate interlinked loops or circles of DNA through the transient introduction of a double strand break, via a process that requires ATP.

1. DNA Topoisomerase I Inhibitors

The first well-characterized inhibitor of type I topoisomerases was camptothecin (CPT). The cytotoxic actions of CPT were shown to be dependent on the action of topoisomerase I (Eng *et al.*, 1988). Topoisomerase I inhibitors stabilize a covalent bond between a tyrosine residue on the protein and the 3′-phosphoryl end of the single strand of DNA it breaks (Hsiang *et al.*, 1985). Formation of cleavable complexes is due to inhibition of the religation phase of the breakage–reunion reaction rather than by promotion of cleavage (Robinson and Osheroff, 1990). These agents kill cells when the unusual ternary complex, comprising CPT:DNA:topoisomerase I, interacts with a DNA replication fork (Zhang *et al.*, 1990).

2. DNA Topoisomerase II Inhibitors

Inhibitors of type II topoisomerase (e.g., eukaryotic DNA topoisomerase IIα) act primarily through the disruption of drug–DNA or drug–enzyme interactions. VP-16 (etoposide) is a semisynthetic derivative of podophyllotoxin capable of acting as a specific poison for topoisomerase II. Epipodophyllotoxins induce cleavable complexes, which sequester double strand cleavage events, without DNA intercalation. Some anticancer DNA intercalators (e.g., doxorubicin and other members of the rhodomycin group of antibiotics) are also able to trap DNA topoisomerase II and generate similar cleavable complexes. The anilinoacridine derivative amsacrine (*m*-AMSA) binds to DNA through weak, reversible intercalation and is one of few intercalators that bind preferentially to AT-rich sequences rather than GC runs. The anthraquinones are a group of synthetic DNA-binding agents, structurally related to the DNA intercalating anthracycline antibiotics. The cytotoxicity of mitoxantrone is attributed to its ability to bind DNA with high affinity by intercalation. This interaction is stabilized by the alkylamino side chains of mitoxantrone, which are cationic at physiological pH and hence can form ionic and hydrogen bonding interactions with the anionic sugar–phosphate backbone of DNA (Lown *et al.*, 1985). Mitoxantrone traps DNA topoisomerase II as a ternary complex with DNA, with evidence of intracellular persistence of the drug and its associated DNA damage (Bowden *et al.*, 1985; Smith *et al.*, 1990, 1992; Fox and Smith, 1995; Roberts *et al.*, 1989).

III. Establishment of Quality Control Parameters

A. Cell/Culture Integrity

1. DNA Fragmentation and Cell Lysis Assay

In general the extracellular free-drug concentrations at which target interactions can be observed are considerably higher than those that are biologically relevant. Thus there is a need to ensure that cell integrity is mainained during the course of analysis. We describe a general method for determining the integrity of cells used in cytometric assays and for assessing the degree of cell loss associated with the use of cytotoxic agents over a prolonged periods.

DNA cleavage (e.g., associated with rapid apoptosis) and the lysis of cells are often seen as acute cellular responses to anticancer drug treatment, particularly if supratoxic concentrations of an agent are used. In determining drug–target interactions in such heterogeneous populations, it is important to evaluate the integrity of cell populations in a manner that provides a balance sheet for the fate of cells. In the case of phase specific agents there may be selective loss of cells and apparent skewing of the age distribution of remaining cells. On the other hand, loss of cells from all phases of the cell cycle without an accompanying arrest may not be apparent. To determine culture integrity we have modified the method described by Kolber *et al.* (1990).

2. Procedure

a. Cell Culture

Exponentially growing cultures of the p53 wild-type B cell lymphoma cell line, DoHH2, were established using standard conditions (cell line kindly supplied by Prof. F. Cotter, LRF Centre for Childhood Leukaemia at the Institute of Child Health, London). The DoHH2 cell line was orignally derived by Dr. Kluin-Nelemans (University Medical Center Leiden; Kluin-Nelemans *et al.*, 1991). Cultures at 2–4×10^6 cells/ml are prelabeled with 1.85×10^{-3} MBq/ml methyl [^{14}C]thymidine (specific activity 1.92 GBq/mmol) for 24 hr, washed twice with warm medium, and then incubated in fresh medium for 2–4 hr prior to addition of drugs.

b. Fractionation

Cells are harvested by centrifugation. At this point aliquots of culture medium can be taken to determine the dpm cell equivalents that cannot be spun down (i.e., lysed cells). Cell pellets are washed with phosphate-buffered saline (PBS) (the washings can also be counted) and suspended in 0.5 ml hypotonic lysis buffer containing 10 mM Tris-HCl (pH 8.0), 1 mM EDTA, and 0.2% Triton X-100. After 30 min of incubation at room temperature, samples are centrifuged at 12,000 g for 30 min, and 0.3 ml supernatant is removed and counted. The remaining pellet (including 0.2 ml equivalent of supernatant) is vortexed and all washed into a scintillation vial. Radioactivity measurements are performed on the supernatant (detergent-soluble low-molecular-weight DNA) and pellet (intact chromatin DNA) by liquid scintillation counting.

c. Analysis and Results

Arithmetic corrections can be made to all values to give dpm values in medium, lysed supernatant, and pellet. DNA fragmentation is calculated as follows: % DNA fragmentation = 100 × (corrected supernatant dpm/total dpm in supernatant + pellet). Accordingly: medium dpm = lysed cells; supernatant dpm = apoptotic/fragmented nuclei cells; pellet dpm = intact cells without fragmented nuclei. Variations on the approach can be used to investigate cell cycle specific effects. Thus [^{14}C]thymidine prelabeled cells may be pulsed using high specific activity [^{3}H]thymidine just prior to cell washing and drug incubation. Using this dual label approach, fragmentation and lysis of cells in S phase at the time of drug treatment may be determined. Figure 1 shows the effects of VP-16 treatment on culture integrity for DoHH2 cells. There is a clear dose- and exposure-dependent increase in the release of radiolabeled DNA into the culture medium (cell lysis). DNA fragmentation (apoptosis) is biphasic with a sensitive phase for drug concentrations up to 0.5 μM VP-16. The results demonstrate the degree of potential loss of cells from drug-treated populations subjected to cytometric analysis alone (Smith *et al.*, 1994a).

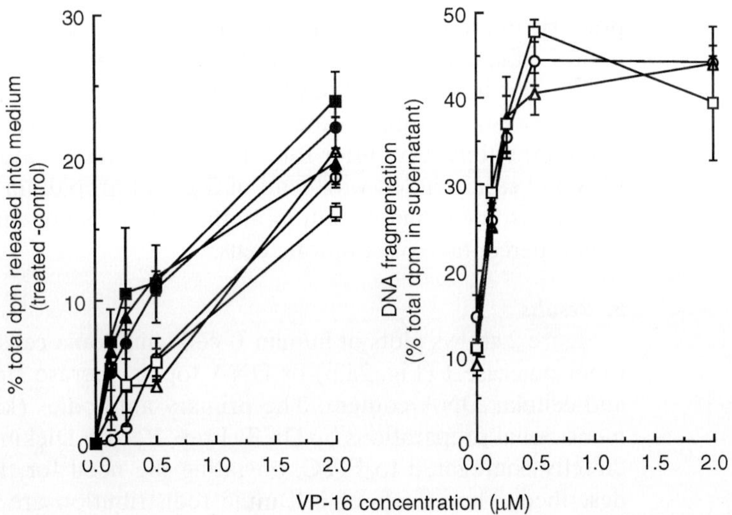

Fig. 1 Effect of VP-16 treatment of DoHH2 cells on culture integrity. (Left) Release of radiolabeled nuclear DNA into culture medium in three independent experiments for 24 hr (○, △, □) or 48 hr (●, ▲, ■) drug exposures. (Right) Fragmentation of nuclear DNA in three independent experiments for a 24-hr drug exposure. Data points are triplicate determinations (± sd).

B. Target Enzyme Availability

The measurement of intracellular enzyme content together with cellular DNA content is particularly pertinent in the case of topoisomerase II since its expression is elevated late in the cell cycle, with targeted destruction in G_1. Thus, drug-induced cell cycle perturbations can generate the impression of enhanced expression because of late cell cycle arrest. The methods used for the detection of any antigens, especially intracellular epitopes, are dependent on the type or location of the antigen, cell type, and fixation protocol. Therefore, several methods of fixation and permeabilization should be attempted. Here we describe a combination of fixatives for the analysis of DNA topoisomerase II expression by flow cytometry, modified from that described previously (Smith and Makinson, 1989; Smith *et al.*, 1994b).

1. Cell Cycle-Related Expression of DNA Topoisomerases

a. Cell Fixation

Approximately $1-5 \times 10^6$ SUD4 cells (a p53 mutant B cell lymphoma cell line supplied by Prof. F. Cotter) are pelleted by centrifugation at 1250 *g* for 10 min at 4°C. The cell pellet is washed once with PBS, loosened by gentle tapping, and incubated for 15 min on ice with 0.25% paraformaldehyde made up in PBS. The cells are pelleted and washed once with PBS. The cells are fixed again overnight with 70% ethanol at −20°C. The fixed cells are rehydrated by washing in PBS. The cells are permeabilized with 0.25% Triton X-100 in PBS for 5 min on ice and rinsed with PBS. The cells are incubated with 100 μl of primary antibody diluted (1:20 to 1:100) in antibody dilution buffer (PBS, 5% nonfat milk, and 0.1% Triton X-100) for 1 hr at room temperature or overnight at 4°C. For indirect antibody staining, cells are rinsed with PBS and incubated with fluorescein isothiocyanate (FITC)-conjugated secondary antibody diluted in antibody dilution buffer for 1 hr at room temperature. Cells are rinsed with PBS and resuspended with 1 ml of 5 μg/ml propidium iodide in PBS and 0.1% RNase A. The ethanol fixation method alone is not suitable for situations with a high percentage of apoptotic cells.

b. Results

Figure 2 shows plots of human B cell lymphoma cells stained for either DNA topoisomerase I (Fig. 2a,b) or DNA topoisomerase IIα (p170 form; Fig. 2b,c), and cellular DNA content. The primary antibodies (kindly supplied as mouse monoclonal preparations by Dr. T. Frey, Becton Dickinson, San Jose, CA) were directly conjugated to FITC, obviating the need for the second antibody step described. The effects of cell cycle redistribution are also shown for cultures exposed to the spindle inhibitor colcemid for 18 hr to generate G_2/M arrest. Expression of topoisomerase I is not cell cycle related, whereas the type II enzyme shows significant enhancement of expression in G_2/M cells (Smith *et al.*, 1994b).

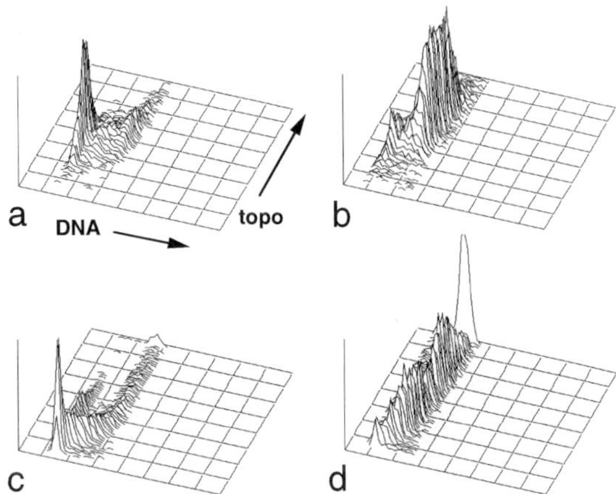

Fig. 2 Flow cytometric frequency distribution plots of DNA topoisomerase protein expression as a function of the cell cycle for a B cell lymphoma cell line (SUD4). (a) Topoisomerase I in asynchronous cultures; (b) topoisomerase I in cultures exposed to 60 ng colcemid/ml for 18 hr; (c) topoisomerase IIα in asynchronous cultures; (d) topoisomerase IIα in cultures exposed to 60 ng colcemid/ml for 18 hr.

2. Drug-Induced Trapping of Topoisomerase II

a. Cell Treatment

Covalent protein–DNA complexes can be precipitated with a high concentration salt solution and measured if the DNA is radiolabeled. The assay was modified from that published by Rowe *et al.* (1986). Briefly, the DNA in logarithmically growing cells (e.g., 2×10^5 cells/ml for a lymphoma suspension culture) is labeled by adding 0.02 μCi/ml[^{14}C]thymidine to the culture medium. Similar conditions are used for attached cells (e.g., human breast tumor cells) except that cells are detached using a standard EDTA-containing buffer. After an overnight incubation, the labeled cells are washed with fresh medium and left to equilibrate under normal culture conditions for 2 hr. Following equilibration, the cells are incubated for an hour with increasing concentrations of drug (e.g., VP-16 at 0–80 μM).

b. Cell Lysis and Protein–DNA Complex Precipitation

One milliliter of drug-treated cells is aliquoted into 1.5-ml tubes and spun at 1500 *g* for 10 min. The supernatants are decanted, and the pellets are washed once with cold PBS and once with solution A (274 mM NaCl, 0.8 mg/ml glucose, 8.3 mM NaHCO$_3$, 2 mM EDTA, pH 8.0). The cells are lysed by incubation with 1.2 ml lysis solution (1.25%, w/v, SDS, 5 mM EDTA, pH 8.0, 0.4 mg/ml calf thymus DNA) at 65°C for 10 min. Two hundred microliters of lysate is mixed with 10 ml of scintillation fluid for scintillation counting. Two hundred and fifty microliters of potassium chloride (650 mM KCl) is added to each tube, and the mixture is vortexed vigorously for 2 min. The protein–DNA complexes are left to precipitate on ice for 15 min. The tubes are centrifuged at 3000 g for 10 min at 4°C, and supernatants are removed. One milliliter of prewarmed (65°C) wash buffer (10 mM Tris-HCl, pH 8.0, 100 mM KCl, 1 mM EDTA, pH 8.0, 0.1 mg/ml calf thymus DNA) is added to the pellets, and tubes are incubated at 65°C to dissolve the precipitate. When all the precipitates have dissolved, the mixtures are allowed to reprecipitate on ice for 15 min. This "precipitate–dissolve" process is repeated twice. The precipitates are finally dissolved in 0.5 ml hot PBS (65°C) and transferred into scintillation vials, and the contents of each tube are washed out with 0.5 ml of PBS. Ten milliliters of scintillation fluid are mixed with the 1 ml of sample in each scintillation vial, and the counts are measured on a scintillation counter.

c. Analysis and Results

The DNA cross-linking index is determined from the percentage of total dpm appearing in the precipitate. Typical results for % total DNA cross-linked for a standard non-multidrug-resistant human B cell lymphoma cell line are as follows: control cells, 3.2 ± 2.0%; doxorubicin-treated cells (10 μM, 1 hr), 6.3 ± 1.8%; VP-16 (10 μM, 1 hr) 18.4 ± 3.7%. Figure 3 shows the differences in cross-linking observed in two human breast cancer cell lines for the drug mitoxantrone. The cross-linking levels reflect both the relative availability of nuclear-located DNA topoisomerase II (data not shown) and nuclear delivery of drug. Comparison of the results of this assay with the flow cytometric measurment of mitoxantrone uptake in intact cells (Fig. 4) shows that the cell line with the highest cross-linking index had reduced whole cell accumulation of mitoxantrone. Such comparisons underline the importance of establishing target availability and drug delivery. Described next are methods to distinguish nuclear binding of drug in a background of whole cell accumulation.

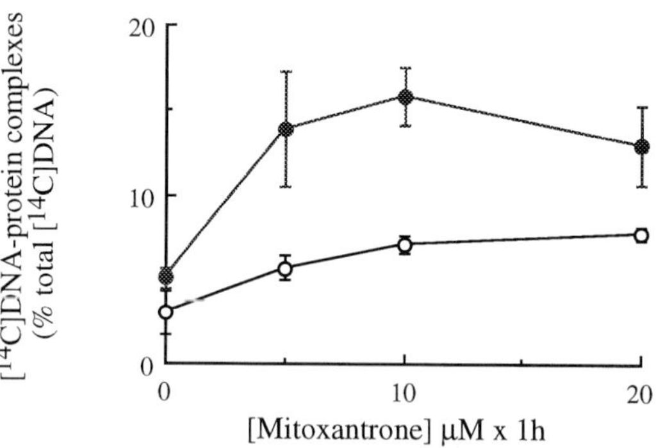

Fig. 3   DNA–protein cross-linking induced by mitoxantrone in human breast tumor cell lines (○) MCF-7 and (●) T-47D.

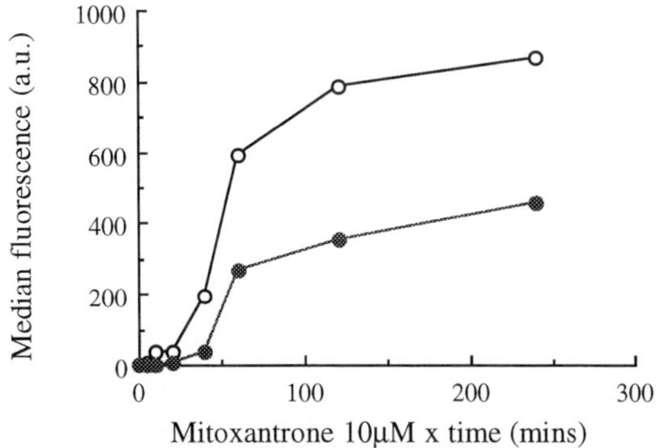

Fig. 4 Kinetics of mitoxantrone uptake, measured by flow cytometry, in human breast tumor cell lines (○) MCF-7 and (●) T-47D using a laser excitation at 647 nm wavelength.

═══ IV. Drug–DNA Interactions

A. Ligand Characteristics

The fluorometric analysis of drug uptake at low concentrations and for long exposure periods is problematic given the complications of drug metabolism, redistribution of cells throughout the cell cycle, and reduction of cell volume/ integrity as a consequence of commitment to apoptosis. Most informative are data derived from acute exposures with immediate or simultaneous analysis, or in the case of efflux experiments after a suitable holding period in drug-free medium. Conventional fluorescence microscopy (Yanowich and Taub, 1983), flow cytometry (Krishan and Ganapathi, 1980), and confocal imaging (Gervasoni *et al.,* 1991) have been used to measure the cellular uptake and subcellular distribution of naturally highly fluorescent drugs such as anthracyclines. In contrast, some drugs, such as the anthraquinone mitoxantrone, demonstrate only weak fluorescence and are more difficult to detect. Mitoxantrone shows weak pH-dependent fluorescence in solution, whereas the mitoxantrone diacid metabolite is essentially nonfluorescent (Bell, 1988). For some agents, nuclear binding is accompanied by fluorescence enhancement or indeed fluorescence quenching, and the investigator should consider such characteristics when attempting to infer stoichiometry from cytometric data.

B. Spectrofluorimetry

Many drugs carry a chromophore suitable for detection by fluorimetry. We have found it informative to use both chemical extraction and cytometric methods to establish drug accumulation characteristics of given cellular system. Often such measurements can be carried out in parallel and allow the investigator to determine whether saturation observed in cytometric assays reflects true limitations on drug accumulation.

Intracellular drug concentrations can be determined using cell lysis and extraction methods. A well-established protocol for anthracyclines has been described previously (Schwartz, 1973). This extraction technique recovers both cytoplasmic and nuclear-bound drug at recovery levels of >90% and can be used to establish differences in anthracycline uptake. Measurements are performed on a fluorimeter, with appropriate detector sensitivity, spectral correction, and with the inclusion of solvent controls. In the case of anthraquinones it is important to enhance the far-red sensitivity of the detector [e.g., use of a Perkin-Elmer (Warrington, UK) LS50B spectrofluorimeter equipped with a Hamamatsu R928 photomultiplier].

In such experiments it is advisable to generate calibration curves for each experiment using spiked samples of intact cells to allow for recovery losses. Care should be taken to consider drug loss in experimental samples through rapid efflux from cells undergoing washing procedures, finite amounts of extracellular drug in the wet volume of cell pellets, binding to macromolecules and sample containers, and drug metabolism. Manipulation procedures will depend on the chemical nature of the agent, but in many cases appropriate extraction procedures have been developed often to support pharmacokinetic studies. A typical result using the Schwartz (1973) method with a standard non-multidrug-resistant human B cell lymphoma cell line exposed to 10 μM doxorubicin (Adriamycin) for 1 hr under normal cell culture conditions is 0.52 $\pm$ 0.24 fmol doxorubicin/cell.

C. Analysis of Drug Uptake by Flow Cytometry

Flow cytometer configurations for mitoxantrone detection (excitation λ_{max} ~640 nm, emission λ_{max} ~680 nm) are given later. For a 488 nm laser-equipped cytometer sufficient activation of intracellular drug occurs to obtain signals for a drug concentration range of 1–20 μM (e.g., detection at FL3 of a standard FACS Vantage system (Becton Dickinson) using a 695 nm long-pass emission filter). Optimal excitation will be achieved using 647 nm wavelength excitation of a krypton–argon laser (Smith *et al.*, 1997). The lower energy output of a helium–neon (633 nm wavelength) laser tends to give a less distinct signal. In all cases it is advisable to use forward light scatter as the master trigger so that all cells are analyzed and any autofluorescence monitored. Figure 4 shows the difference in mitoxantrone uptake between two human breast tumor cell lines assessed using 647 nm wavelength excitation. There are significant differences between the cell lines, exposed as attached cultures, while both show a delay in drug accumulation.

D. Confocal Microscopy

Similar laser requirements, described in the previous section, pertain to the analysis of the subcellular distribution of mitoxantrone by imaging (Smith *et al.*,

1992, 1997). The drug yields both cytoplasmic and nuclear located fluorescence. It is reasonable to assume that only DNA-targeted drug is capable of forming trapped complexes of topoisomerase II on DNA. Cell handling is problematic, and ideally cells should be observed in incubation chambers. Typically this restricts the number of samples that can be observed. A convenient approach is to culture attached cells on sterile glass coverslips and expose them to drug in a multiwell plate. A coverslip is washed twice with PBS containing calcium and magnesium, and the underside wiped across a damp synthetic cloth. The coverslip is then mounted in PBS supported by a ring of petroleum jelly on a microscope slide. This seals the specimen and acts to prevent crushing. A similar approach can be adopted for suspension cultures, where the trapped film restricts cell movement and aids sequential imaging. The sample is observed through the coverslip on an upright or inverted microscope. With practice, the culture-to-observation preparation time is around 30 sec. If a laser scanning microscope is used as the imaging system, care should be taken not to use too high beam powers as this may disturb subcellular drug distribution. For example, cytoplasmic deposits of anthraquinones can absorb photon energy and cause local heating. The result is intracellular release of free drug from points of sequestration, causing a progressive development of nuclear fluorescence. As a rule, successive scans should be analyzed to detect any laser-induced changes in cell fluorescence before images are accumulated or subjected to image filtration. Typical results for mitoxantrone have been described previously (Smith *et al.*, 1997).

E. Flow Cytometric Analysis of Hoechst 33342–DNA Binding

1. Spectral Shift Analysis

The bisbenzimidazole Hoechst dye number 33342 is a minor groove binding agent with DNA topoisomerase disrupting properties. Fluorescence microscopy of Hoechst 33342-stained cells reveals intense nuclear fluorescence. The DNA targeting ability of AT base pair specific minor groove binders has been used in new generations of anticancer agents incorporating the concept of bifunctionality. An example is the agent FCE 24517, which links the alkylating activity of melphalan with the minor groove-binding ability of distamycin (Broggini *et al.*, 1991). Hoechst 33342, being relatively lipophilic, can act as a surrogate agent to analyze drug uptake, targeting, and distribution. In addition, the agent can be used to evaluate multidrug resistance, a subject dealt with elsewhere in this volume. The agent is excited by multiline ultraviolet (UV) lasers, with the emission spectrum of Hoechst 33342-stained DNA extending over a wavelength range of 400 to 500 nm with a significant signal detectable at 600 nm and beyond. This extensive emission range has been used to monitor the ligand binding characteristics in intact cells (Smith *et al.*, 1991). During the time-dependent uptake of the ligand the potential nuclear binding sites are effectively titrated with the initially violet-biased emission spectrum undergoing a shift in emission maximum to longer wavelengths. Consequently, monitoring the time-dependent shift in the emission spectrum provides information on the rate of ligand binding, where the absolute DNA content per cell is no longer a factor (Smith *et al.*, 1991). Two cytometer configurations are given later with results shown for the second.

2. Single Laser Flow Cytometry for Spectral Shift Analysis

Typically cells are resuspended by aspiration and diluted in medium (supplemented with 5–10 mM HEPES, pH 7.4) to 2.5×10^5 cells/ml prior to Hoechst 33342 treatment (1–10 μM, 1–60 min at 37°C) and analysis in a cytometer incorporating a krypton–argon laser tuned to 337 nm wavelength at 200 mW. The optical analysis system at 90° to the intersection of the laser beam with the cell stream includes five dichroic mirrors in series (all Zeiss) with nominal 50% transmission at 390, 420, 460, 510, and 580 nm, respectively. Each dichroic reflects light below the 50% transmission wavelength sequentially into a series of five photomultiplier tubes (PMT), and the last mirror in the series transmits above 580 nm into a sixth PMT. Thus, after the primary filtration due to the dichroic mirrors, PMTs 1 through 6, respectively, receive light in the wavelength bands <390, 390–420, 420–460, 460–510, 510–580, and >580 nm. Additional filtration is then applied. PMT 1 is guarded by a UG11 black glass filter (Melles Griot, Arnhem, Holland) transmitting below 370 nm to analyze 90° scattered light. PMTs 2, 3, 4, and 6 are guarded by narrow-band pass filters centered at 400, 450, 500, and 600 nm each ±5 nm (all Melles Griot) analyzing violet, low blue, blue-green, and red light, respectively. PMT 5 is additionally guarded by a 550 nm long-pass and a 560 nm short-pass filter (both Zeiss, giving a 555 nm ± 5 nm band-pass filter). Forward scatter is analyzed with a solid-state detector. Thus the data sets effectively generate views of the time-dependent changes in the emission spectrum of sequential samples from a given population. Analysis of the increase in fluorescence, with time of Hoechst 33342 exposure, reflects the rate at which intracellular ligand binds to nuclear DNA. Typically there is a rapid increase in fluorescence intensity monitored in the violet region during the first 5 min of ligand exposure for most cell types. Eventually cells show a more predominant increase in the green-red spectral regions as the dye–DNA interactions become equilibrated. Cells that undergo very rapid (<1 min) spectral shifts have damaged membranes and can be excluded from the intact cell analysis or quantified to determine biological response.

3. Dual Laser Flow Cytometry for Spectral Shift Analysis

a. Procedure

The human small cell lung cancer cell line NCI-H69 was grown under standard conditions and exposed to 10 μM Hoechst 33342. The cell line tends to form multicellular aggregates that should be dispersed by aspiration prior to treatment. The stained cells are analyzed using an appropriate dual beam system. The system described here is a FACS Vantage cell system (Becton Dickinson) incorporating a

Coherent Enterprise II laser simultaneously emitting at multiline UV (350–360 nm range) and 488 nm wavelengths with the beams made noncolinear using dichroic separators with a temporal separation of about 25 μsec. Forward light scatter, 90° light scatter, and fluorescence emissions are collected for 1×10^4 cells using the forward light scatter parameter as the master signal from the primary 488 nm beam, while side scatter is collected through a 488/10 nm band-pass filter. The analysis optics are as follows: (i) primary beam-originating signals analyzed at FL1 (FITC filter; barrier filter of 530/30 nm) after transmission at SP610 and SP560 dichroics, or at FL2 (barrier filters of 585/42 or 575/26 nm) after transmission at SP610 and reflection at SP560 dichroics, or at FL3 (barrier filter of LP695 nm) after reflection at a SP610 dichroic; (ii) delayed beam-originating signals analyzed at FL5 (barrier filter of DF424/44 nm) after reflection at a LP640 dichroic or at FL4 (barrier filter of DF675/20 nm) after transmission at the dichroic. Forward and 90° light scatter are analyzed to exclude any cell debris, whereas pulse analysis on the FL5 parameter can be used to exclude doublets. Pulse height parameters are analyzed using CellQuest software (Becton Dickinson).

b. Results

Figure 5 shows the time-dependent changes in fluorescence as Hoechst 33342 stains small cell lung cancer cells (SCLC), with the development of two populations. The SCLC suspension cultures carry a burden of apoptotic and degrading cells together with a significant noncycling population, due to their propensity for spheroid formation. Rapid/instantaneous staining is expected of a population with compromised membrane integrity. Slowly staining, intact, cells should demonstrate a spectral shift which is time-, dye concentration-, and temperature-dependent. The plots show the progression of the two populations and reveal differences in ligand binding in the G_1 compartment of intact cells compared with the remainder of the cell cycle.

4. DNA–Targeting Drug Binding and Fluorescence Quenching

The spectral shifts shown by subpopulations in Fig. 5 at early time points can be analyzed by plotting the ratio of violet and red fluorescence. Such plots are shown in Fig. 6 together with the effects of long-term (VP-16) or short-term (mitoxantrone) drug exposure. In the case of VP-16 the drug does not act to quench Hoechst 33342–DNA fluorescence, and the ratio plots reveal the loss of cells from the intact viable fraction. The presence of a drug that can quench Hoechst 33342:DNA fluorescence results not only in a reduction of signal but also a shift in the ratio distributions to the left. The quench reflects the extent of DNA targeting by mitoxantrone in the two distinguishable populations. This effect is observable at early exposure times, not easily monitored by direct drug fluorescence (Fig. 4). This is a convenient approach to monitoring drug DNA targeting in cells with compromised plasma membrane integrity distinct from

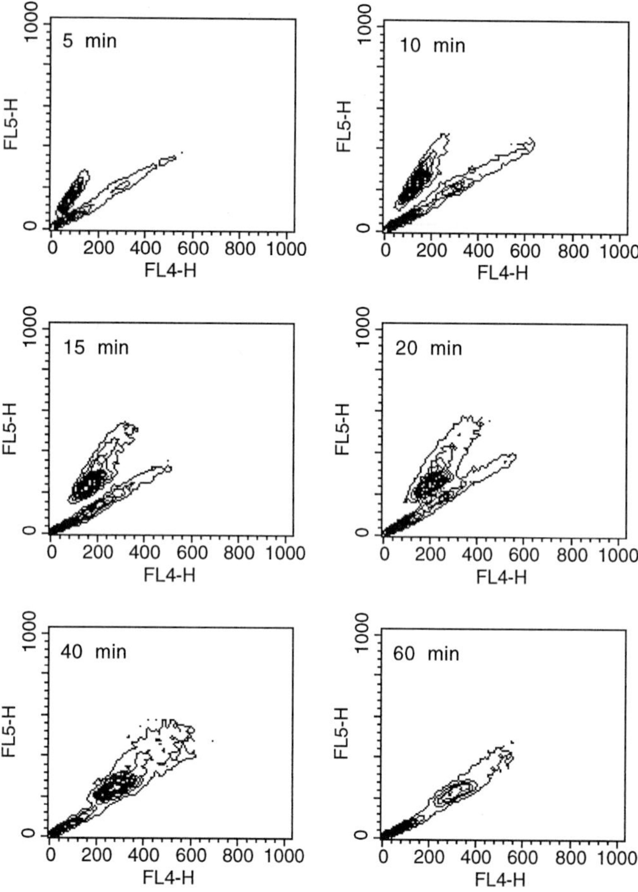

Fig. 5   Flow cytometric contour plots of red (FL4-H) versus violet (FL5-H) fluorescence of Hoechst 33342-stained human small cell lung cancer (NCI-H69) cells as a function of incubation period.

the responses of intact cells. On dual laser systems, it is possible to monitor total cell anthraquinone levels in spectrally shifted populations simultaneously. Since the Hoechst 33342 fluorescence is enhanced greatly on binding to cellular DNA and originates from discrete binding to AT base pairs, the degree of quench can also be used to infer base pair preferences in drug structure–activity studies.

V. Conclusions

There will be an increasing need to devise methods of evaluating drug–target interaction in intact cells. The information gleaned will support drug design,

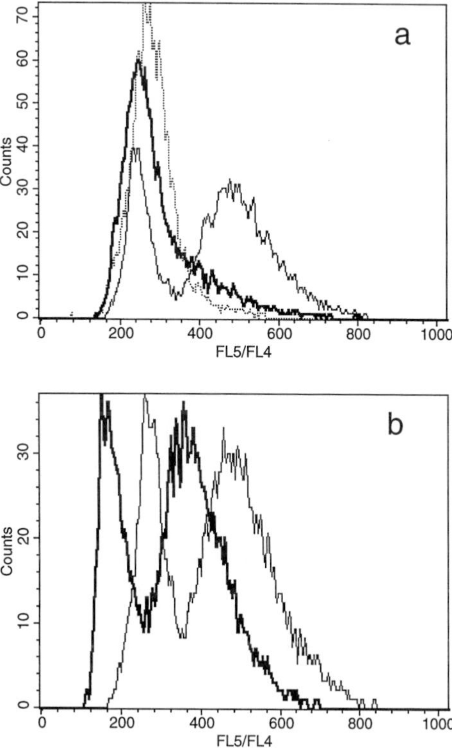

Fig. 6 Flow cytometric frequency distribution plots for the ratio of violet and red fluorescence observed in Hoechst 33342-stained human small cell lung cancer (NCI-H69) cell populations. Plots show the effects of drug treatments on the low ratio population, comprising rapidly staining cells, and on the high ratio population, comprising intact cells: (a) 22-min Hoechst 33342 exposure control (faint line), 22-min Hoechst 33342 exposure for cells pretreated with 2 μM VP-16 for 72 hr (bold line), 60-min Hoechst 33342 exposure (dotted line). (b) 20-min Hoechst 33342 exposure control (faint line), 20-min Hoechst 33342 exposure with a simultaneous exposure to 20 μM mitoxantrone for the last 15-min period (bold line).

while the identification of molecular interactions may identify new targets. In this chapter fluorescence quenching is provided as an example for drug–DNA interactions. Whatever probe or reporter system is used, there is a need to resolve interactions at the biologically important range of <100 Å for the proximity of interacting molecules. An interesting consequence of using changes in fluorescence signals to reveal interactions of molecules is that the resolution is increased beyond that obtainable using direct fluorescence imaging of the partner molecules. In the case of flow cytometry subcellular localization can be inferred, by having a targeted reporter molecule, despite the monitoring of whole cell signals.

An elegant approach is to use fluorescence resonance energy transfer (FRET) as a means of gauging the proximity of two molecules to each other. Importantly, the absorption spectra of the reporter molecules must overlap, and the donor and acceptor transition dipole orientations must be approximately parallel for FRET to be possible. This approach has been successful in a number of experimental systems, and it has been used to monitor interactions of the apoptosis controlling molecules Bax and Bcl-2 through the use of green fluorescent protein (GFP) fusion proteins (Mahajan *et al.*, 1998) in intact cells. In terms of the interaction of anticancer drugs with critical targets in intact cells, the scope for such an approach is currently limited to model systems, despite the potential for drug screening and design. Clearly, fluorimetric methods for analyzing drug–target interactions in viable cells will continue to be applied and refined.

Acknowledgment

The authors acknowledge the support of the U.K. Medical Research Council.

References

Anderson, R. D., and Berger, N. A. (1994). Mutagenicity and carcinogenicity of topoisomerase-interactive agents. *Mutat. Res.* **309,** 109–142.

Bell, D. H. (1988). Characterization of the fluorescence of the antitumour agent, mitoxantrone. *Biochim. Biophys. Acta* **949,** 132–137.

Bowden, G. T., Roberts, R., Alberts, D. S., Peng, Y.-M., and Garcia, D. (1985). Comparative molecular pharmacology in leukaemic L1210 cells of the anthracene anticancer drugs mitoxantrone and bisantrene. *Cancer Res.* **45,** 4915–4920.

Broggini, M., Erba, E., Ponti, M., Geroni, C., Spreafico, F., and D'Incalci, M. (1991). Selective DNA interaction of the novel distamycin derivative FCE 24517. *Cancer Res.* **51,** 199–209.

Chabner, B. A. (1992). Mitotic inhibitors. *Cancer Chemother. Biol. Response Modif.* **13,** 69–74.

Chen, A. Y., and Liu, L. F. (1994). DNA topoisomerases: Essential enzymes and lethal targets. *Annu. Rev. Pharmacol. Toxicol.* **34,** 191–218.

Cummings, J., and Smyth, J. F. (1993). DNA topoisomerase-I and topoisomerase-II as targets for rational design of new anticancer drugs. *Ann. Oncol.* **4,** 533–543.

Eng, W.-K., Faucette, L., Johnson, R. K., and Sternglanz, R. (1988). Evidence that DNA topoisomerase I is necessary for the cytotoxic effects of camptothecin. *Mol. Pharmacol.* **34,** 755–760.

Fox, M. E., and Smith P. J. (1995). Subcellular localisation of the antitumour drug mitoxantrone and the induction of DNA damage in resistant and sensitive human colon carcinoma cells. *Cancer Chemother. Pharmacol.* **35,** 403–410.

Gervasoni, J. E., Jr., Fields, S. Z., Krishna, S., Baker, M. A., Rosado, M., Thurasamy, K., Hindenburg, A. A., and Taub R. N. (1991). Subcellular distribution of daunorubicin in P-glycoprotein-positive and -negative drug-resistant cell lines using laser-assisted confocal microscopy. *Cancer Res.* **51,** 4955–4963.

Hsiang, Y.-H., Hertzberg, R., Hecht, S., and Liu, L. F. (1985). Camptothecin induces protein-linked DNA breaks via mammalian DNA topoisomerase *Int. J. Biol. Chem.* **260,** 14873–14878.

Kluin-Nelemans, H. C., Limpens, J., Meerabux, J., Beverstock, G. C., Jansen, J. H., de Jong, D., and Kluin, P. M. (1991). A new non-Hodgkin's B-cell line (DoHH2) with a chromosomal translocation t(14,18)(q32,q21). *Leukemia* **5,** 221–224.

Kolber, M. A., Broschat, K. O., and Landa-Gonzalez, B. (1990). Cytochalasin B induces cellular DNA fragmentation. *FASEB J.* **4,** 3021–3027.

Krishan, A., and Ganapathi, R. (1980). Laser flow studies on the intracellular fluorescence of anthracyclines. *Cancer Res.* **40,** 3895–3900.

Liu, L. F. (1989). DNA topoisomerase poisons as anti-tumour drugs. *Annu. Rev. Biochem.* **58,** 351–375.

Lown, J. W., Morgan, A. R., Yen, S.-F., Wang, Y. H., and Wilson, W. D. (1985). Characteristics of the binding of the anticancer agents mitoxantrone and ametantrone and related structures to deoxyribonucleic acids. *Biochemistry* **24,** 4028–4035.

Mahajan, N. P., Linder, K., Berry, G., Gordon, G. W., Heim, R., and Herman, B. (1998). Bcl-2 and Bax interactions in mitochondria probed with green fluorescent protein and fluorescence resonance energy transfer. *Nature Biotechnol.* **16,** 547–52.

Mir, L. M., Tounekti, O., and Orlowski, S. (1996). Bleomycin: Revival of an old drug. *General Pharmacol.* **27,** 745–748.

Roberts, R. A., Cress, A. E., and Dalton, W. S. (1989). Persistent intracellular binding of mitoxantrone in a human colon carcinoma cell line. *Biochem. Pharmacol.* **38,** 4283–4290.

Robinson, M. J., and Osheroff, N. (1990). Stabilization of the topoisomerase II DNA cleavage complex by antineoplastic drugs—inhibition of enzyme-mediated DNA religation by 4′-(9-acridinyl-amino)-methane-sulfon-*m*-anisidide. *Biochemistry* **29,** 2511–2515.

Rowe, T. C., Chen, G. L., Hsiang, Y.-H., and Liu, L. F. (1986). DNA damage by antitumor acridines mediated by mammalian DNA topoisomerase II. *Cancer Res.* **46,** 2021–2026.

Schilsky, R. L. (1996). Methotrexate: An effective agent for treating cancer and building careers. The polyglutamate era. *Stem Cells* **14,** 29–32.

Schneider, E., Hsiang, Y. H., and Liu, L. F. (1990). DNA topoisomerases as anticancer drug targets. *Adv. Pharmacol.* **21,** 149–183.

Schwartz, H. S. (1973). A fluorometric assay for daunomycin and adriamycin in animal tissues. *Biochem. Med. J.* **7,** 396–404.

Smith, P. J. (1990). DNA topoisomerase dysfunction: A new goal for antitumor chemotherapy. *BioEssays* **12,** 167–172.

Smith, P. J., and Makinson T. A. (1989). Cellular consequences of overproduction of DNA topoisomerase II in an ataxia-telangiectasia cell line. *Cancer Res.* **49,** 1118–1124.

Smith, P. J., and Soues, S. (1994). Multilevel therapeutic targeting by topoisomerase inhibitors. *Br. J. Cancer* **70,** 47–51.

Smith, P. J., Morgan, S. A., Fox, M. E., and Watson, J. V. (1990). Mitoxantrone-DNA binding and the induction of topoisomerase II associated DNA damage in multi-drug resistant small cell lung cancer cells. *Biochem. Pharmacol.* **40,** 2069–2078.

Smith, P. J., Morgan, S. A., and Watson, J. V. (1991). Detection of multidrug resistance and quantification of responses of human tumour cells to cytotoxic agents using flow cytometric spectral shift analysis of Hoechst 33342–DNA fluorescence. *Cancer Chemother Pharmacol.* **27,** 445–450.

Smith, P. J., Sykes, H. R., Fox, M. E., and Furlong, I. J. (1992). Subcellular distribution of the anticancer drug mitoxantrone in human and drug-resistant murine cells analyzed by flow cytometry and confocal microscopy and its relationship to the induction of DNA damage. *Cancer Res.* **52,** 1–9.

Smith, P. J., Rackstraw, C., and Cotter, F. (1994a). DNA fragmentation as a consequence of cell cycle traverse in doxorubicin and idarubicin treated human lymphoma cells. *Ann. Haematol.* **69,** 7–11.

Smith, P. J., Soues, S., Gottlieb, T., Falk, S. J., Watson, J. V., Osborne, R. J., and Bleehen, N. M. (1994b). Etoposide-induced cell cycle delay and arrest-dependent modulation of DNA topoisomerase II in small cell lung cancer cells. *Br. J. Cancer* **70,** 914–921.

Smith, P. J., Desnoyers, R., Patterson, L. H., and Watson, J. V. (1997). Flow cytometric analysis and confocal imaging of anticancer alkylaminoanthraquinones and their *N*-oxides in intact human cells using 647 nm krypton laser excitation. *Cytometry* **27,** 43–53.

Yanowich, S., and Taub, R. N. (1983). Differences in daunomycin retention in sensitive and resistant P388 leukemic cells as determined by digitized video fluorescence microscopy. *Cancer Res.* **41,** 4167–4171.

Zhang, H., D'Arpa, P., and Liu, L. F. (1990). A model for tumour cell killing by topoisomerase poisons. *Cancer Cells* **2,** 23–27.

CHAPTER 40

Monitoring of Cellular Resistance to Cancer Chemotherapy: Drug Retention and Efflux

Awtar Krishan

Division of Experimental Therapeutics
Radiation Oncology Department
University of Miami School of Medicine
Miami, Florida 33136

I. Introduction and Background
II. Applications
III. Cell Lines, Efflux, Multiple Drug Resistance Drugs, and Blockers
    A. Cell Lines
    B. Multiple Drug Resistance Drugs
    C. Efflux Blockers
IV. Staining Protocols
V. Critical Aspects
VI. Controls, Standards, and Instruments
VII. Results and Conclusions
    References

I. Introduction and Background

Several earlier studies suggest that cellular resistance to cancer chemotherapeutic agents, such as alkaloids and antibiotics (multiple drug resistance, MDR), is related to their rapid efflux from the intracellular environment (see reviews by Ling, 1992; Gottesman, 1993; Biedler, 1994; Sikic *et al.*, 1997). On the basis of these pioneering studies, the phenomenon of MDR was recognized as a phenotypic characteristic that makes cells (inherently or after exposure to drugs) resistant to a variety of unrelated natural products. MDR can be seen in cells exposed to alkaloids (e.g., vinblastine, colchicine, taxol, podophyllotoxins) and antibiotics (e.g., adriamycin, doxorubicin).

One of the most important discoveries for understanding MDR was the identification of a 170-kilodalton P-glycoprotein and the MDR_1 gene. Three of the well-studied drug resistance related transport proteins are the 170-kilodalton P-glycoprotein (Juliano and Ling, 1976), the multidrug resistance associated protein MRP (Cole *et al.*, 1992), and the lung resistance related protein LRP (Izquierdo *et al.*, 1996). Other studies have identified several other drug transporter genes (and their protein products) that seem to accompany *de novo* or intrinsic drug resistance to a variety of drugs and carcinogens (Ross *et al.*, 1997). The reduced cellular drug retention or compartmentalization in drug-resistant cells could be accomplished by direct drug efflux (e.g., P-glycoprotein), by efflux of the drug–glutathione conjugate (e.g., MRP Jedlitschky *et al.*, 1996), or by controlling the nucleocytoplasmic transport through the nuclear pore vault proteins (e.g., LRP). Besides direct drug efflux, which results in reduced cellular retention, another transport related mechanism of drug resistance is seen in cells made resistant to drugs such as methotrexate or cisplatin that are not considered to be the substrates for the P-glycoprotein pump. Gifford *et al.* (1998) have shown that methotrexate resistant cells show the MDR phenotype, indicating the presence of a P-glycoprotein-mediated methotrexate transport mechanism. Their data support the hypothesis that in cells with defective carrier protein, methotrexate can be a substrate for the P-glycoprotein. Shen *et al.* (1998) reported that in two cisplatin resistant cell lines, which were also resistant to methotrexate, there was reduced expression of the folate binding methotrexate transport protein accompanied by the selective loss of several other specific binding proteins.

Thus, on the basis of our current knowledge it is evident that besides the three well-known and well-studied drug resistance associated proteins, several other proteins that modulate cellular drug uptake, intracellular distribution, and drug efflux may be involved in drug resistance.

Methods for identification of MDR include molecular procedures for gene expression (mRNA) *in situ* polymerase chain reaction (PCR) methods (Thomas *et al.*, 1994), use of monoclonal antibodies by immunocytochemistry (Kartner *et al.*, 1985; Chan *et al.*, 1990), and flow cytometry (FCM) (Krishan *et al.*, 1991; Feller *et al.*, 1995). However, some of these methods cannot determine the heterogeneity of MDR marker expression or determine if the efflux protein is functional and active in reducing cellular drug retention. In several cases, phosphorylation of the P-glycoprotein (Wielinga *et al.*, 1997; Ramachandran *et al.*, 1998), cell cycle stage, and proliferation (Krishan and Bourguignon, 1984; Ramachandran *et al.*, 1995) can modulate the functionality of the efflux pump. This can result in discordance between the presence of the efflux proteins and the functional activity of the pump (Xie *et al.*, 1995). Similarly, mutations could alter the specificity or activity of the efflux proteins, resulting in altered drug efflux (Gros *et al.*, 1991).

Analytical methods such as high-pressure liquid chromatography and spectro-fluorometry can be used for monitoring of cellular drug retention and efflux, but of necessity these methods are slow, need large samples, and cannot measure drug retention in single cells. The flow cytometric functional assays that basically monitor the retention of a fluorescent drug and compare the total cellular drug content of the resistant versus sensitive cells, or of resistant cells with or without coincubation with an efflux blocker, overcome some of these obstacles. Laser FCM offers a unique tool for monitoring fluorescent antitumor drug retention and its modulation in tumor cells (Krishan and Ganapathi, 1980; Durand and Olive, 1981). Besides its rapidity, the laser FCM method can identify heterogeneity of drug retention (Krishan, 1995; Krishan et al., 1987) as well as allow for sorting of subpopulations for further biochemical or morphological character-ization.

Anthracyclines such as doxorubicin and daunomycin are important cancer chemotherapeutic agents. Most of the anthracyclines are fluorescent and can be excited with the 488-nm laser line from an argon ion laser (Krishan and Gana-pathi, 1980). Thus, this method can be used for rapid monitoring of anthracycline transport and retention in drug-resistant and -sensitive tumor cells. Cellular resistance to some of the clinically important anthracyclines has been suggested to be due to rapid drug efflux. Thus, drugs that block anthracycline efflux and thereby enhance retention can reduce cellular resistance to anthracyclines (Tsuruo et al., 1981; Ganapathi et al., 1981; Krishan et al., 1985a, Slater et al., 1986). Several unrelated drugs, such as calcium channel blockers (e.g., verapamil) and phenothiazines (e.g., trifluoperazine), will inhibit anthracycline efflux from the resistant cells and thereby render them drug sensitive. Similarly, reduced cellular retention of the vital DNA dye Hoechst 33342 (Krishan, 1987) and the calcium indicator dyes (Hollo et al., 1994) in refractory cells may also be related to rapid efflux. Drugs such as phenothiazines or calcium channel blockers can block Hoechst 33342 efflux and make it possible to generate DNA distribution histograms from living cells, which were heretofore difficult to stain with this vital dye (Krishan, 1987).

The use of efflux blockers has been advocated for chemotherapy of human malignancies (Dalton et al., 1995; Sikic et al., 1997). It would be useful if before the administration of anthracyclines and efflux blocking agents tumor cells could be screened in vitro for their anthracycline retention and efflux characteristics. As shown in the following sections, laser FCM can be used to monitor anthracycline fluorescence in tumor cells and to study the effect of drug efflux blocking agents. We have used this method to monitor anthracycline retention and its modulation by phenothiazines and amphotericin B in P388 and doxorubicin-resistant cells (Krishan and Bourguignon, 1984; Krishan et al., 1985a,b). Some of our data show that the effect of efflux blockers on doxorubicin retention is cell cycle proliferation related (Krishan et al., 1985a), and often one can identify subpopulations based on their differential response to efflux blockers (Krishan et al., 1987). From these studies it follows that modulation of drug transport and thereby cellular resistance by different efflux blockers may not be uniform in a population but may be selective and confined to only certain types of cells and subpopulations.

II. Applications

The rapid laser flow cytometric method can be used for monitoring of the following:

1. Cellular transport of fluorescent drugs (e.g., anthracyclines, rhodamine, Indo-1AM, Calcein, Hoechst 33342).
2. Effect of drugs that either increase drug influx (e.g., amphotericin B) or enhance retention by blocking efflux (verapamil, phenothiazines, cyclosporins, tamoxifen).
3. Effect of efflux blocker combinations on drug retention.
4. Tumor cell heterogeneity in drug retention and response to efflux blockers.
5. Selection of ideal efflux blockers and protocols for possible clinical use.
6. Rapid identification and sorting of cells that have efflux as a major mechanism of drug resistance.

III. Cell Lines, Efflux, Multiple Drug Resistance Drugs, and Blockers

A. Cell Lines

Before one can analyze a clinical specimen for the presence of drug efflux, it is important to use an indicator cell line to set up the instrument, monitor cellular drug retention with and without efflux blockers, and in general to serve as a control and reference for comparative studies. Several well-characterized murine and human cell lines with MDR and drug efflux are available. We have used the murine leukemic line P388 and its doxorubicin-resistant subline P388/ADR or cloned drug-resistant P388/R84 cell line for most of our early work. These cell lines have short doubling times (16–18 hr) and are easy to maintain as suspension cultures and *in vivo* as ascites or subcutaneous implants. The P388/R84 cell line has multifactorial drug resistance involving drug efflux, a detoxification mechanism, and a DNA damage/repair mechanism (Maniar *et al.*, 1988; Deffie *et al.*, 1988; Nair *et al.*, 1990). P388/R84 has stable drug resistance and does not need rechallenge with doxorubicin to maintain the MDR phenotype. The human T cell line CCRF-CEM and its vinblastine-resistant cell line CCRF-CEM$_{VLB100}$ is another excellent pair for the study of drug retention and efflux (Beck, 1983). These CEM cell lines are tetraploid [although the original CCRF-CEM cell line was diploid (Krishan *et al.*, 1969)] and easy to grow in cell suspensions. A minor disadvantage is that the CEM$_{VLB100}$ needs to be grown in the presence of vinblastine to restore the efflux mechanisms. The paired human colon carcinoma cell lines SW620 and SW620/AD300 are excellent models, as efflux is the major and possibly the only mechanism involved in their resistance to doxorubicin (Leibovitz *et al.*, 1976). The SW620 cell lines grow as monolayers with a 24 hr doubling time and can be readily grown in athymic mice as xenografts.

B. Multiple Drug Resistance Drugs

Fluorochromes, which can be excited by either a high-pressure mercury arc source or by a laser in a FCM, are ideal for the study of drug retention and efflux. Fluorochromes commonly used for the study of retention and efflux include doxorubicin (Adriamycin, NSC-123127, Adria Labs, Columbus, OH), daunomycin (NSC-821151, Calbiochem, San Diego, CA), Hoechst 33342 (Calbiochem), rhodamine 123 (Calbiochem), Indo-1AM, Sy-38, Sy-3150 (Frey *et al.*, 1995), and Bodipy-verapamil (Haugland, 1996). Doxorubicin and daunomycin are important cancer chemotherapeutic antibiotics. Cellular retention of doxorubicin is slow (2–3 hr to reach peak cellular content), whereas daunomycin is more lipophilic and its cellular transport rapid (15–30 min). Both of these antibiotics quench their fluorescence on binding to DNA and are excitable by the 488-nm argon ion laser line. Rhodamine 123 is brighter than doxorubicin or daunomycin, and most of the cellular fluorescence is localized on mitochondria. Hoechst 33342 and Hoechst 33258 are DNA binding fluorochromes, which need ultraviolet light (UV) excitation for analysis. The reader is referred to the *Molecular Probes' Handbook of Fluorescent Probes and Research Chemicals* (Haugland, 1996) for discussion of other fluorochromes that can be used for the monitoring of drug retention and efflux.

C. Efflux Blockers

Some of the well-known drug efflux blockers include calcium channel active drugs (e.g., verapamil), phenothiazines (e.g., prochlorperazine), quinine, cyclosporins, dipyridamole, tamoxifen, as well as monoclonal antibodies against the P-glycoprotein (Tsuruo *et al.*, 1981; Ganapathi *et al.*, 1981; Krishan and Bourguignon, 1984; Slater *et al.*, 1986; Shalinsky *et al.*, 1990; Trump *et al.*, 1992; Leonessa *et al.*, 1994; Dalton *et al.*, 1995; Sikic *et al.*, 1997). Other studies suggest that efflux blocker combination may have synergistic effects on drug retention and thereby reduce toxicity that may be caused by the use of higher concentrations of the individual efflux blockers (Stein, 1997).

IV. Staining Protocols

For evaluating drug retention and efflux in a cell line or in tumor cells disassociated from solid tumors or isolated from ascites or pleural fluid, a typical protocol requires the following materials and procedures:

1. Suspension cultures of murine leukemic P388 Adriamycin-resistant cell line P388/R84 (Nair *et al.*, 1990) or vinblastine resistant CEM_{VLB100} (Beck, 1983).
2. Single cell suspensions prepared from bone marrow aspirates, ascites, or pleural fluid or after enzymatic digestion of solid tumor samples.
3. Fluorochromes: daunorubicin or rhodamine 123.
4. Efflux blockers: prochlorperazine (Compazine, Smith, Kline and Beecham Labs, Carolina, Puerto Rico), verapamil (Calan, Searle Pharmaceutical, Chicago, IL), or dipyridamole (Persantin).

Suspension cultures of the P388 cell lines grown in RPMI 1640 medium supplemented with 10% heat-inactivated fetal bovine serum (FBS), penicillin, and streptomycin are used for calibration and as controls. In soft agar assays, the ID_{50} (drug dose needed to cause 50% growth inhibition) for the P388 and P388/R84 cells is 0.0875 and 8.4 μM of doxorubicin, respectively.

Human tumor cells are recovered by centrifugation of pleural fluid (lung cancer), bone marrow or peripheral blood (leukemia), ascites (ovarian, breast), or other solid tumor material after enzymatic digestion. Filtration through nylon mesh (40 μm) is used to remove clumps. The supernatant fluid is aspirated (after centrifugation), and the cell pellets are resuspended and washed in Ca^{2+}- and Mg^{2+}-free phosphate-buffered saline (PBS). After centrifugation, the pelleted cells are resuspended in fresh tissue culture medium supplemented with 10% heat-inactivated FBS. To remove erythrocytes from bone marrow aspirates and peripheral blood samples, specimens are diluted with Ca^{2+}- and Mg^{2+}-free PBS and centrifuged over a 70% performed Percoll gradient (Pharmacia, Piscataway, NJ). Mononuclear cells are recovered and washed before incubation with the drugs at 37°C. Cytospin preparations are used for analysis of morphological heterogeneity and for differential counting purposes.

Stock solutions of drugs are prepared in Ca^{2+}- and Mg^{2+}-free Hanks' balanced salt solution (HBSS). Fresh dilutions of the drugs are prepared in normal saline before each experiment.

For generation of two-parameter dot plots (scattergrams) based on cellular drug fluorescence and length of incubation (time), cell suspensions are directly mixed with the drug-containing medium in the sampling cuvette of the flow cytometer maintained at 37°C. Final drug concentrations used are 1–3 μM doxorubicin or daunomycin, or 0.1 μM rhodamine 123. For efflux blocking experiments, cells are incubated with or without the addition of prochlorperazine (20 μM) or verapamil (10 μM). Samples can be run after 30–60 min of incubation.

V. Critical Aspects

Several parameters related to specimen preparation and instrumentation can cause artifacts and lead to generation of erroneous data. Special consideration should be given to the following factors:

1. Several anthracyclines quench their fluorescence on binding to DNA and other target molecules. It is important to keep this in mind and use other nonquenching fluorochromes for critical experiments.

2. pH can have a major effect on drug fluorescence (Alabaster *et al.*, 1989) either by shifting the excitation maxima or by altering drug transport and retention.

3. Some efflux blockers may precipitate or bind to glass or rubber stoppers. Proper precautions should be taken to avoid these artifacts.

4. High fluorochrome concentrations can overcome efflux, and high levels of some efflux blockers can damage cell membranes and thus increase cellular drug retention.

5. Two parameter histograms showing light scatter and drug fluorescence are preferable than single parameter histograms of cellular drug fluorescence. The multiparametric scattergrams allow for the identification of dead cells and other subpopulations for electronic gating.

6. Dead cells (cells with damaged cell membranes) will rapidly stain with the fluorochromes and give erroneous results. Forward and/or right angle light scatter should be used to isolate and identify these cells in multiparameter dot plots.

7. We have advocated the use of propidium iodide as a stain to distinguish the dead cells from live cells with high drug fluorescence (Krishan *et al.*, 1997).

8. Coated dichroic filters used in the flow cytometer can often with age develop pinholes, which may result in light leaks. Similarly, certain commercially available filters are notorious for generating autofluorescence when excited with high laser excitation.

VI. Controls, Standards, and Instruments

We regularly use Adriamycin-resistant P388/R84 cells coincubated with similar (to that of the test sample) drug concentrations under identical conditions as controls for each experiment. Once the photomultiplier high voltage, laser power, and amplifications are optimized and set, all samples are analyzed without altering any of these parameters. In general, drug-sensitive P388 cells (ID_{50} 0.0875 μM), after incubation with 1.75–3.5 μM of doxorubicin for 30–60 min, are 20 times less fluorescent than human diploid nuclei stained with the propidium iodide (PI)–hypotonic citrate method (Krishan, 1975). One report described the use of beads as calibration standards for drug efflux studies (Pallis and Russell, 1998). The use of drug-resistant tissue culture cells for testing the *ex vivo* efflux blocking activity of plasma from patients on drug efflux blocking protocols has been described by Ayesh *et al.* (1996) and Lehnert *et al.* (1996).

For FCM analysis, cell suspensions incubated *in vitro* with anthracyclines or rhodamine 123, with or without the addition of phenothiazines or verapamil, are analyzed for their total cellular fluorescence. Fluorescence emission (>530 nm) and forward and 90° light scatter are collected. A minimum of 10,000 cells are analyzed for each sample and used for generation of gated histograms or dot plots. For generation of two-parameter dot plots based on cellular drug fluorescence and length of incubation (time), cell suspensions are directly mixed with the drug-containing medium in the sampling cuvette of the flow cytometer,

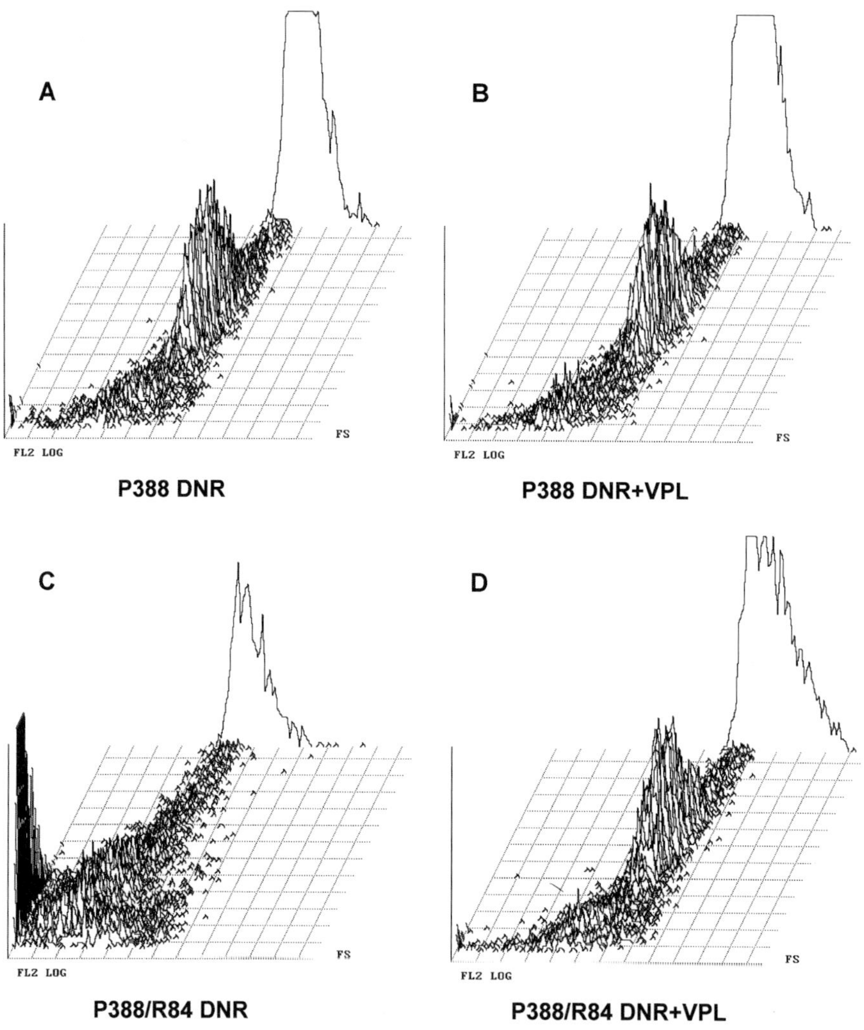

Fig. 1 Uptake and retention of daunorubicin in P388 parental drug-sensitive and P388/R84 drug-resistant cells. In P388 cells (A, B) and in P388/R84 cells incubated with the efflux blocker verapamil (D) a prominent population with high drug retention is seen. In P388/R84 cells incubated with daunorubicin alone, retention is low due to drug efflux (C).

maintained at 37°C. Time as a parameter is available in software from the Coulter and the Becton Dickinson software packages. For efflux blocking experiments, P388/R84 cells are incubated with doxorubicin (3.5 μM), for 120 min at 37°C, with or without the addition of prochlorperazine (25 μM) or verapamil (10 μM).

A Coulter XL or Elite flow cytometer (Coulter Beckman, Miami, FL) or a Becton Dickinson FACScan/FACStar analyzer/cell sorter (San Jose, CA) equipped with an argon laser is used to measure forward angle and 90° light scatter and fluorescence. Fluorescence versus time is used as a parameter in drug uptake studies.

VII. Results and Conclusions

In this section, cellular retention of fluorochromes with or without the presence of efflux blockers is illustrated in a variety of cell lines and clinical specimens. Multiparameter histograms in Fig. 1A,C are of Adriamycin-sensitive and -resistant P388/R84 cells, respectively, incubated with 2 μM of daunorubicin or daunorubicin and 10 μM of verapamil (Fig. 1B,D) for 30 min. Forward angle light scatter (ordinate) and drug fluorescence (abscissa) in log scale are shown. P388/R84 cells have four- to sixfold less drug content (owing to rapid efflux)

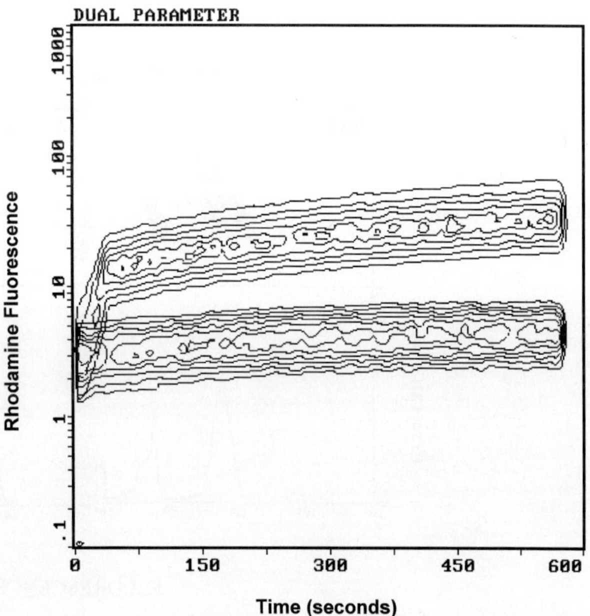

Fig. 2   Appearance of cellular rhodamine 123 fluorescence in P388/R84 cells incubated with drug alone (lower contour plot) or in the presence of the efflux blocker (verapamil).

than the parental drug-sensitive cells. In the presence of an efflux blocker (e.g., verapamil at 10 μM), drug retention is enhanced (Fig. 1D) and results in cellular drug retention profiles similar to that of the drug sensitive (noneffluxing) cells (Fig. 1A,B).

Figure 2 shows contour plots (rhodamine fluorescence versus time) of P388/R84 cells incubated with rhodamine 123 (0.1 μg/ml) and 10 μM of verapamil for 10 min. In cells incubated with the fluorochrome and the efflux blocker verapamil (top contour), the appearance of drug fluorescence is rapid and after

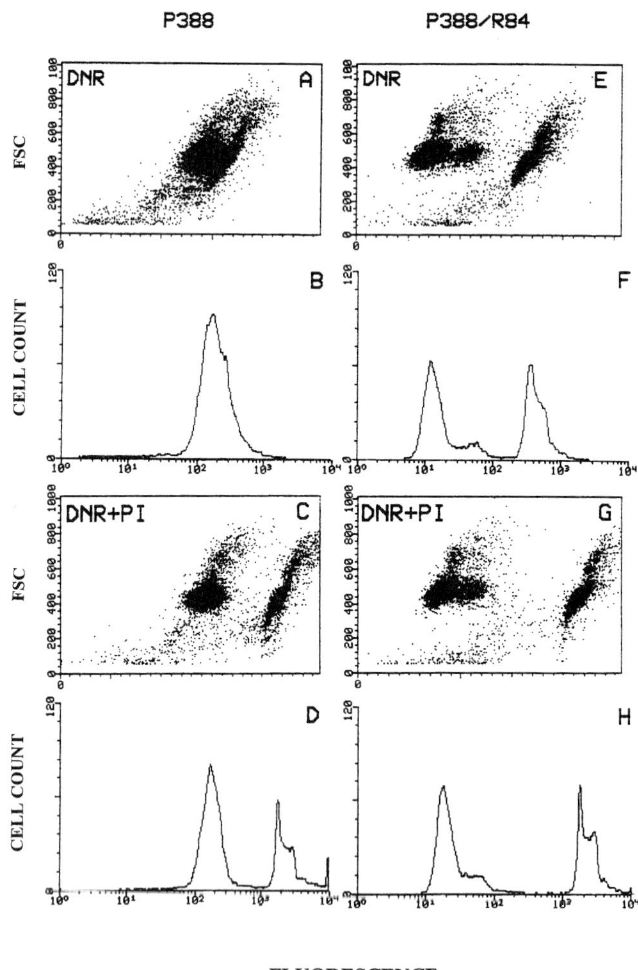

Fig. 3   Use of isotonic propidium iodide for identification of cells with permeable cell membranes or dead cells (which are brightly stained by this dye) that can be differentiated from live cells with high fluorochrome retention. From Krishan *et al.* (1997).

10 min nearly a log higher than that of cells incubated without the efflux blocker (bottom contour).

Figure 3 illustrates the use of propidium iodide (in isotonic saline) for discriminating between damaged cells with permeable membranes and cells, that have high drug retention due to the absence of an efflux pump. Figure 3A,E shows P388 and P388/R84 cells incubated with daunorubicin. Several subpopulations based on forward angle scatter and cellular fluorescence can be seen. In the histogram in Fig. 3E, the presence of two distinct populations is seen in P388/R84 cells. In P388 cells, the addition of isotonic PI results in separation of two distinct populations based on drug fluorescence (Fig. 3C,D). In the P388/R84 cells, the subpopulation with the higher drug fluorescence seen in Fig. 3E also increases fluorescence in the presence of PI by nearly a log and thus can be recognized as consisting of the dead or membrane permeable cells.

In Fig. 4, we have compared cellular retention of doxorubicin (AdR) and daunorubicin (DnR) in human leukemic blasts from the peripheral blood of a patient and incubated with the fluorochromes alone or in the presence of efflux blockers (chlorpromazine, CpZ, Fig. 4C,D) or verapamil (VpL, Fig. 4E,F). A

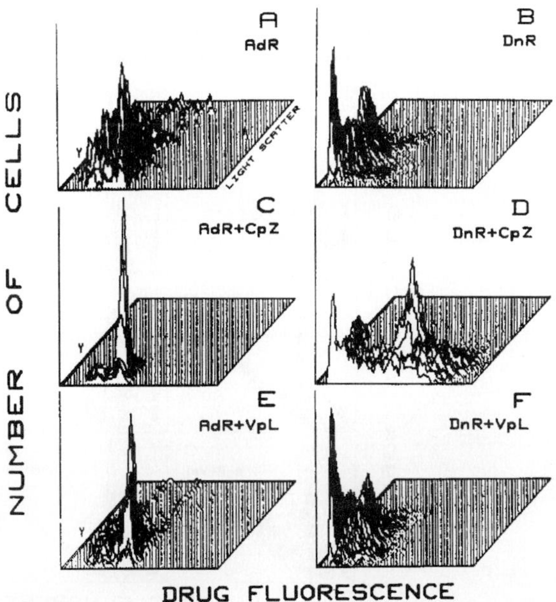

Fig. 4 Cellular retention of doxorubicin (AdR) and daunorubicin (DnR) in human leukemic blasts with or without the presence of efflux blockers chlorpromazine (CpZ) or verapamil (VpL). Note both CpZ and VpL enhanced cellular retention by blocking AdR efflux (C, E), whereas in cells incubated with DnR and VpL (F) no enhancement was seen. From Patterns of anthracycline retention modulation in human tumor cells. A. Krishan, K. S. Sridhar, E. Davila, C. Vogel, and W. Sternheim; *Cytometry*. Copyright © 1987 John Wiley & Sons, Inc. Reprinted by permission of Wiley-Liss, Inc., a subsidiary of John Wiley & Sons, Inc.

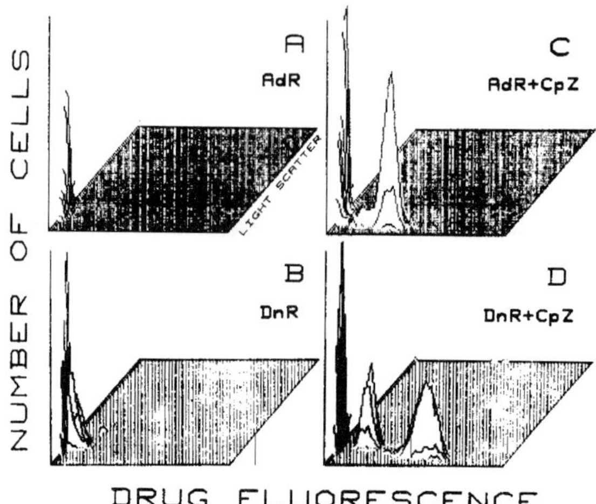

Fig. 5 In leukemia cells from a patient, Adriamycin and daunorubicin retention was low (A, B) and in the presence of CpZ resulted in the emergence of a population with high retention (C). In cells incubated with DnR and CpZ (D), three subpopulations with low, medium, and high retention can be identified. From Patterns of anthracycline retention modulation in human tumor cells. A. Krishan, K. S. Sridhar, E. Davila, C. Vogel, and W. Sternheim; *Cytometry.* Copyright© 1987 John Wiley & Sons, Inc. Reprinted by permission of Wiley-Liss, Inc., a subsidiary of John Wiley & Sons, Inc.

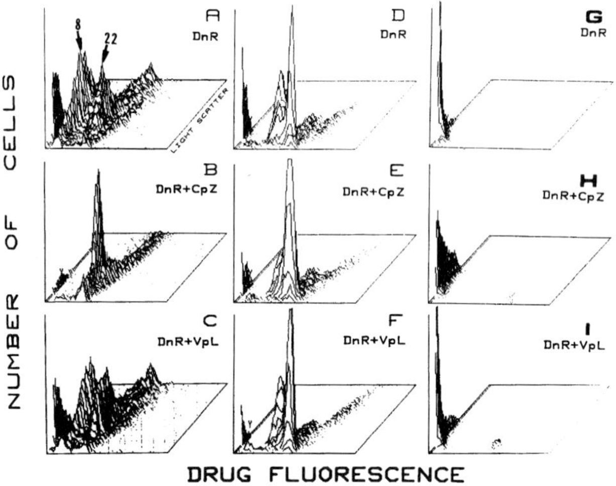

Fig. 6 Daunorubicin retention in tumor cells from the pleural fluid of a lung cancer patient. Note CpZ (B) but not VpL (C) enhanced drug retention, resulting in the emergence of a single population with high drug fluorescence. In samples retrieved after 1 month of therapy, drug retention was low (G), and efflux blockers did not increase drug retention (H, I). From Patterns of anthracycline retention modulation in human tumor cells. A. Krishan, K. S. Sridhar, E. Davila, C. Vogel, and W. Sternheim; *Cytometry.* Copyright© 1987 John Wiley & Sons, Inc. Reprinted by permission of Wiley-Liss, Inc., a subsidiary of John Wiley & Sons, Inc.

comparison of Fig. 4A with Fig. 4C and Fig. 4E shows that both of these efflux blockers had a similar effect in enhancing the drug retention of the heterogeneous population seen in Fig. 4A. In contrast, in cells incubated with daunomycin (Fig. 4B), chlorpromazine, but not verapamil enhanced drug retention (Fig. 4 D,F).

In Fig. 5 leukemia cells from the peripheral blood of a patient show that although in cells incubated with Adriamycin a single population with high retention was seen (Fig. 5C), in cells incubated with daunorubicin and chlorpromazine two distinct populations are recognizable (Fig. 5D).

In Figs. 6 and 7 daunorubicin retention in tumor cells from two lung cancer patients (pleural effusion) incubated with or without the efflux blockers chlorpromazine or verapamil are shown. Heterogeneity of drug retention was seen in cells incubated with daunorubicin alone (Fig. 6A). In Fig. 6B,C chlorpromazine but not verapamil increased drug retention, with the emergence of a single fluorescent population. In tumor cells from the second patient (Fig. 7), however, efflux blockers did not enhance drug retention (Fig. 7B,C). In both of these patients treatment with doxorubicin protocols led to the disappearance of cells with high drug retention and emergence of cells with low retention (Figs. 6G–I and 7D–I), which could not be increased by incubation with the efflux blockers (Figs. 6H,I and Fig. 7E,F).

Figure 8 shows fluorescence of single cells isolated by enzymatic digestion from a pharyngeal carcinoma and lymph node of a head and neck tumor patient

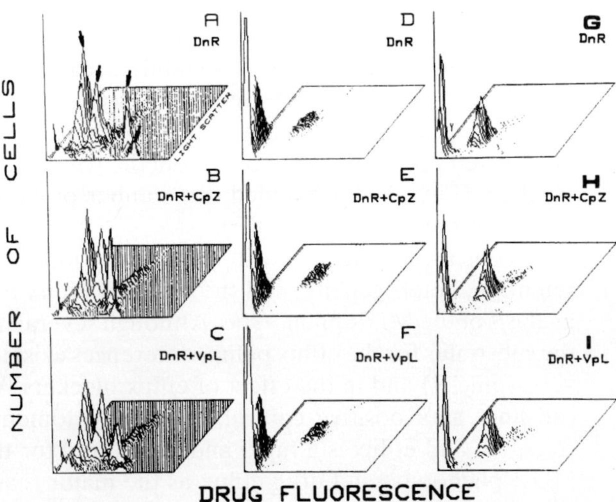

Fig. 7 Daunorubicin retention in tumor cells from a lung cancer patient. Efflux blockers did not enhance drug retention in any of the samples analyzed. From Patterns of anthracycline retention modulation in human tumor cells. A. Krishan, K. S. Sridhar, E. Davila, C. Vogel, and W. Sternheim; *Cytometry*. Copyright © 1987 John Wiley & Sons, Inc. Reprinted by permission of Wiley-Liss, Inc., a subsidiary of John Wiley & Sons, Inc.

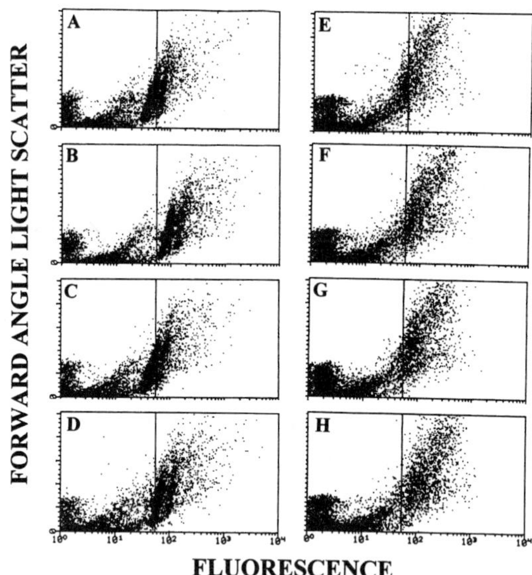

Fig. 8 Forward angle and fluorescence scattergrams of cells from a head and neck tumor and lymph node of a patient incubated with Sy 38 (A, E) with or without efflux blockers prochlorperazine (B, F), dipyridamole (C, G) or verapamil (D, H). Note an increase in the number of gated cells in scattergrams B, D, and F–H.

incubated with fluorochromes (SY 38) and efflux blockers. In Fig. 8B (prochlorperazine) and Fig. 8D (verapamil), the percentage of positive cells (beyond the vertical gate) was increased from 14 (in Fig. 8A) to 39 and 31 (Fig. 8B,D), respectively, by the efflux blockers. Dipyridamole (Fig. 8C) was not so effective (17% gated positive). In the cells from the lymph node, incubation with the efflux blockers (Fig. 8F–H) doubled the number of gated positive cells as compared to that of the control (Fig. 8E).

In conclusion, the flow cytometric functional assay for monitoring of drug retention, heterogeneity, and the effect of efflux blockers is a rapid method for analysis of the MDR phenotype. Although several fluorochromes are available as the substrates for the efflux pump, differences exist in their binding (mitochondria versus nuclei) and in the effect of efflux blockers. With the proper use of MDR cell lines as a positive control, the flow cytometric functional assay for drug retention and efflux is a rapid and useful tool for the detection of cells with the MDR phenotype and drug efflux as the major reason for their resistance.

References

Alabaster, O., Woods, T., Ortiz-Sanchez, V., and Jahangeer, S. (1989). Influence of microenvironmental pH on adriamycin resistance. *Cancer Res.* **49,** 5638–5643.

Ayesh, S., Lyubimov, E., Algour, N., and Stein, W. D. (1996). Reversal of P-glycoprotein is greatly reduced by the presence of plasma but can be monitored by an *ex vivo* clinical assay. *Anti-Cancer Drugs* **7**, 678–686.

Beck, W. T. (1983). Vinca alkaloid-resistant phenotype in cultured human leukemic lymphoblasts. *Cancer Treat. Rep.* **10**, 875–882.

Biedler, J. L. (1994). Drug resistance: Genotype versus phenotype. Thirty-second G. H. A. Clowes Memorial Award Lecture. *Cancer Res.* **54**, 666–678.

Chan, H. S., Thorner, P. S., Haddad, G., and Ling, V. (1990). Immunohistochemical detection of P glycoprotein: Prognostic correlation in soft tissue sarcoma of childhood. *J. Clin. Oncol.* **8**, 689–704.

Cole, S. P. C., Bhardwaj, G., Gerlach, J. H., Mackie, J. E., Grant, C. E., Almquist, K. C., Stewart, A. J., Kurz, E. U., Duncan, A. M. V., and Deeley, R. G. (1992). Overexpression of a transporter gene in a multidrug-resistant human lung cancer cell line. *Science* **258**, 1650–1654.

Dalton, W. S., Crowley, J. J., Salmon, S. S., Grogan, T. M., Laufman, L. R., Weiss, G. R., and Bonnet, J. D. (1995). A phase III randomized study of oral verapamil as a chemosensitizer to reverse drug resistance in patients with refractory myeloma: A Southwest Oncology Group study. *Cancer* **75**, 815–820.

Deffie, A. M., Alam, T., Seneviratne, C., Beenken, S. W., Batra, J. K., Shea, T. C., Henner, W. D., and Goldenberg, G. J. (1988). Multifactorial resistance to adriamycin: Relationship of DNA repair, glutathione transferase activity, drug efflux and P-glycoprotein in cloned cell lines of adriamycin-sensitive and -resistant P388 leukemia. *Cancer Res.* **48**, 3595–3602.

Durand, R. and Olive, P. (1981). Flow cytometric studies of intracellular adriamycin in single cells *in vitro. Cancer Res.* **41**, 3489–3494.

Feller, N., Kuiper, C. M., Lankelma, J., Ruhdal, J. K., Scheper, R. J., Pinedo, H. M., and Broxterman, H. J. (1995). Functional detection of MDR1/p170 and MRP/p190-mediated multidrug resistance in tumour cells by flow cytometry. *Br. J. Cancer* **72**, 543–549.

Frey, T., Yue, S., and Haugland, R. P. (1995). Dyes providing increased sensitivity in flow-cytometric dye-efflux assays for multidrug resistance. *Cytometry* **20**, 218–227.

Ganapathi, R., Grabowski, D., Rouse, W., and Riegler, F. (1981). Differential effect of the calmodulin inhibitor trifluoperazine on cellular accumulation, retention, and cytotoxicity of anthracyclines in doxorubicin (adriamycin)-resistant P388 mouse leukemia cells. *Cancer Res.* **44**, 5056–5061.

Gifford, A. J., Kavallaris, M., Madafiglio, J., Matherly, L. H., Stewart, B. W., Haber, M., and Norris, M. D. (1998). P-glycoprotein-mediated methotrexate resistance in CCRF-CEM sublines deficient in methotrexate accumulation due to a point mutation in the reduced folate carrier gene. *Int. J. Cancer* **78**, 176–181.

Gottesman, M. M. (1993). How cancer cells evade chemotherapy: Sixteenth Richard and Hinda Rosenthal Foundation Award Lecture. *Cancer Res.* **53**, 747–754.

Gros, P., Dhir, R., Croop, J., and Talbot, F. (1991). A single amino acid substitution strongly modulates the activity and substrate specificity of the mouse mdr1 and mdr3 drug efflux pumps. *Proc. Natl. Acad. Sci. U.S.A.* **88**, 7289–7293.

Haugland, R. P. (1996). "Molecular Probes' Handbook of Fluorescent Probes and Research Chemicals," 6th Ed. Molecular Probes, Eugene, Oregon.

Hollo, Z., Homola, L., Davis, C. W., and Sakardi, B. (1994). Calcein accumulation as a fluorometric functional assay of the multidrug transporter. *Biochim. Biophys. Acta* **1191**, 384–388.

Izquierdo, M. A., Scheffer, G. L., Flens, M. J., Giaccone, G., Broxterman, H. J., Meijer, C. J. L. M., van der Valk, P., and Scheper, R. J. (1996). Broad distribution of the multidrug resistance-related vault lung protein in normal human tissues and tumors. *Am. J. Pathol.* **148**, 877–887.

Jedlitschky, G., Leier, I., Buchholz, U., Barnouin, K., Kurz, G., and Keppler, D. (1996). Transport of glutathione, glucuronate, and sulfate conjugates by the MRP gene-encoded conjugate export pump. *Cancer Res.* **56**, 988–994.

Juliano, R. L., and Ling, V. (1976). A surface glycoprotein modulating drug permeability in Chinese hamster ovary cell mutants. *Biochim. Biophys. Acta* **455**, 152–162.

Kartner, N., Evernden-Porelle, D., Bradley, G., and Ling, V. (1985). Detection of P-glycoprotein in multidrug resistant cell lines by monoclonal antibodies. *Nature* **316**, 820–823.

Krishan, A. (1975). Rapid flow cytofluorometric analysis of mammalian cell cycle by propidium iodide staining. *J. Cell Biol.* **66,** 188–193.

Krishan, A. (1987). Effect of drug efflux blockers on vital staining of cellular DNA with Hoechst 33342. *Cytometry* **8,** 642–645.

Krishan, A. (1995). Heterogeneity of anthracycline retention and response to efflux blockers in human tumors. *Cytometry* **21,** 72–75.

Krishan, A., and Bourguignon, L. Y. W. (1984). Cell cycle related phenothiazine effects on adriamycin transport. *Cell Biol. Int. Rep.* **8,** 449–457.

Krishan, A., and Ganapathi, R. (1980). Laser flow cytometric studies on intracellular fluorescence of anthracyclines. *Cancer Res.* **40,** 3895–3900.

Krishan, A., Raychaudhuri, R., and Flowers, A. (1969). Karyotype studies on human leukemic lymphoblasts in vitro and as serial transplants in neonatal Syrian hamsters. *J. Natl. Cancer Inst.* **43,** 1203–1214, 1969.

Krishan, A., Sauerteig, A., and Gordon, K. (1985a). Effect of amphotericin B on adriamycin transport in P388 cells. *Cancer Res.* **45,** 4097–4102.

Krishan, A., Sauerteig, A., and Wellham, L. (1985b). Flow cytometric studies on modulation of cellular adriamycin retention by phenothiazines. *Cancer Res.* **45,** 1046–1051.

Krishan, A., Sridhar, K. S., Davila, E., Vogel, C., and Sternheim, W. (1987). Patterns of anthracycline retention modulation in human tumor cells. *Cytometry* **8,** 306–314.

Krishan, A., Sauerteig, A., and Stein, J. H. (1991). A comparison of three commercially available antibodies for flow cytometric monitoring of P-glycoprotein expression in tumor cells. *Cytometry* **12,** 731–742.

Krishan, A., Sauerteig, A., Andritsch, I., and Wellham, L. (1997). Flow cytometric analysis of the multiple drug resistance phenotype. *Leukemia* **11,** 1138–1146.

Lehnert, M., de Giuli, R., and Twentyman, P. R. (1996). Sensitive and rapid bioassay for analysis of P-glycoprotein-inhibiting activity of chemosensitizers in patient serum. *Clin. Cancer Res.* **2,** 403–410.

Leibovitz, A. L., Stinson, J. C., McCombs, W. B., McCoy, C. E., Mazur, K. C., and Mabry, N. D. (1976). Classification of human colorectal adenocarcinoma cell lines. *Cancer Res.* **36,** 4562–4569.

Leonessa, F., Jacobson, M., Boyle, B., Lippman, J., McGarvey, M., and Clarke, R. (1994). Reversal of P-glycoprotein-mediated multidrug resistance by pure anti-oestrogens and novel tamoxifen derivatives. *Biochem. Pharmacol.* **48,** 277–285.

Ling, V. (1992). P-glycoprotein and resistance to anticancer drugs. *Cancer* **69,** 2603–2609.

Maniar, N., Krishan, A., Israel, M., and Samy, T. S. A. (1988). Anthracycline-induced DNA breaks and resealing in doxorubicin-resistant murine leukemic P388 cells. *Biochem. Pharmacol.* **37,** 1763–1772.

Nair, S., Singh, S. V., Samy, T. S., and Krishan, A. (1990). Anthracycline resistance in murine leukemic P388 cells: Role of drug efflux and glutathione related enzymes. *Biochem. Pharmacol.* **39,** 723–728.

Pallis, M., and Russell, N. H. (1998). Functional multidrug resistance in acute myeloblastic leukemia: A standardized flow cytometric assay for intracellular daunorubicin accumulation. *Br. J. Haematol.* **100,** 194–197.

Ramachandran, C., Mead, D., Wellham, L., Sauerteig, A., and Krishan, A. (1995). Expression of drug resistance associated mdr-1, GSTπ, and topoisomerase II genes during cell cycle. *Biochem. Pharmacol.* **49,** 545–552.

Ramachandran, C., Kunikane, H., You, W., and Krishan, A. (1998). Phorbol ester induced P-glycoprotein phosphorylation and functionality in HTB-123 human breast cancer cell line. *Biochem. Pharmacol.* **56,** 709–718.

Ross, D. D., Gao, Y., Yang, W., Leszyk, J., Shively, J., and Doyle, L. A. (1997). The 95-kilodalton membrane glycoprotein overexpressed in novel multidrug-resistant breast cancer cells is NCA, the nonspecific cross-reacting antigen of carcinoembryonic antigen. *Cancer Res.* **57,** 5460–5464.

Shalinsky, D. R., Andreeff, M., and Howell, S. B. (1990). Modulation of drug sensitivity by dipyridamole in multidrug resistant tumor cells in vitro. *Cancer Res.* **50,** 7537–7543.

Shen, D., Pastan, I., and Gottesman, M. M. (1998). Cross-resistance to methotrexate and metals in human cisplatin-resistant cell lines results from a pleiotropic defect in accumulation of these compounds associated with reduced plasma membrane binding proteins. *Cancer Res.* **58,** 268–275.

Sikic, B. I., Fisher, G. A., Lum, B. L., Halsey, J., Beketic-Oreskovic, L., and Chen, G. (1997). Modulation and prevention of multidrug resistance by inhibitors of P-glycoprotein. *Cancer Chemother. Pharmacol.* **40**(Suppl), S13–S19.

Slater, L. M., Sweet, P., Stupecky, M., Wetzel, M. W., and Gupta, S. (1986). Cyclosporin A corrects daunorubicin resistance in Ehrlich ascites carcinoma. *Br. J. Cancer* **54**, 235–238.

Stein, W. D. (1997). Kinetics of the multidrug transporter (P-glycoprotein) and its reversal. *Physiol. Rev.* **77**, 545–590.

Thomas, G. A., Barrand, M. A., Stewart, S., Rabbitts, P. H., Williams, E. D., and Twentyman, P. R. (1994). Expression of the multidrug resistance-associated protein (MRP) gene in human lung tumours and normal tissue as determined by *in situ* hybridization. *Eur. J. Cancer* **30A**, 1705–1709.

Trump, D. L., Smith, D. C., Ellis, P. G., Rogers, M. P., Schold, S. C., Winer, E. P., Panella, T. J., Jordan, V. C., and Fine, R. L. (1992). High dose oral tamoxifen, a potential multidrug-resistance-reversal agent: Phase I trial in combination with vinblastine. *J. Natl. Cancer Inst.* **84**, 1811–1816.

Tsuruo, T., Iida, H., Tsukagoshi, S., and Sakurai, Y. (1981). Overcoming of vincristine resistance in P388 leukemia *in vivo* and *in vitro* through enhanced cytoxicity of vincristine and vinblastine by verapamil. *Cancer Res.* **41**, 1967–1972.

Wielinga, P. R., Heijn, M., Broxterman, H. J., and Lankelma, J. (1997). P-Glycoprotein-independent decrease in drug accumulation by phorbol ester treatment of tumor cells. *Biochem. Pharmacol.* **54**, 791–799.

Xie, X. Y., Robb, D., Chow, S., and Hedley, D. W. (1995). Disconcordant P-glycoprotein antigen expression and transport function in acute myeloid leukemia. *Leukemia* **9**, 1882–1887.

CHAPTER 41

Resistance of Tumor Cells to Chemo- and Radiotherapy Modulated by the Three-Dimensional Architecture of Solid Tumors and Spheroids

Ralph E. Durand and Peggy L. Olive

Medical Biophysics Department
British Columbia Cancer Research Centre
Vancouver, British Columbia, Canada V5Z 1L3

I. Introduction and Historical Perspective
    A. Are Cells in Tumors Different Than in Tissue Culture?
    B. Complex Culture Systems: A Means to an End
    C. Complexity: An End in Itself
II. Background
    A. Spheroids and Other Three-Dimensional Culture Systems
    B. Cell Sorting from Multicell Systems
III. Methods
IV. Results and Discussion
    A. The Contact Effect
    B. Chemotherapy in Multicell Systems
    C. Kinetic Resistance in Multifraction Exposures
V. Conclusions and Future Directions
    References

I. Introduction and Historical Perspective

A. Are Cells in Tumors Different Than in Tissue Culture?

From the title of this chapter, it might seem obvious that the three-dimensional architecture of tumors and select tumor models is well established as a modulator

METHODS IN CELL BIOLOGY, VOL. 64

of tumor cell response to therapy. Certainly that is true to a degree. The fact that solid tumors do not maintain exponential cell growth was recognized more than 65 years ago (Winsor, 1932; Mayneord, 1932), and the fundamental concepts of cell loss (Mendelsohn *et al.*, 1960) and growth fraction (Mendelsohn, 1960) characteristic of solid tumors were established four decades ago. Similarly, the displacement of a large fraction of solid tumor cells away from the vasculature has been correlated with hypoxia and radioresistance for nearly half a century (Thomlinson and Gray, 1955).

Conversely, a significant effort continues to be expended on the development of predictive assays for tumor chemosensitivity and radiosensitivity, both of which are largely based on the response of biopsied tumor cells growing and tested *in vitro*. The implicit assumption in such studies, carried to an extreme by the current National Institutes of Health (NIH) drug discovery program that emphasizes the differential responses of panels of cultured tumor cell lines, is that the susceptibility of an isolated tumor cell is predictive for that of the same cell growing in a three-dimensional tumor milieu.

B. Complex Culture Systems: A Means to an End

The validity of the assumption that isolated cells are predictive for cells in tumors has seldom been directly tested. More often, detailed characterizations of new tumor models have resulted in the recognition of the role of the cellular microenvironment as a potent modifier of response to cytotoxic agents. An obvious example was the introduction of multicell spheroids as a tumor model in the early 1970s (Inch *et al.*, 1970; Sutherland *et al.*, 1971), where the histology and growth of the spheroids was shown to resemble that of nodular carcinomas in rodents and micrometastases in humans. Early studies confirmed, as expected, that a radioresistant, hypoxic subpopulation of cells developed in large spheroids (Sutherland *et al.*, 1970). Subsequent studies, however, were necessary to solidify the rather unexpected observation that even the aerobic cells were more radioresistant when exposed in the intact spheroid than when irradiated as individual cells after spheroid disaggregation (Durand and Sutherland, 1972).

Spheroids and, more recently, alternative three-dimensional culture systems have proved to be extremely versatile and informative systems for studying the interplay of architecture, environment, and intercellular contact as modulators of antitumor therapies. Consequently, spheroids and other culture systems that exhibit multidimensional complexity while retaining the ease of control and manipulation inherent to *in vitro* systems will be examined in detail (Section II, A), with particular emphasis placed on the characteristics of each culture system that are most exploitable for specific mechanistic studies.

C. Complexity: An End in Itself

Like tumors, three-dimensional culture systems have a fundamental property that is a mixed blessing: cellular and environmental heterogeneity. Although there is clearly an occasional role for binary (yes/no) studies such as cure versus tumor recurrence, studies in which cells respond to a given treatment and why are generally of more interest. To address such questions, however, it is necessary to be able to individually recover and study the different cell subpopulations that compose the tumor or model system.

While acknowledging that other investigators have often made more substantial contributions than we have in development of such techniques, we will limit the current discussion to our own studies. A first and obvious attempt to collect subpopulations of cells from spheroids involved the use of differential trypsinization, where cells from the superficial layers were released (and collected) prior to those from the spheroid core (Sutherland and Durand, 1976). Alternatively, various drug-based differential toxicity approaches were used in an attempt to selectively deplete unwanted subpopulations (Durand and Sutherland, 1973; Sutherland and Durand, 1976). Our first attempts at sorting cell subpopulations were based on cell size using a sedimentation approach. This had the advantages of higher throughput and increased reproducibility (Durand, 1975), yet it was somewhat unsatisfactory since cell size is largely determined by position in the cell cycle.

A quantum leap in our cell selection capabilities resulted from flow cytometry techniques using minimally toxic fluorescent dyes that exhibited a penetration gradient into spheroids (and later, into tumors in animals). Importantly, a few dyes were suitably retained in the monodispersed cells after spheroid/tumor disaggregation. Given the importance of this technology in shaping many of the conclusions drawn in our studies, its strengths and limitations will subsequently be reviewed in detail (see Section II,B).

II. Background

A. Spheroids and Other Three-Dimensional Culture Systems

As has previously been mentioned, the historical use of multicell culture systems can largely be categorized as a means to an end, rather than an end in itself. For example, some of the earliest uses of multicell cultures were in the fields of embryonic cell aggregation (Moscona and Moscona, 1966) and malignancy assays *in vitro* (McAllister *et al.,* 1967). Semisolid culture matrices have had a long history of use for the formation of local multicellular colonies. Such soft agar or methylcellulose cultures have found numerous applications, especially for hematopoietic stem cell assays *in vitro,* but the inability to easily retrieve the three-dimensional colonies from their supporting structure limits the subsequent analyses that can be undertaken. Consequently, that method of cell propagation will not be considered further here.

Even the first culture system generally recognized as an *in vitro* tumor model, multicell spheroids in suspension cultures (Sutherland *et al.,* 1971; Sutherland and Durand, 1976), actually were not designed or developed for that purpose.

Rather, the original intention was to grow single cells in culture; it was only when the resulting clusters were characterized for their growth and histological properties that their potential as a tumor model was recognized (Inch *et al.,* 1970).

The propensity of many cell lines to spontaneously aggregate and grow into multicell clusters when placed in a gyratory shaker or, more commonly, in magnetically stirred liquid suspension cultures has popularized this system. The individual clusters or spheroids in stirred suspension cultures are easily observed, and they can be collected when the stirring is discontinued, allowing the spheroids to settle to the bottom of the container. In fact, it was the ease of growing such cultures, and of retrieving the individual cells by enzymatic and/or mechanical treatments, that led to experimental therapy studies and ultimately the recognition that cellular response *in situ* differed from that of cells grown as monolayers. Even spheroid cells treated after disaggregation responded differently than when in the intact spheroid.

Once the potential advantages of multicell cultures began to be appreciated, the approach changed: investigators started to deliberately grow their favorite cells as three-dimensional structures. The first and one of the more general techniques developed for growing spheroids from cells that would not spontaneously aggregate in stirred suspensions was the liquid overlay method (Carlsson, 1977; Yuhas *et al.,* 1977). Here, three-dimensional growth is initiated by placing the cells into medium above a nonadherent layer (e.g., agar or agarose). Once the cell clusters have grown to sufficient sizes, they are typically transferred into stirred suspension cultures for subsequent bulk analyses or into multiwell plates when individual spheroids are to be observed. Although obviously more labor intensive and less amenable to high-volume culturing than the stirred suspension method, the liquid overlay approach has the distinct advantage that virtually any proliferating cells can be induced to grow as multicell spheroids.

More recently, a number of quite specialized culture systems have been developed for three-dimensional cell growth; most of these can be envisaged as intermediate between monolayers and suspension spheroids insofar as growth starts on a solid substrate. Examples include artificial capillaries or hollow fiber systems (Wiemann *et al.,* 1987; Casciari *et al.,* 1994) and microcarriers (Nilsson, 1988; Rasey *et al.,* 1996). A new model system that has emerged, and is of particular interest for studies of drug delivery to avascular tissue, is cell multilayers growing on a permeable support (Cowan *et al.,* 1996; Minchinton *et al.,* 1997).

B. Cell Sorting from Multicell Systems

Although some investigations can profitably use the response of an entire spheroid as an end point, we take the position that it makes little sense to acknowledge that three-dimensional systems contain a multitude of different microenvironments, but then to simply look at net response without knowing which cells respond or why. Rather, we prefer to undertake reconstruction experiments, in which the response of all cell subpopulations is independently observed, and then combined to give the overall picture.

In essence, that approach has two major prerequisites. First, one must be able to physically isolate and recover cells from different areas of the spheroid (or other three-dimensional system) with reasonable speed, accuracy, and reproducibility. Second, the recovery procedure itself must not interfere with the response being studied. We have found that the use of slowly penetrating but well-retained fluorescent stains, coupled with fluorescence-activated cell sorting, generally meets those requirements.

As part of the evolution of our cell sorting techniques, we evaluated numerous candidate fluorescent stains including DNA-, protein-, or thiol-binding agents, mitochondrial function probes, and/or membrane potential probes. Some membrane potential probes, including several carbocyanine derivatives, can be used effectively in culture systems where cells can be disaggregated by enzymatic means (Olive and Durand, 1987), but they are subject to redistribution in tumors when cutting or mincing the tissue is required to produce single cells. We still find a Hoechst dye, Hoechst 33342, to be the most versatile stain, coupling good intracellular retention with a wide dynamic range of staining through spheroids (Durand, 1982, 1994), multilayers (Minchinton *et al.,* 1997), and even tumors in murine hosts (Durand, 1991a).

Despite the very favorable staining characteristics of Hoechst 33342, it does have some liabilities with regard to the second of our prerequisites: it is cytotoxic at sufficient exposure concentrations, and also vasoactive. Neither is typically a concern at the concentrations used for tumor cell selection *in vivo,* and cytotoxicity can be avoided by lowering the exposure *in vitro.* To further substantiate that position, Table I shows a compilation of cytotoxicity data. As indicated in Table I, toxicity is observed at quite different stain concentrations for many different cell types. However, when the data are analyzed using the common denominator of intracellular stain concentration, toxicity is typically seen only if staining is sufficient to increase fluorescence to more than 50 times background. Unlike the situation when Hoechst 33342 is used at high concentrations for stoichiometric DNA staining, few if any cells recovered from three-dimensional systems approach that intensity level in typical positional sorting studies.

III. Methods

Cell sorting in our laboratory is performed with a Becton Dickinson dual laser FACS 440, or in cases where instrument sensitivity is more critical than sorting purity, a three-laser Coulter Elite ESP. Hoechst 33342 (purchased from Sigma, St. Louis, MO) is prepared in saline and filter sterilized. For spheroids, 0.2–2.0 μM is added directly to the culture medium for 15–20 min; in animals, 0.5–50 μg/g is injected (or infused by Harvard pump) via the lateral tail vein. Stain dose and the time and duration of infusion are dependent on the specific

Table I
Hoechst 33342 Cytotoxicity

Reference	Cell type	Hoechst 33342 (μM)	Fluorescence ($\times$ unstained)	Survival[a] (%)
Arndt-Jovin and Jovin (1977)	B8	10	Not given	>90
Pallavicini *et al.* (1979)	KHT	20	Not given	40
Durand and Olive (1982)	V79	10	40×	100
		100	100×	80
Fried *et al.* (1982)	Hela	10	Not given	100
	SK-DHL2			
		10	Not given	2
Smith and Anderson (1984)	HT29	20	Not given	10
Olive *et al.* (1985)	SCCVII	0.1	5×	100
		1	20×	100
		10	50×	90
Siemann and Keng (1986)	KHT	5	Not given	50
Young and Hill (1989)	KHT	0.01	10×	100
		0.1	100×	10
		1	Not given	~1
Morgan *et al.* (1990)	MRI	5	90×	
	1215R	10	90×	10
	EMT6	10	90×	8
	CR	10	40×	40
	AR	10	30×	80
	VR	10	80×	50
Tenforde *et al.* (1990)	R-1	5	Not given	100
Yelian and Dukelow (1992)	Embryo	5	Not given	~50
Watkins *et al.* (1996)	Spermatozoa	90	Not given	~50
		900	Not given	<1
Zhang and Kiechle (1997)	Myocytes	15	Not given	~80

[a] Although toxicity is observed at quite different exposures in different cells, all reports suggest significant toxicity only when the stain intensity exceeds background by about 50-fold.

goals and cell type used in individual experiments; higher doses provide better resolution for cell sorting, but they can be both cytostatic (Durand and Olive, 1982) and cytotoxic (Table I) and may be vasoactive in animals (Smith *et al.,* 1988; Trotter *et al.,* 1990). An additional complication in animals is the nonconstancy of tumor blood flow (Chaplin *et al.,* 1985, 1987; Durand and LePard, 1995). Posttreatment staining provides information on transient changes in blood flow in the interval between radiation and staining; stain administration during irradiation indicates functional tumor blood flow (Chaplin *et al.,* 1985; Olive *et al.,* 1985). Longer infusion times presumably provide a more representative picture of average tumor perfusion.

Animals are sacrificed by cervical dislocation 15–20 min after Hoechst 33342 injection, or immediately after dye infusion, and the tumor rapidly excised, washed with cold saline (4°C) to remove excess stain and inhibit stain redistribution, minced, and then incubated with an enzyme cocktail of trypsin, DNase, and collagenase for 10–40 min (dependent on tumor type) at 37°C to produce single cells. Spheroid processing is simpler, as a mild trypsin cocktail followed by mechanical agitation rapidly produces single cells (Sutherland and Durand, 1976; Durand, 1986a). After sorting, cells are analyzed as required, or cultured using conventional techniques to determine clonogenicity.

Cell suspensions are maintained at 4°C during the sorting procedures (typically <1 hr) to retain viability and inhibit stain redistribution. The primary laser is operated at 400 mW at 488 nm and the ultraviolet (UV) laser at 40 mW. Hoechst 33342 emission is monitored through a 449 ± 10 nm band-pass filter. Cells are recognized by the forward scatter signal; for tumors from mice, addition of fluorescein isothiocyanate (FITC)-conjugated goat anti-mouse immunoglobulin G (IgG) (Olive, 1989) allows discrimination against many of the host cells that inevitably are present in both xenograft cell suspensions and syngeneic tumors. Eliminating host cells in this manner provides a more accurate tumor cell recovery and count, and consequently a higher and more reproducible cloning efficiency.

After initial gating, the concurrent signals from the Hoechst 33342 stain and the 90° scatter signal are processed through matched logarithmic amplifiers, and the ratio of these signals used to generate a stain concentration profile that is integrated to define windows of equal cell numbers, or, if desired, windows representing cells from shells of essentially equal thickness inward into spheroids, or outward from tumor blood vessels (Durand *et al.*, 1990). Use of the ratioed signals corrects for cell size, and thus provides a better estimate of intracellular stain concentration for increased sort accuracy. Our software automates this procedure for as many as 100 sorted fractions; 10 are typically collected (Durand, 1986a; Durand *et al.*, 1990). A stained but untreated cell population from each batch of spheroids or tumors is processed similarly to define the plating efficiency of each sorted fraction; cytotoxicity of the Hoechst 33342 is monitored with dose-escalation studies.

IV. Results and Discussion

Clearly, understanding all implications of the three-dimensional architecture of tumors is our ultimate goal. However, a number of practical and ethical issues limit the types of studies that can be performed, even with tumors grown in experimental animals. Consequently, *in vitro* systems like spheroids have played a major role in understanding many of the factors that influence response to antitumor therapies. Several of these will now be reviewed.

A. The Contact Effect

1. Ionizing Radiation

The contact effect was first described in 1972 (Durand and Sutherland, 1972), where Chinese hamster V79-171b cells, when grown in suspension culture for only 24 hr, formed spheroids containing about 10–25 cells that were more resistant to killing by ionizing radiation than cells grown as monolayers. The increase in radiation resistance was confined largely to the shoulder of the radiation survival curve, implicating differences in the cellular capacity to sustain and repair damage. Subsequent studies indicated that spheroids showed fewer radiation-induced mutations than monolayers exposed to the same dose (Durand and Olive, 1979; Olive and Durand, 1981). Although cell cycle kinetics are unchanged in 1-day-old spheroids compared to monolayers (Durand, 1976, 1990a), the radiation-induced G_2 block, which is believed to allow cells time to repair before entering mitosis, is significantly shorter for small spheroids than for monolayers (Durand, 1984; Sweigert and Alpen, 1987).

The contact effect is of fundamental interest to radiobiologists and radiotherapists because small survival differences in the low-dose region of the survival curve are magnified during a conventional course of fractionated radiotherapy. Tumor cells displaying a wider survival curve shoulder would therefore be more difficult to eliminate by ionizing radiation. From the analysis of results with 39 different cell lines, Freyer (1992) estimated that approximately 40% of human and rodent cells display resistance to radiation when grown in close contact with other cells. Although not a universal finding, cells removed directly from several murine tumors often survive irradiation better than those allowed to grow as monolayers for several hours prior to irradiation (Dawson *et al.*, 1973; Hill *et al.*, 1979). A contact effect, which decays over several hours when cells are left in culture, has been observed for kidney (Jen *et al.*, 1991), and several other normal tissues similarly exhibit a related phenomenon called *in situ* repair (Gould and Clifton, 1979; Mulcahy *et al.*, 1980).

2. Chemotherapy and Other Agents

Spheroids, even very small ones, often contain cells more resistant than monolayers to a variety of chemotherapeutic agents (Nederman, 1984; Sutherland, 1988; Durand, 1990a; Olive *et al.*, 1991a; Kobayashi *et al.*, 1993; Hoffman, 1994). Likewise, a contact effect akin to that seen for ionizing radiation has been observed for several other anticancer agents. Hyperthermia produced less cell killing in 1-day-old V79 and EMT6 spheroids than in monolayers (Durand, 1978; Dobrucki and Bleehen, 1985; Wigle and Sutherland, 1985), an effect that could not be explained by differences in heating technique, cell cycle distribution, or nutrient availability. Small spheroids (100 μm in diameter) are about twice as resistant to photodynamic therapy (PDT), even when reduced to single cells prior to light exposure (West, 1989). Additional factors further increase PDT resistance in larger spheroids (West and Moore, 1992). Ultrasound treatments are more toxic to single cells than to small spheroids, although dissociation of spheroids prior to treatment eliminated this difference, suggesting that a different mechanism may be operating (Sacks *et al.*, 1981).

Figure 1 shows the contact effect for V79 cells exposed to doxorubicin after growth as monolayers versus 1-day-old spheroids. To ensure that drug delivery was not limiting in the spheroids, they were reduced to a single cell suspension before drug exposure. Also, to ensure consistency of handling, and to provide additional information on the intracellular doxorubicin content (information not shown here), the exposed cells were sorted into 10 fractions of decreasing doxorubicin fluorescence (fraction 1 was the most fluorescent 10% of the population). Note that the cells grown as spheroids were considerably more resistant than those grown and exposed as monolayers. Actually, however, the more relevant comparison is between the lower and upper curves, since doxorubicin uptake into monolayer-grown cells is increased when those cells are exposed in a stirred suspension culture. Further information interrelating exposures and actual intracellular drug levels achieved and required for cytotoxicity under the various growth and exposure conditions has previously been published (Durand and Olive, 1981).

3. Proposed Mechanisms for the Contact Effect

Explanations for the contact effect have been evolving since its original description. Three somewhat interrelated possibilities are (1) intercellular communication that enhances cell recovery after radiation or other damage, (2) a change in nuclear shape and DNA packaging leading to more efficient DNA repair, or

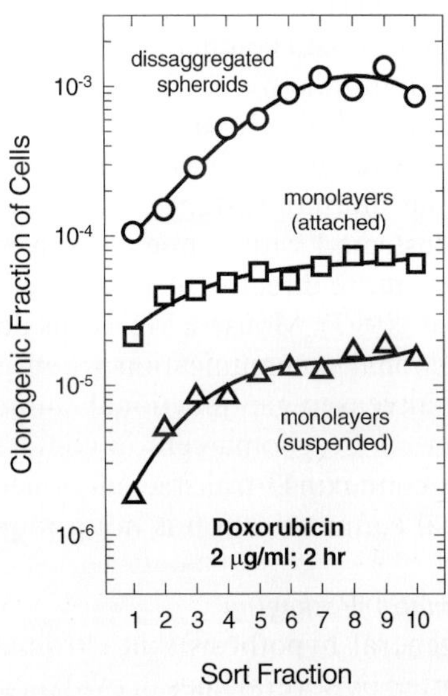

Fig. 1 Survival of Chinese hamster V79 cells grown as spheroids for 1 day, or as monolayers, and then exposed to doxorubicin. Each sample was sorted into 10 equal fractions on the basis of the intracellular doxorubicin content (fluorescence), with fraction 1 being the most fluorescent cells.

(3) a change in cell shape and gene expression resulting in changes in damage recognition, signaling, or repair.

a. Intercellular Communication

Resistance of spheroids to radiation and some drugs was initially thought to be a result of growth with close three-dimensional cell contact (Durand and Sutherland, 1972). Molecules involved in DNA repair might be interchanged via gap junctions, thus allowing doomed cells to enhance the recovery of others. Gap junctional reciprocity has been used to describe the ability of healthy cells to nourish neighboring sick or deprived cells, thus increasing the recovery potential of the population as a whole (Holder *et al.*, 1993). Many transformed cells show fewer gap junctions than their corresponding normal tissues, and as malignancy progresses and metastatic potential increases, a decrease in intercellular communication is often observed (Nicolson, 1987; Yamaski and Katoh, 1988; Yamaski, 1990). If gap junctional communication is important in the contact effect, then the degree of malignancy or even metastatic potential could determine the degree of contact resistance.

Conversely, it was noted even in the initial paper (Durand and Sutherland, 1972) that spheroid cells, separated from each other before irradiation, only gradually lose their radiation resistance. Therefore gap junctional communication, during or after the cytotoxic event, does not appear to be directly involved in the contact effect. Similarly, in a panel of seven human non-small-cell lung cancer cell lines, no correlation was observed between doxorubicin resistance and junctional communication (Bradley *et al.*, 1990). Wigle and Sutherland (1985) addressed cell–cell contact effects on damage by hyperthermia. They found that spheroid multiplicity in the range of 1.16 to 76 cells/spheroid had no effect on cell survival after hyperthermia, and concluded that a physiological change associated with growth in suspension, and not intercellular contact, was responsible for the thermoresistance of EMT6 spheroids. More recently, two cell lines were transfected with connexin43, a major gap junctional protein, in order to determine more directly the role of cell–cell communication in the contact effect (Luo *et al.*, 1997). Mouse EMT6 cells do not display a contact effect or significant gap junctional communication when grown as spheroids; transfection of connexin43 increased gap junctional communication but did not result in a contact effect. In rat C6 glioma cells, a contact effect was observed in both parent cell line and connexin43-transfected clones. Together, these data indicate that gap junctional communication is not a requirement for the contact effect.

b. Chromatin Packaging

The general hypothesis that chromatin structure may influence DNA repair is supported by several lines of evidence: (1) internucleosomal regions and matrix associated DNA are differentially sensitive to X-ray and UV damage (Smerdon, 1991), (2) damage to transcribed sequences is repaired more rapidly (Bohr *et al.*, 1987), (3) induction of strand breaks by radiation is generally greater in transcriptionally active DNA (Chiu *et al.*, 1982), and (4) changes in DNA structure through the cell cycle are accompanied by changes in radiosensitivity (Terasima and Tolmach, 1963). Although these effects are well documented, there may also be a more subtle influence of higher order chromatin structure.

Why would the organization of chromatin affect repair of damage? One theory suggests that the lethal effects of radiation might result from an incorrect configuration of the DNA after repair (Cook, 1973). The idea that specific structural proteins may be involved in maintaining DNA integrity during repair has been slowly developing since the discovery of the nuclear protein matrix (Berezney and Coffey, 1976), demonstration of its role in replication and transcription, and the presence of specific and stable attachment sites (Dijkwell and Hamlin, 1988). Although it is possible that the components of the nuclear matrix may differ for cells of monolayers and spheroids, our results tend to support the hypothesis that the stability of this structure, rather than specific chemical constituents, may be the critical feature in terms of cellular sensitivity to some cytotoxic agents (Olive, 1992). Chromatin structure differences between monolayers and spheroids (Olive and Durand, 1994) were demonstrated using the acridine orange staining protocol developed by Darzynkiewicz and colleagues (1985).

c. Cell Shape and Gene Expression

An important difference between cells growing in a three-dimensional system versus a monolayer is the lack of cell attachment to a solid substrate in the former. This in turn results in quite different shapes of the cell and its nucleus, which has been shown to play a critical role in DNA replication, gene expression and differentiation in anchorage dependent cells (Folkman and Moscona, 1978; Ben-Ze'ev *et al.*, 1980; Ben-Ze'ev, 1991). Several authors have invoked a role for cell shape in the resistance of V79 spheroids to ionizing radiation (Sutherland, 1988; Olive *et al.*, 1991b; Reddy and Lange, 1991; Olive and MacPhail, 1992). Similarly, Dobrucki and Bleehen (1985) suggested that the round cell shape of spheroid cells might increase resistance to heat, perhaps due to alterations in the physical properties of the membrane.

The methods by which cells signal the presence of DNA damage to repair enzymes are incompletely understood. However, the fact that the ataxia telangiectasia gene, associated with radiation sensitivity, appears to code for a signaling molecule involved in damage recognition is intriguing, and may offer an important clue as to the mechanism for the contact effect. Cell signaling processes may be altered when cells are grown as spheroids, and these changes could affect the efficiency of damage recognition or repair.

Damage induction can also be altered in spheroids relative to monolayers through processes that may involve changes in cell signaling pathways as well. An essential nuclear enzyme involved in replication, chromosome segregation, and topological changes in DNA conformation, topoisomerase II, was examined in V79 spheroids. Although the amount and activity of topoisomerase II was similar in monolayers and outer spheroid cells, there was a dramatic decrease in topoisomerase II phosphorylation in spheroids (Luo *et al.*, 1998). This change in phosphorylation was correlated with the 10-fold resistance of the outer cells of spheroids to killing and DNA damage by the topoisomerase II inhibitors etoposide and amsacrine (*m*-AMSA) (Olive *et al.*, 1993). Why topoisomerase II would be less phosphorylated in spheroids is not yet known. Interestingly, tyrosine phosphatases were found to be increased in the outer cells of spheroids relative to monolayers (Mansbridge *et al.*, 1992).

B. Chemotherapy in Multicell Systems

1. Patterns of Drug Activity in Spheroids

Exposure of cells to chemotherapeutic drugs *in situ* in a tumor or spheroid usually produces a response quite different from that of the same cells exposed *in vitro* after disaggregation. The extent and significance of this phenomenon are illustrated in Fig. 2 for doxorubicin exposure of cells *in situ* in large spheroids, compared to the response of the same cell subpopulations after disaggregation. In both cases, fluorescence-activated cell sorting with Hoechst 33342 as previously described (Section III) was used (sort fraction 1 represents the outer 10% of the spheroid). For comparison, the data of Fig. 1 (for the same cell type and expo-

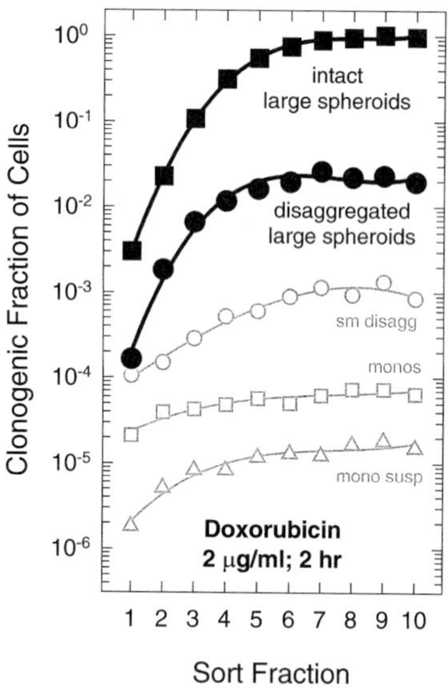

Fig. 2 Survival of cells from ~750 µm Chinese hamster V79 spheroids, exposed to doxorubicin before (squares) or after (circles) being sorted on the basis of Hoechst 33342 diffusion into the spheroids. Fraction 1 represents the 10% outermost cells. For reference, the curves from Fig. 1 have been reproduced in gray; note that similar survival was seen for the disaggregated outer cells (fraction 1) of large spheroids and the comparable (rapidly cycling) cells throughout small spheroids.

sures) have been reproduced. Note that the exposure environment can modify cytotoxicity by several orders of magnitude for the same cell type. Although other drugs typically do not produce such dramatic differences, two generalizations can nonetheless be made. First, except for the nitrosoureas, every drug class that we have studied has shown differences in activity in the intact spheroid compared with cells from disaggregated spheroids exposed under monodispersed conditions. Second, most drugs studied to date have shown considerable variation in activity through intact spheroids, and all show at least some differences.

The pattern of drug resistance in large spheroids shown in Fig. 2 is similar to that described in a number of studies (West *et al.*, 1980; Kwok and Twentyman, 1985; Kerr *et al.*, 1987; Carlsson and Nederman, 1989). Due to a combination of drug delivery and cycle-specific processes (Durand, 1986b, 1991b), one might expect that the external, rapidly cycling cells of spheroids would be more sensitive to most chemotherapeutic agents than the internal, quiescent cells. However, not all agents behave this way. In fact, mitomycin C and other bioreductive drugs, and perhaps somewhat surprisingly, the nitrosoureas, are preferentially toxic to the inner, noncycling cells of spheroids (Durand, 1990b; Durand and Olive, 1992). Since clinical chemotherapy combinations are typically chosen based on non-cross-resistance, studies like these add yet another dimension to the standard clinical definition of cross-resistance. Such data provide a more objective rationale for choosing and applying chemotherapy combinations, and for altering combinations as the status of the tumor/spheroid changes during (or in response to) therapy.

2. Delivery of Antitumor Agents

a. Physical and Physiological Considerations

A continuing problem with treatment of solid tumors, which is likely to become even more acute in the new age of molecular and gene therapies, is delivery of the therapeutic agent at adequate concentrations to all tumor cells. Clearly, for any blood-borne agent, a number of barriers must be traversed, including the vascular endothelium, multiple layers of avascular tumor cells, and even the high interstitial pressure of the tumor (Jain, 1991a,b; Helmlinger *et al.*, 1997). We (Durand, 1981, 1986b), like others (Sutherland *et al.*, 1979; Erlichman and Vidgen, 1984; Kerr and Kaye, 1987; Carlsson and Nederman, 1989), have consequently touted spheroids as a relevant model for evaluating drug penetration. In fact, the rationale for chemosensitivity testing in single cell systems has been openly questioned (Sutherland, 1988; Hoffman, 1994).

Clearly, antitumor agents cannot be effective if they do not reach their targets. However, several factors suggest, at least to us, that delivery is not a major problem for the cancer chemotherapy drugs in common use today. First, given the radial structure of tumor cords where the number of cells increases with the square of distance from the capillary, it is difficult to imagine a tumor responding to treatment if therapeutically effective drug levels do not reach even the most distant cells. Experimental results with radiolabeled drugs are consistent with that expectation (Durand, 1989). Second, virtually all demonstrations of drug delivery problems in spheroids and other three-dimensional systems have used the anthracyclines. As discussed later, those drugs have rather unique characteristics that influence both their reactivity and longevity in tissues.

b. Slow Drug Penetration: A Therapeutic Advantage?

The very large gradient of activity seen for doxorubicin through large spheroids in Fig. 2 would be consistent with poor drug delivery, and since the compound is itself fluorescent, intracellular drug levels can be directly estimated using flow or static cytometry techniques (Krishan and Ganapathi, 1979; Durand, 1981; Wartenberg *et al.*, 1998). These have generally been consistent with the observed levels of cytotoxicity (Durand, 1981). However, such observations in spheroids lead to a major contradiction: only a small subpopulation (the external cells) is sensitive to the drug. Consequently, the apparent penetration problem for doxorubicin suggests that it should have little if any clinical activity. Like others, we originally explained this inconsistency by the fact that clinical treatment is not delivered as a single administration of the drug, and multiple exposures would be more effective. Although that explanation is partially correct, detailed studies of doxorubicin uptake and retention in multifraction experiments have provided further insight (Durand, 1990c).

The problem with most studies reporting slow penetration of doxorubicin into multicell models is that only relatively short times are typically studied. We found a much different response when spheroids were exposed to the drug for a typical 2-hr period, washed free of drug, and left intact for variable times rather than being immediately disaggregated for cytometry and clonogenicity analyses (Fig. 3). Again, these data are superimposed on those shown in the previous figures for the same cells and drug exposure time. Doxorubicin, which had clearly penetrated quite slowly into the spheroid after 2 hr (upper curve), can be inferred to have redistributed in the subsequent 24 hr after drug removal from the growth medium. This slow transport also applied to the outward diffusion of accumulated drug (Durand, 1990c).

Clearly, it is a very different situation for a cell to remain *in situ* after drug exposure as opposed to being immersed in a sea of non-drug-containing medium; the effective drug exposure time is greatly increased *in situ*. In either case, the cell may efficiently pump out the drug, but in a tissuelike situation, this may simply transport the drug into a neighboring cell or produce a high local extracellular concentration favoring diffusion back into the cell. In the even more extreme scenario, it is perhaps possible that the drug is pumped directly into adjacent cells, resulting in a futile role for the P-glycoprotein molecule in cells of solid tumors.

3. Clinical Implications

Two generalizations emerge from our studies. First, extrapolating from well-defined exposures in a tissue culture environment to that in a tumor is fraught

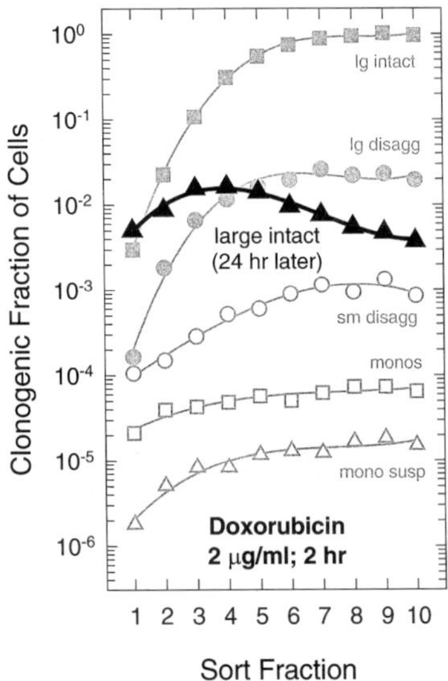

Fig. 3 Survival of cells from the same V79 spheroids as those in Fig. 2, but where the cell sorting was performed 24 hr after terminating the doxorubicin exposure. Note the markedly different response, consistent with the observed redistribution of doxorubicin when the spheroids remained intact (Durand, 1990c). Other curves (gray) are reproduced from the previous figures for reference.

with difficulty in terms of both drug dosage and exposure time. Second and even more important, but perhaps less obvious, is the underlying message in Figs. 1–3. Remembering that the same cell line, drug concentration, and exposure time was used for each curve, the obvious question arises: were these cells sensitive or resistant to doxorubicin? This question directly addresses the validity and utility of *in vitro* chemosensitivity testing and the risks in prescribing clinical therapy based solely on *in vitro* assays.

C. Kinetic Resistance in Multifraction Exposures

1. Solid Tumor Growth

Cancer is, ultimately, a problem of cell growth. Aspects of tumor growth are used to characterize the disease and its response to treatment: a rapidly growing tumor is called aggressive, tumors that decrease in size during therapy are called responsive while those that continue to grow are termed resistant, and success or failure of treatment is defined by the absence or presence of regrowth after therapy. Given that tumors are so commonly characterized in terms of their growth characteristics, it seems surprising to us and others (Preisler and Venugopal, 1995) that their altered growth regulation is seldom recognized as a potential limitation to the multifraction therapy typically employed in the clinic.

2. Spheroids for Multifraction Studies

Spheroids growing in stirred suspensions can be easily (though not inexpensively) maintained, and at larger sizes their relatively slow growth provides a quasi-steady-state system ideal for multifraction experiments with cytotoxic agents. A major advantage of the spheroid system is that large numbers of identical spheroids can be propagated, thus making it feasible and informative to determine the number of viable cells per spheroid before and after every treatment in a multifraction regimen. In this way, changes in response (development of resistance) can be identified, and importantly, quantified.

The advantages of this approach are illustrated in Fig. 4, where spheroids were repeatedly treated with etoposide. This drug was chosen because of its preferential cytotoxicity to proliferating cells due to effects on topoisomerase II, and also because it is a substrate for P-glycoprotein and therefore subject to development of multidrug resistance. Interestingly, the spheroids in Fig. 4 responded to etoposide in a manner reminiscent of many tumors in patients: an initial, rapid, and dose-dependent decrease of the number of viable cells per spheroid was seen. At the lower dose, regrowth occurred in spite of continued treatment; even the higher dose appeared to show progressively diminishing activity. In the clinic, this is inevitably described as emergence of drug resistance.

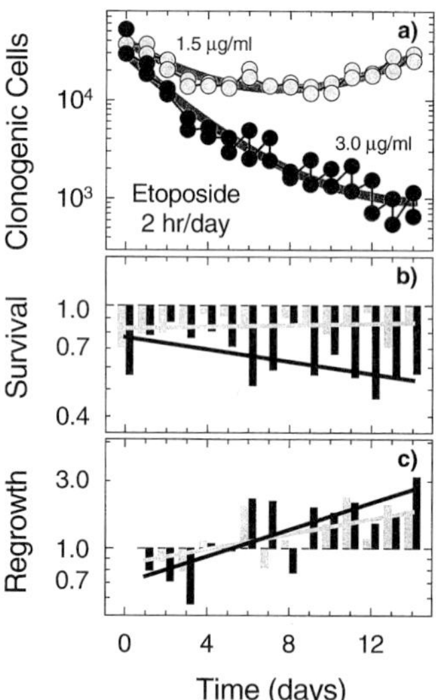

Fig. 4 Consecutive samples from V79 spheroids during daily exposures to etoposide, with assays performed before and after each treatment. (a) Number of viable cells per spheroid at each determination. (b) Cell kill and (c) regrowth measured for each fraction. The lines in (b) and (c) are linear regressions showing the trend of each data set. Note the apparent "emergence of drug resistance," which in fact was due entirely to "accelerated repopulation."

Unlike the case in the clinic, however, we were able to monitor the actual cell kill and regrowth for each treatment, as plotted in Fig. 4b,c. Although the nature of the assays and the inherent problems of spheroid sampling lead to somewhat noisy data on a point by point basis, the trends shown by the regression lines in the lower panels are indisputable. The data in Fig. 4b conclusively show that the apparent resistance was not the result of a change in intrinsic sensitivity to etoposide. At the lower dose, no change in resistance was seen, whereas at the higher dose, the cell kill per treatment actually increased (the cells became progressively more sensitive, not more resistant). Conversely, in both cases, regrowth increased with time. This regrowth was, ultimately, sufficiently rapid to produce the apparent resistance.

These observations raise intriguing questions for cancer therapy. Clearly, they indicate that it is possible to demonstrate drug resistance even though individual cell sensitivity is unchanged. The key variable is the decreased growth fraction due to the three-dimensional growth, and the associated development of a quiescent but potentially clonogenic cell population. If these noncycling tumor cells are recruited back into cycle by depopulation, then the growth fraction will increase (in the spheroids shown in Fig. 4, from about 25–30% to virtually 100% for the higher dose) and may well be able to compensate for treatment-induced cell kill.

If a similar phenomenon occurs in human tumors, one must question whether the drug resistance seen in the clinic has a genetic or perhaps a kinetic basis (Durand and Vanderbyl, 1989; Durand, 1990a, 1991b, 1993; Olive and Durand, 1994). For the former, we need better drugs and treatment strategies. If, however, any part of the resistance of human tumors is due to changing cell kinetics, then the cancer chemotherapy drugs already available are probably adequate, if used appropriately, to overcome that problem.

An intriguing extension of that argument is that cytostatic agents could find increased usage in dynamic systems such as solid tumors with high cell loss. This point is also illustrated in Fig. 4c, where the negative regrowth initially observed simply reflects the spontaneous cell loss processes ongoing in spheroids prior to a sufficient increase in the growth fraction. Similar conclusions have resulted from our initial studies with multifraction irradiation of human tumor xenograft systems (Sham and Durand, 1999).

V. Conclusions and Future Directions

While compelling data are emerging to suggest that cells in a three-dimensional growth state can have a markedly different sensitivity to cancer chemotherapy and/or radiotherapy, practical and ethical issues limit the extent and quality of data that can be produced with tumors *in vivo*. Consequently, model systems remain necessary, and multicell spheroids *in vitro* have proved to be one of the more versatile and informative systems for examination of questions relating to tumor resistance to drugs and radiation. Three mechanistic observations relevant to drug and radiation resistance that have emerged from the study of spheroids are the contact effect, microenvironmental resistance, and regrowth resistance as have been reviewed here.

Our data suggest obvious extensions for the future. Verification of the observations made in model systems will continue to be required in experimental tumors and, wherever possible, in the clinic. We would also argue that an increased emphasis on evaluation of (new) therapies in clinically relevant, multifraction exposures is mandatory. Lest we forget, any three-dimensional system is dynamic, and the local microenvironment of a cell in a solid tumor not only is different from that of a cell in a petri dish, but is also subject to change by treatment.

Acknowledgments

This work was supported by grants from the National Cancer Institute and the Medical Research Council of Canada, and the U.S. National Cancer Institute through the U.S. Public Health Service.

References

Arndt-Jovin, D. J., and Jovin, T. M. (1977). Analysis and sorting of living cells according to deoxyribonucleic acid content. *J. Histochem. Cytochem.* **7**, 585–589.

Ben-Ze'ev, A. (1991). Animal cell shape changes and gene expression. *BioEssays* **13**, 207–211.

Ben-Ze'ev, A., Farmer, S. R., and Penman, S. (1980). Protein synthesis requires cell surface contact while nuclear events respond to cell shape in anchorage dependent fibroblasts. *Cell* **21**, 365–372.

Berezney, R., and Coffey, D. S. (1976). The nuclear protein matrix: Isolation, structure and functions. *Adv. Enz. Regul.* **14**, 63–100.

Bohr, V. A., Phillips, D. H., and Hanawalt, P. C. (1987). Heterogeneous DNA damage and repair in the mammalian genome. *Cancer Res.* **47**, 6426–6436.

Bradley, C., Freshney, R. I., and Pitts, J. D. (1990). Relationship between intercellular communication and doxorubicin resistance in non-small cell lung cancer. *Br. J. Cancer* **62**, 512.

Carlsson, J. (1977). A proliferation gradient in three-dimensional colonies of cultured human glioma cells. *Int. J. Cancer* **20**, 129–136.

Carlsson, J., and Nederman, T. (1989). Tumour spheroid technology in cancer therapy research. *Eur. J. Cancer Clin. Oncol.* **25**, 1127–1133.

Casciari, J. J., Hollingshead, M. G., Alley, M. C., Mayo, J. G., Malspeis, L., Miyauchi, S., Grever, M. R., and Weinstein, J. N. (1994). Growth and chemotherapeutic response of cells in a hollow-fiber *in vitro* solid tumor model. *J. Natl. Cancer Inst.* **86**, 1846–1852.

Chaplin, D. J., Durand, R. E., and Olive, P. L. (1985). Cell selection from a murine tumour using the fluorescent probe Hoechst 33342. *Br. J. Cancer* **51**, 569–572.

Chaplin, D. J., Olive, P. L., and Durand, R. E. (1987). Intermittent blood flow in a murine tumor: Radiobiological effects. *Cancer Res.* **47**, 597–601.

Chiu, S. M., Oleinick, N. L., Friedman, L. R., and Stambrook, P. J. (1982). Hypersensitivity of DNA in transcriptionally active chromatin to ionizing radiation. *Biochim. Biophys. Acta* **699**, 15–21.

Cook, P. R. (1973). Hypothesis on differentiation and the inheritance of gene superstructure. *Nature* **245**, 23–25.

Cowan, D. S., Hicks, K. O., and Wilson, W. R. (1996). Multicellular membranes as an *in vitro* model for extravascular diffusion in tumours. *Br. J. Cancer* **74**, S28–S31.

Darzynkiewicz, Z., Traganos, F., Kapuscinski, J., and Melamed, M. R. (1985). Denaturation and condensation of DNA in situ induced by acridine orange in relation to chromatin changes during growth and differentiation of Friend erythroleukemia cells. *Cytometry* **6**, 195–207.

Dawson, K. B., Madoc-Jones, H., Mauro, F., and Peacock, J. H. (1973). Studies on the radiobiology of a rat sarcoma treated *in situ* and assayed *in vitro*. *Br. J. Cancer* **9**, 59–69.

Dijkwell, P. A., and Hamlin, J. L. (1988). Matrix attachment regions are positioned near replication initiation sites, genes and an interamplicon junction in the amplified dihydrofolate reductase domain of Chinese hamster ovary cells. *Mol. Cell. Biol.* **12**, 5398–5409.

Dobrucki, J., and Bleehen, N. M. (1985). Cell–cell contact affects cellular sensitivity to hyperthermia. *Br. J. Cancer* **52**, 849–855.

Durand, R. E. (1975). Isolation of cell populations from *in vitro* tumor models according to sedimentation velocity. *Cancer Res.* **35**, 1295–1300.

Durand, R. E. (1976). Cell cycle kinetics in an *in vitro* tumor model. *Cell Tissue Kinet.* **9**, 403–412.

Durand, R. E. (1978). Effects of hyperthermia on the cycling, non-cycling, and hypoxic cells of irradiated and unirradiated multicell spheroids. *Radiat. Res.* **75**, 373–384.

Durand, R. E. (1981). Flow cytometry studies of intracellular adriamycin in multicell spheroids *in vitro*. *Cancer Res.* **41**, 3495–3498.

Durand, R. E. (1982). Use of Hoechst-33342 for cell selection from multicell systems. *J. Histochem. Cytochem.* **30**, 117–122.

Durand, R. E. (1984). Repair during multifraction exposures: Spheroids vs. monolayers. *Br. J. Cancer* **49**, 203–206.

Durand, R. E. (1986a). Use of a cell sorter for assays of cell clonogenicity. *Cancer Res.* **46**, 2775–2778.

Durand, R. E. (1986b). Chemosensitivity testing in V79 spheroids: Drug delivery and cellular microenvironment. *J. Natl. Cancer Inst.* **77**, 247–252.

Durand, R. E. (1989). Distribution and activity of antineoplastic drugs in a tumor model. *J. Natl. Cancer Inst.* **81**, 146–152.

Durand, R. E. (1990a). Multicell spheroids as a model for cell kinetics studies. *Cell Tissue Kinet.* **23**, 141–159.

Durand, R. E. (1990b). Cisplatin and CCNU synergism in spheroid cell subpopulations. *Br. J. Cancer* **62**, 947–953.

Durand, R. E. (1990c). Slow penetration of anthracyclines into spheroids and tumors: A therapeutic advantage? *Cancer Chemother. Pharmacol.* **26**, 198–204.

Durand, R. E. (1991a). The influence of microenvironmental factors on the activity of radiation and drugs. *Int. J. Radiat. Oncol. Biol. Phys.* **20**, 253–258.

Durand, R. E. (1991b). Effects of drug distribution and cellular microenvironment on the interaction of cancer chemotherapeutic agents. *In* "Synergism and Antagonism in Chemotherapy" (T. C. Chou and D. C. Rideout, eds.), pp. 659–688. Academic Press, San Diego.

Durand, R. E. (1993). Cell kinetics and repopulation during multifraction irradiation of spheroids: Implications for clinical radiotherapy. *Semin. Radiat. Oncol.* **3**, 105–114.

Durand, R. E. (1994). Contributions of flow cytometry to studies with multicell spheroids. *In* "Flow Cytometry, Methods in Cell Biology" (Z. Darzynkiewicz, J. P. Robinson, and H. A. Crissman, eds.), pp. 405–422. Academic Press, San Diego.

Durand, R. E., and LePard, N. E. (1995). Contribution of transient blood flow to tumour hypoxia in mice. *Acta Oncol.* **34**, 317–324.

Durand, R. E., and Olive, P. L. (1979). Radiation-induced DNA damage in V79 spheroids and monolayers. *Radiat. Res.* **78**, 50–60.

Durand, R. E., and Olive, P. L. (1981). Flow cytometry studies of intracellular adriamycin in single cells *in vitro*. *Cancer Res.* **41**, 3489–3494.

Durand, R. E., and Olive, P. L. (1982). Cytotoxicity, mutagenicity and DNA damage by Hoechst-33342. *J. Histochem. Cytochem.* **30**, 111–116.

Durand, R. E., and Olive, P. L. (1992). Evaluation of bioreductive drugs in multicell spheroids. *Int. J. Radiat. Oncol. Biol. Phys.* **22**, 689–692.

Durand, R. E., and Sutherland, R. M. (1972). Effects of intercellular contact on repair of radiation damage. *Exp. Cell Res.* **71**, 75–80.

Durand, R. E., and Sutherland, R. M. (1973). Growth and radiation survival characteristics of V79-171B Chinese hamster cells: A possible influence of intercellular contact. *Radiat. Res.* **56**, 513–527.

Durand, R. E., and Vanderbyl, S. L. (1989). Tumor resistance to therapy: A genetic or kinetic problem? *Cancer Commun.* **1,** 277–283.

Durand, R. E., Chaplin, D. J., and Olive, P. L. (1990). Cell sorting with Hoechst or carbocyanine dyes as perfusion probes in spheroids and tumors. *In* "Methods in Cell Biology" (Z. Darzynkiewicz and H. A. Crissman, eds.), Vol. 33, pp. 509–518. Academic Press, San Diego.

Erlichman, C., and Vidgen, D. (1984). Cytotoxicity of doxorubicin in MGH-U1 cells grown as monolayer cultures, spheroids and xenografts in immune-deprived mice. *Cancer Res.* **44,** 5469–5475.

Folkman, J., and Moscona, A. (1978). Role of cell shape in growth control. *Nature* **273,** 345–349.

Freyer, J. P. (1992). Spheroids in radiobiology. *In* "Multicell Spheroids" (R. Bjerkvig, ed.), pp. 217–275. CRC Press, Boca Raton, Florida.

Fried, J., Doblin, J., Takamoto, S., Perez, A., Hansen, H., and Clarkson, B. (1982). Effects of Hoechst 33342 on survival and growth of two tumor cell lines and on hematopoietically normal bone marrow cells. *Cytometry* **3,** 42–47.

Gould, M. N., and Clifton, K. H. (1979). Evidence for a unique *in situ* component of the repair of radiation damage. *Radiat. Res.* **77,** 149–155.

Helmlinger, G., Yuan, F., Dellian, M., and Jain, R. K. (1997). Interstitial pH and pO_2 gradients in solid tumors *in vivo:* High-resolution measurements reveal a lack of correlation. *Nat. Med.* **3,** 177–182.

Hill, R. P., Ng, R., Warren, B. F., and Bush, R. S. (1979). The effect of intercellular contact on radiation sensitivity of KHT sarcoma cells. *Radiat. Res.* **77,** 182–192.

Hoffman, R. M. (1994). The three-dimensional question: Can clinically relevant tumor drug resistance be measured *in vitro? Cancer Metastasis Rev.* **13,** 169–173.

Holder, J. W., Elmore, E., and Barrett, J. C. (1993). Gap junction function and cancer. *Cancer Res.* **53,** 3475–3485.

Inch, W. R., McCredie, J. A., and Sutherland, R. M. (1970). Growth of nodular carcinomas in rodents compared with multi-cell spheroids in tissue culture. *Growth* **34,** 271–282.

Jain, R. K. (1991a). Haemodynamic and transport barriers to the treatment of solid tumours. *Int. J. Radiat. Biol.* **60,** 85–100.

Jain, R. K. (1991b). Therapeutic implications of tumor physiology. *Curr. Opin. Oncol.* **3,** 1105–1108.

Jen, Y., West, C. M. L., and Hendry, J. H. (1991). The lower radiosensitivity of mouse kidney cells irradiated *in vivo* than *in vitro:* A cell contact effect phenomenon. *Int. J. Radiat. Oncol. Biol. Phys.* **20,** 1243–1248.

Kerr, D. J., and Kaye, S. B. (1987). Aspects of cytotoxic drug penetration, with particular reference to anthracyclines: A Review. *Cancer Chemother. Pharmacol.* **19,** 1–11.

Kerr, D. J., Wheldon, T. E., Kerr, A. M., and Kaye, S. B. (1987). *In vitro* chemosensitivity testing using the multicellular tumor spheroid model. *Cancer Drug Del.* **4,** 63–73.

Kobayashi, H., Man, S., Graham, C. H., Kapitain, S. J., Teicher, B. A., and Kerbel, R. S. (1993). Acquired multicellular-mediated resistance to alkylating agents in cancer. *Proc. Natl. Acad. Sci. U.S.A.* **90,** 3294–3298.

Krishan, A., and Ganapathi, R. (1979). Laser flow cytometry and cancer chemotherapy: Detection of intracellular anthracyclines by flow cytometry. *J. Histochem. Cytochem.* **27,** 1655–1656.

Kwok, T. T., and Twentyman, P. R. (1985). The relationship between tumour geometry and the response of tumour cells to cytotoxic drugs—an *in vitro* study using EMT6 multicellular spheroids. *Int. J. Cancer* **35,** 675–682.

Luo, C., MacPhail, S., Dougherty, G. J., Naus, C. C., and Olive, P. L. (1997). Radiation response of connexin43-transfected cells in relation to the "contact effect". *Exp. Cell Res.* **234,** 225–232.

Luo, C., Johnston, P. J., MacPhail, S., Banáth, J. P., Oloumi, A., and Olive, P. L. (1998). Cell fusion studies to examine the mechanism for etoposide resistance in Chinese hamster V79 spheroids. *Exp. Cell Res.* **243,** 282–289.

McAllister, R. M., Reed, G., and Huebner, R. J. (1967). Colonial growth in agar of cells derived from adenovirus-induced hamster tumours. *J. Natl. Cancer Inst.* **39,** 43–47.

Mansbridge, J. N., Knuchel, R., Knapp, A. M., and Sutherland, R. M. (1992). Importance of tyrosine phosphatases in the effects of cell–cell contact and microenvironments on EGF stimulated tyrosine phosphorylation. *J. Cell Physiol.* **151,** 433–442.

Mayneord, W. V. (1932). On a law of growth of Jensen's rat sarcoma. *Am. J. Cancer* **16,** 841–846.

Mendelsohn, M. L. (1960). The growth fraction: A new concept applied to tumors. *Science* **132,** 1496–1496.

Mendelsohn, M. L., Dohan, J., and Moore, J. (1960). Autoradiographic analysis of cell proliferation in spontaneous breast cancer of C3H mouse. I. Typical cell cycle and timing of DNA synthesis. *J. Natl. Cancer Inst.* **25,** 477–484.

Minchinton, A. I., Wendt, K. R., Clow, K. A., and Fryer, K. H. (1997). Multilayers of cells growing on a permeable support. An *in vitro* tumour model. *Acta Oncol.* **36,** 13–16.

Morgan, S. A., Watson, J. V., Twentyman, P. R., and Smith, P. J. (1990). Reduced nuclear binding of a DNA minor groove ligand (Hoechst 33342) and its impact on cytotoxicity in drug resistant murine cell lines. *Br. J. Cancer* **62,** 959–965.

Moscona, A. A., and Moscona, M. H. (1966). Aggregation of embryonic cells in a serum-free medium and its inhibition at suboptimal temperatures. *Exp. Cell Res.* **41,** 697–702.

Mulcahy, R. T., Gould, M. N., and Clifton, K. H. (1980). The survival of thyroid cells: *In vivo* irradiation and *in situ* repair. *Radiat. Res.* **84,** 523–528.

Nederman, T. (1984). Effects of vinblastin and 5-fluorouracil on human glioma and thyroid cancer cell monolayers and spheroids. *Cancer Res.* **44,** 254–258.

Nicolson, G. (1987). Tumor cell instability, diversification and progression to the metastatic phenotype: From oncogene to oncofetal expression. *Cancer Res.* **46,** 1473–1487.

Nilsson, K. (1988). Microcarrier cell culture. *Biotechnol. Genet. Eng. Rev.* **6,** 403–439.

Olive, P. L. (1989). Distribution, oxygenation, and clonogenicity of macrophages in a murine tumor. *Cancer Commun.* **1,** 93–100.

Olive, P. L. (1992). DNA organization affects cellular radiosensitivity and detection of initial DNA strand breaks. *Int. J. Radiat. Biol.* **62,** 389–396.

Olive, P. L., and Durand, R. E. (1981). Uptake and mutagenicity of AF-2 in Chinese hamster V79 spheroids under aerobic and hypoxic conditions. *Environ. Mutagen.* **3,** 659–670.

Olive, P. L., and Durand, R. E. (1987). Characterization of a carbocyanine derivative as a fluorescent penetration probe. *Cytometry* **8,** 571–575.

Olive, P. L., and Durand, R. E. (1994). Drug and radiation resistance in spheroids: Cell contact and kinetics. *Cancer Metastasis Rev.* **13,** 121–138.

Olive, P. L., and MacPhail, S. (1992). Radiation-induced DNA unwinding is influenced by cell shape and trypsin. *Radiat. Res.* **130,** 241–248.

Olive, P. L., Chaplin, D. J., and Durand, R. E. (1985). Pharmacokinetics, binding and distribution of Hoechst 33342 in spheroids and murine tumours. *Br. J. Cancer* **52,** 739–746.

Olive, P. L., Durand, R. E., Banáth, J. P., and Evans, H. H. (1991a). Etoposide sensitivity and topoisomerase II activity in Chinese hamster V79 monolayers and small spheroids. *Int. J. Radiat. Biol.* **60,** 453–466.

Olive, P. L., Vanderbyl, S. L., and MacPhail, S. (1991b). Resistance to DNA denaturation in irradiated Chinese hamster V79 fibroblasts is linked to cell shape. *Exp. Cell Res.* **193,** 339–345.

Olive, P. L., Banáth, J. P., and Evans, H. H. (1993). Cell killing and DNA damage by etoposide in Chinese hamster V79 monolayers and spheroids: Influence of growth kinetics, growth environment and DNA packaging. *Br. J. Cancer* **67,** 522–530.

Pallavicini, M. G., LaLande, M. E., Miller, R. G., and Hill, R. P. (1979). Cell cycle distribution of chronically hypoxic cells and determination of the clonogenic potential of cells accumulated in G_2 + M phases after irradiation of a solid tumor *in vivo*. *Cancer Res.* **39,** 1891–1897.

Preisler, H. D., and Venugopal, P. (1995). Regrowth resistance in cancer: Why has it been largely ignored? *Cell Prolif.* **28,** 347–356.

Rasey, J. S., Cornwell, M. M., Maurer, B. J., Boyles, D. J., Hofstrand, P., Chin, L., and Cerveny, C. (1996). Growth and radiation response of cells grown in macroporous gelatin microcarriers (CultiSpher-G). *Br. J. Cancer* **74,** S78–S81.

Reddy, N. M. S., and Lange, C. S. (1991). Serum, trypsin, and cell shape but not cell-to-cell contact influence the X-ray sensitivity of Chinese hamster V79 cells in monolayers and in spheroids. *Radiat. Res.* **127,** 30–35.

Sacks, P. G., Miller, M. W., and Sutherland, R. M. (1981). Influences of growth conditions and cell–cell contact on responses of tumor cells to ultrasound. *Radiat. Res.* **87**, 175–186.

Sham, E., and Durand, R. E. (1999). Repopulation characteristics and cell kinetic parameters resulting from multifraction irradiation of xenograft tumors in SCID mice. *Int. J. Radiat. Oncol. Biol. Phys.*

Siemann, D. W., and Keng, P. C. (1986). Cell cycle specific toxicity of the Hoechst 33342 stain in untreated or irradiated murine tumor cells. *Cancer Res.* **46**, 3556–3559.

Smerdon, M. J. (1991). DNA repair and the role of chromatin structure. *Curr. Opin. Cell Biol.* **3**, 422–428.

Smith, K. A., Hill, S. A., Begg, A. C., and Denekamp, J. (1988). Validation of the fluorescent dye Hoechst 33342 as a vascular space marker in tumours. *Br. J. Cancer* **57**, 247–253.

Smith, P. J., and Anderson, C. O. (1984). Modification of the radiation sensitivity of human tumour cells by a bis-benzimidazole derivative. *Int. J. Radiat. Biol.* **46**, 331–344.

Sutherland, R. M. (1988). Cell and environment interactions in tumor microregions: The multicell spheroid model. *Science* **240**, 177–184.

Sutherland, R. M., and Durand, R. E. (1976). Radiation response of multicell spheroids—an *in vitro* tumour model. *Curr. Top. Radiat. Res. Q.* **11**, 87–139.

Sutherland, R. M., Inch, W. R., McCredie, J. A., and Kruuv, J. (1970). A multi-component radiation survival curve using an *in vitro* tumour model. *Int. J. Radiat. Biol.* **18**, 491–495.

Sutherland, R. M., McCredie, J. A., and Inch, W. R. (1971). Growth of multicell spheroids in tissue culture as a model of nodular carcinomas. *J. Natl. Cancer Inst.* **46**, 113–120.

Sutherland, R. M., Eddy, H. A., Bareham, B., Reich, K., and Vanantwerp, D. (1979). Resistance to adriamycin in multicellular spheroids. *Int. J. Radiat. Oncol. Biol. Phys.* **5**, 1225–1230.

Sweigert, S. E., and Alpen, E. L. (1987). Radiation-induced division delay in 9L spheroid versus monolayer cells. *Radiat. Res.* **110**, 473–478.

Tenforde, T. S., Kavanau, K. S., Afzal, S. M., and Curtis, S. B. (1990). Host cell cytotoxicity, cellular repopulation dynamics, and phase-specific cell survival in X-irradiated rat rhabdomyosarcoma tumors. *Radiat. Res.* **123**, 32–43.

Terasima, T., and Tolmach, L. J. (1963). Variations in several responses of HeLa cells to x-irradiation during the division cycle. *Biophys. J.* **3**, 11–33.

Thomlinson, R. H., and Gray, L. H. (1955). The histological structure of some human lung cancers and the possible implications for radiotherapy. *Br. J. Cancer* **9**, 539–549.

Trotter, M. J., Olive, P. L., and Chaplin, D. J. (1990). Effect of vascular marker Hoechst 33342 on tumour perfusion and cardiovascular function in the mouse. *Br. J. Cancer* **62**, 903–908.

Wartenberg, M., Hescheler, J., Acker, H., Diedershagen, H., and Sauer, H. (1998). Doxorubicin distribution in multicellular prostate cancer spheroids evaluated by confocal laser scanning microscopy and the "optical probe technique". *Cytometry* **31**, 137–145.

Watkins, A. M., Chan, P. J., Kalugdan, T. H., Patton, W. C., Jacobson, J. D., and King, A. (1996). Analysis of the flow cytometer stain Hoechst 33342 on human spermatozoa. *Mol. Hum. Reprod.* **2**, 709–712.

West, C. M. L. (1989). Size-dependent resistance of human tumour spheroids to photodynamic treatment. *Br. J. Cancer* **59**, 510–514.

West, C. M. L., and Moore, J. V. (1992). Mechanisms behind the resistance of spheroids to photodynamic treatment: A flow cytometry study. *Photochem. Photobiol.* **55**, 425–430.

West, G. W., Weichselbaum, R., and Little, J. B. (1980). Limited penetration of methotrexate into human osteosarcoma spheroids as a proposed model for solid tumor resistance to adjuvant chemotherapy. *Cancer Res.* **40**, 3665–3668.

Wiemann, M. C., Dexter, D. L., and Calabresi, P. (1987). Artificial capillary culture studies of human tumor cell growth, differentiation, and marker production. *J. Natl. Cancer Inst.* **79**, 93–102.

Wigle, J., and Sutherland, R. M. (1985). Increased thermotolerance developed during growth of small multicellular spheroids. *J. Cell. Physiol.* **122**, 281–289.

Winsor, C. P. (1932). The Gompertz curve as a growth curve. *Proc. Natl. Acad. Sci. U.S.A.* **18**, 1–7.

Yamaski, H. (1990). Gap junctional intercellular communication and carcinogenesis. *Carcinogenesis* **11**, 1051–1058.

Yamaski, H., and Katoh, H. (1988). Further evidence for the involvement of gap-junctional intercellular communication in induction and maintenance of transformed foci in BALB/c 3T3 cells. *Cancer Res.* **48,** 3490–3495.

Yelian, F. D., and Dukelow, W. R. (1992). Effects of a DNA-specific fluorochrome, Hoechst 33258, on mouse sperm motility and fertilizing capacity. *Andrologia* **24,** 167–170.

Young, S. D., and Hill, R. P. (1989). Radiation sensitivity of tumour cells stained *in vitro* or *in vivo* with the bisbenzimide fluorochrome Hoechst 33342. *Br. J. Cancer* **60,** 715–721.

Yuhas, J. M., Li, A. P., Martinez, A. O., and Ladman, A. J. (1977). A simplified method for production and growth of multicellular tumor spheroids. *Cancer Res.* **37,** 3639–3643.

Zhang, X., and Kiechle, F. L. (1997). Hoechst 33342-induced apoptosis in BC3H-1 myocytes. *Ann. Clin. Lab. Sci.* **27,** 260–275.

CHAPTER 42

Analysis of DNA Damage in Individual Cells

Peggy L. Olive, Ralph E. Durand, Judit P. Banáth, and Peter J. Johnston
Medical Biophysics Department
British Columbia Cancer Research Centre
Vancouver, British Columbia, Canada V5Z 1L3

I. Introduction
II. The Development of the Single Cell Gel Electrophoresis/Comet Assay
III. Comet Preparation and Analysis
    A. Comet Preparation
    B. Analysis of Comet Images
IV. Types of Damage Detected by the Comet Assay
V. Using DNA Damage to Predict Cell Survival in Complex Systems
VI. Future Directions
    References

I. Introduction

DNA is the most common intracellular target for the action of agents used to treat malignant disease, and it is an essential target for mutagenic agents. Measuring damage to cellular DNA is a relatively straightforward process using techniques developed over the last 30 years (Olive, 1998). However, tumors and some normal tissues are heterogeneous, and the average amount of DNA damage in a population of cells will not adequately predict for response. It is the undamaged or minimally damaged cells, which may constitute only a small fraction of the tumor, that are likely to cause tumor recurrence. Conversely, after low level exposure to genotoxic agents, it is the cells with damaged DNA that are more likely to develop mutations.

METHODS IN CELL BIOLOGY, VOL. 64

In 1984, Ostling and Johanson developed a microgel electrophoresis method that allowed detection of DNA strand breaks in individual tumor cells. Prior to this, Rydberg and Johanson (1978) had described a way of observing strand breaks in single cells based on differential lysis of irradiated cells in alkali. Later this technique was adapted for flow cytometry by first encapsulating cells in agarose beads before irradiation and alkaline lysis, then staining with acridine orange (Rydberg, 1984). Subsequent methods employed static or flow cytometry to evaluate damage produced in "nucleoids" composed of cells extracted with high salt and anionic detergents (Roti Roti and Wright, 1987; Vaughan *et al.*, 1993). However, the method introduced by Ostling and Johnason offered important advantages over earlier single-cell methods in terms of sensitivity and reliability. Electrophoresis of damaged DNA greatly improved the sensitivity for detecting small amounts of DNA damage. In addition, because cells were stabilized in agarose before lysis, a long delay was acceptable between slide preparation and analysis of DNA damage.

II. The Development of the Single Cell Gel Electrophoresis/ Comet Assay

There have been several reviews on the development and applications of the comet assay in genetic toxicology and oncology (see McKelvey-Martin *et al.*, 1993; Fairbairn *et al.*, 1995; Collins *et al.*, 1997a,b; Olive, 1999). Following the introduction of the microgel method by Ostling and Johanson in 1984, subsequent work improved on sensitivity and reproducibility of this method. An alkaline single-cell gel electrophoresis method was developed by Singh and colleagues (1988) to measure radiation-induced DNA single-strand breaks. Alkali allowed the DNA duplex to denature and unwind, and single strands of DNA could then migrate independently. Image analysis was first applied to a different version of the alkaline method called the comet assay based on the obvious appearance of damaged cells after electrophoresis (Olive *et al.*, 1990). The terms single-cell gel electrophoresis (SCGE) and comet assay are now used interchangeably. Figure 1 shows digitized images of two comets prepared from Chinese hamster V79 cells exposed to ionizing radiation. These complex three-dimensional images are first reduced to two dimensions for analysis, and then given a single numeric descriptor, so that some information is inevitably lost in describing DNA damage.

The comet or single-cell gel electrophoresis method can detect, either directly or indirectly, a variety of DNA lesions, including single-strand breaks, double-strand breaks, interstrand cross-links, and damage to the DNA bases. Apoptotic cells can be easily identified since highly fragmented DNA in apoptotic cells migrates efficiently (Olive *et al.*, 1993; Fairbairn *et al.*, 1996). Because so few cells are required for this method, fluorescence-activated cell sorting becomes a

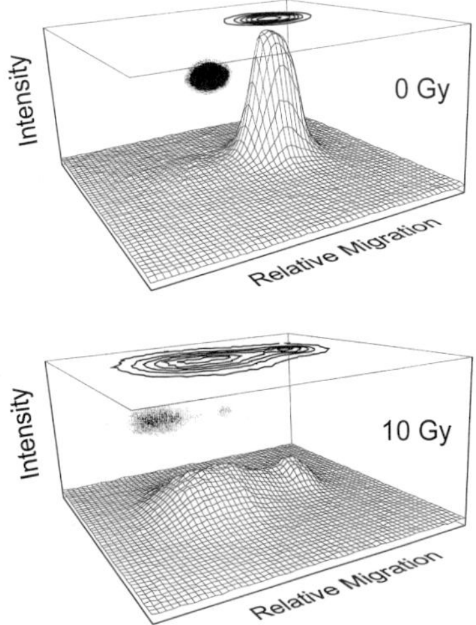

Fig. 1 Appearance of two irradiated TK6 lymphoblast cells in the alkaline comet assay using three-dimensional displays, digitized images (back projection), and contour plots (upper projection). The contours on the upper projections represent percentage of peak fluorescence.

practical way to provide purified, populations for analysis of DNA damage (e.g., Durand and Olive, 1997; Luo *et al.*, 1998). Comets can be stained with antibodies against incorporated bromodeoxyuridine (BrdUrd) (Olive and Banáth, 1992) or against specific DNA adducts (Buschfort *et al.*, 1997; Sauvaigo *et al.*, 1998). Comets can be examined using fluorescent probes for telomeres (Santos *et al.*, 1997) or for specific gene sequences (McKelvey-Martin *et al.*, 1998). Figure 2 shows the appearance of comets prepared from SiHa human cervical carcinoma cells. Cells were exposed to 50 Gy, then allowed to repair damage for 1 hr before conducting the alkaline comet assay. The cell with residual damage, equivalent to about 5 Gy, is easily discriminated from the cell that was not irradiated (Fig. 2b,c). These comets were stained with a fluorescent probe for Cot-1 DNA, a repetitive DNA sequence associated with the centromeres of human chromosomes (Fig. 2a). Similar percentages of Cot-1-labeled DNA and propidium iodide-labeled DNA were observed in the comet tail.

Much of the popularity of the comet assay is due to the ability of this method to measure DNA damage in virtually any single cell. A more unique aspect of this method is the ability to detect heterogeneity in response within different cells of a single population (Ostling and Johanson, 1987; Olive *et al.*, 1998). This information can be used in a variety of ways, for example, to identify drug-resistant or apoptotic cells within a population or to examine the response of cells of different ploidy or in different phases of the cell cycle.

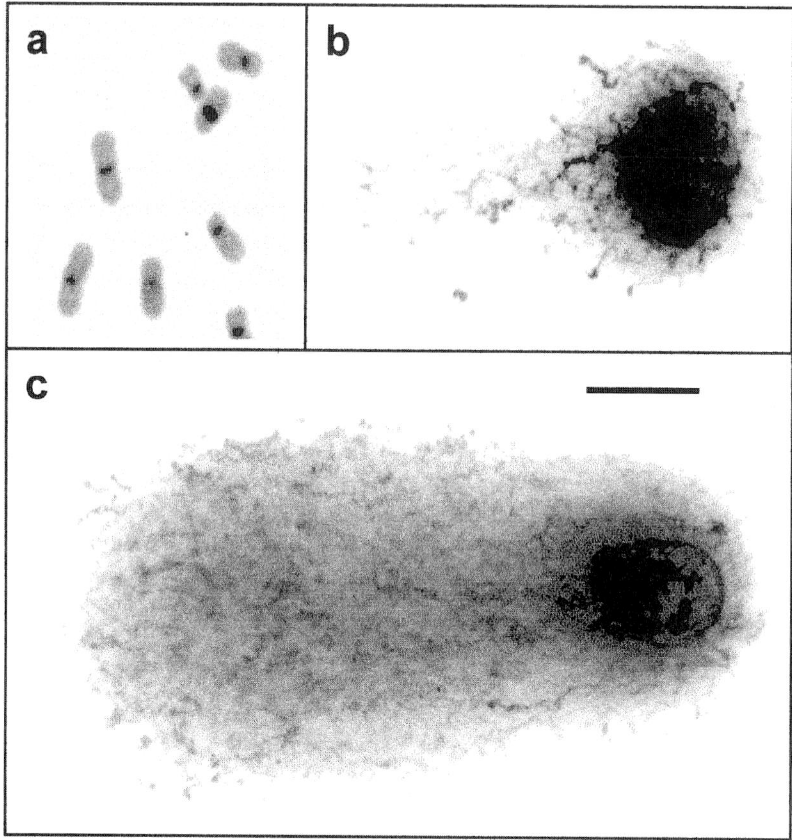

Fig. 2. Comet analysis using fluorescence *in situ* hybridization (FISH) probes. After performing the alkaline comet assay, SiHa human cervical carcinoma DNA was stained using a FISH probe for Cot-1 repetitive DNA that stains centromeric DNA (a). Digitized images of SiHa cervical carcinoma cells exposed to 0 Gy (b) or 50 Gy followed by a 1-hr repair period (c) are shown. The magnification bar (c) represents 10 μm.

III. Comet Preparation and Analysis

A. Comet Preparation

The general steps for comet preparation are as follows:

1. Cells (about 15,000) are embedded in a small amount (0.15 to 1.5 ml) of 0.75% low gelling temperature agarose and spread on a microscope slide that has been precoated with agarose to promote attachment.

2. Slides are placed in a lysis solution, usually containing 1–2.5 M NaCl and ionic and nonionic detergents (sarkosyl or sodium dodecyl sulfate, and

Triton X-100). Lysis solutions may also contain proteinase K, EDTA, and DMSO. For the neutral method, incubation at 50°C can be used to promote lysis and DNA diffusion. For the alkali method, the pH should be greater than 12.3 for rapid denaturation and unwinding. Alkali can be included in the lysis solution, or it can be added to the electrophoresis buffer. Lysis time can vary from a few minutes to overnight in a cold room.

3. Slides are equilibrated in electrophoresis buffer, electrophoresed in a horizontal electrophoresis chamber at low voltage for a few minutes (often at 0.6 V/cm for about 20 min), and finally stained with a fluorescent DNA binding dye. Silver staining provides a nonfluorescent alternative (Cerda *et al.*, 1997).

There are many versions of the alkaline comet method (Fairbairn *et al.*, 1995), although the end result is similar in terms of the appearance of the comets. In our efforts to reproduce different published protocols for alkaline comet preparation, we have observed no more than a factor of 2 difference in sensitivity of mammalian cells to ionizing radiation. A potentially critical point, however, is whether the salt concentration within the agarose on the slide is different from the salt concentration in the electrophoresis buffer at the time of electrophoresis. Ideally, these should be identical to prevent anomalous patterns of DNA migration within slides and problems with reproducibility between experiments (Fairbairn *et al.*, 1995). Other areas of potential problems are as follows:

1. Precoating slides with a small amount (100 μl) of agarose, and allowing it to dry before adding the agarose containing the cells, is commonly used to prevent slippage of the agarose from the slide during the various steps. Fully frosted slides (also precoated) are essential for the neutral method when using a higher temperature (50°C) during lysis.

2. Once DNA is exposed to alkali, it must not be allowed to renature until after electrophoresis. Renaturing promotes tangling and prevents efficient migration during electrophoresis. Exposure to alkali is far more effective in promoting DNA unwinding when 1–2 M NaCl is present (Rydberg, 1975).

3. Addition of up to 1 mg/ml proteinase K to the lysis solution may be necessary to reveal protein-linked DNA strand breaks produced by some drugs.

4. We routinely use propidium iodide or ethidium bromide for DNA staining. SYBR green, YOYO, and TOTO (all available from Molecular Probes, Eugene, OR) are also very efficient but more costly (Singh *et al.*, 1994). DAPI (4',6-diamidino-2-phenylindole) and Hoechst 33258 bleach too rapidly under ultraviolet (UV) excitation.

5. Addition of 2–10% DMSO to the agarose and/or lysing solution is necessary to prevent damage to neighboring nucleated cells when red blood cells are present in the slide and are lysed using the alkaline method.

6. After staining, slides can be air dried or fixed in alcohol and stored for future analysis. Fewer cells are required if slides are to be dried (no more than 10,000 per slide), and drying samples on fully frosted slides is not recommended.

At the time of analysis, we add about 1 ml agarose to the slide to reduce background fluorescence.

B. Analysis of Comet Images

Figure 2c shows the typical appearance of a single cell containing about 5000 single-strand breaks after alkaline lysis and electrophoresis. Large molecular weight DNA remains in the original position of the cell casting at the time of embedding. This DNA is presumably greater than 5 Mb in size, and does not appear to migrate during electrophoresis. Smaller fragments can migrate through the agarose to form the tail of the comet. Since only a few minutes are required for migration (longer electrophoresis would pull DNA out of the field of view), there is no opportunity for fragments to separate according to size.

Our software was initially written to use hardware functions for an image board that is no longer available. Many laboratories have developed their own comet analysis programs, and there are now several commercial sources. NIH Image is also available on the Internet. Several features used to describe comet images have become standard: percent DNA in the tail, tail moment (variously defined), and tail length. These end points are illustrated in Fig. 3 for mammalian

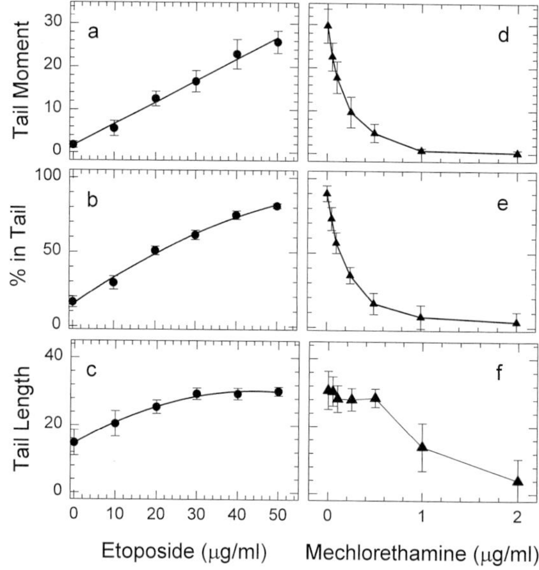

Fig. 3 Dose–response relationships using the alkaline comet assay. (a–c) A subline of Chinese hamster V79 cells resistant to killing by etoposide were examined for DNA strand breaks after a 1-hr exposure to etoposide. The means and standard deviations for three independent experiments are shown. (d–f) TK6 human lymphoblasts were exposed to the cross-linking agent mechlorethamine (nitrogen mustard) for 1 hr and then exposed at 0°C to 15 Gy X rays. The means and standard deviations for 100 individual comets per dose point, from one representative experiment, are shown.

cells exposed to the topoisomerase II inhibitor, etoposide, or to a bifunctional alkylating agent, nitrogen mustard. Note that the standard deviations for three independent experiments in Fig. 3a–c are relatively small, indicating the excellent reproducibility of the method. Tail moment (the product of percent DNA in the comet tail and the distance between the means of the head and tail DNA distributions) is linear over a large dose range and shows a high signal-to-noise ratio (Fig. 3a). However, since the value of tail moment includes a measure of distance, it is dependent on both the hardware (microscope, camera, image board) and software, and is not immediately comparable between laboratories. The percentage of DNA in the comet tail (Fig. 3b) should be less variable between experiments and between laboratories. However, this end point is not as sensitive as tail moment at low doses, and it still requires a discrimination between comet head and tail. Tail length increases over the low dose range but then remains fairly constant (Fig. 3c). The maximum distance of migration is determined primarily by the electrophoresis voltage and time. Even apoptotic cells, where the average fragment size can be 50 kb, show the same maximum tail length as cells with an average fragment size of 500 kb. DNA pieces smaller than about 40 kb appear to be lost during alkaline lysis and electrophoresis.

Other considerations in image analysis are as follows:

1. Various types of cameras (image intensifying and integrating) and fluorescence microscopes can be used. Filter sets should be chosen that are appropriate for the DNA fluorochrome, and uniform illumination of the field of view is critical. Camera output should be linear with intensity.

2. The camera image should not be saturated; otherwise, DNA content, percent DNA in tail, and tail moment cannot be accurately determined. Data in Fig. 5 show that in spite of significant heterogeneity in DNA damage (as indicated by the range in tail moments), DNA content remains constant if appropriate gains and thresholding are used. This is a good indication that the images are not saturated and that the field is uniformly illuminated.

3. Once a specific fluorescence light intensity (for variable intensity power sources) or camera gain/integration time is chosen, all slides for that experiment should be collected on that setting.

4. Depending on the specific software, the speed of image collection can be quite variable (<60 to >600 comets/hr). For some of our applications that require analysis of up to 1000 images per slide, speed of collection is an important consideration. We therefore analyze images in real time (<2 sec/image) and do not archive the raw images. The rate-limiting step is finding and focusing on the next comet image. Debris and overlapping images are potential sources of error in automatic collection of comets.

5. To date, methods have not been developed to standardize either preparation or analysis of comets (Bocker *et al.*, 1997). Intercomparison of image analysis systems or comparisons of results between laboratories using different methods

may be possible by using ionizing radiation to produce a given number of strand breaks (i.e., Gy-equivalent damage). To use this approach, comet preparation and analysis methods must be reproducible, and the dose–response relationship must be known.

IV. Types of Damage Detected by the Comet Assay

Under alkaline conditions, the comet assay detects DNA single-strand breaks, double-strand breaks, and alkali labile lesions. Sensitivity for detecting these lesions is as good or better than 5 cGy. Assuming about 1000 single-strand breaks per cell per Gy (Rydberg, 1980; Powell and McMillan, 1990), and random distribution of damage, this is equivalent to about 50 single-strand breaks per cell. Of course, the sensitivity for detecting induced damage is dependent on the amount of background damage. This in turn is dependent on the proportion of cells in the sample that are actively replicating their DNA. Replication forks appear as single-strand breaks in the presence of alkali (Rydberg, 1975) (Fig. 5), and each S phase cell may contain the equivalent of 2000 or more breaks (Olive and Banáth, 1993a).

Under neutral lysis and electrophoresis conditions, the method primarily detects DNA double-strand breaks (Olive *et al.*, 1991). The original Ostling and Johanson method was also performed with neutral conditions, but it was able to detect single-strand breaks because these breaks relaxed supercoiled DNA and enhanced its migration during electrophoresis (Ostling and Johanson, 1984, 1987). Therefore, at low doses, single-strand breaks are preferentially detected even under neutral conditions. However, once all of the DNA supercoils are relaxed, only double-strand breaks will allow DNA to migrate under neutral conditions (Olive and Johnston, 1997). The sensitivity of the neutral method is similar to that of the alkaline method in that about 50 double-strand breaks (produced by about 2 Gy) can be detected in nonreplicating cells. In this assay, replication bubbles inhibit DNA migration, reducing the sensitivity for detecting damage in S phase cells. It is possible to convert tail moment to number of breaks per cell by allowing cells to incorporate ^{125}I-labeled dUrd into DNA, and by assuming that one decay of this isotope produces one double-strand break (Olive and Banáth, 1993b).

DNA base damage can be identified using enzymes such as endonuclease III and fapy-DNA glycosylase (fpg) that create single-strand breaks at sites of a variety of oxidative base damages (Wallace, 1997). To detect base damage using the comet assay, cells are embedded in agarose and then lysed in 1% Triton X-100 and 2.5 *M* NaCl (Collins *et al.*, 1993, 1996). After rinsing, base damage detecting enzymes (now available from Trevigen, Gaithersburg, MD) are pipetted onto slides and covered with coverslips. After 20 min at 37°C, coverslips are removed, and slides continue onto alkaline lysis and electrophoresis. Enzyme sensitive sites can be detected with excellent sensitivity using the alkaline comet assay (Collins *et al.*, 1996; Evans *et al.*, 1995). In addition, artifacts associated with some chemical methods of base damage detection can be avoided by using the comet assay (Cadet *et al.*, 1997).

Figure 4 shows the response of cells exposed to the bioreductive drug tirapazamine, now in Phase III clinical trials as a hypoxic cell cytotoxin and chemopotentiator. Tirapazamine causes DNA single- and double-strand breaks (Olive, 1995a), but it also creates base damage that is recognized by endonuclease III and fpg. Figure 4a–c shows the strand breaks induced in aerobic SiHa cervical carcinoma cells exposed to tirapazamine. Note that S phase cells show more DNA damage as the result of the natural breaks at active replication sites. Figure 4d–f shows the additional strand breaks that become evident after treatment of permeabilized, dehistonized cells (i.e., cells treated with 0.5% Triton X-100 and 2 M NaCl) to endo III and fpg. These enzymes improve detection of tirapazamine-induced DNA damage by three- to fourfold. There is also the suggestion that tirapazamine produces more base damage in G_1 than in G_2 phase cells. Hypoxic SiHa cells show similar effects but at 30-fold lower drug doses.

Drugs that produce interstrand cross-links prevent DNA from migrating, even when the DNA contains significant numbers of single-strand breaks (Olive, 1995b; Pfuhler and Wolf, 1996). To determine if cross-links are present, cells can be exposed to a strand-breaking agent such as ionizing radiation. Observing fewer single-strand breaks than expected is an indication that interstrand cross-links are present. No DNA damage is evident when TK6 lymphoblasts were exposed to mechlorethamine (nitrogen mustard) for 30 min. However, Fig. 3d–f shows the response of TK6 cells exposed to mechlorethamine for 30 min followed immediately by irradiation with 15 Gy. The tail moment was initially high, but as the mechlorethamine concentration increased and interstrand cross-links increased, fewer radiation-induced strand breaks were detected. After high drug doses, DNA actually migrated less efficiently in mechlorethamine-treated than

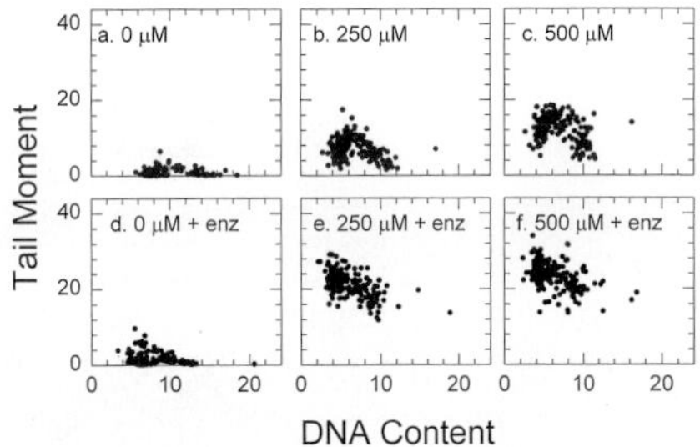

Fig. 4 Detection of base damage in SiHa human cervical carcinoma cells exposed to tirapazamine. Cells were incubated for 1 hr with tirapazamine under aerobic conditions. They were then analyzed for DNA damage using the alkaline comet assay (a–c). Alternatively, cells from the same population were permeabilized as described by Collins *et al.* (1996) and incubated for 20 min with 85 ng endo III and fpg. Slides were then processed using the alkaline comet method (d–f). Each symbol shows the response of a single cell.

in untreated cells. In Fig. 3d–f, the standard deviation shows the range of responses for 100 comets. The relatively small range indicated that cells within this population were responding in a similar fashion, and therefore showed a similar number of cross-links. This is not always the case, however, and Fig. 5 shows results when Chinese hamster V79 cells were grown as multicell spheroids and then incubated for 30 min with the bioreductive drug RSU-1069. This drug also creates interstrand cross-links, but only in the internal, hypoxic cells of spheroids (Olive, 1995b). Cross-links were not apparent after treatment with the drug alone (Fig. 5c), but they became visible after exposure to 10 Gy. Under the treatment conditions used in this experiment, approximately 50% of the cells failed to show radiation-induced strand breaks (Fig. 5d, box) and can therefore be considered to be hypoxic.

V. Using DNA Damage to Predict Cell Survival in Complex Systems

The relation between biological end points of response (killing, mutation, chromosome aberrations) and DNA damage varies for each agent. For agents that produce strand breaks, base damage, or cross-links, there is generally a tail moment threshold above which (or below in the case of cross-linking agents) cells begin to die. This threshold can be determined using model systems (e.g., spheroids and murine tumors) in which tail moment can be compared to surviving fraction for cells from the same sample. Once this threshold is known, it may be used to predict the fraction of cells likely to survive treatment (Olive *et al.*, 1996; Olive and Banáth, 1997). If the threshold can be shown to be largely

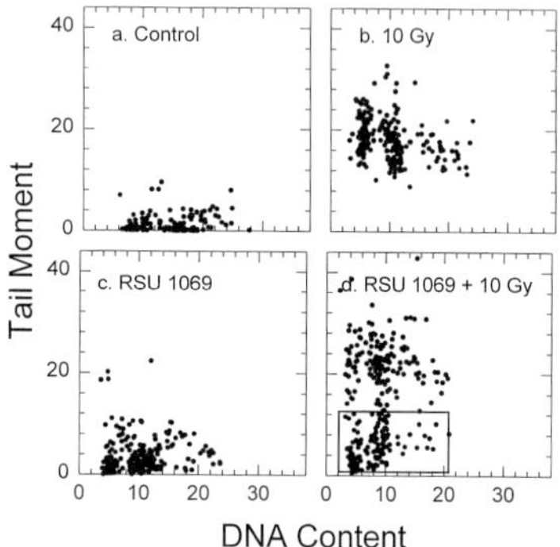

Fig. 5 Detection of interstrand cross-links in Chinese hamster V79 spheroids. Spheroids (600 μm in diameter) were equilibrated with 5% oxygen and exposed for 30 min to 0 or 50 μg/ml RSU 1069, then reduced to single cells using trypsin. Single cells were exposed to 0 or 10 Gy X rays. Single cells were embedded in agarose and analyzed using the alkaline comet assay. The box in (d) indicates the population of (hypoxic) cells with cross-links.

independent of cell type, this information could be particularly useful for estimating damage to nonclonogenic normal cells.

Figure 6 shows the response of Chinese hamster V79 lung fibroblasts exposed for 30 min to several DNA damaging agents (irradiated cells were exposed on ice). The amount of cell killing for a given number of strand breaks varies over several orders of magnitude. Clearly the chemical nature of the DNA damage and the ability of the V79 cells to repair this damage must differ enormously. An added complication is that strand breaks can represent the actions of repair enzymes. Therefore, damage measured at a specific time during drug incubation is the sum of damage induction and repair processes (Fortini *et al.*, 1996; Collins *et al.*, 1997b). Finally, the toxicity of many agents cannot be predicted based only on DNA damage. For example, ionizing radiation produces the same initial number of strand breaks in all (well-oxygenated) cells, regardless of their sensitivity to killing by radiation (e.g., Olive *et al.*, 1994). The dotted line in Fig. 6 shows the response of the radiosensitive lymphoblast cell line TK6 to X rays. Although the same number of breaks per gray were produced in V79 cells and TK6 cells, V79 cells are clearly much more proficient in repair.

For certain drugs, the average amount of DNA damage produced in a population of cells can be a good indicator of toxicity. However, in a heterogeneous cell mixture, a small fraction of nonresponding (drug-resistant) cells can go unnoticed. Similarly, a slow rate of rejoining of strand breaks can be the result of the presence of a small population of degrading cells, or a slow rejoining rate of all of the cells. The comet assay allows a distinction between these different

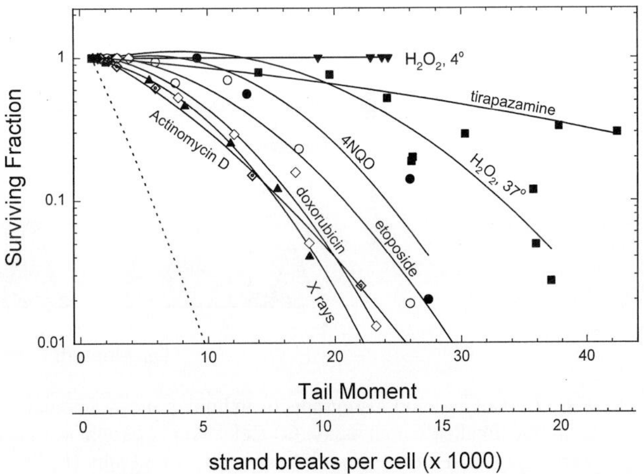

Fig. 6. Comparison between tail moment and cell survival (clonogenicity) in Chinese hamster V79 monolayers exposed for 30 min to various DNA damaging agents. The same cells were used for both assays. The dotted line shows the response of TK6 cells to X rays (revised from Olive and Johnston, 1997).

possibilities. As examples, Fig. 7 shows histograms from tumor biopsies containing etoposide-resistant cells (Fig. 7a) or necrotic/apoptotic cells (Fig. 7b). In Fig. 7c, a lung cancer metastasis initially contained 32% hypoxic tumor cells, determined by fitting two normal curves to the tail moment histogram. Hypoxic cells are distinguished from well-oxygenated cells based on the fact that hypoxic cells are about three times more resistant to DNA damage and cell killing than aerobic cells (Olive and Durand, 1992; Hu *et al.*, 1995). After treatment with nicotinamide to improve tumor blood flow (McLaren *et al.*, 1997), there appears to be a reduction in hypoxic fraction in this tumor (Fig. 7d). Of some practical importance, a single fine needle aspirate biopsy provides sufficient cells for the comet assay.

VI. Future Directions

Comet preparation and analysis should become more consistent between laboratories as optimum methods (simple, cost-effective, and reproducible) are developed and adopted. It is also likely that software will continue to undergo further improvements. Information on comet shape, the location of specific DNA sequences and their response to cytotoxic treatments, or the strength of association of DNA with the nuclear matrix may require new methods of comet analysis. Use of pulsed fields during electrophoresis (e.g., transverse alternating fields) could provide some information on DNA sizes, which is currently lacking. Com-

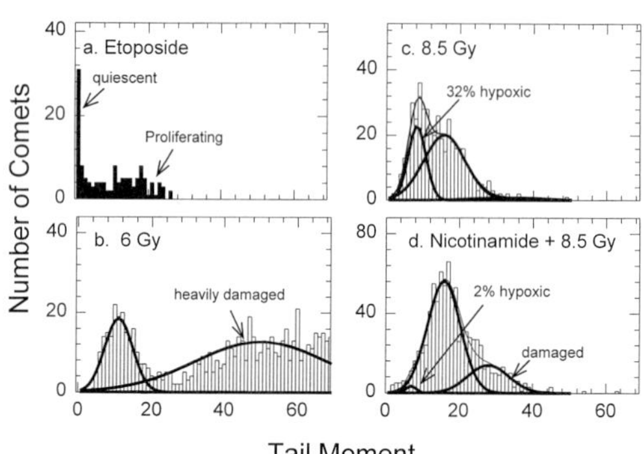

Fig. 7 Tail moment histograms from human tumor fine needle aspirate biopsies (FNAB) analyzed using the alkaline comet assay. (a) Cells from a patient with a squamous cell carcinoma of the neck were exposed *in vitro* to 5 μg/ml etoposide for 30 min. (b) Presence of a large population of heavily damaged (probably necrotic) cells from a patient with a head and neck cancer exposed to 6 Gy. (c) Cells taken immediately following 8.5 Gy to an adenocarcinoma of the bronchus metastatic to the skin. (d) Cells from the same tumor taken after treatment with nicotinamide and radiation (McLaren *et al.*, 1997). Note the decrease in tumor hypoxia and the presence of more damaged cells from the first dose of radiation.

bining antibody and FISH detection methods with the comet assay should provide important information on the relation between DNA adducts, DNA damage, and repair.

Application of the comet method to clinical samples has been shown to be feasible and could provide useful insights into degree of resistance, time to develop resistance, and reversal of resistance by chemical modifiers (e.g., Huang *et al.*, 1998). The method has already been used to identify those patients whose tumors contain radioresistant hypoxic cells (McLaren *et al.*, 1997). The current limitation of this particular application is that the tumor must be exposed to 4 Gy or more in order to reliably identify the hypoxic cells, and the clinical dose is usually 2 Gy. Alternatively, tumors containing hypoxic cells sensitive to tirapazamine can be identified using the comet assay at the start of treatment (J. M. Brown, personal communication, 1999). Routine application of this method could provide a basis for limiting exposure to those patients likely to benefit from this drug.

The majority of the applications of the comet method have been in the area of genetic toxicology. Standardization of the method is even more important in this field. Initial evaluation of chemicals will require comparison of comet results with more established end points (e.g., micronuclei, chromosome aberrations) since strand breaks, per se, do not necessarily lead to permanent genetic damage (Speit *et al.*, 1996; Wagner *et al.*, 1998). Analysis of the genotoxic effects of mixtures of chemicals will provide a much greater challenge.

Acknowledgments

This work was supported by National Institutes of Health Grant CA37879 and by the National Cancer Institute of Canada with funds provided by the Canadian Cancer Society.

References

Bocker, W., Bauch, T., Muller, W. U., and Streffer, C. (1997). Image analysis of comet assay measurements. *Int. J. Radiat. Biol.* **72,** 449–460.

Buschfort, C., Muller, M. R., Seeber, S., Rajewsky, M. F., and Thomale, J. (1997). DNA escision repair profiles of normal and leukemic human lymphocytes: Functional analysis at the single-cell level. *Cancer Res.* **57,** 651–658.

Cadet, J., Douki, T., and Ravanat, J. L. (1997). Artifacts associated with the measurement of oxidized DNA bases. *Environ. Health Perspect.* **105,** 1034–1039.

Cerda, H., Delincee, H., Haine, H., and Rupp, H. (1997). The DNA 'comet assay' as a rapid screening technique to control irradiated food. *Mutat. Res.* **375,** 167–181.

Collins, A. R., Duthie, S. J., and Dobson V. L. (1993). Direct enzymic detection of endogenous oxidative base damage in human lymphocyte DNA. *Carcinogenesis,* **14,** 1733–1735.

Collins, A. R., Dusinska, M., Gedik, C. M., and Stetina, R. (1996). Oxidative damage to DNA: Do we have a reliable biomarker? *Environ. Health Perspect.* **104,** 465–469.

Collins, A. R., Dusinska, M., Franklin, M., Somorovska, M., Petrovska, H., Duthie, S., Fillion, L., Panayiotidis, M., Raslova, K., and Vaughan, N. (1997a). Comet assay in human biomonitoring studies: Reliability, validation, and applications. *Environ. Mol. Mutagen.* **30,** 139–146.

Collins, A. R., Dobson, V. L., Dusinska, M., Kennedy, G., and Stetina R. (1997b). The comet assay: What can it really tell us? *Mutat. Res.* **375,** 183–193.

Durand, R. E., and Olive, P. L. (1997). Physiologic and cytotoxic effects of tirapazamine in tumor-bearing mice. *Radiat. Oncol. Invest.* **5,** 213–219.

Evans, M. D., Podmore, I. D., Daly, G. J., Perrett, D., Lunec, J., and Herbert, K. E. (1995). Detection of purine lesions in cellular DNA using single cell gel electrophoresis with Fpg protein. *Biochem. Soc. Trans.* **23,** 434S.

Fairbairn, D. W., Olive, P. L., and O'Neill, K. L. (1995). The comet assay: A comprehensive review. *Mutat. Res.* **339,** 37–59.

Fairbairn, D. W., Walburger, D. K., Fairbairn, J. J., and O'Neill, K. L. (1996). Key morphologic changes and DNA strand breaks in human lymphoid cells: Discriminating apoptosis from necrosis. *Scanning* **18,** 407–416.

Fortini, P., Raspaglio, G., Falchi, M., and Dogliotti E. (1996). Analysis of DNA alkylation damage and repair in mammalian cells by the comet assay. *Mutagenesis* **11,** 169–175.

Hu, Q., Kavanagh, M., Newcombe, D., and Hill, R. P. (1995). Detection of hypoxic fractions in murine tumours by comet assay: Comparison with other techniques. *Radiat. Res.* **144,** 266–275.

Huang, P., Olive, P. L., and Durand, R. E. (1998). Use of the comet assay for assessment of drug resistance and its modulation in vivo. *Br. J. Cancer* **77,** 412–416.

Luo, C., Johnston, P. J., MacPhail, S. H., Banáth, J. P., Oloumi, A., and Olive, P. L. (1998). Cell fusion studies to examine the mechanism for etoposide resistance in Chinese hamster V79 spheroids. *Exp. Cell Res.* **243,** 282–289.

McKelvey-Martin, V. J., Green, M. H. L., Schmezer, P., Pool-Zobel, B. L., De Meo, M. P., and Collins, A. (1993). The single cell gel electrophoresis assay (comet assay): A European review. *Mutat. Res.* **288,** 47–63.

McKelvey-Martin, V. J., Ho, E. T., McKeown, S. R., Johnston, S. R., McCarthy, P. J., Rajab, N. F., and Downes, C. S. (1998). Emerging applications of the single cell gel electrophoresis (Comet) assay. I. Management of invasive transitional cell human bladder carcinoma. II. Fluorescent in situ hybridization Comets for the identification of damaged and repaired DNA sequences in individuals. *Mutagenesis* **13,** 1–8.

McLaren, D. B., Pickles, T., Thomson, T., and Olive, P. L. (1997). Impact of nicotinamide on human tumour hypoxic fraction measured using the comet assay. *Radiother. Oncol.* **45,** 175–182.

Olive, P. L. (1995a). Detection of hypoxia by measurement of DNA damage in individual cells from spheroids and murine tumours exposed to bioreductive drugs. I. Tirapazamine. *Br. J. Cancer* **71,** 529–536.

Olive, P. L. (1995b). Detection of hypoxia by measurement of DNA damage in individual cells from spheroids and murine tumours exposed to the bioreductive drugs. II. RSU-1069. *Br. J. Cancer* **71,** 537–542.

Olive, P. L. (1998). Molecular assays for DNA damage. *In* "DNA Damage and Repair, Volume 2: DNA Repair in Higher Eukaryotes" (M. F. Hoekstra and J. A. Nickoloff, eds.), pp. 539–557. Humana Press, Totawa, New Jersey.

Olive, P. L. (1999). DNA damage and repair in individual cells: Applications of the comet assay in radiobiology. *Int. J. Radiat. Biol.* **75,** 395–405.

Olive, P. L., and Banáth, J. P. (1992). Growth fraction measured using the comet assay. *Cell Prolif.* **25,** 447–457.

Olive, P. L., and Banáth, J. P. (1993a). Induction and rejoining of radiation induced DNA single-strand breaks: "Tail moment" as a function of position in the cell cycle. *Mutat. Res. (DNA Repair)* **294,** 275–283.

Olive, P. L., and Banáth, J. P. (1993b). Detection of DNA double-strand breaks through the cell cycle after exposure to X-rays, bleomycin, etoposide and ^{125}IdUrd. *Int. J. Radiat. Biol.* **64,** 349–358.

Olive, P. L., and Banáth, J. P. (1997). Multicell spheroid response to drugs predicted with the comet assay. *Cancer Res.* **57,** 5528–5533.

Olive, P. L., and Durand, R. E. (1992). Detection of hypoxic cells in a murine tumor with the use of the comet assay. *J. Natl. Cancer Inst.* **85,** 707–711.

Olive, P. L., and Johnston, P. J. (1997). DNA damage from oxidants: Lesion complexity and chromatin organization. *Oncol. Res.* **9,** 287–294.

Olive, P. L., Banáth, J. P., and Durand, R. E. (1990). Heterogeneity in radiation-induced DNA damage and repair in tumour and normal cells measured using the "comet" assay. *Radiat. Res.* **122,** 69–72.

Olive, P. L., Wlodek, D., and Banáth, J. P. (1991). DNA double-strand breaks measured in individual cells subjected to gel electrophoresis. *Cancer Res.* **51,** 4671–4676.

Olive, P. L., Frazer, G., and Banáth, J. P. (1993). Radiation-induced apoptosis measured in TK6 human B lymphoblast cells using the comet assay. *Radiat. Res.* **136,** 130–136.

Olive, P. L., MacPhail, S. H., and Banáth, J. P. (1994). Lack of correlation between DNA double-strand break induction/rejoining and radiosensitivity in six human tumor cell lines. *Cancer Res.* **54,** 3939–3946.

Olive, P. L., Vikse, C. M., and Banáth, J. P. (1996). Use of the comet assay to identify cells sensitive to tirapazamine in multicell spheroids and tumours in mice. *Cancer Res.* **56,** 4460–4463.

Olive, P. L., Johnston, P. J., Banáth, J. P., and Durand, R. E. (1998). The comet assay: A new method to examine heterogeneity associated with solid tumours. *Nat. Med.* **4,** 103–105.

Ostling, O., and Johanson, K. J. (1984). Microelectrophoretic study of radiation-induced DNA damage in individual mammalian cells. *Biochem. Biophys. Res. Commun.* **123,** 291–298.

Ostling, O., and Johanson, K. J. (1987). Bleomycin, in contrast to γ irradiation, induces extreme variation of DNA strand breakage from cell to cell. *Int. J. Radiat. Biol.* **52,** 683–691.

Powell, S., and McMillan, T. J. (1990). DNA damage and repair following treatment with ionizing radiation. *Radiother. Oncol.* **19,** 95–108.

Pfuhler, S., and Wolf, H. U. (1996). Detection of DNA-crosslinking agents with the alkaline comet assay. *Environ. Mol. Mutagen.* **27,** 196–201.

Roti Roti, J. L., and Wright, W. D. (1987). Visualization of DNA loops in nucleoids from HeLa cells: Assays for DNA damage and repair. *Cytometry* **8,** 461–467.

Rydberg, B. (1975). DNA unwinding in alkali applied to the study of DNA replication in mammalian cells. *FEBS Lett.* **54,** 196–200.

Rydberg, B. (1980). Detection of induced DNA strand breaks with improved sensitivity in human cells. *Radiat. Res.* **81,** 492–495.

Rydberg, B. (1984). Detection of DNA strand breaks in single cells using flow cytometry *Int. J. Radiat. Biol.* **46,** 521–527.

Rydberg, B., and Johanson, K. J. (1978). Estimation of DNA strand breaks in single mammalian cells. *In* "DNA Repair Mechanisms" (P. C. Hanawalt, E. C. Friedberg, and C. F. Fox, eds.), pp. 465–468. Academic Press, New York.

Santos, S. J., Singh, N. P., and Natarajan, A. T. (1997). Fluorescence in situ hybridization with comets. *Exp. Cell Res.* **232,** 407–411.

Sauvaigo, S., Serres, C., Signorini, N., Emonet, N., Richard, M. J., and Cadet J. (1998). Use of the single-cell gel electrophoresis assay for the immunofluorescent detection of specific DNA damage. *Anal. Biochem.* **592,** 1–7.

Singh, N. P., McCoy, M. T., Tice, R. R., and Schneider, E. L. (1988). A simple technique for quantitation of low levels of DNA damage in individual cells. *Exp. Cell Res.* **175,** 184–191.

Singh, N. P., Stephens, R. E., and Schneider, E. L. (1994). Modifications of alkaline microgel electrophoresis for sensitive detection of DNA damage. *Int. J. Radiat. Biol.* **66,** 23–28.

Speit, G., Hanelt, S., Helbig, R., Seidel, A., and Hartmann, W. (1996). Detection of DNA effects in human cells with the comet assay and their relevance for mutagenesis. *Toxicol. Lett.* **88,** 91–98.

Vaughan, A. T., Anderson, P., Wallace, D. M., Beaney, R. P., and Lymch, T. H. (1993). Local control of T2/3 transitional cell carcinoma of bladder is correlated to differences in DNA supercoiling: Evidence for two discrete tumor populations. *Cancer Res.* **53,** 2300–2303.

Wallace, S. S. (1997). Oxidative damage to DNA and its repair. *In* "Oxidative Stress and the Molecular Biology of Antioxidant Defenses" (J. G. Scandalios, ed.), pp. 49–90. Cold Spring Harbor Laboratory, Cold Spring Harbor, New York.

Wagner, E. D., Rayburn, A. L., Anderson, D., and Plewa, M. J. (1998). Calibration of the single cell gel electrophoresis assay, flow cytometry analysis and forward mutation in Chinese hamster ovary cells. *Mutagenesis* **13,** 81–84.

CHAPTER 43

Cytometric Methods to Analyze Ionizing-Radiation Effects

William D. Wright, Isabelle Lagroye, Peng Zhang, Robert S. Malyapa, and Joseph L. Roti Roti

Mallinckrodt Institute of Radiology
Section of Cancer Biology
Washington University School of Medicine
St. Louis, Missouri 63108, USA

I. Introduction
II. Applications
III. Methods for Measuring DNA Damage on a Cell-by-Cell Basis
    A. Alkaline and Neutral Comet Assays
    B. Fluorescent Halo Assay
    C. Halo–Comet Method
IV. Summary
    References

I. Introduction

The biological effects of ionizing radiation are studied for two principal goals: (1) to better understand and improve the effectiveness of radiation therapy for cancer treatment and (2) to assess the environmental hazards associated with radiation exposure. In the former case, important goals are to better understand the mechanisms of cell killing and the reasons for radiosensitivity. In the latter case the goal is to understand the mechanisms of mutation and neoplastic transformation. Radiation-induced DNA damage is thought to be the main cause of cell killing, neoplastic transformation, and mutation. For these reasons, much of the studies of ionizing radiation effects involves the study of DNA damage and its repair.

Early methods to measure DNA damage were either biophysical or chemical methods conducted on bulk samples, which would give a single value for the entire cell population used to produce the sample. The advantage of cytometric methods is that they usually can provide quantified data on a cell-by-cell basis. In the past several years an increasing number of cytometric techniques for the measurement of DNA damage have become available. One of the first is the single-cell gel electrophoresis or comet assay, pioneered by Ostling and Johanson (1984) and refined by several other groups, including Singh *et al.* (1988) and Olive *et al.* (1990). This assay utilized automated and semiautomated image analysis. A latter development utilized *in situ* DNA supercoiling ability as an assay for DNA damage. With this method, known as the fluorescent halo assay, Roti Roti and Wright (1987) utilized either flow cytometry (Milner *et al.*, 1993) or image analysis to detect the extent of DNA supercoiling. A more recent method developed by Rhee *et al.* (1998) combines the halo assay with the comet assay. In this chapter we will describe these assays and discuss applications for each of them.

II. Applications

The goals previously described present the research worker with different criteria for the application of these assays. Specifically, as a measure of DNA damage following low dose exposures, one attempts to maximize the sensitivity of the assay so that DNA damage can be found following the lowest exposure. Studies have pushed the alkaline comet assay to sensitivities below the centigray level (Singh *et al.*, 1994, 1995; Malyapa *et al.*, 1998). The observations that DNA damage can be detected at these low levels are surprising based on theoretical calculations of the number of DNA single-strand breaks expected following these doses. Two possible explanations are that there is a predominance of alkali-labile damage induced by radiation compared with frank single-strand DNA breaks or that low levels of DNA damage affects the DNA unwinding step as well as DNA migration. Thus, further study with, and of, these assays should provide new information regarding the induction of DNA damage and the behavior of nuclear DNA under the assay conditions.

In terms of studies related to radiation oncology, the goals are to delineate the mechanisms of radiation effects and to identify cell types that are radiosensitive. Tumor cells, as well as normal tissues, respond to radiation treatment with a broad spectrum of degrees of sensitivity. Most radiosensitive cells display this phenotype due to defective DNA repair mechanisms. However, nuclear DNA is highly organized spatially within the nucleus by periodic attachments to the nuclear matrix, and evidence has been accumulating showing that the nature of this association can influence the way in which a cell copes with DNA damage (for review, see Roti Roti *et al.*, 1993). For example, the fluorescent halo assay (described in Section III,B) measures the DNA supercoiling ability of cell lines expressing altered radiosensitivity in a context that reflects nuclear DNA organization (Roti Roti and Wright, 1987). The CHO cell lines 4364 (wild type), XR-1 double-strand break (dsb) (repair-deficient, radiosensitive), and XR-122 (a radioresistant variant of XR-1 bearing human chromosome 5) were studied as a model system for radiosensitivity due to repair defects (Malyapa *et al.*, 1994). Several rat embryonal cell (REC) lines, primary (wild type), myc (primary transfected with c-*myc*), ras (primary/H-*ras*), and the myc/ras contransfectant, were studied as a model system for oncogene-mediated radioresistance (Malyapa *et al.*, 1996a). Analysis of the ability of DNA loop domains to undergo supercoiling changes in the presence of radiation-induced DNA damage by this assay revealed that in all cases DNA loop rewindability was reduced as cells became increasingly radiosensitive. Thus, data from the halo assay, that is, inhibition of DNA loop rewinding, appear to correlate with radiosensitivity in cells rendered radiosensitive by very different mechanisms. Although more work is needed, this approach appears promising to identify a certain type of cellular radioresistance. Thus, the methods described in this chapter have a variety of applications in the studies of the cellular effects of ionizing radiation.

III. Methods for Measuring DNA Damage on a Cell-by-Cell Basis

A. Alkaline and Neutral Comet Assays

The alkaline comet assay is one of the most sensitive methods to measure DNA damage. We describe the method developed by Singh *et al.* (1988) and the method developed by Olive *et al.* (1990).

All versions of the comet assay take advantage of the fact that, under appropriate lysis conditions, damaged DNA, having a radiation dose-dependent reduction in molecular weight, will migrate away from the residual nuclear mass toward the anode in an electric field. The extent of this migration is proportional to the amount and type of damage. Where appropriate, digestion of the agarose-embedded residual nuclear structures with proteinase K can reveal the presence of DNA–protein cross-links by releasing such constraints on damaged DNA and restoring the expected electrophoretic migration.

1. The Alkaline Method Developed by Singh *et al.*

The version presented next is the one most recently described in the literature (Singh *et al.*, 1994).

a. Materials

1. Agarose solution: 0.5% low-melting point agarose (Amresco, Solon, OH) in 0.9% NaCl. Melt agarose by microwaving in a beaker for 30 sec; shake

and repeat boiling until thoroughly dissolved. Adjust to the desired volume by adding 0.9% NaCl and store in a 41°–42°C water bath.

2. Lysis buffer: 2.5 M NaCl, 1% sodium N-lauroyl sarcosinate, 100 mM disodium EDTA, 10 mM Tris base, pH 10. This solution can be stored at room temperature. At the time of the experiment, add 1% Triton X-100 and readjust the pH to 10.

3. Electrophoresis buffer: 300 mM NaOH, 10 mM tetrasodium EDTA, 2% DMSO, 0.1% 8-hydroxyquinoline, pH 13.

4. DNA precipitation solution: 1:9 (v:v) 10 M ammonium acetate:100% ethanol.

5. Prestain solution: 5% sucrose, 10 mM NaH$_2$PO$_4$. This solution can be stored at 4°C. Add 5% DMSO at the time of use.

6. Staining solution: The DNA specific cyanine dye YOYO-1 (Molecular Probes, Eugene, OR) stock solution is prepared at a final concentration of 1 mM in DMSO. YOYO-1 staining solution is 1 μM YOYO in 5% DMSO (in distilled water). All YOYO-1 solutions are stored in the dark.

b. Considerations

The comet assay or single-cell gel electrophoresis in alkaline conditions is currently one of the most sensitive assays available to detect alkali-labile DNA damage in single cells (base damage, single-strand breaks, alkali-labile sites, and bound DNA protein cross-links). Two major versions have been described in the literature: one by Singh *et al.* (1988) and the other by Olive *et al.* (1992). The main difference between the methods described is that controls show comet tails in Singh's version (presumably due to electrophoresis duration and conditions, which could elongate DNA loops) in contrast to those in Olive's version in which controls show no comet tails. Although the sensitivity of both versions has been found quite similar (Singh *et al.*, 1994; Malyapa *et al.*, 1998), one can think that the presence of DNA migration in controls would allow the detection of potential protein–DNA cross-links (visualized as a reduction in DNA migration). However, Singh *et al.* (1994) usually add proteinase K in order to more accurately measure DNA damage. Finally, using any version of the alkaline comet assay, cross-linking agents can be efficiently detected as they impede the migration of DNA induced by a genotoxic agent (Olive *et al.*, 1998).

2. The Neutral Method Developed by Singh *et al.*

a. Materials

The materials are as for the alkaline method by Singh *et al.* (1994), except for the electrophoresis buffer: 100 mM Tris, 300 mM sodium acetate, pH 9.

Under neutral conditions, the comet assay allows the measurement of the amount of DNA double-strand breaks, and only variations from the alkaline comet assay will be described here. Because of the neutral pH, removal of nuclear proteins takes place as an additional step.

In all the following procedures each step, without exception, needs to be performed in minimum indirect light. The cell suspension is usually prepared in a saline solution such as phosphate-buffered saline (PBS) or Hanks' balanced salt solution. After treatment, samples from animal tissues or cell culture containing a suspected DNA damage-inducing agent either have to be prepared as fast as possible and kept on ice to prevent any DNA repair. Antioxidants such as N-t-butyl-phenylnitrone (200 mM) can be added at this time. In the case of monolayer cell cultures, a prolonged trypsinization step should be avoided (e.g., a 2-min trypsinization is optimal for C3H 10T$_{1/2}$ cells). However, enzymatic dissociation is preferred to mechanical scraping to ensure the production of a single cell suspension. Cell suspensions should be gently handled, and pipettes with large apertures should be used.

b. Procedures

Three-layer agarose slides are prepared as follows: the first layer is prepared by pipetting 50 μl of agarose solution onto a fully frosted slide and covering it with a cover glass. After the agarose solidifies while on ice for 1 min, the cover glasses are removed, and the layer is allowed to air dry (this dried agarose provides firm attachment for subsequent layers). These slides can be stored for many months. In the next step, 5 μl of the cell suspension (10^5 to 10^6 cells/ml) is mixed with 50 μl of agarose, layered onto the slide, covered with a cover glass, and put on ice as before. The cover glasses are removed, and 200 μl of agarose solution is placed on top of the microgel as the third layer and then covered as described above. After 1 min on ice, the cover glasses are removed and the slides are immersed in cold lysis buffer for a minimum of 1 hr. Where protein/DNA cross-links are suspected, a treatment with DNase-free proteinase K (1 mg/ml) in lysis solution without detergents can be performed at 37°C for 2 hr. The slides are then transferred into an electrophoresis chamber (Hoefer He 100 Super Sub, Amersham Pharmacia Biotech, Piscataway, NJ), which can be modified as previously described (Singh *et al.*, 1994) and immersed in electrophoresis buffer (~1 liter) for 20 min at room temperature to accomplish DNA unwinding. Electrophoresis is then performed at 0.4 V/cm (12 V, 225–250 mA) for 1 hr in the dark. Recriculation of the buffer (~50 ml/min) is recommended for a uniform distribution of salts. Thereafter, DNA is precipitated by immersion in ammonium acetate solution for 30 min at room temperature. Dehydration is achieved by placing the slides in 100% ethanol for a minimum of 2 hr, and slides are subsequently rehydrated for 5 min in 75% ethanol (to prevent cracking of the dehydrated microgels during air drying). Once dried at room temperature, the slides can be stored indefinitely. At the time of the analysis, 50 μl of prestain solution is pipetted onto the microgel and covered with a cover glass for 5 min. After the cover glass is removed, 50 μl of YOYO-1 staining solution is put onto the microscope slide and covered as before to stain the DNA.

For the detection of DNA double-strand breaks, the microgel preparation and cell lysis procedures are the same as in the alkaline comet assay. After lysis, the slides are treated first with RNase A (10 μg/ml in lysis solution without detergents) at 37°C for 2 hr and then with proteinase K as described for the alkaline comet assay (1 mg/ml, 37°C, 2 hr). The DNA is then allowed to unwind for 20 min in electrophoresis buffer, and the electrophoresis is performed at 0.4 V/cm (12 V, 100 mA) for 1 hr in the dark with buffer recirculation. Procedures for drying the agarose microgel and staining the DNA are identical to the alkaline version of the assay, described earlier.

3. The Alkaline (Comet) Method Developed by Olive *et al.* (1992)

a. Materials

1. Agarose solution: 1% of low-melting point agarose (SeaKem Gold, FMC, Rockland, ME) in PBS. Agarose solution is prepared and stored as described earlier.
2. Lysis buffer: 1.2 M NaCl, 30 mM NaOH, 0.1% sodium N-lauroyl sarcosinate, 0.1 M Na$_2$EDTA, pH 12.0. The buffer is prepared at the time of the experiment and cooled to 4°C before use.
3. Electrophoresis buffer: 30 mM NaOH, 2 mM EDTA (acid), pH 12.5. The buffer is kept at 4°C until use.
4. Staining solution: Propidium iodide (PI) stock solution (Sigma, St. Louis, MO) is prepared at a final concentration of 5 mg/ml in distilled water. PI staining solution is 2.5 μg/ml PI in 0.1 M NaCl prepared after thoroughly mixing the PI stock solution. All PI solutions are stored in the dark.

b. Procedures

The slides are prepared as follows: 1.5 ml of agarose is added to 0.5 ml of cell suspension ($\sim$3 $\times$ 10^4 cells/ml) in a tube. Approximately one milliliter of this mixture is quickly layered onto a labeled fully frosted microscope slide, covered with a coverslip (22 $\times$ 60 mm^2), and placed on ice for 1–2 min before continuing. After the cover glasses are removed, the slides are placed into a container of cold lysis buffer for at least 1 hr. If needed, they can be stored in the dark at 4°C in the lysis buffer for several days. To remove salts that impede DNA migration, the slides are first washed with cold electrophoresis buffer (once for 5 min, once for 20 min). They are then placed in a horizontal electrophoresis chamber as described earlier, covered with electrophoresis buffer (900 ml), and allowed to unwind for 40 min in the dark at room temperature. The electrophoresis is performed for 25 min at 0.6 V/cm (18V, 35–45 mA). A constant volume of buffer allows the amperage to be reproducible from experiment to experiment. To obtain a uniform migration among the slides, it is recommended to put only one row of slides in the electrophoresis chamber. For the same reason the use of a modified Hoefer He100 (Amersham/Pharmacia) electrophoresis unit with two anodes and two cathodes on each side is advised. Also, always using the same power supply avoids variations between experiments. After electrophoresis, the slides are rinsed in distilled water, placed in PI staining solution for 15 min, and rinsed again in distilled water. To avoid the diffusion of the PI stain in the agarose, it is recommended to stain one set of three to four slides at a time (the other slides are maintained in distilled water). Observation of the slides must be performed immediately after staining. When frosted slides are used, transferring the agarose layer to a regular microscopic slide can improve observation of the comets by decreasing background light.

4. The Neutral Method Developed by Olive *et al.* (1992)

a. Materials

Materials are as in the previous section, except for the changes below.

1. Lysis buffer: 30 mM EDTA, 0.5% sodium dodecyl sulfate (SDS), pH 8.0.
2. Electrophoresis buffer: 90 mM Tris, 90 mM boric acid, 2 mM EDTA, pH 8.5.

b. Procedures

Once the agarose is solidified, slides are submersed in lysis buffer at 50°C for 4 hr and then washed in electrophoresis buffer for 2–16 hr. Electrophoresis is performed at 0.55 V/cm for 25 min. Slides are then rinsed and stained for 1 hr in PI staining solution.

c. Instrumentation (Both Olive Methods)

Images of comets are visualized using an inverted fluorescence microscope, usually attached to a camera and an eyepiece micrometer or an image analysis system. Slides are illuminated with green light (510–560 nm) from a 100-W mercury lamp when stained with PI. In the case of YOYO-1 staining, a filter combination for fluorescein isothiocyanate (excitation at 490 nm, emission at 515 nm) is used.

The parameters determined are mainly the comet length and the normalized comet moment. The comet length, which provides a measure of DNA migration, can either take into account the total distance between both edges of the comet (the whole nucleus area is included) or consider the distance between the center of the head and the distal edge of the comet. The normalized comet moment is defined as the sum of the amount of DNA (based on fluorescence intensity) at different points along the axis of migration, multiplied by the distance from the center of the head and weighted by the total amount of DNA (see Kent *et al.*, 1995). Komet (Kinetic Imaging, Liverpool, U.K.) and Optimas (Data Cell, Maidenhead, U.K.) can be used for the measurement of these or more parameters. We currently use a comet macro program designed by Computer Imaging

Applications (Madison, WI) running on an Optimas image analysis platform (Media Cybernetics, Silver Spring, MD).

As described earlier, the Singh *et al.* (1992) version of the assay is optimally designed for manual analysis (micrometer), although semiautomated analysis can be performed, and it allows measurement of comet length only.

5. Results

For all versions of the comet assay described here, comet lengths and normalized comet moments are conventionally displayed as frequency histograms that provide a mean value for each parameter as well as a population distribution for each experimental condition. A typical example of comet frequency distributions is shown in Fig. 1. This experiment examines initial ionizing radiation-induced DNA damage in two CHO cell lines of differing intrinsic radiosensitivity. XRC-421 is a variant of the radiosensitive cell line XR-1 (Stamato *et al.*, 1983) that has been rendered radioresistant through transfection with the human DNA repair gene *XRCC4* (Li *et al.*, 1995), while EBV-1 are XR-1 cells transfected with the Epstein–Barr virus transformation vector alone and remain radiosensitive. These data indicate that the degree of initial damage induction does not account for differential radiosensitivity in these cell lines.

B. Fluorescent Halo Assay

Of the assays for DNA damage and repair described here, only the fluorescent halo assay and that derived from it (the comet–halo assay) detect the expression of radiation-induced DNA damage that reflects the recipient cells' intrinsic radiosensitivity. The specific nature of this parameter is unknown, but the decreased ability of nucleoid DNA from radiosensitive cells to rewind in the presence of DNA damage has been observed in all intrinsically radiosensitive cell lines studied so far—over 18 in all.

1. Imaged Based Cytometry Method

a. Materials

1. Propidium iodide (PI) stock solution: 5mg/ml PI in deionized water. Store in a light-tight container at room temperature.

2. Dye/lysis solutions: 1.95 M NaCl, 20 mM EDTA, 20 mM Tris base (pH 8.0), 0.5% Triton X-100. This formulation has been optimized for use with HeLa cells. For use with other cell lines see Section III,B,1,c. To produce a standard unwinding/rewinding PI response curve, lysis solutions are supplemented with PI stock solution to final concentrations of 1, 4, 10, 15, 20, 70, and 100 μg/ml. These solutions are stable for several months when stored in the dark at 4°C.

b. Procedures and Instrumentation

Single cell suspensions obtained by brief (2–3 min) trypsinization of monolayer cultures are adjusted to 2.3×10^5 cells/ml with ice-cold PBS and exposed to graded doses of ionizing radiation. Immediately following radiation exposure, 250 μl of the cell suspension is placed in each well of a four well tissue culture chamber slide (Nalge Nunc International, Naperville, IL). An equal volume of each dye/lysis buffer solution is then added, and the cells are allowed to lyse for 5 min at room temperature before visualization using an inverted microscope equipped with epifluorescence optics with 510–560 nm excitation. Fields are selected for image capture using care not to expose the samples to excess light exposure because the heat generated by such exposure may cause sample damage. Digitized images are stored for each PI concentration, and particle diameters are subsequently obtained using an AutoHalo macro program (Computer Imaging Applications) running on an Optimas image analysis platform (Media Cybernetics).

c. Special Considerations

Cell lysis conditions must frequently be optimized according to the cell type under study. Such optimization most commonly involves alterations in the NaCl and Triton X-100 concentrations in the dye/lysis buffer formulations. Such optimizations are performed by constructing a matrix of NaCl/Triton X-100 ratio combinations using sample integrity over time as an end point. Overly aggressive lysis conditions usually result in a rapid loss of sample internal structure accompanied by a loss of DNA loop rewinding ability due to the disruption of DNA–matrix attachment regions. Under ideal lysis conditions, a well-defined brightly staining core surrounded by a more weakly staining DNA halo persists from 5 to 30 min of lysis time with the majority of particles settling to the slide bottom, providing numerous sedentary objects within a stationary focal plane. Examples of alternate lysis conditions for other cell types determined empirically are as follows: for CHO cells, 3.0 M NaCl, 20 mM EDTA, 20 mM Tris base (pH 8.0), 0.025% Triton X-100; for murine lymphoma (L5178Y) cells, 3.0 M NaCl, 20 mM EDTA, 50 mM Tris base (pH 8.0), 0.012% Triton X-100 (Malyapa *et al.,* 1994, 1996b).

d. Results

The results from a typical experiment demonstrating ionizing radiation dose-dependent inhibition of DNA loop rewinding is shown in Fig. 2. DNA damage is expressed as the decreased ability of DNA loops to undergo PI-driven supercoil rewinding at concentrations above 15 μg/ml between irradiated and control nucleoids. The determination of this parameter is shown schematically in Fig. 3. An example of the correlation of excess halo diameter (EHD) with intrinsic cellular radiosensitivity is shown in Fig. 4. Here, two groups of CHO-derived cell lines [xrs-5 and XR-1 (radiosensitive) and CHO.K1, 4364, and XR-122 (radioresistant)] are compared for clonogenic survival and EHD as a function of radiation dose. The results clearly show an inverse relationship between radiation

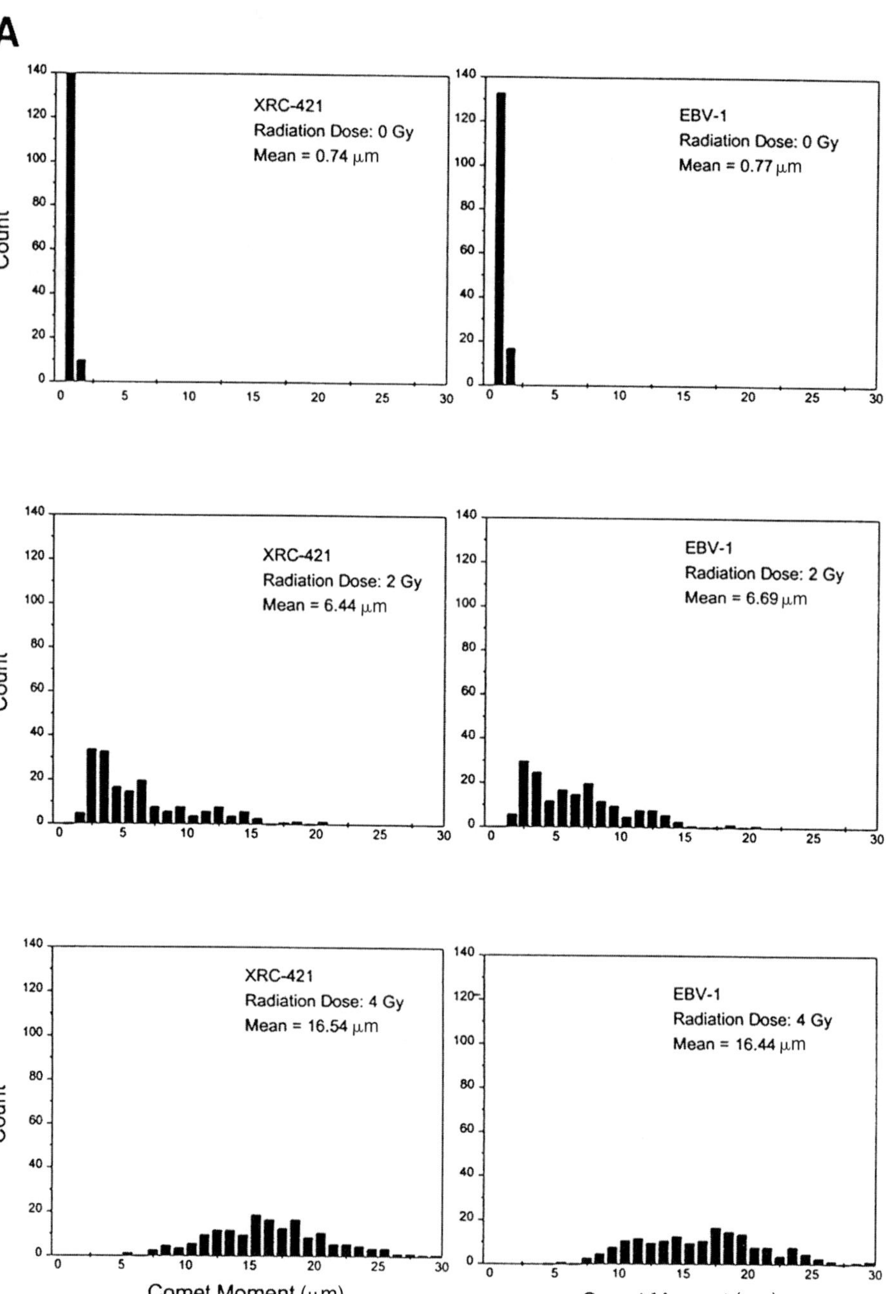

Fig. 1 Dose–response analysis of ionizing radiation-induced DNA damage in CHO cells using the alkaline comet assay according to Olive *et al.* (1992). (A) Frequency distributions of comet moment in radioresistant (XRC-421) and radiosensitive (EBV-1) cells following irradiation with 0, 2, or 4 Gy. The exposure dose and mean of the frequency distributions are indicated. Each data set represents images from 150 comets. (B) Frequency distribution of comet length derived from the same images as in (A).

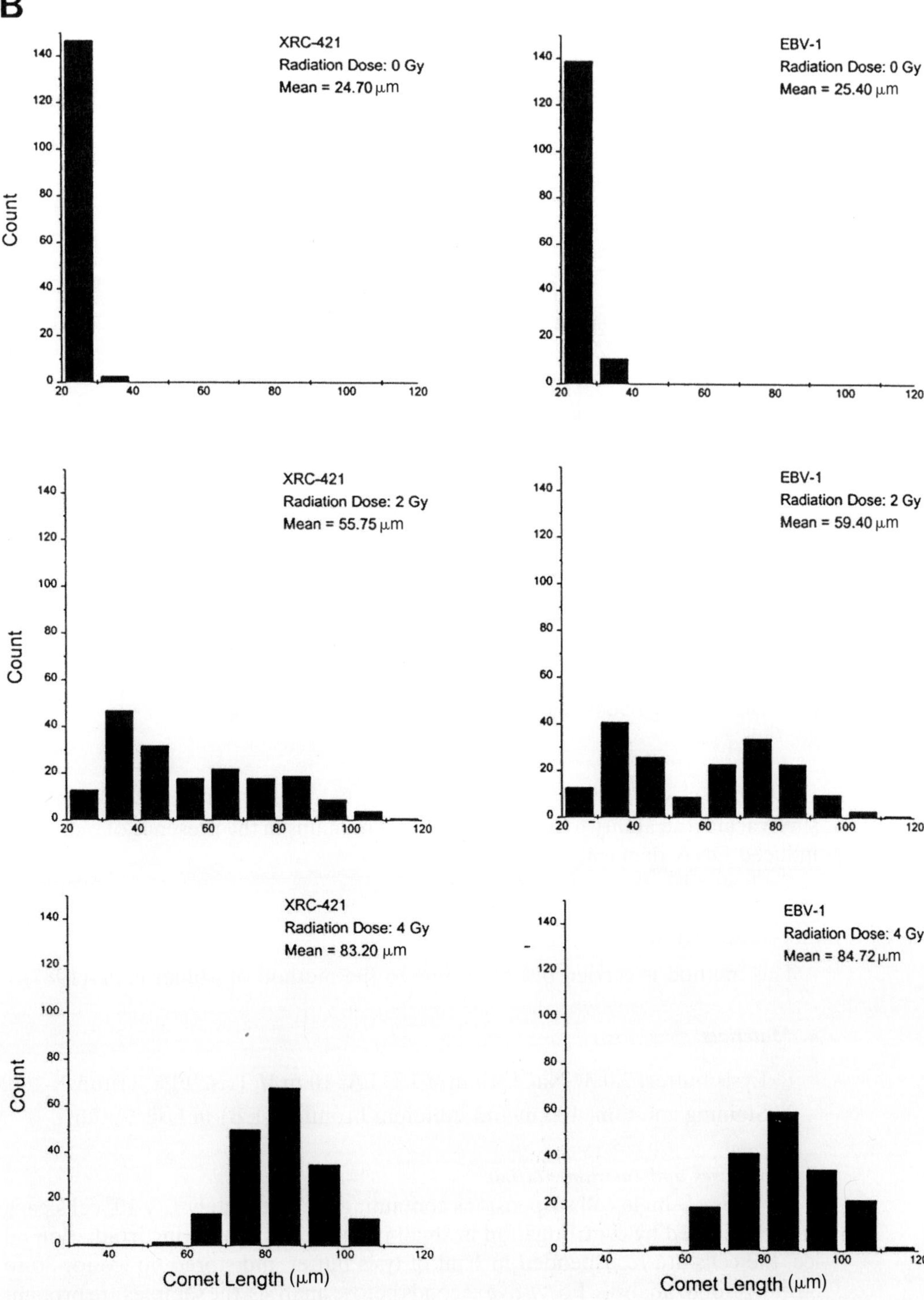

Fig. 1 (*continued*).

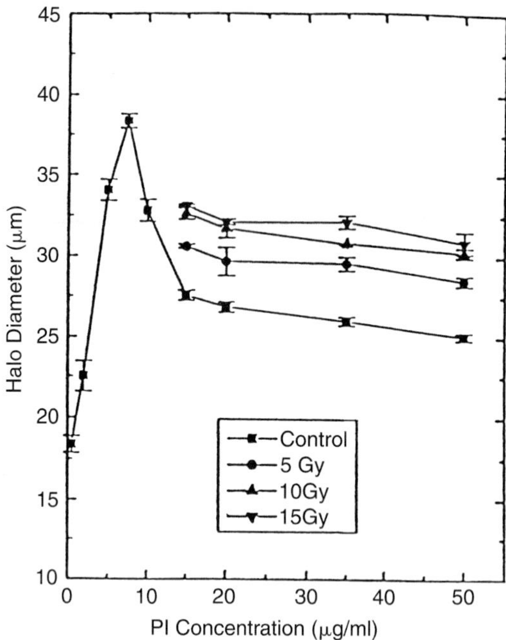

Fig. 2 DNA loop unwinding and rewinding as a function of PI concentration in nucleoids from unirradiated control CHO cells or from cells exposed to the indicated doses of γ-radiation. The plotted points represent mean values obtained from three to four independent experiments, and the error bars represent 1 SEM. Reproduced with permission from Malyapa *et al.* (1994).

survival and the ability to rewind DNA loop domains in the presence of radiation-induced DNA damage.

2. Flow Cytometry-Based Method

This method is carried out according to the method of Milner *et al.* (1987).

a. Materials

1. Lysis buffer: 2.0 *M* NaCl, 10 m*M* EDTA, 10 m*M* Tris, 0.1% Triton X-100.
2. Staining solution: 100 mg/ml ethidium bromide (EB) in lysis buffer.

b. Procedures and Instrumentation

Samples of single cell suspensions containing approximately 1×10^6 cells each are sedimented by centrifugation in small plastic tubes. Following irradiation on ice, the cells are resuspended in 1 ml of lysis buffer and stored on ice for 30 to 50 min prior to analysis. Forty-five seconds before analysis, the samples are brought

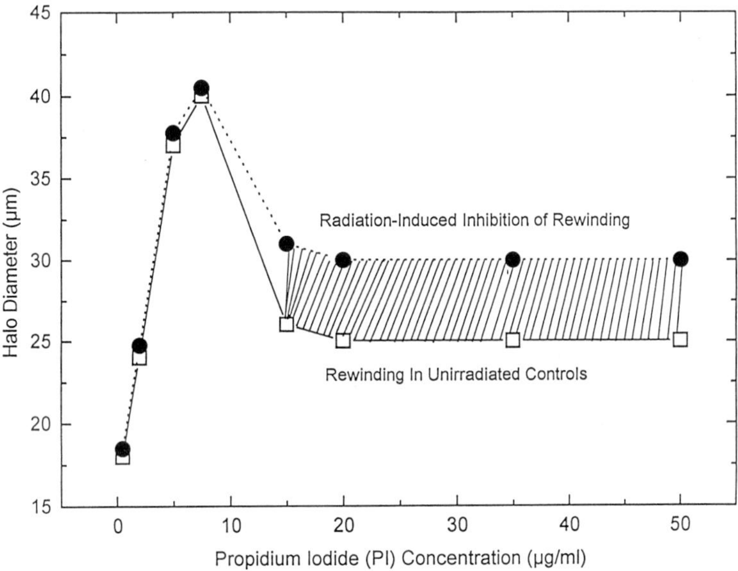

Fig. 3   Schematic representation of the calculation of excess halo diameter from data such as that presented in Fig. 2. The PI-dependent loop unwinding and rewinding curve in control unirradiated in nucleoids is depicted by the open squares, whereas that showing radiation-induced inhibition of rewinding is represented by the solid circles. The shaded region indicates the DNA damage-induced excess halo diameter. The data points are for illustration purposes and were not generated experimentally.

to 0, 2, 10, 20, or 50 mg/ml EB by addition of staining solution. After the 45-sec interval for dye/sample equilibration, the samples are analyzed by flow cytometry for red fluorescence and forward angle light scatter. A standard flow cytometric device set up to detect red fluorescence and forward angle light scatter, and fitted with a 60-μm diameter orifice is used for data generation. Data acquisition and analysis are achieved through an interface to a personal computer (PC) with resident flow cytometry analysis software. Raw data are conventionally plotted as fluorescence or scatter (y axis) versus channel number (x axis). EB and radiation dose responses may then be expressed as a mean channel number of either parameter (fluorescence as scatter) versus experimental condition (EB concentration or radiation dose. Using this technique these researchers have demonstrated differential DNA loop rewinding in several human tumor cell lines expressing altered radiosensitivity with the more radiosensitive cell lines displaying reduced ability to rewind damaged DNA loops (Lynch *et al.*, 1991; Milner *et al.*, 1993).

C. Halo–Comet Method

A method has been developed that combines features of the fluorescent halo and comet assays (Rhee *et al.*, 1998). In this assay PI-induced nucleoid halos are

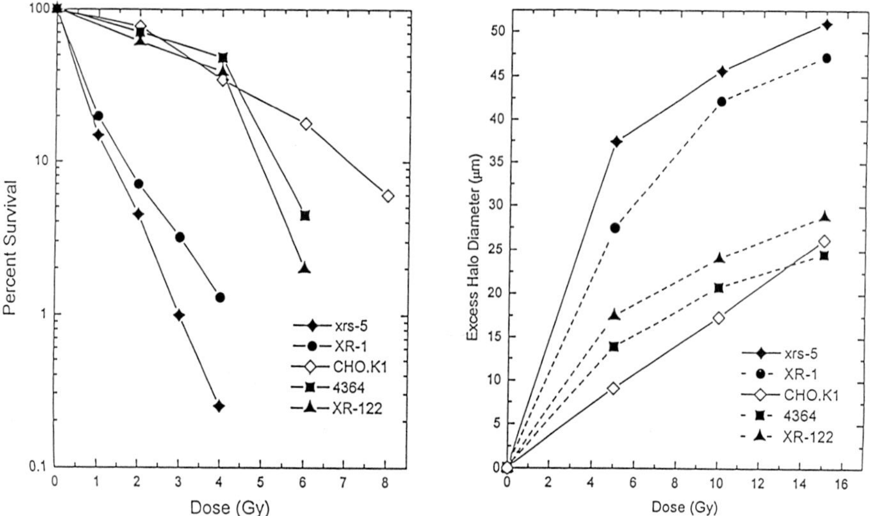

Fig. 4 The relationship of clonogenic survival and excess halo diameter (EHD) in a group of CHO cells of differing radiosensitivity. Percent survival (left) and EHD are plotted as functions of radiation dose. The symbols are defined in each graph. Results are from a typical experiment. Reproduced with permission from Malyapa *et al.* (1995).

produced in cells embedded in agarose. The resultant halo structure is then subjected to an electric field that rolls the DNA toward the cathode. Under these conditions, Rhee *et al.* (1998) report assay sensitivity to as low as 0.5 Gy of ionizing radiation and the detection of radiation-induced alterations in nucleoid DNA organization, which persist up to 6 days following postirradiation repair. This latter observation may represent the presence of global genomic structural alterations that remain long after frank radiation-induced strand breaks are repaired.

a. Materials

1. Lysis buffer: 2.0 M NaCl, 10 mM EDTA, 2mM Tris, pH 8.0, 0.5% Triton X-100. This solution may be stored indefinitely at room temperature.
2. 1% low-gelling-temperature agarose solution: low-gelling temperature agarose (Seakam Gold, FMC, Rockland, ME) is dissolved in PBS by microwave heating to a final concentration of 1% and maintained at 43°C.
3. Propidium iodide (PI) stock solution: PI (Sigma) is dissolved to a final concentration of 5 mg/ml in deionized water and stirred overnight in a light-tight container.
4. Propidium iodide–lysis solutions: the following dye–lysis solution is made by adding an appropriate amount of PI stock to lysis buffer, the final

concentration of PI being 6 μg/ml. This solution is stored at 4°C in a light-tight container.

5. Electrophoresis buffer (TBE): 2 mM EDTA, 90 mM Tris, pH 8.0, 90 mM boric acid.

6. Software: Optimas image analysis software with comet macro program can be used to acquire and analyze the halo–comet images. The system is precalibrated using a stage micrometer.

b. Procedures

Conditions for the optimal extent of cell lysis and electrophoresis that result in halo–comet structure may vary among different cell types. The following method applies to CHO cells grown as monolayer cultures in Dulbecco's modified essential medium (DMEM) supplemented with 10% fetal bovine serum. Cells are trypsinized with 0.05% trypsin/0.02% EDTA for 3 min and suspended in PBS. After being centrifuged at 1000 rpm for 5 min, cells are resuspended at a concentration of 2–4 $\times$ 10^4 cells/ml in PBS. Then 0.2 ml of the cell suspension is mixed with 0.6 ml of 1% low-gelling-temperature agarose in PBS maintained at 43°C, and the mixture is immediately layered onto a microscope glass slide and covered with a glass coverslip. The slides are kept on ice for 1 min to allow the agarose to gel. The glass coverslips are removed and slides are transferred to a Coplin jar containing lysis–dye buffer. Cell lysis is carried out in the dark at room temperature for 30 min. Then the slides are rinsed for several times and kept in deionized water for 1 hr. The slides are then transferred to the horizontal bed of a modified Hoefer electrophoresis unit. Electrophoresis buffer (TBE) is added into the chamber to a level about 5 mm above the agarose gel on the slides. Electrophoresis is carried out at 1.8 V/cm for 25 min. After electrophoresis, the slides are rinsed in water and visualized with an Olympus IX-70 inverted fluorescence microscope using a 20 $\times$ objective with 520–540 nm excitation from a 100-W mercury lamp. Images are taken in the 24-bit mode with a cooled color charge-coupled device (CCD) camera (Optronics, Goleta, CA), using the same exposure settings throughout the experiment, and digitized using a flashpoint pixel frame grabber. The CCD camera exposure is set such that there is no saturation in the intensity of the head of the comet. Digitized images are stored, and each image is analyzed for comet length and normalized comet moment using the Optimas image analysis software and the comet macro program. One comet is analyzed at a time. Threshold settings are selected manually for each comet to encompass the whole comet by including all pixels above the background. Once the threshold is set, the entire analysis is done by the computer with no additional input from the operator.

The results from a typical set of experiments are presented in Fig. 5. As described in the fluorescent halo assay, at certain PI concentrations (5–7.5 μg/ml) the negatively supercoiled DNA loops of the nucleoids can be relaxed, and this unwinding process is not affected by ionizing radiation. When cells are

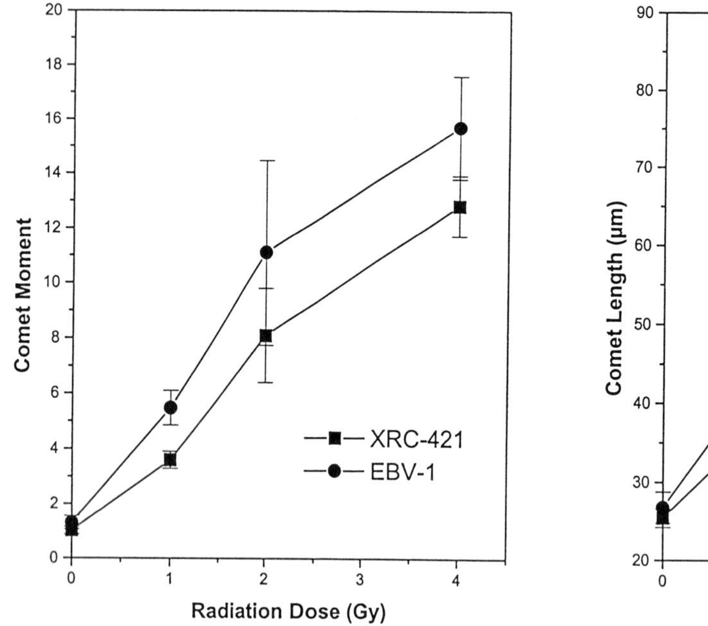
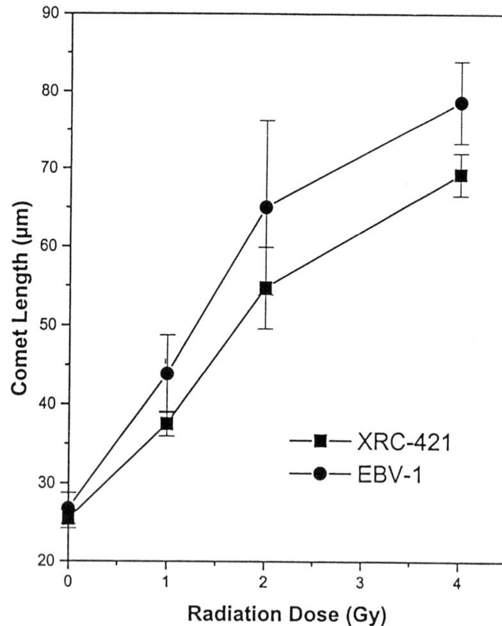

Fig. 5 Induction of damage in DNA organization by ionizing radiation determined using the halo–comet technique. EBV-1 (solid circles) and XRC-421 (solid squares) cells were irradiated with the indicated doses of ^{137}Cs γ-rays and embedded in an agarose gel. Cells were lysed, and supercoiled DNA loops were relaxed by exposing to high salt solution containing Triton X-100 and 6 μg/ml PI. The nucleoids with unwound DNA loops were subjected to electrophoresis. Normalized comet moment (left) and comet length (right) were used as parameters to indicate damage to DNA organization. The DNA organization of EBV-1 cells expressed a greater degree of damage than that of XRC-421 cells after ionizing radiation as measured by either parameter. The data points represent the means of three independent experiments, with the error bars being ±1 SEM.

embedded in an agarose gel and exposed to high salt lysis buffer that contains 6 μg/ml PI, nucleoids are formed with relaxed DNA loops extruding through the nuclear lamina. When subjected to electrophoresis, DNA loops are stretched toward the anode and form a tail. This tailed structure is called a halo–comet to indicate migration of DNA loops. The halo–comet images are quantified in terms of tail moment and tail length as described in the comet assay.

The type of DNA damage that is determined by this assay is not very clear. Since no rewinding process of DNA loops and no separation of DNA double strands (pH value is maintained between 8-0 and 8.5) are involved in this assay, it is most likely that the amount of stretched DNA is dependent on the strength of DNA–nuclear matrix association with the amount of DNA single-strand breaks present in the nucleoids. Cells with different radiosensitivity were studied in the experiment shown in Fig. 5. The DNA organization of EBV-1, the radiosensitive cell line, showed a greater degree of damage than that of XRC-421 cells after ionizing radiation.

IV. Summary

Four cytometric assays for the assessment of radiation-induced DNA damage in individual cells are presented. Two of these, the alkaline and neutral comet assays, are useful for the detection of DNA damage due to very low radiation doses and promise to be useful for the quantitation of genomic damage after clinically or environmentally relevant exposures. The other two, the halo and halo–comet assays, reveal aspects of chromatin structure in the presence of DNA damage that reflect differences in intrinsic cellular radiosensitivity. Further development of these assays used alone, or in combination, should eventually lead to the definition of readily measurable cytometric parameters that will be useful as predictive markers for cellular responses to DNA damaging agents.

Acknowledgments

The authors thank Ms. Kathy Bles for manuscript preparation. This work was supported by National Institutes of Health Grant PO1 CA75556 and by the Motorla Corporation.

References

Kent, C. R. H., Eady, J. J., Ross, G. M., and Steel, G. G. (1995). The comet moment as a measure of DNA damage in the comet assay. *Int. J. Radiat. Biol.* **67,** 655–660.

Li, Z., Otevrel, T., Gao, Y., Cheng, H.-L., Seed, B., Stamato, T. D., Taccioli, G. E., and Alt, F. W. (1995). The XRCC4 gene encodes a novel protein involved in DNA double-strand break repair and V(D)J recombination. *Cell* **83,** 1079–1089.

Lynch, T. S., Anderson, P., Wallace, D. M., Kondratowicz, G. M., Beaney, R. P., and Vaughan, A. T. (1991). A correlation between nuclear supercoiling and the response of patients with bladder cancer to radiotherapy. *Brit. J. Cancer* **64,** 867–871.

Malyapa, R. S., Wright, W. D., and Roti Roti, J. L. (1994). Radiosensitivity correlates with changes in DNA supercoiling and nucleoid protein content in cells of three Chinese hamster cell lines. *Radiat. Res.* **140,** 312–320.

Malyapa, R. S., Wright, W. D., and Roti Roti, J. L. (1995). Possible role(s) of nuclear matrix and DNA loop organization in fixation or repair of DNA double-strand breaks. *In* "Radiation Damage in DNA Structure/Function Relationships at Early Times" (A. F. Fuciarelli and J. D. Zimbrick, eds.), pp. 409–418. Battelle Press, Columbus, OH.

Malyapa, R. S., Wright, W. D., Taylor, Y. C., and Roti Roti, J. L. (1996a). DNA supercoiling changes and nuclear matrix-associated proteins. Possible role in oncogene-mediated radioresistance. *Int. J. Radiat. Oncol. Biol. Phys.* **35,** 963–973.

Malyapa, R. S., Wright W. D., and Roti Roti, J. L. (1996b). DNA supercoiling changes and nucleoid protein composition in a group of L5178Y cells of varying radiosensitivity. *Radiat. Res.* **145,** 239–242.

Malyapa, R. S., Bi, C., Ahern, E. W., and Roti Roti, J. L. (1998). Detection of DNA damage by the alkaline comet assay after exposure to low-dose γ radiation. *Radiat. Res.* **149,** 396–400.

Milner, A. E., Vaughan, A. T. M., and Clark, I. P. (1987). Measurement of DNA damage in mammalian cells using flow cytometry. *Radiat. Res.* **110,** 108–117.

Milner, A. E., Gordon, D. J., Turner, B. M., and Vaughan, A. T. M. (1993). A correlation between DNA-nuclear matrix binding and relative radiosensitivity in two human squamous cell carcinoma cell lines. *Int. J. Radiat. Biol.* **63,** 13–20.

Olive, P. L., Banath, J. P., and Durand, R. E. (1990). Heterogeneity in radiation-induced DNA damage and repair in tumour and normal cells measured using the 'comet' assay. *Radiat. Res.* **122,** 86–94.

Olive, P. L., Wlodek, D., Durand, R. E., and Banath, J. P. (1992). Factors influencing DNA migration from individual cells subjected to gel electrophoresis. *Exp. Cell Res.* **198,** 259–267.

Olive, P. L., Johnston, P. J., Banáth, J. P., and Durand, R. E. (1998). The comet assay: A new method to examine heterogeneity associated with solid tumors. *Nat. Med.* **4,** 103–105.

Ostling, O., and Johanson, K. J. (1984). Microelectrophoretic study of radiation-induced DNA damages in individual cells. *Biochem. Biophys. Res. Commun.* **123,** 291–298.

Rhee, J. G., Liu, J., and Suntharalingam, J. (1998). Halo-comet assay. Armed Forces Radiobiology Research Institute Worshop (1998). Proceedings: Triage of irradiated personnel. *AFRRI Special Publ.* **98,** D5–D9.

Roti Roti, J. L., and Wright, W. D. (1987). Visualization of DNA loops in nucleoids from HeLa cells: Assays for DNA damage and repair. *Cytometry* **8,** 461–467.

Roti Roti, J. L., Wright, W. D., and Taylor, Y. C. (1993). DNA loop structure and radiation response. *Adv. Radiat. Biol.* **17,** 227–259.

Singh, N. P., McCoy, M. T., Tice, R. R., and Schneider, E. L. (1988). A simple method for quantitation of low levels of DNA damage in individual cells. *Exp. Cell Res.* **175,** 184–191.

Singh, N. P., Stephens, R. E., and Schneider, E. L. (1994). Modifications of alkaline microgel electrophoresis for sensitive detection of DNA damage. *Int. J. Radiat. Biol.* **66,** 23–28.

Singh, N. P., Graham, M. M., Singh, V., and Khan, A. (1995). Induction of DNA single-strand breaks in human lymphocytes by low doses of γ-rays. *Int. J. Radiat. Biol.* **68,** 563–569.

Stamato, T. D., Weinstein, R., Giacci, A., and Mackenzie, L. (1983). Isolation of cell cycle-dependent γ ray-sensitive Chinese hamster ovary cell. *Somat. Cell Genet.* **9,** 165–173.

CHAPTER 44

Cytometric Methods to Analyze Thermal Effects

**Robert P. VanderWaal, Ryuji Higashikubo, Mai Xu,
Douglas R. Spitz, William D. Wright,
and Joseph L. Roti Roti**

Mallinckrodt Institute of Radiology
Radiation Oncology Center
Section of Cancer Biology
Washington University School of Medicine
St. Louis, Missouri 63108 USA

  I. Introduction
 II. Application
III. Methods to Measure Nuclear and Nuclear Matrix Protein Content
      A. Nuclear Protein Content by Flow Cytometry
      B. Nuclear Matrix Protein Content by Flow Cytometry
      C. Nuclear Matrix Stability Assay Following Heat Shock Using Flow Cytometry
 IV. Identification of Altered DNA Replication Patterns Following Heat Shock
  V. Nuclear Localization of hsp70
      A. Flow Cytometric Detection of Nuclear hsp/hsc70
      B. Nuclear Localization of hsp70 by *in Situ* Immunofluorescence
 VI. Prooxidant Measurement
VII. Results
      A. Nuclear Protein Content
      B. Pattern of DNA Replication
      C. hsp/hsc70
      D. Prooxidant Production
    References

I. Introduction

The study of thermal effects on cells is of critical interest to at least three diverse fields: (1) studies of the regulation of gene expression in the response(s) of cells to environmental stress; (2) the application of hyperthermia as a single or combined modality in the treatment of cancer (Urano and Douple, 1988, 1989); and (3) the identification of thermal artifacts (or effects) in studies of the biological effects of physical agents such as radiofrequency radiation and ultrasound (Miller and Ziskin, 1989). In these studies thermal effects have been detected at temperatures as low as 39°C and up to 47°C (Henle and Roti Roti, 1988; A. Laszlo, unpublished, 1996). For most cell lines the study of temperatures above 48°–50°C are not biologically relevant owing to rapid cytotoxicity and/or massive protein denaturation (Wright *et al.*, 1998). For the most part, the effects described in this chapter are generally studied following (or during) heat shocks at temperatures between 41° and 46°C.

Hyperthermia in the 41°–45°C range causes a plethora of effects on most cellular organelles and functions. These effects range from the interaction of the plasma membrane with the extracellular matrix, to protein aggregation with the nuclear matrix (Roti Roti and Laszlo, 1988; Laszlo, 1992; Kampinga, 1993). Functional changes can be seen in cytosol and the nucleoplasm. In this sense most, if not all, of the methods described in this book can be used to detect differences between heat-shocked and normal cells. The challenge for the workers in this field is to find out those changes that contribute mechanistically to the effects of interest. Thus, we have selected parameters that reflect thermal cytotoxicity uniquely, as far as we know, distinguish between heat-sensitive and heat-resistant cells, are markers for transient heat resistance, thermotolerance, and those that show interrelationships between thermal stress and oxidative stress. The last parameter is included since there are a number of studies that suggest that both the oxidative and thermal stress responses, while not identical, share some overlapping features. The methods described in this chapter, therefore, should have a wide variety of applications in the study of cells and their responses to environmental stress, particularly heat shock.

II. Application

As stated earlier, several cytometric methods have been used to assay the effects of hyperthermia on cells. The first assay, the amount of protein that coisolates with nuclei or nuclear matrices, measures a parameter that correlates with numerous heat effects on nuclear function such as inhibition of DNA repair, DNA synthesis, transcription, and processing of hnRNA. This parameter correlates very well with cell death in many cell lines, particularly those that do not die by apoptosis. Because the mode of cell death induced by nuclear protein aggregation appears to involve progression through S phase, we will describe methods to determine the pattern of DNA replication factories which changes in a time-dependent manner. The next parameter will be the expression of heat-shock proteins whose nuclear localization and delocalization correlates with the development of thermotolerance and heat resistance. Finally, we will describe a method to measure intracellular prooxidant capacity. One of the effects that can be induced by the exposure of cells to hyperthermia is an imbalance between prooxidant and antioxidant status either by a reduced antioxidant function or an increased prooxidant function, or both. An increased prooxidant level may, in turn, lead to a myriad of biochemical alterations that disrupt normal cellular processes and even lead to cell death.

III. Methods to Measure Nuclear and Nuclear Matrix Protein Content

A. Nuclear Protein Content by Flow Cytometry

Two methods for isolating nuclei have been described in the previous volume of this series (Higashikubo *et al.*, 1990). We describe here another method that works very well for cultured cells that grow on monolayer. All reagents are from Sigma, St. Louis, MO.

1. Reagents

1. Spinner salt solution, pH 7.0; 270–300 mOs/kg; per 1ml double-distilled water (ddH$_2$O): 0.4 mg KCl, 2.2 mg NaHCO$_3$, 0.1 mg MgSO$_4$, 1.4 mg NaH$_2$PO$_4$, 6.8 mg NaCl, and 1.0 mg D-glucose.
2. Nuclear isolation buffer: 0.5% Nonidet P-40, 50 mM Tris base/Tris-HCl (pH 7.4), 10 mM EDTA, 50 mM NaCl.
3. Bicarbonate buffer (pH 8.1): 0.03 M NaCl, 0.03 M NaHCO$_3$.
4. Staining solution: (a) 1 mg/ml RNase A in ddH$_2$O (boil for 5 min before each use), (b) 3 μg/ml fluorescein isothiocyanate (FITC) in bicarbonate buffer, and (c) 70 μg/ml propidium iodide (PI) in bicarbonate buffer.

2. Nuclear Isolation and Staining Procedure

1. Detach cells by trypsinization or other enzymatic methods and neutralize by the addition of complete medium. Make sure cells are well monodispersed. Centrifuge at 150 g and decant medium. Adjust the cell number to 1–1.5 $\times$ 10^6 per sample.
2. Wash twice with Spinner salt solution, each followed by centrifugation at 150 g.
3. Resuspend in 5 ml of nuclear isolation buffer. Keep on ice. The duration for this treatment depends on the cell type. For example, Chinese hamster ovary K1 cells need 5 min, while some cell lines of epithelial origin may need longer treatment.

4. Collect nuclei by centrifugation at 900 g.

5. Wash twice with bicarbonate buffer. Resuspend in 400 μl of the same buffer.

6. Add 50 μl of RNase A solution and incubate for 30 min at room temperature.

7. Add 50 μl of FITC stock solution and 500 μl of PI solution. Stain for at least 1 hr before analysis.

3. Special Considerations

1. The FITC fluorescence intensity of nuclei is sensitive to the concentration of nuclei and pH of the final suspension. The concentration of nuclei in the final staining solution should not exceed 1.5×10^6 per ml. The isolated nuclei should be washed clean of cellular debris and organelles that may cause pH change of the final staining solution.

2. The stained nuclei should be examined under a fluorescence microscope to ensure that they are free of cytoplasmic contamination. However, the loss of cellular particles during the nuclear isolation procedure and the subsequent washes should be kept at a minimum, for example, no more than 20%. A low nuclear yield indicates that the isolation procedure is too harsh and may have extracted some nuclear proteins.

4. Instruments

1. Both dyes can be excited with the 488 nm line of an argon laser. Green FITC fluorescence is detected through a 535 nm band-pass filter, whereas red PI fluorescence can be detected through a 640 nm long-pass filter.

2. Very strong FITC signals, unlike those from immunofluorescence staining, may necessitate fluorescence compensation on the red fluorescence channel.

B. Nuclear Matrix Protein Content by Flow Cytometry

The nuclear matrix is perhaps the most heat labile structure in eukaryotic cells. As the result of denaturation, soluble nuclear proteins bind to the nuclear matrix altering its structure (vanderWaal *et al.*, 1996). One result of this protein binding is that nuclei isolated from heat-shocked cells (see earlier) have a higher protein content, compared to controls (Warters and Roti Roti, 1982). As another result of this protein binding, many nuclear functions are compromised. These include DNA replication, and RNA transcription and processing (Higashikubo and Roti Roti, 1993).

1. Reagents

1. HeLa cells in tissue culture flasks with watertight caps. Alternatively, the flasks can be sealed with Parafilm during the heat-shock steps. Another

possibility is to use HeLa cells grown in suspension and transfer the culture to an appropriately sized centrifuge tube prior to immersion.

2. Medium prewarmed to the temperature selected for the heat shock experiment.

3. Trypsin-EDTA.

4. 0.01 M phosphate-buffered saline (PBS) (pH 7.4): For 1 liter of 10× stock solution, 80 g NaCl, 2 g KCl, 11.5 g $Na_2HPO_4 \cdot 7H_2O$, 2 g KH_2PO_4.

5. Digestion Buffer: 10 mM PIPES, pH 6.8, 50 mM NaCl, 300 mM sucrose, 3 mM $MgCl_2$, 1 mM EGTA, 0.5% (v/v) Triton X-100, 4 mM vanadyl riboside complex (add just before use), 1.2 mM phenylmethylsulfonyl fluoride (PMSF) (add just before use).

6. 5 mg/ml deoxyribonuclease I (DNase I) stock: Add 1 ml 50% glycerol to 5 mg DNase I (Worthington Biochemical, Freehold, NJ, DPRF grade), store at $-20°C$.

7. 1 M NH_4SO_4 in water.

8. Cytoskeleton buffer: 0.292 g NaCl in 100 ml digestion buffer.

9. FITC stain: 3 μg FITC (Calbiochem, La Jolla, CA) per ml in digestion buffer.

2. Nuclear Matrix Isolation and Staining Procedure

1. Heat shock cells by submerging flasks in a water bath set at the desired temperature and incubate.

2. Immediately after removal from the water bath, aspirate medium, and replace with trypsin/EDTA. Keep at room temperature for 3 min.

3. Centrifuge cells at 235 g for 5 min.

4. Wash cells with PBS at 4°C and centrifuge.

5. Extract cells with cytoskeleton buffer for 3 min at 4°C at a density of 4×10^6 cells per ml. Centrifuge for 5 min at 940 g.

6. Resuspend pellet in digestion buffer, using same volume as in Step 5. Add DNase I to a final concentration of 150 μg DNase I per ml of digestion buffer. Incubate 30 min at room temperature.

7. Add 1 M NH_4SO_4 stock to cell suspension to 250 mM final concentration, incubate 5 min at room temperature, and centrifuge for 5 min at 940 g.

8. Resuspend pellet with 0.9 ml digestion buffer, add 0.1 ml stock FITC stain, and incubate at 4°C for 1 hr to overnight.

9. Analyze nuclear matrices by flow cytometry (FCM).

3. Instruments

For flow cytometry using a 488 nm excitation line from an argon laser, FITC fluorescence is detected through a 535 nm band-pass filter.

C. Nuclear Matrix Stability Assay Following Heat Shock Using Flow Cytometry

Evidence is accumulating which suggests that the nuclear matrix and its associated functions are a primary target for the effects of heat shock (for review, see Roti Roti *et al.*, 1998). This subnuclear organelle is composed mostly of RNA and protein. Evidence for the structural role of RNA in the matrix resides in the fact that the matrix fragments as a result of RNase treatment (for review, see Nickerson *et al.*, 1990). In contrast, the nuclear matrices from heated cells are stable following exhaustive digestion with RNase (Wright *et al.*, 1989).

1. Reagents

1. Exponentially growing HeLa cells in suspension with Joklik modified minimal essential medium (MEM), supplemented with 3.5% calf serum and 3.5% fetal bovine serum.
2. 0.01 M PBS (pH 7.4): For 1 liter of 10× stock solution, 80 g NaCl, 2 g KCl, 11.5 g $Na_2HPO_4 \cdot 7H_2O$, 2 g KH_2PO_4.
3. Lysis buffer: 1% Triton X-100, 80 mM NaCl, 10 mM EDTA, pH 7.2.
4. TMNP buffer: 10 mM Tris, pH 7.4, 5 mM Mg_2Cl, 0.1 mM PMSF.
5. 5 mg/ml DNase I stock: Add 1 ml 50% glycerol to 5 mg DNase I (Worthington Biochemical, DPRF grade), store at $-20°C$.
6. High salt buffer: 3 M NaCl, 10 mM EDTA, 10 mM Tris, pH 9.0, 20 mM β-mercaptoethanol.
7. RNase A: 1 mg/ml in TMNP, boil for 5 min before use.
8. Staining solution: TMNP buffer with 3 mg/ml FITC and 35 μl/ml PI (Sigma, St. Louis, MO).

2. Procedures

1. Collect HeLa cells growing in suspension by centrifugation, then resuspend in prewarmed 45°C medium at 10 times their original concentration, then submerse in a 45°C water bath.
2. Place cells in an ice bath for 2 min.
3. Centrifuge at 240 g for 5 min.
4. Wash three times with PBS by resuspending the pellet, and centrifuging at 240 g.
5. Wash pellet three times in lysis buffer by resuspending the pellet, and centrifuging at 1000 g.
6. Wash pellet twice with TMNP buffer, by resuspending the pellet, and centrifuging at 1000 g.

7. Resuspend the final pellet in TMNP to a concentration of 2×10^6 cell nuclei per ml.

8. Add DNase I to a final concentration of 130 μg/ml. Incubate for 2 hr at 37°C.

9. Centrifuge 5 min at 1000 g.

10. Resuspend pellet to original volume with TMNP, and centrifuge again.

11. Resuspend pellet in 1/8 starting volume of TMNP and add 5 volumes of high salt buffer. Incubate 10 min at room temperature.

12. Centrifuge as in Step 9.

13. Wash twice using original volume of TMNP.

14. Digest the nuclear matrices with RNase by resuspending 2–3×10^6 particles in 200 μl TMNP. Separate samples are digested with a final RNase concentration of 0.001, 0.01, 0.1, 1, 10, and 100 μg/ml RNase. Digestion is carried out at 0°C for 30 min.

15. Collect nuclear matrices by centrifugation at 1000 g for 5 min.

16. Resuspend pellet in the staining solution, and incubate on ice for at least 1 hr.

17. Analyze matrices by FCM, collecting FITC fluorescence, light scatter, and PI fluorescence.

3. Special Considerations

1. The differential stability between nuclear matrices from control and heated cells can cause yield differences, which, if large enough, can alter dye binding characteristics. Therefore, it is important to count the matrices recovered with a Coulter counter or a hemocytometer as a means to compute yield differences to adjust the number of particles in each sample. Alternatively, cell sorting can be used to ensure that the same number of matrices are compared in samples from control and heated cells.

2. Nuclear matrices from rodent cell lines tend to be less stable than those from human cell lines.

3. This protocol can be adapted for cells growing in monolayer.

4. Instruments

1. Nuclear matrices from HeLa cells retain enough double-stranded RNA to be detected with PI. Thus, the excitation and detection configuration described earlier can be used. However, fluorescence compensation is a must because the PI signal will be weak.

2. Even when both the PI and FITC signals become weak, the nuclear matrices can be detected by forward light scatter (unless they have become dissociated).

IV. Identification of Altered DNA Replication Patterns Following Heat Shock

DNA replication is organized temporally and spatially within the nuclei during S phase (Ma *et al.*, 1998). To study this process following heat shock, aphidicolin (APH) is used to synchronize HeLa cells at the G_1/S phase border. Cells are given heat shock (i.e., 15–30 min at 45°C) and then allowed to move through S phase. Periodically during the recovery from heat shock, cells are pulse-labeled with bromodeoxyuridine (BrdU), harvested, and the morphology of DNA replication determined using immunohistochemical methods. Our choice of primary antibody permits the immunostaining of incorporated BrdU without the harsh DNA denaturation steps with HCl or NaOH required by other procedures that may alter the morphology of DNA replication factories (Taragi *et al.*, 1993).

1. Reagents

1. HeLa cells cultured glass coverslips in 25-mm tissue culture dishes, in F10 medium supplemented with 10% calf serum. One dish with several coverslips are required for each time point.
2. 30 μM APH (10× stock): 1 mg APH is dissolved in a small volume of DMSO, then brought to 98 ml with F10 medium. The stock solution can be divided into small volumes and stored at −20°C for several months. The stock solution is diluted 10-fold with complete F-10 medium (i.e., supplemented with 10% calf serum) immediately before use.
3. Fixing solution: Histochoice (Amresco, Solon, OH) fixative supplemented with 5% acetic acid and 0.5% Triton X-100.
4. 0.01 M PBS (pH 7.4): For 1 liter of 10× stock solution, 80 g NaCl, 2 g KCl, 11.5 g $Na_2HPO_4 \cdot 7H_2O$, 2 g KH_2PO_4.
5. 1 mM BrdU (100×).
6. Blocking reagent: 1% bovine serum albumin, 10% normal goat serum in PBS.
7. Primary antibody: Anti-bromodeoxyuridine monoclonal antibody mouse plus nuclease (Amersham, Arlington Heights, IL, RPN 202).
8. Secondary antibody: Sheep anti-mouse immunoglobulin G (IgG), Cy-3 conjugate (Sigma), diluted 1:1000 with blocking reagent.
9. FITC staining solution: 10 μg/ml FITC in PBS.

2. Procedures

1. Culture cells in the APH containing medium for 15 to 20 hr at 37°C to align them at the G_1/S border.
2. Remove APH by three quick washes with prewarmed complete F10 medium. Replace the medium and return the cells to a 37°C incubator. Allow cells to recover from the APH block for approximately 30 min prior to exposure to heat shock. Without this recovery time, the cells fail to resume DNA replication.
3. For heating, seal the culture dishes with Parafilm, and submerge in a 45°C water bath for 15 or 30 min. At the completion of heat exposure, return culture dishes to a 37°C incubator.
4. At each time point, label cells for 5 min by adding 10 μl of BrdU solution per milliliter of medium and wash with PBS. BrdU labeling is conducted immediately after heat shock and every 1–2 hr thereafter.
5. Fix cells by submerging coverslips in the fixative solution for 30 min.
6. Following fixation replace the fixative with PBS. The cells can be stored in PBS for several days at 4°C.
7. Remove the coverslips from the PBS solution.
8. Place 100 μl blocking reagent on the cells, and incubate cells for 15 min.
9. Aspirate the blocking reagent and replace with 50 μl primary antibody for 1 hr.
10. Aspirate the primary antibody off the cells and quickly wash with blocking reagent three times, followed by three additional washes of 5 min each.
11. Apply diluted secondary antibody to the cells and incubate for 30 min.
12. Remove secondary antibody and wash the cells six times with blocking reagent as in Step 9, above.
13. Counterstain by applying FITC staining solution for 5 min, then rinse well with PBS.
14. For mounting on glass slides, place 7 μl of PBS on the slide, and gently place the coverslip, cell side down on the PBS. Seal the coverslip with fingernail polish. The slides can be stored in this condition for several months at 4°C. Alternatively, the mounting medium, described in the next section, can be used.
15. Visualize nuclear sites of DNA replication under a fluorescence microscope using a rhodamine filter cube.

3. Special Considerations

1. All steps are carried out at room temperature, except as noted otherwise.
2. Careful attention should be paid to ensure that the cells do not dry out at any time.

4. Instrumentation

A fluorescent microscope equipped with a rhodamine and FITC filter cubes.

V. Nuclear Localization of hsp70

A. Flow Cytometric Detection of Nuclear hsp/hsc70

1. Level of hsp/hsc 70 Associated with Nuclei

Nuclei from heated and control cells are isolated with the procedures described earlier. Nuclei can be used immediately, or they can be fixed in 70% cold ethanol or methanol. It should be noted, however, that fixing nuclei in alcohol results in excessive clumping of nuclear particles, especially if they are not clean nuclei and have numerous cytoplasmic tabs remaining.

2. Reagents

1. Primary antibody: Mouse anti-hsp70 antibody clone C92F3A-5 (diluted 1:100 in PBS).
2. Secondary antibody: Goat anti-mouse IgG, FITC conjugate (diluted 1:100 in PBS).
3. RNase A: 100 μg/ml in ddH$_2$O, boil for 5 min before use.

3. Procedures

1. Wash isolated nuclei (fixed or unfixed) with PBS.
2. Label with the primary antibody for 1 hr at room temperature.
3. Wash twice with PBS.
4. Label with the secondary antibody for 1 hr at room temperature.
5. Wash twice with PBS.
6. Treat with RNase A for 30 min at room temperature.
7. Stain with 25 μg/ml PI for 1 hr.

4. Special Considerations

The assay can be applied to isolated nuclear matrix. Because the nuclear matrix is more fragile than the nucleus, more care should be paid in its handling as described earlier.

5. Instruments

Flow cytometry can be set up for routine dual parameter PI/FITC data acquisition as described for nuclei (earlier).

B. Nuclear Localization of hsp70 by *in Situ* Immunofluorescence

Although the FCM methods provide a rapid measurement of nuclear hsc/hsp70 amounts, this method provides a visualization of intranuclear hsp70 localization and will allow a comparison of relative amounts in the cytoplasm and nucleus. Both methods provide unique information and complement each other well.

1. Reagents

1. 0.01 M PBS (pH 7.4): For 1 liter of 10× stock solution, 80 g NaCl, 2 g KCl, 11.5 g $Na_2HPO_4 \cdot 7H_2O$, 2 g KH_2PO_4.
2. 3.7% formaldehyde fixative solution: for 100 ml 3.7% formaldehyde solution, 10 ml 37% formaldehyde, 0.2 ml Triton X-100, and 89.8 ml 0.01 M PBS.
3. Absolute acetone: stored at $-20°C$.
4. Blocking solution: 0.01 M PBS with 10% goat serum.
5. Mouse anti-hsp70 monoclonal antibody (StressGen, Victoria, Canada, product SPA-810). Working concentration is 1:80 and diluted in blocking buffer.
6. FITC-conjugated goat anti-mouse antibody (Becton Dickinson, San Jose, CA, product 347580). Working concentration is 1:100 and diluted in blocking buffer.
7. Mounting medium: 2.4 g Mowiol 4-88 (Calbiochem 475904), 6 g glycerol (analytical grade), 6 ml doubly distilled H_2O, 12 ml 0.2 M Tris buffer, pH 8.5, mixed for 4 hr on a shaker; then the solution is incubated in water bath at 50°C for 10 min with occasional stirring to dissolve the Mowiol 4-88. Centrifuge the mixture at 5000 g for 15 min, divide the supernatant into aliquots, and store in $-20°C$ (Heimer and Taylor, 1974).

2. Staining Procedure

1. Culture cells on the coverslips in the tissue culture dish for 48 hr, then shift to 41.1° from 37°C for different periods of time.
2. Wash in three changes of PBS at room temperature for 5 min each.
3. Fix cells with fixative solution for 15 min.
4. Wash in three changes of PBS at room temperature for 5 min each.
5. Incubate the cells with cold absolute acetone (keep at $-20°C$) for 10 min for permeabilization at room temperature.
6. Wash in three changes of PBS at room temperature for 5 min each.
7. Incubate cells with blocking solution for 30 min.
8. Remove the blocking solution, add mouse anti-hsp70 antibody, and incubate for 1 hr at 37°C.

9. Wash in three changes of PBS at room temperature for 5 min each.

10. Wash in three changes of cold PBS (4°C) for 5 min each.

11. Incubate the cells with FITC-conjugated goat anti-mouse antibody for 40 min at room temperature.

12. Wash in three changes of room temperature PBS for 5 min each.

13. Place a small drop of mounting medium with a needle on the slide and mount the coverslip on it.

14. Keep the slides in dark and observe under a fluorescence microscope.

3. Special Considerations

1. The cell number seeded in the dish containing coverslips has to be optimized considering the growth rate of cells and area of the dish bottom. Too many or too few cells will cause problems in the observation of the fluorescent cells. To achieve an optimal condition, we used 35×10 mm tissue culture dishes and placed four coverslips in each dish along with 2 ml medium containing 5×10^4 cells.

2. The washing step must be very gentle. When using the 24-well plate with one coverslip per well, wash each well with 1 ml of PBS for each washing step. Cells, especially heat sensitive cells, are more easily detached from the coverslip after heat treatment.

4. Instruments

A fluorescent microscope equipped with an FITC fluorescence cube.

VI. Prooxidant Measurement

Much indirect evidence has suggested that heat stress and oxidative stress have some commonalties. We report here a method to directly measure the oxidative potential of the intracellular milieu following heat shock.

1. Reagents

Labeling solution: 5-(and-6)-carboxy-2',7'dichlorodihydrofluorescein diacetate (carboxy-H_2DCFDA): dissolved in DMSO and diluted to 10 μg/ml in PBS (pH 7.2) containing 1 mg/ml glucose.

2. Staining Procedure

1. Suspend monodispersed cells (1 to 3×10^6) in 15 ml of labeling solution and incubate for 30 min at 37°C.

2. Centrifuge and remove labeling solution, and resuspend cells in 1 ml of fresh labeling solution and keep on ice.

3. Special Considerations

Three steps are involved for the probe to be used as an indicator of cellular oxidative activity. They are (1) the transport through the cellular membrane into the cell, (2) the cleavage of ester groups by intracellular esterases to deter its leakage out of the cell, and (3) the oxidation of the probe to become a fluorescent molecule. Hyperthermia is known to affect membrane permeability. To ensure that changes in fluorescence intensity is due to changes in oxidative activity and not due to other cellular properties, it is necessary to use an appropriate control. To this end the use of a probe such as 5-(and 6-)carboxy-2′,7′-dichlorofluorescein diacetate (carboxy-DCFDA), which does not require oxidation to become fluorescent but retains other characteristics of 5-(and 6-)carboxy-2′,7′-dichlorodihydrofluorescein diacetate, is recommended as a parallel control.

In some cases the fluorescence intensity from the oxidation of the probe is very weak. Therefore, it is also recommended that the background autofluorescence should be monitored at each time point and subtracted to obtain net fluorescence intensity.

4. Instruments

The fluorescent dyes are excited with the 488 nm line of an argon laser. Fluorescence is detected through a 525 nm band-pass filter.

VII. Results

A. Nuclear Protein Content

The amount of protein that coisolates with isolated nuclei from HeLa cells exposed to varying temperature is shown in Fig. 1. The increase in the amount of nuclear protein is dependent on exposure temperature and time, especially at the temperatures of 43°C or above. In many cell lines, the amount of protein correlates with cell killing. At 42°C or lower the increase is seen to plateau. The increase ceased immediately on the removal of elevated temperature, and the excess proteins are removed during incubation at 37°C. As the excess nuclear proteins are removed, macromolecular synthesis resumes; for example, when the nuclear protein content is reduced to 1.25 times that of the control the rate of DNA synthesis recovers to 25% of control (Higashikubo and Roti Roti, 1993).

B. Pattern of DNA Replication

Three distinct patterns of DNA replication factories are evident as mammalian cells progress through S phase (Fig. 2). Type I occurs early to mid S phase, as

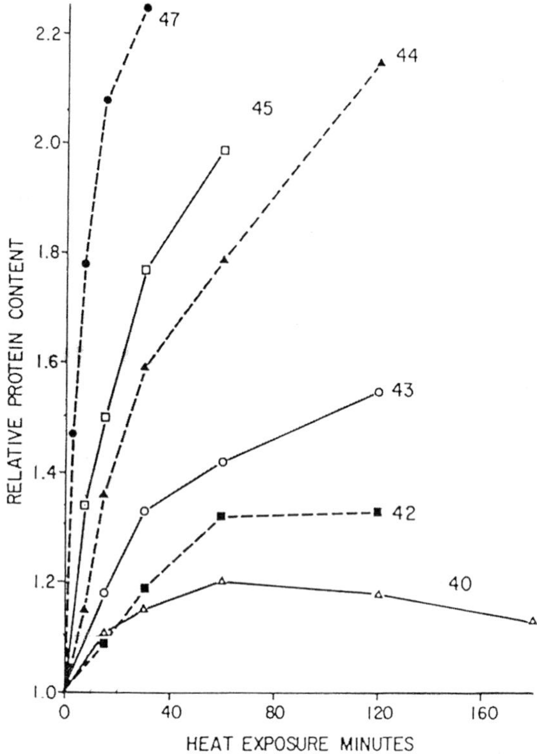

Fig. 1 Effects of heat on nuclear protein content. HeLa cells were exposed to the various time–temperature combinations indicated on the graph, and their nuclei were isolated and stained for DNA and protein content according to published methods (Roti Roti *et al.*, 1982). The relative nuclear protein content (control = 1) is plotted versus heating time for each temperature. The plotted points represent the mean of at least three repeated experiments; error bars have been omitted for clarity. Reproduced with permission from Roti Roti and Laszlo, 1988.

the cells are replicating euchromatin. Numerous small foci are evident over regions of euchromatin. Initiation of heterochromatin synthesis starting in mid S phase results in a Type II pattern where the replication factories are concentrated around the nuclear periphery, nucleoli, and other heterochromatin domains. During late S phase a Type III DNA replication pattern is seen where factories become less numerous, are larger, and are found within the heterochromatin regions. Following heat shock, the distinctive Type II pattern is not observed. Instead, an altered pattern (Type II/III) appears, characterized by substantial DNA replication in heterochromatin regions. Dual labeling will be needed to resolve Type II and Type III DNA replication patterns in heated cells. This technique is currently under development.

DNA Replication Morphology

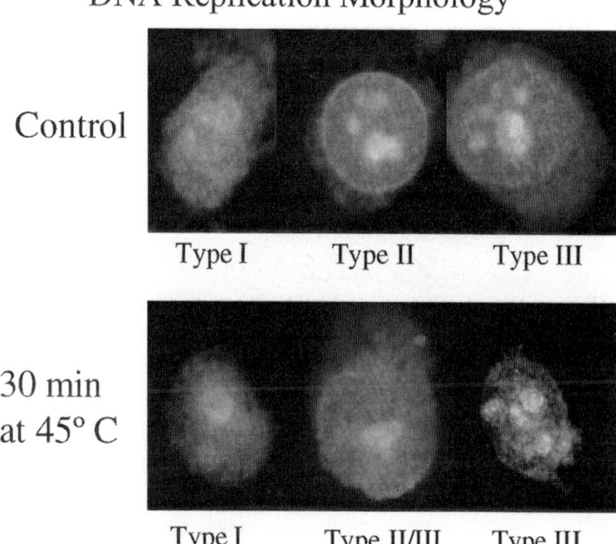

Fig. 2 DNA replication patterns observed in control and heat-shocked HeLa cells. Type I replication pattern replicates euchromatin, leaving heterochromatic regions (nucleoli) bright green. Types II and III (and unresolved II and III following heat shock) are involved in heterochromatin replication, resulting in the yellow nucleolar region (red plus green). Photographed at 1000 × magnification. (See color plates.)

C. hsp/hsc70

Expression of hsp70 associated with nuclei and nuclear matrices following exposure of Chinese hamster ovary cells to heat shock of 45°C for 30 min is detected by flow cytometry and shown in Fig. 3. Heat-induced increases in the association of the proteins are clearly indicated in both cases. Although cell-cycle dependent DNA content was preserved in nuclei, DNA content distribution is amorphous in nuclear matrices following exhaustive DNase digestion.

The localization of hsp70 is a very useful parameter. Following heat shock, hsp70 is found in the nucleus. The rate of delocalization from the nucleus after heat shock correlates with the thermal resistance of the cell. In heat-resistant cells the delocalization is faster, whereas in heat-sensitive cells, it is slower (Ohtsuka and Laszlo, 1992), although hsp70 exhibits nuclear localization, the lack of nucleolar staining appears to correlate with survival for continuous heating at 41.1°C (see Fig. 4) (Xu *et al.*, 1998).

D. Prooxidant Production

The fluorescence intensity should be corrected by subtracting background and normalizing to that of control unheated cells. Prooxidant production is expected

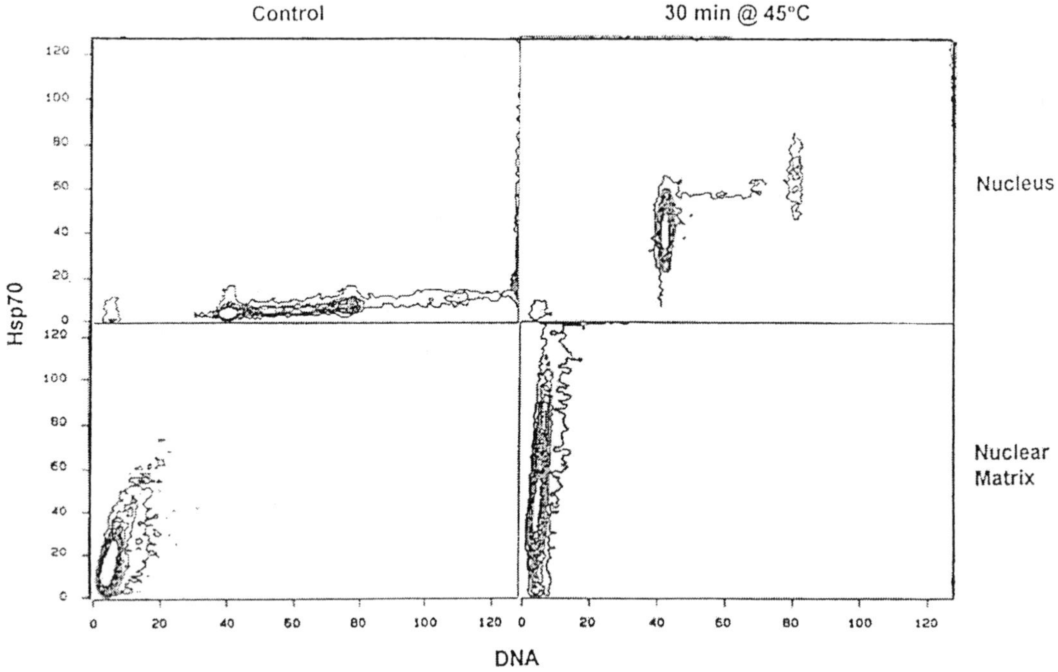

Fig. 3 Expression of hsp70 in nuclei (top) and in nuclear matrix (bottom) following exposure of cells to heat shock of 30 min at 45°C (right) compared to unheated control (left).

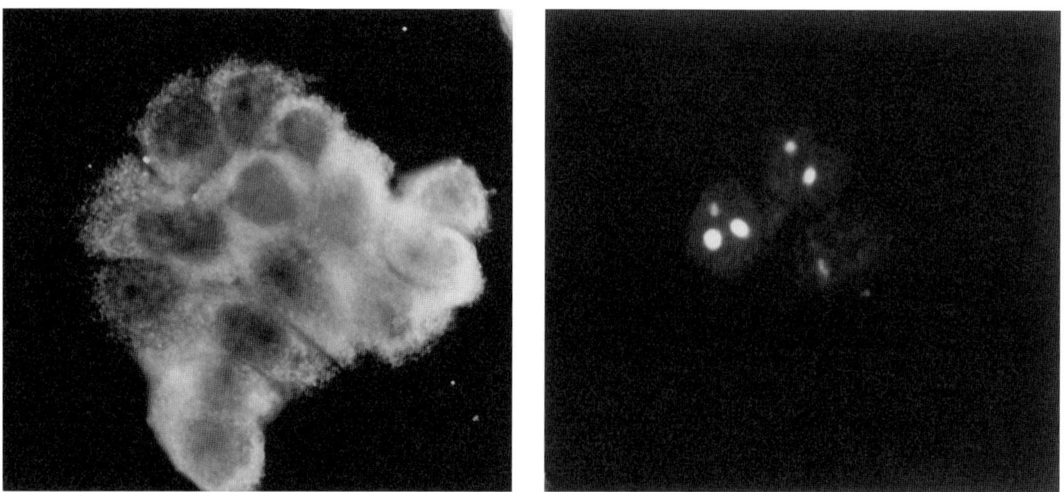

Fig. 4 (Right) Intensive cytoplasmic immunocytochemical staining pattern of hsp70 from NSY42129 cells heated at 41.1°C for 16 hr. (Left) Nucleolar punctate staining pattern of hsp70 from transformed fibroblast (TF) cells heated at 41.1°C for 16 hr. Photographed at 1000 × magnification.

to vary depending on heating protocol, that is, temperature and heating duration, as well as cell types. In most cases, however, we found that prooxidant production decreases initially on heat treatment and recovers to and surpasses the control level during the postheat recovery following acute heat shock (i.e., 45°C for 15 min) or the adjustment period during long duration moderate hyperthermia (i.e., 24 hr at 41.5°C).

References

Heimer, C. V., and Taylor, C. E. (1974). Improved mountant for immunofluorescence preparations. *J. Clin. Pathol.* **27,** 254–256.

Henle, K. J., and Roti Roti, J. L. (1988). Response of cultured mammalian cells to hyperthermia. *In* "Hyperthermia and Oncology, Thermal Effects on Cells and Tissues" (M. Urano and E. Douple, eds.), Vol. 1, pp. 57–82. VSP, Utrecht, The Netherlands.

Higashikubo, R., and Roti Roti, J. L. (1993). Alterations in nuclear protein mass and macromolecular synthesis following heat shock. *Radiat. Res.* **134,** 193–201.

Higashikubo, R., Wright, W. D., and Roti Roti, J. L. (1990). Flow cytometric methods for studying isolated nuclei: DNA accessibility to DNase I and protein–DNA content. *In* "Methods in Cell Biology" (Z. Darzynkiewicz and H. A. Crissman, eds.), Vol. 33, pp. 325–336. Academic Press, New York.

Kampinga, H. H. (1993). Thermotolerance in mammalian cells: Protein denaturation and aggregation, and stress proteins. *J. Cell Sci.* **104,** 11–17.

Laszlo, A. (1992). The effects of hyperthermia on mammalian cell structure and function. *J. Cell. Prolif.* **25,** 59–87.

Ma, H., Sanmarabandu, S., Devdhan, R. S., Acharya, R., Cheng, P.-C., Meng, C., and Berezney, R. (1998). Spatial and temporal dynamics of DNA replication sites in mammalian cells. *J. Cell Biol.* **143,** 1415–1425.

Miller, M. W., and Ziskin, M. C. (1989). Biological consequences of hyperthermia. *Ultrasound Med. Biol.* **15,** 707–722.

Nickerson, J. A., He, D., Fey, E. G., and Penman, S. (1990). The nuclear matrix. *In* "The Eukaryotic Nucleus" (P. R. Strauss and S. H. Wilson, eds.), Vol. 2, pp. 763–782. The Telford Press, Caldwell, NJ.

Ohtsuka, K., and Laszlo, A. (1992). The relationship between hsp70 localization and heat resistance. *Exp. Cell Res.* **202,** 507–518.

Roti Roti, J.L., and Laszlo, A. (1988). The effects of hyperthermia on cellular macromolecules. *In* "Hyperthermia and Oncology, Thermal Effects on Cells and Tissues" (M. Urano and E. Douple, eds.), Vol. 1, pp. 13–56. VSP, Utrecht, The Netherlands.

Roti Roti, J. L., Higashikubo, R., Blair, O. C., and Uygur, N. (1982). Cell-cycle position and nuclear protein content. *Cytometry* **3,** 91–96.

Roti Roti, J. L., Wright, W. D., and VanderWaal, R. (1997). The nuclear matrix: A target for heat shock effects and a determinant for stress response. *Critical Rev. Eukaryotic Gene Expression* **7,** 343–360.

Taragi, S., McFadden, M. L., Humphreys, R. E., Woda, B. A., and Sairenji, T. (1993). Detection of 5-bromo-2-deoxyuridine (BrdUrd) incorporation with monoclonal anti-BrdUrd antibody after deoxyribonuclease treatment. *Cytometry* **14,** 640–648.

Urano, M., and Douple, E. (eds.). (1988). Thermal effects on cells and tissues. *In* "Hyperthermia and Oncology," Vol. 1. VSP, Utrecht, The Netherlands.

Urano, M., and Douple, E. (eds.). (1989). Biology of the thermal potentiation of radiotherapy. *In* "Hyperthermia and Oncology," Vol. 2. VSP, Utrecht, The Netherlands.

VanderWaal, R., Thampy, G., Wright, W. D., and Roti Roti, J. L. (1996). Heat-induced modifications in the association of specific proteins with the nuclear matrix. *Radiat. Res.* **145,** 746–753.

Warters, R. L., and Roti Roti, J. L. (1982). Hyperthermia and the cell nucleus. *Radiat. Res.* **92,** 458–462.

Wright, N. T., Chen, S. S., and Humphrey, J. D. (1998). Time-temperature equivalence of heat-induced changes in cells and proteins. *Trans. ASME* **120,** 2–26.

Wright, W. D., Higashikubo, R., and Roti Roti, J. L. (1989). Flow cytometric studies of the nuclear matrix. *Cytometry* **10,** 303–311.

Xu, M., Wright, W. D., Higashikubo, R., and Roti Roti, J. L. (1998). Intracellular distribution of hsp70 during long duration moderate hyperthermia. *Int. J. Hyperthermia* **14,** 211–225.

PART X

Clinical Oncology

Multiparameter Data Acquisition and Analysis of Leukocytes by Flow Cytometry

Carleton C. Stewart and Sigrid J. Stewart

Laboratory of Flow Cytometry
Roswell Park Cancer Institute
Buffalo, New York 14263

  I. Introduction
 II. Correlated List Mode Data
III. Verification of Instrument Performance
IV. Optical Filtration and Spectral Compensation
        A. For Two- or Three-Color Compensation
        B. For Four-Color Compensation
 V. Multiparameter Data Analysis
VI. Summary
    References

I. Introduction

Multiparameter flow cytometry can be used to define the unique repertoire of epitopes on cells using forward scatter (FSC), side scatter (SSC), and several antibodies labeled with a different colored fluorochrome (Stewart, 1994, 1996; Stewart and Stewart, 1997, 1999). No other analysis system can obtain the amount of correlated information as rapidly on large numbers of cells while also providing the capability of sorting the unique population. For collecting multiparameter data it is essential that the instrument has been properly set up and evaluated. As simple as this is to do, the verification is very often improperly or incompletely performed. The result is that data generated from excellent cell preparations are poorly and improperly acquired.

By using more than one antibody simultaneously it is possible to perform single antibody histogram analysis or multiparameter histogram analysis. The simplest is single antibody histogram analysis in which each of the individual fluorescence histograms is treated as single parameter data. Data analysis is rapid because only the percentage of positive cells for each marker is considered. This method of analysis, however, does not provide any correlation among the different parameters that could be measured.

Multiparameter analysis is more complex and time consuming because populations that are labeled with more than one antibody must be individually identified. Each file must initially be analyzed separately to obtain the correct analysis strategy, but then subsequent files, containing the same antibody combinations, can be analyzed automatically using macros that reproduce the original strategy. A macro is a list of computer instructions for processing information repetitively the same way. Multiple antibodies provide the opportunity to resolve unique cell populations that cannot otherwise be identified. These populations may be most important in understanding normal and disease processes.

In this chapter we will describe the rationale and method for proper instrument quality control. Strategies for analyzing properly acquired multiparameter data will be discussed. Whereas the various aspects of analysis will be illustrated using human peripheral blood, the principles discussed can be applied to any type of cell population. Because storage of large data sets is no longer expensive, we recommend acquiring all data ungated. One can never retrospectively evaluate data that was not acquired.

II. Correlated List Mode Data

For purposes of discussion, we use "acquisition" to mean obtaining the cellular data, using a flow cytometer. We use the word "analysis" to mean the analysis of that data. List mode data should always be acquired when performing multiparameter specimen analysis so that retrospective data analysis can be performed on all events in the sample tube. A flow cytometer measures particles, which are often called events. It is the analysis process and subsequent data interpretation that will determine whether the events are cells. A simple correlated list mode file is illustrated in Table I.

Suppose we wish to acquire forward scatter (FSC), which reflects cell size, and one color of fluorescence. When cells are analyzed, the measured values are recorded sequentially in the computer as each event is interrogated at the intersection of the laser beam and cell stream. For the first cell, the first number in the list is FSC; the second number is fluorescence. The second cell produces the third and fourth numbers in the list, the third cell produces the fifth and sixth numbers, and so forth until all the cells have been measured. In this example, if 10,000 cells had been measured there would be 20,000 numbers in the list in exactly the same order as they were acquired. The first and every third number thereafter would be FSC; the second and every fourth number thereafter would be fluorescence. Thus, the computer can reanalyze measured parameters for

Table I
Correlated List Mode Data File

Data list	Simple list mode data file explanation
200	Forward scatter channel # for cell #1
400	Fluorescence channel # for cell #1
100	Forward scatter channel # for cell #2
700	Fluorescence channel # for cell #2
200	Forward scatter channel # for cell #3
500	Fluorescence channel # for cell #3
$FSCn$	Forward scatter channel # for cell #n
Fn	Fluorescence channel # for cell #n

each cell, over and over again, in the same order that the cells were the first measured by the flow cytometer.

By selecting regions utilizing one set of parameters that contain the values for a particular cell population, the location of these cells in the other parameters can be found and displayed. The display could be a simple histogram or bivariate plot of one parameter versus the other.

An example of list mode data analysis and gating is shown in Fig. 1 for human peripheral blood mononuclear cells obtained from a patient who had been infused with interleukin 2 and stained with fluoresceinated CD4 (FL-CD4). The purpose of this example is to show the importance of including all lymphocytes in the analysis region and the importance of list mode data for performing this task. In general, any single property of a uniform group of cells is distributed in a Gaussian fashion, that is, either a normal or log normal distribution. As shown in Fig. 1 in the top left plot, there are three overlapping normal distributions of cells with different FSC properties. The FSC measurement reflects cell size, and events to the left represent the smallest and those to the right represent the largest cells. We establish a gating region, R1, containing the smallest cells and display the fluorescence histogram by processing the list mode file for all the fluorescence measurements associated with cells meeting this size criterion. The gated data are shown in the upper right histogram. The cells in the left peak of the fluorescence histogram are the unstained cells that have a low innate fluorescence called autofluorescence. A population of brightly stained cells is contained in the right peak, and a very small population of cells between the two major peaks is dimly stained with CD4. Note the Gaussian nature of all three of the populations.

By processing the list mode file for cells meeting the size criterion for the cells in region R2, the fluorescence histogram, in the lower left plot of Fig. 1, shows most cells are dimly stained with CD4. However, a few brightly stained cells, similar to those found when region A was used, are also found in this region. Finally, there is a small population of very large cells in region R3 (Fig. 1, upper left). When their fluorescence histogram is displayed (Fig. 1, lower right), most

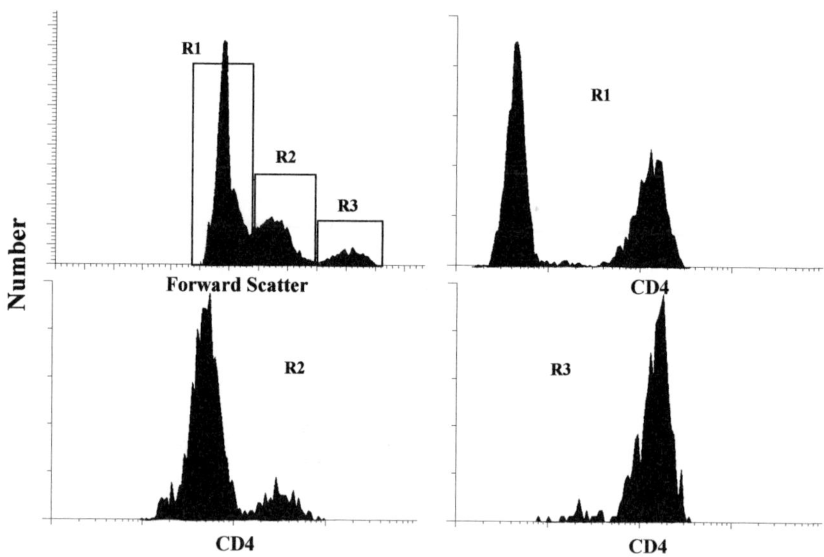

Fig. 1 Single-color analysis. Human cells were isolated from blood using neutrophil isolation medium (NIM, Cardinal Associates, Santa Fe, NM). The cells were adjusted to 20×10^6 cells/ml in phosphate buffered saline containing 0.1% sodium azide and protease and nuclease free 0.5% bovine serum albumin (PAB), and 50 μl was used for labeling. The mononuclear cells were incubated 10 min with 30 μg mouse IgG, then for 15 min with fluoresceinated CD4, washed, and analyzed. Using forward light scatter, region R1 was selected to include small cells, region R2 intermediate cells, and region R3 large cells. The CD4 staining characteristics of the cells in each of these regions is shown in the appropriately labeled histograms.

of these cells are brightly stained with CD4 characteristic of the positive cells found in region R1, but region R3 also contains some cells with the intensity of those in region R2. Although this population of large lymphocytes is not a prominent one in normal individuals, it may be frequently found in sick patients. If the common practice of acquiring only the small cells, found in region R1, had been employed, the population of large cells would not have been found. (This underscores the reason data should always be acquired ungated.)

When the cells in each of these populations are sorted and stained with Wrights blood stain, the cells in region R1 are lymphoid, those in region R2, with intermediate fluorescence intensity, are monocytes, and in those in region R3, with bright fluorescence intensity, are large lymphocytes. It is better to acquire more data for retrospective analysis than to never acquire it at all.

A bivariate histogram is created by the projections of two univariate histograms into two-parameter space, as shown in Fig. 2. There are three common methods for visualizing the correlated events in a bivariate plot, namely, contour plots, density plots, and dot plots. Contour plots provide a series of isometric lines that are intended to show the number of cells at selectable levels. Contour plots are useful when there is a high frequency of cells in a population in two-parameter

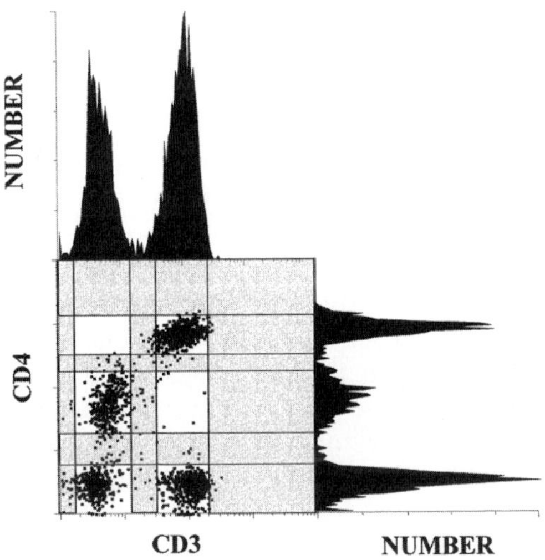

Fig. 2 Bivariate dot plot to provide correlation of parameters. When the univariate histograms displayed in Fig. 1 for CD3 and CD4 are arranged and parallel markers describing the position of each Gaussian peak are drawn across the bivariate space, the x and y coordinates of each event produce clusters of dots that form a population of cells. In this example, four distinct populations, $CD3^-CD4^+$, $CD3^+CD4^+$, $CD3^+CD4^-$, and $CD3^-CD4^-$, are clearly resolved. Their number can then be computed to provide quantitative information on the frequency of each subset. (See color plates.)

space, but they are useless for displaying populations containing a low frequency of cells. Density plots are like contour plots except different colors are used to display different cell numbers. Populations containing small numbers of cells are readily visualized while using color to retain the event contours. For dot plots, each cell is plotted as a dot. The greater the size of any given population, the more dots are plotted so that eventually the dots may be plotted on top of one another. This causes a loss in the ability to see differences in the frequency of cells among populations with high cells numbers (an advantage of contour and density plots). This method is best for displaying populations that contain few cells. This process is performed to provide a bivariate view of each combination of parameters evaluated. For multiparameter data where more than one bivariate display is necessary, dot plots are the preferred method because assigning a specific color to each population can identify its location in each bivariate display. This is called color gating. All bivariate plots used in this chapter are color gated dot plots.

III. Verification of Instrument Performance

There are two types of flow cytometers; those whose laser beams require alignment and those whose laser beams' alignment is fixed. The rationale and procedures to be described do not include, but depend on, proper laser beam alignment. The verification methods will, however, clearly indicate whether alignment has been properly adjusted.

Microspheres (or beads) have historically been the primary standard for instrument performance verification. They have three major advantages over any other kind of material: they are consistent in their properties over time, they are relatively inexpensive, and they are easy to use. However, they have one major disadvantage: they often do not behave like cells. For example, because their index of refraction is so different from that of a cell, their forward and side scatter properties are totally different from cells of exactly the same volume.

The first step in the verification process is to evaluate overall instrument performance. Beads that have a very low coefficient of variation (CV) and that will produce detectable fluorescence in every detector should be used. An acceptable range should be established for the mean channel fluorescence value and CV for each parameter. This is shown in Fig. 3 for the daily performance of a FACScan.

RULE: ESTABLISHED INSTRUMENT SETTINGS SHOULD NOT BE CHANGED TO BRING BEADS INTO THE ACCEPTABLE RANGE.

As shown in Fig. 3, the instrument is so stable that the same settings are valid from day to day unless there is a problem. Changing the settings may simply

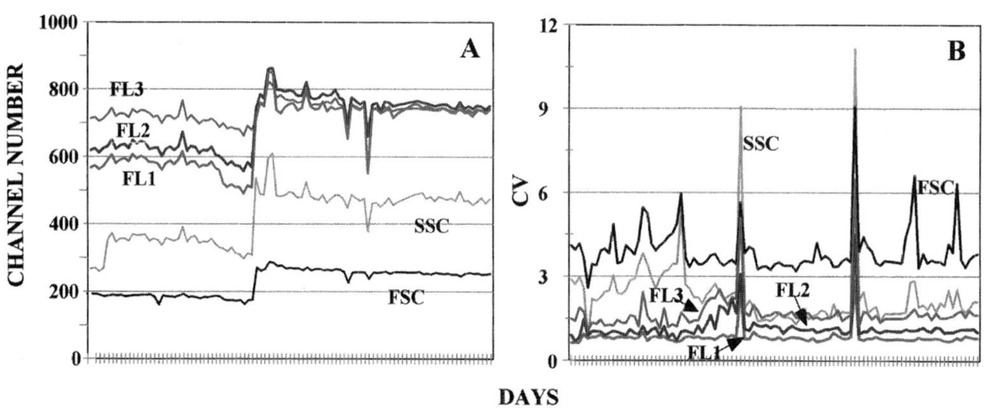

Fig. 3 Performance verification of a FACScan. Microspheres (DNA check) were used daily to measure the median channel for each parameter (A). The threshold channel number was allowed to vary ±10% for acceptable instrument performance per median; channel value depended on the lot of beads as shown by their abrupt change when the lot changed, indicating that the lot to lot reproducibility was far worse than the instrument's reproducibility. The coefficient of variation (CV) was established at less than 3% for all parameters except FSC and SSC, which were 6% (B). Instrument failure shown as spikes required maintenance to reestablish performance prior to use. At no time were instrument settings changed to provide compliance; instead, the reason for noncompliance was sought and repaired. (See color plates.)

cover up a problem, and this is not an acceptable practice; instead, the problem needs to be diagnosed and fixed. Acute deterioration is usually caused by debris in the flow cell that distorts the hydrodynamic flow pattern. This can be corrected by cleaning using the manufacturer's suggested procedure or your own trusted method. Chronic deterioration in the optical system is detected by an upward trend of the CV over a long period of time. This is caused by deterioration of the gel used to couple the objective lens to the flow cell (when present) or by dirty optics. This can be prevented by routine preventive maintenance. Both chronic and acute failure can occur as a result of the deterioration of laser beam power. This can be corrected by getting a new laser or plasma tube.

Once the instrument has passed its initial verification with beads, spectral compensation is usually performed. This is such an important process that we will devote an entire section of this chapter to a discussion of it. There are several bead-based systems for performing this task, some of which are instrument specific, such as adjustment of laser beam timing in a FACSCalibur. The reasons for this will become clear in Section IV on compensation. Thus, bead compensation is useful for initial instrument evaluation, but may be inadequate for the final analysis.

For final instrument verification, relevant cell based surrogates are recommended. The selection of the appropriate cell surrogate will depend on the instrument's use. For example, if the instrument will be used mainly with hematopoietic cells, lymphocytes represent the best surrogate. These can be from blood or animal organs such as mouse spleen cells. If the main use will be to make measurements on cell lines, an appropriate one can be selected as the surrogate.

RULE: AN APPROPRIATE CELLULAR SURROGATE MUST BE PREPARED FOR VERIFICATION OF ALL PARAMETERS.

There are also stabilized cells available, which behave much like their viable counterparts, that can provide a convenient surrogate to verify instrument performance. The rest of the instrument performance verification will be described in Section IV.

RULE: A CHANGE IN SETTINGS REQUIRES A COMPLETE REVERIFICATION OF ALL PARAMETERS.

Once the instrument has been verified, settings must not be changed to accommodate cells that may differ significantly from the surrogate. If settings are changed, the cells that lead to the change can be appropriately stained with bright antibodies for readjustment of all parameters and compensation. Failure to perform this step will result in erroneous data that are certain to be misinterpreted.

IV. Optical Filtration and Spectral Compensation

Before considering the use of more than one antibody, the problem of overlapping fluorescence spectra of fluorochromes used to label monoclonal antibodies needs to be addressed. As shown in Fig. 4, when excited at 488 nm, the emission from fluorescein (FL) overlaps considerably with that of phycoerythrin (PE), whereas phycoerythrin emission overlaps fluorescein emission to a lesser extent. PETR (phycoerythrin–Texas Red tandem) also overlaps the PECY5 (PE/CY5 tandem) (or peridinin chlorophyll protein, PerCP) emission spectrum, and PE photon leakage from this construct is high due to less efficient energy transfer. When excited at 635 nm, CY5 and allophycocyanin (APC) are not distinguishable from one another in their emission distributions. Although not yet a commonly used fluorochrome, APC-CY7 or CY7 may be a useful reagent for increased spectral separation (Beavis and Pennline, 1996). Antibodies conjugated with this fluorochrome are available from Caltag (Burlingame, CA). The emission from one fluorochrome can be optimized to reduce but not eliminate unwanted fluorescence overlap by using optical filtration. Table II lists the optical filters commonly used for fluorescence detection when the fluorochromes are excited by spatially separated lasers operating at 488 and 635 nm. PECY5, PerCP, and APC

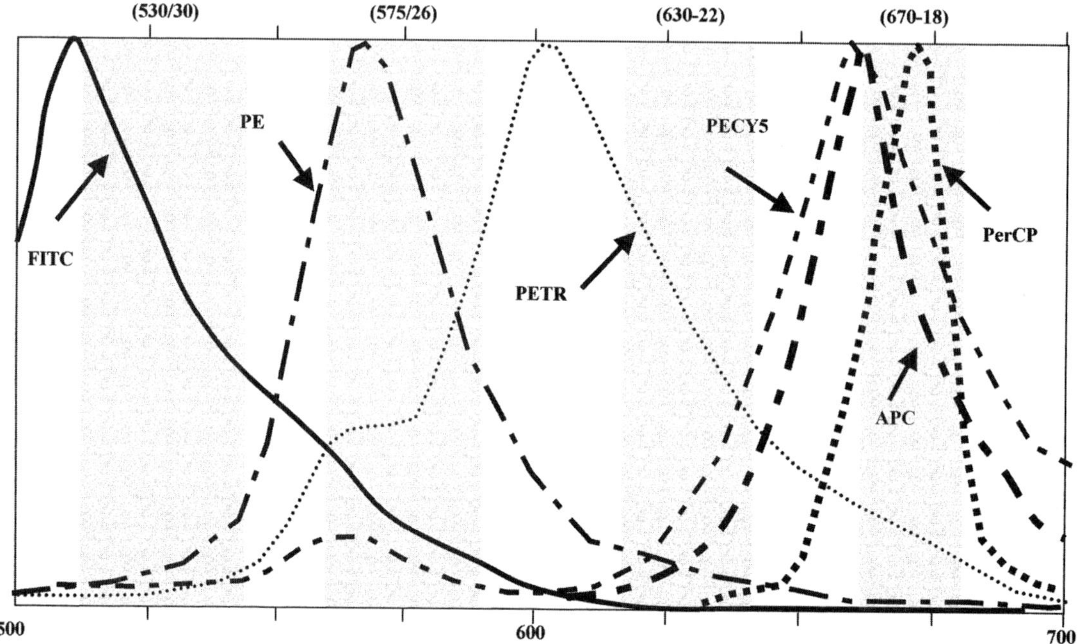

Fig. 4 Emission spectra from commonly used fluorochromes. For each fluorochrome, the relative photon frequency (intensity) as a function of wavelength (energy) is shown. The commonly used band-pass filter system is also shown for each fluorochrome. For the tandem complexes, the emission spectrum is composed of a PE component and the acceptor component (TR or CY5). The height of the PE component, identical to the PE emission spectrum, is greater for Texas Red than for CY5. Note also there is virtually no spectral separation possible between CY5, PerCP, and APC. The band-pass filter range usually used is shown by the stippled bars. (See color plates.)

Table II
Filters Commonly Used for
Immunofluorescence Measurements

Fluorochrome	Dichroic mirror (nm)	Band-pass filter (nm)
488 nm excitation		
Fluorescein	SP 560	530–30
Phycoerythrin	SP 560	575–26
PETR	SP 610	630–22
PECY5	SP 640	670–18
PerCP	SP 640	670–18
635 nm excitation		
APC	SP 660	670–18
CY5	SP 660	670–18
CY7	SP 730	780–50

are all excited at 635 nm, but APC is not excited at 488 nm. Thus, APC or CY5 emission can be time resolved from the others.

Optical filters are identified by the wavelengths they process. Long-pass (LP) filters will transmit wavelengths that are longer than the wavelength provided, whereas short-pass (SP) filters transmit shorter wavelengths. For a band-pass (BP) filter, the first number refers to the wavelength that is transmitted at or near 100%, and the second number refers to the bandwidth (50% transmission). Thus, 530–30 means that the filter transmits 50% of light from 515 to 545 nm (30-nm bandwidth) and the maximum transmission is at 530 nm. Dichroic mirrors (DM) are used to both transmit and reflect selective wavelengths of light along the optical path. An LP dichroic mirror transmits the longer and reflects the shorter designated wavelengths whereas a SP dichroic mirror does the opposite. An LP560 dichroic mirror transmits wavelengths longer than 560 nm (e.g., PE fluorescence) and reflects shorter wavelengths [e.g., fluorescein isothiocyanate (FITC) fluorescence]. By placing the correct dichroic mirrors and filters in the optical path, the appropriate colors can be focused on designated photomultiplier tube detectors. In this way, each colored fluorochrome is detected. Unfortunately, due to overlapping spectra, optical filtration is not completely effective in resolving a specific fluorochrome.

Because the optical filters cannot remove all the overlap from the unwanted fluorochrome, electronic subtraction, called spectral compensation, is used to eliminate the remaining overlapping fluorescence from unwanted fluorochromes. Figure 5 shows a schematic diagram of how this is accomplished. The signal from the fluorescence 1 (FL1) photomultiplier tube is routed directly into the positive side of the fluorescence 1 operational amplifier (OPAMP). This device is used to subtract one signal from another. The output of this OPAMP is routed through an adjustable circuit into the negative side of the fluorescence 2 (FL2) OPAMP. Similarly, the fluorescence 2 photomultiplier signal is routed into the positive

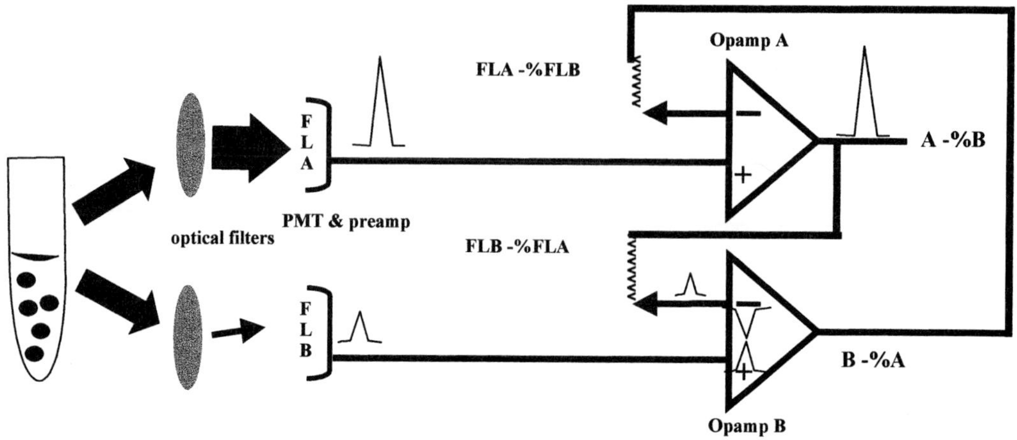

Fig. 5 Electronic compensation. A simplified representation of the circuit for electronic compensation is shown. In this example, cells are stained with the fluorochrome detected by FLA, namely, FITC. The optical filters select for the appropriate wavelength to reduce some but not all the unwanted fluorescence. FLA has a high detection efficiency for the fluorochrome detected by A, so the signal is high, whereas FLB efficiency is low and the output is also low. These signals are routed to OPAMPs A and B, which are devices that invert and add voltages. The output signal from OPAMP A is routed through an adjustable circuit so the input can be made the same height as that detected by FLB. It is inverted by OPAMP B, where it is added to the signal from FLB, thereby canceling it.

side of its OPAMP and its output, via an adjustable circuit, into the negative side of the fluorescence 1 OPAMP. By adjusting the amount of signal fed into the negative side of the appropriate OPAMP from particles labeled with a single fluorochrome, the unwanted fluorescence can be removed. For more than two fluorochromes several compensation networks can be ganged together to achieve the correct compensation. A properly compensated instrument will provide good resolution of single- or double-labeled cells (Fig. 6).

Improper compensation can lead to false positive or false negative cell populations. This is shown in Fig. 7 where microspheres of different FITC fluorescence intensities are shown uncompensated (Fig. 7A), partially compensated (Fig. 7B), and fully compensated (Fig. 7C). Any stained cells that are brighter than the beads used for compensation (in this case the third brightest) will be undercompensated (Fig. 7B), producing a hockey stick pattern and potential misinterpretation of the results.

RULE: ALWAYS USE CELLS STAINED WITH THE BRIGHTEST CONJUGATED ANTIBODY FOR VERIFICATION OF COMPENSATION.

Although microspheres are often used initially for compensation adjustment, stained cells should also be used to verify the final settings. There are two reasons for this. First, commercial beads are often not as bright as some antibodies used

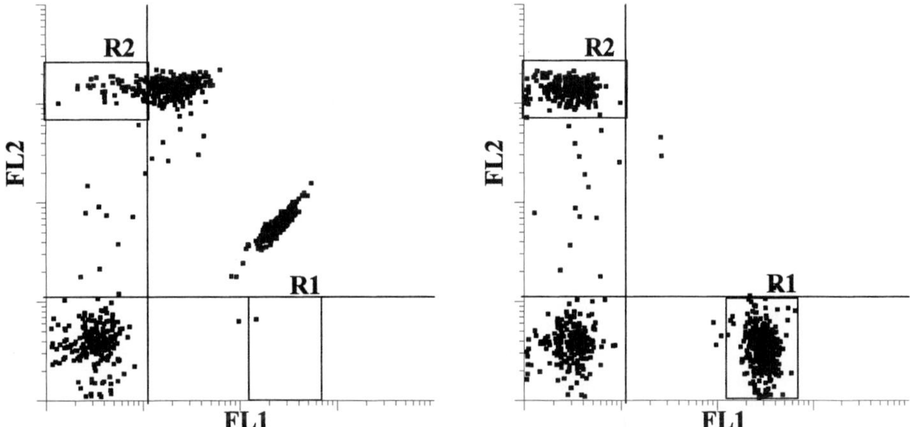

Fig. 6 The bivariate distribution of uncompensated (left) and properly compensated (right) cells are shown. The PMT high voltage is adjusted so that the unlabeled cells (autofluorescence) are in the lower left corner. The fluorescence 1 boundary limit is shown by the line parallel to the x-axis and the fluorescence 2 boundary limit by the line parallel to the y-axis. Events beyond these boundaries represent spectral overlap. Regions R1 and R2 are established so that correct compensation can be readily achieved in a reproducible fashion from time to time.

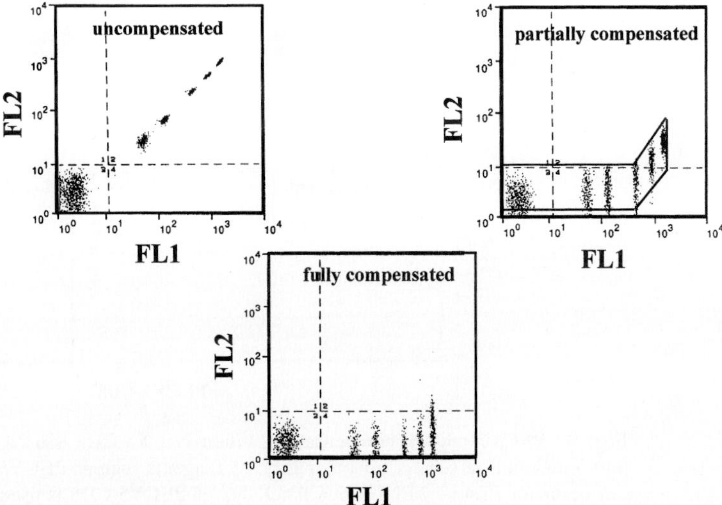

Fig. 7 Compensation and intensity. Microspheres with various levels of FITC fluorescence (FL1) were acquired uncompensated and compensated using the beads with the third level of fluorescence or with the highest level of fluorescence. When partially compensated, the brighter beads are under-compensated, giving a hockey stick pattern.

to stain cells. Second, when some fluorochromes, such as the tandems, are used, there are no beads available to properly adjust compensation.

Figure 8 illustrates the problem that is encountered when a properly compensated instrument for one PECY5 conjugated antibody is used to measure several other PECY5 conjugated antibodies. Three different situations occur. First, a properly compensated instrument for one PECY5 antibody (Fig. 8A) is under-compensated for a different antibody (Fig. 8B). Second, a properly compensated

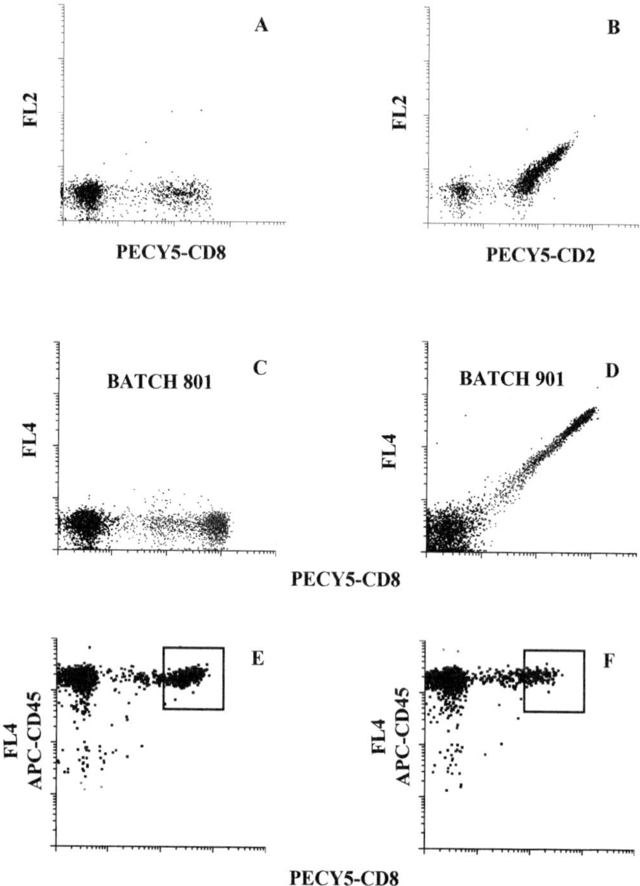

Fig. 8 PECY5 reagent compensation. When PECY5-CD8 is used for compensation, the FL2 − %FL3 may not be correct for other PECY5 reagents such as PECY5-CD2 (B). A similar situation can occur for FL4 − %FL3, when batch 801 of PECY5-CD8 is used for compensation (C). Batch 901 is not properly compensated (D). Another problem occurs when a bright APC reagent such as CD45 is combined with a PECY5 reagent such as CD8. When properly compensated (E), there is good separation between the CD8 positive and negative cells. The tendency, however, is to overcompensate FL3 − %FL4 (F), where the events shift to the left and fewer are in the box, causing the PECY5 antibody to appear dimmer or completely absent.

instrument for one PECY5 antibody in the CY5 APC channel (Fig. 8C) is undercompensated for a different batch of the same antibody (Fig. 8D). Third, CY5 fluorescence (Fig. 8E) can be markedly decreased (Fig. 8F) because of overcompensation of APC (FL3 − %FL4). Thus, each PECY5 conjugated antibody requires its own unique compensation. This creates a logistical problem in acquiring sample data.

There are three approaches that can be used to solve all of these problems. One could always use the same PECY5 conjugated antibody. For example, CD45 is often used as a gating reagent (Stelzer *et al.*, 1993). Such an approach, however, limits the investigator's flexibility in multicolor flow cytometry. Another approach would be to use biotinylated antibodies so that a single PECY5 avidin is used throughout. This approach provides the investigator the flexibility of using any antibody, but it requires a second processing step.

Finally, in order to use several antibodies directly conjugated with PECY5, software compensation can be used that contains the unique compensation matrix for each PECY5 conjugated antibody. Whenever antibody is used, its unique values are applied at the time of data analysis by the software. For software that can provide this function on data acquisition (e.g., Beckman Coulter XL, Hialeah, FL), the unique values are applied at that time. Otherwise, uncompensated data can be collected and compensation performed at the time of data analysis (e.g., with WinList, Topsham ME; or FLO JO, San Diego, CA).

We will perform compensation for up to four separate colors. Compensation standards are prepared for each color using normal blood or another source of cells commonly used in the laboratory. The cells are divided into separate tubes and then stained with the conjugated antibodies. See Chapter 11 of Volume 63 of this series for the methodology for staining cells with single-color reagents.

A. For Two- or Three-Color Compensation

1. Stain cells in separate tubes with a FITC-antibody, PE-antibody, and PECY5 (or PerCP) antibody. Prepare a separate tube of unstained cells.

2. Refer to Fig. 9 and create the bivariate histograms FSC versus SSC, FL1 versus FL2, and FL3 versus FL2 using the instrument's computer. Using unstained cells, acquire 2000 events and draw a region, R1, around the lymphocyte cluster (Fig. 9A). Create quadrant markers set at the $x = 10$ and $y = 10$ for both fluorescence bivariate histograms. Reacquire cells, adjusting the photomultiplier tube (PMT) high voltage so that the cells are in the lower right quadrant, as shown in Fig. 9B,C.

3. Using cells stained with the FITC antibody, adjust FL2 − %FL1 so none of the cells are on the positive side of the quadrant marker that is parallel to the *x*-axis in the bivariate histogram of FL1 versus FL2. Draw a region, R2, around the cluster (Fig. 9B).

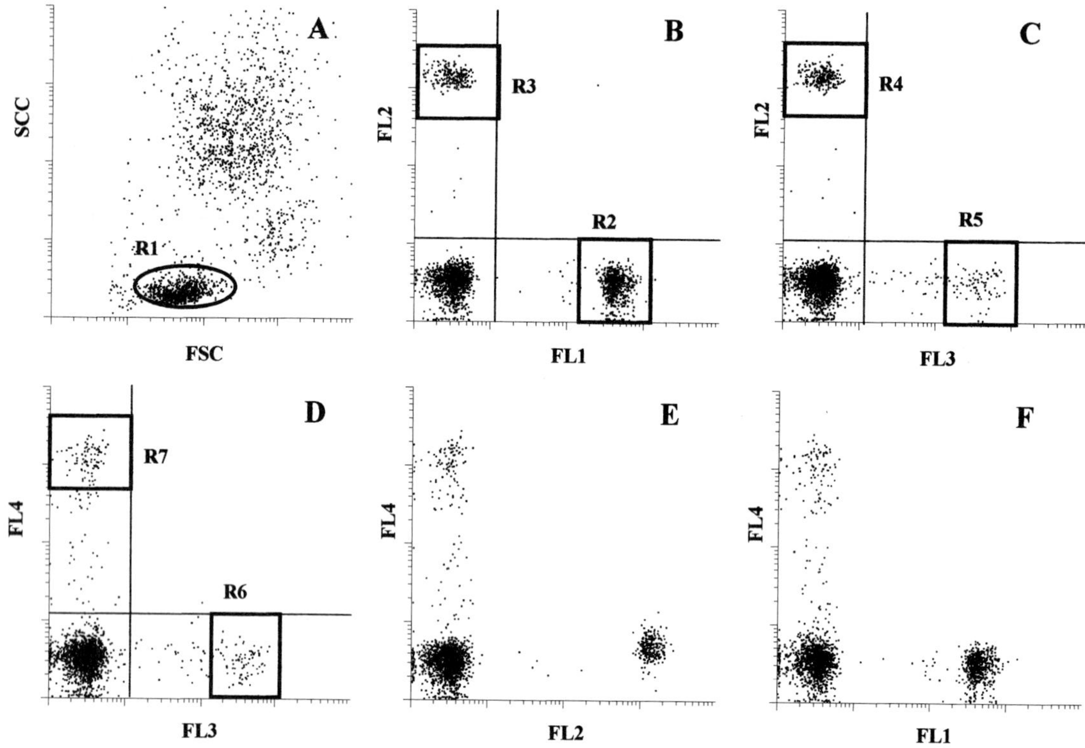

Fig. 9 Electronic compensation for up to four colors. Human blood was stained with FITC-CD45, PE-CD4, PECY5-CD8, and ECD-CD8 for single-laser or APC-CD45 for dual-laser instruments, in separate tubes. Although it is recommended that the generation of the initial compensation settings should be produced using separate tubes, the cells can be mixed together and acquired as a reference file that will contain the instrument settings for daily comparisons and verification of compensation. The same reference file is created for either single or dual laser excitation.

4. Using cells stained with the PE antibody, adjust FL1 − %FL2 so none of the cells are on the positive side of the quadrant marker that is parallel to the *y*-axis in the bivariate histogram of FL1 versus FL2. Draw a region, R3, around the cluster (Fig. 9B). Two-color compensation is complete.

5. For three-color compensation, adjust FL3 − %FL2 so none of the cells are on the positive side of the quadrant marker that is parallel to the *y*-axis in the bivariate histogram of FL3 versus FL2. Draw a region, R4, around the cluster (Fig. 9C).

6. Using cells stained with PECY5, adjust FL2 − %FL3 so none of the cells are on the positive side of the quadrant marker that is parallel to the *x*-axis in the bivariate histogram of FL3 versus FL2. Draw a region, R5, around the cluster (Fig. 9C).

If different PECY5 conjugated antibodies are used that require different compensation, it is necessary to collect an uncompensated list mode file of 2000 events for each and use software compensation described later.

B. For Four-Color Compensation

1. Perform Steps 1–4 for three-color compensation. Add to the display FL3 versus FL4, FL2 versus FL4, and FL1 versus FL4 (Fig. 9D–F), add quadrant markers at $x = 10$ and $y = 10$, and adjust FL4 high voltage so unstained cells are inside the lower left quadrant.

2. Using cells stained with PECY5-CD8, adjust FL4 − %FL3 so none of the cells are on the positive side of the quadrant marker that is parallel to the x-axis in the FL3 versus FL4 bivariate histogram. Draw a region, R6, around them (Fig. 9D). Replace the tube with the one containing ECD-CD8 (for single laser) or APC-CD45 (for dual laser), and adjust FL3 − %FL4 so that none of the cells are on the positive side of the quadrant marker that is parallel to the y-axis in the bivariate histogram of FL3 versus FL4. Draw a region, R7, around the positive cells (Fig. 9D). This completes the initial instrument compensation adjustments for four colors.

Because PE, Texas Red, CY5 (or PerCP), and APC will all have overlapping spectra (Fig. 2) in a four-color combination, there will be intercompensation network interactions that negate the use of compensation standards stained with a single-color fluorochrome as an effective means for compensation. The reasons for this have been described in detail elsewhere (Stewart and Stewart, 1999). After compensating the instrument using single-color stained cells, it is necessary to verify that there is no overcompensation.

1. For single laser excitation, stain cells with an antibody combination that is mutually exclusive for PE and PETR but costains all cells with a dim PECY5 fluorescence, for example, PE-CD4, ECD-CD8, and PECY5-CD2. (See Fig. 10A.)
2. Adjust compensation as necessary as shown in Fig. 10B.
3. For dual laser excitation, stain cells with an antibody combination that is mutually exclusive for PE and PECY5 but costains all cells with APC, for example, PE-CD4, PECY5-CD8, and APC-CD45.
4. Adjust, as necessary, FL3 − %FL2, so that cell populations are mutually exclusive, as shown in Fig. 10, and adjust FL3 − %FL4 for maximum separation of the coexpressing population from the single FL4$^+$ expressing cells (Fig. 10D).

Having generated a template that provides the positions of compensated cells, individual tubes no longer need to be run separately. Instead, the single-stained cells can be pooled and acquired. The position of each cluster must appear in

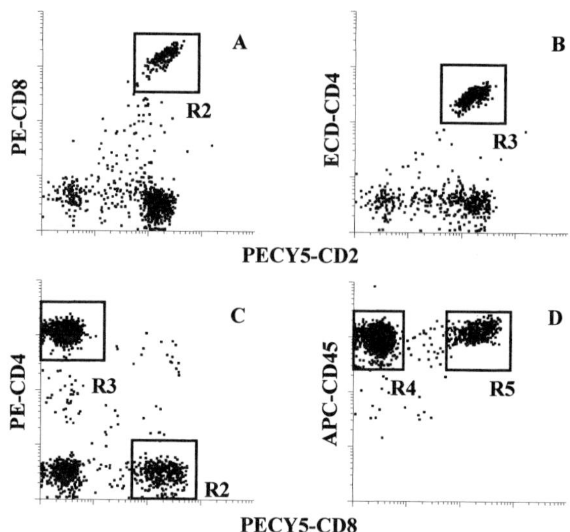

Fig. 10 The top row shows the verification for single laser excitation. It is important that both CD8 bright (A, R2) and CD4 bright (B, R3) positive cells are also CD2 positive. R1 is used to gate on lymphocytes (Fig. 9). The bottom row shows the verification combination for dual laser excitation. It is important that CD4 (R3) and CD8 (R2) do not appear to be coexpressed (C) and that CD8 (R5) is maximally separated from CD45 (R4).

the appropriate regions (as defined in Figs. 9 and 10). For four-color compensation the appropriate combination of four antibodies is also evaluated, and the stained cells must appear in the appropriate regions.

Compensation files generated for each compensation standard are saved and provide a continuous log of instrument performance. The two sets can be compared to verify that the instrument is performing in an identical fashion from time to time. If, after acquisition, the previous compensation standard does not compare favorably with its reference file (generated on the previous day), then there is an instrument problem. If the previous compensation standard looks like it did before, but the new mixture looks poor, then the new compensation standard is faulty and should be prepared again.

It is essential that the instrument itself be in proper alignment when these files are created. In a properly aligned instrument, there should be no changes in settings over time; if there are, it is likely there is an instrument problem. Using both microspheres (available from several sources) and the cellular compensation mixture described here, instrument performance can be monitored daily, trends established, and deterioration in instrument or sample preparation performance detected (see Fig. 3). A lymphocyte (or other low autofluorescent cell) gate is used for acquisition to ensure the best homogeneity of each measurement.

The use of a single large cellular compensation mixture that can be used over a period of time as a compensation standard is not recommended because fixation results in a slowly increasing autofluorescence with time, as shown in Fig. 11. The increase in cellular autofluorescence can be documented by comparing old (7–14 days) cellular compensation mixtures with freshly prepared ones to evaluate instrument performance. By preparing the standards daily, this problem does not occur.

As shown in Fig. 8, the PECY5 conjugated antibodies vary in their compensation requirements. This problem can be easily corrected using software compensa-

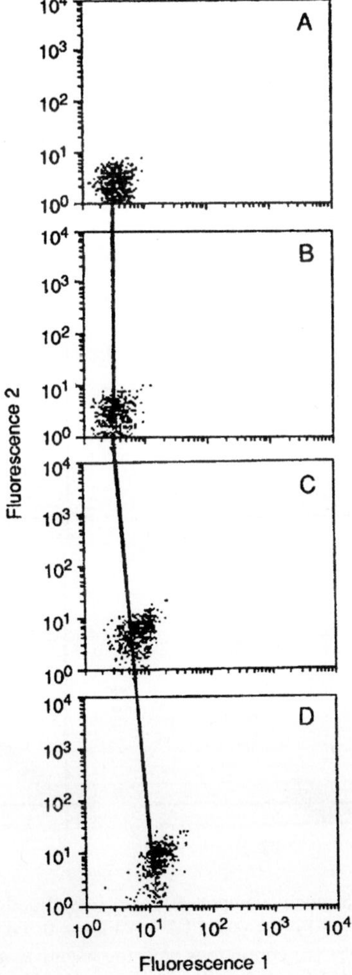

Fig. 11 Storage of lysed whole blood in 2% ultrapure formaldehyde. Human blood was lysed and fixed in 2% ultrapure formaldehyde. The same sample was analyzed (A) 1 hr, (B) 7 days, (C) 14 days, and (D) 21 days after fixing, using the same instrument settings. There is a fivefold increase in the mean autofluorescence from day 7 and day 21.

tion. One could compensate all parameters using software or only those that are associated with the PECY5 reagent. We have chosen the latter approach.

1. Using cells stained with FITC-CD45, adjust FL2 − %FL1 on the instrument.
2. Using cells stained with PE-CD4, adjust FL1 − %FL2, FL2 − %FL3 = 0, and FL3 − %FL2.
3. Using cells stained with both APC-CD45 and PECY5-CD8, adjust FL3 − %FL4 so that there is maximum separation between the two clusters. This will ensure the instrument is not overcompensated (FL4 − %FL3 = 0).
4. Using cells stained for each properly titered PECY5 conjugated antibody, acquire 2000 uncompensated (FL2 − %FL3 = 0, FL4 − %FL3 = 0) events. This process is repeated every time a new batch of the same reagent or a different reagent is obtained.
5. Using WinList, display all bivariate histogram combinations and apply trace lines (Fig. 12). Adjust them to compensate FL2 − %FL3 and FL4 − %FL3. Make sure all other trace lines are set parallel to the *x*- or *y*-axis, as shown

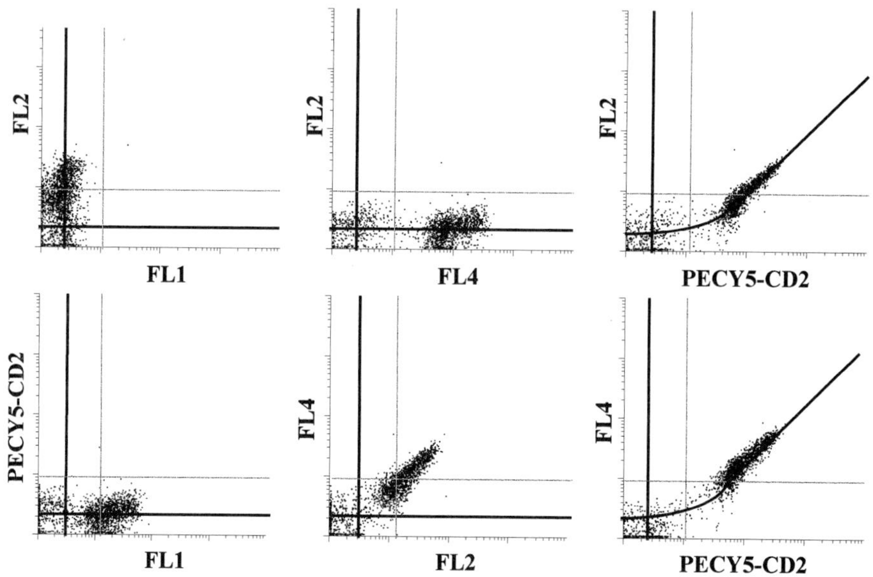

Fig. 12 Software compensation process. Blood was stained with PECY5-CD2 and the data acquired with FL3 − %FL2 = 0 or FL3 − %FL4 = 0. To perform compensation in WinList, adjust the trace lines so they are continuous with the negative and positive cells in views FL3 versus FL2 and FL3 versus FL4. Make sure all other trace lines are parallel to their respective axes, because these parameters have been compensated electronically. This process is performed for each PECY5 reagent. The unique matrix is saved as ÇDxxxbbb.cmp, where CDxxx is the antibody field and bbb the batch field.

in Fig. 12, because these parameters have been compensated using the instrument.

6. Save the compensation matrix. We use the file name convention CDxxxbbb. cmp where xxx is the CD or clone designation field for the antibody, bbb is the batch field, and .cmp stands for compensation.

The compensation matrix can be applied manually every time any antibody combination is used containing the particular PECY5 conjugated antibody. Figure 13 shows three commonly used PECY5 antibodies that require different compensation. The matrix can be incorporated into a macro where it will be automatically applied every time the combination is used.

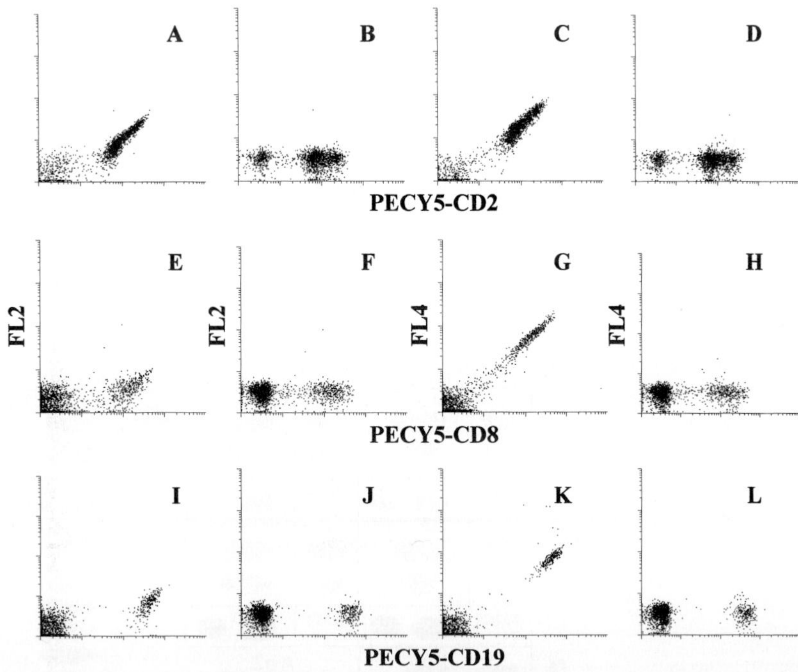

Fig. 13 Software compensation of PECY5 antibodies. Each PECY5 reagent requires unique compensation. Blood was stained with either PECY5-CD2, -CD8, or -CD19 and the data were acquired with FL3 − %FL2 = 0 (A, E, I) or FL3 − %FL4 = 0 (C, G, K). The reagents were compensated using WinList (B, D, F, H, J, L), and then the matrix was saved as CDxxxbbb.cmp, where CDxxx stands for any CD marker or provides a field for a non-CD designated clone, bbb provides a field for the batch number, and .cmp stands for compensation. PECY5-CD2 (A, C), -CD8 (E, G), and -CD19 (I, K) were acquired uncompensated. These were compensated in software and shown in Fig. 11. The amount of compensation for CD2 was FL2 − %FL3 = 8%, FL4 − %FL3 = 21%. The amount of compensation for CD8 was FL2 − %FL3 = 1%, FL4 − %FL3 = 50%. The amount of compensation for CD19 was FL2 − %FL3 = 1%, FL4 − %FL3 = 20%.

V. Multiparameter Data Analysis

The customary approach to data analysis is the marker method. For a bivariate histogram, four quadrants are created from which negative, single-positive, and double-positive cells for each antibody can be readily determined. When more than two colors of fluorescence are used, multiple bivariate plots are required. These are summarized in Fig. 14. There are 8 or 16 possible binary combinations that might be present when three- or four-color data are analyzed. As the number of antibodies increases, the number of binary combinations increases by 2^n, where n is the number of antibodies. It has been customary to first plot FSC versus SSC, form a gate around the desired cluster, such as lymphocytes, and then gate the fluorescence histogram on them. The problem with this approach, however, is that all the desired cells may not be included in the gate, especially in sick patients. Therefore, healthy individuals may not always be good surrogates of sick individuals. This has led to the realization that a process called cell gating may be a much better approach. In this strategy, a specific antibody to the desired cells is used to resolve them. For example, CD45 for leukocytes (Stelzer *et al.*, 1993), CD3 for T cells (Mandy *et al.*, 1992), CD19 for B cells, CD56 and not CD3 for NK cells (CDC, 1997), CD14 for monocytes, and CD34 for progenitor cells (Sutherland *et al.*, 1996) are a few examples. In this approach, side scatter, in linear or log units, versus the fluorescent antibody is used to identify the desired cell. A region can be drawn around positive cells, and that is then applied to other displays of the scatter or fluorescence parameters as a gate. The

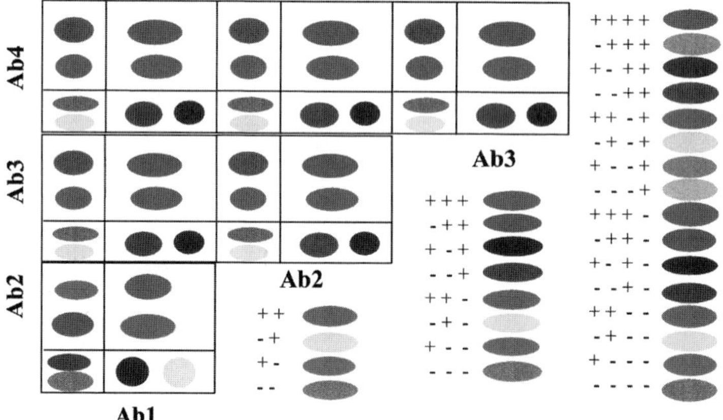

Fig. 14 Bivariate plots for binary combinations of multiparameter data. For two-parameter fluorescence measurements, there are four possible binary combinations, and each is assigned a different color. These same colors along with four additional colors are assigned to the eight binary combinations for three-parameter fluorescence measurements. When four colors are used, there are eight additional binary combinations for a total of 16, so we repeat the eight colors again. This process is continued for five parameters, which produces 32 binary combinations, etc. (See color plates.)

advantage of using this approach is that only one region for each antibody is required, and any one of them or a Boolean combination can be used as a cell gate.

We have incorporated this feature into a new generic approach to data analysis, which can easily be applied to any number of fluorescence parameters. This approach becomes increasingly easier as the number of analysis parameters increases, as shown in Fig. 15 for a combination used to determine the lymphocyte subsets of T cells, NK cells, and B cells in human blood (see also Table III).

1. Display SSC versus the fluorescence parameter for the number of fluorescence parameters (n) measured and draw a region around positive events or the desired cluster of cells, as shown in Fig. 15A for four-color data. An isotype control or autofluorescence control could be used to define negative events.

2. Next the FSC versus SSC display for the cells of interest is obtained by gating on the positive events in each SSC versus antibody view (R1 to R4). This would provide information on the scatter properties of the positive cells, which might actually provide for the resolution of previously unknown populations. Another region, R5, can be applied to enumerate a specific population in this view. Note that granulocytes contaminate the regions R1–R3, and these can be eliminated by either combining them with R4 or R5 in a Boolean expression.

3. Display all fluorescence combinations, as shown in Fig. 15B. These bivariate displays can be gated on any desired combination of fluorescence and scatter parameters. In this example, we have used CD45$^+$ cells (R4), exhibiting the scatter characteristics of mononuclear cells (R5), as the gate (R4 and R5). Clearly,

Table III
Frequency of Cell Populations Resolved by Four-Color Analysis

Binary combination						
CD3	CD5 6	CD1 9	Boolean expression	Color	%	Cell type
−	−	−	R4 and not (R1 or R2 or R3)	Black	7.9	Monocytes
+	−	−	R4 and [R1 and not (R2 or R3)]	Yellow	59	NK cells and monocyte subset
−	+	−	R4 and [R2 and not (R1 or R3)]	Cyan	12	NK cells
+	+	−	R4 and [R1 and R2 and not R3]	Green	11	Cytotoxic T cells
−	−	+	R4 and [R3 and not (R1 or R2)]	Brown	9.4	B cells
+	−	+	R4 and [R1 and R3 and not R2]	Blue	<0.1	None, dead cells, and debris
−	+	+	R4 and [R2 and R3 and not R1]	Violet	<0.1	None, dead cells, and debris
+	+	+	R4 and [R1 and R2 and R3]	Red	<0.1	None, dead cells, and debris

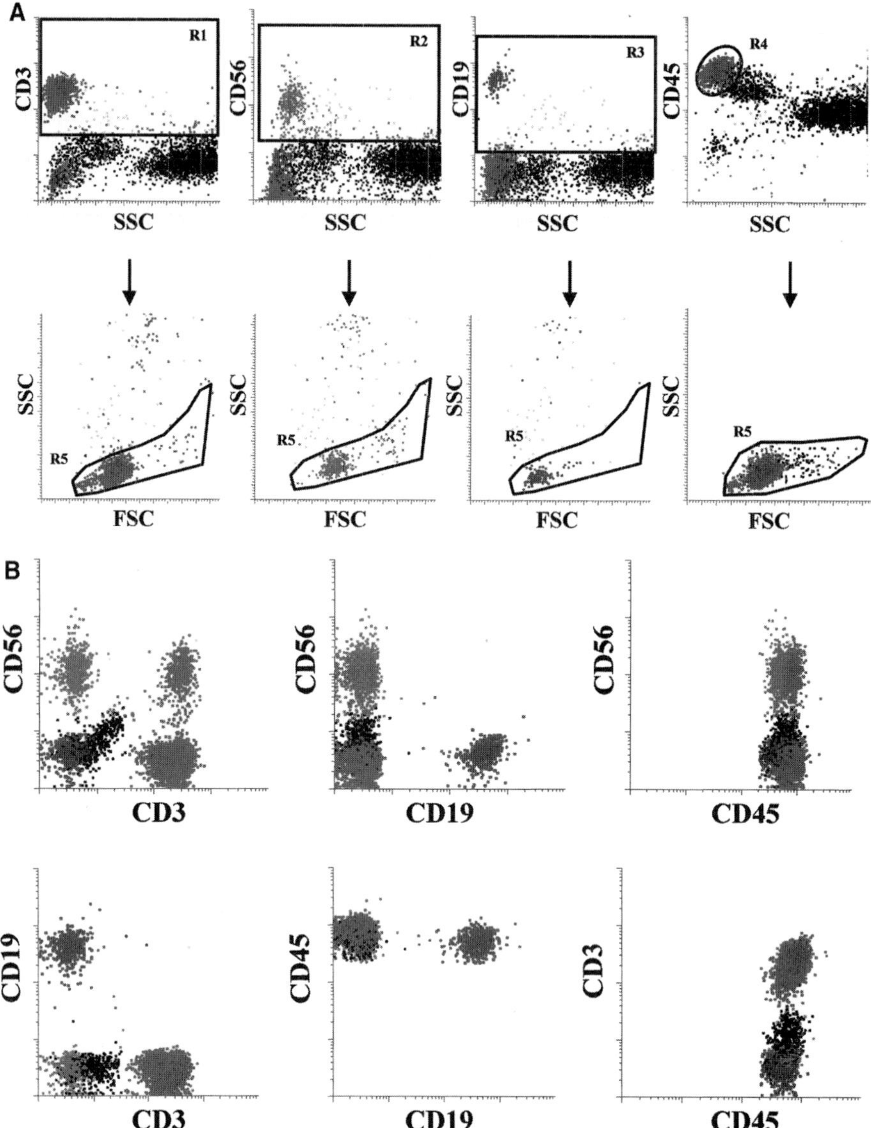

Fig. 15 Multiparameter data analysis. To illustrate the strategy for evaluating multiparameter data analysis, human blood was stained with the antibody combination FITC-CD3, PE-CD56, PECY5-CD19, and APC-CD45. A bivariate plot of SSC versus each fluorescence parameter is displayed in the top row (A). A region (R1–R4) is drawn on the desired population of positive cells or to all events above a negative control, as shown. Next the FSC versus SSC for each selected cluster is displayed by gating them on each of the regions. Another region, R5, can be drawn to further define the desired population (A, bottom row). Finally, the bivariate plots of each antibody combination are displayed (B). For four-color data, there are six bivariate combinations. The frequency and color of each cluster of events defined by a Boolean expression are shown in Table III. In this example, the gate R4 and R5 that defines CD45$^+$ mononuclear cells was used. Because there are no CD45$^-$ events defined by the gate used, these values have not been included in Table III. (See color plates.)

any region could be drawn in the fluorescence or scatter display for gating providing unlimited flexibility for data analysis.

In order to visualize each of the subsets of cells in all the bivariate combinations, color gating is used by assigning a color to each binary combination. Our strategy is to assign the color to the binary combination representing all possible phenotypes. Therefore, in the three-color combination $+++$, red is assigned. Thus, all cells stained with all three antibodies (whatever they might be) will be colored red in the bivariate histograms.

Color gating, however, is not useful for five or more parameters because most individuals cannot resolve more than 11 or 12 colors (e.g., red, yellow, orange, green, blue, violet, brown, gold, cyan, magenta, gray, and black). Although shades can be made between any two pairs, we are not readily able to perceive them in a colored figure. Thus, a five-color experiment that would generate 32 binary combinations using 32 different colors would be visually useless. To solve this visualization problem we repeat the same colors in increments of up to 12. For example, four-color would be represented by 8 + 8, five-color by 12 + 12 + 8. Selecting one or more of the fluorescence parameters as a (sequential) gate performs three separate analyses. Insight into the purpose of the antibody combinations is required to choose the proper color gating antibodies and their Boolean combination.

Although there is no minimization of the increasing complexity of multiparameter analysis produced by correlated list mode data, this strategy provides for a simple generic approach for its analysis and visualization. Thus, data become increasingly complex, but the analysis remains simple. We call this process cell gating because only the cells that meet the criteria selected can pass through the gate for display and numerical evaluation.

VI. Summary

Each laboratory has to establish its own experience base and standard operating procedures. The intent of this discussion has been to illustrate the procedures that will lead to good flow cytometry data acquisition and analysis and to illustrate problematic areas. The most important rule of all is to recognize when there is a problem and find the correct solution. It is hoped the information provided herein will be of help in the recognition process.

References

Beavis, A. J., and Pennline, K. J. (1996). Allo-7: A new fluorescent tandem dye for use in flow cytometry. *Cytometry* **24**(4), 390–395.

Center for Disease Control and Prevention (CDC) (1997). Revised guidelines for performing CD4[+] T cell determinations in persons infected with human immunodeficiency virus (HIV). *MMWR* **46**, 1–29.

Mandy, F. F., Bergeron, M., Recktenwald, D., and Izaguirre, C. A. (1992). A simultaneous three-color T cell subsets analysis with single laser flow cytometers using T cell gating protocol. *J. Immunol. Methods* **156,** 151–162.

Stelzer, G. T., Shults, K. E., and Loken, M. R. (1993). CD45 gating for routine flow cytometric analysis of human bone marrow specimens. *In* "Clinical Flow Cytometry" (A. Landay, K. Ault, K. Bauer, and P. Rabinovitch, eds.), Vol. 677, pp. 265–280. Ann. NY Acad. Sci.

Stewart, C. C. (1994). Multiparameter flow cytometry. *In* "Immunochemistry" (C. J. van Oss, ed.), Chap. 32, pp. 849–866. Dekker, New York.

Stewart, C. C. (1996). Clinical applications of multiparameter flow cytometry. *In* "Haematology 1996, Education Programme of the 26th Congress of the International Society of Haematology" (J. R. McArthur, S. H. Lee, J. E. Wong, and Y. W. Ong, eds.). The International Society of Haematology, Singapore.

Stewart, C. C., and Stewart, S. J. (1997). Immunophenotyping. *In* "Current Protocols in Cytometry" (J. P. Robinson, Z. Darzynkiewicz, P. Dean, L. Dressler, P. Rabinovitch, C. Stewart, H. Tanke, and L. Wheeless, eds.), pp. 6.2.1–6.2.15. Wiley, New York.

Stewart, C. C., and Stewart, S. J. (1999). Four-color compensation. *Commun. Clin. Cytometry* **38,** 161–175.

Sutherland, D. R., Anderson, L. M., Keeney, M., Nayar, R., and Chin-Yee, I. (1996). The ISHAGE guidelines for CD34[+] cell determination by flow cytometry. *J. Hematother.* **5,** 213–226.

CHAPTER 46

Immunophenotyping of Hematological Malignancies by Laser Scanning Cytometry

Richard J. Clatch

Department of Pathology
Highland Park Hospital
Highland Park, Illinois 60035

I. Immunophenotyping of Leukemia and Lymphoma: General Considerations
II. Description of Laser Scanning Cytometry
III. Immunophenotyping by Laser Scanning Cytometry
    A. Method Development and Description
    B. Example Results
    C. Advantages/Disadvantages
IV. Extensions of the Method
    A. Relocalization for Light Microscopy and Fluorescence *in Situ* Hybridization
    B. Multipass and Multiplex Systems
V. Conclusions
    References

I. Immunophenotyping of Leukemia and Lymphoma: General Considerations

Immunophenotypic analysis of hematologic specimens is a useful laboratory adjunct to surgical pathology and cytology to confirm or further characterize diagnoses of leukemia or lymphoma (Duque and Braylan, 1991; Knowles *et al.*, 1992; Sun, 1993; Willman, 1992). In a generic sense, immunophenotypic analysis can pertain to any type of tumor or tissue; but for the purposes of this chapter, discussion will be confined to hematologic and lymphoreticular specimens. Methods of immunophenotyping vary most significantly in the type of detection system used to ascertain the occurrence of antigen–antibody binding, and the instrumentation used to observe or quantify that binding. The most common detection systems used for immunophenotyping are fluorescence based and enzymatic–histochemical based. Instrumentation ranges from flow cytometry (FCM), to laser scanning cytometry (LSCM), to epifluorescence microscopy, to immunohistochemistry (IHC), and bright field light microscopy. Before describing the techniques of laser scanning cytometric immunophenotyping, we will first examine certain technical aspects of the various existing methodologies, so that, for his or her own purposes, the reader may choose the method best suited to the clinical or research purpose at hand.

The optimal method should be capable of assessing multiple antigens simultaneously. Simultaneous assessment of at least two and, better yet, even three or four antigens on individual cells permits definitive identification of different populations and subpopulations of cells within a particular specimen. For example, in quantitative analysis of helper and suppressor T cells, it is mandatory to have first defined the population of T lymphocytes in general, apart from all other cell types. Similarly, this is true in the assessment of κ and λ immunoglobulin light chains on B lymphocytes, which is currently by far the most useful means of evaluating B cell clonality. Most immunofluorescence systems, such as immunophenotyping by FCM or LSCM, can easily assess three or four antigens simultaneously.

As an alternative to immunofluorescence systems, immunohistochemical means of immunophenotyping hematologic or lymphoreticular specimens suffer in that at most two antigens can be assayed simultaneously. Moreover, in current clinical practice, IHC is almost always a single antigen assay, making conclusions regarding the interrelationship of different antigens incumbent on the microscopist based on cytologic and architectural morphology.

It is also important to clinical immunophenotyping that reliable antibodies are available specific for the antigens to be tested. Moreover, antibodies must be available for those antigens in whatever state (fresh, fixed, dried, etc.) any particular method demands. In this regard, methods designed to work on fresh unfixed and undried cells have a distinct advantage owing simply to the overwhelming abundance of available antibodies. With time, as antibody development progresses, there may someday be available as many antibodies to formalin fixed CD antigens, as there are antibodies to fresh CD antigens today.

Another important consideration in designing or choosing a method of clinical immunophenotyping is the ability to assess small specimens. In the realm of hematopathology, fine needle aspiration biopsies of lymph nodes and other masses, hypocellular body fluids, hypocellular bone marrows, biopsies of skin, and endoscopic biopsies of gastrointestinal or mucosal tissues are becoming more and more common. Methods that require millions of cells for analysis are simply inadequate for many of these specimens. Moreover, because of the interpretive limitations of cytology alone, it is precisely with such limited specimens that immunophenotypic analysis can often prove to be of greatest diagnostic value (Clatch and Walloch, 1997; Hanson, 1994; Pitts and Weiss, 1992).

One additional and potentially very important attribute of certain clinical immunophenotyping methods is the ability of the method to correlate antigen expression with cytologic and/or architectural morphology. This is the realm in which IHC truly shines; and, at least in regard to correlation with cytologic morphology, laser scanning cytometric immunophenotyping is also particularly advantageous. Although subtle, lymph nodes do have architecture, and this is extremely important in the assessment of potential lymphoid neoplasms. Tissue architecture is also extremely important in evaluating lymphoid infiltrates in nonlymphoid organ systems such as the skin, gastrointestinal tract, and other sites. Unfortunately, aside from frozen sections, good light microscopic assessment of tissue morphology requires formalin fixation and paraffin embedding which can often mask or destroy important hematologic antigens. This limitation can only be overcome through the use of frozen sections or with time as antibodies specific for formalin fixed antigens become available. In clinical situations where correlation of individual cell immunophenotype and architectural morphology is imperative, IHC is the only practically viable solution. As compared with FCM, LSCM does offer the ability to correlate antigen expression with individual cell cytologic morphology. But as of yet, LSCM has not been adapted to tissue applications for clinical purposes. Thus, simultaneous multiparameter immuno- phenotyping on tissue sections stained for light microscopic examination for a broad array of antigens (as are available for flow cytometric and other laser scanning cytometric applications) remains impossible.

Another important characteristic of immunophenotyping methods is the sensi- tivity of the method in detecting weakly expressed antigens or subtle differences in antigen density expression between different populations of cells. In this arena, immunofluorescence and automation are invaluable. Immunofluorescence systems, as compared with immunohistochemical systems, are generally linear because (at least for direct systems) there is no amplification built into the detection means. That is, fluorescence intensity can be stoichiometrically related to antigen density. However, immunofluorescence requires the use of complex instrumentation to excite the fluorochromes, and the fluorescence intensity and spectral characteristics of those fluorochromes are often outside the limited detection capabilities of the naked human eye. IHC generally results in accumula- tion of an easily visible chromogen overlying the positive tissues; but, because enzymes are utilized as a means of amplifying the signal, the stoichiometric relationship is lost, and subtle distinctions in antigen density are often not pos- sible.

Additional considerations in evaluating clinical immunophenotyping methods include issues such as simplicity, time, and cost. All three technologies (IHC, FCM, and LSCM) have evolved to become quite simple and reliable, but there are significant cost differences. LSCM and FCM suffer from the need to purchase expensive capital equipment, whereas the capital equipment necessary for auto- mated IHC is significantly less costly. Once capital equipment has been pur- chased, reagent costs become the driving economic factor; and reagent usage for most laser scanning cytometric methods is diminutive compared to those of either FCM or IHC.

The discussions in the subsequent sections of this chapter will be confined to laser scanning cytometric immunophenotyping techniques, which can vary quite dramatically as the clinical or research needs demand.

II. Description of Laser Scanning Cytometry

Laser scanning cytometry (LSCM) is a new laboratory technology closest in similarity to FCM. In both LSCM and FCM, cells treated with one or more fluorescent reagents are interrogated by a laser beam, with light scatter and emitted fluorescence at multiple wavelengths rapidly measured to determine multiple features of the cells. The principal difference between the two technologies is that for LSCM the cells are located on a solid surface such as a glass slide, whereas for FCM the cells are in a fluid suspension. A laser scanning cytometer (LSC) block diagram is shown in Chapter 3 of this volume. The detailed description of the instrument and the principles of cell measurements are also presented in Chapter 3. The reader is also referred to the original papers by Kamentsky and Kamentsky (1991) and Kamentsky *et al.* (1997a,b).

III. Immunophenotyping by Laser Scanning Cytometry

A. Method Development and Description

In designing a method of immunophenotyping for LSCM, certain fundamental issues must be addressed. Issues to be thoughtfully considered include the following:

How to do the immunofluorescence reactions?
How to get the specimen on the slide?
How to contour?
How to gate?
How to minimize compensation requirements?
How to stain for light microscopy and relocalization?

Many of these issues are interrelated, but each will be discussed in turn with the goal of elucidating the rationale for the development of our preferred method of immunophenotyping and to highlight other possibilities. In order to decide how the immunofluorescence reactions should be performed, one must examine the nature of the specimen to be tested. Most important, is the specimen fresh, or is it dried, fixed, or paraffin-embedded? Fresh specimens, or suspensions thereof, offer the tremendous advantage that antigenicity is virtually undisturbed, making the antigen–antibody reactions reliable without antigen retrieval techniques or amplification. Even so, fresh specimens might be reacted with antibodies in a number of different ways. Cell suspensions may be reacted with antibodies in test tubes, microtiter wells, or reaction chambers built into slides. Of course, different physical systems will have different consequent advantages and disadvantages.

There is a wide variety of different types of dried or fixed specimens. They may include cytospin preparations, touch and smear preparations, paraffin-embedded tissue sections, or frozen sections. Depending on the exact preparation, specific antigens will remain intact to a greater or lesser degree, and the investigator must account for this variability. In most cases, some type of antigen retrieval technique and/or amplification of the signal will be required. Additionally, particularly for hematologic specimens, the number of antibodies available for dried or fixed antigens is significantly limited as compared with antibodies directed against fresh antigens. On the other hand, dried or fixed specimens do potentially offer advantages compared with fresh cell suspensions because, at least for tissue sections, architectural tissue relationships are not necessarily lost. Nevertheless, for the purposes of developing a clinical method of immunophenotyping hematologic specimens by LSCM, a decision was made (for our preferred method to be described later in this chapter) that specimens would be assayed fresh. That is, suspensions of fresh unfixed cells in saline would be reacted with antibodies in a fashion closely analogous to conventional FCM techniques.

Presently in our clinical laboratory, specimens are initially purified using methods similar to those for flow cytometric immunophenotyping. Clinical specimens may include peripheral blood, bone marrow aspirates, body fluids, solid tissue biopsies, and fine needle aspiration biopsies. The goal for all specimens is to obtain a suspension of dissociated leukocytes with relatively few contaminating red blood cells. Red blood cells can be removed from bloody specimens using either Ficoll-Hypaque density centrifugation or ammonium chloride lysis techniques. Solid tissue biopsies can be mechanically disaggregated by pressing the specimen through a 200-μm stainless steel mesh with the plunger of a syringe. Unless they are particularly bloody, body fluid and fine needle aspiration biopsy specimens often require virtually no preparation aside from counting the leukocytes present and making an appropriate dilution.

Because of the nature of LSCM, the specimen must eventually be placed on a solid surface, such as a glass slide, for analysis. Even if the immunofluorescence reactions are performed on suspensions of fresh unfixed cells in test tubes, the cells must eventually be placed on a slide. This might be accomplished in several ways, some of which result in subsequent drying of the cells, potentially compromising pH or chemically sensitive fluorochromes as well as the integrity of the antigen–antibody complexes themselves. Cytospin, touch, or smear preparations offer some advantages in that individual cells are fixed in location on the slide to be analyzed, and methods for subsequent light microscopic staining are well established. However, we have found that drying and/or fixing cells, even after the immunofluorescence reactions have already occurred, significantly compromises the fluorescence signals from many particularly large and complex fluorochromes such as phycoerythrin (PE) which are widely utilized in flow cytometric applications. Therefore, we preferred to develop methods wherein the specimen was always maintained in an aqueous environment, not only for the immunofluorescence reactions, but also afterwards during specimen analysis.

Once a specimen has been reacted with antibodies and is located on a slide or other solid surface, the means of cell contouring needs to be defined. As described earlier, cell contouring is a critical step in any form of laser scanning cytometric data collection. Contouring must be based on one of the fluorescence parameters or on the scatter parameter. Fluorescence might reflect binding of a fluorochrome-conjugated antibody (i.e., immunofluorescence); or it might reflect binding of a fluorochrome that, by its chemical nature, binds to some cellular constituent. A great many laser scanning cytometric applications use red fluorescence from the nucleic acid binding dye propidium iodide (PI) as the contouring parameter (Clatch *et al.*, 1997; Darzynkiewicz *et al.*, 1998; Furuya *et al.*, 1997; Gorczyca *et al.*, 1997a,b; Kamentsky *et al.*, 1997a,b; Li and Darzynkiewicz, 1995; Li *et al.*, 1995, 1996; Martin-Reay *et al.*, 1994; Numa *et al.*, 1996; Reeve and Rew, 1997; Sasaki *et al.*, 1996). For most applications, PI fluorescence is an excellent choice because all nucleated cells can be intensely stained making the contouring operation very robust and simple. Also, when used in combination with RNase, PI stoichiometrically binds DNA, making analysis of cell ploidy and proliferation possible. Working with tissue culture and cytospin preparations, other investigators have shown that it is possible to differentiate leukocyte subsets one from another, and to differentiate cells in different phases of the cell cycle, using only PI-staining and laser scanning cytometric analysis with PI-based contouring (Bedner *et al.*, 1997; Gorczyca *et al.*, 1996; Kakino *et al.*, 1996; Kawasaki *et al.*, 1997; Luther and Kamentsky, 1996).

However, for the purposes of immunophenotyping hematologic specimens, PI-based contouring has a great many drawbacks. First and foremost, PI is a very bright and broadly fluorescent fluorochrome, dominating the orange, red, and even the long red wavelengths. It is quite simple to perform one-color immunophenotyping using a single fluorescein isothiocyanate- (FITC-) conjugated antibody in combination with PI fluorescence as the contouring parameter. However, additional antibodies conjugated to fluorochromes such as PE, PE/Cyanin 5 (PE/Cy5), and others would be totally useless in such a system. Therefore, if one's goal is to perform multiparameter (two- or three-color) immunophenotyping of hematologic specimens, PI fluorescence cannot be used as the contouring parameter. Furthermore, PI requires permeabilization of the cells, and in some cases this may also affect antigenicity.

Immunofluorescence might also be used as the contouring parameter provided that the antibody chosen for contouring reliably binds all leukocyte subsets of interest. Anti-CD45 seems the most likely candidate for most applications, although pan-T cell, pan-B cell, and other antibodies might also be used in specific circumstances. Indeed, we have used CD3, CD14, CD15, CD20, and CD45 conjugated to FITC, PE, or PE/Cy5 in several experimental circumstances with excellent results. However, for routine purposes in a clinical immunophenotyping laboratory, what seems most desirable is a parameter that reliably contours all cells regardless of their type without obstructing any fluorescence wavelength.

Although there are certain limitations as described later, light scatter provides such a contouring parameter [Chapter 3 (Fig. 1) of this volume]. Using light scatter, all cells within a specimen regardless of their type can be reliably contoured and good cell size determinations made. Limitations and difficulties of using light scatter as the contouring parameter, which we have found manageable, are described next. First, the use of light scatter requires that the specimen be located on a transparent solid surface such as a glass slide because the light scatter detector measures laser light scattered in the forward direction through the slide itself. For strictly fluorescence applications, it would be possible to use opaque solid surfaces although such would also prohibit bright field microscopy. Second, the use of forward scatter dictates that the specimen be overlaid by a cover glass to control the refractive interface as the laser strikes the specimen. Third, unlike the fluorescence signals which have been purposefully designed to be focus insensitive, good light scatter-based contouring is quite sensitive to focus making it imperative that the user assures that the stage and slide are level (orthogonal to the laser beam). Fourth, because scatter is a simple physical property of all cells and many particles, undesired events such as red blood cells and platelets may also be contoured. Under most circumstances, these undesired events can be effectively gated out during data analysis, but if the number of these events is overwhelming relative to the number of leukocytes, data acquisition may be compromised or impossible without additional specimen purification to remove the undesired constituents.

At this juncture, it is most useful to describe the first method of immunophenotyping hematologic specimens that we developed, as well as the subsequently developed and preferred method that we now use in our clinical laboratory. Having thoughtfully considered the issues regarding the immunofluorescence reactions, the necessity of a glass slide, and the constraints of contouring, we designed our first method as follows. Suspensions of lymphoreticular or other hematologic cells were divided into aliquots in 500-μl Eppendorf test tubes and reacted with FITC-, PE-, and PE/Cy5-conjugated monoclonal antibodies. The cells were subsequently washed by centrifugation, resuspended in an appropriate volume of phosphate-buffered saline (PBS), pipetted onto glass slides, and coverslipped. Such ''drop preparations'' of immunostained lymphoreticular cells were analyzed using the LSC, contouring on light scatter, to generate immunofluorescence scattergrams directly comparable to those of FCM. We published a systematic comparison of this laser scanning cytometric drop preparation method with conventional FCM on 71 consecutive clinical hematologic specimens (Clatch *et al.*, 1996). Although successful, this method did not take full advantage of the capabilities of the LSC, mostly because the lymphoreticular cells were not fixed in specific locations on the slide. Rather, the cells were simply "floating" under the cover glass, and attempts to remove the cover glass and stain the cells for light microscopic examination were completely unsuccessful because the cells washed away or changed location. Clearly what was necessary was an entirely different means of affixing the cells to the slide, without the complications of drying or fixation.

Therefore, we designed a specialized chamber slide as shown in Fig. 1. These chamber slides are integral to our current and preferred method of immunophenotyping, as well as several extensions of the method to be described later in this chapter. Each slide has 12 individual chambers measuring 2.7 cm $\times$ 2 mm $\times$ 260 μm, with each chamber thereby defining a volume of 14 μl. In order to optimally use these chamber slides, it is necessary that (after disaggregation and red blood cell removal as may be necessary) specimens are optimally diluted in PBS without serum or protein additives at an approximate concentration of 4000 cells/μl or less. Depending on the original cellularity of the specimen, dilutions may be made as low as 400 cells/μl without compromising the number of antibodies to be tested. If desired, viability can be assessed using trypan blue dye exclusion. All of the 12 chambers are individually filled with 16 μl (slightly overfilling each chamber to allow for some drying) of the diluted and purified cell suspension (equating to 6400–64,000 cells/chamber). The dry chambers fill very easily due to capillary action. The slide is then stored for 15 min at room temperature in a humidified chamber. During this time, the leukocytes settle by gravity and adhere firmly by an electrostatic interaction with the bottom glass surface of the chambers. Thereafter, various monoclonal antibody combinations can be pipetted into each of the chambers, displacing the original fluid by slightly inclining the slide and absorbing the effluent with a paper towel or other absorbent material. To allow for some dead space and intrachamber mixing, we routinely use 18 μl of prediluted antibody combinations for each chamber. The antibodies we use are triple combinations of FITC, PE, and PE/Cy5 conjugates diluted 1:10 in PBS. In this way it is possible to react a single specimen with 36 different antibodies on a single slide. However, for gating and other comparative purposes, we have built some redundancy of the antibodies tested into our typical lymphoma or leukemia panels as shown in Fig. 2. The cells adherent within the chambers of the slide are allowed to react with the antibody combinations for 30 min at 4°C in a humidified chamber. Then, the chambers of the slide are all

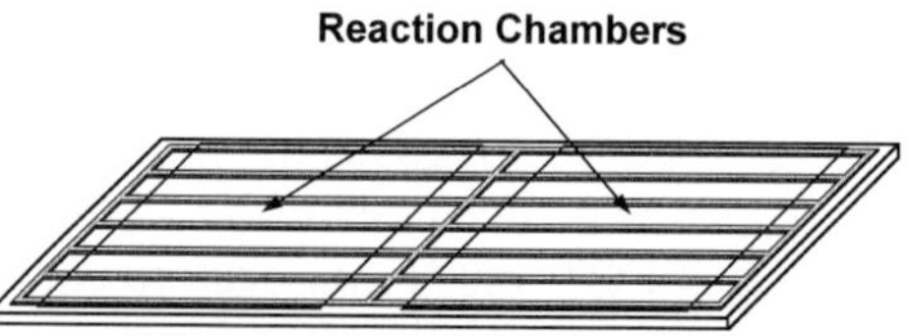

Fig. 1   Immunophenotyping chamber slide

Lymphoma / CLL Panel

FITC	PE	PE-Cy5
CD3	CD20	CD45
HLA-DR	CD20	CD45
CD5	CD19	CD45
CD10	CD19	CD45
Kappa	CD19	CD45
Lambda	CD19	CD45
CD19	CD23	CD45
FMC7	CD25	CD45
CD8	CD4	CD45
CD7	CD56	CD45
CD16	CD2	CD45
CD11c	CD14	CD45

Acute Leukemia Panel

FITC	PE	PE-Cy5
CD3	CD20	CD45
CD10	CD19	CD45
Kappa	CD19	CD45
Lambda	CD19	CD45
CD7	CD20	CD45
HLA-DR	CD34	CD45
CD13	CD34	CD45
CD13	CD34	CD45
CD11c	CD34	CD45
CD14	CD34	CD45
CD41	CD34	CD45
TdT	CD34	CD45

Fig. 2   Antibody panels used for clinical immunophenotyping.

washed simultaneously by passing 200 μl of PBS through each end of the chamber slide twice, 1 min apart.

The slide is then placed onto the stage of the LSC, which has been programmed to automatically scan a rectangular area within the center of each of the chambers. Individual cells are contoured based on light scatter. Green, orange, and long red fluorescence are subsequently assessed for each cell with all results stored in a single list mode data file. We currently scan only approximately 10% of the entire area of each chamber. Data acquisition for all 12 chambers takes a total of approximately 25 min, with the data for each chamber including between approximately 600 and 6000 cells depending on cellularity. As we often do for extremely hypocellular specimens, enlarging the scan area within each chamber provides a means of easily increasing the number of cells assayed. Table I shows in outline form our current method of clinical immunophenotyping that we recommend for LSCM (Clatch *et al.*, 1998).

Data analysis, gating, and compensation are accomplished within the WinCyte software in a fashion directly analogous to FCM. For gating purposes, we have

Table I
Method of Immunophenotyping by
Laser–Scanning Cytometry

Purify specimen
   Lyse red blood cells/disaggregate solid tissues
Load specimen into slide
   6400–64,000 cells in 16 μl per chamber
Add prediluted antibody combinations
   FITC-, PE-, and PE/Cy5-conjugated antibodies
Wash with PBS
Analyze on LSC (contour on light scatter)

found it most practical to include CD45-PE/Cy5 as one component of all the antibody combinations. So, our routine method of analyzing clinical cases includes first displaying all cells from all 12 chambers in a single scattergram interrelating cell size and CD45 expression. Lymphocytes, neutrophils, monocytes, maturing myeloid cells, and blasts are normally well separated in this scattergram, allowing only cells of interest to be gated and passed to other scattergrams (Fig. 3). The gated cells are next displayed in a scattergram showing their locations on the slide (possible because the x and y positions are part of the list mode data file). This effectively provides a map of the chamber slide itself (Fig. 4). The 12 rectilinear regions created in this cell location scattergram allow data from each of the chambers to be passed to one of 12 individual scattergrams reflecting binding of the FITC- and PE-conjugated antibodies that are unique for each chamber.

Figure 5 shows, for each of two cases, three scattergrams representing binding of basic lymphocyte-specific antibodies. The case shown in the top three scattergrams comprises benign T and B lymphocytes obtained from a reactive lymph node. The bottom three scattergrams also show a lymph node with mixture of T and B lymphocytes. But, in this case the B cells monotypically express κ as compared with λ surface membrane immunoglobulin light chains, indicative of a clonal B-lineage lymphoproliferative disorder such as non-Hodgkin's lymphoma.

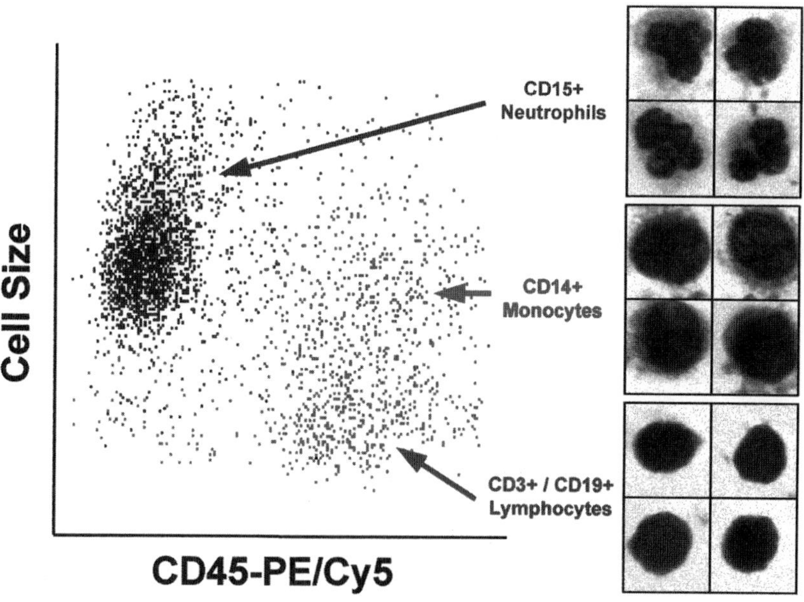

Fig. 3   Initial gating based on cell size and CD45 positivity. (See color plates.)

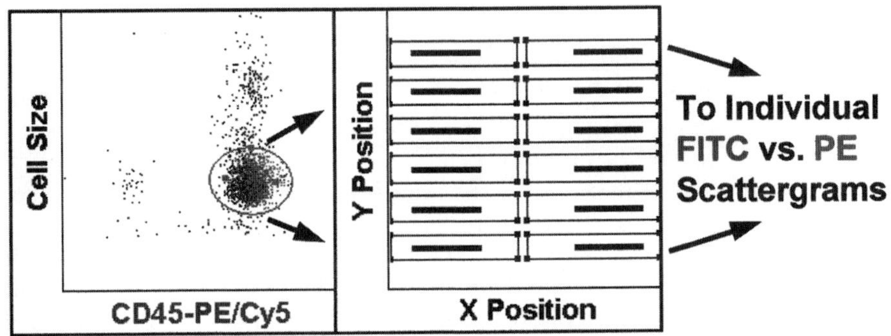

Fig. 4 Subsequent gating based on cell position. (See color plates.)

Polyclonal B cells (Benign)

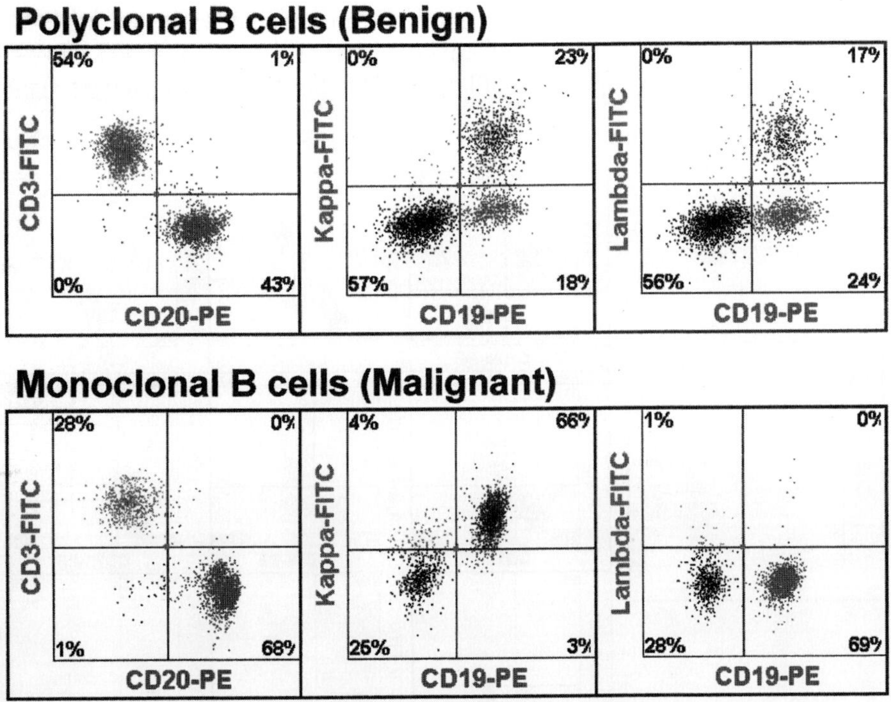

Fig. 5 Example results (top) benign lymph node with polyclonal B cells (bottom) malignant lymph node with monoclonal B cells. (See color plates.)

B. Example Results

Example immunophenotyping results from two additional clinical cases are shown in Figs. 6 and 7. Figure 6 shows the results from a fine needle aspiration biopsy of a left cervical lymph node from a 41-year-old woman who presented to medical attention complaining of lymphadenopathy of several weeks duration. The lymph node was easily palpable and tender, and a fine needle aspiration biopsy was performed using a 25-gauge hypodermic needle and no anesthesia. Approximately 350,000 total lymphoid cells were obtained and processed routinely for the possibility of lymphoma using the antibody panel shown in Fig. 2. As shown in Fig. 6, there was a single well-defined population of small to medium-sized moderately to strongly CD45 positive cells that were gated (as shown on the left) and subsequently displayed within the 12 separate FITC versus PE scattergrams (shown to the right). In this example the colorization of each of the FITC versus PE scattergrams is based solely on the quadrants drawn for the individual scattergrams with negative cells colorized black, FITC positive cells colorized green, PE positive cells colorized red, and dual FITC positive and PE positive cells colorized blue. The colorization within the cell size versus CD45-PE/Cy5 scattergram is a reflection of that within the CD3-FITC versus CD20-PE scattergram effectively showing T cells in green and B cells in red. The minor population of larger CD45 positive cells seen in the cell size versus CD45

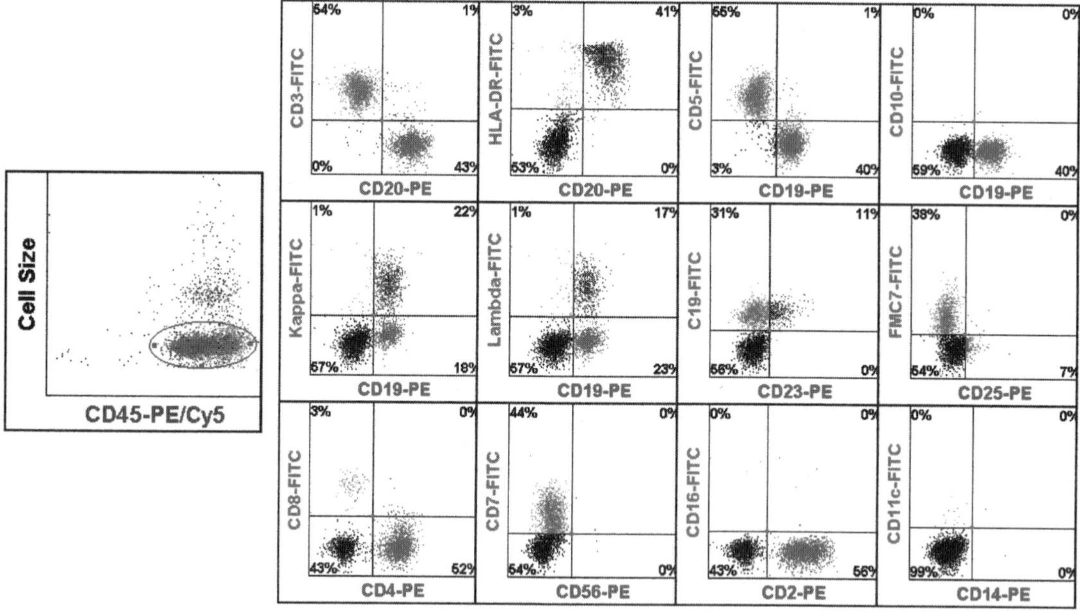

Fig. 6 Example clinical case 1: Fine needle aspiration biopsy of a benign lymph node. (See color plates.)

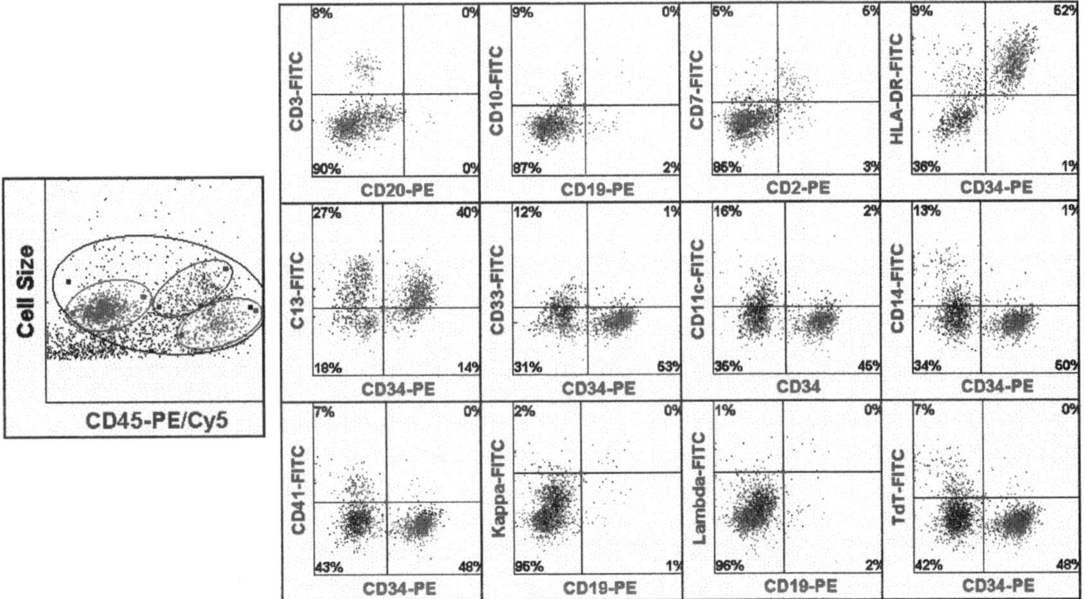

Fig. 7   Example clinical case 2: Peripheral blood with acute myelogenous leukemia. (See color plates.)

scattergram is composed of doublets. The immunophenotyping results in this case are clearly benign. There is a mixed population of T and B lymphocytes with a T:B cell ratio of approximately 1.25:1. The T cells correctly express all pan-T cell antigens and are appropriately divided into CD4+ helper and CD8+ suppressor subpopulations with an increased Th:Ts ratio of approximately 17:1. The B cells are also antigenically normal without aberrant expression of CD5 or CD10, and there is good polyclonal expression of both κ and λ surface membrane immunoglobulin light chains. Combined with the cytologic assessment of the Wright–Giemsa and Papanicolaou stained smears, these immunophenotypic findings strongly supported a benign reactive cause for the patient's lymphadenopathy that spontaneously resolved several weeks after the biopsy.

The second example case shown in Fig. 7 is a peripheral blood specimen with acute myelogenous leukemia. The specimen was obtained from a 37-year-old woman presenting with purpura. A complete blood count showed a marked leukocytosis with a white blood cell count of 35,000/μl with a manual differential showing approximately 50% morphologically undifferentiated blasts. The specimen was treated with ammonium chloride to remove red blood cells and processed for immunophenotyping using the acute leukemia panel shown in Fig. 2. The gating in this case is slightly more complex. The cell size versus CD45 scattergram shows at least three well-defined populations of cells. By experience one can surmise that the smaller more brightly CD45 positive cells (colorized green) are benign lymphocytes. The larger moderately CD45 positive cells (colorized blue) are most likely mature and maturing myeloid and/or monocytoid cells. The relatively heavy population of medium sized weakly CD45 positive cells (colorized red) are most likely the blasts. In this example the colorization is more conventional in that the colors are defined by the bit maps drawn within the cell size versus CD45 gating scattergram and are transposed into the various FITC versus PE scattergrams. The immunophenotyping data clearly show that the aforementioned hypotheses regarding the three cell populations are indeed correct. There is a relatively minor population of lymphocytes that are mostly T cells (CD2, CD3, and CD7 positive) and also a significant population of mature and maturing myelomonocytoid cells that variably and weakly express CD10 and HLA-DR, and moderately to strongly express CD11c and CD13. The blasts in this case strongly express HLA-DR and the human progenitor cell antigen CD34. They are moderately positive for the myeloid antigen CD13, but are negative for other myeloid, monocytoid, megakaryocytoid, and lymphoid antigens. The diagnosis of acute myeloblastic leukemia, subtype M2 in the FAB categorization, was substantiated by cytochemical special stains showing myeloperoxidase and Sudan black positivity in some of the blast cells. Appropriate chemotherapy was begun. In our own and several other LSCM laboratories, clinical immunophenotyping of this type has been performed with considerable success, with cases analyzed numbering in the thousands at present.

C. Advantages/Disadvantages

A brief comparison of FCM and LSCM for the purpose of clinical immunophenotyping is shown in Table II. FCM clearly has the advantage of speed and is the method of choice in cases in which hundreds of thousand of cells need to be counted. Also, because of the addition of a side angle light scatter detector, gating of complex cell populations is more precise by FCM.

Laser scanning cytometry is highly suited to the analysis of very small specimens such as fine needle aspiration biopsies and hypocellular body fluids. The method

Table II

Comparison of Flow Cytometric and Laser Scanning Cytometric Immunophenotyping

Flow Cytometry
    more precise cell gating with side angle light scatter
    counts more cells faster
Laser Scanning Cytometry
    complete analysis with only 64,000 total cells
    simplified methodology
    antibody usage reduced by >80%
    relocalization enables direct correlation of results, cytology, and FISH

is remarkably simple to perform, and antibody usage is reduced by greater than 80% compared with flow cytometric methods. And, as discussed in detail below, with LSCM it is possible to relocalize individual cells after immunophenotyping, enabling direct correlation of antigenic characteristics, cytology, and other cell features that may be apparent after special staining procedures such as fluorescence *in situ* hybridization (FISH).

IV. Extensions of the Method

A. Relocalization for Light Microscopy and Fluorescence *in Situ* Hybridization

Relocalization is a relatively simple process that can be done at any time following the immunophenotyping procedure from minutes to months after analysis. Following data acquisition, the slide can be dismantled, and the specimen fixed as desired. Fig. 8 shows a dismantled immunophenotyping chamber slide with rows of cells remaining adherent to the slide. In certain clinical cases, we have found it very useful to fix and stain the slides for light microscopic examination, and then to relocalize individual cells based on their immunofluorescence (antigenic) characteristics. This relocalization is accomplished automatically by the LSC on placement of the stained slide onto the microscope stage.

An interesting clinical case in which relocalization proved to be a useful undertaking is shown in Figs. 9–11. The specimen is a fine needle aspiration biopsy of a right parotid mass from a 13-year-old boy. The most significant immunophenotyping results are shown in Fig. 9. There are three distinct populations of lymphoid cells within the specimen. The first population is one of T lymphocytes and constitutes approximately 30% of all lymphoid cells in the specimen. The T cells correctly express the pan-T cell antigens CD3 and CD5, and are appropriately divided into CD4 positive helper and CD8 positive suppressor subpopulations with Th:Ts ratio of approximately 2:1. There is also a population of benign B lymphocytes constituting approximately 50% of the specimen. These B cells moderately express the pan-B cell antigens CD19 and CD20, do not aberrantly express CD5 or CD10, and show good polyclonal expression of both κ and λ

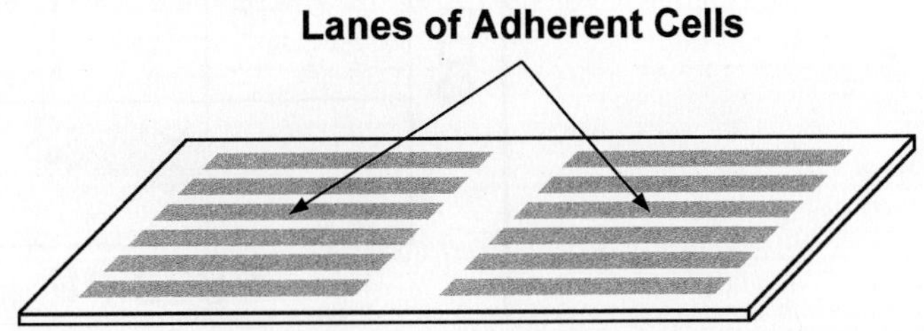

Fig. 8 Dismantled immunophenotyping chamber slide with adherent cells.

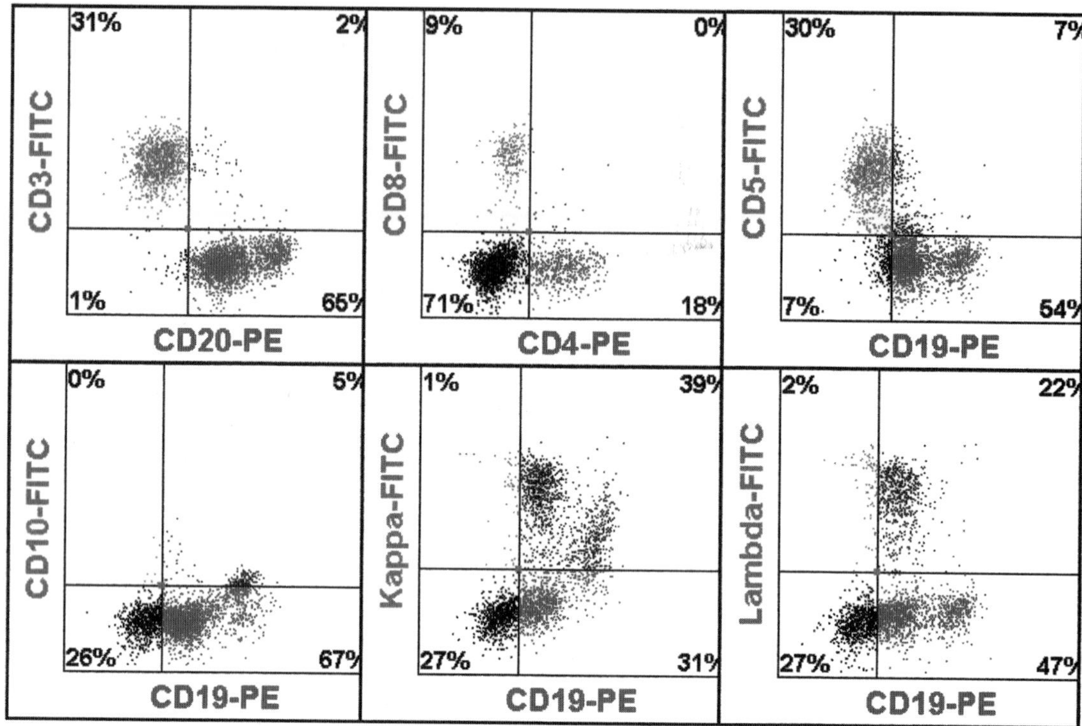

Fig. 9 Example clinical case 3: Fine needle aspiration biopsy of a parotid mass with Burkitt's lymphoma. (See color plates.)

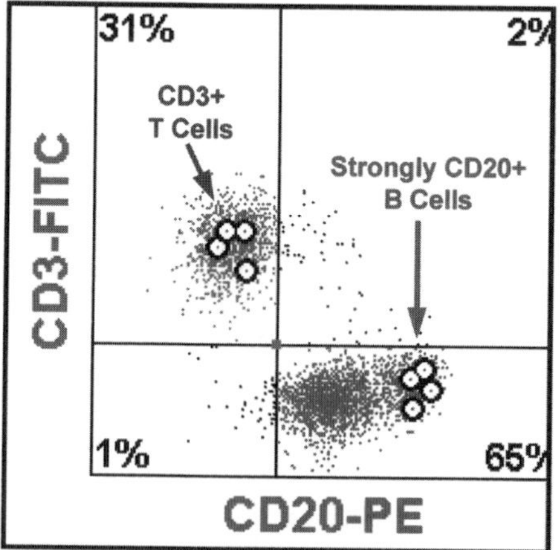

Fig. 10 Example clinical case 3: Selection of cells for relocalization. (See color plates.)

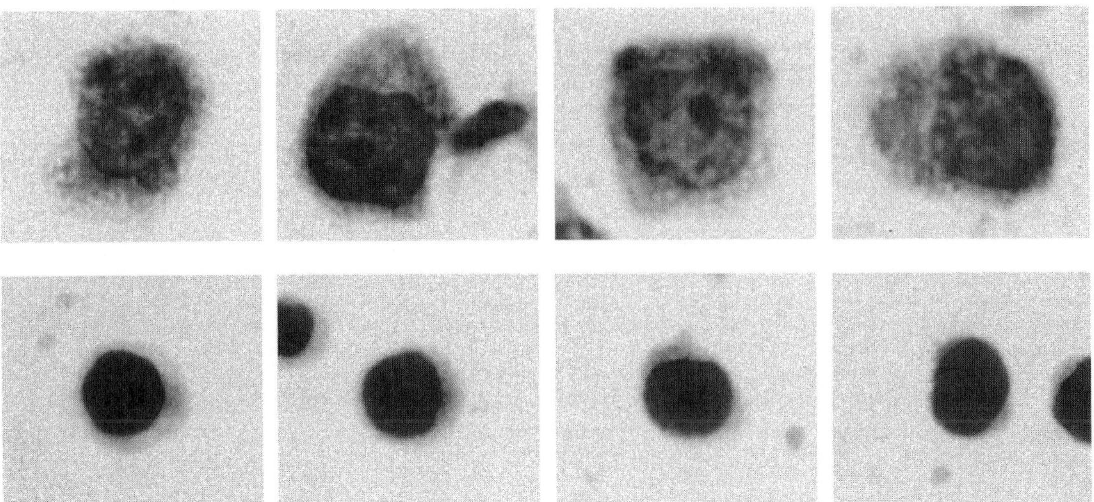

Fig. 11   Example clinical case 3: Relocalization for light microscopy. (See color plates.)

surface membrane immunoglobulin light chains. Last, there is a relatively minor population of clonal B cells that more strongly express CD19 and CD20, weakly and aberrantly express CD10, and monotypically express κ as compared with λ surface membrane immunoglobulin light chains. After immunophenotyping, the slide was dismantled and stained with a Wright–Giemsa stain. Four of the strongly CD20 positive clonal B cells and four of the CD3 positive T cells were chosen for relocalization as shown in Fig. 10. The light microscopic images of these eight cells captured through the CCD camera of the LSC are shown in Fig. 11. The pathologic diagnosis based on the morphology of the FNA biopsy smears and the immunophenotyping results was that of Burkitt's lymphoma. Appropriate chemotherapy was begun.

After taking photographs or capturing video images, the slides can be destained and processed for FISH to demonstrate known or suspected chromosomal abnormalities. Table III shows the entire method we developed, including immunophenotyping and relocalization for light and epifluorescence microscopy. Two interesting cases with known chromosomal abnormalities were chosen for relocalization for light microscopy and FISH with the results shown in Figs. 12-15 and 17-20. Both cases were first subjected to full panel immunophenotyping and secondarily stained for light microscopy. Individual cells were then relocalized and photographed under light microscopy based on antigenic characteristics. The slides were then hybridized *in situ* with fluorescent probes for the cytogenetic abnormalities in question, and the same cells again relocalized and photographed under epifluorescence microscopy.

The first of these two cases is a peripheral blood specimen from a 67-year-old man with a marked lymphocytosis. Immunophenotypic analysis showed a clonal

Table III

Method of Immunophenotyping with Relocalization for Light Microscopy and Fluorescence *in Situ* Hybridization by Laser Scanning Cytometry

Purify specimen
    lyse red blood cells/disaggregate solid tissues
Load specimen into slide
    6400–64,000 cells in 16 μl per chamber
Add prediluted antibody combinations
    FITC-, PE-, and PE/Cy5-conjugated antibodies
Wash with PBS
Analyze on LSC (contour on light scatter)
Stain slide for light microscopy
    PAP, Wright-Giemsa, Hematoxylin and Eosin, etc.
Relocalize individual cells for light microscopy based on immunophenotype
Destain and fix with MeOH then Carnoy's
Perform fluorescence in-situ hybridization (FISH)
Relocalize the same individual cells for epifluorescence microscopy

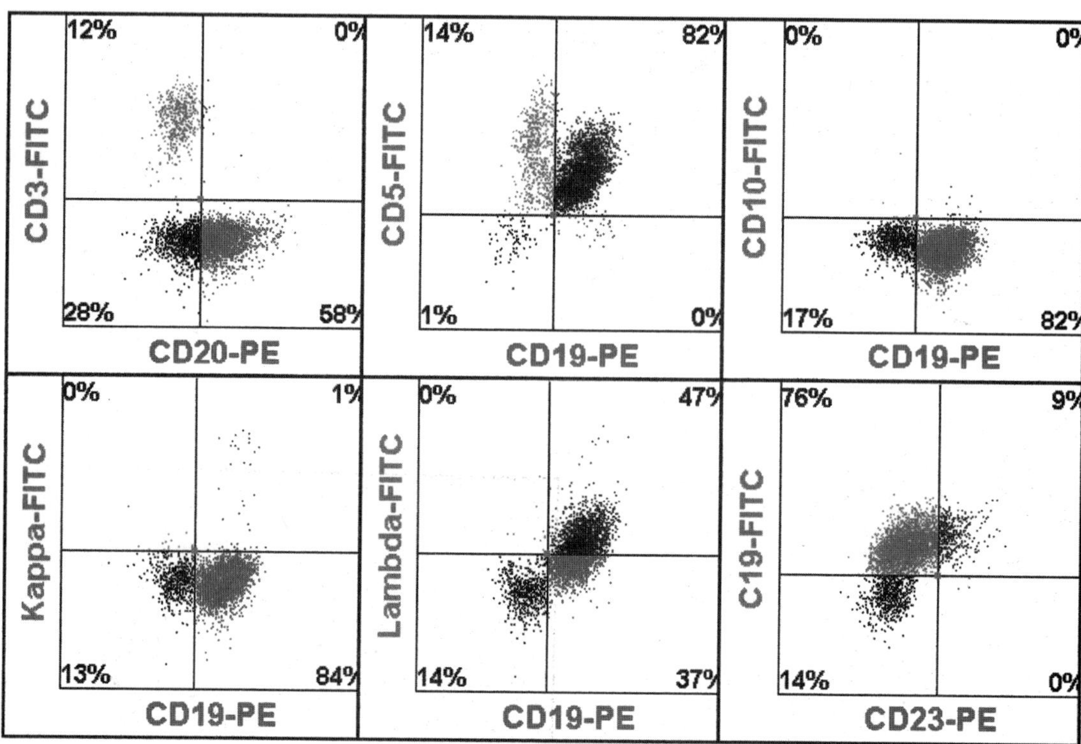

Fig. 12 Example clinical case 4: Peripheral blood with chronic lymphocytic leukemia. (See color plates.)

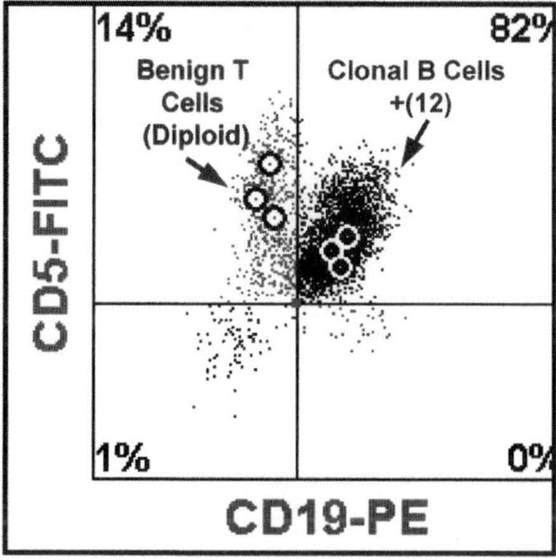

Fig. 13 Example clinical case 4: Selection of cells for relocalization. (See color plates.)

population of B lymphocytes with a pattern of antigenicity highly characteristic of chronic lymphocytic leukemia (Fig. 12). The B cells had diminished expression of CD20, strong and aberrant expression of CD5, weak but monotypic expression of λ as compared with κ surface membrane immunoglobulin light chains, and

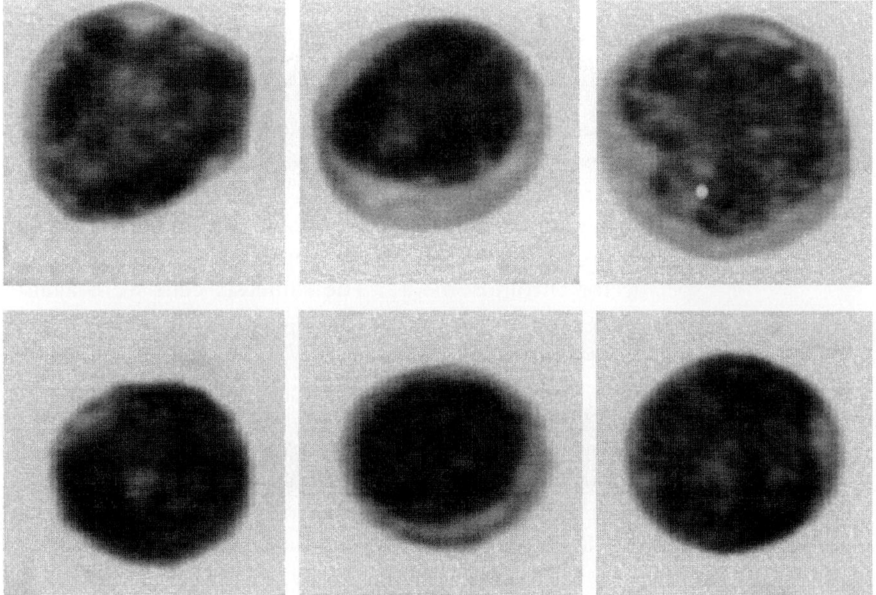

Fig. 14 Example clinical case 4: Relocalization for light microscopy. (See color plates.)

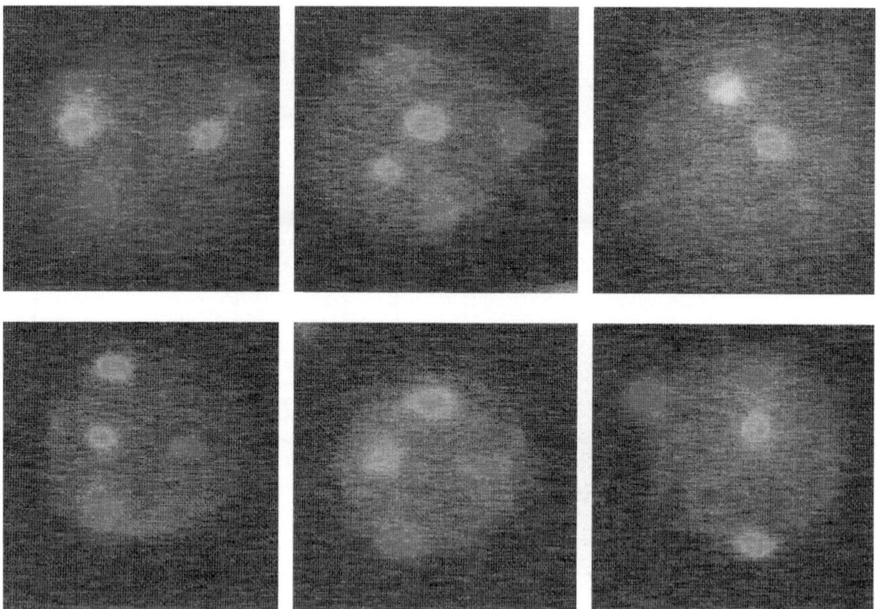

Fig. 15 Example clinical case 4: Relocalization for epifluorescence microscopy, FISH. (See color plates.)

preserved expression of CD23. Based on the CD5-FITC versus CD19-PE scattergram, three T lymphocytes and three B lymphocytes were selected for relocalization (Fig. 13) The Wright–Giemsa stained B cells and T cells are shown in the top and bottom rows of Fig. 14, respectively. As might be expected, the morphologic differences between the clonal malignant B cells and the benign T cells are relatively subtle.

The slide was subsequently destained and hybridized with a spectrum green-labeled probe specific for chromosome 3 and a spectrum orange-labeled probe specific for chromosome 12. The same six cells were again relocalized in this case using epifluorescence microscopy with the results shown in Fig. 15. The T cells in the bottom row all have two green probe spots and two red probe spots indicating that they are normal diploid cells. The three B cells in the top row all have two green probe spots and three red probe spots, indicating that they are positive for the trisomy (12) chromosomal abnormality that is relatively common in B-lineage chronic lymphocytic leukemia.

The epifluorescence images shown in Fig. 15 have been artificially enhanced by a simple computerized method outlined in Fig. 16. The individual green and red fluorescence images from a single cell are first amplified and then each broken into their individual green and red components. The red component of the red epifluorescence image is combined with the green component of the green

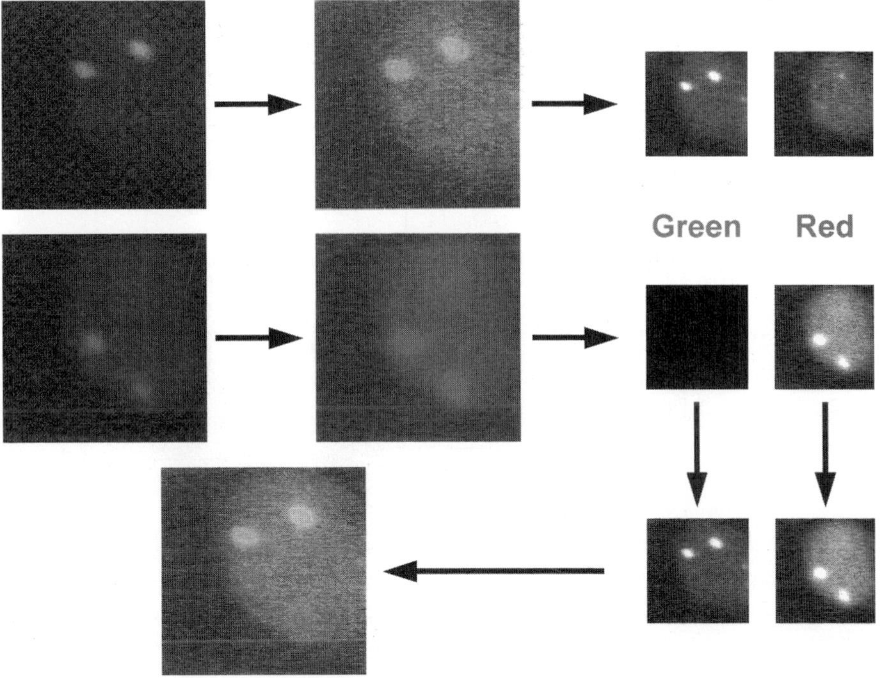

Fig. 16   Enhancement of epifluorescence video images of FISH probe spots. (See color plates.)

epifluorescence image to yield a single image in which both the red and green probe spots are clearly visible without the expense of complex optical systems.

A second example case is shown in Fig. 17–20. The specimen is a bone marrow aspiration biopsy from a 48-year-old woman with a known history of chronic myelogenous leukemia. On complaining of increasing fatigue, peripheral blood was initially drawn and shown to contain rare blast cells. A bone marrow aspiration and biopsy was then performed. The most relevant scattergrams of the immunophenotypic analysis performed on the bone marrow aspirate are shown in Fig. 17. The cell size versus CD45-PE/Cy5 scattergram clearly shows four populations of cells. The large population of ungated relatively small CD45 negative cells are mostly nucleated erythroid cells that have escaped the ammonium chloride lysis procedure. The three other populations of colorized cells are gated to be included in the FITC versus PE immunophenotyping scattergrams shown at the bottom of the Fig.. The small strongly CD45 positive cells (colorized green) are a nearly even mixture of T and B lymphocytes. The larger moderately CD45 positive cells (colorized blue) are mature and/or maturing myeloid cells as evidenced by the variable but strong CD13 positivity. Lastly, the population of relative large but only weakly CD45 positive cells (colorized red) consists of myeloblasts. The myeloblasts constitute between 6 and 8% of all nonerythroid

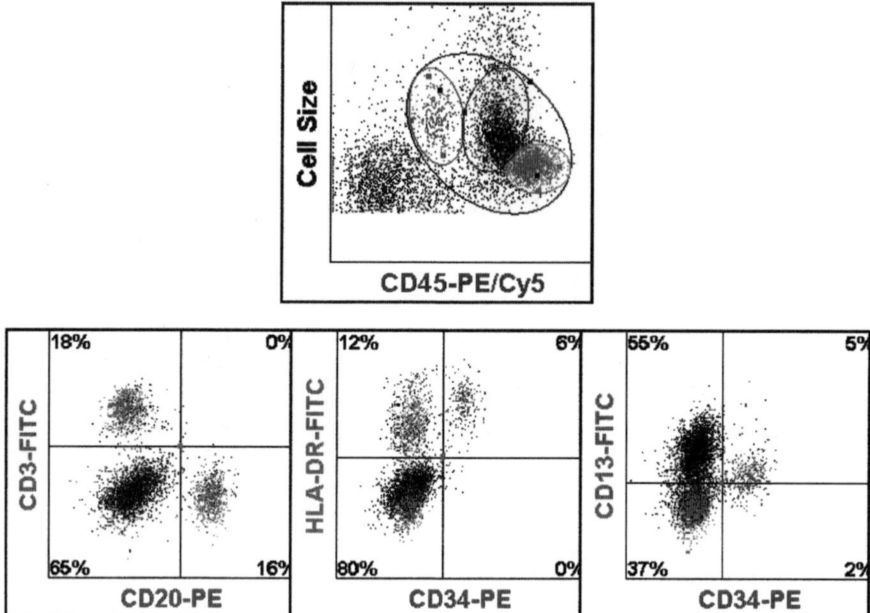

Fig. 17 Example clinical case 5: Bone marrow aspirate with chronic myelogenous leukemia in blast crisis. (See color plates.)

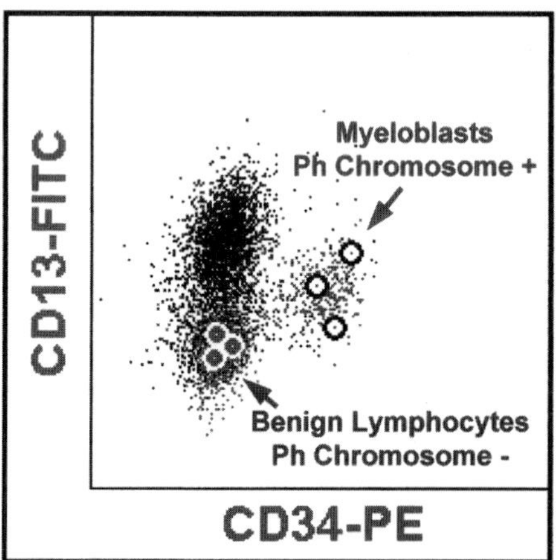

Fig. 18 Example clinical case 5: Selection of cells for relocalization. (See color plates.)

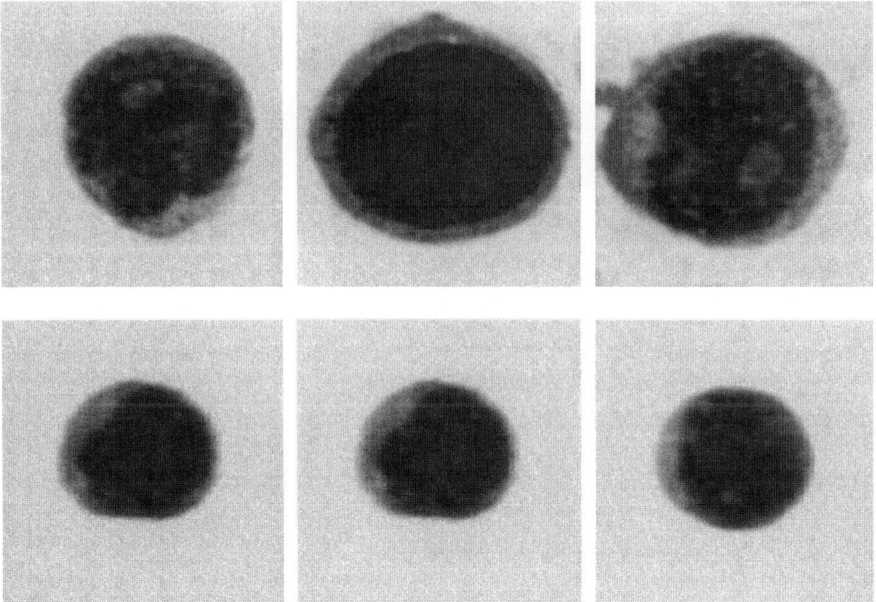

Fig. 19 Example clinical case 5: Relocalization for light microscopy. (See color plates.)

cells based on this analysis, and moderately express the human progenitor cell antigen CD34, strongly express HLA-DR, and weakly to moderately express the myeloid antigen CD13.

The patient had a known t(9;22) translocation (Philadelphia chromosome), and so the case provided an opportunity to demonstrate the ability of the LSC to relocalize individual cells for light and epifluorescence microscopy. Based on the CD13 versus CD34 scattergram (Fig. 18), three myeloblasts and three benign lymphocytes were chosen for relocalization after staining for light microscopy. The Wright–Giemsa stained cells are shown in Fig. 19 with the myeloblasts in the top row and the lymphocytes in the bottom row. After destaining and further fixation, the slide was reacted with a spectrum green probe for chromosome 8 and a spectrum orange probe for chromosome 22, and the same six cells relocalized under epifluorescence (Fig. 20). The blasts in the top row clearly have colocalized green and red probe spots indicative of the t(9;22) translocation, whereas the lymphocytes in the bottom row do not. Interestingly, the acid fixation necessary for FISH has resulted in the apparent enlargement of the lymphocyte images in the bottom row.

These capabilities of LSCM are of interest not only because of their novelty, but more so for their potential usefulness in medical research and clinical diagnosis. Because individual cells can be automatically relocalized, it is simple to correlate subtle variabilities in antigen expression within a population of tumor cells with

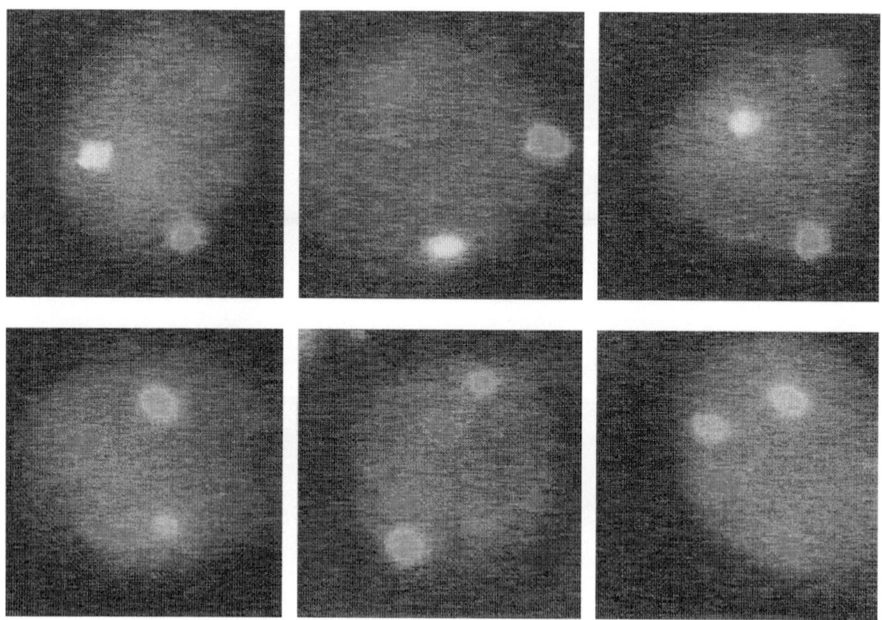

Fig. 20 Example clinical case 5: Relocalization for epifluorescence microscopy, FISH. (See color plates.)

any potentially corresponding cytogenetic or morphologic variability. The study of tumor heterogeneity is now considered essential to understanding tumor evolution and evaluating potential therapies, and should benefit from this technology. For clinical purposes, the method may be most applicable to the detection of residual disease after chemotherapy or radiation therapy. Without complex sorting techniques, it is relatively simple to determine, for example, whether the small percentage of CD34 positive blasts that may be present the bone marrow of a patient following induction chemotherapy represents normal regenerating marrow precursors or residual acute leukemia. To enhance sensitivity while retaining the capabilities of full-panel immunophenotyping and relocalization for light and epifluorescence microscopy, it would also be relatively simple to substitute *in situ* PCR in place of the FISH procedure. Such work is now in progress.

B. Multipass and Multiplex Systems

As described earlier, three-color immunophenotypic analysis of hematologic specimens is easily performed using the LSC. This can be accomplished in a single scan pass using the argon laser and a combination of antibodies bound to three fluorochromes such as FITC, PE, and PE/Cy5. An alternate fluorochrome combination such as PE, PE/Texas Red conjugate (PE/TR), and PE/Cy5 works equally well, but with somewhat greater compensation requirements. Regardless of the fluorochrome combination chosen, in order to achieve simultaneous three-color immunophenotypic analysis with a single laser and a single scan pass, it is necessary to rely on light scatter-based contouring of the cells in the specimen. This is true because of the wide fluorescence emission spectral characteristics of PI and many other DNA binding dyes that are often used for contouring purposes in LSCM applications.

Because, for most LSCM applications, the locations of individual cells within a specimen remain fixed, it is a simple matter to scan the same specimen more than once using the same laser and different experimental conditions, or even using a different laser. Algorithms built into the software of the LSC enable merging of individual computer data files created using different experimental conditions or lasers, such that all the separate fluorescence and physical measurements obtained can be simultaneously interrelated during data analysis. This process of merging separate data files collected for a single specimen at different times is designated as a multipass system. The LSC also has a built-in multiplex capability that enables different lasers to be used in the collection of a single data file. For multiplex systems, corresponding fields of raw uncontoured data (768,000 pixels) are sequentially collected with each of the two lasers one at a time, and then merged in real time in a fashion transparent to the LSC user. Thus, armed with an LSC with both helium neon (HeNe) and argon lasers and the WinCyte software, we sought to expand beyond three colors our ability to perform immunophenotypic analysis.

The HeNe laser is optimally suited to excite the fluorochrome Cy5. This fluorochrome is available conjugated to anti-mouse immunoglobulin antibodies, and therefore is an optimal secondary immunofluorescent reagent to use with the HeNe laser. Unfortunately, few if any other fluorochromes appropriate for use with the HeNe laser (most desirably with longer Stoke's shifts) are commercially available. Nevertheless, because Cy5 alone does not excite with the argon laser, multiplexing or a simple two scan pass system (one pass with the HeNe laser and another with the argon laser) makes four-color immunophenotypic analysis possible. Cellular contouring for both scan passes would be based on light scatter.

The LSC has four fluorescence photomultiplier tubes (PMTs) and one light scatter detector. However, because of the design of the LSC's electronics, data collection can simultaneously occur from only four of the total of these five input sources. Since one of these must be utilized for the light scatter-based contouring parameter, in any individual scan pass it is only possible to collect three fluorescence parameters simultaneously. But, as described earier, there are at least four fluorochromes (commercially available bound to antibodies) with sufficiently distinct emission spectral characteristics that work very well with the argon laser. Specifically, these are FITC, PE, PE/TR, and PE/Cy5. Therefore, because of this limitation in the number of available input channels, in order to collect fluorescent

Laser	Pass #1 HeNe	Pass #2 Argon	Pass #3 Argon
PMT #1	Orange —	Orange PE, Ab #2 (CD4)	Green FITC, Ab #5 (CD8)
PMT #2	Red —	Red PE/TR, Ab #3 (CD19)	Red* PI, DNA Content
PMT #3	Far Red Cy5, Ab #1 (CD45)	Far Red PE/Cy5, Ab #4 (CD3)	—
PMT #4	Scatter*	Scatter*	—

Fig. 21 Multipass system of five-color immunophenotyping plus DNA content analysis by laser scanning cytometry.

data for all four of these fluorochromes, it is necessary to make two separate passes with the argon laser, each scan pass with a different optical configuration defining the PMTs.

Although at first glance cumbersome, the necessity of this second scan pass with the argon laser can be put to further advantage by using the second pass to also collect fluorescence data defining cellular DNA content. This can be simply accomplished by reacting PI with the specimen between the two argon laser scan passes. The PI fluorescence is so bright that residual fluorescence of PE, PE/TR, or PE/Cy5 that may remain present is inconsequential. Thus, a three scan pass system as shown in Fig. 21 accomplishes five-color immunophenotyping plus DNA content analysis using only a single specimen aliquot.

The results from an actual experiment using this three scan pass system are shown in Fig. 22 (Clatch and Foreman, 1998). Peripheral blood leukocytes were purified by lysing the red blood cells with ammonium chloride. The leukocytes were then placed into a single chamber of an immunophenotyping chamber slide and simultaneously reacted with a combination of five different fluorochrome-conjugated antibodies: CD45-Cy5, CD3-PE/Cy5, CD19-PE/TR, CD4-PE, and

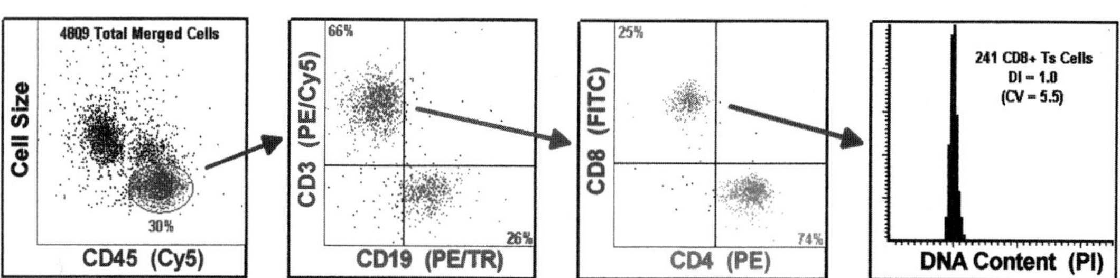

Fig. 22 Five-color immunophenotyping plus DNA content analysis of peripheral blood lymphocytes. (See color plates.)

CD8-FITC. After washing unbound antibody from the chamber, the slide was scanned twice, first with the HeNe laser and second with the argon laser. Then the cells in the chamber were permeabilized with 70% ethanol and subsequently reacted with PI (50 μg/ml) with RNase in PBS. After again washing unbound reagents from the chamber, and after changing the optical filters defining the PMTs, the slide was scanned again with the argon laser. The individual data files from these three scan passes were merged in the WinCyte software, and the data was analyzed and displayed as shown in Fig. 22. Beginning with a total of 4809 merged peripheral blood leukocytes, 1448 (30%) lymphocytes were gated based on CD45 positivity. Of the lymphocytes, 957 (66%) were CD3$^+$ T cells, and 241 (25%) of the T cells were CD8$^+$ suppressor T cells. As to be expected, the gated T suppressor cells were shown to be diploid with a DNA index of 1.0.

One advantage of LSCM relative to FCM, that was at first unrecognized and discovered during the course of these experiments, is the ability to minimize the unwanted artifacts of spectral crossover between fluorescence channels that is typically addressed through mathematical or instrument compensation. Because the optical design of the LSC is similar to that of modern flow cytometers, compensation requirements and methods are similar for flow cytometric applications and single laser scan pass laser scanning cytometric applications. In the three scan pass system that has just been described for five-color immunophenotyping plus DNA content analysis by LSCM, spectral crossover is particularly problematic due to orange light from the bright PE fluorochrome contaminating (crossing over into) the red PMT channel used for the relatively dim PE/TR fluorochrome. In flow cytometers, such problems can only be addressed through mathematical or instrument compensation or by using multiple lasers and different fluorochromes. In LSCM, because the same cells within an individual specimen can be scanned over and over again at will, a problem of this kind can be solved simply by resorting to an additional scan pass, the data from which can be merged with that from former and subsequent passes.

For example, since there is no compensation problem relative to the Cy5 and PE/TR fluorochromes (because different lasers are used to excite them), the antibodies bound to these two fluorochromes can be added simultaneously to the specimen to begin the experiment. A first scan pass with the HeNe laser collects immunofluorescence data relative to the Cy5-conjugated antibody. A second scan pass with the argon laser collects immunofluorescence data relative to the PE/TR-conjugated antibody impervious to PE-induced spectral crossover because PE is not present in the specimen at this time. After a subsequent reaction with FITC-, PE-, and PE/Cy5-conjugated antibodies (for which compensation requirements are significant but easy to manage), a third scan pass with the argon laser is made. Because the PE/TR fluorochrome is relatively dim compared with PE and PE/Cy5, there is little if any problem with spectral crossover in this direction. After this third scan pass, the specimen is reacted with PI and RNase. A fourth scan pass with the argon laser effects DNA content analysis. This four scan pass system also eliminates any requirement to compensate FITC and PI, as is necessary with the three scan pass system described earlier.

Although these experimental techniques are somewhat tedious and time-consuming, they are not technically demanding and serve well to highlight the potential capabilities of the LSCM technology for any number of scientific applications requiring the analysis of multiple cellular constituents on a single population of cells. Of course, after analysis the specimen can be stained for light or epifluorescence microscopy as desired, and individual cells meeting user-defined fluorescent or physical parameters can be individually relocalized for cytologic examination.

V. Conclusions

Modern diagnostic categorization of hematologic neoplasms is increasingly dependent on special laboratory methods to characterize the antigenicity or phenotype of the cells in question. Currently, this is most often accomplished by one of two methods: FCM or IHC. IHC is invaluable in cases in which it is necessary to correlate tissue architecture with antigenic features. FCM precludes correlation of immunophenotype and tissue architecture, but allows innumerable antigens to be tested because the antigen–antibody reactions occur prior to fixation or drying and because so many antibodies are commercially available. LSCM is a practical alternative method of immunophenotyping hematologic specimens with some advantages relative to FCM and IHC. Most important of these are the adaptability of the technology to very small clinical specimens and the capability to relocalize individual cells for microscopic examination after immunophenotyping. We have extended laser scanning cytometric immunophenotyping to include sequential light microscopy and epifluorescence microscopy for FISH. We have also adapted the technology to multipass systems, greatly increasing the potential number of antigens that can be sequentially assayed and simultaneously reducing compensation problems. LSCM provides many new research opportunities in hematology and molecular and cell biology, and has immediate practical relevance for clinical immunophenotyping of small specimens.

References

Bedner, E., Burfeind, P., Gorczyca, W., Melamed, M. R., and Darzynkiewicz, Z. (1997). Laser scanning cytometry distinguishes lymphocytes, monocytes, and granulocytes by differences in their chromatin structure. *Cytometry* **29**, 191–196.

Clatch, R. J. and Walloch, J. L. (1997). Multiparameter immunophenotypic analysis of fine needle aspiration biopsies and other hematologic specimens by laser scanning cytometry. *Acta Cytol.* **41**, 109–122.

Clatch, R. J., Walloch, J. L., Zutter, M. M., and Kamentsky, L. A. (1996). Immunophenotypic analysis of hematologic malignancy by laser scanning cytometry. *Am. J. Clin. Pathol.* **105**, 744–755.

Clatch, R. J., Walloch, J. L., Foreman, J. R., and Kamentsky, L. A. (1997). Multiparameter analysis of DNA content and cytokeratin expression in breast carcinoma by laser scanning cytometry. *Arch Pathol. Lab. Med.* **121,** 585–592.

Clatch, R. J., Foreman, J. R., and Walloch, J. L. (1998). Simplified immunophenotyping by laser scanning cytometry. *Commun. Clin. Cytometry,* **34,** 3–16.

Clatch, R. J. and Foreman, J. R. (1998). Five-color immunophenotyping plus DNA content analysis by laser scanning cytometry. *Commun. Clin. Cytometry* **34,** 36–38.

Darzynkiewicz, Z., Bedner, E., Traganos, F., and Murakami, T. (1998). Critical aspects in the analysis of apoptosis and necrosis. *Hum Cell.* **11,** 3–12.

Duque, R. E. and Braylan, R. C. (1991). Applications of flow cytometry to diagnostic hematopathology. *In* "Diagnostic Flow Cytometry" (J. S. Coon and R. S. Weinstein, eds.), pp. 89–102. Williams & Wilkins, Baltimore.

Furuya, T., Kamada, T., Murakami, T., Kurose, A., and Sasaki, K. (1997). Laser scanning cytometry allows detection of cell death with morphologic features of apoptosis in cells stained with PI. *Cytometry* **29,** 173–177.

Gorczyca, W., Melamed, M. R., and Darzynkiewicz, Z. (1996). Laser scanning cytometer (LSC) analysis of fraction of labelled mitoses (FLM). *Cell Prolif.* **29,** 539–547.

Gorczyca, W., Darzynkiewicz, Z., and Melamed, M. (1997a). Laser scanning cytometry in pathology of solid tumors. A review. *Acta Cytol.* **41,** 98–108.

Gorczyca, W., Sarode, V., Juan, G., Melamed, M., and Darzynkiewicz, Z. (1997b). Laser scanning cytometric analysis of Cyclin B1 in primary human malignancies. *Hum. Pathol.* **10,** 457–462.

Hanson, C. A. (1994). Fine-needle aspiration and immunophenotyping: A role in diagnostic hematopathology? *Am. J. Clin. Pathol.* **101,** 555–556.

Juan, G., and Darzynkiewicz, Z. (1998). Detection of cyclins in individual cells by flow and laser scanning cytometry. *Methods Mol. Biol.* **91,** 67–75.

Kakino, S., Sasaki K., Kurose, A., and Ito, H. (1996). Intracellular localization of cyclin B1 during the cell cycle in glioma cells. *Cytometry* **24,** 49–54.

Kamentsky, L. A., and Kamentsky, L. D. (1991). Microscope-based multiparameter laser scanning cytometer yielding data comparable to flow cytometry data. *Cytometry* **12,** 381–387.

Kamentsky, L. A., Burger, D. E., Gershman, R. J., Kamentsky, L. D., and Luther, E. (1997). Slide-based laser scanning cytometry. *Acta Cytol.* **41,** 123–143.

Kamentsky, L. A., Kamentsky, L. D., Fletcher, J. A., Kurose, A., and Sasaki, K. (1997). Methods for automated multiparameter analysis of fluorescence in situ hybridized specimens with a laser scanning cytometer. *Cytometry* **27,** 117–125.

Kawasaki, M., Sasaki, K., Satoh, T., Kurose, A., Kamada, T., Furuya, T., Murakami, T., and Todoroki, T. (1997). Laser scanning cytometry (LSC) allows detailed analysis of the cell cycle in PI stained human fibroblasts (TIG-7). *Cell Prolif.* **30,** 139–147.

Knowles, D. M., Chadburn, A., and Inghirami, G. (1992). Immunophenotypic markers useful in the diagnosis and classification of hematopoietic neoplasms. *In* "Neoplastic Hematopathology" (D. M. Knowles, ed.), pp. 73–167. Williams & Wilkins, Baltimore.

Li, X. and Darzynkiewicz, Z. (1995). Labelling DNA strand breaks with BrdUTP. Detection of apoptosis and cell proliferation. *Cell Prolif.* **28,** 571–579.

Li, X., Traganos, F., Melamed, M. R., and Darzynkiewicz, Z. (1995). Single-step procedure for labeling DNA strand breaks with fluorescein- or BODIPY-conjugated deoxynucleotides: Detection of apoptosis and bromodeoxyuridine incorporation. *Cytometry* **20,** 172–180.

Li, X., Melamed, M. R., and Darzynkiewicz, Z. (1996). Detection of apoptosis and DNA replication by differential labeling of DNA strand breaks with fluorochromes of different color. *Exp. Cell Res.* **222,** 28–37.

Luther, E. and Kamentsky, L. A. (1996). Resolution of mitotic cells using laser scanning cytometry. *Cytometry* **23,** 272–278.

Martin-Reay, D. G., Kamentsky, L. A., Weinberg, D. S., Hollister, K. A., and Cibas, E. S. (1994). Evaluation of a new slide-based laser scanning cytometer for DNA analysis of tumors: Comparison with flow cytometry and image analysis. *Am. J. Clin. Pathol.* **102,** 432–438.

Numa, Y., Matsudaira, K., Tsukazaki, H., Kawamoto, K., Sato, T., and Kiyomatsu, Y. (1996). Analysis of cell nuclei and the quantity of chromosomal DNA by laser scanning cytometer (LSC). *Hum. Cell.* **9,** 237–243.

Pitts, W. C. and Weiss, L. M. (1992). The role of fine needle aspiration biopsy in diagnosis and management of hematopoietic neoplasms. *In* "Neoplastic Hematopathology" (D. M. Knowles, ed.), pp. 385–405. Williams & Wilkins, Baltimore.

Reeve L. and Rew, D. A. (1997). New technology in the analytical cell sciences: The laser scanning cytometer. *Eur. J. Surg. Oncol.* **23,** 445–50.

Sasaki, K., Kurose, A., Miura, Y., Sato, T., and Ikeda, E. (1996). DNA ploidy analysis by laser scanning cytometry (LSC) in colorectal cancers and comparison with flow cytometry. *Cytometry* **23,** 106–109.

Sun, T. (1993). "Color Atlas—Text of Flow Cytometric Analysis of Hematologic Neoplasms." Igaku Shoin, New York.

Willman, C. L. (1992). Flow cytometric analysis of hematologic specimens. *In* "Neoplastic Hematopathology" (D. M. Knowles, ed.), pp. 169–195. Williams & Wilkins, Baltimore.

CHAPTER 47

Immunophenotyping of Acute Leukemia: Utility of CD45 for Blast Cell Identification

J-P. Vial and F. Lacombe

Laboratory of Hematology
University Hospital Haut-Lévêque
33604 Pessac, France

I. Introduction
II. Background
III. Methods
    A. Sample Processing
    B. Immunofluorescence Staining and Erythrocyte Lysis
    C. Flow Cytometry and Data Analysis
    D. Statistical Methods
IV. Results and Comparison with Other Methods
    A. Identification and Quantification of Bone Marrow Cell Lineages
    B. Typical CD45/SSC Gating Correlates with the FAB Classification of Acute Myeloid Leukemia
    C. Determination of Surface Antigens of Blast Cell Population
V. Critical Aspects of the Methodology
    A. Specimen Handling, Transportation, Storage, and Processing
    B. Instrumental Procedures
    C. Data Analysis
VI. Pitfalls and Misinterpretation of the Data
    A. An Intermediate Cell Population Could Be Found Occasionally between Lymphocytes and Blast Cells
    B. An Intermediate Cell Population Could Be Found Occasionally between Blast Cells and Monocytes in Monocytic Acute Myeloid Leukemia
    C. Some Acute Myeloid Leukemias Show Original Features of CD45 Expression
VII. Future Directions
    A. Acute Leukemia Diagnosis
    B. Acute Leukemia Follow-up

METHODS IN CELL BIOLOGY, VOL. 64

C. Acute Leukemia Prognosis
D. Other Hematological Malignancies
References

I. Introduction

Flow cytometry (FCM) is widely used to phenotype cases of acute leukemia, and is now important in their classification (Foon and Todd, 1986; Neame *et al.*, 1986; Drexler, 1987). Although many teams are experienced in this technology, there is no single accepted method for preparation of samples, or for analyzing, interpreting, and reporting data. When bone marrow (BM) is largely infiltrated by blasts, there is generally no problem to determine the phenotype of leukemic cells. However, in some cases of acute leukemia (especially most of acute myeloid leukemia, AML), a high degree of heterogeneity is the rule (Terstappen *et al.*, 1991), because (1) both leukemic and normal cells are simultaneously present and (2) blast cells themselves can be heterogeneous. Although there is much agreement about the most important lineage-specific markers for myeloid and lymphoid lineages (Bene *et al.*, 1995; Rothe *et al.*, 1996), there is still disagreement in two areas. First, it is not yet decided whether blast cells should be enriched by Ficoll-density gradient centrifugation prior to phenotypic analysis or should be studied in the whole BM, following only red cell lysis, that is, as close to the native state as possible. Although opinions tend to favor the study of lysed whole BM, samples are still frequently treated with Ficoll for cryopreservation and storage. Second, within the malignant blast cell populations, it is not yet agreed whether results should be expressed as percentage of positive cells or as mean fluorescence intensity of all cells of interest. Thus, routine methods of analysis may make it difficult to interpret phenotypic information if no reliable practical discrimination between malignant blasts and normal cell types is used.

II. Background

Multiparametric analysis in flow cytometry refers to the simultaneous measurement and display of four or five different parameters of scatter and fluorescence. With appropriate combinations of antibodies, it is frequently possible to distinguish normal and neoplastic cells, particularly when fluorescence measurements are combined with information from light scatter (Stelzer *et al.*, 1997). Since the first studies (Borowitz *et al.*, 1993; Stelzer *et al.*, 1993), the idea of combining the intensity of CD45 with linear side scatter (SSC) parameters for individualizing leukemic cells to perform comprehensive multicolor immunofluorescence has gained ground. A brief review of the literature on the CD45 gating is presented in Table I.

The systematic use of leukocyte common antigen (LCA, CD45) marker in a multiple ($\geq$3) immunofluorescence (IF) combination is demonstrated here to

Table I
Brief Review of the Literature

Author	Comment
Lacombe *et al.* (1997)	Flow cytometry CD45 gating for immunophenotyping of AML: study of 74 patients at diagnosis (Lacombe *et al.*, 1997).
Rainer *et al.* (1995)	CD45 gating correlates with bone marrow differential: study of 26 AML, 15 ALL,[a] 2 myelodysplasic syndromes, 1 lymphoproliferative disorder, 6 normal patients (Rainer *et al.*, 1995).
Abrahamsen *et al.* (1995)	Flow cytometric assessment of peripheral blood contamination and proliferative activity of human cell populations: study of 12 normal bone marrow aspirates (Abrahamsen *et al.*, 1995).
Borowitz *et al.* (1993)	Use of CD45 and right-angle light scatter to gate on leukemic blasts in three-color analysis: approach applied to 39 cases of acute leukemia and 8 cases of myelodysplasia or myeloproliferative disorders (Borowitz *et al.*, 1993).
Stelzer (1993)	CD45 gating for routine flow cytometric analysis of normal human bone marrow specimens (Stelzer *et al.*, 1993).

[a] ALL, acute lymphoblastic leukemia.

achieve a reliable and highly reproducible discrimination between the blast cell populations and the normal cells, showing that CD45/SSC gating is a simple method to detect blast cells in large series of acute leukemia.

III. Methods

A. Sample Processing

A total of 192 BM samples were consecutively collected from consenting adult patients with unequivocal diagnosis of AML based on morphological and cytochemical criteria. According to the French–American–British (FAB) classification (Bennett *et al.*, 1985) and immunophenotyping (Dubosc-Marchenay *et al.*, 1992; Garand *et al.*, 1994; Borowitz *et al.*, 1997a; Stewart *et al.*, 1997), diagnosis was AML0 ($n = 19$), AML1 ($n = 38$), AML2 ($n = 52$), AML3 ($n = 16$), AML4 ($n = 21$), AML5 ($n = 27$), AML6 ($n = 7$), AML7 ($n = 3$), unclassified ($n = 5$), and biphenotypic acute leukemia ($n = 4$). BM samples (2–3 ml) were collected in EDTA K3 tubes (Vacutainer; Becton Dickinson, Pont de Claix, France) and received in the laboratory within 1 hr of collection. Morphological analysis and differential count (excluding erythroid cells) were performed by two experienced readers on 200 cells on MGG-stained smears.

B. Immunofluorescence Staining and Erythrocyte Lysis

Tricolor IF staining was performed on fresh cells by direct IF as follows. Samples were diluted to 1×10^7 cells/ml with AB-positive pooled plasma, and aliquots of 50 μl (5×10^5 cells in each tube) were incubated for 30 min at room temperature in the dark with 5 μl of both fluorescein isothiocyanate (FITC)- and phycoerythrin (PE)-conjugated antibody (Table II) plus Cy5-PE-conjugated CD45.

AML diagnosis was confirmed by positivity of cytoplasmic markers for myeloperoxidase (MPO-Ag), intracytoplasmic CD13c and CD22c, or the presence of megakaryocytic CD41a, CD61, or erythroid marker glycophorin A. Samples were then treated using the Coulter Immunoprep reagent kit and Multi-Q-Prep system (Coultronics, Margency, France) according to the manufacturer's instructions.

C. Flow Cytometry and Data Analysis

Samples were analyzed on a Coulter XL flow cytometer (Coultronics). Using the 488 nm wavelength excitation, FITC green, PE orange, and Cy5-PE red emission fluorescences were measured through 525, 575, and 675 band-pass filters, respectively. Multiparametric analysis collected five parameters simultaneously: linear forward-angle light scatter (FSC), linear side scatter (SSC), Log FITC,

Table II
Panel of Monoclonal Antibodies Used

CD antigen		Reactivity[a]	Source[b]
CD2	F[a]	T lineage	Coulter
CD3	F	T lineage	Coulter
CD4	F	T lineage	Immunotech
CD5	P	T lineage	Dako
CD7	F	T lineage	Coulter
CD8	P	T lineage	Coulter
CD10	P	B lineage	Dako
CD11b	F	Granulocytes, monocytes, NK cells	Immunotech
CD13	P	Myeloid cells, monocytes	Dako
CD14	F	Monocytes	Immunotech
CD19	P	B lineage	Dako
CD20	F	B lineage	Coulter
CD22	F	B lineage	Dako
CD24	P	B lineage	Immunotech
CD33	P	Myeloid cells, monocytes	Coulter
CD34	F	Hematopoietic precursor cells	Becton Dickinson
CD45	Cy	Leukocyte common antigen	Coulter
CD56	P	NK cells	Immunotech
CD117	P	Hematopoietic progenitor cells	Immunotech
HLA-DR	F	Histocompatibility complex	Immunotech

[a] F = FITC coupled; P = PE coupled; Cy = Cy5PE coupled; NK = natural killer.
[b] Coulter, Coultronics, Margency, France; Immunotech, Marseille, France; Dako, Glostrup, Denmark.

Log PE, and Log Cy5-PE. Levels for positivity were determined according to PE-, FITC-, and Cy5-PE-conjugated isotypic immunoglobulin (Ig) control (IgG1-FITC, IgG1-PE, and IgG1-Cy5PE from Coulter). A first step (gate A) was to exclude debris and fat cells on a FSC/SSC histogram and to record data in list mode (Fig. 1a). List mode data were analyzed using the Coulter System II program. In the FSC/SSC histogram, presumed blast cells were identified by using gate S. In the CD45/SSC histogram four gates were set up: gate L for lymphocytes, M for monocytes, G for granulocytes, and B for blast cells (Fig. 1b). At diagnosis 1×10^4 blast cells were analyzed. CD antigen expression on the blast cells was compared with these two gating procedures, and positivity was classically recorded when more than 20% blasts marked the CD antibody tested.

D. Statistical Methods

Comparison of light microscopic BM examination and flow differential for hematopoietic cell evaluation was performed using Pearson's coefficient of correlation (r). The degree of agreement between the two methods could also be visualized using a plot of the mean reading of the two methods against the difference in reading between the two methods. To determine the degree of concordance of the two methods, we used the intraclass correlation coefficient *RI* (Shrout and Fleiss, 1979; Lee *et al.*, 1989) instead of using Pearson's coefficient of correlation which is just an estimation of trend. It is generally assumed that

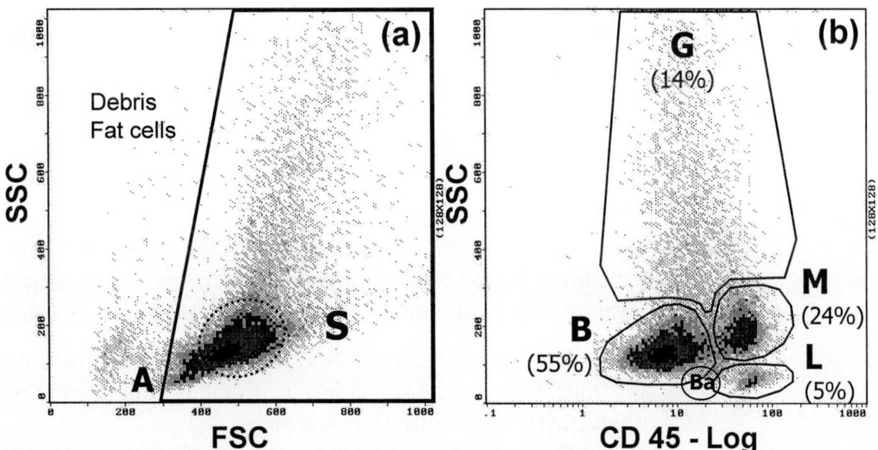

Fig. 1 Flow cytometry bivariate histograms of an AML. (a) FSC/SSC histogram with two gates: gate A to eliminate debris and fat cells, gate S on presumed blast cells. Blast cells partially overlap the gates for normal lymphocytes and monocytes. (b) CD45/SSC histogram of the same AML with four gates: lymphocytes (L), monocytes (M), granulocytes (G), and blast cells (B). Corresponding cell percentages are indicated (the sum is not 100% since rare cells were found outside the gates).

a good agreement between the two methods is achieved when the lower limit of the 95% confidence interval of *RI* is at least 0.75.

==== **IV. Results and Comparison with Other Methods**

A. Identification and Quantification of Bone Marrow Cell Lineages

Figure 1 shows the advantage of the CD45/SSC gating procedure over traditional FSC/SSC for identifying leukemic cells. When the FSC/SSC gate was used without any additional immunological marker, the AML blast cells partially overlapped within the gates of normal lymphocytes and monocytes. Indeed, Fig. 1a shows the difficulty of defining a specific area for blast cells. In contrast, CD45/SSC gating clearly separated the four cell categories (Fig. 1b). Mature lymphocytes presented the highest CD45 fluorescence intensity and the lowest SSC signal (gate L). Mature monocytes expressed almost the same amount of CD45, but were easily distinguished by a higher intensity of SSC signal (gate M). Granulocytic lineage expressed CD45 less than mature lymphocytes, and were distinguished from monocytes by the very high intensity of SSC signal (gate G). Blast cells had the lowest CD45 intensity and a low intensity of SSC signal, thus reflecting features of normal immature bone marrow precursor cells (gate B).

The morphological differential count was compared with the results of CD45/SSC gating procedure according to Pearson's (*r*) and intraclass (*RI*) correlation methods. A strong positive correlation was found for the four cell subpopulations (Fig. 2c–e): blast cells (*r* = 0.98, *RI* = 0.98), granulocytes (*r* = 0.96, *RI* = 0.96), lymphocytes (*r* = 0.97, *RI* = 0.97), and monocytes (*r* = 0.98, *RI* = 0.98). A similar comparison between the morphological blast cell count and the corresponding count performed on the FSC/SSC histogram (Fig. 2a,b) showed a poorer correlation (*r* = 0.70, *RI* = 0.67).

B. Typical CD45/SSC Gating Correlates with the FAB Classification of Acute Myeloid Leukemia

A good correlation was found between the different AML patterns obtained with a primary CD45/SSC gating and the FAB classification. A set of typical CD45/SSC gating is shown on Fig. 3 and is described as follows:

Undifferentiated AML1 (or AML0): The blastic infiltration (gate B) is massive (greater than 90% of the total cells), is replacing the normal hematopoietic cells, and granulocytic differentiation is absent.

Differentiated AML2: The blastic infiltration is moderate (between 30 and 90%) and associated with a granulocytic differentiation (gate G) greater than 10% of the total cells.

Promyelocytic AML3: The typical promyelocytic infiltrate is massive. The high SSC signal makes it difficult to separate the infiltrate from the granulocytic

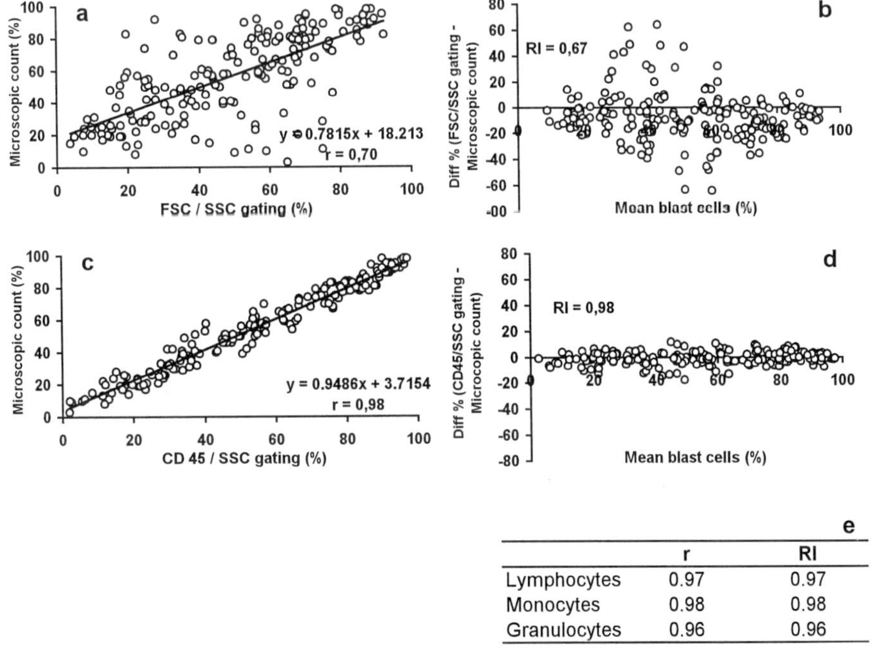

e		
	r	RI
Lymphocytes	0.97	0.97
Monocytes	0.98	0.98
Granulocytes	0.96	0.96

Fig. 2 Correlation between a light microscopic bone marrow cell differential and the flow differential for hematopoietic cell evaluation. Two statistical methods were used: classic Pearson's (*r*) and intraclass (*RI*) coefficient of correlation. (a,b) FSC/SSC gating for blast cell evaluation. (c,d) CD45/SSC gating for blast cells, and (e) for lymphocyte, monocyte, or granulocyte evaluation.

maturation. However, this population appears abnormal considering its percentage.

Myelo-monocytic AML4: As in AML2, AML4 blastic infiltration is moderate and associated with a granulocytic maturation greater than 10% of the total cells. Moreover, a monocytic infiltrate is present and clearly separated from blast cells (AML4, lane 1). Sometimes, an additional population of promonocytes (pM) may be identified morphologically and clearly separated from blast cells (B) and monocytes (M) on the basis of CD45/SSC gating (AML4, lane 2). The gate pM may be drawn between the B and M gates.

Monoblastic AML5: Unlike AML4, the blast cells are massively infiltrating the BM and replacing the normal hematopoietic lineages. Granulocytic differentiation is absent, whereas the percentages of monocytes or promonocytes are increasing with the monocytic maturation.

AML5a: Monoblastic cells (B) are the essential component of the BM infiltration. They are characterized by a SSC signal intensity close to mature monocytes (which represent less than 20% of the infiltration) but distinguished from them by lower expression of CD45.

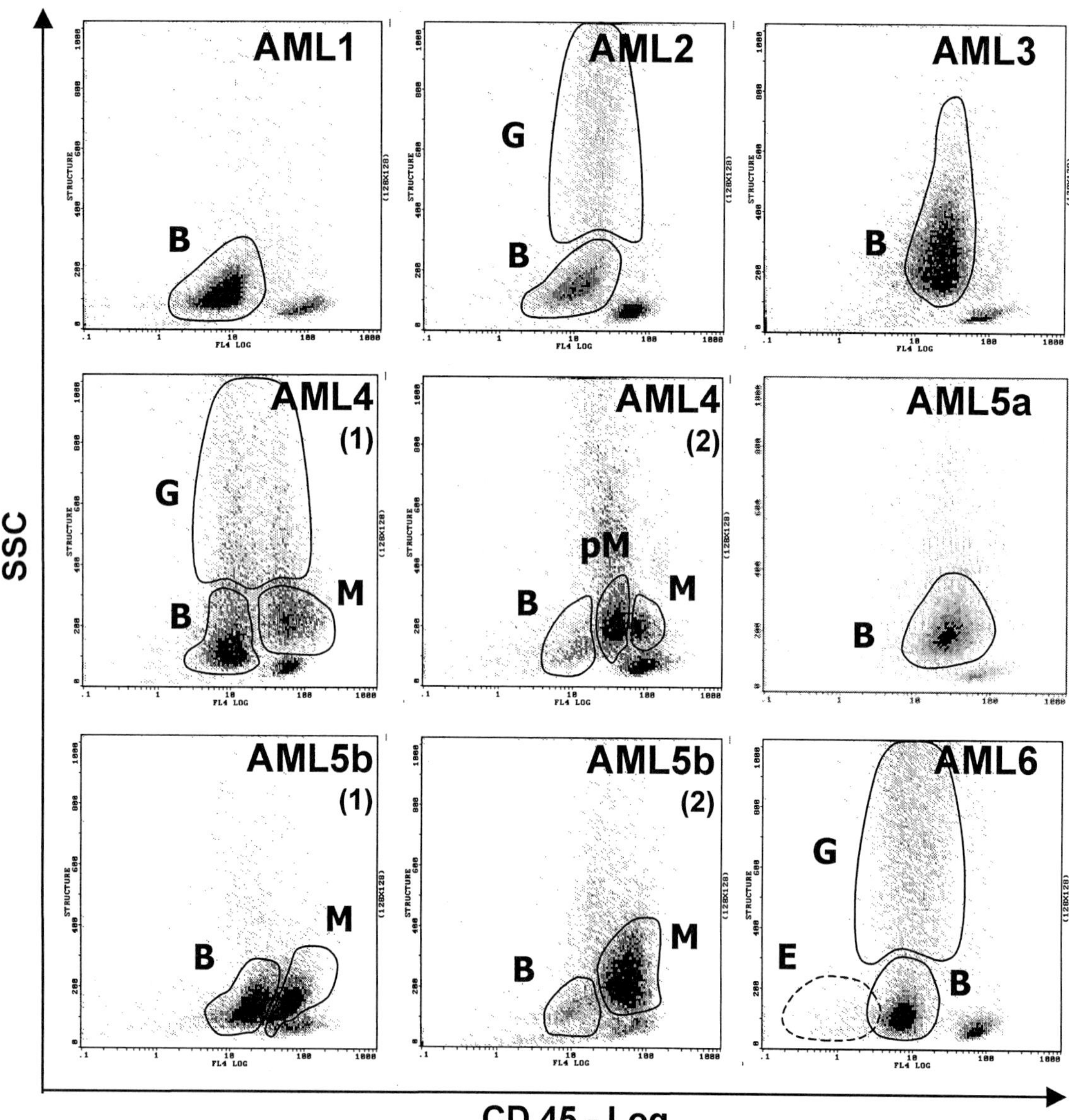

Fig. 3 Typical CD45/SSC gating set of AML cases correlated with FAB classification. Classic gates B, G and M are shown on bivariate histograms of undifferentiated AML1, differentiated AML2, promyelocytic AML3, myelo-monocytic AML4 (charts 1 and 2), monoblastic AML5 without (AML5a) or with monocytic maturation (AML5b, charts 1 and 2), and AML6. In AML4 (chart 2), a pM gate (promonocytes) is indicated between B and M gates. In AML6, remaining erythroid cells (gate E) were found.

AML5b: Monoblastic infiltration (B) is variable because the monocytic maturation (M) is present and higher than 20% of the infiltration. Several patterns illustrate the heterogeneity of this AML according to the degree of monocytic maturation; for example, an AML5b moderately differentiated as shown in lane 1 and a well differentiated AML5b in lane 2.

AML6: The blast cell infiltrate (B) is variable, depending on the granulocytic differentiation (G) and percentage of erythroid cells present in the sample. In this case, the major difficulty is to remove the erythroid cells from the sample without affecting blast cell antigenicity. Indeed, erythroid cells (gate E) may persist and prevent analysis because of their lack of CD45 expression, thus causing an overlap with the blast cells.

C. Determination of Surface Antigens of Blast Cell Population

We used a triple-labeling method with systematic CD45-Cy5PE and varied pairs of FITC- and PE-conjugated antibodies to appreciate the gain of informa-

Table III
Antigenic Expression Discrepancies Between FSC/SSC and CD45/SSC Gating[a]

			Positive cases[c]	
	FP[b]	FN[b]	Number	Antigenic discrepancies (%)
CD2	7.8	0	5	20
CD3	12.1	0	0	0
CD5	10.5	0	1	0
CD7	13.1	1	37	24.3
CD10	0	0	2	0
CD19	3.2	0	10	10
CD20	0	0	0	0
CD22	0.5	0	3	0
CD24	0.7	0	3	0
CD56	2.9	0.7	22	4.5
HLA-DR	3.7	0	149	31.5
CD11b	15.9	0.5	29	55.2
CD13	2.1	5.8	144	34.7
CD14	14.3	0	0	0
CD33	2.6	3.7	153	26.1
CD34	0	4.7	96	40.6
CD117	0	4.3	38	36.8

[a] CD45/SSC gating was considered as the reference procedure to study CD-antigen expression.

[b] Percentages of false positive (FP) and false negative (FN) results given by the FSC/SSC procedure.

[c] For positive cases only (i.e., sample with ≥20% positivity): CD expression discrepancies >15% between FSC/SSC and CD45/SSC gating.

tion in the immunophenotyping of blast cells. The positivity of CD-antigen expression on blast cells was evaluated using either FSC/SSC or CD45/SSC gating. Because CD45/SSC gating determined blast cells better, it was chosen as the reference procedure. If the positivity threshold for each CD-antigen expression was strictly fixed at 20%, false positive (FP) samples would have been recorded in samples that were positive (CD expression $\geq$20%) using FSC/SSC gating, but negative (CD <20%) with CD45/SSC gating. By contrast, false negative (FN) samples appeared negative using FSC/SSC gating but positive with CD45/SSC gating. Table III shows that FP cases were found for lymphoid (CD2, CD3, CD5, CD7, CD19, CD22, CD24), natural killer (CD56), HLA-DR, and monocytic antigens (CD11b, CD13, CD14, CD33). In each case, the false positivity was due to contamination of analysis by lymphocytes or monocytes. On the other hand, false negativity was found for myeloid (CD11b, CD13, CD33), immature (CD34, CD117), and, in a few samples, aberrant cross-lineage antigens (CD7, CD56). Among these FN cases, most were AML with a low percentage of blast cells, that is, with myeloid and/or monocytic maturation (AML2, n = 19, AML4 or AML5, n = 9) or hemodiluted BM samples (n = 9).

Even when FP and FN cases were absent, significant differences were noted in many patients regarding the degree of CD-antigen expression when the FSC/SSC and CD45/SSC gating procedures were compared (Table III). For each positive CD-antigen expression (i.e. blast cells with $\geq$20% positivity), the values of antigen expression were considered to be in agreement when the difference using the two gating procedures was below an arbitrary value of 15%. In these conditions, discrepancies in positive antigenic determination were mostly seen in the myeloid lineage (CD11b, CD13, CD33) and in HLA-DR, but also in immature (CD34, CD117) and aberrant cross-lineage antigens (CD2, CD7, CD19, CD56).

V. Critical Aspects of the Methodology

A. Specimen Handling, Transportation, Storage, and Processing

A low volume of BM has to be briefly aspired to prevent hemodilution of the samples. Collection in EDTA tubes was suitable because heparinized samples were not compatible with the Multi-Q-Prep lysis method.

In general, specimens should be transported to the flow cytometry laboratory, processed, and stained for analysis immediately after collection. Short-term storage (1 hr or less) should be maintained at room temperature (round about 20°C). For prolonged storage (12–24 hr), BM specimens should be maintained at room temperature or, if possible, at a more strictly regulated temperature of 16°C.

The Multi-Q-Prep lysis method worked well only in samples diluted with plasma since the admixture of PBS caused cell aggregation. CD45 staining and CD45/SSC gating did not seem contraindicated on Ficolled-treated samples or thawed samples (Lacombe *et al.*, 1997).

B. Instrumental Procedures

To keep the settings and histograms uniform during the study, the Coulter XL flow cytometer was checked daily with calibration beads (DNA-Check; Coulter). Fluorochrome compensation settings were routinely adjusted using CD4-FITC/ CD8-PE/CD45-Cy5PE staining of whole blood lymphocytes.

C. Data Analysis

Erythrocytes and platelets, which do not express CD45 antigen, are theoretically excluded from the analysis. However, care must be taken with the putative presence of lysis-resistant erythrocytes or platelet clumps that could overlap within the blast cell population. Note that similar interferences could be found with any small debris that weakly bound CD45 causing false positive antigen expression. This emphasizes a first constraint of this method: FSC/SSC pregating is necessary to eliminate debris and fat cells. A second constraint is the corollary of using Multi-Q-Prep or any other lysis method: the impossibility of analyzing immature erythroid cells.

Granulocytes (gate G) can appear horizontally divided in two populations displaying the same SSC signal intensity. However, the multiparametric immuno-phenotypic study (CD16 labeling) made it possible to separate immature (CD45 weak, CD16 negative) and mature granulocytes (CD45 stronger, CD16 positive).

VI. Pitfalls and Misinterpretation of the Data

A. An Intermediate Cell Population Could Be Found Occasionally between Lymphocytes and Blast Cells

These cells display the same low SSC signal intensity, but a moderate CD45 expression (Fig. 1b, gate Ba). They should not be misidentified as blast cells because their immunophenotypic study and cell sorting (data not shown) revealed that this was a basophilic granulocytes population.

B. An Intermediate Cell Population Could Be Found Occasionally between Blast Cells and Monocytes in Monocytic Acute Myeloid Leukemia

Promonocytes (pM) were identified morphologically and were clearly separated from blast cells (B) and monocytes (M) on the basis of CD45/SSC gating (Fig. 3, AML4 chart 2). These cells showed higher SSC signal intensity and CD45 expression than blast cells, but lower intensity than mature monocytes. They should not be misidentified as blast cells. Among the 27 AML5 of the series, pM were morphologically identified in 17 cases (all were AML5b); CD45/SSC gating always clearly separated promonocytes from blast cells and monocytes.

Figure 4 illustrates the monocyte lineage maturation study in AML5 according to CD45/SSC gating (*Oy*-axis: percentage of positive cells for CD11b and CD14; labels: mean fluorescence intensities +/− SD). Blast cells were always negative for CD14, and the percentage of positivity for CD11b varied (blast cells of AML5 did not always bind CD11b: 50% of cases, and degree of binding was variable, data not shown), whereas monocytes were always positive for both markers (CD11b: $p < 0.001$; CD14: $p < 0.001$). Study of the mean fluorescence intensities of these CD antigens (see labels) confirmed a lack of CD14 on AML5 blast cells and the weaker expression of CD11b and CD14 on promonocytes than on their mature monocyte counterparts (all p were <0.001).

C. Some Acute Myeloid Leukemias Show Original Features of CD45 Expression

Some AML were of original cytological presentation (heterogeneous blast cells) and were unclassifiable in accordance with FAB. Many of these cases displayed original features of CD45/SSC gating like the example shown in Fig. 5. Although in this case the presumed blast cells were identified with classic FSC/SSC gating (see gate S in the insert of chart a), the heterogeneity of this population was revealed using the CD45/SSC histogram (chart a): B1 blast cells with less CD45 expression were clearly separated from B2 blast cells. The study of these different populations (S, B1, or B2: chart c) showed that the blast cells expressed different proportions of CD-antigens according to their CD45 expression. The CD45^{low} B1 population was positive with high fluorescence intensity for the antigens related to myeloid immaturity (HLA-DR, CD34, CD117), contrary to the CD45^{intermediate} B2 which strongly expressed the mature myeloid antigen CD11b. Both blast cell populations were positive for the myeloid CD33 and CD13; the fluorescence intensity of CD33 was greater on mature B2 blast cells, unlike that of the CD13. CD7 expression was noted on the most immature B1 blast cells which caused a relapse a few months later (chart b).

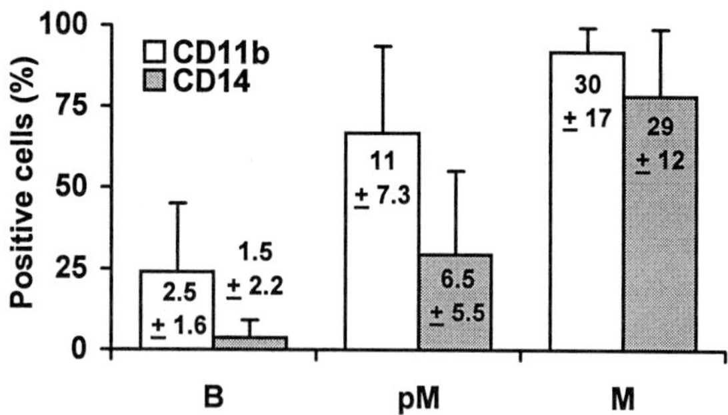

Fig. 4 Immunophenotyping for studying monocyte lineage maturation. Among the 27 AML5 of the series, pM were morphologically identified, clearly separated from B and M on the basis of the CD45/SSC histogram, and immunophenotyped in 17 cases (which were AML5b). *Oy*-axis: percentage of positive cells for CD11b and CD14. Labels: Mean fluorescence intensities +/− SD.

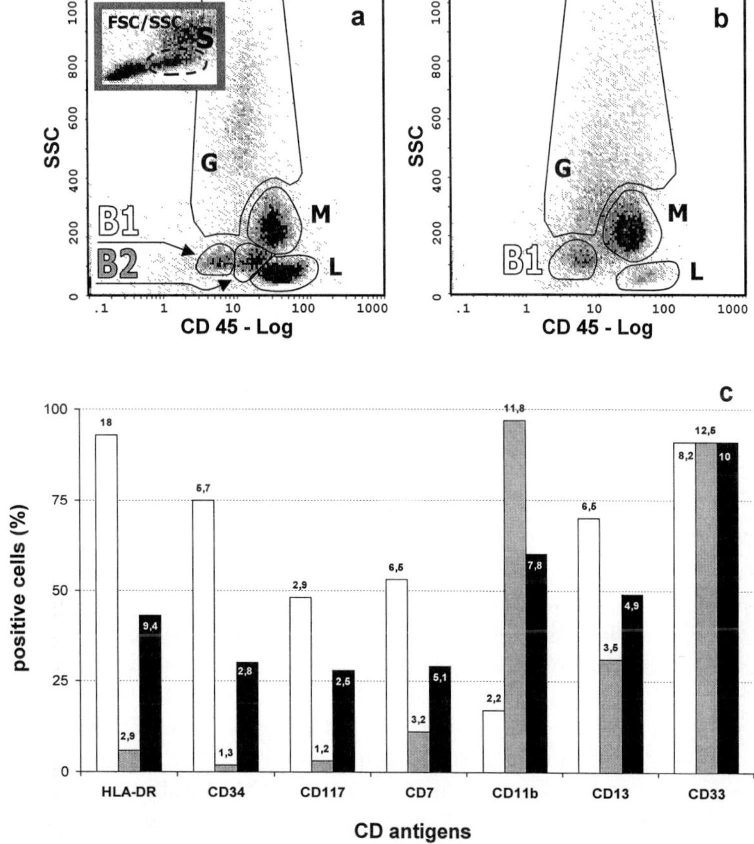

Fig. 5 Flow cytometry histograms of an unclassifiable AML: an example of blast cell heterogeneity studied according to CD45 expression. (a) Although in this case the presumed blast cells were identified using classic FSC/SSC gating (see gate S inserted in lane a), heterogeneity of this population was found only using the CD45/SSC histogram: B1 cells with less CD45 expression were clearly separated from B2 blast cells. (b) Same patient relapsing 3 months later; only B1 blast cells were present at this time. (c) Corresponding percentages of positive cells and mean fluorescence intensity (labels) for the different CD antigens studied. CD antigens related to myeloid immaturity (HLA-DR, CD34, CD117) were present only in the CD45^{low} B1 population (white bars) while CD11b was on the CD45^{intermediate} B2 (gray bars). The myeloid antigens CD33 and CD13 were positive on both blast cell populations; the fluorescence intensity of CD33 was greater on mature B2 blast cells, unlike the CD13. Note the CD7 expression on the most immature B1 blast cells. Classic FSC/SSC gating did not detect this heterogeneity of S blast cells (black bars).

VII. Future Directions

A. Acute Leukemia Diagnosis

CD45/SSC gating may provide a common platform for uniform data processing during the immunophenotyping of AML at diagnosis, because it could be considered as a useful bridge between cytology and immunophenotyping of leuke-mic diseases.

B. Acute Leukemia Follow-up

Right lineage assignment and detection of aberrant phenotypes in blast cells (e.g., CD7, Fig. 5b) could facilitate early diagnosis of relapse and probably the study of minimal residual disease.

The systematic use of CD45 in combination with lineage-specific markers in normal BM analysis could be helpful to refine the antigenic profile of normal myeloblasts that present the same CD45/SSC characteristics as leukemic blasts. Thus, the comparison with myeloblasts could contribute to a reliable definition of blast cell antigenic overexpression, which is recognized as a helpful tool during the immunological follow-up of AML.

C. Acute Leukemia Prognosis

The accurate study of blast cells could optimize the intrinsic prognostic value of other CD-antigens and contribute to understanding some of the mechanisms responsible for the changes detected in AML blast cells during treatment: for example, possible effect of antiblastic drugs on cell differentiation, expression of different proteins and antigens, induction of multidrug resistance or apoptosis, and selection of certain tumor cell clones (Stass *et al.*, 1984; Lacombe *et al.*, 1994; Durrieu *et al.*, 1998). The intensity of CD45 fluorescence itself has proved to be a prognostic factor in acute lymphoblastic leukemia (Borowitz *et al.*, 1997b). In our hands, such differences in CD45 fluorescence intensity were not found in AML, but refined calculation of CD45 fluorescence intensity (Shah *et al.*, 1988) is now necessary to make more solid conclusions.

D. Other Hematological Malignancies

Because this procedure is particularly adapted to the study of low percentages of immature cells in heterogeneous samples without removing the other cells, it could be successfully applied to the study of myelodysplasic syndromes or myeloproliferative disorders. For example, in chronic myelogenous leukemia, easy identification of immature cells and basophilic granulocytes could improve the detection of the acceleration phase.

Although acute lymphoblastic leukemia (ALL) blast cells are generally homo-geneous and massively infiltrating the BM at diagnosis, CD45/SSC gating could be a good mean for follow-up.

Acknowledgments

We gratefully acknowledge Alex Briais, Elisabeth Bascans, and Patrice Dumain for their technical assistance.

References

Abrahamsen, J., Lund-Johansen, F., Laerum, O., Schem, B., Sletvold, O., and Smaaland, R. (1995). Flow cytometric assessment of peripheral blood contamination and proliferative activity of human bone marrow cell populations. *Cytometry* **19,** 77–185.

Bene, M. C., Castoldi, G., Knapp, W., Ludwig, W. D., Matutes, E., Orfao, A., and Van't Veer, M. B. (1995). Proposals for the immunological classification of acute leukemias. European group for the immunological characterization of leukemias (EGIL). *Leukemia* **9,** 1783–1786.

Bennett, J. M., Catowsky, C., Daniel, M. T., Flandrin, G., Galton, D. A. G., Gralnick, H. R., and Sultan, C. (1985). Classification of acute myeloid leukemia. A report of the French–American–British cooperative group. *Ann. Intern. Med.* **103,** 620–624.

Borowitz, M. J., Guenther, K. L., Shults, K. E., and Stelzer, G. T. (1993). Immunophenotyping of acute leukemia by flow cytometric analysis. Use of CD45 and right-angle light scatter to gate on leukemic blasts in three-color analysis. *Am. J. Clin. Pathol.* **100**(5), 534–540.

Borowitz, M. J., Bray, R., Gascoyne, R., Melnick, S., Parker, J. W., Picker, L., and Stetler-Stevenson, M. (1997a). U.S.–Canadian consensus recommendations on the immunophenotypic analysis of hematologic neoplasia by flow cytometry: Data analysis and interpretation. *Cytometry* **30,** 236–244.

Borowitz, M. J., Shuster, J., Carroll, A. J., Nash, M., Look, A. T., Camitta, B., Mahoney, D., Lauer, S. J., and Pullen, D. J. (1997b). Prognostic significance of fluorescence intensity of surface marker expression in childhood B-precursor acute lymphoblastic leukemia. A pediatric Oncology Group Study. *Blood* **89**(11), 3960–3966.

Drexler, H. G. (1987). Classification of acute myeloid leukemias–a comparison of FAB and immunophenotyping. *Leukemia* **1,** 697–705.

Dubosc-Marchenay, N., Lacombe, F., Dumain, P., Marit, G., Montastruc, M., Belloc, F., and Reiffers, J. (1992). Role of blast cell immunophenotyping for the diagnosis and prognosis of acute myeloid leukemia. *Hematol. Oncol.* **10,** 235–249.

Durrieu, F., Belloc, F., Lacoste, L., Dumain, P., Chabrol, J., Dachary-Prigent, J., Morjani, H., Boisseau, M. R., Reiffers, J., Bernard, P., and Lacombe, F. (1998). Caspase activation is an early event in anthracycline-induced apoptosis and allows detection of apoptotic cells before they are ingested by phagocytes. *Exp. Cell Res.* **240,** 165–175.

Foon, K. A., and Todd, R. F. (1986). Immunological classification of leukemia and lymphoma. *Blood* **68,** 1–31.

Garand, R., Béné, M. C., and Geil, A. T. (1994). A new approach of acute lymphoblastic leukemia immunophenotypic classification. *Leukemia Lymphoma* **13**(1), 1–5.

Lacombe, F., Belloc, F., Dumain, P., Puntous, M., Cony Makhoul, P., Saux, M.-C., Bernard, P., Boisseau, M. R., and Reiffers, J. (1994). Detection of cytarabine resistance in patients with acute myelogenous leukemia using flow cytometry. *Blood* **84**(3), 716–723.

Lacombe, F., Durrieu, F., Briais, A., Dumain, P., Belloc, F., Bascans, E., Reiffers, J., Boisseau, M. R., and Bernard, P. (1997). Flow cytometry CD45 gating for immunophenotyping of acute myeloid leukemia. *Leukemia* **11**(11), 1878–1886.

Lee, J., Koh, D., and Ong, C. N. (1989). Statistical evaluation of agreement between two methods for measuring a quantitative variable. *Comput. Biol. Med.* **19,** 61–70.

Neame, P., Soamboosrup, P., Browman, G., Meyer, R., Benger, A., Wilson, W., Walker, I., Seed, N., and McBride, J. (1986). Classifying acute leukemia by immunophenotyping: A combined FAB-immunologic classification of AML. *Blood* **68,** 1355–1362.

Rainer, R. O., Hodges, L., and Seltzer, G. T. (1995). CD 45 gating correlates with bone marrow differential. *Cytometry* **22**(2), 139–145.

Rothe, G., and Schmitz, G., (1996). Consensus protocol for the flow cytometric immunophenotyping of hematopoietic malignancies. *Leukemia* **10,** 877–895.

Shah, V. O., Civin, C. I., and Loken, M. R. (1988). Flow cytometric analysis of bone marrow: IV. Differential quantitative expression of T-200 common leucocyte antigen during normal hematopoiesis. *J. Immunol.* **140,** 1861–1867.

Shrout, P. E., and Fleiss, J. L. (1979). Intraclass correlations: Uses in assessing rater reliability. *Psychol. Bull.* **86,** 320–428.

Stass, S., Mirro, J., Melvin, S., Pui, C. H., Murphy, S. B., and Williams, D. (1984). Lineage switch in acute leukemia. *Blood* **64,** 701–708.

Stelzer, G. T., Shults, K. E., and Loken, M. R. (1993). CD45 gating for routine flow cytometric analysis of human bone marrow specimens. *Ann. N.Y. Acad. Sci.* **677,** 265–280.

Stelzer, G. T., Marti, G., Hurley, A., McCoy, P., Lovett, E. J., and Schwartz, A. (1997). U.S.–Canadian consensus recommendations on the immunophenotypic analysis of hematologic neoplasia by flow cytometry: Standardization and validation of laboratory procedures. *Cytometry* **30,** 214–230.

Stewart, C., Behm, F., Carey, J., Cornbleet, J., Duque, R., Hudnall, S., Hurtubise, P., Loken, M., Tubbs, R., and Wormsley, S. (1997). U.S.-Canadian consensus recommendations on the immunophenotypic analysis of hematologic neoplasia by flow cytometry: Selection of antibody combinations. *Cytometry* **30,** 231–235.

Terstappen, L. W. M. M., Safford, M., Könemann, S., Loken, M. R., Zurlutter, D., Büchner, T., Hiddemann, W., and Wörmann, B. (1991). Flow cytometric characterization of acute myeloid leukemia. Part II. Phenotypic heterogeneity at diagnosis. Leukemia **5,** 757–767.

CHAPTER 48

Cell Proliferation Markers in Human Solid Tumors: Assessing Their Impact in Clinical Oncology

Maria Grazia Daidone, Aurora Costa, and Rosella Silvestrini
Istituto Nazionale per lo Studio e la Cura dei Tumori
20133 Milan, Italy

I. Introduction
II. Proliferation Markers
    A. Mitosis: Mitotic Index or Mitotic Count
    B. S Phase: Incorporation of DNA Precursors
    C. S Phase: Quantification of Nuclear DNA Content
    D. Growth Fraction: Determination of Enzymes, Antigens, or Structural Alterations
III. Biological Studies
IV. Clinical Studies
    A. Proliferation Markers and Prognosis
    B. Proliferation Markers and Response to Systemic Treatments
V. Conclusions on First Generation Translational Studies with Proliferation Markers
References

I. Introduction

Proliferation represents a fundamental biological process. It is involved in determining growth and maintenance of tissue homeostasis. The proliferative activity of a tissue can be considered as the resultant of a complex and dynamic equilibrium among the cell subpopulations which compose it and which progress under the control of regulatory elements through the four consecutive phases of the cell cycle, $G_1 \rightarrow S \rightarrow G_2 \rightarrow M$, in which cytoplasmic proteins, organelles, and RNA are synthesized (G_1, G_2 phases), DNA is replicated (S phase), and cells undergo mitosis (M phase), or remain in a state of quiescence (G_0). The activity of cell cycle regulators is modulated according to stimulatory or inhibitory growth signals and is subjected to strict control in normal cells, whereas in cancer cells a variable degree of independence from such stimuli seems to occur. In cancer, the activation of oncogenes, likely accomplished by the inactivation of one or more tumor suppressor genes, is responsible for the induction of stimulatory signals as well as for the disruption of checkpoints that ensure an orderly progression of cells in the cell cycle (Clurman and Roberts, 1995; Gillett and Barnes, 1998). Thus, the study of proliferative activity of neoplastic tissues contributed and still contributes to improve our knowledge of tumor biology.

Any tumor contains cell subpopulations characterized by proliferative conditions singly reproducing the kinetic behavior of the different normal tissues. Such subpopulations are represented by: (a) proliferating cells; (b) temporarily quiescent cells that still maintain a proliferative capability and eventually may join the cycling population following suitable stimuli; and (c) cells that permanently exit from the cell cycle, and undergo the genetically controlled process of programmed cell death (apoptosis), or necrosis. The growth of a tumor is the net result of cell gain by proliferative activity, which is directly related to the growth fraction (i.e., the proportion of cells committed to the cycle), and cell loss by apoptosis or necrosis. The relative importance of cell gain and cell loss can be modified following clinical treatments, but may also vary spontaneously as a function of tumor age or physiological status (nutrients, hypoxic conditions, etc.).

Initially, cell kinetic studies have been devoted to investigate growth patterns of several systemic and solid experimental tumors and to establish rules about tumor growth. In parallel, investigation on experimental tumors has identified and classified antitumor drugs according to their action on nonproliferating or proliferating cells and, for the latter, with a specific reference to the different cell cycle phases. A major statement emerging from studies on experimental tumors was the relevance of cell proliferation as an indicator of tumor aggressiveness and of response to chemical and physical agents. In experimental studies, proliferative activity and sensitivity to drugs or radiation have consistently proved to be associated events, since a direct relation between the rate of proliferative activity and sensitivity to chemical and physical agents has been consistently observed.

However, clinical practice soon showed many of the hypotheses derived from experimental research to be far from true. Major discrepancies between human and animal tumors can be summarized as due to a large biological heterogeneity among tumors of different histotypes as well as within each histotype and to a lack of a general relation between cell proliferation and response to systemic treatment. Such findings provided support to the belief that experimental tumors, chemically induced or implanted in animals and generally characterized by a limited metastasizing potential, cannot be considered representative of the complex biology of human tumors. Thus, considering the biological intertumor heterogeneity, more recently an increasing effort has been made to obtain as much cell kinetic information as possible from individual clinical lesions to improve the knowledge on tumor biology and potential aggressiveness and to provide clinicians data on consecutive series of cases to assess the clinical utility of proliferation markers.

To make cell kinetics determination feasible on consecutive series of patients, investigators have addressed their attention to specific aspects of the complex phenomenon of proliferation and growth. In particular, kinetic characterization has focused on the proliferating cell compartment, that is, on cells that transit through the four phases of the cell cycle, that are generally responsible for tumor growth, and that are most susceptible to the action of therapeutic agents.

Biologists and pathologists have used several approaches to determine the cell proliferative fraction, according to their professional background (Barnes and Gillett, 1995). Such approaches are based on different rationales, employ different methods of evaluation, such as morphometric, immunocytochemical, cytometric, or autoradiographic detection, and each of them has inherent advantages and disadvantages. They are intended to analyze and quantify either the whole proliferating fraction or discrete fractions of cells in specific cell cycle phases, mainly in the S phase. Methodologically, common requirements for clinically useful proliferation markers should be defined by technical–biological effectiveness, in terms of their ability to describe a specific biological phenomenon and to provide results that are informative, reliable, accurate, reproducible within and among the different laboratories at a reasonable cost, and obtainable easily and quickly when needed for clinical decision making.

II. Proliferation Markers

Notwithstanding the general initial skepticism for studies identifying approaches able to adequately reflect the potential proliferative activity of a cell population by a determination carried out on a tumor sample and mainly at only one time in tumor life, which generally corresponds to surgical intervention for diagnosis and/or tumor removal, proliferation indices (assessed in a number of ways including morphometry, immunocytochemistry, flow cytometry, and incorporation techniques) have become widely accepted determinants of prognosis and, more recently, of response to specific treatments in several human solid tumor types (Table I). In this chapter, following a brief description of the several approaches used to evaluate the proliferative activity of human solid tumors (Silvestrini *et al.,* 1995a), we will focus on results from studies in which the relevance on clinical outcome of S-phase markers has been investigated in breast, colorectal, and lung cancers. Those tumor types account for the major causes of death from cancer in Western countries and, owing to their high incidence, frequently represent the objective of translational research.

Table I
Prognostic and Predictive Relevance of Cell Proliferation in Human Solid Tumors

| | | Retrospective studies | | | |
| | | Treatment[a] response | | | |
	Prognosis	CT	RT	ET	Prospective use
Breast	Yes	Yes	?[b]	Yes	Yes
Head and neck	Yes	Yes	Yes		Yes
Lung (NSCLC)[c]	Yes		Yes		
Stomach	Yes				
Colorectum	Yes	Yes	Yes		
Liver	Yes				
Ovary	Yes	Yes?[d]			
Uterine cervix	Yes	Yes	Yes		
Kidney	Yes				
Bladder	Yes	Yes			
Neuroblastoma	Yes	Yes			
Melanoma	Yes				

[a] CT, chemotherapy; RT, radiotherapy; ET, endocrine therapy.
[b] ? indicates data insufficient to draw any conclusion.
[c] NSCLC, non-small cell lung cancer.
[d] Yes? indicates a trend not firmly established because of insufficient studies.

A. Mitosis: Mitotic Index or Mitotic Count

Quantification of cells in the mitotic phase is currently expressed as the number of mitoses per 100 cells or per 10 high-power fields (HPF). These indices, which have long been employed as diagnostic and prognostic tools in the study of tumor pathology, are an important component of all histologic grading systems and are routinely reported by many pathologists. However, although an increased mitotic activity is a frequent finding in aggressive tumors, the validity of these measurements as markers of tumor proliferative activity remains controversial (Quinn and Wright, 1990). Counting mitotic figures represents a simple and highly feasible approach, which, however, does not take into account differences in cell number and size among tumors, and can be affected by a number of factors, including type of fixative, suboptimal or inadequate fixation, nonstandardized section thickness, as well as differences in defining the HPF area among laboratories, that markedly affect the interstudy comparability of results. In addition, besides the poor resolution of the measurement, due to the relatively short time of the M phase (40–60 min) compared to the duration of the entire cell cycle (40–50 hr), the mitotic phase may be of highly variable length, and metaphase arrest may occur in some tumors. There are several cytometric methods, however, that can be used to estimate mitotic index (see Chapter 14).

B. S Phase: Incorporation of DNA Precursors

This approach is based on the use of labeled pyrimidine bases, such as ^{3}H-thymidine (Silvestrini and SICCAB, 1991), or halogenated analogs, such as bromodeoxyuridine (BUdR) and iododeoxyuridine (Gratzner, 1982), which are specifically incorporated into DNA during the S phase. Such measurements involve autoradiographic or immunohistochemical techniques and can be performed on histologic sections or on cytological samples obtained from surgical or biopsy specimens. Quantification of S-phase cells is expressed as the percentage of DNA precursor-incorporating cells over the total number of tumor cells. The main advantages of these approaches are a high feasibility and, as *in situ* procedures, the possibility to discriminate tumor from nontumor cells and to assess tissue heterogeneity. The limitation to a widespread use of these methodologies is the requirement of fresh tumor material, with a sufficient number of viable cells. This constraint has been partially overcome by the availability of kits for ^{3}H-thymidine labeling index (TLI) determination (distributed by Euroframe, Asti, Italy) and from BUdR (distributed by Amersham, Buckinghamshire, England), which guarantee the standardization of the first methodological steps and facilitate the performance even in peripheral institutions. Additionally, a quality control of TLI determination has been ongoing in Italy since 1988 to assure the reproducibility of subjective evaluations, a problem common to all microscopic quantitative determinations (Silvestrini, 1991). The TLI, which is generally considered a troublesome approach, has the advantage of not being influenced by type and time of fixation, of insuring a reliable detection of the labeled cells after long preservation of paraffin blocks, and of offering a clear-cut positive image of reduced silver grains and a persistent record.

C. S Phase: Quantification of Nuclear DNA Content

This approach, which generally provides information on total DNA content and gross genomic abnormalities, can be used to quantify the cells in the different cell cycle phases, and in particular in the S phase (Dean *et al.,* 1982). The determination is based on the knowledge that S-phase cells have a variable DNA content ranging from the presynthetic phase $G_{0/1}$ ($2n$) to the postsynthetic G_2 phase ($4n$). The utilization of dyes that specifically bind DNA [propidium iodide, ethidium bromide, mitramycin, 4,6-diamidino-2-phenylindole (DAPI), acridine orange, Hoechst 33258] allows a quantitation of nuclear DNA content by flow cytometry on isolated nuclei or cell suspensions, or by image (static) cytometry on cytohistological specimens. The results of both approaches is a frequency histogram of DNA content, which is a representation of the cell cycle, in which the fraction of cells in the different phases can be quantified from DNA histograms by computerized cell cycle analysis. Besides the S-phase cell fraction, also the fraction of cells in the $S+G_2M$ phases is considered by some authors as a proliferation index that defines the proportion of cells in the cell cycle excluding those in the $G_{0/1}$ phase. The most diffuse approach for the evaluation of the S-phase cell fraction is flow cytometry (FCM-S), whose main advantage is that a potentially objective evaluation can be made in a short time on a large number of cells obtained from different types of specimens (surgical, biopsy or fine needle aspirates, effusions, and bone marrow aspirates). Such an automated technique received a major impetus in the late 1980s, with the development of procedures to perform flow cytometry in solid tumors using material from formalin-fixed paraffin-embedded blocks (Hedley, 1989) or from frozen tumor specimens. The use of frozen material, maintained in liquid nitrogen or at $-70°C$ until processed, represents an easy and reproducible approach to obtain information on specimens stored for some time and accrued from different centers. The feasibility of FCM-S, which is potentially high, and the quality of results, are markedly affected by several methodological factors. For reproducibility of results and comparison among centers, a standardization of assay methodologies, cell cycle analysis techniques and cutpoints for classifying and interpreting FCM-S from DNA histograms, and a quality control are therefore necessary (Frierson, 1991). The main limit of the FCM-S evaluation, which is common to all the non-*in situ* techniques, is the impossibility to discriminate tumor from nontumor cells. Additional constraints derive from assessability, which is around 75% in solid tumors and which is highly reduced or null in multiploid tumors for the presence of overlapped DNA histograms.

D. Growth Fraction: Determination of Enzymes, Antigens, or Structural Alterations

These approaches [determination of enzymes (DNA polymerase α, thymidine kinase), antigens (Ki-67/MIB-1, KiS1, cyclin-PCNA), or structural alterations (AgNOR, i.e., argyrophilic nucleolar organizer regions)] (Cattoretti *et al.*, 1992; Galand and Degraef, 1989; Gerdes *et al.*, 1984; Howell, 1982; Mushika *et al.*, 1988; Nelson and Schiffer, 1973; Simonsson *et al.*, 1985), which in some instances represent the natural evolution and the integration of morphologic and functional determinations, should provide information on the overall fraction of proliferating cells, that is, the growth fraction of the tumor. However, particularly for solid tumors, the assessment of sensitivity, specificity, and reproducibility of these last approaches is ongoing (Silvestrini *et al.*, 1995a). Available information on their clinical relevance, although interesting, is still controversial and indicates the necessity of a methodological verification and standardization through quality control assessments (Van Diest *et al.*, 1998).

Among proliferation-associated antigens, Ki-67 was regarded as the most reliable marker of proliferating cells. The expression of Ki-67 antigen, detectable by Ki-67 antibody, was first identified in phytohemagglutinin-stimulated lymphocytes and described as putatively expressed only by proliferating cells (Gerdes *et al.*, 1984). The main limitation to its widespread use was its detection by Ki-67 antibody in acetone-fixed frozen sections. However, the more recent availability of a series of reagents—the MIB antibodies, which recognize Ki-67 proliferation antigen and can be used on formalin-fixed paraffin-embedded material following antigen retrieval with pretreatment in a microwave oven or in a pressure cooker (Cattoretti *et al.*, 1992), overcomes the constraint due to the need of frozen specimens. Such a finding opens interesting perspectives for an assessment of the clinical utility of evaluating MIB-1 immunoreactivity on consecutive series of cancers, once assay methodology, fixation, and criteria for the quantification and interpretation of results are accurately standardized.

III. Biological Studies

A general aim of basic studies on human tumors was to define whether and to what extent the rate of tumor cell proliferation is related to the most important conventional pathologic factors such as extension of the disease at diagnosis, that is, at time of clinical detection. Available information shows that cell proliferation of primary breast (Amadori and Silvestrini, 1998; Wenger and Clark, 1998), colon (Costa *et al.*, 1992), and non-small cell lung cancers (NSCLC) (Dalquen *et al.*, 1998) is only weakly related to clinical and pathologic stage, and such a relation is generally observed in large case series in which minimal differences in mean or median TLI or FCM-S within stage subsets reached statistical significance. Overall, such a finding should suggest that proliferative activity is an independent variable, not indicative of the preclinical tumor life. Only for breast cancers is the S-phase cell fraction significantly related to histologic and nuclear grading, with a positive correlation between increased FCM-S or proportion of precursor incorporating cells and worse tumor grade (Amadori *et al.*, 1991; Meyer *et al.*, 1986; Wenger and Clark, 1998). Anyway, the results consistently showed a wide variability in cell kinetics, with overlapped ranges of proliferation indices within each grade subsets. Conversely, a significant, although weak, relation with histologic type has been generally observed in breast and NSCLC, regardless of the proliferation index used to assess the S-phase cell fraction. In fact, in breast cancer, intraductal or infiltrating lobular carcinomas, and medullary carcinomas generally show a lower and higher proliferative activity, respectively, as compared to invasive carcinomas (Amadori *et al.*, 1991; Meyer *et al.*, 1986). However, such a relation is not paralleled by a better long-term prognosis or lower rates of lymph nodal metastasis for patients with lobular breast cancer or by worst long-term prognosis or higher rates of lymph node metastasis for patients with medullary breast cancer compared to patients with other breast histotypes. In NSCLC proliferative activity is significantly higher in patients with squamous cell carcinoma than in those presenting with adenocarcinoma (Costa *et al.*, 1996; Dalquen *et al.*, 1998). Such an association indirectly results in a different prognosis for the two histotypes, owing to a higher clinical radiosensitivity of squamous cell carcinomas compared to adenocarcinomas.

Several studies have analyzed the relation between cell kinetics and biological features common or specific for different tumor types. A general association between ploidy (i.e., nuclear DNA content) and cell kinetics has been reported for all investigated tumor types (Costa *et al.*, 1992; Kallioniemi *et al.*, 1988; Wenger and Clark, 1998) when the proliferative activity was defined as the FCM-S phase. In fact, aneuploid tumors are characterized by a higher S-phase value than diploid tumors. Conversely, only a weak relation, or no relation at all, was observed when TLI was considered (Silvestrini *et al.*, 1993a). This finding represents indirect evidence, which emerged also from other basic and clinical studies (Costa *et al.*, 1992; Meyer and Coplin, 1988; Silvestrini *et al.*, 1995a), that the two cell kinetic variables which evaluate the fraction of S-phase cells through a phenomenological detection (FCM-S) or through an active incorporation (TLI) provide different biological information and thus can be only roughly considered as alternatives.

As regards to biological variables that are tumor-type specific, several studies have shown a relation between S-phase cell fraction and estrogen (ER) or progesterone receptor (PgR) levels in about two-thirds of breast cancers (Amadori and Silvestrini, 1998; Dressler *et al.*, 1988; Meyer *et al.*, 1986; Wenger and Clark, 1998). In particular, a high proliferative activity is associated with the lack of ER in about 50% of cases, whereas a low proliferative activity is associated with the presence of ER in about 70% of cases. This correlation is less evident for PgR. In colorectal cancer, cell proliferation has been analyzed in relation to CEA, a marker whose prognostic relevance is still controversial but whose usefulness in monitoring tumor progression seems to be widely accepted. No relation was evidenced between the S-phase cell fraction, as expressed by TLI or FCM-S, and CEA serum or tissue levels determined at surgery in previously untreated patients (Costa *et al.*, 1992).

IV. Clinical Studies

Translational research on human tumors has developed along two main paths: (1) to search for markers to use as a complement to clinicopathologic staging in order to identify patients destined to relapse with progression independent of treatment, and (2) to predict patients likely to be responding to or developing resistance to a specific treatment (Tumor Marker Expert Panel Members, 1996). At present, the contribution of the different cell proliferation indices has been defined or is still under investigation for the identification of patients at high risk of relapse or death (who need aggressive treatments) and for patients with an indolent disease (who are potentially curable by local–regional treatment alone).

The interest of investigators involved in translational studies has been directed to analyze the role of proliferation indices as indicators of response to chemical and physical treatments. Preliminary evidence, which confirms previous studies on advanced tumors, has prospected a potential relevance to cell proliferation, mainly in breast cancer (Amadori and Silvestrini, 1998; Gardin *et al.*, 1994; Hietanen *et al.*, 1995; Stål and Nordenskjold, 1994). However, most of these studies have been retrospectively performed in adjuvant settings in which the advantage of a long-term follow-up could be counterbalanced by a marked heterogeneity in technical and analytical procedures for biomarker determination. More recently, the determination of cell proliferation has been prospectively planned within the context of adjuvant and neoadjuvant treatment protocols, in which the evaluation of utility of biological information accounted for a secondary objective of the clinical study. Although such studies have not been specifically designed to test the predictivity of proliferation indices, it is likely that they will improve the quality of information on the predictive accuracy and provide a definite evaluation of the clinical utility of a cell kinetic characterization.

A. Proliferation Markers and Prognosis

Much effort has been devoted to investigate whether proliferative activity of a tumor cell population, which is independent of important prognostic factors such as clinical or pathologic stage, could provide clinically relevant information even in the presence of other prognostic factors. Cell proliferation has generally emerged as an important indicator of clinical outcome in several tumor types (Table I) (Silvestrini, 1994a; Tubiana and Courdi, 1989; Van Diest *et al.*, 1998), and results indicate a worse clinical outcome for patients with rapidly proliferating tumors than for those with slowly proliferating tumors, in keeping with the hypothesis of a different metastatic potential for tumors with different proliferative activity. This is a common finding even for tumors with a different natural history and impact of conventional prognostic factors. Some contrasting results are probably due to heterogeneity of case series, in terms of stage and treatment, and to differences in sample size, but also to lack of methodological standardization and quality controls, and to variability in result analysis and in cutpoints used for classifying and interpreting S-phase results. This evidence emphasizes the importance to activate and maintain quality control programs which are ongoing for some cell kinetic variables (Silvestrini and SICCAB, 1991; Silvestrini, 1994b), and to establish guidelines for comparison of the results (Altman and Lyman, 1998; McGuire, 1991).

In particular, the prognostic relevance of cell kinetic variables should first be investigated in phase I exploratory studies (hypothesis generating) on series of patients possibly treated with local–regional therapy alone until relapse (i.e., out of any interference by systemic therapy on subclinical metastases) and adequate for size and follow-up. Interlaboratory consistency of the results needs to be verified, and the relative prognostic contribution of investigated variables evaluated by multivariate analysis, including pathologic and biological consolidated prognostic factors in phase II exploratory studies (Altman and Lyman, 1998). The newly proposed prognostic variables should then be validated in confirmatory phase III studies and their clinical impact assessed for the usefulness in making a therapeutic decision in the management of individual patients only when such criteria are fulfilled. In these phases, all the possible common causes of variation among studies investigating prognostic factors, including sample size, ancillary variables, differences in assay techniques, cutpoints, subsets of patients, study end-points, and statistical approaches, should be taken into consideration. Moreover, besides clinicobiological effectiveness and usefulness, defined as the ability of the marker to describe a biological process and its impact on clinical outcome when it influences the choice of therapy, even laboratory effectiveness (in terms of presence of quality assurance programs and methods comparison for any analyte) should be considered to improve the diagnostic armamentarium in oncology (Gion *et al.*, 1999).

Breast cancer, even at an early stage, is a heterogeneous disease, in which the presence of several molecular mechanisms affecting tumor growth, progression and metastatic potential, and the tendency to acquire drug resistance limit the prognostic value of TNM. Early studies on proliferation markers, independently carried out in the early 1980s using incorporation techniques by Italian (Gentili *et al.*, 1981; Silvestrini *et al.*, 1986), French (Tubiana *et al.*, 1984), and American groups (Meyer *et al.*, 1983) on patients with stage I tumors or at all the stages but treated with local–regional treatment alone (surgery and/or radiotherapy), showed that the relapse rate is at least two times higher for women with rapidly proliferating tumors than for women with slowly proliferating tumors and death rate was from two to more than four times higher, probably as a function of treatments given after relapse (Table II). These results have been confirmed in more recent studies on large series and /or with a longer follow-up by the same (Meyer and Province, 1988; Silvestrini *et al.*, 1989, 1995b; Tubiana *et al.*, 1987) or other groups (Héry *et al.*, 1987; Paradiso *et al.*, 1991) for patients with operable or advanced breast cancer treated with local–regional therapy alone, regardless of whether TLI was used either as a continuous or as a dichotomous variable and regardless of the length of follow-up (Silvestrini *et al.*, 1997). The only discordant results were ascribed by the authors (Cooke *et al.*, 1992) to the lack of quality control of TLI determination. Multiple regression analyses have shown that TLI maintains its independent prognostic role as an indicator of relapse and death also in the presence of important factors such as tumor size, DNA ploidy, p53 and bcl-2 expression, ER and PgR status, and histologic or nuclear grade (Meyer and Province, 1988; Silvestrini *et al.*, 1989, 1993a, 1994), and of other cell kinetic variables (Silvestrini *et al.*, 1993a; Rudas *et al.*, 1994). Moreover, analysis on the same series of patients showed that TLI is a further prognostic discriminant within subsets with a favorable prognosis, that is, among patients with small, diploid, or ER^+ tumors. Such findings support the hypothesis that the contribution of a strong prognostic marker, such as cell proliferative activity, could be further potentiated by considering it in association with weaker biological or pathologic prognostic factors that are independent or only partially related. In fact, a gradual increase in risk of relapse was observed from tumors presenting with two favorable (low TLI and small size, euploidy or presence of ER) to those with two unfavorable prognostic factors (high TLI and large size, aneuploidy or absence of ER). In order to provide even more clinically useful information, techniques for tree-structured regression analysis, which represents a formal

Table II
Evaluation of the Prognostic Value of S-Phase Fraction in Patients with Stage I Breast Cancer

Authors (year)	Marker	Follow-up (years)	Relapse (%)			Death (%)		
			Low S	High S	p value	Low S	High S	p value
Tubiana et al.(1984)[a]	TLI	10	25	58	<0.02	30	57	<0.02
Tubiana et al. (1987)	TLI	14	25	70	<0.01	30	60	<0.01
Héry et al. (1987)	TLI	8	17	44	0.02	0	64	<0.001
Meyer et al. (1983)	TLI	4	15	50	0.0001			
Meyer and Province (1988)	TLI	5	20	40	<0.001	11	35	0.001
Gentili et al. (1981)	TLI	4	0	67	<0.0005			
Silvestrini et al. (1986)	TLI	5	15	38	0.0002	6	18	0.005
Silvestrini et al. (1989)	TLI	6	22	40	<0.0001	5	18	<0.0001
Silvestrini et al. (1995b)	TLI	8	30	43	$<10^{-6}$			
Paradiso et al. (1991)	TLI	5	15	30	0.03	9	22	0.05
Cooke et al. (1992)	TLI	10	38	50	NS			
Silvestrini et al. (1993a)	FCM-S	4	23	23	NS			
Muss et al. (1989)	FCM-S	5	22	33	NS	13	30	0.04
Sigurdsson et al. (1990)	FCM-S	4	11	31	0.01	1	22	0.01
Clark et al. (1989)[b]	FCM-S	5	10	30	0.01	10	15	0.01
Clark et al. (1992)[c]	FCM-S	5	15	32	0.0004			NS
O'Reilly et al. (1990a)[d]	FCM-S	5	22	48	0.006			
Fisher et al. (1991)	FCM-S	10	54	73	0.003	40	60	0.005
Stanton et al. (1992)	FCM-S	10				36	48	NS

[a] Also included a few patients with stage II and III tumors, given only local–regional treatment.
[b] Only on diploid tumors.
[c] Only on ER$^+$, small tumors.
[d] Only on aneuploid tumors.

method for optimal categorization of continuous variables such as TLI, were used to identify subgroups at different risk of locoregional relapse or distant metastasis (Silvestrini *et al.*, 1995b). As regards local recurrence, patients with slowly proliferating tumors showed the lowest risk regardless of age, similar to that observed for older patients with rapidly proliferating tumors. The risk was more than double for younger patients with rapidly proliferating tumors, whereas size and steroid receptor status were not selected by the model as important predictors of local–regional recurrence. For distant metastases, TLI was an important discriminant within the subset of 1–2 cm tumors, which represents a large fraction of all breast cancers, and the risk of relapse for slowly proliferating lesions of this size subset was similar to that observed for tumors smaller than 1 cm.

Due to the requirement of fresh and adequate material and for the time-consuming microscopic analysis for ^{3}H-thymidine or bromodeoxyuridine labeling index determination, FCM evaluation of the S-phase cell fraction has been intensively pursued. In breast cancers not all the results obtained from comparable series of node-negative tumors have ascribed a prognostic role for this cell kinetic variable (Table II). In fact, according to different studies, the prognostic relevance was observed in diploid (Clark *et al.*, 1989) and only successively in aneuploid tumors (Clark *et al.*, 1992), only in aneuploid tumors (O'Reilly *et al.*, 1990a), or on the overall series (Fisher *et al.*, 1991; Sigurdsson *et al.*, 1990), whereas some studies failed to detect any relevance even when using several models to quantify the S-phase cell fraction (Muss *et al.*, 1989; Silvestrini *et al.*, 1993a; Stanton *et al.*, 1992). Such discrepancies could be attributed to case series selection owing to the exclusion of DNA plots with a high variation coefficient of the $G_{0/1}$ peak and debris background, to the inability of many models to quantify S-phase cell fraction for tumors with more than one cell subpopulation, or to the variability among laboratories regarding assay methodologies and result interpretation. However, the prognostic value of FCM-S, even though moderate and declining over time (Hilsenbeck *et al.*, 1998), clearly emerged from an overall review of 1987–1997 literature on flow cytometric measurements (Wenger and Clark, 1998). In fact, high FCM-S was associated with decreased relapse-free survival in 21 of 24 studies in univariate analysis, and in 17 of the 21 studies in which multivariate analysis was performed. Similarly, high FCM-S was associated with decreased overall survival in 18 of 20 studies in univariate analysis, and in 14 of the 19 studies in multivariate analysis. In addition, prognostic information provided by FCM-S was also independent of those provided by other kinetic variables, such as MIB-1 (Brown *et al.*, 1996).

In some instances, however, information present in the literature does not allow one to separately analyze the role of cell proliferation as a function of tumor stage or treatment. In general, as clearly shown by colorectal cancer (Table III) and NSCLC (Table IV), even in these clinical situations cell proliferation eventually emerged as an indicator of survival, notwithstanding the impact of therapy on the natural history. This finding, which has been reported for the majority of tumor types regardless of stage and treatment (Schipper *et al.*, 1998; Silvestrini, 1994a; Tubiana and Courdi, 1989; Van Diest *et al.*, 1998), indicates that the original biological aggressiveness, which could be reasonably represented by cell kinetics, generally prevails over treatment response as regards to long-term clinical outcome. Moreover, as an indicator of long-term relapse-free or overall survival, cell kinetics maintains its significance even in the presence of prognostic information provided by variables that are generally relevant for all neoplasms (i.e., patient age, pathologic stage, grade of differentiation) or specific for some histotypes (i.e., histology, hormone receptors). However, these positive findings do not allow one to neglect publication bias similar to those affecting meta-analysis of prognostic factor studies (Altman and Lyman, 1998) or to overcome the need for a cautious interpretation of clinical outcome as a function of cell kinetics owing to the dual, and in some instances opposite, role of this marker that indicates biological aggressiveness but also, possibly, susceptibility to specific treatments.

For colorectal cancer, as for other solid tumors, surgical resection represents the best treatment option, but it does not prevent the risk of relapse, which

Table III

Evaluation of the Prognostic Value of S-Phase Fraction in Patients with Primary Colorectal Cancer

Authors (year)	Dukes' stage	Follow-up (years)	Death (%)		p value
			Low S	High S	
Bauer et al. (1987)	A–D[a]	5	33	40	0.003
	A,B	5	18	57	0.019
Scott et al. (1987)	A–D[b]	4	10	40	0.03
Quirke et al. (1987)	A–D[b]	5	30	60	0.002
Auvinen et al. (1994)	A–D[a]	10	49	67	0.05
Cascinu et al. (1998)	A	5[c]	10	28	0.05
Pietra et al. (1996)	A–C	2.5	15	62	0.0001
Cosimelli et al. (1998)	A–C	5[c]	38	43	NS
Witzig et al. (1991)	B_2,C	5	25	48	0.001
Ahnen et al. (1998)	B,C	7	48	47	NS
Schutte et al. (1987)	C	5	32	67	0.01
Harlow et al. (1991)	C[b]	5	20	60	0.02

[a] Colon cancer.
[b] Rectal cancer.
[c] Relapse (%).

Table IV

Evaluation of the Prognostic Value of S-Phase Fraction in Patients with Primary Non-Small Cell Lung Cancer

Authors (years)	Histology	Stage	Marker	Follow-up (years)	Death (%)		p value
					Low S	High S	
Alama et al. (1990)	All	I-III	TLI	3	35	78	0.04
Volm et al. (1985b)	All	I-III	TLI	3	52	78	0.04
Matturri et al. (1994)	All	I-III	TLI	4	41	89	0.01
Silvestrini et al. (1991b)	All	I	TLI	5	21	59	0.04
Costa et al. (1996)	AC[a]	II	TLI	3[b]	25	70	0.02
		III	TLI	3[b]	50	85	0.0001
Ten Velde et al. (1988)	All	I–III	FCM-S	2	75	84	0.04
	SCC[c]	I–III	FCM-S	2	70	100	0.02
Volm et al. (1985a)	All	I–III	FCM-S	3	58	90	0.01
	SCC[c]	I–III	FCM-S	5	42	80	0.05
Dalquen et al. (1998)	All	I–III	FCM-S	10	79	80	NS
Filderman et al. (1992)	All	I	FCM-S	5	7	79	0.0001

[a] Adenocarcinoma.
[b] Relapse (%).
[c] Squamous cell carcinoma.

depends on degree of local invasion, spread to regional lymph nodes, and tumor site and differentiation (Johnston and Allegra, 1995). Alterations in cell proliferating compartments have been shown to be the early event of colon mucosa transformation (Bosman, 1995; Gilliland *et al.*, 1996), and an increase in proliferating or S-phase cell fraction has been generally shown to be related to poor prognosis. Most studies on the cell proliferative status of colorectal cancer have analyzed the distribution of cells in the different cell cycle phases by FCM. Results have generally shown the prognostic relevance of cell proliferation indices (FCM-S or FCM-S+G_2M), regardless of the mathematical method used to quantify the cell cycle fractions, and their independence of stage in predicting clinical outcome (Table III). A better prognosis was generally observed (Ahnen *et al.*, 1998; Cosimelli *et al.*, 1998; Pietra *et al.*, 1996), for patients with slowly proliferating than for those with rapidly proliferating lesions on the overall population of colon (Auvinen *et al.*, 1994; Bauer *et al.*, 1987) or rectal cancers (Quirke *et al.*, 1987; Scott *et al.*, 1987) and within stage subsets (Harlow *et al.*, 1991; Schutte *et al.*, 1987; Witzig *et al.*, 1991), and such findings held true even in multivariate analysis (Quirke *et al.*, 1987). In a more recent study, Cascinu *et al.* (1998) observed an association between high FCM-S and poor prognosis also in patients with stage A disease, and emphasized the need for adjuvant chemotherapy for patients classified at a poor prognosis only on the basis of tumor proliferation.

Also for resectable NSCLC, surgery represents the treatment of choice. In fact, adjuvant treatments including chemotherapy and radiotherapy have been extensively used, but no definitive conclusions have been reached about their effectiveness (Le Chevalier *et al.*, 1991; Marino *et al.*, 1994). The S-phase cell fraction—defined as TLI or FCM-S—has been consistently shown to be a prognostic factor in series including all stages of NSCLC (Table IV), regardless of histology (Alama *et al.*, 1990; Matturri *et al.*, 1994; Ten Velde *et al.*, 1988; Volm *et al.*, 1985a,b) as well as in the subgroup of patients with squamous cell carcinoma (Ten Velde *et al.*, 1988; Volm *et al.*, 1985a). In stages II–III NSCLC patients, TLI has an additive prognostic role for adenocarcinoma but not for squamous cell carcinoma (Costa *et al.*, 1996), thus suggesting that cell proliferation does not provide prognostic information in patients with operable locally advanced disease and favorable histotype. Studies performed by Filderman *et al.* (1992) and by our group (Silvestrini *et al.*, 1991b) demonstrated that FCM-S and TLI provide significant prognostic information and identify subgroups with different risk of relapse among patients with stage I NSCLC. However, Dalquen *et al.* (1998) in a substantial series of NSCLC patients failed to observe any relation between FCM-S and prognosis, in contrast with previous findings, probably because of the markedly stringent criteria adopted by these authors for interpretation and analysis of DNA histograms.

B. Proliferation Markers and Response to Systemic Treatments

The search of biological predictive factors, that is, markers able to identify patients more or less likely to benefit from therapy, has received a renewed emphasis. However, this field of translational research is more difficult to be investigated compared to that of prognostic factors (Henderson and Patek, 1998). In fact, the ideal study to assess the role of any biological variable as a predictor of response to a specific treatment should request the prospective evaluation of the investigated marker within the context of a randomized clinical study specifically designed to analyze biomarker predictivity, or to compare systemic with local–regional therapies. However, the few prospective, high-powered studies that address the issue of biomarker utility are still ongoing, and present results only rarely derive from therapeutic clinical trials with follow-up information already available, even including the biological characterization as an ancillary study. In most of the reports from which present information on the association between proliferative activity and treatment response is derived, tumor specimens have been collected for a variety of reasons and were available for determining proliferation indices without any a priori study design.

Proliferative activity is the typical biomarker that may be both prognostic and predictive. However, as for most prognostic factors, present data related to proliferative activity are insufficient to draw firm conclusions regarding its predictive role in choosing either endocrine or chemotherapy, but only suggestive of a relation that should be further investigated on independent adjuvant settings and analyzed with techniques appropriately designed to determine the clinical utility of biomarkers. In the two successive sections, available information on the relation between proliferation and treatment response in advanced and adjuvant settings is reported and discussed for breast cancer, the tumor type for which such an aspect of translational studies has been and is deeply investigated.

1. Chemotherapy

An overview of basic and clinical information shows no correlation between median TLI values for different tumor types and sensitivity to chemical therapy (Amadori and Silvestrini, 1998). This finding should indicate the lack of a direct relation in human tumors between cell proliferation and response to drugs, which, conversely, has been observed for experimental tumors. Natural drug resistance or inadequate treatment in terms of type and intensity, might be equally responsible for the absence of such a relation. However, emerging evidence from retrospective studies should suggest specific quantitative interaction in subgroups of patients with potentially sensitive tumors adequately treated.

The retrospective analysis of the relation between cell kinetic variables and response to clinical treatment was carried out on advanced breast cancer even taking advantage of the Tumor Marker Utility Grading System-Plus proposed by Hayes *et al.* (1998), in which the relative predictive strength of proliferative activity can be estimated and expressed in terms of benefit ratio. A relation between TLI and objective clinical response was observed as a function of the type of polychemotherapy used (Table V). In fact, a similar response rate, with a benefit ratio of 1, was observed following Adriamycin and vincristine (AV) treatment in patients with slowly or rapidly proliferating tumors (Silvestrini *et*

Maria Grazia Daidone et al.

Table V

Evaluation of the Predictive Value on Treatment Response of S-Phase Fraction in Patients with Advanced Breast Cancer

Authors (year)	Number of cases	Marker	Treatment	Drugs	Objective clinical response Advantage for	Benefit ratio[a]	p value
Silvestrini *et al.* (1987)	52	TLI	CT[b]	AV[c]	None	1.0	NS
Sulkes *et al.* (1979)	25	TLI	CT	FAC±V[d]	High S	4.6	0.01
Amadori *et al.* (1997)	76	TLI	CT	CMF,[e] FAC	High S	3.1	0.01
Remvikos *et al.* (1989)	60	FCM-S	CT	FAC	High S	1.9	0.004
Meyer and Lee (1980)	20	TLI	ET[f]	Various	Low S	6.0	0.05
Paradiso *et al.* (1990)[g]	29	TLI	ET	Tamoxifen	Low S	1.9	0.04
Amadori *et al.* (1994)[g]	43	TLI	ET	Tamoxifen	Low S	2.9	0.003

[a] The benefit ratio is defined as the likelihood of achieving objective clinical response (i.e., tumor reduction $\geq 50\%$) from chemotherapy or endocrine therapy for patients with tumors belonging to the biological category for which a beneficial outcome after treatment was observed (Hayes *et al.*, 1998). The benefit ratio can be calculated by dividing the response rate in those patients who fall into the favorable category (high S following chemotherapy or low S following endocrine therapy) by that for those who fall into the unfavorable category (low S following chemotherapy or high S following endocrine therapy). In particular, from the study of Remvikos *et al.* (1989), a benefit ratio following chemotherapy equal to 1.9 has been obtained by dividing the 89% response rate of patients presenting with rapidly proliferating tumors by the 46% response rate of patients presenting with slowly proliferating tumors. Similarly, from the study of Meyer and Lee (1980), a benefit ratio following endocrine therapy equal to 6.0 has been obtained by dividing the 60% response rate of patients presenting with slowly proliferating tumors by the 10% response rate of patients presenting with rapidly proliferating tumors.

[b] CT, chemotherapy.

[c] AV, Adriamycin + vincristine.

[d] FAC±V, 5-fluorouracil + Adriamycin + cyclophosphamide ± vincristine.

[e] CMF, cyclophosphamide + metothrexate + 5-fluorouracil.

[f] ET, endocrine therapy.

[g] Including only ER$^+$ tumors.

al., 1987). Conversely, in advanced tumors treated with 5-fluorouracil, Adriamycin, and cyclophosphamide (FAC), with or without vincristine, the clinical response was more than four times higher for rapidly proliferating tumors than for slowly proliferating tumors (Sulkes *et al.*, 1979). These data have been confirmed on a series of patients with locally advanced or metastatic breast cancer treated with FAC or cyclophosphamide, methotrexate, and 5-fluorouracil (CMF) (Amadori *et al.*, 1997; Remvikos *et al.*, 1989). Such findings should indicate that metastatic and locally advanced rapidly proliferating tumors benefit from intensive polychemotherapy including S-phase specific agents. Conversely, in slowly proliferating breast cancers, an intensive treatment does not add any therapeutic benefit compared to AV, which can be considered a monochemotherapy, since experimental and clinical results have shown that these two drugs are cross-resistant. The major benefit of polychemotherapy including S-phase specific drugs for patients with rapidly proliferating tumors, which has emerged from retrospective analyses on advanced tumors, has also been confirmed in adjuvant settings for node-positive resectable cancers (Hietanen *et al.*, 1995; O'Reilly *et al.*, 1990b; Stål and Nordenskjold, 1994). In addition to the type of drugs included in the clinical treatment, dose intensity also proved to be an important determinant of treatment efficacy in rapidly proliferating tumors (Kute *et al.*, 1995). This evidence emerged even from a series of patients with node-positive tumor, for whom the high biological aggressiveness associated with rapid cell proliferation was reduced when full dose chemotherapy was administered, whereas it dramatically emerged when less intensive doses were given (Daidone *et al.*, 1989). In an innovative clinical protocol on node-negative, high-risk ER-negative tumors, patients were randomized to receive surgery alone or surgery plus adjuvant CMF (Bonadonna *et al.*, 1986). A retrospective analysis as a function of TLI in patients from the adjuvant compared to those from the control arm showed a major benefit in relapse-free survival for women with rapidly proliferating tumors and a moderate advantage for those with slowly proliferating tumors (Zambetti *et al.*, 1992).

More recently, ncoadjuvant chemotherapy protocols became available as an innovative approach in translational studies, in which they could represent an ideal model to evaluate the clinical impact of a biological investigation, to analyze the predictivity of the biological variables on the different clinical end points, and to monitor, at the cellular and molecular level, treatment effect by sequential determinations of biomarkers within a single tumor, in the presence of only intralesional heterogeneity (Daidone *et al.*, 1999). Overall, proliferative activity appears as a biomarker which can provide information on tumor biological changes and which is sensitive enough to reflect at a cellular level the biological downstaging induced by treatment (Daidone *et al.*, 1990; Gardin *et al.*, 1994), with direct implications on long-term follow-up. In fact, significant changes after treatment were mainly observed in markers of proliferation, and generally consisted in a reduction of proliferative activity (Baldini *et al.*, 1996; Briffod *et al.*, 1995; Collecchi *et al.*, 1998; Daidone *et al.*, 1995). Tumor shrinkage proved to be less frequent in patients presenting before treatment with slowly proliferating tumors (Baldini *et al.*, 1996; Briffod *et al.*, 1995; Chevillard *et al.*, 1996; Collecchi *et al.*, 1998; Pierga *et al.*, 1997; Rozan *et al.*, 1998), whereas a favorable long-term clinical outcome was generally, although not univocally (Rozan *et al.*, 1998), observed for patients with posttreatment indolent tumors, that is slowly proliferating (Collecchi *et al.*, 1998; Remvikos *et al.*, 1997), with wild-type TP53 (Aas *et al.*, 1996), weakly or not expressing p53 (Aas *et al.*, 1996; Honkoop *et al.*, 1998), or HER2 (MacGrogan *et al.*, 1996).

The outcome from retrospective analyses in which cell proliferation appears to provide both prognostic and predictive information prompted the activation of two prospective multicentric clinical trials in Italy on node-negative operable breast cancers in which patients that were candidates for adjuvant chemotherapy were selected on the basis of TLI (Amadori *et al.*, 1994; Paradiso *et al.*, 1993). These randomized studies, which can be considered as the phase III confirmatory studies described by Altman and Lyman (1998), represent an important challenge to prospectively verify the possibility of modulating treatment intensity on the basis of biological indicators of risk and of tailoring clinical treatment on the basis of biological putative predictors of response to different types of therapy, and to start a definite assessment of the clinical usefulness of determining cell proliferation in human solid tumors. Results from the first of these prospective studies provide convincing evidence in favor of a survival advantage of CMF treatment for patients with rapidly proliferating tumors (Amadori *et al.*, 2000).

2. Endocrine Therapy

With the methodology proposed by Hayes *et al.* (1998) to assess the benefit ratio, already applied to chemotherapy results, proliferative activity has been also investigated as a predictor of response to hormonal treatment (Table V). Studies carried out on patients with advanced breast cancer shows a higher response rate to various endocrine treatments for slowly proliferating tumors than for rapidly proliferating tumors (Meyer and Lee, 1980). In two subsequent studies on patients with ER^+ metastatic or advanced disease, mainly treated with tamoxifen, the response rate was still significantly higher (two to three times) in slowly proliferating tumors than in rapidly proliferating tumors (Amadori *et al.*, 1994; Paradiso *et al.*, 1990). Such findings, which held true regardless of PgR status, clearly and consistently show that slowly proliferating metastatic ER^+ breast cancers benefit from tamoxifen, whereas in rapidly proliferating tumors response to endocrine treatment was moderate, even if tumor lesions are ER^+, that is, traditionally considered hormone responsive. Therefore, the presence of a rapidly proliferating and likely ER-negative cell population is a limiting factor for response to endocrine therapy and may be responsible for the failure of some ER^+ lesions to take advantage of hormonal treatment.

The evidence for a major benefit from endocrine therapy for slowly proliferating tumors has been confirmed also in adjuvant settings (Silvestrini *et al.*, 1993b; Wenger and Clark, 1998). In particular, in a large series of postmenopausal patients with ER^+ tumors treated with tamoxifen for at least 1 year, 6-year relapse-free survival was about two times lower in patients with slowly rather than rapidly proliferating tumors, regardless of extent of nodal involvement, tumor size, PgR, p53, and bcl-2 expression (Silvestrini *et al.*, 1996). These data were confirmed in elderly patients (Daidone *et al.*, 2000), for whom, besides a low proliferative rate, absence or low levels of p53 expression and bcl-2 overexpression were indicative of a favorable clinical outcome.

All these data have been obtained from retrospective clinical analyses, and prospective studies are needed to confirm them and to define whether the improved clinical outcome for patients with ER^+, slowly proliferating breast cancer, is due to their natural indolence or to a peculiar susceptibility to endocrine treatment.

V. Conclusions on First Generation Translational Studies with Proliferation Markers

In neoplastic cell populations, proliferative activity represents a functional characteristic, which is expressed in the absence or in the partial lack of control mechanisms, and proved to be related to biological and clinical aggressiveness in different tumor types. However, notwithstanding, these findings become available for the diverse proliferation indices on substantial case series from different tumor types during the 1990s, the validation process to which information provided by proliferation indices is submitted, as well as the critical assessment of the clinical effectiveness and usefulness are still ongoing. In fact, in 1996 the Expert Panel Members of the American Society of Clinical Oncology (ASCO), while proposing guidelines for the use of tumor markers in breast and colorectal cancers, did not ascribe to measurements of proliferative activity a validated role in clinical practice. The final report (Tumor Marker Expert Panel Members, 1996) stated that present data are insufficient to recommend the routine use of S-phase determination (mainly by FCM) for assigning patients to prognostic groupings, for selecting the type of adjuvant therapy as well as for selecting among different treatment options of metastatic disease. However, "these guidelines are intended for use in the care of patients outside of clinical trials." Within clinical trials, conversely, the ASCO Guidelines Panel recognized the potential usefulness of cell proliferation information, although they suggested standardization of assay techniques and planning of true confirmatory studies, possibly prospective, in well-defined low-risk breast cancer populations.

Such an area for future research is already ongoing, since cell kinetic characterization is prospectively planned in phase III confirmatory studies, included in those activated in the late 1980s in Italy (Amadori *et al.*, 1994; Paradiso *et al.*, 1993). Such research protocols are aimed to validate an a priori hypothesis attempting to use proliferation indices to discriminate between patients at high and low risk of disease progression or death and to evaluate with particular emphasis by the methodology of randomized treatment protocols and evidence-based medicine whether (and which) subset of patients classified on the basis of tumor kinetic features is likely to benefit from treatment. Once the clinical utility of a cell kinetic characterization is assessed, studies should be focused on the development of models in which proliferation indices will be introduced and combined with other clinically relevant variables in an attempt to maximize the ability to predict outcome for groups or for individual patients.

In parallel with translational studies in which first-generation proliferation indices were investigated as prognosticators or predictors of treatment response, a considerable effort has been and is being addressed to elicit mechanisms involved in the control and regulation of the cell cycle, to better understand tumorigenesis and to identify novel therapeutic targets. The identification of positive and negative regulators of the cell cycle engine, including cyclins, cyclin-dependent kinases (CDKs) and CDK inhibitors (CKIs), and the recent availability of antibodies raised against them and suitable to detect their expression in paraffin-embedded archival specimens of clinical tumors favored the proliferation of studies in which these markers have been investigated for prognostic and therapeutic purposes. As expected, in clinical tumors, proteins involved in driving cell proliferation are frequently overexpressed, whereas those which restrain cell proliferation, such as the classes of CKIs belonging to the Kip/Cip family (p21, p27, p57) as well as those comprised in the INK4 family (p15, p16, p18, and p19) are frequently inactivated. Cyclin overexpression and/or CKI underexpression are generally, although not univocally, associated with poor prognosis and aggressive phenotype. However, translational studies involving such new generation of proliferation-related proteins are extremely complex. In fact, the absolute number of positive and negative modulators of cell cycle checkpoints, the presence of complex interplays among them, and, most importantly, the possibility that cancer cells could partially compensate for the deregulated expression of one of these proteins (Gillett and Barnes, 1998) raise some doubts about the possibility to understand the network of alterations determining an abnormal regulation of the cell cycle and to investigate their role on tumor progression by studying individual factors in clinical specimens. It is conceivable that a multifactorial evaluation of proteins controlling G_1 and/or G_2 checkpoints is more informative than considering each factor one at a time, even though it requires the use of complex statistical approaches to investigate the interrelation among variables. In addition, also these variables should undergo the translational validation process described by Altman and Lyman (1998), which includes exploratory and confirmatory phase I, II, or III studies.

Acknowledgments

Supported in part by grants from the Italian Association for Cancer Research (AIRC), the National Research Council (CNR), and the Italian Health Ministry. We thank B. Canova for editorial assistance, and R. Motta, L. Ventura, R. Erdas, and G. Abolafio for their skilled technical collaboration.

References

Aas, T., Børrresen, A. L., Geisler, S., Smith Sorensen, B., Johnsen, H., Varhaug, J. E., Akslen, L. A., and Lonning, P. E. (1996). Specific P53 mutations are associated with de novo resistance to doxorubicin in breast cancer patients. *Nat. Med.* **2,** 811–814.

Ahnen, D. J., Feigl, P., Quan, G., Fenoglio-Preiser, C., Lovato, L. C., Bunn, P. A., Jr., Stemmerman, G., Wells, J. D., Macdonald, J. S., and Meyskens, F. L., Jr. (1998). Ki-ras mutation and p53 overexpression predict the clinical behavior of colorectal cancer: A Southwest Oncology Group Study. *Cancer Res.* **58,** 1149–1158.

Alama, A., Costantini, M., Repetto, L., Cone, P. F., Serrano, J., Nicolin, A., Barbieri, F., Ardizzoni, A., and Bruzzi, P. (1990). Thymidine labelling index as prognostic factor in resected non-small cell lung cancer. *Eur. J. Cancer* **26,** 622–625.

Altman, D. G., and Lyman, G. H. (1998). Methodological challenges in the evaluation of prognostic factors in breast cancer. *Br. Cancer Res. Treat.* **52,** 289–303.

Amadori, D., and Silvestrini, R. (1998). Prognostic and predictive value of thymidine labelling index in breast cancer. *Breast Cancer Res. Treat.* **51,** 267–281.

Amadori, D., Bonaguri, C., Nanni, O., Gentilini, P., Lundi, N., Zoli, W., Riccobon, A., Vio, A., and Saragoni, A. (1991). Cell kinetics and hormonal features in relation to pathological stage in breast cancer. *Breast Cancer Res. Treat.* **18,** 19–25.

Amadori, D., Volpi, A., Callea, A., Amaducci, L., Morgagni, S., Magni, E., and Nanni, O. (1994). Clinical relevance of cell kinetics in breast cancer. *Ann. N.Y. Acad. Sci.* **698,** 186–192.

Amadori, D., Volpi, A., Maltoni, R., Nanni, O., Amaducci, L., Amadori, A., Giunchi, D. C., Vio, A., Saragoni, A., and Silvestrini, R. (1997). Cell proliferation as a predictor of response to chemotherapy in metastatic breast cancer: A prospective study. *Breast Cancer Res. Treat.* **43,** 7–14.

Amadori, D., Nanni, O., Marangolo, M., Pacini, P., Ravaioli, A., Rossi, A., Gambi, A., Catalano, G., Perroni, D., Scarpi, E., Casadei-Giunchi, D., Tienghi, A., Becciolini, A., and Volpi, A. (2000). DFS advantage of adjuvant CMF in node-negative rapidly proliferating breast cancer patients: A randomised multicentre study. *J. Clin. Oncol.* in press.

Auvinen, A., Isola, J., Visakorpi, T., Koivula, T., Virtanen, S., and Hakama, M. (1994). Overexpression of p53 and long-term survival in colon carcinoma. *Br. J. Cancer* **70,** 293–296.

Baldini, E., Giannessi, P. G., Collecchi, P., Naccarato, A. G., Passoni, A., Bevilacqua, G., and Conte, P. F. (1996). Effects of primary chemotherapy on proliferative activity, IGF-1R and bcl2 expression in locally advanced breast cancer (Meeting abstract). *Proc. Annu. Meet. Am. Soc. Clin. Oncol.* **15,** A139.

Barnes, D. M., and Gillett, C. E. (1995). Determination of cell proliferation. *J. Clin. Pathol. Mol. Pathol.* **48,** M2–M5.

Bauer, K. D., Lincoln, S. T., Vera-Roman, J. M., Wallemark, C. B., Chmiel, J. S., Madurski, M. L., Murad, T., and Scarpelli, D. G. (1987). Prognostic implications of proliferative activity and DNA aneuploidy in colonic adenocarcinomas. *Lab. Invest.* **57,** 329–335.

Bonadonna, G., Valagussa, P., Tancini, G., Rossi, A., Brambilla, C., Zambetti, M., Bignami, P., Di Fronzo, G., and Silvestrini, R. (1986). Current status of Milan adjuvant chemotherapy trials for node-positive and node-negative breast cancers. *NCI Monogr.* **1,** 45–49.

Bosman, F. T. (1995). Prognostic value of pathological characteristics of colorectal cancer. *Eur. J. Cancer* **31A,** 1216–1221.

Briffod, M., Tubiana-Hulin, M., Spyratos, F., Hacène, K., Pallud, C., Mayras, C., and Rouessé, J. (1995). Fine-needle cytopunctures for early prediction of tumor response to preoperative chemotherapy in 94 operable breast carcinomas (Meeting abstract). *Proc. Annu. Meet. Am. Soc. Clin. Oncol.* **14,** A261.

Brown, R. W., Allred, D. C., Clark, G. M., Osborne, C. K., and Hilsenbeck, S. G. (1996). Prognostic value of Ki-67 compared to S-phase fraction in axillary node-negative breast cancer. *Clin. Cancer Res.* **2,** 585–592.

Cascinu, S., Ligi, M., Graziano, F., Del Ferro, E., Valentini, M., Grianti, C., Bartolucci, M., and Catalano, G. (1998). S-phase fraction can predict event free survival in patients with pT2-T3N0M0 colorectal carcinoma. Implications for adjuvant chemotherapy. *Cancer* **83,** 1081–1085.

Cattoretti, G., Becker, M. H. G., Key, G., Duchrow, M., Schluter, C., Galle, J., and Gerdes, J. (1992). Monoclonal antibodies against recombinant parts of the Ki-67 antigen (MIB 1 and MIB 3) detect proliferating cells in microwave-processed formalin-fixed paraffin sections. *J. Pathol.* **168,** 357–363.

Chevillard, S., Pouillart, P., Beldjord, C., Asselain, B., Beuzeboc, P., Magdelenat, H., and Viel, P. (1996). Sequential assessment of multidrug resistance phenotype and measurement of S-phase fraction as predictive markers of breast cancer response to neoadjuvant chemotherapy. *Cancer* **77,** 292–300.

Clark, G. M., Dressler, L. G., Owens, M. A., Pound, G., Oldaker, T., and McGuire, W. L. (1989). Prediction of relapse or survival in patients with node-negative breast cancer by DNA flow cytometry. *N. Engl. J. Med.* **320,** 627–633.

Clark, G. M., Mathieu, M. C., Owens, M. A., Dressler, L. G., Eudey, L., Tormey, D. C., Osborne, C. K., Gilchrist, K. W., Mansour, E. G., and Abeloff, M. D. (1992). Prognostic significance of S-phase fraction in good-risk, node-negative breast cancer patients. *J. Clin. Oncol.* **10,** 428–432.

Clurman, B. E., and Roberts, J. M. (1995). Cell cycle and cancer. *J. Natl. Cancer Inst.* **87,** 1499.

Collecchi, P., Baldini, E., Giannessi, P., Naccarato, A. G., Passoni, A., Gardin, G., Roncella, M., Evangelista, G., Bevilacqua, G., Conte, P. F., *et al.* (1998). Primary chemotherapy in locally advanced breast cancer (LABC): Effects on tumour proliferative activity, bcl-2 expression and the relationship between tumour regression and biological markers. *Eur. J. Cancer* **34,** 1701–1704.

Cooke, T. G., Stanton, P. D., Winstanley, J., Murray, G. D., Croton, R., Holt, S., and George, W. D. (1992). Long term prognostic significance of thymidine labeling index in primary breast cancer. *Eur. J. Cancer* **28,** 424–426.

Cosimelli, M., D'Agnano, I., Tedesco, M., D'Angelo, C., Botti, C., Giannarelli, D., Vasselli, S., Cavaliere, F., Zupi, G., and Cavaliere, R. (1998). The role of multiploidy as unfavourable prognostic variable in colorectal cancer. *Anticancer Res.* **18,** 1957–1966.

Costa, A., Faranda, A., Scalmati, A., Quagliuolo, V., Colella, G., Ponz, de Leon, M., and Silvestrini, R. (1992). Autoradiographic and flow-cytometric assessment of cell proliferation in primary colorectal cancer: Relationship to DNA ploidy and clinico-pathological features. *Int. J. Cancer* **50,** 719–723.

Costa, A., Silvestrini, R., Mochen, C., Lequaglie, C., Boracchi, P., Faranda, A., Vessecchia, G., and Ravasi, G. (1996). P53 expression, DNA ploidy and S-phase cell fraction in operable locally advanced non-small cell lung cancer. *Br. J. Cancer* **73,** 914–919.

Daidone, M. G., Silvestrini, R., Canova, S., Valagussa, P., and Bonadonna, G. (1989). Tumor cell kinetics and course of node-positive (N+) breast cancer. *Proc. Am. Soc. Clin. Oncol.* **8,** 24.

Daidone, M. G., Silvestrini, R., Valentinis, B., Ferrari, L., and Bartoli, C. (1990). Changes in cell kinetics induced by primary chemotherapy in breast cancer. *Int. J. Cancer* **47,** 380–383.

Daidone, M. G., Silvestrini, R., Luisi, A., Mastore, M., Benini, E., Veneroni, S., Brambilla, C., Ferrari, L., Greco, M., Andreola, S., and Veronesi, U. (1995). Changes in biological markers after primary chemotherapy for breast cancers. *Int. J. Cancer* **61,** 301–305.

Daidone, M. G., Veneroni, S., Benini, E., Tomasic, G., Coradini, D., Brambilla, C., Ferrari, L., and Silvestrini, R. (1999). Biological markers and changes induced in their profiles following primary chemotherapy: Relevance on short- and long-term clinical outcome. *In* "Primary Medical Therapy for Breast Cancer" (A. Howell and M. Dowsett, eds.), pp. 53–72. ESO Scientific Updates, Milan.

Daidone, M. G., Luisi, A., Martelli, G., Veneroni, S., Tomasic, G., De Palo, G., and Silvestrini, R. (2000). Biomarkers and outcome after tamoxifen treatment in node-positive breast cancers from elderly women. *Br. J. Cancer* **82,** 270–277.

Dalquen, P., Moch, H., Feichter, G., Lehmann, M., Soler, M., Stulz, P., Jordan, P., Torhorst, J., Mihatsch, M. J., and Sauter, G. (1998). DNA aneuploidy, S-phase fraction, nuclear p53 positivity, and survival in non-small cell lung carcinoma. *Virchows Arch.* **431,** 173–179.

Dean, P. N., Gray, J. W., and Dolbeare, F. A. (1982). The analysis and interpretation of DNA distributions measured by flow cytometry. *Cytometry* **3,** 188–195.

Dressler, L. G., Seamer, L. C., Owens, M. A., Clark, G. M., and McGuire, W. L. (1988). DNA flow cytometry and prognostic factors in 1331 frozen breast cancer specimens. *Cancer* **61,** 420–427.

Filderman, A. E., Silvestri, G. A., Gatsonis, C., Luthringer, D. J., Honig, J., and Flynn, S. D. (1992). Prognostic significance of tumour proliferative fraction and DNA content in stage I non-small cell lung cancer. *Am. Rev. Resp. Dis.* **146,** 707–710.

Fisher, B., Gunduz, N., Costantino, J., Fisher, E. R., Edmond, C., Mamounas, E. P., and Siderits, R. (1991). DNA flow cytometric analysis of primary operable breast cancer. *Cancer* **68,** 1465–1469.

Frierson, H. F. (1991). The need for improvement in flow cytometric analysis of ploidy and S-phase fraction. *Am. J. Clin. Pathol.* **96,** 439–441.

Galand, P., and Degraef, C. (1989). Cyclin/PCNA immunostaining as an alternative to tritiated thymidine pulse labeling for marking S phase cells in paraffin sections from animal and human tissues. *Cell Tissue Kinet.* **22,** 383–392.

Gardin, G., Alama, A., Rosso, R., Campora, E., Rapetto, L., Pronzato, P., Merlini, P., Naso, C., Camoriano, A., and Meazza, R. (1994). Relationship of variations in tumor cell kinetics induced by primary chemotherapy to tumor regression and prognosis in locally advanced breast cancer. *Breast Cancer Res. Treat.* **32,** 311–318.

Gentili, C., Sanfilippo, O., and Silvestrini, R. (1981). Cell proliferation in relation to clinical features and relapse in breast cancers. *Cancer* **48,** 974–979.

Gerdes, J., Lemke, H., Baisch, H., Wacker, H. H., Schwab, U., and Stein, H. (1984). Cell cycle analysis of a cell proliferation-associated human nuclear antigen defined by the monoclonal antibody Ki-67. *J. Immunol.* **133,** 1710–1715.

Gillett, C. E., and Barnes, D. M. (1998). Cell cycle. *J. Clin. Pathol. Mol. Pathol.* **51,** 310–316.

Gilliland, R., Williamson, K. E., Wilson, R. H., Anderson, N. H., and Hamilton, P. W. (1996). Colorectal cell kinetics. *Br. J. Surg.* **83,** 739–749.

Gion, M., Boracchi, P., Biganzoli, E., and Daidone, M. G. (1999). A guide for reviewing submitted manuscripts (and indications for the design of translational research studies on biomarkers). *Int. J. Biol. Markers* **13,** 123–133.

Gratzner, H. G. (1982). Monoclonal antibody to 5-bromo-and 5-iododeoxyuridine: A new reagent for detection of DNA replication. *Science* **218,** 474–476.

Harlow, S. P., Eriksen, B. L., Poggensee, L., Chmiel, J. S., Scarpelli, D. G., Murad, T., and Bauer, K. D. (1991). Prognostic implications of proliferative activity and DNA aneuploidy in Astler-Coller Dukes stage C colonic adenocarcinomas. *Cancer Res.* **51,** 2403–2409.

Hayes, D. F., Trock, B., and Harris, A. L. (1998). Assessing the clinical impact of prognostic factors: When is "statistically significant" clinically useful? *Br. Cancer Res. Treat.* **52,** 305–319.

Hedley, D. W. (1989). Flow cytometry using paraffin-embedded tissue: Five years on. *Cytometry* **10,** 229–241.

Henderson, I. C., and Patek, A. J. (1998). The relationship between prognostic and predictive factors in the management of breast cancer. *Br. Cancer Res. Treat.* **52,** 261–288.

Héry, M., Gioanni, J., Lalanne, C. M., Namer, M., and Courdi, A. (1987). The DNA labeling index: A prognostic factor in node-negative breast cancer. *Breast Cancer Res. Treat.* **9,** 207–212.

Hietanen, P., Blomqvist, C., Wasenius, V. M., Niskanen, E., Franssila, K., and Nordling, S. (1995). Do DNA ploidy and S-phase fraction in primary tumor predict the response to chemotherapy in metastatic breast cancer? *Br. J. Cancer* **71,** 1029–1032.

Hilsenbeck, S. G., Ravdin, P. M., de Moor, C. A., Chamnen, G. C., Osborne, C. K., and Clark, G. M. (1998). Time-dependence of hazard ratios for prognostic factors in primary breast cancer. *Br. Cancer Res. Treat.* **52,** 227–237.

Honkoop, A. H., van Diest, P. J., de Jong, J. S., Linn, S. C., Giaccone, G., Hoekman, K., Wagstaff, J., and Pinedo, H. M. (1998). Prognostic role of clinical, pathological and biological characteristics in patients with locally advanced breast cancer. *Br. J. Cancer* **77,** 621–626.

Howell, W. M. (1982). Selective staining of nucleolar organizer regions (NORs). *In* "The Cell Nucleus" (H. Bush and L. Tothblum, eds.), pp. 89–142. Academic Press, New York.

Johnston, P. G., and Allegra, C. J. (1995). Colorectal cancer biology: Clinical implications. *Semin. Oncol.* **22,** 418–432.

Kallioniemi, O., Blanco, G., Alavaikko, M., Hietanen, T., Mattila, J., Lauslahti, K., Lehtinen, M., and Koivula, T. (1988). Improving the prognostic value of DNA flow cytometry in breast cancer by combining DNA index and S-phase fraction. *Cancer* **62,** 2183–2190.

Kute, T. E., Quadri, Y., Muss, H., Zbieranski, N., Cirrincione, C., Berry, D. A., Barcos, M., Thor, A. P., Liu, E., Koerner, F., *et al.* (1995). Flow cytometry in node-positive breast cancer: Cancer and Leukemia Group B protocol 8869. *Cytometry* **22,** 297–306.

Le Chevalier, T., Arriagada, R., Quoix, E., Ruffie, P., Martin, M., Tarayre, M., Douillard, J. Y., and Laplanche, A. (1991). Radiotherapy alone versus combined chemotherapy and radiotherapy in nonresectable non-small cell lung cancer: First analysis of a randomized trial in 353 patients. *J. Natl. Cancer Inst.* **83,** 417–423.

MacGrogan, G., Mauriac, L., Durand, M., Bonichon, F., Trojani, M., de Mascarel, I., and Coindre, J. M. (1996). Primary chemotherapy in breast invasive carcinoma: Predictive value of the immunohistochemical detection of hormonal receptors, p53, c-erbB-2, Mib1, pS2 and GST. *Br. J. Cancer* **74,** 1458–1465.

McGuire, W. L. (1991). Breast cancer prognostic factors: Evaluation guidelines. *J. Natl. Cancer Inst.* **83,** 154–155.

Marino, P., Pampallona, S., Preatoni, A., Cantoni, A., and Invernizzi, F. (1994). Chemotherapy versus supportive care in advanced non-small cell lung cancer. Results of a meta-analysis of the literature. *Chest* **106,** 861–865.

Matturri, L., Lavezzi, A. M., Grignani, F., Salomoni, G., and Roviaro, G. C. (1994). The prognostic value of cell proliferation in non-small cell lung cancer assessed with tritiated thymidine and anti-PCNA antibodies. *Eur. J. Cancer* **30A,** 1397–1398.

Meyer, J. S., and Coplin, M. D. (1988). Thymidine labeling index, flow cytometric S-phase. Measurement, and DNA index in human tumors. *Am. J. Clin. Pathol.* **89,** 586–589.

Meyer, J. S., and Lee, J. Y. (1980). Relationships of S-phase fraction of breast carcinoma in relapse to duration of remission, estrogen receptor content, therapeutic responsiveness, and duration of survival. *Cancer Res.* **40,** 1890–1896.

Meyer, J. S., and Province, M. (1988). Proliferative index of breast carcinoma by thymidine labeling: Prognostic power independent of stage, estrogen and progesterone receptors. *Breast Cancer Res. Treat.* **12,** 191–199.

Meyer, J. S., Friedman, E., and McCrote, M. M. (1983). Prediction of early course of breast carcinoma by thymidine labeling. *Cancer* **51,** 1879–1886.

Meyer, J. S., Prey, M. U., and Babcock, D. S. (1986). Breast carcinoma cell kinetics, morphology, stage and host characteristics. *Lab. Invest.* **54,** 41–51.

Mushika, M., Miwa, T., Suzuoki, Y., Hayashi, K., Masaki, S., and Kaneda, T. (1988). Detection of proliferative cells in dysplasia, carcinoma in situ, and invasive carcinoma of the uterine cervix by monoclonal antibody against DNA polymerase α. *Cancer* **61,** 1182–1186.

Muss, H. B., Kute, T. E., Case, L. D., Smith, L. R., Booher, C., Long, R., Kammire, L., Gregory, B., and Brockschmidt, J. K. (1989). The relation of flow cytomety to clinical and biological characteristics in women with node negative primary breast cancer. *Cancer* **64,** 1894–1900.

Nelson, J. S., and Schiffer, L. M. (1973). Autoradiographic detection of DNA polymerase containing nuclei in sarcoma 180 ascites cells. *Cell Tissue Kinet.* **6,** 45–54.

O'Reilly, S. M., Camplejohn, R. S., Barnes, D. M., Millis, R. R., Rubens, R. D., and Richards, M. A. (1990a). Node-negative breast cancer: Prognostic subgroups defined by tumor size and flow cytometry. *J. Clin. Oncol.* **8,** 2040–2045.

O'Reilly, S. M., Camplejohn, R. S., Millis, R. R., Rubens, R. D., and Richards, M. A. (1990b). Proliferative activity, histological grade and benefit from adjuvant chemotherapy in node positive breast cancer. *Eur. J. Cancer* **26,** 1035–1038.

Paradiso, A., Tommasi, S., Mangia, A., Lorusso, V., Simone, G., and De Lena, M. (1990). Tumor proliferative activity, progesterone receptor status, estrogen receptor level, and clinical outcome of estrogen receptor-positive advanced breast cancer. *Cancer Res.* **50,** 2958–2962.

Paradiso, A., Mangia, A., and Picciariello, M. S. (1991). Fattori prognostici nel carcinoma della mammella operabile N-: Attività proliferativa e caratteristiche clinico patologiche. *Folia Oncol.* **14,** 247–252.

Paradiso, A., Mangia, A., Barletta, A., Catino, A. M., Giannuzzi, A., Schittulli, F., Radogna, N., Longo, S., Palmieri, D., and Marzullo, F. (1993). Randomized clinical trial of adjuvant chemotherapy in patients with node negative, fast proliferating breast cancer drug. *Drugs* **45,** 68–74.

Pierga, J. Y., Lainé-Bidron, C., Beuzeboc, P., De Cremoux, P., Pouillart, P., and Magdelenat, H. (1997). Plasminogen activator inhibitor-1 (PAI-1) is not related to response to neoadjuvant chemotherapy in breast cancer. *Br. J. Cancer* **76,** 537–540.

Pietra, N., Sarli, L., Sansebastiano, G., Jotti, G. S., and Peracchia, A. (1996). Prognostic value of ploidy, cell proliferation kinetics, and conventional clinicopathologic criteria in patients with colorectal carcinoma. *Dis. Colon Rectum* **39,** 494–503.

Quinn, C. M., and Wright, N. A. (1990). The clinical assessment of proliferation and growth in human tumours: Evaluation of methods and applications as prognostic variables. *J. Pathol.* **160,** 93–102.

Quirke, P., Dixon, M. F., Clayden, A. D., Durdey, P., Dyson, J. E., Williams, N. S., and Bird, C. C. (1987). Prognostic significance of DNA aneuploidy and cell proliferation in rectal adenocarcinomas. *J. Pathol.* **151,** 285–291.

Remvikos, Y., Beuzeboc, P., Zajdela, A., Voillemot, N., Magdelenat, H., and Pouillart, P. (1989). Correlation of pretreatment proliferative activity of breast cancer with the response to cytotoxic chemotherapy. *J. Natl. Cancer Inst.* **81,** 1383–1387.

Remvikos, Y., Mosseri, V., Asselain, B., Fourquet, A., Voillemot, N., Magdelenat, H., and Pouillart, P. (1997). S-phase fractions of breast cancer predict overall and post-relapse survival. *Eur. J. Cancer* **33,** 581–586.

Rozan, S., Vincent-Salomon, A., Zafrani, B., Validire, P., De Cremoux, P., Bernoux, A., Nieruchalski, M., Fourquet, A., Clough, K., Dieras, V., and Pouillart, P. (1998). No significant predictive value of c-erbB-2 or p53 expression regarding sensitivity to primary chemotherapy or radiotherapy in breast cancer. *Int. J. Cancer (Pred. Oncol.)* **79,** 27–33.

Rudas, M., Gnant, M. F., Mittlbock, M., Neumayer, R., Kummer, A., Jakesz, R., Reiner, G., and Reiner, A. (1994). Thymidine labeling index and Ki-67 growth fraction in breast cancer: Comparison and correlation with prognosis. *Breast Cancer Res. Treat.* **32,** 165–175.

Schipper, D. L., Wagenmans, M. J. M., Peters, W. H .M., and Wagener, D. J. (1998). Significance of cell proliferation measurement in gastric cancer. *Eur. J. Cancer* **34,** 781–790.

Schutte, B., Reynders, M. M. J., Wiggers, T., Arends, J. W., Volovics, L., Bosman, F. T., and Blijham, A. M. (1987). Retrospective analysis of the prognostic significance of DNA content and proliferative activity in large bowel carcinoma. *Cancer Res.* **47,** 5494–5496.

Scott, N. A., Rainwater, L. M., Wieand, H. S., Weiland, L. H., Pemberton, J. H., Beart, R. W., Jr., and Lieber, M. M. (1987). The relative prognostic value of flow cytometric DNA analysis and conventional clinicopathologic criteria in patients with operable rectal carcinoma. *Dis. Colon Rectum* **30,** 513–520.

Sigurdsson, H., Baldetorp, B., Borg, A., Dalberg, M., Ferno, M., Killander, D., Olsson, H., and Ranstam, J. (1990). Indicators of prognosis in node-negative breast cancer. *N. Engl. J. Med.* **322,** 1045–1049.

Silvestrini, R. (1994a). Cell kinetics: Prognostic and therapeutic implications in human tumor. *Cell Prolif.* **27,** 579–596.

Silvestrini, R. (on behalf of the SICCAB Group for Quality Control of Cell Kinetic Determinations) (1994b). Quality control for the evaluation of the S-phase fraction by flow cytometry: a multicentric study. *Cytometry (Comm. Clin. Cytometry)* **18,** 11–16.

Silvestrini, R., and the SICCAB Group for Quality Control of Cell Kinetic Determination (1991). Feasibility and reproducibility of the 3H-dT labeling index in breast cancer. *Cell Prolif.* **24,** 437–445.

Silvestrini, R., Daidone, M. G., Di Fronzo, G., Morabito, A., Valagussa, P., and Bonadonna, G. (1986). Prognostic implication of labelling index versus estrogen receptors and tumor size in node-negative breast cancer. *Breast Cancer Res. Treat.* **7,** 161–169.

Silvestrini, R., Daidone, M. G., Valagussa, P., Salvadori, B., Rovini, D., and Bonadonna, G. (1987). Cell kinetics and prognosis in locally advanced breast cancer. *Cancer Treat. Rep.* **71,** 375–379.

Silvestrini, R., Daidone, M. G., Valagussa, P., Di Fronzo, G., Mezzanotte, G., and Bonadonna, G. (1989). Cell kinetics as a prognostic indicator in node-negative breast cancer. *Eur. J. Cancer Clin. Oncol.* **25,** 1165–1171.

Silvestrini, R., Muscolino, G., Costa, A., Lequaglie, C., Veneroni, S., Mezzanotte, G., and Ravasi, G. (1991). Could cell kinetics be a predictor of prognosis in non-small cell lung cancer? *Lung Cancer* **7,** 165–170.

Silvestrini, R., Daidone, M. G., Del Bino, G., Mastore, M., Di Fronzo, G., and Boracchi, P. (1993a). Prognostic significance of proliferative activity and ploidy in node-negative breast cancer. *Ann. Oncol.* **4,** 213–219.

Silvestrini, R., Daidone, M. G., Mastore, M., Di Fronzo, G., Coradini, D., Boracchi, P., Squicciarini, P., Salvadori, B., and Veronesi, U. (1993b). Cell kinetics as a predictive factor in node-positive breast cancer treated with adjuvant hormone therapy. *J. Clin. Oncol.* **11,** 1150–1155.

Silvestrini, R., Veneroni, S., Daidone, M. G., Benini, E., Boracchi, P., Mezzetti, M., Di Fronzo, G., Rilke, F., and Veronesi, U. (1994). The bcl-2 protein: A prognostic indicator strongly related to p53 protein in lymph node-negative breast cancer patients. *J. Natl. Cancer Inst.* **86,** 499–504.

Silvestrini R., Daidone, M. G., and Costa, A. (1995a). Determination of the proliferative fraction in human tumors. *In* "Cell Growth and Apoptosis. A Pratical Approach" (G. P. Studzinski, ed.), pp. 59–77. IRL Oxford Univ. Press, Oxford.

Silvestrini, R., Daidone, M. G., Luisi, A., Boracchi, P., Mezzetti, M., Di Fronzo, G., Andreola, S., and Veronesi, U. (1995b). Biologic and clinico-pathologic factors as indicators of specific relapse types in node-negative breast cancer. *J. Clin. Oncol.* **13,** 697–704.

Silvestrini, R., Benini, E., Veneroni, S., Daidone, M. G., Tomasic, G., Squicciarini, P., and Salvadori, B., (1996). P53 and Bcl-2 expression correlates with clinical outcome in a series of node-positive breast cancer patients. *J. Clin. Oncol.* **14,** 1604–1610.

Silvestrini, R., Daidone, M. G., Luisi, A., Boracchi, P., Mezzetti, M., Di Fronzo, G., Andreola, S., Salvadori, B., and Veronesi, U. (1997). Cell proliferation in 3800 node-negative breast cancers: Consistency over time of biological and clinical information provided by 3H-thymidine labelling index. *Int J. Cancer (Pred. Oncol.)* **74,** 122–127.

Simonsson, B., Killander, C. F. R., Brenning, G., Kallander C. F., Ahre, A., and Gronowitz, J. S. (1985). Evaluation of serum deoxythymidine kinase as a marker in multiple myeloma. *Br. J. Hematol.* **61,** 215–218.

Stål, O., and Nordenskjold, B. (1994). S-phase fraction and survival benefit from adjuvant chemotherapy and radiotherapy of breast cancer. *Br. J. Cancer* **70,** 1258–1263.

Stanton, P. D., Cooke, T. G., Oakes, S. J., Winstanley, J., Holt, S., George, W. D., and Murray, G. D. (1992). Lack of prognostic significance of DNA ploidy and S phase fraction in breast cancer. *Br. J. Cancer* **66,** 925–929.

Sulkes, A., Livingstone, R. B., and Murphy, W. K. (1979). Tritiated thymidine labeling index and response in human breast cancer. *J. Natl. Cancer Inst.* **62,** 513–515.

Ten Velde, G. P. M., Schutte, B., Vermeulen, A., Volovics, A., Reynders, M. M., and Blijham, G. H. (1988). Flow cytometric analysis of DNA ploidy level in paraffin-embedded tissue of non-small-cell lung cancer. *Eur. J. Cancer Clin. Oncol.* **24,** 455–460.

Tubiana, M., and Courdi, A. (1989). Cell proliferation kinetics in human solid tumors: Relation to probability of metastatic dissemination and long-term survival. *Radiother. Oncol.* **15,** 1–18.

Tubiana, M., Pejovic, M. H., Chavaudra, G., Contesso, G., and Malaise, E. P. (1984). The long term prognostic significance of the thymidine labeling index in breast cancer. *Int J. Cancer* **33,** 441–445.

Tubiana, M., Pejovic, M. H., Koscielny, S., Chavaudra, N., and Malaise, E. (1987). Growth rate, kinetics of tumor cell proliferation and long-term outcome in human breast cancer. *Int. J. Cancer* **44,** 17–22.

Tumor Marker Expert Panel Members (1996). Clinical practice guidelines for the use of tumor markers in breast and colorectal cancer. *J. Clin. Oncol.* **14,** 2843–2877.

Van Diest, P. J., Brugal, G., and Baak, J. P. A. (1998). Proliferation markers in tumours: Interpretation and clinical value. *J. Clin. Pathol.* **51,** 716–724.

Volm, M., Drings, P., Mattern, J., Sonka, J., Vogt-Moykopf, I., and Wayss, K. (1985a). Prognostic significance of DNA patterns and resistance-predictive tests in non-small cell lung carcinoma. *Cancer* **56,** 1396–1403.

Volm, M., Matter, J., Sonka, J., Vogt-Schaden, M., and Wayss, K. (1985b). DNA distribution in non-small-cell lung carcinomas and its relationship to clinical behaviour. *Cytometry* **6,** 348–356.

Wenger, C. R., and Clark, G. M. (1998). S-phase fraction and breast cancer—a decade of experience. *Br. Cancer Res. Treat.* **51,** 255–265.

Witzig, T. E., Loprinzi, C. L., Gonchoroff, N. J., Reiman, H. M., Cha, S. S., Wieand, H. S., Katzmann, J. A., Paulsen, J. K., and Moertel, C. G. (1991). DNA ploidy and cell kinetic measurements as predictors of recurrence and survival in stages B2 and C colorectal adenocarcinoma. *Cancer* **68,** 879–888.

Zambetti, M., Bonadonna, G., Valagussa, P., Daidone, M. G., Coradini, D., Bignami, P., Contesso, G., and Silvestrini, R. (1992). Adjuvant CMF for node-negative and estrogen receptor-negative breast cancer. *NCI Monogr.* **11,** 77–83.

Detection of Minimal Residual Disease

Andrzej Deptala[*] and Sharon P. Mayer[†]

[*]Brander Cancer Research Institute
The New York Medical College
Hawthorne, New York 10532; and USA
Department of Hematology, Oncology and Internal Medicine
Warsaw Medical University
02-097 Warsaw, Poland

[†]Departments of Pediatrics and Pathology
New York Medical College
Valhalla, New York 10595

I. Introduction
II. Tissue Sources to Detect Minimal Residual Disease
III. Methods to Detect Minimal Residual Disease
    A. Morphology
    B. Immunophenotyping
    C. Cell Culture
    D. Cytogenetics
    E. Flow Cytometry
    F. Polymerase Chain Reaction
    G. Sensitivity and Applicability
IV. Technical Problems
V. Concluding Remarks
    References

I. Introduction

Despite improvements in risk-adapted therapy for patients with hematological as well as solid tumor malignancies, relapse appears inevitable for many. Conventional methods to define complete remission (CR) relying on light microscopy examination and/or on sophisticated imaging techniques cannot reach the appropriate sensitivity to detect or to discriminate a malignant cell. In other words, conventional methods do not recognize the malignancy until the time of relapse, and patients who may have residual cancer cells are managed like those with no residual malignancy at all.

Minimal residual disease (MRD) can be defined as the presence of a small number of cancer cells, in hematological malignancies usually below 10^{10} cells (that amount approximately corresponds to 1 g of the total cancer cell burden in a human body). This small number of malignant cells is below the level of detection by conventional diagnostic techniques in patients who have survived a treatment that had induced a complete remission of the malignancy.

The ability to distinguish residual malignant cells among a normal population of cells is inherently dependent on the presence of identifiable cellular or subcellular features specific to the malignant clone. Characteristic chromosomal abnormalities and clone-specific gene rearrangements have been utilized as markers for assessing the MDR burden, by cytogenetic or molecular analysis. Unique combinations of surface, cytoplasmic, or nuclear molecules that are expressed by the malignant clone and are not present in normal cell population allow multiparameter flow-cytometric detection of MRD. Theoretical requirements for detection of MRD are shown in Table I.

The study of remission status is an active area of research, whereby the monitoring of MRD during all stages of treatment and the prognostic significance thereof are being investigated. Results from numerous studies seem promising: most researchers agree that an increase in levels of residual malignant cells is the strongest predictor of relapse of a cancer (Cavé *et al.*, 1998; Coustan-Smith *et al.*, 1998; Deptala *et al.*, 1995a). However, certain contrasting results have emerged. For example, the absence of residual cells does not always guarantee cancer-free survival (Ito *et al.*, 1993), and the presence of a certain amount of residual malignant cells may be compatible with long-term survival (Roberts *et al.*, 1997). In order to elucidate the cause of these conflicts, adoption of universally accepted and standardized detection methods may be necessary. In this chapter, a comprehensive overview of the most current technology in the advancement of defining MRD is provided. Because the most significant studies have been

Table I

Theoretical Requirements for Detection of Minimal Residual Disease

Disease-related	Technique-related
Marker:	Test:
1. Present in all or majority of malignant cells	1. Sensitive
2. Stable	2. Specific
3. Individual/unique	3. Repetitious
	4. Quantitative
	5. Performed on a compartment that is representative for cancer population

performed on hematological malignancies, we also focus on MRD in acute and chronic leukemias and some lymphoproliferative disorders.

II. Tissue Sources to Detect Minimal Residual Disease

In both acute and chronic leukemias as well as in multiple myeloma, bone marrow (BM) is the primary and most significant compartment for MRD analysis. Peripheral blood (PB) may also be a suitable tissue source to detect MRD, but when seeking MRD in the blood, one has to keep in mind that leukemic abnormalities in PB only reflect and follow the transformation event in the BM. Therefore, false-negative results may be obtained more often when the detection of residual leukemia is performed from the blood. However, there are exceptions, discussed later, that favor PB instead of BM for MRD detection in leukemias. Extramedullary disease is not a common feature of acute leukemia [e.g., primary involvement of the central nervous system (CNS) in acute leukemias is about 5%]. Bone marrow relapse usually precedes an extramedullary relapse of the disease. However, detection of MRD in isolated measurable "sanctuary" sites can be a sign that unmeasurable bone marrow MRD exists. In these particular cases, monitoring of residual disease in all of the tissues in which leukemia involvement is possible must be performed.

Non-Hodgkin's lymphoma (NHL) is primarily a disease of lymph nodes, so lymph node biopsy seems to be a method of choice to detect MRD. However, lymph node biopsy is usually performed at the time of initial diagnosis or at relapse and very seldom when a patient is in CR. Therefore, BM and PB samples provide easily available tissue sources to detect MRD in NHL. NHL type and clinical stage determine the likelihood of BM infiltration by lymphoma cells. In general, the higher the grade of the tumor and the more advanced the stage of NHL, the more likely the BM and/or PB are involved. However, in some subtypes [e.g., anaplastic large-cell lymphoma (ALCL) or T-cell lymphoblastic lymphoma (LL)], at the diagnosis, BM can be spared from the lymphomatic infiltration, or a relapse can occur only at nodal sites with no evidence of MRD in other tissues. In those cases MRD detection has to be performed by lymph node biopsy. In NHL, there is also higher probability of extranodal and extramedullary involvement than is observed in a leukemia.

In solid tumor malignancies, reliable tissue sources for MRD detection are BM and PB, especially in epithelial-derived tumors (e.g., breast carcinoma, prostate carcinoma).

Contrary to homogeneity in a cancer infiltration of primary tissue, anatomic distribution of residual disease is very irregular. Residual cells form so-called clusters or foci of MRD. Therefore, it is important to examine a sufficient quantity of a specimen for the detection of a residual disease. It has also been suggested to take samples from numerous sites of suspect distinct locations of MRD.

III. Methods to Detect Minimal Residual Disease

A summary of the methods used in detection of MRD is given in Table II.

A. Morphology

In the acute leukemias and in chronic myelogenous leukemia (CML), the disease is considered to be in CR when less than 5% of the cells in the bone marrow are blasts. Because blasts (myeloblasts, lymphoblasts, monoblasts, etc.) are components of normal hematopoiesis, they are unidentifiable from leukemic blasts. In rare instances, however, the leukemic blasts are associated with a unique morphological feature that is readily detectable. For example, the discrimination of residual malignant cells can be achieved at a sensitivity of 1% in patients with acute lymphoblastic leukemia (ALL) L3 or with acute myeloid leukemia (AML) M2/M3 with blasts containing Auer rods. Because of the limitations of the light microscope, sensitivity greater than 1% in leukemias as well as other cancers is not readily attainable. Due to this deficiency in sensitivity and specificity, morphology has extremely limited application in the monitoring of MRD (Campana and Pui, 1995).

B. Immunophenotyping

Immunophenotyping is based on a very specific reaction between the antigen and antibody. Antibodies have a high affinity to one [monoclonal antibody (MAb)] or to several [polyclonal antibody (PolAb)] epitopes of a particular antigen. In immunophenotyping, the antigen is a protein that is localized either in the plasma membrane (surface marker, s) or in the cytoplasm (cytoplasmic

Table II
Methods Used for Detection of Minimal Residual Disease

Detection at cellular level
   A. Morphology
   B. Immunophenotyping
   C. Cell culture
Detection at chromosome level
   D. Cytogenetics
     1. Conventional cytogenetics
     2. Prematurely condensed chromosomes (PCC)
     3. Fluorescence *in situ* hybridization (FISH)
   E. Flow cytometry
     1. Flow karyotyping
     2. DNA index
Detection at molecular level
   F. Polymerase chain reaction (PCR)

marker, c) of the cell, and it is usually involved in the normal physiological functions of that cell. For hematological malignancies, surface markers called CD, which means cluster of differentiation or cluster of designation, are the most commonly used for immunophenotyping. According to the Sixth International Workshop on Human Leukocyte Differentiation Antigens held in Kobe, Japan, 1996, there are 166 antigens (or epitopes of the antigen) identifiable by appropriate antibodies (Kishimoto *et al.*, 1997). There are also many other antigens not classed as the CDs [e.g., terminal deoxynucleotidyltransferase (TdT), IgM-μ chain of immunoglobulin (Ig), p53, Bcl-2] that are utilized in the detection of MRD.

Stewart and Stewart (1994a,b) have provided a comprehensive overview of sample preparation, staining procedures, and multiparameter analysis of cells. Therefore, only methodological considerations applicable for the detection of MRD are revealed in this chapter.

Various detection methodologies are available to assess antigen–antibody binding, such as immunofluorescence or immunocytochemistry [alkaline phosphatase antialkaline phosphatase (APAAP), peroxidase antiperoxidase (PAP), and Immunogold]; however, for MRD studies, immunofluorescence is certainly the method of choice. Altough very sophisticated fluorescence microscopes with phase-contrast attachments are still utilized, the use of such instruments is time-consuming, arbitrary, and nonquantitative. Consequently, nearly all laboratories that perform MRD analysis by immunophenotyping favor the flow cytometer. In fact, modern flow cytometry allows reproducible and rapid acquisition of a large number of cells, and it is capable of achieving a quantitative, multicolor, and multiparameter analysis (Darzynkiewicz *et al.*, 1994).

Development of commercially available reagents for the fixation and permeabilization of cell membranes has standardized the flow cytometric identification of cytoplasmic antigens in cell suspensions. A second advantage of these solutions in comparison to older permeabilization/fixation procedures is that cell morphology is preserved without changing the scatter characteristics. In our hands, and according to the literature (Groeneveld *et al.*, 1996), a reagent, Fix & Perm (An der Grub, Vienna, Austria; Caltag, Burlingame, CA), and IntraPrep (Coulter/Immunotech, Miami, FL) give the optimum detection sensitivity of intracellular markers such as CD3, CD22, immunoglobulin μ, κ, and λ chains, TdT, and myeloperoxidase (MPO). Other reagents, for instance, PermeaFix (Ortho Diagnostic System, Raritan, NJ), OptiLyse B (Immunotech, Marseilles, France), and FACS Brand (Becton Dickinson, San Jose, CA), can also provide sufficient staining of these proteins; however, they are not suitable for MPO labeling. In addition, OptiLyse B and FACS Brand do not give reliable results for κ, λ, and cCD3 staining (Knapp *et al.*, 1994).

The laser scanning cytometer (LSC) is well suited for multicolor and multiparameter analysis and can potentially be utilized in MRD detection (for a review of LSC, see Chapters 46, 50, and 51 of this volume and Chapter 3 of Volume 63 of this series). LSC has a unique characteristic that enables analysis of an individual cell within a fixed location of the specimen. Therefore, it is possible to scan the same slide multiple times, under different experimental conditions, using the same or a different laser. Thus, an individual cell can be stained with different antibody-conjugated fluorochromes and can be repeatedly scanned. Equally important, LSC allows the capability to correlate antigen expression with tissue architecture, which could be useful in the assessment of tissue involvement in patients, for example, with NHL or solid tumor malignancies. Indeed, immunophenotypic analysis combined with examination of tissue architecture by LSC could provide an extra benefit by enabling the detection of residual disease in fresh or frozen sections obtained from lymph nodes, skin, soft tissue, liver, or spleen (Chapter 46 of this volume).

Almost all markers expressed on leukemia or lymphoma cells are also present on normal hemato- and lymphopoietic cells. For example, CD10 (CALLA, the common acute lymphoblastic leukemia antigen) is expressed on normal B-cell progenitors, bone marrow stromal cells, and neutrophils. CD10 can even be normally found on cells from different tissues such as kidney, gastrointestinal tract, bile duct, and brain. CD19 and CD33, which are identified on most leukemic B lymphoblasts or myeloblasts, are expressed on normal B-lymphoid or myeloid precursors, respectively (Deptala, 1995). Thus, under most conditions a single marker is not suitable for MRD detection and quantitation. Some exceptions exist, though, as follows:

1. Detection of TdT, CD10, or CD34 in the PB. Less than 0.1% of normal cells in PB are TdT$^+$ or CD10$^+$, and less than 0.5% are CD34$^+$; however, after chemotherapy or during granulocyte macrophage colony stimulating factor (GM-CSF)/granulocyte colony stimulating factor (G-CSF) treatment, the percentage of normal TdT$^+$, CD10$^+$, or CD34$^+$ cells can escalate.

2. Detection of CD34 or TdT in cerebrospinal fluid (CSF). Even the detection of one cell carrying these markers is suggestive of MRD with CNS involvement; however, any contamination of the CSF by blood cells during the puncture procedure precludes a definitive result.

3. Detection of leukemia-associated proteins such as those formed from specific gene fusion events as a result of a chromosomal translocation [e.g., t(9;22)(q34;q11) = BCR-ABL, t(1;19)(q23;p13) = PBX1-E2A, t(15;17)(q11-22;q21) = PML-RARα]. In theory, utilization of these leukemia-specific markers has enormous potential; however, only two antibodies have been developed, one that recognizes the PBX1-E2A chimera protein (Sang *et al.*, 1997) and one ("7.1") that is specific to cells bearing 11q23 chromosomal abnormalities (Behm *et al.*, 1996).

4. Quantitative differences in the expression of CD10, CD34, etc. (Lavabre-Bertrand *et al.*, 1994; Porwitt-MacDonald *et al.*, 1996). Each leukemic cell from hyperdiploid B-lineage ALL expresses greater than 3×10^4 CD10 molecules, which is 10 times more than in normal B-lymphoid precursors. Because the lymphoid precursors in normal fetal BM and fetal liver can express up to $5 \times$

10^4 CD10 molecules/cell, it is necessary to combine analysis of CD10 expression with a second maker such as a chromosomal abnormality.

Discrimination between a residual leukemic cell and a normal myelo-, mono-, or lymphoblast improves as the number of different parameters that are simultaneously examined increases. In general, flow cytometric discrimination between various cell populations is based on differences in scatter [forward light scatter (FLS) and side scatter (SSC)] characteristics, the existence of unique combinations of certain antigens, and heterogeneity in the expression of differentiation-associated antigens. Because the BM contains cell populations with overlapping sizes (FLS) and levels of morphologic complexity (SS), analysis based on scatter characteristics is incapable of distinguishing normal from leukemic cells on its own merit. This tactic is suitable for MRD detection when it is combined with surface antigen-specific and/or leukemia-specific markers. Certain antigen combinations characterizing malignant clones that are not observed (or are at an extremely low quantity) in normal cells in normal tissues are known as leukemia-associated phenotypes or aberrant phenotypes (Campana and Pui, 1995; van Dongen *et al.*, 1993). These atypical phenotypes are essential in monitoring MRD. Because leukemic cells do not follow the same differentiation pathways that normal cells do, very often in leukemogenesis certain differentiation-associated antigens arise simultaneously. This heterogeneous expression of the differentiation-associated antigens is specified as asynchronous differentiation or asynchronous phenotype (Greaves *et al.*, 1986). Normal multipotential progenitor cells may reveal asynchronous differentiation, but the amount of antigen expression is lower than in leukemic cells, and during the normal differentiation process, the asynchronous phenotype consequently disappears. Thus, the utilization of an asynchronous phenotype for monitoring of MRD has been valuable (Deptala, 1995).

A panel of antibodies conjugated to two, three, or four different fluorochromes can recognize leukemia-associated phenotypes or asynchronous phenotypes. In particular, antibodies conjugated to fluorescein isothiocyanate (FITC), phycoerythrin (PE), peridinim chlorophyll protein (PerCP), and PE-Cyanine 5 tandem conjugate (PECy5) have been successfully used. A standard, single-laser (e.g., argon laser, 488 nm) flow cytometer can perform up to three-color analysis for MRD detection, but a dual-laser flow cytometer (e.g., argon laser + HeNe laser, 633 nm) that can detect another fluorochrome, for example, allophycocyanin (APC), enables four-color analysis and perhaps increases sensitivity and specificity of MRD detection. To our knowledge, no appropriate studies comparing three-color and four-color analysis have been performed.

Flow cytometry is able to detect one target cell in 10^6 to 10^7 cells (Campana and Pui, 1995; Gross *et al.*, 1995). This sensitivity can be achieved by removing common sources of false-positive events, including nonspecific immunofluorescence, autofluorescence, background particles from previous experiments, and breaking events during acquisition (Gross *et al.*, 1993). However, when analyzing tissue samples, a more feasible sensitivity has to be postulated. For practical applications, if analysis is performed on only aberrant or asynchronous phenotypes using the multicolor/multiparameter analysis, a sensitivity of one residual cell in 10^4 to 10^5 normal cells is expected (Campana, 1994). Such high sensitivity is mandatory to obtain reliable results, especially if application to clinical management is desired. Thus, markers that provide lower levels of sensitivity are not recommended for the detection of residual disease. To follow MRD, detailed knowledge about the immunophenotype of the leukemic cells at the time of diagnosis is required, because it influences the selection of the suitable markers. If the immunophenotypic characteristics of the leukemia or lymphoma are not known from the beginning, we do not recommend this method for detection of the residual disease. Immunophenotyping then becomes very expensive and also time-consuming, and its results may eventually fail to recognize MRD.

1. T–Cell Acute Lymphoblastic Leukemia

More than 95% of T-ALL cells express nuclear TdT in combination with T-cell markers such as CD2, cCD3, CD5, and CD7. Many of them express additional T-cell markers such as CD1, CD3, CD4, and/or CD8. The cCD3/TdT (also CD3/TdT) and CD5/TdT aberrant phenotypes are observed in healthy individuals only in the thymus during normal thymocyte development but never in the normal BM or PB. T-cell marker$^+$/TdT$^+$ cells can occur in normal BM and PB only if they express the CD2 and/or CD7 antigens but neither cCD3, CD3, CD5, nor CD1, CD4, CD8 (van Dongen *et al.*, 1993). Thus, in T-ALL the double-staining technique is good enough to detect MRD in almost all cases. This approach (CD5/TdT, cCD3/TdT) also contributes the best sensitivity that can be obtained for MRD detection by immunophenotyping: 10^{-5} (i.e., one leukemic cell among 100,000 normal cells) (Campana, 1994; Deptala *et al.*, 1995b).

2. B–Lineage Acute Lymphoblastic Leukemia

In B-lineage ALL, including CD10$^+$ or CD10$^-$ B-precursor ALL, pre-B ALL, and B-ALL, MRD detection is much more complicated, because aberrant phenotypes have been demonstrated in only 3–50% of the cases, depending on the age of the patient (infant, child, adolescent, adult) (Jennings and Foon, 1997). An additional confounding factor is that these phenotypes, when examined by double-color staining techniques, are detectable in normal BM, and consequently the sensitivity (becomes $>10^{-3}$) and reliability of the analysis are decreased (Campana and Pui, 1995; Deptala, 1995; Drach *et al.*, 1992). The only reasonable pairs for MRD detection seem to be CD19/PBX1-E2A-protein in pre-B ALL and CD19/7.1 in B-precursor ALL. Hence, most laboratories perform triple- or quadruple-color analysis to detect leukemia-associated phenotypes and simultaneously utilize common lymphoid/progenitor markers such CD10 or TdT and/or CD19 and CD34, in combination with CD21, cIgM, CD13, CD15, CD33, CD56, and CD65 (Coustan-Smith *et al.*, 1998; Macedo *et al.*, 1995). However, it is feasible to monitor MRD in only about 20% of B-lineage ALL cases. Another informative aberrant phenotype is CD19/CD10/KOR-SA3544 (or CD19/CD34/CD10/KOR-SA3544). The KOR-SA3544 monoclonal antibody reacts to Philadelphia-chromosome positive cells (Ph$^+$) with high sensitivity and allows the detection of MRD in up to 5% of pediatric and 30% of adult ALL patients (Mori *et al.*, 1995). Because this antibody also recognizes myeloid cells, it must be used in combination with lymphoid-associated markers.

Detection of asynchronous phenotypes in B-lineage ALL requires a three- or four-color labeling procedure to monitor MRD. Using a similar pattern of lymphoid-associated antigens (CD10, TdT, CD19, CD34) with differentiation-associated markers (CD45, CD45RA/RB, CD44, CD40, CD38, CD22, CD20, etc.), it is possible to recognize residual cells with appropriate specificity and high sensitivity (Weir *et al.*, 1999). Quantitative multiparameter flow cytometry can also distinguish between normal and malignant B-cell precursors. Normal B-cell precursors usually harbor a significantly higher number of TdT ($>100 \times 10^3$) and a lower number of CD10 ($<50 \times 10^3$) and CD19 ($<10 \times 10^3$) molecules per cell than B-lineage ALL blasts ($<100 \times 10^3$, $>50 \times 10^3$, and $>10 \times 10^3$ molecules per cell, respectively) (Farahat *et al.*, 1995). Although the significance of this phenomenon for the monitoring of residual disease is still unclear, this approach is helpful in the distinction between regenerating normal precursors and reoccurrence of leukemic blasts.

Table III gives the example of a flow cytometry procedure used for MRD detection in B-precursor CD10$^-$ ALL. This procedure is suitable for the detection of residual disease in AML and other hematological as well as solid tumor malignancies.

Figure 1 shows a graphic representation of the flow cytometric protocol. Figure 1A shows MRD detected by triple staining and gating for CD19$^+$/CD34$^+$/CD15$^+$ cells. Figure 1B shows the same analysis at the time of relapse.

Table III

An Example of a Flow Cytometric Protocol for Detecting Minimal Residual Disease in Acute Lymphoblastic Leukemia

A. Leukemic cells at diagnosis are CD19$^+$/CD34$^+$/CD15$^+$.

B. Mononuclear cells are stained with CD34 PerCP, CD19 PE, and CD15 FITC. In parallel tube isotype-matched IgG is used as a negative control for CD15.

  1. Collect 5000–10000 cells.

  2. In FLS versus SSC dot plot draw a gate including lymphoblasts and lymphocytes (region 1, R1).

  3. In PE versus SSC dot plot draw a gate including all lymphoid and CD19$^+$ cells (region 2, R2).

  4. Collect as many events as possible (at least 20,000 cells) that fulfill the criteria defined by R1 and R2

  5. Plot the CD19$^+$/CD34$^+$ cells versus CD15 FITC and analyze for the expression of CD15. Most of the cells are CD15$^+$ and should be negative with the IgG FITC.

C. The light scatter and immunophenotypic features of the residual cells should correspond to those parameters of lymphoblasts determined at diagnosis.

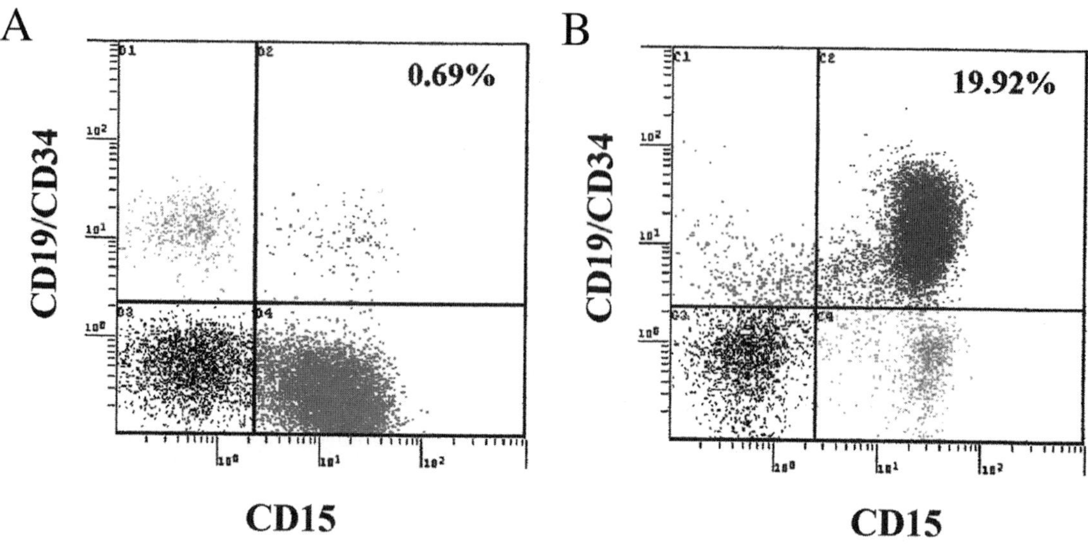

Fig. 1 CD19 PE/CD34 PerCP versus CD15 FITC bivariate distributions of bone marrow cells of B-precursor ALL at time of complete clinical remission (A) and relapse (B). Percentages indicate residual leukemic cells. (See color plates.)

3. Acute Myeloid Leukemia

Acute myeloid leukemia is subclassified according to the French–American–British (FAB) criteria as revised by the National Cancer Institute-sponsored workshop that recognizes eight major groups: M0, M1, M2, M3, M4, M5, M6, and M7 (Jennings and Foon, 1997). According to the common immunophenotype, only five main subgroups can be distinguished:

1. Myeloblastic (includes M0, M1, M2)
2. Promyelocytic (includes M3, M3 variant)
3. Monoblastic (includes M4, M5)
4. Erythromyeloblastic (includes M6)
5. Megakaryoblastic (includes M7)

The immunophenotype of AML blasts is more heterogeneous than in ALL, and a larger panel of antibodies is required for MRD detection. Unusual expression of antigens occurs in about 70–85% of AML cases. Because several atypical phenotypes are commonly observed at the time of the diagnosis, the complexity of the immunophenotype can confuse the selection of the most suitable antigens for the monitoring of the residual disease (Wörmann *et al.*, 1993).

For MRD detection in AML, van Dongen *et al.* (1993) reported that double-color staining techniques could identify unique pairs of antigens such as CD13$^+$/TdT$^+$, CD33+/TdT$^+$, CD65$^+$/TdT$^+$, CD15$^+$/TdT$^+$, and CD14$^+$/TdT$^+$. According to Macedo *et al.* (1995), other phenotypes are detectable at relatively low incidences ($<1 \times 10^{-3}$) in normal BM, including $CD34^+/CD14^+$, $CD34^+/CD65^+$, $CD34^+/CD15^+/HLADR^-$, $CD34^+/CD20^+$, $CD34^+/CD22^+$, $CD34^+/CD3^+$, $CD34^+/CD56^+$, $CD34^+/CD38^-/HLADR^+$, $CD34^+/CD38^-/HLADR^-$, $CD34^+/CD33^-/HLADR^-$, $CD34^+/CD13^-/HLADR^+$, $CD34^+/CD71^+/HLADR^-$, $CD34^+/CD61^+/HLADR^-$, and $CD34^+/CD41^+/HLADR^-$. According to Campana (1994) and our observations, it is not obvious whether all of these unusual phenotypes can be suitable for sensitive ($\leq 10^{-4}$) detection of residual disease, because they represent apparently quantitative changes in the expression of the antigens. The only exception is the combination of the $CD34^+/CD56^+$ pair, which became one of the most useful in high sensitive detection of residual disease (Wörmann *et al.*, 1993). In our experience, the combination of $CD33^+/CD3^+$ also allows detection of residual disease at a high sensitivity level (in normal BM, $<0.05\%$ of the cells are double positive). Presumably this phenotype is a rare event, because it was the only case of biphenotypic AML in the study (Deptala *et al.*, 1995b) and other cases have not been reported in the literature. Another currently published pair that can be valuable for MRD detection is $CD34^+/CD87^+$ (Lanza *et al.*, 1998). Up to 70–80% of AML is CD87 positive, but the antigen is observed in less than 0.2% of normal $CD34^+$ cells. In general, for MRD detection in AML, it is better to use a multicolor staining technique (at least three color; probably four color is superior) that includes a pattern of common myeloid/progenitor antigens (e.g., CD13 and/or CD33 and CD34 and/or CD117) with an another aberrant or asynchronous marker (e.g., CD19, CD2, CD7, CD11b, CD14, CD15, CD38, CD65, HLA-DR).

4. Lymphoproliferative Disorders

Chronic lymphoproliferative disorders (CLD) include chronic lymphocytic leukemia (CLL), hairy cell leukemia (HCL), prolymphocytic leukemia (PLL), multiple myeloma (MM), large lymphocytic leukemia (LGL), and low-grade NHLs. They are usually indolent diseases, so the strategy of clinical maintenance is different from that utilized in acute leukemias. There is apparently no necessity to eradicate the malignant clone, unless the patient has undergone bone marrow transplantation. Therefore, there are few data about detection of MRD in these diseases. Besides, in CLD as well as in intermediate/high-grade NHL, malignant cells often do not display aberrant phenotypes. Therefore, polymerase chain reaction (PCR) analysis that recognizes clone-specific Ig/T-cell receptor (TcR) gene rearrangements or characteristic molecular abnormalities is the method of choice for MRD detection in these cases. If one uses immunophenotyping as the detection technique, the same strategy that is utilized for the study of residual acute leukemia is applied; however, certain modifications are highlighted later.

Chronic lymphocytic leukemia is the most common adult leukemia in Europe and North America. The predominant cell type is a small lymphocyte with well-established immunophenotype and is usually B-cell associated. Three-color staining techniques provide high sensitivity of detection for the residual disease.

Cells falling into the lymphocyte gate are plotted for CD5 versus CD20 expression. Then, dual-positive cells with weak CD20 are gated and plotted for both surface immunoglobulin κ and λ light chains versus CD20. Clonal κ^+ or λ^+ cells for the CD5$^+$/CD20$^+$ cells indicates MRD. The same strategy is suitable for low-grade small cell lymphocytic lymphoma (SCLL) (Jennings and Foon, 1997).

In another type of chronic B-cell leukemia, HCL, residual disease can be monitored with phenotypes CD22$^+$/CD11c$^+$/CD25$^+$ or CD103$^+$/CD11c$^+$/CD25$^+$, which are the most specific for this malignancy.

Multiple myeloma has typically been difficult to measure by flow cytometry, because there are few strong specific single-color markers, and because of the focal nature of bone marrow involvement. However, multicolor staining can recognize plasma cells that express bright CD38$^+$ and dim CD45$^+$. Thus, when gating for cells that have intermediate SSC with intermediate to high FLS and express low CD45$^+$/bright CD38$^+$ and when plotting them for clonal intracellular κ^+ or λ^+ (or cytoplasm IgG or IgA or IgD or IgE), one can identify the pure population of residual myeloma cells. Another immunophenotype suitable for MRD detection in MM is CD45$^{\mathrm{dim}+}$/CD38^{++}/CD56$^+$ (Jennings and Foon, 1997).

Chronic T-cell leukemias [T-CLL, T-PLL, T-LGL, natural killer (NK)-LGL, adult T-cell leukemia/lymphoma (ATLL)] are characterized by a post-thymic T-cell immunophenotype. Although atypical immunophenotypes are common in this group (e.g., loss of pan-T-cell antigens: CD2, CD7, cCD3), it is not clear if they are reliable for the sensitive detection of residual disease. Typically only the quantity of the particular antigen varies. Therefore, TcR gene rearrangement analysis by PCR is superior and is recommended for MRD detection in these diseases.

Only a few types of NHL immunophenotyping are suitable to detect MRD. Marker combinations usually represent quantitative changes that may not allow recognition of residual disease with high sensitivity. MRD detection in SCLL is described earlier. In another low-grade NHL, mantle cell lymphoma (MCL), it is possible to recognize MRD if lymphoma cells express CD5 (most of the cases do). The aberrant phenotypes are as follows: CD20$^+$/CD5$^+$/λ^+, CD20$^+$/CD5$^+$/μ^+, or CD20$^+$/CD5$^+$/CD23$^-$/λ^+. Among high-grade NHL, only ALCL can be monitored by immunophenotyping. Most of the cases are CD30$^+$ and express activation antigens such as CD25, Cdw70, CD71, and HLA-DR. However, this is a controversial entity that can display either T-cell, B-cell, or non-T, non-B phenotype. Of these, the T-cell phenotype is the most frequent, especially in childhood. Possible aberrant phenotypes are as follows: CD30$^+$/CD2$^+$/CD25$^+$/CD71$^+$, CD30$^+$/CD2$^+$/CD71$^+$/HLA-DR$^+$ or CD30$^+$/CD19$^+$/CD25$^+$/CD71$^+$, CD30$^+$/CD19$^+$/CD25$^+$/Cdw70$^+$ (Jennings and Foon, 1997).

5. Solid Tumor Malignancies

In solid tumor malignancies, especially in epithelial-derived tumors (e.g., breast cancer), the MAb targets cellular expression of epithelial cell-associated cytokeratins (usually CK18 or CK19). Dual-color staining techniques that detect cytokeratin$^+$/CD45$^-$ cells combined with the unique FLS properties of carcinoma cells allow flow cytometry to distinguish between residual tumor cells and normal cells in PB or BM. In addition, the detection of CK-18/oncoprotein p185^{erbB2}, CK-18/p120, or CK-18/Ki-67 double-positive cells in bone marrows obtained from patients with breast, gastric, and colon cancers prove marrow micrometastases (Ross, 1998; Pelkey *et al.*, 1996; Schoenfeld *et al.*, 1997). Staining for CD56 in bone marrow may allow the detection of invasive neuroblastoma (Nagai *et al.*, 1994), especially when it is combined with CD45. CD56$^+$/CD45$^-$ cells indicate marrow involvement.

C. Cell Culture

Cell culture techniques have been utilized for the identification of various hematopoietic progenitors that form cell colonies [CFU (colony forming unit) or CFC (colony forming cells)] in semisolid or liquid medium. For example, leukemic cells have the ability to make colonies, so called CFU-L (some of the epithelial-derived tumors, e.g., breast carcinoma, can also form colonies under appropriate conditions). The rationale of the cell culture method for MRD detection is to selectively grow cancer cells while suppressing the growth of the normal CFU (Estrov *et al.*, 1986, 1994). Despite substantial progress in culturing technology, it is still rather difficult to obtain a suitable, pure cancer cell population necessary for optimal sensitivity of MRD detection. Thus, false-positive results may originate from the normal CFU proliferating over CFU-L under conditions that should support only the leukemic cell growth. Moreover, cell culture methodologies require an additional step [e.g., PCR, fluorescence *in situ* hybridization (FISH), immunophenotyping] in order to demonstrate the clonality of each individual CFU. As a result of these cumbersome techniques, the sensitivity can differ from one culture assay to another, and thus it is very difficult to establish a clinically useful cutoff level of MRD from which to estimate the probability of relapse. In addition, a combination of these potential pitfalls surrounding cell culture methodologies could preclude an accurate analysis (Campana and Pui, 1995).

Other important reasons described reveal the inadequacy of cell culture in MRD detection:

1. Low applicability. Less than 30% of human leukemias can be cultured *in vitro*.

2. Lack of standardization in culture conditions. Various laboratories use different culture conditions, often relying on poorly defined components (e.g., supernatant obtained from phytohemagglutinin (PHA)-stimulated lymphocytes, human plasma, animal sera) that make the results difficult to compare.

3. Low numbers of CFU-L (0.01–1% of the total leukemic clone) that can be recognized with colony assays. The low quantity of the cells may not represent the total cancer population and can make an outcome of the disease unpredictable.

D. Cytogenetics

Cytogenetic analysis reveals that the majority of hematological as well as solid-tumor malignancies have chromosomal abnormalities. For example, 75–90% of ALL cases and 60–90% of AML cases show clonal qualitative, that is, structural chromosome changes such as translocation and inversion as well as quantitative cytogenetic aberrations such as monosomy and trisomy (Walker *et al.*, 1994). The most common chromosomal aberrations with their accompanying genotypic abnormalities that can be used as targets for karyotype or molecular detection of MRD are listed in Table IV.

1. Conventional Cytogenetics

Traditional cytogenetics utilizes conventional banding techniques of metaphase chromosomes. Success of this method depends on the number of metaphases (usually ≥25 is required) that can be analyzed and on the proliferative rate of cancer cells (Freireich *et al.*, 1993). For example, in ALL, the analysis is more difficult than in AML, because a fewer number of blasts enter mitosis; in solid tumors, obtaining mitotic cells is very often unsuccessful (Drexler *et al.*, 1995).

Although traditional cytogenetics identifies important cancer-specific markers, it suffers from several limitations that make karyotyping inappropriate for MRD detection:

1. Inability to recognize chromosomal alterations less than 5–10 Mb of DNA, which determines the low sensitivity of detection, which is similar to morphology
2. Suitable only to the mitotic cell
3. Culture conditions that may select cells proliferating well only *in vitro* but not representative for the total cancer population
4. Lack of correlation between cellular morphology or phenotype and genotype
5. Time-consuming, labor-intensive technique

2. Prematurely Condensed Chromosomes

The prematurely condensed chromosomes (PCC) method is the only modification of traditional cytogenetics that enables the analysis of karyotype in interphase (nondividing) cells (Hittelman *et al.*, 1990). Owing to the limitations just listed, this technique is also not suitable for MRD detection.

3. Fluorescence *in Situ* Hybridization

For a comprehensive review of the methodologies employed in FISH, see Fischer *et al.* (1994). Fluorescence *in situ* hybridization is essentially a molecular

Table IV
Karyotype Aberrations and Genotype Abnormalities That Can Be Used as Targets for Cytogenetic and/or Molecular Detection of Minimal Residual Disease in Hematological Malignancies

Karyotype aberration	Genetic abnormality	Disease[a]
del 1p32	SIL-TAL1	T-ALL
t (1;7)(p32;q35)	TAL1-TCRβ	T-ALL
t (1;7)(p34;q34)	LCK-TCRβ	T-ALL
t (1;14)(p32;q11)	TAL1-TCRδ	T-ALL
t (1;19)(q23;p13)	PBX1-E2A	Pre-B-ALL
t (2;8)(p12;q24)	MYC-IgLκ	B-ALL
t (3;21)(q26;q22)	EVI1-AML1(CBFα), EVI1-EAP, EVI1-MDS1	AML, CML-blastic phase, MDS-CMML
t (4;11)(q21;q23)	AF4-MLL(ALL1)	ALL, AML
t (5;12)(q33;p13)	PDGFβ-TEL	MDS
t (5;14)(q31;q32)	IL3-IgH	B-precursor-ALL
t (5;17)(q31;q11)	NPM-RARα	AML M3
t (6;9)(p23;q34)	DEK-CAN	AML, AUL, CML, MDS
t (6;11)(v;q23)	AFG-MLL(ALL1)	AML
t (7;9)(q34;q32)	TCRβ-TAN	T-ALL
t (7;10)(q35;q24)	TCRβ-HOX11	T-ALL
t (7;11)(q35;p13)	TCRβ-TTG2	T-ALL
t (7;11)(p15;p15)	NUP98-HOXA9	AML
t (7;19)(q35;p13)	TCRβ-LYL1	T-ALL
t (8;14)(q24;q11)	MYC-TCRα	T-ALL
t (8;14)(q24;q11)	PVY1-TCRδ	T-ALL
t (8;14)(q24;q32)	MYC-IgH	B-ALL
t (8;21)(q22;q22)	ETO-AML1(CBFα)	AML M2
t (8;22)(q24;q11)	MYC-IgLλ	B-ALL
del 9q34	SET-CAN	AUL
t (9;11)(p21-22;q23)	AF9-MLL(ALL1)	AML M5
t (9;22)(q34;q11)	BCR-ABL	CML, ALL, AML
t (10;14)(q24;q11)	HOX11-TCRδ	T-ALL
t (11;14)(p13;q11)	TTG2-TCRδ	T-ALL
t (11;14)(p15;q11)	TTG1-TCRδ	T-ALL
t (11;14)(q13;q32)	BCL1-IgH	CLL
t (11;17)(q23;q21)	PLZF-RARα	AML M3
t (11;19)(q23;p13)	MLL(ALL1)-ENL	ALL, AML
t (12;21)(p13;q11)	TEL-AML1(CBFα)	ALL, AML
t (12;22)(p13;q11)	TEL-ABL	ALL, AML, CML
t (14;19)(q32;q13.1)	IgH-BCL3	CLL
t (15;17)(q11-22;q21)	PLZF-RARα	AML M3
inv (16)(p13q22)/t (16;16)	CBFβ-MYH11	AML M4 eoz
t (17;19)(q22;p13)	HLF-E2A	Pre-B-ALL

[a] ALL, Acute lymphoblastic leukemia; AML, acute myeloid leukemia; AUL, acute undifferentiated leukemia; CLL, chronic lymphocytic leukemia; CML, chronic myelogenous leukemia; CMML, chronic myelomonocytic leukemia; MDS, myelodysplastic syndrome.

hybridization technique. A distinguishing feature between this and other hybridization methods such as Southern blotting and PCR is that the target DNA is detected inside an intact cell. This is beneficial to the investigator who may want to compare the results to the morphological or immunological phenotype. A substantial advantage of FISH over conventional cytogenetics is that information can be obtained from interphase cells. In effect, cells with a low mitotic index can be analyzed without the need for cell culture and its inherent risk of overpopulation by normal cells. The molecular target is most often a specific gross chromosomal abnormality. For example, numerical aberrations are identified by DNA probes specific to the centromeric portions of the chromosomes, and structural chromosomal abnormalities are detected by specific patterns generated by hybridization of DNA probes specific to both sides of a translocation, inversion, or deleted segments. When utilizing FISH to detect chromosomal abnormalities, the size of the cDNA probe is restricted to greater than 1 kb, mainly because probes smaller than this compromise the sensitivity of FISH analysis when the copy number of the target nucleic acid is fewer than 10.

a. Diagnostic Screening

Because FISH analysis is performed on specific cytogenetic aberrations and because of the vast numbers of abnormalities that have been characterized in malignancies (refer to Table IV), it is generally more practical to perform a conventional cytogenetic analysis when screening diagnostic samples. Some exceptions do exist, however. For instance, when a particular abnormality is strongly associated with a certain malignancy, such as the Philadelphia chromosome, t(9;22), in CML, or the bcl2 translocation, t(14;18), in follicular center cell lymphoma (FCCL), FISH can be used as a diagnostic tool. FISH is also useful in cases when the diagnostic samples are inadequate for conventional cytogenetic analysis. For example, cells from CLL, which usually have a low mitotic index, can be screened for trisomy 12 using a whole-chromosome paint probe (Anastasi, 1993).

b. Monitoring Minimal Residual Disease

For the detection of residual disease, FISH can be performed on cells in the metaphase and interphase stages of the cell cycle. However, because it is rare to observe more than 20 to 25 metaphases in a given specimen, standard metaphase FISH analysis is not recommended. Hypermetaphase FISH, which is a FISH technique performed after the introduction to the cell culture of a mitotic arresting agent (e.g., colcemid), significantly increases the number of cells that can be analyzed. However, because the cells from some malignancies (e.g., blastic phase Ph^+ CML) have a higher mitotic index than normal cells, hypermetaphase FISH analysis could result in an overestimation of the malignant burden. Conversely, in ALL, most of the leukemic cells remain in interphase *in vitro*, and metaphase FISH could result in an underestimation of the residual leukemic burden. Therefore in some cases of monitoring MRD, interphase FISH provides more accurate results and is the method of choice for most laboratories. However, interphase FISH has an inherent risk of "counting" cells of the malignant clone that have somehow reverted or are dying, and it should be used with caution.

The sensitivity of FISH improves as the number of markers used increases by reducing the occurrence of false positives. For instance, when using a single marker, positive signals can be generated in approximately 4% of normal cells by an apparent juxtaposition of the signals (e.g., locus-specific probes directed against the *BCR* and *ABL* genes) caused by colocalization of the chromosomes in the interphase nuclei. Additionally, normal diploid control specimens with "false trisomy" and "false monosomy" are routinely detected in about 2% and 6–11% of the cells, respectively. Use of a triple probe/multicolor system that involves labeling three probes with separate fluorochromes has reduced the incidence of false positives of detecting Ph^+ CML to 0.065–0.27% (Sinclair *et al.*, 1997). Other methods to increase sensitivity include the utilization of probes that span the break points and the use of a combination of translocation-specific probes and whole chromosome painting (sensitivity of 1×10^{-3}) (Gray *et al.*, 1990).

In ALL patients with high hyperdiploid clones (>50 chromosomes), a sensitivity of 10^{-4} can be achieved when interphase FISH specific for the hyperdiploid clone is performed (Kasprzyk and Secker-Walker, 1997). In this approach, up to three α satellite probes specific to the centromeric regions of the extra chromosomes are simultaneously utilized. A report has demonstrated the usefulness of combining FISH analysis with immunophenotyping in the quantitation of MRD in AML and MDS (Engel *et al.*, 1999).

Owing to advances in molecular techniques in the detection and quantitation of MRD, PCR has become a simpler and more sensitive assay. The use of genotypic abnormalities as markers for MRD detection is discussed in more detail later in Section III,F.

E. Flow Cytometry

1. Flow Karyotyping

Gray *et al.* (1990) give a comprehensive overview of flow cytomeric measurement of isolated chromosomes. The chromosomes usually are stained with two fluorescent dyes, for example, Hoechst 33258 (HO) and chromomycin A3 (CA3). HO binds preferentially to adenine–thymine-rich DNA, whereas CA3 stains guanine–cytosine base pairs. The flow cytometer has to be equipped with two lasers: an ultraviolet (UV) laser that first selectively excites HO (at 351 plus 363 nm) and a blue light laser, subsequently exciting CA3 (at 458 nm). The advantage of the method is the ability to analyze hundreds to thousands of chromosomes per second. The bivariate fluorescence distribution (CA3 versus HO), called flow karyotype, shows numerous peaks, each made by chromosomes with similar dye contents. The mean of the peak is defined by the DNA content and DNA base configuration of the corresponding chromosome type. Flow karyotyping, in theory, should have a better sensitivity than conventional cytogenetics because the technique allows the recognition of chromosomal alterations in about 2 Mb of DNA. However, there are disadvantages that make this method irrelevant for MRD detection:

1. Suitable only to the mitotic cell, as culture may select cells that proliferate well *in vitro* but are not representative of the total cancer population
2. Inability to identify chromosomes 9, 10, 11, and 12
3. High background owing to nuclei fragmentation and isolation of chromosomes that apparently makes the sensitivity similar to that of conventional cytogenetics

2. DNA Index

The DNA index method describes the quantity of cellular DNA assayed by flow cytometry, where the DNA index (DI) is the ratio of the G_1 peak position of examined cells to the G_1 peak position of a known diploid standard (Darzynkiewicz *et al.*, 1994). In diploid cells DI equals 1.0. Malignant cells can have various quantities of DNA, and the identification of aneuploidy (which occurs approximately in 60% of solid-tumor and 30% of hematological malignancies) using DNA-binding fluorochromes may afford detection of residual disease. For characteristics of the fluorochromes and DNA analysis, see Darzynkiewicz *et al.* (1994). Because of the low sensitivity of detection, this method is also negligible in the monitoring of MRD unless it is combined with immunophenotyping (Nowak *et al.*, 1997). However, certain limitations apply:

1. Paraformaldehyde-based fixatives, which are optimal for preserving membrane antigens, interfere with DNA dye binding and decrease the quality of DNA analysis. Small changes in DNA content, such as near-diploidy, may result in false-negatives.
2. Single ethanol fixation, which is superior for DNA analysis, as well as treatment with nonionic detergents can destroy most surface and some intracellular markers utilized in MRD detection.
3. Multicolor staining for aberrant or asynchronous phenotypes may be inconvenient due to overlapping DNA–dye-associated fluorescence spectra, which can decrease the sensitivity of the detection.

F. Polymerase Chain Reaction

The polymerase chain reaction (PCR) is a powerful *in vitro* molecular technique that allows for an exponential increase in the number of copies of a specific DNA segment by a series of repetitive steps. Each PCR cycle consists of DNA denaturation, annealing of oligonucleotide primers to the DNA template, and polymerization of deoxynucleotides by a DNA polymerase. The oligonucleotide primers, which are necessary to initiate the reaction, are designed to flank the DNA segment of interest on opposite strands of the double-stranded DNA template, with the 3′ ends facing toward each other. Under maximal efficiency, the polymerase is able to amplify 2^n (where n is the number of cycles) of product. PCR can be performed on any single- or double-stranded DNA target, including genomic DNA (native or cut with restriction enzymes), PCR products, or cDNA. In the case of a single-stranded DNA template, the complementary primer will anneal and elongate to synthesize the corresponding DNA strand during the first PCR cycle, and thus generate a double-stranded template for subsequent cycles.

The polymerase chain reaction has been crucial to the advancement of MRD analysis because it allows for direct visualization of DNA with as few as one initial target copy. The importance of PCR is overtly evident because it requires only small amounts of starting material, is extremely sensitive, and can be completed within a short period of time (from a few hours to a few days). The PCR product is usually sufficient to be visible after resolution on an agarose or polyacrylamide gel, followed by staining of the gel with a dye that specifically intercalates between the bases of the DNA and fluoresces with UV illumination. To confirm that the reaction is specific to the desired target, and that the DNA band is not the result of artifactual amplification, the PCR products can be further analyzed by transferring them to a nitrocellulose or nylon membrane (Southern transfer). The transferred DNA is then probed with a radiolabeled or fluorescence-labeled oligonucleotide, cDNA, or restriction digest fragment. Otherwise, the PCR product can be directly spotted onto the membrane for hybridization analysis (dot-blot hybridization). It is essential that the probe is specific to the internal sequence of the desired amplicon and that it does not cross-react with other DNA fragments. The hybridized products can be visualized by standard autoradiography or by scanning with PhosphorImager or Storm (Molecular Dynamics, Sunnyvale, CA), in the case of radiolabeled probes, or with FlourImager or Storm (Molecular Dynamics), for fluorescence applications. Although they are still utilized, radiolabeled probes are not recommended because their use is tedious and expensive. In contrast, fluorescence-labeled probes are less expensive, easier to handle, and more sensitive than radioactive counterparts (Dibenedetto *et al.*, 1997).

1. Characterization of Markers

a. Tumor-Specific Markers

Specific clonal chromosomal translocations, deletions, or inversions are associated with certain types of cancer. Molecular analysis of cytogenetic abnormalities allows the identification of unique DNA sequences that can be used as targets for molecular detection of MRD. This kind of MRD detection utilizes so-called tumor-specific markers. Owing to the abundance of abnormalities that have been identified and molecularly characterized, we will focus on the few most useful markers for the detection of MRD. Table IV lists the most commonly studied genotype abnormalities for the monitoring of residual malignant cells. Most of these tumor-specific markers can be characterized either at diagnosis or relapse by classic cytogenetic methods as well as FISH, Southern transfer, reverse transcriptase (RT)–PCR, or PCR. Although most translocations result in a gross chromosomal abnormality that can be identified by karyotyping, some translocations [e.g., the t(12;21)] are usually cryptic to such analysis (Sawyers, 1998). Thus, screening for the TEL/AML1 fusion product of the t(12;21) translocation must be performed by molecular methods such as RT–PCR and Southern transfer. Most translocations involve a large junction comprising long stretches of unknown intronic sequences and are not easily amenable to DNA amplification. When using molecular methods to screen for these translocations, the mRNA must first be transcribed into cDNA by utilizing an enzyme called reverse transcriptase, followed by amplification of the cDNA using primers specific to the chimeric transcript.

Virtually all translocations, deletions, and inversions for which the interrupted genes have been molecularly defined are good candidates for leukemia-specific markers. For instance, detection of the inversion of chromosome 16 [inv(16)(p13q22)] and the translocation t(16;16)(p13;q22) in patients with AML-M4Eo can be identified by RT–PCR with primers specific to the CBFB/MYH11 fusion transcripts (Tobal *et al.*, 1995). In order to prevent unwanted RNA degradation, it is important to process the diagnostic sample within 1 hr. If the specimen is BM or PB, separate the mononuclear cells from the red blood cells by density-gradient centrifugation, then either freeze the cells at $-80°C$ in 10% DMSO or start the extraction process. We obtain a suitable RNA yield and quality by using the RNAzol B RNA isolation kit (Tel-Test, Inc., Friendswood, TX). Because certain abnormalities are associated with specific malignancies, a panel of primers can be used in multiple RT–PCRs as a screening process for the most common aberrations. After the PCR, it is important to sequence the products in order to confirm the presence of the aberration. For a description of direct sequencing of PCR products, see later.

b. Clone-Specific Markers

One of the earliest events during B-cell differentiation is the rearrangement of the genes that code for the immunoglobulin heavy (μ) chain (IgH), the protein product of which makes up part of the surface immunoglobulin receptor later in B-cell development. T lymphocytes, like their B-cell counterparts, also undergo gene rearrangements early in differentiation. The genes that code for the T-cell receptor (TcR) code for two types of receptors, the $\alpha–\beta$ and the $\gamma–\delta$ receptor. For both the IgH and TcR β and δ, the exons that code for the variable portion are termed variable (V), diversity (D), and joining (J) segments. Because each cell retains its unique gene rearrangement during clonal expansion, the rearrangement can be used as a marker to probe for the persistence of malignancies that are of B-lineage and T-cell origin. However, these rearrangements are not directly linked to the transformation process and consequently are used as surrogate markers for the malignant clone. This kind of MRD detection utilizes so-called clone-specific markers.

Up to 95, 54, 55, and 33% of B-lineage ALL patients contain a detectable clonal IgH, TcR δ, TcR γ, and TcR β gene rearrangements, respectively (Campana and Pui, 1995). In T-cell ALL, rearrangements of the TcR δ and TcR γ genes occurs in up to 95% of the cases. Of the δ rearrangements in T-cell ALL, Vδ1–Jδ1, Vδ2–Jδ1, Vδ3–Jδ1, and Dδ2–Jδ2 are the most common, occurring in up to 70% of the cases (Cole-Sinclair *et al.*, 1994; Campana and Pui, 1995). Clonal antigen receptor gene rearrangements are not limited to the acute leukemias but are found in many malignancies of the B-lineage and in T-cell malignancies.

Methods to detect clonal IgH and TcR gene rearrangements in B- and T-cell malignancies include Southern blotting and enzymatic amplification. DNA analysis of these gene rearrangements by Southern blotting, though sensitive enough (1–5%) to identify clonal rearrangements, is labor-intensive, time-consuming, and not practical for MRD analysis. Enzymatic amplification of the genes by PCR is less costly and simpler to perform. It also allows for subsequent direct sequencing and design of clone-specific oligonucleotides.

Because the majority of normal B and T cells contain rearranged antigen receptor genes, it is necessary that the sample used to characterize the rearrangement of the malignant clone is taken at the time of diagnosis or relapse. The optimal source of material from which to analyze for clonal gene rearrangements is fresh or cryopreserved mononuclear cells that were isolated from a bone marrow aspirate by density-gradient centrifugation. BM smears of stained or unstained slides are acceptable. Our yield in characterizing clonal rearrangements from isolated mononuclear cells and unstained bone marrow smears taken from pre-B ALL patients are 90 and 77%, respectively. DNA can be extracted by standard phenol–chloroform extraction and ethanol precipitation, or by using one of the many commercially available kits.

A multitude of oligonucleotide primers are available that are specific to the consensus portions of the complementarity determining regions I and III of the IgH, as well as of the δ and γ TcR gene rearrangements, and PCR analysis has become rather standardized. The PCR products are concentrated using Microcon 30 filters (Amicon, Beverly, MA) and then run on a high percentage low melting point agarose (such as Nusieve, FMC BioProducts, Rockland, ME) gel. The gel is stained with ethidium bromide, and the DNA bands are visualized using a standard UV light box. Figure 2 shows the PCR products of the IgH and TcR (Vδ2–Dδ3 and Dδ2–Dδ3) rearrangements from the diagnostic DNA of a pre-B ALL patient. The presence of one or more distinct bands indicates a clonal expansion. In contrast, the presence of a DNA smear (or sometimes a blank lane) in the expected location represents amplification of a polyclonal population. This particular patient has a monoclonal IgH gene rearrangement and a monoclonal Vδ2–Dδ3 gene rearrangement.

Most samples do not need to be cloned before sequence analysis, but rather can be directly sequenced from the PCR products. This is achieved by excising the band of interest from the agarose gel using a sterile scalpel and eluting the DNA from the gel slice by centrifugation through a Microcon 30 filter and gel

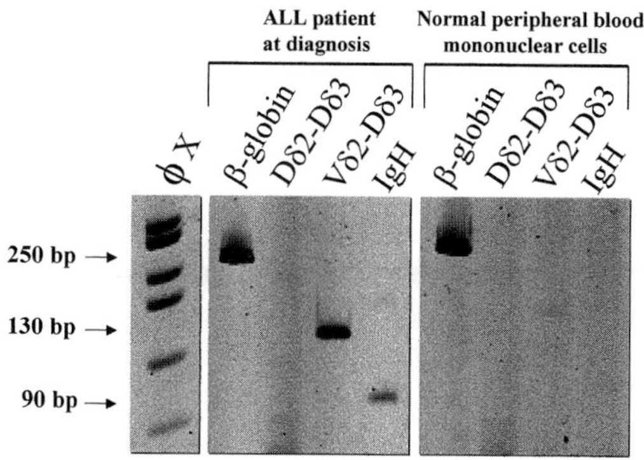

Fig. 2 PCR amplification of a pre-B ALL patient at diagnosis reveals monoclonal bands in the IgH and Vδ2–Dδ3 lanes, whereas no monoclonal rearrangements are evident in normal PB.

nebulizer (Amicon). The eluted DNA can then be sequenced in both directions using each of the PCR primers, which can be performed at a core facility or by one of many available commercial sequencing facilities.

Over 25% of the pediatric ALL patients whose diagnostic samples harbor a detectable, clonal IgH gene rearrangement exhibit multiple bands from the PCR products. In the past, if the bands ran too close together, the only option for sequencing was to clone the products and perform up to 10 sequencing reactions (Yamada *et al.*, 1990). Not only is cloning expensive, it considerably increases the chances for contamination of subsequent PCRs. If the bands are too close to each other to cleanly cut with a scalpel, stab the bands with an autoclaved toothpick, then dip the toothpick into a premixed PCR tube, and reamplify the bands in separate reactions. Occasionally, two bands in the PCR product will comigrate on an agarose gel. Add 1 U of Resolver Gold (Novagen, Madison, WI) per milliliter of gel solution before casting the gel. Resolver Gold is an intercalating agent that preferentially binds to A- and T-rich regions of DNA and retards the movement of DNA fragments that are A+T rich.

2. Technical Considerations of PCR Approaches for Minimal Residual Diseases

Chromosomal translocations that have been defined at the molecular level are good candidates for amplification by PCR. Amplification of the *bcr–abl* gene fusion transcript in CML patients was the first example of RT–PCR amplification to detect MRD by a leukemia-specific translocation break point. The types of techniques currently used to detect and quantitate malignant cells by chromosomal translocations as markers are quite varied. Quantitation of RNA transcripts can be complicated. The largest difficulties are associated with unknown half-lives that can vary for each type of transcript. Therefore, successful amplification of cDNA from a housekeeping gene can, at best, provide only an approximation of the integrity of the transcript being quantitated. In addition, RT–PCR does not quantify the number of positive malignant cells in the sample, but rather the number of transcripts. Thus, for translocations whose gene expressions are low, whether due to inherent or drug-related phenomena, the malignant burden could be underestimated. In fact, fewer than 1000 PML–retinoic acid receptor (RARα) molecules may be present in 1 μg of total RNA extracted from the bone marrow of acute promyelocytic leukemia (APL) patients, derived from approximately 10^6 blasts. One approach to improve overall sensitivity in this particular case is to expose the blasts *in vitro* to interferon-α (IFNα) prior to RNA extraction and subsequent PCR (Yokota *et al.*, 1991; Westbrook *et al.*, 1992).

A disadvantage to utilizing the antigen receptor gene rearrangements for MRD monitoring is the potential for clonal evolution. There is evidence of clonal instability in almost 10% of the TcR δ and γ rearrangements in T-cell ALL, and in 31% of the IgH gene rearrangements in pre-B ALL. However, if the diagnostic sample is screened for both the IgH and TcR gene rearrangements, the overall rate of false-negative relapse prediction can be reduced to less than 5% (Steward *et al.*, 1994).

Enzymatic amplification of the IgH and TcR gene rearrangements to monitor the persistence of MRD can be very sensitive, but it has traditionally been relatively labor-intensive and time-consuming. Previous methods involved amplification of the gene rearrangement using a set of primers that was specific to consensus regions flanking the unique rearrangement of DNA segments, followed by Southern transfer of the PCR products and subsequent hybridization using radioactive-labeled probes specific to the leukemic clone. Because this methodology cannot easily be utilized in a large-scale clinical trial, emphasis has been placed on reducing the time, expertise, and expenses that are necessary to obtain reliable results.

Quantitation of the number of copies of a DNA target in a sample has been achieved by utilizing various, rather ingenious, methods. However, the molecules being examined are the products of numerous cycles of exponential amplification, where the variables and levels of efficiency at each cycle cannot be precisely determined. Therefore in the world of molecular biology, when a DNA target is reported to have been ''quantitated'' by PCR, it is understood that the target was at best semiquantitated. The more useful reports of quantitative PCR analysis state a ''most probable range'' of the level of starting copies, and include the calculated error associated with the results.

a. Fingerprinting

The fingerprinting approach involves PCR amplification of the IgH and TcR gene rearrangements using primers specific to the consensus regions; however, rather than hybridize the products to a clone-specific probe, the pattern of the consensus PCR product is analyzed. Specifically, the PCR products are allowed to migrate on a large polyacrylamide or agarose gel, and if a DNA band is present and dominant over the background levels of polyclonal amplification products, and is the same size as that identified at the time of diagnosis, it is considered to be a positive result (Chim *et al.*, 1996). A similar procedure has been developed whereby one of the primers is 5' end labeled with a green fluorochrome such as HEX, JOE, or TET. The PCR products are then loaded onto a 6% denaturing polyacrylamide gel, electrophoresed on an Applied Biosystems DNA sequencer, and analyzed using Genescan analysis software. The gel image is converted to an electrophoretogram, which consists of multiple peaks, the heights of which are directly proportional to the intensity of the fluorescence on the gel. A specific pattern or "fingerprint" identical to that expected is considered a positive result. Although this approach is simpler, less expensive, and more expedient than other methods, its sensitivity at best reaches only 10^{-3}, and it is not quantitative. However, it has the potential to identify emerging clones that are independent or have evolved from the original malignant clone (Evans *et al.*, 1998).

b. Clone–Specific PCR

DNA amplification using oligonucleotide primers that are specific to unique, clone-specific portions of the IgH or TcR sequence is as sensitive as, and simpler than, the labor-intensive Southern blotting and radioisotope labeling methods. A heminested PCR can be performed whereby the PCR products of consensus primers for the TcR γ and δ rearrangements in T-ALL patients are subjected to a second round of PCR. The heminested, second-round PCR is performed using a fluorescence 5' end-labeled primer spanning the unique junction of the malignant clone and one consensus primer from the first PCR step. Overall, fluorescence heminested PCR is more sensitive than standard Southern blot hybridization to consensus primer-amplified DNA (Dibenedetto *et al.*, 1997).

The use of this PCR technique alone does not confer the opportunity to quantitate the percentage of clone-specific cells in the sample. Essentially, when a heminested PCR is performed, the PCR products, which are already exponentially amplified, are subjected to a second round of PCR. The intensity of the DNA band may not be directly proportional to the number of starting copies, because there is a plateau effect whereby the number of copies increases very little during the last few cycles. In addition, the reamplification procedure introduces a major source of potential contamination and is difficult to apply to the clinical setting.

c. RNase Protection Assay

An RNase protection assay has been developed for the detection of TcR γ genes in T-cell disease and the IgH gene rearrangement in B-cell malignancies (Veelken *et al.*, 1991; Kurokawa *et al.*, 1996). In this procedure, PCR amplification is performed on the test samples using consensus primers for the IgH or TcR gene. The PCR products are then transcribed into RNA using T7 RNA polymerase, which is followed by digestion of the genomic template DNA with DNase I. The transcripts are hybridized with radiolabeled (or, theoretically, fluorochrome-labeled) RNA probes prepared from clonal PCR products of the malignant clone. The mismatched nucleotides are subsequently removed by a digestion step with RNase A, and the resultant RNA is resolved on a polyacrylamide gel containing urea, followed by image analysis (autoradiograph or phosphorimaging). This approach is convenient because it does not require molecular characterization of the diagnostic clonal rearrangement or synthesis of clone-specific primers or probes. This assay has high sensitivity (10^{-5}), but it is more susceptible to false negatives due to clonal evolution because the clone-specific probes span the V–N–D junction, the most unstable portion of the rearrangement.

d. Competitive PCR Assay

The simultaneous addition of competitor DNA to the native DNA template of interest prior to PCR is currently the most widely used quantitative PCR assay. In this method, a plasmid construct is engineered to introduce an extra DNA sequence between the primer sites used for PCR. Both the competitor and native DNA templates compete for the same PCR reagents, including the primers. A titration is performed that includes a constant amount of sample DNA but varying amounts of competitor. The equivalence point, the point at which the intensity of the two products are the same, allows the estimation of the number of native DNA copies that are present in the sample. This technique has proved useful for quantitating the number cDNA products of *bcr–abl* transcripts in CML patients (Cross *et al.*, 1993; Lin *et al.*, 1996), for which a sensitivity of 10^{-5} was achieved.

The competitive PCR assay is not without problems. An additional standard must be included to confirm that the specific native DNA (or mRNA) remained intact through sample processing. In an attempt to avoid potential underestimation of the *bcr–abl* transcript target in CML patients, a known amount of the competitor was introduced to the sample before RNA purification (Moravcová *et al.*, 1998). In this way, both the DNA target of interest and internal standard were influenced in the same way throughout the procedure from RNA extraction to the analysis of products. Although this procedure cannot account for RNA damage of the specific target before sample processing, if the samples are systematically processed, say, within 1 hr, the quality of the samples could be normalized throughout the project.

e. Limiting Dilution

The quantitation of DNA by PCR can be tricky because the molecules that are estimated are the products of a reaction for which there is no means to truly observe and account for the dynamic interactions that occur. It has been demonstrated numerous times that sample-to-sample variability, with all other things being equal, results in drastic differences in the amount of product generated. One way to prevent this variability is to adjust the concentration of the DNA such that there are only two possible end results, positive or negative.

The limiting dilution method can be applied to any PCR/RT–PCR detectable marker (Ouspenskaia *et al.*, 1995, Sandoval *et al.*, 2000). It involves a stepwise series of dilutions of the sample into a "negative control" DNA sample, the amplification product of which is negative. When we utilize this approach, 1 μg of DNA (equivalent to 100,000 cells) is subjected to a single round of PCR in each of 10 separate reactions (Mayer *et al.,* 1999). Because each reaction is designed to give a positive result when at least one copy is present, it is considered to be an all-or-none reaction. In other words, the absence of target yields no PCR product, and the presence of one or more copies yields a single distinct band of the expected molecular weight. By performing a series of dilutions, the number of targets in the sample is reduced, until only a fraction of the 10 replicates are positive. From there, the number of target copies in the original stock can be calculated.

In order to carry out this approach successfully, there must be strict adherence to the rules governing clone-specific primer design. The clone-specific primers must be designed to hybridize to a unique region of the rearrangement such that no other normally rearranged lymphocyte can be amplified by it. In our laboratory, we design clone-specific primers to span the D–N–J junction of the IgH gene rearrangement to amplify with a universal V primer. For the Vδ2–Dδ3 and Dδ2–Dδ3 TcR gene rearrangements, we design the clone-specific primers to span the V–N–D and D–N–D junctions, respectively, to amplify with a universal downstream primer from the D segment. Figure 3 shows a schematic representation of the gene rearrangements and the respective primer orientation. The V segments are in green, the D segments are in red, and the J segments are in blue. Notice that the 3′ end of the clone-specific primers span the rearrangements at a unique junction of the segments.

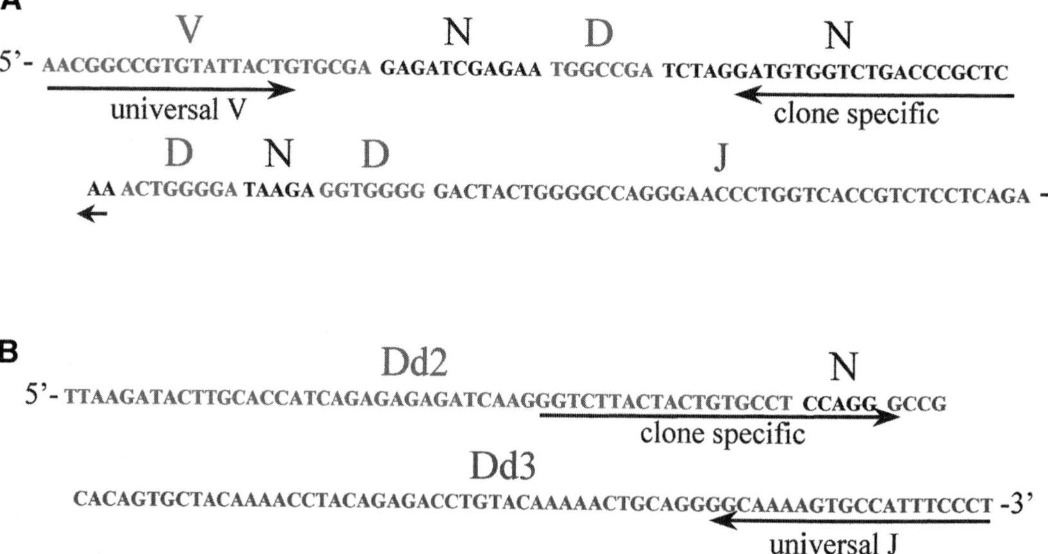

Fig. 3 Sequences of IgH (A) and Vδ2–Dδ3 (B) gene rearrangements from a pediatric pre-B ALL patient at diagnosis. Primers for clone-specific PCR are indicated. (See color plates.)

We generally do not include an internal control in our enzymatic reactions. We and others have found that the internal control reduces the overall sensitivity of the reaction, most likely by competing for the enzyme, dNTPs, and primers. The all-or-none mechanism precludes the need to quantitate the amount of target in each reaction and thus obviates the need for comparisons to a simultaneous control. We do, however, confirm the DNA integrity and absence of inhibitors by simultaneously amplifying the β-globin housekeeping gene in a separate reaction.

Figure 4 shows an example of the results from a pre-B pediatric ALL patient at week 10 of diagnosis. Five of the ten samples are positive, and after interpolating to the model curve, the malignant burden is calculated to be 6.6×10^{-6}.

f. Real-Time Quantitative PCR

Nucleic acid amplification techniques are now starting to be semiautomated, which allows for a more rapid and standardized assessment of MRD status. One novel automated approach to monitoring MRD is referred to as real-time automated PCR. The Taqman system (Perkin–Elmer Applied Biosystems, Foster City, CA) takes advantage of an oligonucleotide, with both a reporter and a quencher dye attached, that anneals to the template DNA between the primers. As the polymerase extends the primer, the $5'$ nuclease activity causes the reporter dye to be separated from the quencher dye, which results in the generation of a fluorescent signal. The change in fluorescence intensity is monitored in the PCR tube by a microplate fluorescence scanner that is connected to a computer for analysis. High set-up costs for optimizing the conditions for each fluorogenic probe diminishes the usefulness of the technique in monitoring MRD with IgH and TcR rearrangements as markers; however, this system could be useful for detection of transcripts from translocations. In fact, its usefulness has already been demonstrated in MRD detection in Ph[+] CML (Mensink *et al.*, 1998), in AML/ETO-associated AML (Marucci *et al.*, 1998), and on IgH and TcR gene rearrangements in ALL (Pongers-Willemse *et al.*, 1998) patients.

A more versatile system using the LightCycler (Roche Molecular Biochemicals, Indianapolis, IN) can be used with hybridization probes and by direct detection of clone-specific PCR products (Nakao *et al.*, 2000). The LightCycler system is feasible

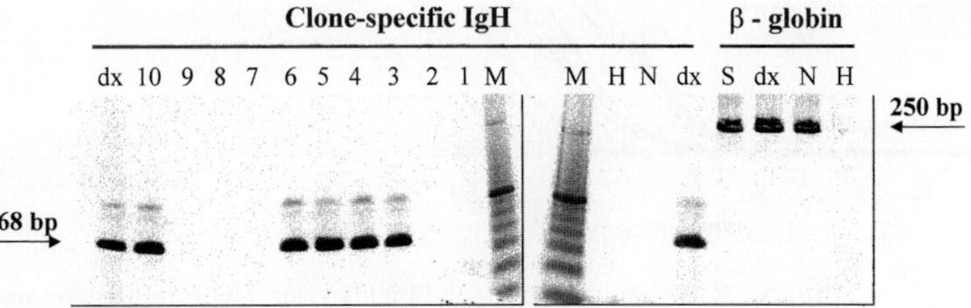

Fig. 4   A 10-replicate clone-specific PCR assay on a pre-B ALL patient 10 months after diagnosis yields five positive reactions. This corresponds to a leukemic burden of 6×10^{-6}. Lane dx, diagnostic sample; H, water; PB, peripheral blood; S, sample.

for monitoring IgH and TCR rearrangements. In fact, comparison studies between Limiting Dilution PCR and the LightCycler in our laboratory show that the Light-Cycler is accurate and rapid, albeit less sensitive (2×10^{-5}) than the limiting dilution method. Figure 5 shows an example of a LightCycler standard curve. Dilutions of a diagnostic bone marrow sample from a pre-B ALL patient were tested using IgH clone-specific primers, and the points at which the fluorescence crossed the noise band correlated with the number of added template copies.

3. Monitoring MRD by Molecular Analysis

a. Chronic Myeloid Leukemia

At least 95% of CML patients have a leukemia-specific translocation, t(9;22), resulting in fusion of the ABL and BCR genes. This translocation is also identified in 50% of adult ALL and 3–5% pediatric ALL. At the molecular level, there appear to be two distinct areas in which the translocation break points occur. In CML, the BCR gene break point usually occurs in the major break point cluster region (M-bcr), and in two-thirds of ALLs containing the BCR-ABL translocation, the BCR gene break point falls in the minor break point cluster region. There is evidence that some normal hematopoietic cells might express BCR-ABL (Bose *et al.*, 1998). Although this could limit the sensitivity of the assay with this particular marker, a quantitative assessment of the samples should prevent false-positive conclusions. Despite the fact that quantitation of transcripts by RT–PCR is difficult, numerous attempts have been made to estimate the level of expression by performing competitive RT–PCR. For instance, by adding a natural competitor (e.g., cells expressing the fusion gene) to the BM sample

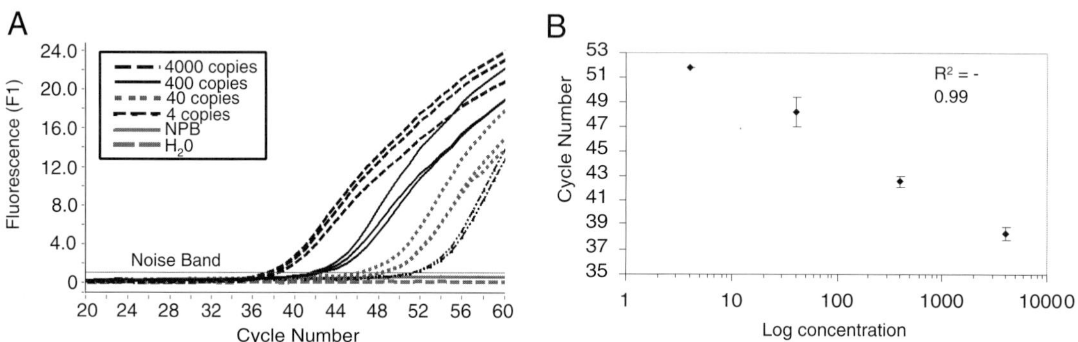

Fig. 5 Real-time PCR using the LightCycler. (A) Real-time fluorescence readings versus the cycle number. Samples that contained a higher starting copy number amplified at an earlier cycle number. (B) Standard curve calculated for the LightCycler assay. The point at which the fluorescence increased beyond the noise band was plotted as cycle number versus concentration for each sample. Linearity was maintained from 4000 copies to 4 copies.

prior to RNA extraction, one can account for variables in RNA extraction and amplification efficiency. In addition, if the competitor fusion gene expresses a variant transcript that is different in size from the target cDNA, one can easily compare the band intensities. Using this method, a sensitivity as high as 10^{-5} can be obtained (Moravcová *et al.*, 1998).

b. Acute Lymphoblastic Leukemia

The most common translocation occurring in pediatric pre-B ALL, characterized in 20–25% of the cases is fusion of the TEL-AML1 genes (also known as ETV6-CBFα2), is t(12;21). It is also found in 2% of adult ALL. This cytogenetic–cryptic translocation fuses the helix–loop–helix domain of the TEL gene to the DNA-binding and transactivating domains of the AML1 gene. A sensitivity of only 10^{-4} can be obtained by a seminested, two-step PCR approach (Satake *et al.*, 1997). Residual disease can be quantitated using the TEL-AML1 fusion as a marker by use of an internal standard (Nakao *et al.*, 1996). This competitor can be derived from a plasmid vector construct that contains either a long TEL-AML1 amplicon or a short one. PCR can then be carried out on different mixtures of both to determine the equivalence point. Quantification of unknown remission samples is made possible by comparisons of the relative amplification to that of the standard.

c. Acute Promyelocytic Leukemia

The most common translocation characterized in AML-M3 is t(15;17), which results in the fusion of the PML and RARα genes. Until recently, detection of residual leukemia, mediated by the RT-PCR detection of the PML-RARα transcript, lacked the sensitivity required to be useful. In effect, even if there are several serial negative samples, residual disease could be as high as one malignant cell in 1000 normal cells. This confirms that monitoring MRD by assessing the amount of mRNA is sometimes inaccurate, because there can be a variable level of transcript in each cell, and it is difficult to account for the unknown half-life of a particular transcript. One group of investigators attempted to improve the sensitivity by increasing expression of PML-RARα in the malignant cells. The cells were incubated in INFα prior to RNA extraction, and the sensitivity improved by one order of magnitude (Seale *et al.*, 1996). Tobal and Liu Yin (1998) attributed the low sensitivity of detection for the PML-RARα transcript to suboptimal reverse transcription conditions. They developed a "hot-start" RT–PCR method with increased sensitivity at the level of two leukemic cells in 1×10^{-6} cells.

d. Acute Myelomonocytic Leukemia

Acute myelomonocytic leukemia with bone marrow eosinophilia (AML-M4Eo) is strongly correlated with the inv(16) and the t(16;16) rearrangements, both of which result in the fusion of the CBFβ and MYH 11 genes (Liu *et al.*, 1993; Sawyers, 1998). Positivity by qualitative RT–PCR of this gene fusion has no prognostic value, thereby necessitating development of a quantification assay. A competitor titration assay has been developed to quantitate the CBFβ/MYH

11 transcripts in AML patients in CR (Evans *et al.*, 1997). The assay utilized an internal competitor artificially prepared by a site-specific directed PCR reaction using a plasmid as a template, and it achieved a sensitivity of 1×10^{-4}.

e. Non–Hodgkin's Lymphoma

For the monitoring of residual disease in NHL patients, the Ig/TcR gene rearrangements are the most frequently utilized markers. However, in specific subtypes of NHL, certain chromosomal translocations and gene rearrangements are common. In the lymphoid malignancies, the translocations often involve a juxtaposition of the antigen receptor gene on chromosome 14 and a protooncogene. Some of these translocations are suitable for MRD detection by PCR of the genomic DNA because the break point region clusters within a small region, such as for t(11;14) and t(14;18). Either the major break point region (MBR) or minor break point region (mbr) can be sequenced in t(14;18)-positive NHL patients. Oligonucleotide junction-specific probes can then be designed for hybridization to PCR-amplified CR samples, allowing a sensitivity of 1×10^{-6} (Galoin *et al.*, 1996).

f. Solid Tumor Malignancies

RT–PCR is the most widely used method of detecting residual disease in solid tumors. Otherwise, immunocytochemistry is an acceptable technique if RT–PCR is not available. For example, mRNA from the prostate-specific antigen (PSA) and the prostate-specific membrane antigen (PSMA) genes in PB samples of patients with prostate cancer can be amplified and analyzed by standard gel electrophoresis. However, PSA and PSMA are not cancer-specific, but rather are normal genes strongly associated with prostate cancer, and low levels of these genes are expressed by brain, salivary glands, and small intestine. Thus, the sensitivity of detection is dependent on the level of expression on noncancer cells (Pelkey *et al.*, 1996; Ross, 1998).

Cytokeratins 18 and 19, epithelial-cell associated cytokeratins, have been utilized in the monitoring of MRD in breast cancer and other epithelial-derived cancers (Schoenfeld *et al.*, 1997; Ross, 1998). Also, amplification of the surfactant proteins A, C, and D has been utilized to monitor the presence of non-small-cell lung cancer cells in lymph nodes (Pelkey *et al.*, 1996). For neuroblastoma, neuroendocrine protein gene product 9.5 and tyrosine hydroxylase mRNA have been assessed in the PB and BM (Pelkey *et al.*, 1996).

G. Sensitivity and Applicability

Sensitivity and applicability of the methods are given in Table V.

IV. Technical Problems

Potential technical difficulties that are associated with particular MRD detection methods are discussed in the appropriate sections of this chapter, and Table VI summarizes the most common pitfalls in MRD detection.

Table V
Sensitivity and Applicability of Methods for Detecting of Residual Leukemia in Bone Marrow Samples

Disease	Method	Sensitivity	Applicability (%)
ALL	Immunophenotyping	10^{-4}–10^{-5a}	5–50[b]/90–95[a]
	Conventional cytogenetics	$>10^{-1}$	5–25
	FISH	10^{-1}–10^{-4}	5–35
	PCR, translocations	10^{-4}–10^{-6}	5–30
	PCR, IgH gene rearrangements	10^{-3}–10^{-6}	$>90^{b}$
	PCR, TcR γ/δ gene rearrangements	10^{-3}–10^{-6}	40–60
AML	Immunophenotyping	10^{-4}–10^{-5a}	5–40
	Conventional cytogenetics	$>10^{-1}$	40–60
	FISH	10^{-1}–10^{-4}	50–70
	PCR, translocations	10^{-4}–10^{-6}	20–40

[a] T-ALL.
[b] B-lineage ALL.

Table VI
Potential Sources of Error in Detection of Minimal Residual Disease

Method	False-negative	False-positive
All	1. Few number of cells in a sample; 2. Heterogeneous distribution of MRD clusters	
FISH	1. Weak hybridization signal	1. Natural occurrence of aneuploidy in normal cells; 2. Artifactual colocalization of probes
PCR, translocations	1. Degradation of DNA or RNA, lack of mRNA expression 2. Inhibitors of *Taq* polymerase (e.g., hemoglobin, heparin) 3. Irrelevant probes 4. Inappropriate hybridization conditions	1. Contamination with DNA or RNA from other samples; 2. Contamination with PCR products
PCR, IgH /TcR gene rearrangements	1–4. As for PCR, translocations 5. Noncomplete rearrangement or gene deletion 6. Clonal change in rearrangement (usually in a relapse of, e.g., leukemia)	1 and 2. As for PCR, translocations 3. Cross hybridization with similar sequences derived from normal lymphocytes
Cell culture	1. Low proliferative activity of cancer cells 2. Apoptosis of cancer cells due to deficiency in stromal support	1. Analysis of colonies formed from noncancer cells
Immunophenotyping	1. Phenotype switch (usually in a relapse of, e.g., leukemia)	1. Nonspecific antibody binding; 2. Cancer-associated phenotypes expressed in normal cells

V. Concluding Remarks

   Comparison of the current MRD techniques demonstrates the wide applicability to analyzing varying markers. Utilization of PCR and immunophenotyping seem the most promising in monitoring residual disease. Although each of these methods is valuable, they have weak points that should not be dismissed. However, where one methodology is lacking, the other seems to excel. For example, PCR, though highly sensitive, cannot discriminate viable from dying cells. In contrast, immunophenotyping can distinguish live cells. In addition, although immunophenotyping is more quantitative, it is not usually as sensitive as PCR. Thus, an approach to monitoring residual disease that utilizes both methodologies simultaneously would maximize the benefits of reliable detection.

References

Anastasi, J. (1993). Fluorescence in situ hybridization in leukemia. Applications in diagnosis, subclassification, and monitoring the response to therapy. *Ann. N.Y. Acad. Sci.* **677,** 214–224.

Behm, F. G., Smith, F. O., Raimondi, S. C., Pui, C. H., and Bernstein, I. D. (1996). Human homologue of the rat chondroitin sulfate proteoglycan NG2, detected by monoclonal antibody 7.1, identifies childhood acute lymphoblastic leukemias with t(4;11)(q21;q23) or t(11;19)(q23;p13) and MLL gene rearrangements. *Blood* **87,** 1134–1139.

Bose, S., Deninger, M., Gora-Tybor, J., Goldman, J. M., and Melo, J. V. (1998). The presence of typical and atypical BCR-ABL fusion genes in leukocytes of normal individuals: Biologic significance and implications for the assessment of minimal residual disease. *Blood* **92,** 3362–3367.

Campana, D. (1994). Applications of cytometry to study acute leukemia: In vitro determination of drug sensitivity and detection of minimal residual disease. *Cytometry* **18,** 68–74.

Campana, D. and Pui, C. H. (1995). Detection of minimal residual disease in acute leukemia: Methodologic advances and clinical significance. *Blood* **85,** 1416–1434.

Cavé, H., van der Werff ten Bosch, J., Suciu, S., Guidal, C., Waterkeyn, C., Otten, J., Bakkus, M., Thielemans, K., Grandchamp, B., and Vilmer, E. (1998). Clinical significance of minimal residual disease in childhood acute lymphoblastic leukemia. European Organization for Research and Treatment of Cancer- Childhood Leukemia Cooperative Group. *N. Engl. J. Med.* **339,** 591–598.

Chim, J. C, Coyle, L. A, Yaxley, J. C., Cole-Sinclair, M. F., Cannell, P. K., Hoffbrand, V. A., and Foroni, L. (1996). The use of IgH fingerprinting and ASO-dependent PCR for the investigation of residual disease (MRD) in ALL. *Br. J. Haematol.* **92,** 104–115.

Cole-Sinclair, M., Foroni, L., and Hoffbrand, V. A. (1994). Genetic changes: Relevance for diagnosis and detection of minimal residual disease in acute lymphoblastic leukemia. *Bailliére's Clin. Haematol.* **7,** 183–233.

Coustan-Smith, E., Behm, F. G., Sanchez, J., Boyett, J. M., Hancock, M. L., Raimondi, S. C., Rubnitz, J. E., Rivera, G. K., Sandlund, J. T., Pui, C. H., and Campana, D. (1998). Immunological detection of minimal residual disease in children with acute lymphoblastic leukemia. *Lancet* **351,** 550–554.

Cross, N. C., Feng, L., Chase, A., Bungey, J., Hughes, T. P., and Goldman, J. M. (1993). Competitive polymerase chain reaction to estimate the number of BCR-ABL transcripts in chronic myeloid leukemia patients after bone marrow transplantation. *Blood* **82,** 1929–1936.

Darzynkiewicz, Z., Robinson, J. P., and Crissman, H. A. (eds.) (1994). "Methods in Cell Biology: Flow Cytometry" 2nd Ed., Parts A and B, Vols. 41 and 42. Academic Press, San Diego.

Deptala, A. (1995). The significance of immunophenotyping in detection of minimal residual disease in acute leukemias. Doctoral thesis, Warsaw Medical University, Warsaw, Poland (in Polish).

Deptala, A., Widzyńska, I., and Kuratowska, Z. (1995a): Clinical significance and prognosis in acute leukemias because of the detection of minimal residual disease. *Acta Haematol. Pol.* **26,** 413–420 (in Polish).

Deptala, A., Widzyńska, I., and Kuratowska, Z. (1995b). Immunophenotyping of bone marrow cells detects minimal residual disease in acute leukemias. *Acta Haematol. Pol.* **26,** 403–411 (in Polish).

Dibenedetto, S. P., Lo Nigro, L., Mayer, S. P., Rovera, G., and Schiliro, G. (1997). Detectable molecular residual disease at the beginning of maintenance therapy indicates poor outcome in children with T-cell acute lymphoblastic leukemia. *Blood* **90,** 1226–1232.

Drach, J., Drach, D., Glassl, H., Gattringer, G., and Huber, H. (1992). Flow cytometric determination of atypical antigen expression in acute leukemia for the study of minimal residual disease. *Cytometry* **13,** 893–901.

Drexler, H. G., Borkhardt, A., and Janssen, J. W. (1995). Detection of chromosomal translocations in leukemia–lymphoma cells by polymerase chain reaction. *Leukemia–Lymphoma* **19,** 359–380.

Engel, H., Drach, J., Keyhani, A., Jiang, S., Van, N. T., Kimmel, M., Sanchez-Williams, G., Goodacre, A., and Andreef, M. (1999). Quantitation of minimal residual disease in acute myelogenous leukemia and myelodysplastic syndromes in complete remission by molecular cytogenetics of progenitor cells. *Leukemia* **13,** 568–577.

Estrov, Z., Grunberger, T., Dubé, I. D., Wang, Y.-P., and Freedman, M. H. (1986). Detection of residual acute lymphoblastic leukemia cells in cultures of bone marrow obtained during remission. *N. Engl. J. Med.* **315,** 538–542.

Estrov, Z., Ouspenskaia, M. V., Felix, E. A., McClain, K. L., Lee, M.-S., Harris, D., Pinkel, D. P., and Zipf, T. F. (1994). Persistence of self-renewing leukemia cell progenitors during remission in children with B-precursor acute lymphoblastic leukemia. *Leukemia* **8,** 46–51.

Evans, P. A. S., Short, M. A., Jack, A. S., Norfolk, D. R., Child, J. A., Shiach, C. R., Davies, F., Tobal, K., Liu Yin, J. A., and Morgan, G. J. (1997). Detection and quantitation of the CBFβ/MYH11 transcripts associated with the inv(16) in presentation and follow-up samples from patients with AML. *Leukemia* **11,** 364–369.

Evans, P. A. S., Short, M. A., Owen, R. G., Jack, A. S., Forsyth, P. D., Shiach, C. R., Kinsey, S., and Morgan, G. J. (1998). Residual disease detection using fluorescent polymerase chain reaction at 20 weeks of therapy predicts clinical outcome in childhood acute lymphoblastic leukemia. *J. Clin. Onco.* **16,** 3616–3627.

Farahat, N., Lens, D., Zomas, A., Morilla, R., Matutes, E., and Catovsky, D. (1995). Quantitative flow cytometry can distinguish between normal and leukaemic B-cell precursors. *Br. J. Haematol.* **91,** 640–646.

Fischer, H., Hindkjaer, J., Pedersen, S., Koch, J., Brandt, C., and Kølvraa, S. (1994). Primed *in situ* labeling (PRINS) and fluorescence *in situ* hybridization (FISH). *In* "Methods in Cell Biology: Flow Cytometry" (Z. Darzynkiewicz, J. P. Robinson, and H. A. Crissman, eds.), 2nd Ed., Part B, Vol. 42, pp. 72–93. Academic Press, San Diego.

Freireich, E. J., Cork, A., Stass, S. A., McCredie, K. B., Keating, M. J., Estey, E. H., Kantarjian, H. M., and Trujillo, J.M. (1993). Cytogenetics for detection of minimal residual disease in acute myelogenous leukemia. *Leukemia* **7,** 736–741.

Galoin, S., al Saati, T., Schlaifer, D., Huynh, A., Attal, M., and Delsol, G. (1996). Oligonucleotide clonospecific probes directed against the junctional sequence of t(14;18): A new tool for the assessment of minimal residual disease in follicular lymphomas. *Br. J. Haematol.* **4,** 676–684.

Gray, J. W., Kuo, W. L., Liang, J., Pinkel, D., van den Engh, G., Trask, B., Tkachuk, D., Waldman, F., and Westbrook C. (1990). Analytical approaches to detection and characterization of disease-linked chromosome aberration. *Bone Marrow Transpl.* **6**(Suppl. 1), 14–19.

Greaves, M. F., Chan, L. C., Furley, A. J. W., Watt, S. M., and Molgaard, H. V. (1986). Lineage promiscuity in hemopoietic differentiation and leukemia. *Blood* **67,** 1–11.

Groeneveld, K., te Marvelde, J. G., van den Beemd, M. W. M., Hooijkaas, H., and van Dongen, J. J. M. (1996). Flow cytometric detection of intracellular antigens for immunophenotyping of normal and malignant leukocytes. *Leukemia* **10,** 1383–1389.

Gross, H. J., Verwer, B., Houck, D., and Recktenwald, D. (1993). Detection of rare cells at a frequency of one per million by flow cytometry. *Cytometry* **14,** 519–526.

Gross, H. J., Verwer, B., Houck, D., Hoffman, R. A., and Recktenwald, D. (1995). Model study detecting breast cancer cells in peripheral blood mononuclear cells at frequencies as low as $10^{(-7)}$. *Proc. Natl. Acad. Sci. U.S.A.* **92,** 537–541.

Hittelman, W. N., Tigaud, J. D., Estey, E., and Vadhan-Raj, S. (1990). Premature chromosome condensation in the study of minimal residual disease. *Bone Marrow Transpl.* **6**(Suppl. 1), 9–13.

Ito, Y., Wasserman, R., Galili, N., Reichard, B. A., Shane, S., Lange, B., and Rovera, G. (1993). Molecular residual disease status at the end of chemotherapy fails to predict subsequent relapse in children with B-lineage acute lymphoblastic leukemia. *J. Clin. Oncol.* **11,** 546–553.

Jennings, C. D. and Foon, K. A. (1997). Recent advances in flow cytometry: Application to the diagnosis of hematologic malignancy. *Blood* **90,** 2863–2892.

Kasprzyk, A. and Secker-Walker, L. M. (1997). Increased sensitivity of minimal residual disease detection by interphase FISH in acute lymphoblastic leukemia with hyperdiploidy. *Leukemia* **11,** 429–435.

Kishimoto, T., Goyert, S., Kikutani, H., Mason, D., Miyasaka, M., Moretta, L., Ohno, T., Okumura, K., Shaw, S., Springer, T. A., Sugamura K., Sugawara, H., von dem Borne, A. E. G. K., and Zola, H. (1997). Update: New CD antigens, 1996. *Tissue Antigens* **49,** 287–288.

Knapp, W., Strobl, H., and Majdic, O. (1994). Flow cytometric analysis of cell-surface and intracellular antigens in leukemia diagnosis. *Cytometry* **18,** 187–198.

Kurokawa, T., Kinoshita, T., Ito, T., Saito, H., and Hotta, T. (1996). Detection of minimal residual disease B cell lymphoma by a PCR-mediated RNase protection assay. *Leukemia* **10,** 1222–1231.

Lanza, F., Castoldi, G. L., Castagnari, B., Todd, R. F., Moretti, S., Spisani, S., Latorraca, A., Focarile, E., Roberti, M. G., and Traniello, S. (1998). Expression and functional role of urokinase-type plasminogen activator receptor in normal and acute leukaemic cells. *Br. J. Haematol.* **103,** 110–123.

Lavabre-Bertrand, T., Janossy, G., Ivory, K., Peters, R., Secker-Walker, L., and Porwitt-MacDonald, A. (1994). Leukemia-associated changes identified by quantitative flow cytometry: I. CD10 expression. *Cytometry* **18,** 209–217.

Lin, F., van Rhee, F., Goldman, J. M., and Cross, N. C. (1996). Kinetics of increasing BCR-ABL transcript numbers in chronic myeloid leukemia patients who relapse after bone marrow transplantation. *Blood* **87,** 4473–4478.

Liu, P., Tarle, S. A., Hajra, A., Claxton, D. F., Marlton, P., Freedman, M., Siciliano, M. J., and Collins, F. S. (1993). Fusion between transcription factor CBFβ/PEBP2β and a myosin heavy chain in acute myeloid leukemia. *Science* **261,** 1041–1044.

Macedo, A., Orfao, A., Gonzalez, M., Vidriales, M. B., Lopez-Berges, M. C., Martinez, A., and San Miguel, J. F. (1995). Immunological detection of blast cell subpopulations in acute myeloblastic leukemia at diagnosis: Implications for minimal residual disease studies. *Leukemia* **9,** 993–998.

Macedo, A., Orfao, A., Ciudad, J., Gonzalez, M., Vidriales, B., Lopez-Berges, M. C., Martinez, A., Landolfi, C., Canizo, C., and San Miguel, J. F. (1995). Phenotypic analysis of CD34 subpopulations in normal human bone marrow and its application for the detection of minimal residual disease. *Leukemia* **9,** 1896–1901.

Marcucci, G., Livak, K. J., Bi, W., Strout, M. P., Bloomfield, C. D., and Caligiuri, M. A. (1998). Detection of minimal residual disease in patients with AML1/ETO-associated acute myeloid leukemia using a novel quantitative reverse transcription polymerase chain reaction assay. *Leukemia* **12,** 1482–1489.

Mayer, S. P., Giamelli, J., Sandoval, C., Roach, A. S., Ozkaynak, M. F., Tugal, O., Rovera, G., and Jayabose, D. (1999). Quantitation of leukemia clone-specific antigen gene rearrangements by a single-step PCR and fluorescence-based detection method. *Leukemia* **13,** 1843–1852.

Mensink, E., van de Locht, A., Schattenberg, A., Linders, E., Schaap, N., Geurts van Kessel, A., and De Witte, T. (1998). Quantitation of minimal residual disease in Philadelphia chromosome positive chronic myeloid leukemia patients using real-time quantitative RT–PCR. *Br. J. Haematol.* **102,** 768–774.

Moravcová, J., Lukasova, M., Stary, J., and Haskovec, C. (1998). Simple competitive two-step RT–PCR assay to monitor minimal residual disease in CML patients after bone marrow transplantation. *Leukemia* **12,** 1303–1312.

Mori, T., Sugita, K., Suzuki, T., Okazaki, T., Manabe, A., Hosoya, R., Mizutani, S., Kinoshita, A., and Nakazawa, S. (1995). A novel monoclonal antibody, KOR-SA3544 which reacts to Philadelphia chromosome-positive acute lymphoblastic leukemia cells with high sensitivity. *Leukemia* **9,** 1233–1239.

Nagai, J., Kigasawa, H., Tomioka, K., Koga, N., Nishihira, H., and Nagao, T. (1994). Immunocytochemical detection of bone marrow-invasive neuroblastoma cells. *Eur. J. Haematol.* **53,** 74–77.

Nakao, M., Yokota, S., Horiike, S., Taniwaki, M., Kashima, K., Sonoda, Y., Koizumi, S., Takaue, Y., Matsushita, T., Fujimoto, T., and Misawa, S. (1996). Detection and quantification of TEL/AML1 fusion transcripts by polymerase chain reaction in childhood acute lymphoblastic leukemia. *Leukemia* **10,** 1463–1470.

Nakao, M., Janssen, J. W., Flohr, T., and Bartram, C. R. (2000). Rapid and reliable quantification of minimal residual disease in acute lymphoblastic leukemia using rearranged immunoglobulin and T-cell receptor loci by LightCycler technology. *Cancer Res.* **60,** 3281–3289.

Nowak, R., Oelschlaegel, U., Schuler, U., Zengler, H., Hofmann, R., Ehninger, G., and Andreeff, M. (1997). Sensitivity of combined DNA/immunophenotype flow cytometry for the detection of low levels of aneuploid lymphoblastic leukemia cells in bone marrow. *Cytometry* **30,** 47–53.

Ouspenskaia, M. V., Johnston, D. A., Roberts, W. M., Estrov, Z., and Zipf, T. F. (1995). Accurate quantitation of residual B-precursor acute lymphoblastic leukemia by limiting dilution and a PCR-based detection system: A description of the methods and principles involved. *Leukemia* **9,** 321–328.

Pelkey, T. J., Frieson, H. F., and Bruns, D. E. (1996). Molecular and immunological detection of circulating tumor cells and micrometastases from solid tumors. *Clin. Chem.* **42,** 1369–1381.

Pongers-Willemse, M. J., Verhagen, O. J. H. M., Tibbe, G. J. M., Wijkhuijs, A. J. M., de Haas, V., Roovers, E., and van der Schoot, C. E. (1998). Real-time quantitative PCR for the detection of minimal residual disease in acute lymphoblastic leukemia using junctional regional specific TaqMan probes. *Leukemia* **12,** 2006–2014.

Porwitt-MacDonald, A. T., Janossy, G., Ivory, K., Swirsky, D., Peters, R., Wheatly, K., Walker, H., Turker, A., Goldstone, A. H., and Burnett, A. (1996). Leukemia-associated changes identified by quantitative flow cytometry: IV. CD34 overexpression in acute myelogenous leukemia M2 with t(8;21). *Blood* **87,** 1162–1169.

Roberts, W. M., Estrov, Z., Ouspenskaia, M. V., Johnston, D. A., McClain, K. L., and Zipf, T. F. (1997). Measurement of residual leukemia during remission in childhood acute lymphoblastic leukemia. *N. Engl. J. Med.* **336,** 317–323.

Ross, A. A. (1998). Minimal residual disease in solid tumor malignancies: A review. *J. Hematother.* **7,** 9–18.

Sandoval, C., Mayer, S. P., Giamelli, J., Farley, T., Ozkaynak, M. F., Tugal, O., and Jayabose, S. (2000). Cytogenetic abnormalities during clinical, immunophenotypic, and molecular remission in pediatric acute lymphoblastic leukemia. *Cancer Genet. Cytogenet.* **118,** 9–13.

Sang, B. C., Shi, L., Dias, P., Liu, L., Wei, J., Wang, Z. X., Monell, C. R., Behm, F., and Gruenwald, S. (1997). Monoclonal antibodies specific to the acute lymphoblastic leukemia t(1;19)-associated E2A/pbx1 chimeric protein: Characterization and diagnostic utility. *Blood* **89,** 2909–2914.

Satake, N., Kobayashi, H., Tsunematsu, Y., Kawasaki, H., Horikoshi, Y., Koizumi, S., and Kaneko, Y. (1997). Minimal residual disease with TEL-AML1 fusion transcript in childhood acute lymphoblastic leukaemia with t(12;21). *Br. J. Haematol.* **97,** 607–611.

Sawyers, C. L. (1998). Molecular abnormalities in myeloid leukemias and myelodysplastic syndromes. *Leukemia Res.* **22,** 1113–1122.

Schoenfeld, A., Kruger, K. H., Gomm, J., Sinnett, H. D., Gazet, J. C., Sacks, N., Bender, H. G., Luqmani, Y., and Coombes, R. C. (1997). The detection of micrometastases in the peripheral blood and bone marrow of patients with breast cancer using immunohistochemistry and reverse transcriptase polymerase chain reaction for keratin 19. *Eur. J. Cancer* **33,** 854–861.

Seale, R. C., Varma, S., Swirsky, D. M., Pandolfi, P.-P., Goldman, J. M., and Cross, N. C. P. (1996). Quantification of PML-RARα transcripts in acute promyelocytic leukaemia: Explanation for the lack of sensitivity of RT–PCR for the detection of minimal residual disease and induction of the leukaemia-specific mRNA by α interferon. *Br. J. Haematol.* **95,** 95–101.

Sinclair, P. B., Green, A. R., Grace, C., and Nacheva, E. P. (1997). Improved sensitivity of BCR-ABL detection: A triple-probe three-color fluorescence in situ hybridization system. *Blood* **90,** 1395–1402.

Steward, C. G., Goulden, N. J., Katz, F., Baines, D., Martin, P. G., Langlands, K., Potter, J. M., Chessells, J. M., and Oakhill, A. (1994). A polymerase chain reaction study of the stability of Ig heavy-chain and T-cell receptor δ gene rearrangements between presentation and relapse of childhood B-lineage acute lymphoblastic leukemia. *Blood* **83,** 1355–1362.

Stewart, C. C. and Stewart, S. J. (1994a). Cell preparation for the identification of leukocytes. *In* "Methods in Cell Biology: Flow Cytometry" (Z. Darzynkiewicz, J. P. Robinson, and H. A. Crissman, eds.), 2nd Ed., Part A, Vol. 41. Academic Press, San Diego.

Stewart, C. C. and Stewart, S. J. (1994b). Multiparameter analysis of leukocytes by flow cytometry. *In* "Methods in Cell Biology: Flow Cytometry" (Z. Darzynkiewick, J. P. Robinson, and H. A. Crissman, eds.), 2nd Ed., Part A, Vol. 41. Academic Press, San Diego.

Tobal, K. and Liu Yin, J. A. (1998). RT–PCR method with increased sensitivity shows persistence of PML-RARA fusion transcripts in patients in long-term remission of APL. *Leukemia* **12,** 1349–1354.

Tobal, K., Johnson, P. R., Saunders, M. J., Harrison, C. J., and Liu Yin, J. A. (1995). Detection of CBFB/MYH11 transcripts in patients with inversion and other abnormalities of chromosome 16 at presentation and remission. *Br. J. Haematol.* **91,** 104–108.

van Dongen, J. J. M., Breit, T. M., Adriaansen, H. J., Beishuizen, A., and Hooijkaas, H. (1993). Immunophenotypic and immunogenotypic detection of minimal residual disease in acute lymphoblastic leukemia. *Recent Results Cancer Res., Recent Adv. Cell Biol. Acute Leukemia* **131,** 157–184.

Veelken, H., Tycko, B., and Sklar, J. (1991). Sensitive detection of clonal antigen receptor gene rearrangements for the diagnosis and monitoring of lymphoid neoplasms by a polymerase chain reaction-mediated ribonuclease protection assay. *Blood* **78,** 1318–1326.

Walker, H., Smith, F. J., and Betts, D. R. (1994). Cytogenetics in acute myeloid leukaemia. *Blood Rev.* **8,** 30–36.

Weir, E. G., Cowan, K., LeBeau, P., and Borowitz, M. J. (1999). A limited antibody panel can distinguish B-precursor acute lymphoblastic leukemia from normal B precursors with four color flow cytometry: Implications for residual disease detection. *Leukemia* **13,** 558–567.

Westbrook, C. A., Hooberman, L. H., Spino, C., Dodge, R. K., Larsen, R. A., Davey, F., Wurster-Hill, D. H., Sobol, R. E., Schiffer, C., and Bloomfield, C. D. (1992). Clinical significance of the BCR-ABL fusion gene in adult acute lymphoblastic leukemia: A cancer and leukemia group B study (8762). *Blood* **80,** 2983–2990.

Wörmann, B., Safford, M., Könemann, S., Büchner, T., Hiddemann, W., and Terstappen L. W. M. M. (1993). Detection of aberrant antigen expression in acute myeloid leukemia by multiparameter flow cytometry. *Recent Results Cancer Res., Recent Adv. Cell Biol. Acute Leukemia* **131,** 185–196.

Yamada, M., Wasserman, R., Lange, B., Reichard, B. A., Womer, R. B., and Rovera, G. (1990). Minimal residual disease in childhood B-lineage lymphoblastic leukemia. Persistence of leukemic cells during the first 18 months of treatment. *N. Engl. J. Med.* **323,** 448–455.

Yokota, S., Hansen-Hagge, T. E., Ludwig, W. D., Reiter, A., Raghavachar, A., Kleihauer, E., and Bartram, C. R. (1991). Use of polymerase chain reactions to monitor minimal residual disease in acute lymphoblastic leukemia patients. *Blood* **77,** 331–339.

CHAPTER 50

Analysis of Human Tumors by Laser Scanning Cytometry

Wojciech Gorczyca,* Andrzej Deptala,[†] Elżbieta Bedner,[‡] Xun Li,*, Myron R. Melamed,* and Zbigniew Darzynkiewicz****

Department of Pathology
New York Medical College
Valhalla, New York 10595

[†] Department of Hematology, Oncology and Internal Medicine
Warsaw Medical University
02-097 Warsaw, Poland

[‡] Department of Pathology
Pomeranian School of Medicine
Szczecin, Poland

** Brander Cancer Research Institute
New York Medical College
Hawthorne, New York 10532

I. Introduction
II. Analysis of Cellular DNA Content by Laser Scanning Cytometry
    A. Materials
    B. Procedure
    C. Results
III. Analysis of Apoptosis by Laser Scanning Cytometry
    A. Materials
    B. Procedure
    C. Results
IV. Analysis of Proliferation Associated Antigens by Laser Scanning Cytometry
    A. Materials
    B. Procedures
    C. Results
V. Analysis of Estrogen Receptors by Laser Scanning Cytometry
    A. Materials
    B. Procedure

METHODS IN CELL BIOLOGY, VOL. 64

        C. Results
   VI. Analysis of Transcription Factors by Laser Scanning Cytometry
        A. Materials
        B. Procedure
        C. Results
  VII. Measurement of Nucleoli Using Laser Scanning Cytometry Fluorescence *in Situ*
        Hybridization Protocol
        A. Materials
        B. Procedure
        C. Results
        References

I. Introduction

The laser scanning cytometer (LSC) is a microscope-based cytofluorometer that combines advantages of flow cytometry and image analysis (Darzynkiewicz *et al.*, 1999; Kamentsky and Kamentsky, 1991; Kamentsky *et al.*, 1997; see Chapter 3 of Volume 63, this series). Fluorescence of individual cells is measured rapidly by LSC, with sensitivity and accuracy comparable to that of flow cytometry (FC). The unique feature of LSC, which distinguishes it from FC, is that the specimen is interrogated by a single or two laser beams while located on a microscope slide rather than in suspension. Furthermore, the position of the slide is monitored by computer-interfaced sensors and is recorded together with fluorescence data. Fluorescence emission signals are separated optically according to wavelength and are directed to photomultipliers for measurement and to a CCD camera for imaging. This allows for (1) correlation of fluorescence data with the morphology of individual cells as viewed by fluorescence or light absorption microscopy; (2) sequential measurements of the same cells over time or after sequential staining with different dyes and integration of the results into a single file recorded in list mode, making use of the "merge" attribute of LSC; and (3) possibility of additional measurement of the same cells in the future, following archival storage of the specimen.

The following attributes can be measured and recorded by LSC for each analyzed cell in a given cellular specimen: (1) the integrated fluorescence intensity over the integration contour at a given wavelength band, (2) the value of maximal pixel within the measured area, (3) the perimeter of the contour, (4) fluorescence intensity at the selected wavelength band, integrated over the area defined by the peripheral contour (torus) that is located around the primary integration contour, (5) the position of the measured object on X and Y coordinates of the slide, and (6) the computer clock time at the moment of measurement. The first four measurements in addition to providing information on total fluorescence intensity, integrated either over the whole cell or a selected portion (e.g., nucleus), reveal fluorescence distributions of selected cell components in relation to cell morphology. In addition, the translocation of a measured component can be followed, from cytoplasm to nucleus, for example (Deptala *et al.*, 1998). The *XY* coordinates not only permit relocating cells of interest for visual inspection or secondary analysis, but also reveal tumor architecture when tissue sections or imprints are measured. Finally, the time parameter makes it possible to measure time resolved events such as enzyme kinetics or fluorochrome uptake rates for individual cells (Bedner *et al.*, 1998).

This chapter is focused on the clinical applications of LSC for tumor analysis. Several analytical methods, frequently used in FC for tumor diagnosis and prognosis, have been adapted to LSC and are presented here. The analytical power of LSC combined with its unique advantages, in particular to correlate measurements with visual examination of the corresponding cell, are of special value for pathologists.

II. Analysis of Cellular DNA Content by Laser Scanning Cytometry

Analysis of cellular DNA content in human malignancies became one of the most frequently used applications of flow cytometry. In many solid and hematologic tumors DNA ploidy bears prognostic implications (Koss *et al.*, 1989). Data similar to those obtained by FC can be generated by LSC using a variety of fluorochromes, excitable either by argon ion or HeNe laser (Darzynkiewicz *et al.*, 1999; Gorczyca *et al.*, 1997a; Kamentsky and Kamentsky, 1991; Kamentsky *et al.*, 1997). Although cell analysis by LSC is slower than that by FC, it offers a major advantage in correlating the analyzed parameters with cell morphology. Thus, the DNA ploidy of individual cells can be correlated with cell morphology, cell classification, or any other parameter of interest. Moreover, the slides can be restained (e.g., with hematoxylin–eosin, Wright–Giemsa, or classic cytochemical or immunocytochemical techniques) and the cells visualized for specific classification. In addition to relative DNA content in tumors (DNA index, DI) and distribution of cells in the cell cycle (e.g., S-phase fraction), DNA staining is used in most LSC applications as a contouring parameter ("counterstaining"; see Kamentsky and Kamentsky, 1991; Kamentsky *et al.*, 1997). This allows one to identify the cell nucleus and thus separately measure nuclear versus cytoplasmic fluorescence.

A large number of DNA specific fluorochromes with different spectral properties is currently available (Haugland, 1994). The most commonly used with LSC are propidium iodide (PI), 7-amino-actinomycin D (7-AAD), and ethidium bromide (EB). The choice of a specific fluorochrome for DNA staining is dictated primarily by the emission spectrum of other fluorochrome(s) used in the experiment, in order to diminish or eliminate spectral overlap.

A. Materials

Stock solution of PI or 7-AAD (Molecular Probes, Eugene, OR) can be prepared in distilled water, at the dye concentration of 1 mg/ml, and stored at 4°C in the dark.

Stock solution of DNase-free RNase A (Sigma Chemical, St. Louis, MO) is prepared by dissolving RNase A in distilled water (5 mg/ml) and boiling for 3–5 min. Aliquots can be stored at −20°C.

Staining solution of PI: Add stock solutions of PI and RNase A to PBS to obtain the desired PI concentration (see later) and a final concentration of RNase A of 100 μg/ml. This solution should be prepared fresh.

B. Procedure

1. The tissues can be disintegrated mechanically (see Visscher and Crissman, 1994). Briefly, place the resected fragment of tissue on gauze and with a surgical blade trim fat and necrotic areas, if distinctly visible. Place the specimen in a petri dish containing 5–10 ml of phosphate-buffered saline (PBS) containing 1% (w/v) bovine serum albumin (BSA; Sigma). Using forceps and a surgical blade gently scrape the tissue surface until the suspension becomes turbid. Then bisect the tissue to provide a new surface area and repeat scraping. Filter the suspension through an 80-μm mesh metal sieve, transfer into a 15-ml conical centrifuge tube, and centrifuge at 300 g for 5 min.

Cells from a fine needle aspirate specimen can be disaggregated by vortexing the aspirate, suspended in a small volume (0.5 ml) of PBS containing 1% (w/v) BSA in an Eppendorf tube. The cells also can be prepared and stained in suspension, as described in protocols designed for FC. The cells are then transferred onto slides for measurement by LSC. However, because the cells are not attached to the slide in those preparations, relocating them (e.g., for morphologic identification) is problematic.

2. Resuspend the cells in PBS containing 1% (w/v) BSA to have approximately 60,000–80,000 cells per milliliter. Transfer 300 μl (~20,000 cells) of this suspension into a Cytospin chamber (Shandon Scientific, Pittsburgh, PA) and spin in a cytocentrifuge (Shandon) at 1000 rpm for 6 min. Alternatively, if the tumor has little stroma and cells can be easily isolated, the "touch" specimens can be prepared by pressing the dry surface of a microscope slide onto the freshly cut surface of the tissue. "Smear" films also can be used, prepared from dense tumor cell suspensions in a manner similar to that of blood smear films made in hematologic laboratories. The morphology of the smeared tumor cells may not always be satisfactory; in general, cell morphology is best preserved in the Cytospin preparations. For preparation of paraffin-embedded material, see Leers *et al.* (1999).

3. Without allowing the Cytospins (or touch preparations or smear films) to completely dry, fix the slides in 80% ethanol for at least 30 min, then rinse in PBS twice for 5 min. To ensure optimal cell density on slides it is advisable to

check the slides under the microscope prior to staining. If the specimens are overcrowded with cells or too sparse, the cell density in suspensions used for centrifugation can then be adjusted accordingly.

4. Immerse the slides in a Coplin jar containing the staining solution of PI. The final concentration of PI may vary between 5 and 20 μg/ml; lower PI concentrations should be used if the cells are to be counterstained with other, weaker fluorochromes [e.g., green (FITC, fluorescein isothrocyanate) and orange (PE, phycoerythrin) fluorescences are expected to be weak]. Keep slides immersed in PI solution for 20–30 min at room temperature in the dark. Although for best results whole slides should be immersed in the staining solution, it is also possible to deposit small volumes (0.2–0.5 ml) of PI solution onto the slide over the area containing the cells.

5. Mount the cells under a coverslip by adding a drop of a solution containing 90% glycerol in PBS in which PI was dissolved at the same concentration as in the staining solution (5–20 μg/ml). Alternatively, a commercial antifade mounting medium (e.g., Vectashield, Vector Laboratories, Burlingame, CA) can be used. Preventing the specimen from drying during measurement improves the quality of the results (i.e., it ensures better accuracy of DNA analysis and lower coefficient of variation of mean DNA content of G_1 population). The slides should be kept in the dark until measurement on LSC. PI can be replaced with 7-AAD at similar concentration and incubation time. No RNase is needed when using 7-AAD.

6. Measure cell fluorescence by LSC using the red sensor [photomultiplier tube (PMT) settings 18–24] for contouring. Adjust contouring threshold based on scan data display. Gate the cells of interest based on red maximal pixel versus red area scattergram. The obtained DNA data can be transferred and analyzed by Multicycle software.

C. Results

Figure 1 presents DNA content frequency histograms obtained by LSC analysis of two solid human tumors. The cells were disaggregated mechanically, deposited on slides by cytocentrifugation, stained with PI (5 μg/ml) as described in the protocol earlier, and their PI fluorescence (integrated value) was measured by LSC. The differences in DNA ploidy (hyperdiploid tumor in Fig. 1a and hypodiploid in Fig. 1b) and in cell cycle distribution between these tumors are apparent. Control diploid cells (e.g., blood lymphocytes or nontumor normal tissue) can be used to estimate the DI of tumor cells. The DNA index is calculated as the ratio of mean or modal channel number of DNA fluorescence of aneuploid G_0/G_1 cells to mean or modal channel of DNA fluorescence of diploid G_0/G_1 cells (see Dressler and Seamer, 1994).

III. Analysis of Apoptosis by Laser Scanning Cytometry

There are numerous flow cytometric methods for the analysis of apoptosis (for reviews, see Bedner *et al.*, 1999; Darzynkiewicz *et al.*, 1997). The most

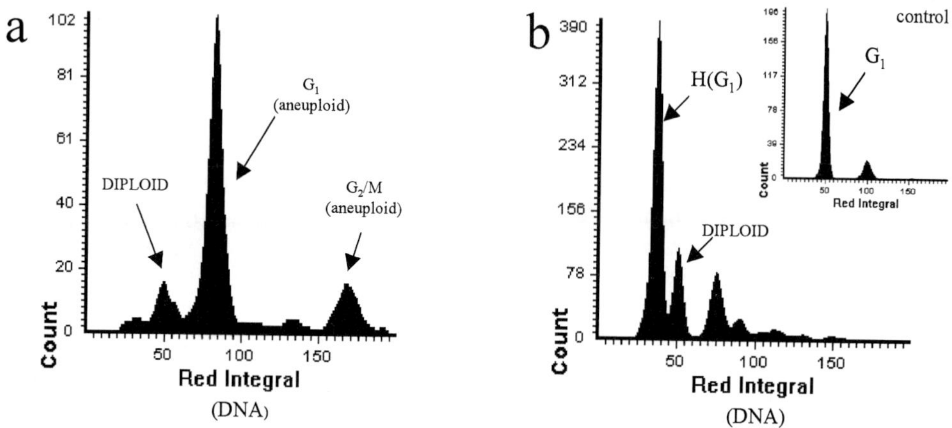

Fig. 1 DNA histograms of aneuploid uterine carcinosarcoma (a) and hypodiploid chromophobe renal cell carcinoma (b) processed by mechanical disintegration, stained with propidium iodide, and analyzed with LSC.

frequently used cytometric assays are based on the detection of DNA fragmentation resulting from the activation of endonuclease, which is considered to be a hallmark of apoptosis (Arends *et al.*, 1990). One of the methods relies on measurement of cellular DNA content (as described) and identification of a "subdiploid" (sub-G_1) peak that represents cells with fractional DNA content (Fig. 2). Loss of cellular DNA from apoptotic cells is the result of extraction of fragmented, small molecular weight DNA during rinsing and staining procedures. Some loss may also be due to shedding of apoptotic bodies that contain fragments of nuclei. It is essential in this procedure that cells are fixed in ethanol; the small size DNA fragments are not retained within the cells following ethanol fixation and leak out during subsequent washings (Gong *et al.*, 1994).

Another widely used technique, which was developed in our laboratory, relies on the detection of fragmented DNA by indirect or direct labeling of 3' hydroxyl termini in DNA breaks with fluorochrome-labeled nucleotides (TdT assay; "*in-situ* end-labeling," or TUNEL) (Gorczyca *et al.*, 1992, 1993, 1998a,b; Gorczyca, 1999; Li and Darzynkiewicz, 1995; Li *et al.*, 1995). Fixation of the cells by a cross-linking fixative (such as formaldehyde) prevents loss of DNA during the staining and labeling procedure and is essential in this procedure. Direct labeling is simpler and involves deoxynucleotides that are directly conjugated with a fluorochrome (Li *et al.*, 1995). Indirect labeling utilizes biotin- or digoxigenin-conjugated deoxynucleotides that subsequently are detected by fluorochrome-conjugated avidin or antibody, respectively (Gorczyca *et al.*, 1992, 1993, 1998a,b; Gorczyca, 1999). The most sensitive assay is based on labeling DNA breaks with BrdUTP, which is in turn detected with fluorochrome-conjugated anti-BrdU antibody (Li and Darzynkiewicz, 1995).

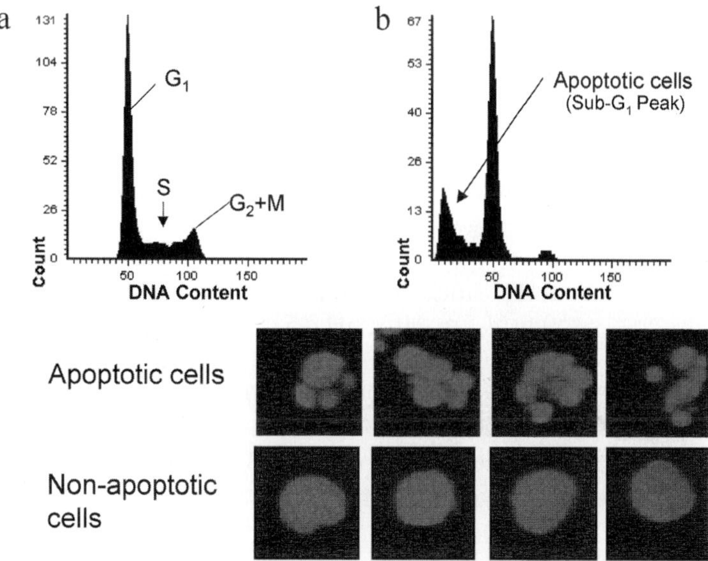

Fig. 2   DNA frequency histograms of HL-60 human leukemic cells, untreated (a) and treated with camptothecin to induce apoptosis (b). Note the subdiploid peak in treated cells reflecting apoptosis. Also shown are LSC computer images of apoptotic and nonapoptotic cells. Propidium iodide staining, 20 × objective. (See color plates.)

Essentially all techniques originally developed for FC can be adapted to LSC (for review, see Gorczyca *et al.*, 1998a,b; Bedner *et al.*, 1999; Darzynkiewicz *et al.*, 1999). As mentioned, LSC not only measures the fluorescence of a cell, but records its position on the slide; all cells with positive fluorescence can be relocated and examined by direct visual microscopy or indirectly by image analysis. This feature of LSC is of particular importance in analysis of apoptosis because changes in cell morphology are still considered the gold standard in identification of apoptotic cells.

A. Materials

Labeling of apoptotic cells can be performed using commercially available apoptosis detection kits (e.g., *In situ* cell detection system, Boehringer Mannheim, Indianapolis, IN; APO-DIRECT, or APO-BRDU, both from Phoenix Flow Systems, San Diego, CA, also offered by Alexis Biochemicals, San Diego, CA, and PharMingen, San Diego, CA). When using these kits follow the protocols provided by the vendors.

DNA strand breaks also can be labeled using original reagents according to the published methods (Gorczyca *et al.*, 1992; Li and Darzynkiewicz, 1995). The following reagents are needed to label DNA strand breaks with BrdUTP:

First fixative: 1% methanol-free formaldehyde (available from Polysciences, Warrington, PA) in PBS, pH 7.4. Prepare fresh before use.

Second fixative: 70% ethanol.

The TdT reaction buffer (5 × concentrated) contains 1 M potassium (or sodium) cacodylate, 125 mM Tris-HCl, pH 6.6, 1.25 mg/ml bovine serum albumin (BSA), 10 mM cobalt chloride ($CoCl_2$), and 25 units in 1 μl TdT in storage buffer. The buffer, TdT, and $CoCl_2$ are available from Boehringer Mannheim.

BrdUTP stock solution: 2 mM BrdUTP (Sigma) (100 nmol in 50 μl) in 50 mM Tris-HCl, pH 7.5.

FITC-conjugated anti BrdU monoclonal antibody (MAb) solution (per 100-μl of PBS) contains 0.3 μg of anti-BrdU-FITC conjugated MAb (available from Becton Dickinson, San Jose, CA), 0.3% Triton X-100, and 1% BSA.

Rinsing buffer: dissolve, in PBS, 0.1% (v/v) Triton X-100 and 5 mg/ml BSA.

PI staining buffer: dissolve in PBS: PI to 5 μg/ml and DNase-free RNase A to 100 μg/ml.

B. Procedure

1. Add 300 μl of cell suspension in tissue culture medium (with serum) containing approximately 20,000 cells into a Cytospin chamber (e.g., Shandon Scientific). Spin in a Cytocentrifuge at 1000 rpm for 6 min. Alternatively, the cells may be electrostatically attached to the slides. To be attached electrostatically the cells while still in suspension have to be first rinsed in serum-free PBS and then resuspended again in serum-free PBS. A drop of such suspension (~20,000–50,000 cells) should then be placed within the well (~1 × 1 cm) made with hydrophobic pencil (Shandon) on a clean and dry (prewashed in 100% ethanol) microscope slide. The slide is left horizontally, at room temperature and at 100% humidity, for 15 min. PBS is then removed by vacuum suction or Pasteur pipette.

2. Without allowing the Cytospin chambers (or electrostatically attached cells) to completely dry, prefix cells by immersing the slide in a Coplin jar containing 1% formaldehyde in PBS, for 15 min on ice.

3. Transfer the slide to a Coplin jar with 70% ethanol and keep for at least 1 hr; the cells can be stored in ethanol several days at 4°C.

4. Transfer the slide to Coplin jars filled with PBS. Keep for 10 min at room temperature.

5. Remove the slide, blot excess PBS with filter paper, and deposit over the cytospin area a 50-μl aliquot of the solution containing the following:

10 μl of the reaction buffer

2.0 μl of BrdUTP stock solution

0.5 μl (12.5 units) of TdT in storage buffer

5 μl of $CoCl_2$ solution

33.5 μl distilled water

6. Cover the Cytospin area with a small piece (2.5 × 1.0 cm) of thin polyethylene foil to prevent drying. Incubation of the horizontally kept slide should be carried out at 100% humidity, within a closed box to prevent drying at any step of the reaction. Incubate cells in this solution for 40 min at 37°C. Alternatively, incubation can be carried at 22°–24°C overnight.

7. After incubation remove the BrdU labeling solution with Pasteur or vacuum suction pipette and rinse the cells with excess of PBS.

8. Deposit over the Cytospin area 100 μl of FITC conjugated anti-BrdU MAb solution, cover with polyethylene film, and incubate at room temperature for 1 hr or at 4°C overnight.

9. Rinse the cells with PBS and transfer the slide into a Coplin jar containing the staining solution of PI. Keep slides immersed in PI solution for 20–30 min at room temperature in the dark. For best results the whole slide should be immersed in the staining solution, but it is also possible to deposit a small volume (0.2–0.5 ml) of PI solution onto the slide over the area containing the cells.

10. Negative controls should be processed identically through Steps 1–9 except that TdT is replaced by the same aliquots of 1% BSA in PBS.

11. Measure green (FITC) and red (PI) fluorescence by LSC. Contouring should be based on red fluorescence.

C. Results

Figure 3 shows bivariate distributions (scatterplots) of DNA strand breaks versus DNA content of HL-60 cells untreated (Fig. 3a) and, to induce apoptosis, treated with DNA topoisomerase I inhibitor camptothecin (CPT) (Fig. 3b). As

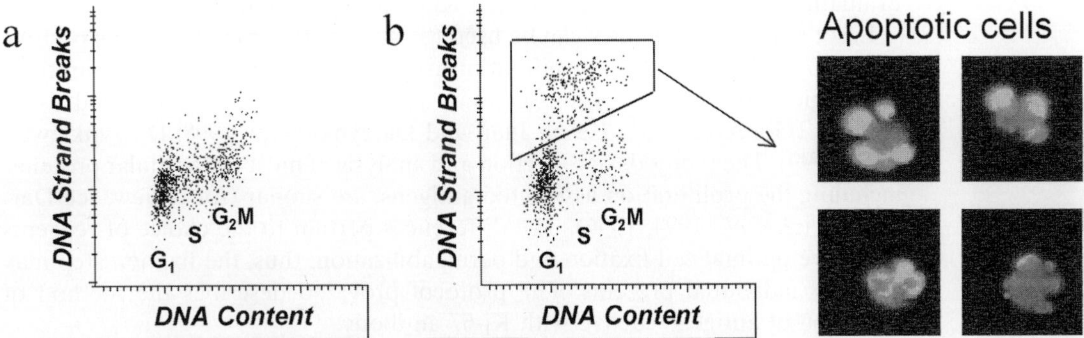

Fig. 3   DNA strand labeling by TdT assays (TUNEL method) in HL-60 cells treated with camptothecin to induce apoptosis (b). In contrast to control (nontreated) cells (a), apoptotic cells have high values of green fluorescence (b, box), reflecting the incorporation of FITC-labeled nucleotide. When relocated on a LSC computer screen, the positively labeled cells displayed characteristic features of apoptosis (right). (See color plates.)

is apparent, apoptotic cells can be identified on the basis of the presence of DNA strand breaks. Because the cells were prefixed in formaldehyde, which is a cross-linking fixative, no significant loss of fragmented DNA of apoptotic cells occurred during the staining procedure. Therefore, the cell cycle distribution of both apoptotic and nonapoptotic cell populations can be estimated from their DNA content. This method is particularly suitable to reveal the cell cycle phase specificity of apoptosis (Gorczyca *et al.*, 1993), and the data show that primarily S phase cells undergo apoptosis in the CPT-treated culture.

Visual inspection of the presumed apoptotic cells, to identify the presence of DNA strand breaks, is possible by LSC and allows one to confirm the apoptotic mode of death by morphology. This feature is useful to distinguish genuine apoptotic cells from "false positive" apoptotic cells. The latter may be macrophages, monocytes, or other bystander cells that engulfed apoptotic bodies containing fragmented DNA (Gorczyca *et al.*, 1998a,b; Bedner *et al.*, 1999).

IV. Analysis of Proliferation Associated Antigens by Laser Scanning Cytometry

Assays to detect PCNA (proliferating cell nuclear antigen), p120, cyclins, the antigen detected by Ki-67 antibody (Gerdes *et al.*, 1984), and many other proliferation associated proteins immunocytochemically have been widely used in studies assessing the growth fraction or proliferative potential of tumors (Brown *et al.*, 1996; Gerdes *et al.*, 1984; Juan and Darzynkiewicz, 1998; Rew *et al.*, 1998). The proliferative index of the tumor often has prognostic value and may provide clues regarding the optimal strategy for antitumor drug selection. Introduction of Ki-67 MIB-1 antibody reacting with the antigen in paraffin-embedded material provided an opportunity to conduct retrospective studies evaluating the prognostic value of the Ki-67 index in archival specimens.

Laser scanning cytometry can be used to estimate the fraction of cells positive for Ki-67, cyclins, and other proliferation associated antigens in fresh preparations (cytological specimens), frozen sections, and routine paraffin-embedded tissue sections (Gorczyca *et al.*, 1997a; Juan and Darzynkiewicz, 1998; Darzynkiewicz *et al.*, 1999). The methods of detection and analysis of most intracellular proteins, including the proliferation associated antigens, are similar (for review, see Darzynkiewicz *et al.*, 1994, 1996). The differences pertain to the choice of reagents to ensure optimal cell fixation and permeabilization; thus, the fixation step may vary for individual proteins. The protocol provided describes the method of detection of antigen reactive with Ki-67 antibody.

A. Materials

Fixative: 80% ethanol. Optimal fixation may vary depending on the antigen that is being detected. The most common fixatives are (1) 70 or 80% ethanol,

(2) 100% methanol (preferred to detect most cyclins), and (3) solution of 1% methanol-free formaldehyde in PBS; the cells are then often postfixed and permeabilized in 70–80% ethanol or treated with detergents.

Ki-67 MAb (FITC-conjugated or unconjugated, available from DAKO, Carpinteria, CA) is used for tissues fixed in ethanol or methanol. MIB-1 (Coulter/Immunotech, Miami, FL) is used for for paraffin-embedded specimens.

Secondary FITC-conjugated goat anti-mouse antibody (DAKO) for indirect staining (similar to FITC-conjugated primary and secondary antibodies; and phycoerythrin-conjugated antibodies can be used for three-color analysis).

PI staining solution: Dissolve in PBS PI to 5 μg/ml and DNase-free RNase A to 100 μg/ml.

Primary antibodies to cyclins are available from DAKO, PharMingen, Santa Cruz Biotechnology (Santa Cruz, CA), Upstate Biotechnology (Lake Placid, NY), and other sources. Before ordering it is crucial to check that the antibody has already been successfully used in immunocytochemical assays. Many antibodies listed in cataloges of different vendors detect only denatured proteins on immunoblots and are not reactive with the antigen in tissue. It is also imperative to titrate each antibody against the specimen(s) to find the optimal concentration for maximal signal-to-noise (background) ratio.

B. Procedures

1. Ki-67 Analysis in Fresh Cytological Specimens

Ki-67 can be detected immunocytochemically in cells attached to microscope slides, either electrostatically or by cytocentrifugation, as described earlier in this chapter. The cells are fixed, rinsed, and incubated with the Ki-67 antibody while attached to the slide, and counterstained with PI. Alternatively, the cells may be fixed and stained in suspension according to the protocols developed for FC. The suspension of stained cells is then placed on a microscope slide, mounted under a coverslip, and analyzed by LSC. In the latter case, however, the cells are not attached to the slide and may move, so these preparations are not ideal for cell relocation after the initial analysis by LSC. The following protocol describes Ki-67 immunostaining of cells that are attached to the slides.

1. Attach cells to a slide electrostatically, by cytocentrifugation, or by touch or smear preparation, as described earlier in this chapter.

2. Fix the specimen by immersing the slide in a Coplin jar filled with 80% cold ethanol ($-20°C$) for at least 1 hr, rinse in PBS, and to enhance subsequent cell permeability to antibody treat the slide with 0.1% Triton X-100 in PBS for 5 min at 4°C, then rinse twice in PBS.

3. Incubate the cells with 10% BSA ("blocking agent") to suppress subsequent nonspecific binding of antibody.

4. Remove the slide, blot excess PBS/BSA solution with filter paper, and deposit a 50-μl aliquot of the antibody solution in PBS over the Cytospin area. In a pilot experiment titrate the antibody and select its optimal concentration (generally within the range of 0.2–2.0 μg antibody in 1 ml of PBS) such that it gives the maximum intensity of immunofluorescence (plateau level during titration) at a still low level of background fluorescence (the greatest difference between Ki-67-positive and -negative cells).

5. Incubate the cells in the presence of FITC-conjugated mouse Ki-67 MAb for 1 hr at room temperature in the dark.

6. Wash the slides in PBS twice for a total of 5 min, resuspend in 5 μg/ml of PI and 0.1% RNase A in PBS, and incubate at room temperature for 30 min.

7. Mount under a coverslip using 90% glycerol in PBS with PI (5 μg/ml) and analyze by LSC. Use red fluorescence signal for contouring.

Steps 2, 3, and 6 can be carried out in Coplin jars, while incubation with antibody should be done on a horizontally placed slide, covered with a thin layer of polyethylene film, at 100% humidity as described earlier in this chapter for incubation with TdT to label DNA strand breaks (TUNEL reaction).

2. Ki-67 Analysis in Histological Sections of Paraffin–Embedded Tissues

1. Cut sections at 5 μm from paraffin-embedded, formalin-fixed tissue, deparaffinize (Histo-Clear, National Diagnostics, Atlanta, Georgia), rehydrate by passing through decreasing ethanol concentrations (100, 95, 80, and 50%), and immerse in distilled water.

2. Heat the slides in antigen retrieval buffer (Target Retrieval Solution, DAKO) in the microwave oven (750 W) for three cycles of 5 min each.

3. Block the sections with avidin and biotin (both from DAKO) and, after washing in PBS, incubate for 60 min at room temperature with primary Ki-67 antibody (MIB-1) for 20 min. Rinse with PBS twice for a total of 5 min.

4. Incubate sections with biotinylated secondary antibody [rabbit anti-mouse immunoglobulin G (IgG), DAKO] for 30 min at room temperature. Wash in PBS three times for a total of 10 min.

5. Incubate the sections with FITC–avidin (Boehringer Mannheim), diluted 1:100, for 30 min in the dark, and counterstain with a solution of PI containing RNase A, prepared as described earlier in this chapter.

6. Mount the specimen as described earlier in this chapter, and analyze by LSC. Use red fluorescence signal for contouring.

3. Staining for Cyclins

Cyclins are key components of the cell cycle machinery, involved in activation of their partners cyclin-dependent kinases (Cdks). During unperturbed growth of normal cells the timing of expression of several cyclins, particularly D-type cyclins, cyclin E, cyclin A, and cyclin B1, is discontinuous, occurring at discrete and well-defined periods of the cell cycle. The expression of these cyclins provides new cell cycle landmarks. The bivariate analysis of cyclins D1–3, E, A, or B1 versus cellular DNA content can be used to subdivide the cell cycle into several compartments (Darzynkiewicz *et al.*, 1996). The point of cell cycle arrest by antitumor agents can be estimated more precisely in relation to these compartments compared to the traditional subdivision into four phases. It should be stressed, however, that some tumor cell lines show unscheduled expression of cyclins; namely, the G_1 cyclins (D cyclins, cyclin E) are present in G_2 cells, and conversely, cyclin B1, which normally is expressed during G_2, can be detected in G_1 cells.

As mentioned, the technique of immunocytochemical detection of cyclins is similar to that of the antigen detected by Ki-67 antibody. Optimal fixation, however, requires 100% methanol (cyclins E, A, and B1) or methanol-free formaldehyde (1% in PBS, 15–30 min on ice) followed by 70% ethanol (D-type cyclins). The methodology of detection and analysis of cyclins by FC is described by Darzynkiewicz *et al.* (1994).

C. Results

Figure 4 presents Ki-67 expression in a chromophobe renal cell carcinoma (cytological preparations). The hypodiploid population of tumor cells expresses strong reactivity with Ki-67 antibody (Fig. 4b, dashed line). Positive staining with Ki-67 antibody is observed in the nuclei. It is restricted to proliferating cells $(G_1 + S + G_2/M)$, and the percentage of positively stained cells is the percentage

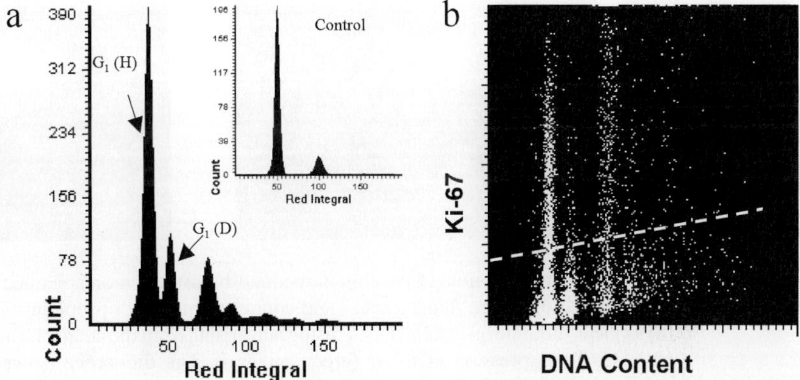

Fig. 4 Renal cell carcinoma, chromophobe type, stained with Ki-67 monoclonal antibody: DNA histogram (a, where H is hypodiploid and D is diploid) and dot plot (b). The hypodiploid cell population (comprising tumor cells) shows high expression of Ki-67 (b). The dashed line presents highest level of green fluorescence with isotypic IgG negative control.

of cells with increased nuclear staining compared to negative controls (where primary antibody was substituted by isotypic immunoglobulin). Cyclins are analyzed as described for Ki-67 (data not shown). Most of the tumors studied by us showed unscheduled expression of cyclin B1; that is, cyclin B1 was not restricted to G_2 + M but was also expressed in G_1 and S phase (see Gorczyca *et al.*, 1997b).

In tissue sections, both fresh ("frozen sections") and paraffin-embedded, the level of expression of Ki-67 is determined by measuring the integrated green (FITC) fluorescence of the tumor cells and subtracting the background fluorescence as measured in a parallel control section stained with FITC-labeled isotypic IgG antibody. The ratio FITC:PI also is determined, to compensate for transected nuclei. The latter analysis offers better discrimination between negative and positive staining in tissue sections (Fig. 5).

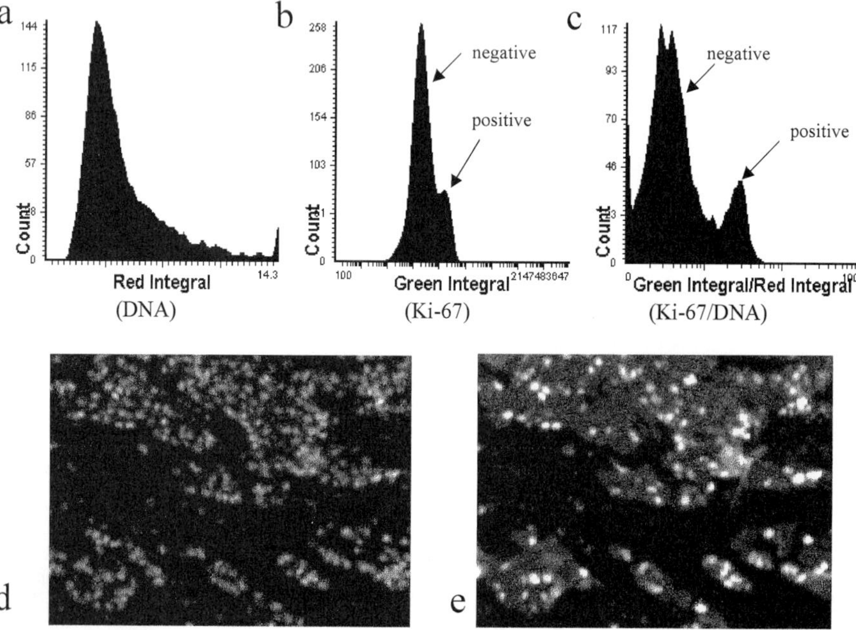

Fig. 5 Routine section of paraffin-embedded breast carcinoma stained with MIB-1 monoclonal antibody (Ki-67, green fluorescence) and counterstained with propidium iodide (red fluorescence). (a) Typical "noninterpretable" DNA histogram arising from the cut nuclei in a tissue section measured by LSC. (b) Expression of Ki-67 (green integral). The difference between negative and positive staining is set individually for each tumor, based on the nonspecific green fluorescence of negative control. The discrimination between negative and positive tumor population is more pronounced when the section is analyzed for the ratio of green versus red fluorescence (c). (d) Red fluorescence (DNA) LSC image of breast carcinoma. (e) Green fluorescence (Ki-67) of the same field.

V. Analysis of Estrogen Receptors by Laser Scanning Cytometry

The introduction of monoclonal antibodies that could recognize steroid receptors in cytological specimens or tissue sections, including routine paraffin-embedded preparations, replaced the older dextran-coated charcoal methods (Allred, 1993; Battifora *et al.*, 1993). Pathologists can easily assess the immunohistochemical staining of estrogen receptors (ER) in sections counterstained with hematoxylin and selectively evaluate only the cancerous component of the specimen. It is apparent that immunostaining is most often heterogeneous; some nuclei stain more intensely than others. Thus, there have been efforts to develop microscope-imaging techniques that would provide sensitive, objective, and reproducible measurements of the receptors on a cell by cell basis by immunohistochemistry (Gorczyca *et al.*, 1998a; Sklarew *et al.*, 1990; Nichols *et al.*, 1996; Remmele and Schicketanz, 1993). LSC allows for that type of measurement of ER in routine histological sections and cytological preparations, and it can display cell by cell measurements in single or dual parameter histograms (Gorczyca *et al.*, 1997a, 1998a).

A. Materials

Primary antibody: mouse monoclonal anti-ER antibody (ER ID5, Immunotech, Westbrook, ME).

Secondary antibody: biotinylated rabbit anti-mouse IgG (DAKO Universal Kit).

FITC–avidin (Boehringer Mannheim).

PI/RNase A solution (described earlier in the chapter).

B. Procedure

This is a protocol for paraffin-embedded tissue sections. For ER staining of cytological specimens for LSC, follow the method described for Ki-67 staining.

1. Cut 5-μm sections, deparaffinize in xylene (or Histo-Clear), and rehydrate in decreasing concentrations of ethanol (100, 95, 80, and 50%); incubate in distilled water for 30 min.

2. Heat the slides for three cycles in a microwave oven in antigen retrieval buffer (e.g., 0.01 *M* citrate buffer, pH 6.0, from BioGenex, San Ramon, CA, or DAKO Target Retrieval Solution).

3. Block the sections with biotin (0.01%; DAKO Biotin Blocking System) and avidin (0.1%; DAKO) for 30 min; wash in PBS twice for a total of 5 min.

4. Incubate the sample with primary antibody diluted 1:50 for 60 min at room temperature.

5. Wash the slides with PBS and incubate with secondary antibody at room temperature for 30 min.

6. Incubate the sections with avidin-FITC diluted 1:100 for 30 min at room temperature in the dark.

7. Counterstain the nuclei with 5 μg/ml PI and 0.1% RNase for 30 min at room temperature.

8. Apply one drop of 90% glycerol in PBS, mount under a cover glass, and analyze by LSC using 20× objective. Use red fluorescence signal for contouring.

C. Results

The tumors positive for ER show characteristic nuclear staining of ER, at least in some of the tumor cells. Receptor expression of individual nuclei was displayed in relative terms by the histogram distribution of integrated FITC fluorescence per cell. The cutoff point between positive and negative fluorescence was determined by negative controls, which were run for each case in parallel with the slide stained for ER (sequential sections from the same paraffin block are used for ER and negative control). Figure 6 presents ER histograms generated by LSC measurement of ER-immunostained tumors superimposed on negative controls.

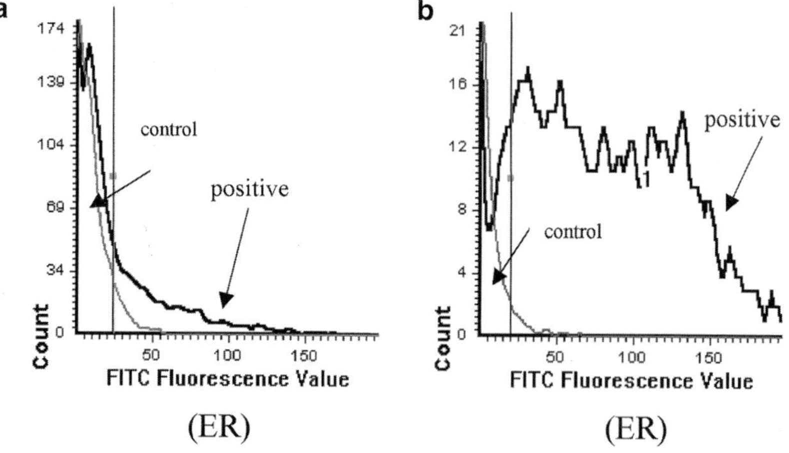

Fig. 6 Histograms of weakly positive (a) and strongly positive (b) breast carcinoma stained for estrogen receptor in routine paraffin-embedded tissue sections. The threshold between negative and positive populations is set individually for each tumor based on the fluorescence of the negative controls. (See color plates.)

VI. Analysis of Transcription Factors by Laser Scanning Cytometry

Nuclear factor κB (NF-κB) belongs to the family of ubiquitous transcriptional activators involved in regulation of immune responses to inflammatory agents, and in controlling cell growth and apoptosis (Baeurle and Baltimore, 1996; Baldwin, 1996). In its inactive form NF-κB remains in the cytoplasm; on activation it is rapidly translocated into the nucleus (Zabel *et al.*, 1993; Baeurle and Baltimore, 1996; Baldwin, 1996). The p53 gene, which belongs to the tumor suppressor gene family, is a key component of the cell cycle and apoptosis regulatory pathways and is essential for the maintenance of genome integrity. Functional inactivation of wild-type p53, which can be caused by mutation, by loss of both alleles, and by its sequestration in the cytoplasm (owing to interaction with other cellular or viral oncogenes), is the most common abnormality in human cancer. LSC offers the possibility not only to analyze the degree of expression of both NF-κB and p53, but also to assess their nuclear versus cytoplasmic localization. Translocation of these and other proteins (e.g., some memebers of the signal transduction pathways) from cytosol to nucleus (or vice versa) can easily be detected and measured by LSC.

A. Materials

Purified mouse IgG3 anti-NF-κB monoclonal antibody (Boehringer Mannheim).

Rabbit polyclonal NF-κB p65 subunit of NF-κB (Santa Cruz Biotechnology).

Antibody to p53: FITC-conjugated mouse IgG2b anti-human p53 (clone DO-7; PharMingen). This antibody recognizes an epitope between amino acids 1 and 45 of all known forms of human p53 and does not cross-react with mouse or rat p53.

Secondary antibodies: FITC-conjugated goat anti-mouse F(ab')$_2$ (DAKO) and FITC-conjugated goat anti-rabbit Ig (Santa Cruz).

PI /RNase A staining solution (as described earlier in this chapter).

B. Procedure

1. Attach cells to microscope slides by spinning in a Cytocentrifuge at 1000 rpm for 6 min, using a Shandon Cytospin (Shandon).

2. Fix the cells in methanol-free 1% formaldehyde in PBS for 15 min on ice, wash with PBS, transfer into 70% ethanol at −20°C, and store for up to 24 hr before subjecting to incubations with antibody.

3. Wash specimen twice with PBS containing 1% BSA and 0.1% sodium azide (PBS–BSA).

4. Incubate the cells with PBS–BSA containing primary antibody for 2 hr at room temperature (2 μg of anti-NF-κB antibody or 1 μg of anti-p53 monoclonal antibody). Use mouse anti-IgG and rabbit anti-IgG as isotype negative controls. For specimens on slides use 20 to 40 μl of antibody solution; when staining cells in suspension use 100 μl of antibody solution.

5. Rinse the cells with PBS–BSA and add FITC-conjugated secondary antibody (goat-anti-mouse Ig diluted 1:20 in PBS–BSA or FITC-conjugated goat-anti-rabbit Ig diluted 1:50). Incubate for 1 hr at room temperature in the dark.

6. Conterstain DNA by adding PI/RNase A staining solution for 30 min at room temperature in the dark.

7. Add 90% glycerol in PBS or Vectashield (Vector Laboratories), mount under coverslips, and analyze at least 5000 cells by LSC using 20× objective.

C. Results

Activation of NF-κB is measured as a function of its translocation from the cytoplasm to the nucleus. Using red fluorescence (nuclear DNA) signal for contouring and measuring the green NF-κB immunofluorescence (integrated value) separately over the nucleus (within the contoured area) and over the cytoplasm (within the torus outside of the contoured area), a ratio of cytoplasmic to nuclear fluorescence is obtained. The change of this ratio is a sensitive parameter of NF-κB activation. Details of the strategy to measure NF-κB translocation or activation of tumor suppressor p53 are presented in our more recent publications (Deptala *et al.*, 1998, 1999; Darzynkiewicz *et al.*, 1999). Examples of LSC analysis of NF-κB are presented in Fig. 7.

VII. Measurement of Nucleoli Using Laser Scanning Cytometry Fluorescence *in Situ* Hybridization Protocol

The nucleolus is a structural and functional organelle of the cell involved in the synthesis of rRNA. It is clearly visible in the routine hematoxylin–eosin preparations, and its size and shape is carefully analyzed by the pathologist in order to differentiate malignant cells from their benign counterpart. Reed–Sternberg cells, malignant melanoma, and clear cell carcinoma of the kidney are just a few examples of tumor cells with prominent nucleoli. Unfortunately, some benign cells (e.g. immunoblasts, regenerating hepatocytes, or activated fibroblasts) may have prominent nucleoli; therefore, additional methods were developed in the past several years in the hope that they will allow for better distinction between malignant and benign characteristics of the nucleolus. One of the most commonly used markers for that purpose is the group of nucleolar proteins

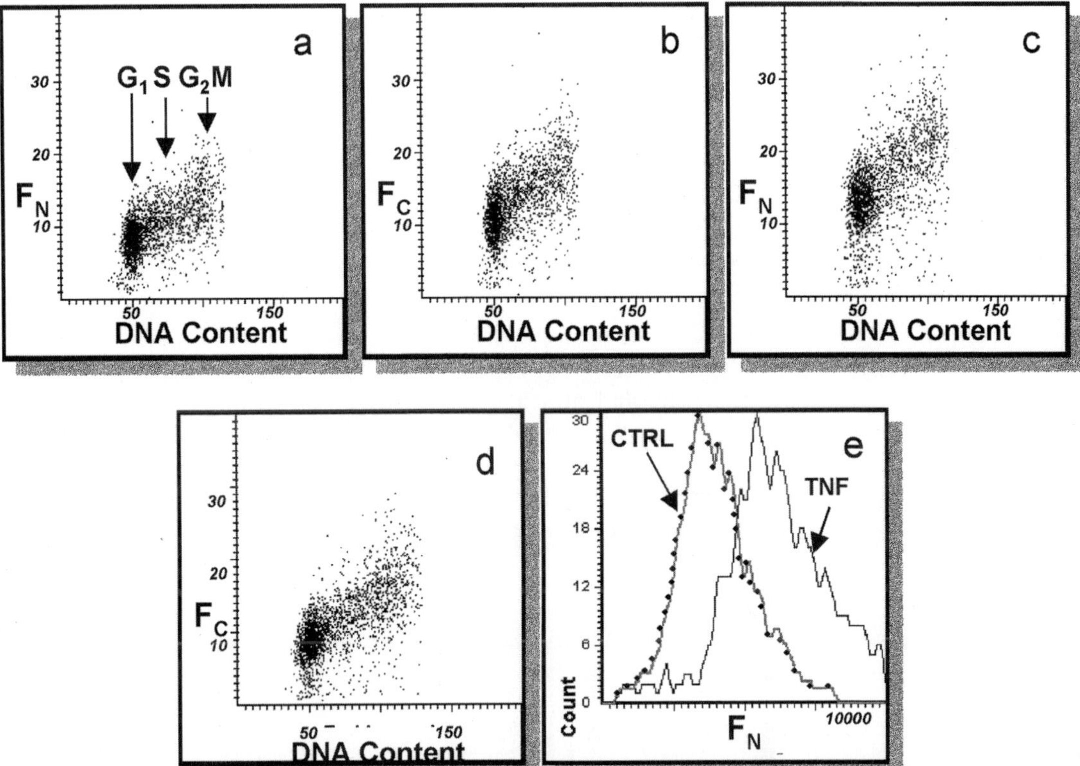

Fig. 7 Analysis of the changes in expression of nuclear factor κB (NF-κB) in U-937 cells treated with 10 ng/ml of tumor necrosis factor β (TNF-β) for 1 hr. (a, b) Scattegrams representing NF-κB located in the nucleus (F_N) and cytoplasm (F_C) versus DNA content prior to the treatment (c, d) Scattergrams for F_N and F_C, respectively, after treatment. (e) Change in F_N after treatment (ctrl is control; TNF, tumor necrosis factor). From Deptala *et al.* (1998). Activation of nuclear factor kappa B (NF-κB) assayed by laser scanning cytometry (LSC). *Cytometry* **33,** 376–382. Copyright © 1998 John Wiley & Sons, Inc. Reprinted by permission of Wiley-Liss, Inc., a subsidiary of John Wiley & Sons, Inc. (See color plates.)

stained by silver methods, called AgNOR (Derenzini *et al.,* 1990, 1995; Sirri *et al.,* 1995, 1997). There is a wealth of evidence in the literature that nucleolar activity is a strong prognostic marker in many malignancies.

The morphometric capabilities of LSC make this instrument ideally suited to analyze cell structure, including measurement of cell organelles. The method described below is based on immunostaining of nucleolar antigen with comercially available monoclonal antibody ("nucleoli," Chemicon International, Temecula, CA). Contouring on DNA-associated red (PI) fluorescence and using the "FISH analysis" feature of LSC, which is included in the standard software package provided with the instrument, we have been able to estimate overall nucleolar mass (represented by green fluorescence integrated over the nucleus) and also to count the number of nucleoli per nucleus.

A. Materials

Mouse anti-human nucleoli monoclonal antibody (Chemicon International); Secondary goat-anti mouse antibody conjugated with FITC (DAKO); PI/RNase A solution (see earlier).

B. Procedure

1. Isolate the cells and attach to slides by cytocentrifugation as described earlier in this chapter. Without allowing the cells to dry, fix in 2% formalin in PBS for 15 min at room temperature, wash in PBS, and resuspend for 3 min in cold ($-20°C$) acetone to make cells permeable. After a final rinse in PBS to remove acetone, the slides can be stored in PBS at 4°C in Coplin jars for up to

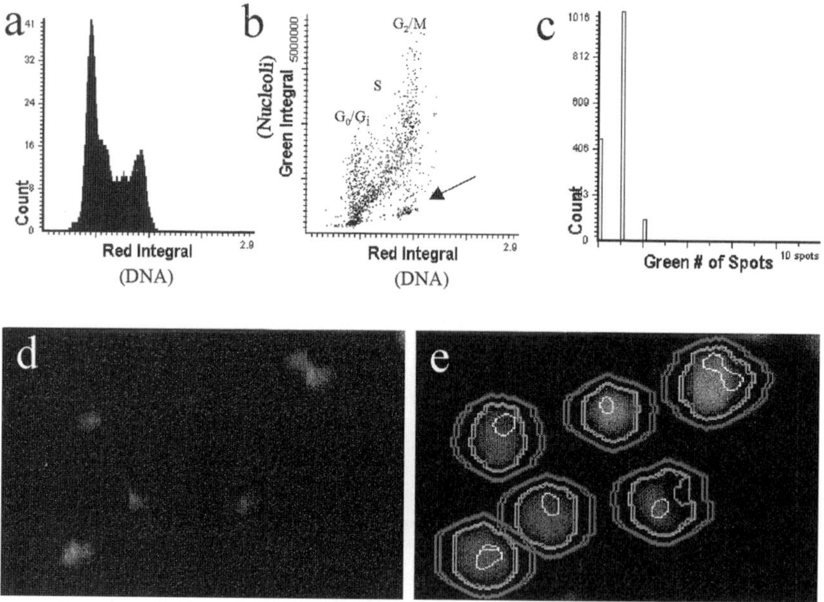

Fig. 8 Human lymphocytes from peripheral blood, stimulated by phytohemagglutinin for 48 hr, stained with anti-nucleolar monoclonal antibody, and analyzed on LSC using the FISH protocol. (a) DNA histogram (PI staining). (b) Values of green ("nucleolar") versus red (DNA) fluorescence (arrow, mitotic cells). (c) Data showing the number of spots (number of nucleoli) per nucleus. Actual LSC images are also shown [(d), green fluorescence (nucleoli); (e), red fluorescence with LSC contouring of the nucleoli]. (See color plates.)

4 hr. Frozen tissue sections from human solid tumors can be fixed and processed the same way.

2. Incubate the specimen with nucleolar antibody at 4°C overnight in a tightly closed box at 100% humidity (in a pilot experiment, titrate the antibody to find out the optimal titer, as described earlier in the chapter).

3. Rinse cells with PBS and incubate with FITC-conjugated anti-mouse antibody at a concentration as specified by the vendor for 30 min at room temperature in the dark.

4. Counterstain cells with PI/RNase solution for 30 min at room temperature in the dark.

C. Results

Figure 8 presents LSC data from the analysis of nucleoli in a culture of mitogen (phytohemagglutinin). stimulated lymphocytes. Cell relocation followed by morphological analysis revealed that the cells with a G_0/G_1 DNA content and minimal nucleolar fluorescence (green integral) were nonstimulated G_0 lymphocytes, whereas the cells with a G_2/M DNA content and minimal nucleolar fluorescence (arrow) were mitotic cells. Likewise, most cells with no nucleoli (zero "FISH" spots) also were G_0 lymphocytes.

References

Allred, D. D. (1993). Should immunohistochemical examination replace biochemical hormone receptor assays in breast cancer? *Am. J. Clin. Pathol.* **99,** 1–2.

Arends, M. J., Morris, R. G., and Wyllie, A. H. (1990). Apoptosis: The role of endonuclease. *Am. J. Pathol.* **136,** 593–608.

Baeurle, P. A., and Baltimore, D. (1996). NF-κB: Ten years after. *Cell* **87,** 13–20.

Baldwin, A. S., Jr. (1996). The NF-κB and IκB proteins: New discoveries and insights. *Ann. Rev. Immunol.* **14,** 649–681.

Battifora, H., Mehta, P., Ahn, C., and Felder, B. (1993). Estrogen receptor immunohistochemical assay in paraffin-embedded tissue: A better gold standard? *Appl. Immunohistochem.* **1,** 39–45.

Bedner, E., Melamed, M. R., and Darzynkiewicz, Z. (1998). Time resolved kinetic reactions measured in individual cells by laser scanning cytometry (LSC). *Cytometry* **33,** 1–9,

Bedner, E., Xun, L., Gorczyca, W., Melamed, M. R., and Darzynkiewicz, Z. (1999). Analysis of apoptosis by laser scanning cytometry. *Cytometry* **35,** 181–195.

Brown, R. W., Allred, D. C., Clark, G. M., Osborne, C. K., and Hilsenbeck, S. G. (1996). Prognostic value of Ki-67 compared to S-phase fraction in axillary node-negative breast cancer. *Clin. Cancer Res.* **2,** 585–592.

Darzynkiewicz, Z., Gong, J., and Traganos, F. (1994). Analysis of DNA content and cyclin protein expression in studies of DNA ploidy, growth fraction, lymphocyte stimulation, and the cell cycle. *In* "Methods in Cell Biology: Flow Cytometry" (Z. Darzynkiewicz, J. P. Robinson, and H. A. Crissman, eds.), 2nd Ed., Part A, Vol. 41. Academic Press, San Diego.

Darzynkiewicz, Z., Gong, J., Juan, G., Ardelt, B., and Traganos, F. (1996). Cytometry of cyclin proteins. *Cytometry* **35,** 1–13.

Darzynkiewicz, Z., Juan, G., Xun, L., Gorczyca, W., Murakami, T., and Traganos, F. (1997). Cytometry in cell necrobiology: Analysis of apoptosis and accidental cell death (necrosis). *Cytometry* **27,** 1–20.

Darzynkiewicz, Z., Bedner, E., Xun, L., Gorczyca, W., and Melamed, M. R. (1999). Laser scanning cytometry: A new instrumentation with many applications. *Exp. Cell Res.* **249**, 1–12.

Deptala, A., Bedner, E., Gorczyca, W., and Darzynkiewicz, Z. (1998). Activation of nuclear factor kappa B (NF-κB) assayed by laser scanning cytometry (LSC). *Cytometry* **33**, 376–382.

Deptala, A., Li, X., Bedner, E., Cheng, W., Traganos, F., and Darzynkiewicz, Z. (1999). Differences in induction of p53. p21$^{\mathrm{WAF1}}$, and apoptosis in relation to cell cycle phase of MCF-7 cells treated with camptothecin. *Int. J. Oncol.* **15**, 861–871.

Derenzini, M., Pession, A., and Trere, D. (1990). Quantity of nucleolar silver-stained proteins is related to proliferating activity in cancer cells. *Lab. Invest.* **63**, 137–140.

Derenzini, M., Sirri, V., Pession, A., and Trere, D., Rousel, P., Ochs, R. L., and Hernandez-Verdun, D. (1995). Quantitative changes of the two major AgNOR proteins, nucleolin and protein B23, related to stimulation of rDNA transcription. *Exp. Cell Res.* **219**, 276–282.

Dressler, L. G., and Seamer, L. C. (1994). Controls, standards, and histogram interpretation in DNA flow cytometry. *In* "Methods Cell Biology: Flow Cytometry" (Z. Darzynkiewicz, J.P. Robinson, and H. A. Crissman, eds.), 2nd Ed., Part A, Vol. 41. Academic Press San Dieg.

Gerdes, J., Lemke, H., Baish, H., Wacker, H. H., Schwab, U., and Stein, H. (1984). Cell cycle analysis of a cell proliferation-associated human nuclear antigen defined by the monoclonal antibody Ki-67. *J. Immunol.* **133**, 1710–1715.

Gong, J., Traganos, F., and Darzynkiewicz, Z. (1994). A selective procedure for DNA extraction from apoptotic cells applicable for gel electrophoresis and flow cytometry. *Anal. Biochem.* **218**, 314–319.

Gorczyca, W. (1999). Cytometric analyses to distinguish death processes. *Endocrine-Rel. Cancer* **6**, 17–19.

Gorczyca, W., Bruno, S., Darzynkiewicz, R. J., Gong, J., and Darzynkiewicz, Z. (1992). DNA strand breaks occurring during apoptosis: Their early in situ detection by the terminal deoxynucleotidyl transferase and nick translation assays and prevention by serine protease inhibitors. *Int. J. Oncol.* **1**, 639–648.

Gorczyca, W., Gong, J., and Darzynkiewicz, Z. (1993). Detection of DNA strand breaks in individual apoptotic cells by the in situ terminal deoxynucleotidyl transferase and nick translation assays. *Cancer Res.* **52**, 1945–1951.

Gorczyca, W., Darzynkiewicz, Z., and Melamed, M. R. (1997a). Laser scanning cytometry in pathology of solid tumors. *Acta Cytol.* **41**, 98–108.

Gorczyca, W., Sarode, V., Juan, G., Melamed, M. R., and Darzynkiewicz, Z. (1997b). Laser scanning cytometric analysis of cyclin B1 in primary human malignancies. *Mod. Pathol.* **10**, 457–462.

Gorczyca, W., Davidian, M., Ghershon, J., Ashikari, R., Darzynkiewicz, Z., and Melamed, M. R. (1998a). Laser scanning cytometry quantification of estrogen receptors in breast cancer. *Anal. Quant. Cytol. Histol.* **20**, 470–476.

Gorczyca, W., Bedner, E., Burfeind, P., Darzynkiewicz, Z., and Melamed, M. R. (1998b). Analysis of apoptosis in solid tumors by laser scanning cytometry. *Mod. Pathol.* **11**, 1052–1058.

Haugland, R. P. (1994). Spectra of Fluorescent dyes used in cytometry. *In* "Methods in Cell Biology: Flow Cytometry" (Z. Darzynkiewicz, J. P. Robinson, and H. A. Crissman, eds.), 2nd Ed., Part B, Vol. 42, pp. 641–663. Academic Press, San Diego.

Juan, G., and Darzynkiewicz, Z. (1998). Detection of cyclins in individual cells by flow and laser scanning cytometry. *Methods Mol. Biol.* **11**, 3–12.

Kamentsky, L. A., and Kamentsky, L. D. (1991). Microscope-based multiparameter laser scanning cytometer yielding data comparable to flow cytometry data. *Cytometry* **12**, 381–387.

Kamentsky, L. A., Burger, D. E., Gershman, R. J., Kamentsky, L. D., and Luther, E. (1997). Slide-based laser scanning cytometry. *Acta Cytol.* **41**, 123–143.

Koss, L. G., Czerniak, B., Hertz, F., and Wersto, R. P. (1989). Flow cytometry measurements of DNA and other cell components in human tumors: A critical appraisal. *Hum. Pathol.* **20**, 528–548.

Leers, M. P. G., Schutte, B., Theunissen, P. H. M. H., Raaekers, F. C. S., and Nap, M. (1999). Heat pretreatment increases resolution in DNA flow cytometry of paraffin-embedded tumor tissue. *Cytometry* **35**, 260–266.

Li, X., and Darzynkiewicz, Z. (1995). Labelling DNA strand breaks with BrdUTP. Detection of apoptosis and cell proliferation. *Cell Prolif.* **28,** 271–279.

Li, X., Traganos, F., Melamed, M. R., and Darzynkiewicz, Z. (1995). A single-step procedure for labeling DNA strand breaks with fluorescein or BODIPY-conjugated deoxynucleotides. detection of apoptosis and bromodeoxyuridine incorporation. *Cytometry* **20,** 172–182.

Nichols, G. E., Fierson, H. F., Boydÿ, J. C., and Hanigan, M. H. (1996). Automated immunohistochemical assay for estrogen receptor status in breast cancer using monoclonal antibody CC4-5 on the Ventana. *Am. J. Clin. Pathol.* **106,** 332–338.

Remmele, W., and Schicketanz, K. H. (1993). Immunohistochemical determination of estrogen and progesterone receptor content in human breast cancer: Computer-assisted image analysis (QIC score) vs. subjective grading (IRS). *Pathol. Res. Pract.* **189,** 862–869.

Rew, D. A., Reeve, L. J., and Wilson, G. D. (1998). Comparison of flow cytometry and laser scanning cytometry for the assay of cell proliferation in human solid tumors. *Cytometry* **33,** 355–361.

Sklarew, R. J., Bodmer, S. C., and Pertschuk, L. P. (1990). Quantitative imaging of immunocytochemical (PAP) estrogen receptor staining patterns in breast cancer sections. *Cytometry* **11,** 359–378.

Visscher, D. W., and Crissman, J. D. (1994). Dissociation of intact cells from tumors and normal tissues. *In* "Methods in Cell Biology: Flow Cytometry" (Z. Darzynkiewicz, J. P. Robinson, and H. A. Crissman, eds.), 2nd Ed., Part A, Vol. 41. Academic Press, San Diego.

Zabel, U., Henkel, T., dos Santos Silva, M., and Bauerle, P. A. (1993). Nuclear uptake control of NF-κB by MAD-3, an IκB protein present in the nucleus. *EMBO J.* **12,** 201–211.

CHAPTER 51

Laser Cytometry of Human Tissues and Tumors: Proliferation and Therapeutic Applications

David A. Rew

Royal South Hants Cancer Centre
Southampton University Hospitals
Southampton SO14 0YG, England

I. Introduction
II. Ploidy and Proliferation in Surgical Oncology
   A. Prognostic Studies
   B. Retrospective Studies
   C. Prospective Studies
III. Cytometric Studies of Proliferation
   A. Ploidy Measurements as Prognostic Markers
   B. S-Phase Fraction Measurements as Prognostic Markers
   C. Cytometry of Other Clinical Tumor Markers
IV. The Halogenated Pyrimidines in Cell Proliferation Research
   A. *In Vitro* Studies of Halogenated Pyrimidine Labeling
   B. *In Vivo* Halogenated Pyrimidine Labeling Indices
   C. Time-Dependent Parameters of Tumor Proliferation
   D. Correlation of Proliferation Data and Biomarker Expression
V. Clinical Studies of Cell Production Rates with Thymidine Analogs
   A. Descriptive Clinical Studies of Dynamic Tumor Proliferation
   B. Descriptive Studies of Cell Production Rates in Nonmalignant Epithelium
   C. Cell Production Rates and Therapy
   D. An Overview of Proliferation Data in Human Tissues and Tumors
   E. Concluding Comments on Cell Proliferation Studies
VI. Further Applications of Cytometry in Clinical Oncology
   A. Laser Scanning Cytometry in Clinical Studies
   B. Cytometric Assays in Cancer Therapy
   C. Cytometric Assays of Fluorescent Cytotoxic Drugs
VII. Conclusions
   References

METHODS IN CELL BIOLOGY, VOL. 64

445

I. Introduction

Treatment strategies for cancers are often inadequate. Surgery, chemotherapy, and radiotherapy all have serious limitations. The clinician thus seeks help from science and technology to achieve better clinical outcomes. The modern tools of cytometry have given us a whole new range of investigative capabilities at the cell and tissue level. Nevertheless, the trained human eye and the brain of the cytologist and histopathologist remain the fastest, most efficient, and most versatile image processor. The conventionally stained histological section contains an immense amount of information about tissue architecture, constitution, cell size, and characteristics. Histology remains the gold standard of clinical diagnosis and prognosis, because of its facility to describe the architecture and the class of tumor, for example, an adenocarcinoma; the grade of the tumor, or its degree of differentiation, usually based on the semiquantitative scale of well, moderately, or poorly differentiated features; and the presence of tumor in metastases. Visual assessment can be enhanced, where appropriate, by techniques such as histochemical staining or automated image analysis. Most descriptive information about tumors is still derived from visualization, either from inspection of the lesion, its local invasion, and its metastases *in situ* or from microscopy to describe the cell and tissue architecture.

Cytometric technologies must thus provide information not otherwise available to the trained observer across a range of applications in clinical oncology. These include the attainment of a precise clinical diagnosis, prognostication, research in the basic sciences, and therapeutic research. We may broadly classify human tumors for cytometric purposes as those presented in suspension, such as in blood or effusions; in homogeneous solid form such as lymphomas, carcinoids, sarcomas, poorly differentiated carcinomas, and melanomas; or in heterogeneous solid form allowing classification on tissue morphology alone, as are most carcinomas. A useful cytometric measurement of outcome must thus complement rather than substitute for tissue morphological criteria. Cytometers can help further classify tumors in those situations where morphology is homogeneous and insufficiently discriminatory. In the case of reticuloendothelial tumors such as lymphomas, cytometric immunophenotyping allows rapid, precise classification of lymphomas and the optimization of treatment (see Chapter 46 of this volume).

Tumors in suspension such as leukemias are relatively easy to study using cytometric instruments. Fresh and archival samples of solid, nonhematological tumors pose considerable problems for cytometric assays. These problems include tissue disaggregation and cell extraction, tumor heterogeneity, and biomarker preservation. Solid tumors must be disaggregated by mechanical or enzymatic techniques. These can introduce cell and epitope damage, and they may yield an unrepresentative sample. All these factors have a significant bearing on the interpretation of cytometric data.

Tumor heterogeneity poses critical and unresolved problems in cytometric data evaluation. Tissue and tumor extracts invariably comprise a complex admixture of cells of many different tumor, stromal, and inflammatory lineages. At the cell, subcellular, and organelle level, there are many similarities between cells and nuclei from tumors and normal tissues. Cell size, granularity, and scatter characteristics are generally heterogeneous, thus ruling out simple classifications based on the standard morphologies and parameters that help characterize hematological lineages. Intratumor heterogeneity of morphology, form, and expression of many biomarkers can be considerable, both within a planar microscope field of tissue section and from site to site within a tumor. This renders measurements based on one or a few samples per tumor unreliable. They may not be representative either of the entire tumor or of the most biologically aggressive regions within it (Rew, 1996). Heterogeneity also varies with time during the growth of each tumor, as for example, heterogeneity of DNA content (Shackney and Shankey, 1995; Shackney et al., 1995).

II. Ploidy and Proliferation in Surgical Oncology

A. Prognostic Studies

Tumor behavior is unpredictable. The analysis of tumor samples for prognostic purposes seeks to detect predictors of future biological behavior, including studies that measure the rate of tumor growth (Steel, 1977). Many biomarkers and tumor characteristics have been studied by cytometric assays in terms of outcome measures, commonly time to local recurrence or death of advanced disease. However, it is of little use to the patient or to the clinician to be able to predict the precise hour of death using the best tools of cytometric science if survival or quality of life cannot be improved.

The best index of the biological aggressiveness of a tumor remains evidence of its spread to regional or distant lymph nodes, liver, or other metastatic site. No measurement of any marker on the primary tumor yet equates as an independent prognostic indicator. Nevertheless, tumors do offer clues as to their future behavior in their morphological appearance, with poorly differentiated tumors tending toward greater aggression. Clinical cytometric studies of prognosis fit two general categories: retrospective and prospective.

B. Retrospective Studies

Retrospective studies are usually conducted on archival series of clinical pathology samples. They offer the considerable advantage of established clinical follow-up of patients, often over many years, and they allow the study of rare and infrequent tumor classes. However, the cytometric study of archival samples can be flawed. The investigator has no control over the sampling and representative nature of the tissue block from the original tumor. Such series are rarely able to address the problem of intratumor heterogeneity, and randomly archived samples may well not be representative of the entire tumor.

C. Prospective Studies

Prospective studies are planned in advance of tissue collection. They have the advantage of allowing the investigator to collect fresh material in optimal conditions and to select specific areas of tumor for study. They have the major disadvantage of duration and unpredictability of specimen availability, the dependence on close liaison with surgeons and oncologists, and the need for collocation of clinical and research facilities.

This chapter offers examples of how laser cytometry has helped our understanding of clinical tumor behavior with examples from the literature and from our own studies. We also consider the constraints to the use of cytometric technologies in cancer medicine, and how laser scanning cytometry in particular promises to advance our knowledge, as with our studies of fluorochromatic, cytotoxic drug uptake in tumors.

III. Cytometric Studies of Proliferation

A. Ploidy Measurements as Prognostic Markers

The cell cycle model (Howard and Pelc, 1951) underpins much cytometric research. Cell cycle related DNA content (ploidy) and the derived S-phase fraction (SPF) have been intensively studied in all classes of clinical tumors, both in prospective and archival clinicopathological series, with respect to prognostic outcome. These studies were aided by techniques for the extraction of cell nuclei from wax-embedded, formalin-fixed archival tissue blocks (Frankfurt *et al.*, 1984, 1986; Hedley, 1989).

Ploidy studies have highlighted the frequency of aneuploidy in clinical tumor series. Aneuploidy is a feature of some 75% of tumors. We are still uncertain whether the gross chromosomal disorder of aneuploidy is a cause or a consequence of malignant change (Rew, 1994). Conventional cytometry fails to detect subtle changes in DNA such as translocations and mutations that may cause cancer. In global terms, ploidy measurements have not proved to be useful independent prognostic indicators in clinical practice, when compared to histological assessment of tumor grade and stage, even in the widely studied model of breast tumors.

B. S-Phase Fraction Measurements as Prognostic Markers

The SPF indicates the proportion of cells or nuclei in the S-phase compartment at the time of measurement. Measurement of the SPF by flow cytometry (FCM) poses a number of problems. Within aneupoid populations, it is difficult to distinguish the S-phase cells of the diploid population within the overlapping aneuploid populations. The interpretation of ploidy and SPF data is further complicated by methodological problems in many series, including different analytical protocols, methods of sample preparation, and instruments (Wheeless *et al.*, 1991). We must thus treat with caution data which purport to demonstrate a correlation between the SPF and clinical behavior of tumors. The SPF is often presented as a surrogate measure of proliferation. However, even where measured accurately using sophisticated deconvolution models, the SPF gives no measure of time, nor of the rate of transition of the cells through the cell cycle. Thus, a tumor with a large S-phase fraction of slowly transiting cells may be much less proliferative than a tumor with a small S-phase fraction of rapidly transiting cells.

C. Cytometry of Other Clinical Tumor Markers

The labeling index (LI) is a generic way to quantify biomarker expression in cell populations. It is the proportion of cells expressing the biomarker above a given threshold in the total population under study. The LI describes a snapshot of the number of labeled cells in a tissue or tumor sample at one time point. It gives no indication of the rate of turnover or transit of the labeled cells in the population.

Cytometry allows quantitation of many cell surface, cytoplasmic, and nuclear proteins and epitopes, for which monoclonal antibodies and fluorescent tags offer accurate identification (Watson, 1992). Such markers include those that are closely associated with key regulatory processes such as proliferation, apoptosis, cell signaling, and oncoprotein function. Many proliferative biomarkers are known to act in the cell cycle, and their expression changes between quiescent and cycling cells, and with the phase of the cell cycle (Quinn and Wright, 1990). Examples include Ki-67, PS1, and the cyclin-dependent kinases. Multiparameter assays that plot DNA content against the expression of a chosen protein marker allow study of its cell cycle related expression, as for example, c-Myc (Rew *et al.*, 1991b) proliferating cell nuclear antigen (PCNA) (Hall *et al.*, 1990; Sawtell *et al.*, 1995), p53 (Rew *et al.*, 1996), and the cyclins (Darzynkiewicz *et al.*, 1996).

IV. The Halogenated Pyrimidines in Cell Proliferation Research

Proliferating cells may also be identified by an exogenous label, usually a thymidine analog. The nonradioactive halogenated pyrimidine (HP) thymidine analogs bromodeoxyuridine (BrdUrd) and iododeoxyuridine (IdUrd) are robust and reliably incorporated into living cells during the S phase. They can be detected by a range of monoclonal antibodies developed in the early 1980s (Gratzner, 1982; Gonchoroff *et al.*, 1985; Gray, 1985; Vanderlaan and Thomas, 1985; Sasaki *et al.*, 1986). These can in turn be identified by fluorochromatically or histochemically labeled probes suitable for cytometric detection and analysis. BrdUrd was originally developed and used as a tumor radiosensitizer in the 1950s, to be given as an adjunct to radiotherapy by intravenous injection in doses of 1 g per day for up to 40 days (Kinsella *et al.*, 1984). Iododeoxyuridine has similar properties to BrdUrd and is used as an alternative for *in vivo* kinetic studies. It is cross-reactive to some anti BrdUrd monoclonal antibodies. The HP analogs can be used in one of three ways to obtain data on the proliferating compartment (Waldman *et al.*, 1988): (1) by *in vitro* incubation of freshly obtained tumor samples (Miwa *et al.*, 1989), (2) by *in vivo* infusion over a period of hours to "saturate" the replicating fraction of cells (these studies have been reviewed in detail by Dolbeare, 1995a,b, 1996), and (3) by *in vivo*, intravenous pulse labeling preoperatively to obtain dynamic indices, of which we have particular experience as reviewed in this chapter.

The incorporation of HPs into DNA raises concerns about mutagenicity. In experimental cell systems, ultraviolet light increases the damage to DNA containing BrdUrd in tumor cells in culture and inhibits cell differentiation (Barrett *et al.*, 1978; Kaufman, 1986; Wright, 1986; Raffel *et al.*, 1988). Findings from these models must be interpreted with caution, as the unusual experimental conditions of the animal or cell model may not translate to the human body. For example, a DNA strand break assay may not take into account damage reversal by the DNA repair enzymes that regulate the integrity of chromosomes. HPs continue to be used in high doses as radiosensitizers, such as for brain tumors, and for the delivery of therapeutic doses of radioiodine to liver metastases (Speth *et al.*, 1989). Photosensitization has been reported with high doses (Fine and Breathnach, 1986). No untoward acute or long-term effects from low dose, single shot labeling have come to light in studies of more than 2500 cancer patients to date.

A. *In Vitro* Studies of Halogenated Pyrimidine Labeling

In vitro cytometric studies of human tumor biopsies by incubation with HPs allow derivation of an S-phase LI (Dolbeare *et al.*, 1983, 1985; Lloveras *et al.*, 1994), but they provide no measurement of the S-phase transit time. Incubation of freshly harvested and viable tissue and tumor cells can be performed in the laboratory without the need for *in vivo* injection or clinical consent. It has been used by many groups over a wide range of tumor and tissue types, as reviewed in detail by Dolbeare (1995a,b). The comparability between the tritiated thymidine and the nonisotopic analogs *in vitro* has been established in studies of human breast tumors (Meyer *et al.*, 1993; Maas *et al.*, 1996) and human colorectal and cervical tumors (Wilson *et al.*, 1985).

B. *In Vivo* Halogenated Pyrimidine Labeling Indices

In vivo labeling offers clear advantages over *in vitro* techniques. When given intravenously (iv) or intraperitoneally (ip) in animal models, the HP label is delivered physiologically to the tumor mass. Over short infusion periods immediately prior to, or at the time of biopsy, the labeled fraction will consist almost entirely of cells in the S phase, and a few that have passed into G_2/M.

The administration of a HP by continuous iv infusion over several hours progressively saturates cycling cells with label, such that, as more and more cells enter and pass through and out of the S phase, the labeled fraction approaches the growth fraction. *In vivo* HP labeling index (S-phase fraction) data have been reported on various classes of tumor (Wilson *et al.*, 1985; Raza *et al.*, 1985), including intracranial gliomas (Hoshino *et al.*, 1985, 1989), bronchial tumors (Tinnemans *et al.*, 1995), meningiomas (Langford *et al.*, 1996), transitional cell carcinoma of the bladder (Popert *et al.*, 1993), renal carcinoma (Larsson *et al.*, 1994), squamous cell carcinomas (SCC) of head and neck (Kotelnikov *et al.*, 1995a,b), and breast tumors (Sasaki *et al.*, 1987; Goodson *et al.*, 1993; Christov *et al.*, 1994).

C. Time–Dependent Parameters of Tumor Proliferation

Time is a key component of biological processes. Tumors grow, regress, stabilize, or evolve with time. Using HPs, proliferation measurements with a time component can be made in clinical tumors. These "dynamic" indices include the S-phase duration (Ts), the cell cycle time (Tc), and the potential doubling time (Tpot). The data from such clinical measurements are reviewed at greater length here.

Continuous prebiopsy infusion denies the option of time-dependent data provided by pulse labeling. A pulse label of an HP and multiparameter flow cytometry can be used to derive time-dependent data from a single biopsy of a tumor following *in vivo* administration by the intraperitoneal or intravenous route (Begg *et al.*, 1985, 1988; Begg, 1989; Terry *et al.*, 1991). HP markers are robust within tumor biopsies, surviving degradation, enzymatic extraction, and acid denaturation during analysis, and surviving long-term storage in ethanol- and in formalin-fixed tissues. It is a remarkable and fortunate observation that these markers survive metabolism, sequestration, and dilution, enter the tumor mass, pass into proliferating cells, and incorporate into S-phase DNA within an hour of intravenous injection.

The technique is illustrated by a study of BrdUrd incorporation after bolus intraperitoneal injection into the human HT29 colorectal tumor grown in a mouse model in Fig. 1. In this model, the G_0/G_1 phase of the principal tumor cell population is centered on channel 40 on the X axis. Within 2 hr, labeled S-phase cells are clearly identified in the G_2M phase. By 5 hr, many labeled cells have appeared in G_0/G_1 of the daughter cell cycle. By 14 hr, the majority of labeled cells have reached this phase. By 29 hr, daughter cells are streaming through the S phase once again. The cell cycle duration is measured by plotting the proportion of labeled cells in the mid S phase against time (Fig. 2). Such serial biopsy is rarely practical in the clinical setting, where single biopsies must usually suffice.

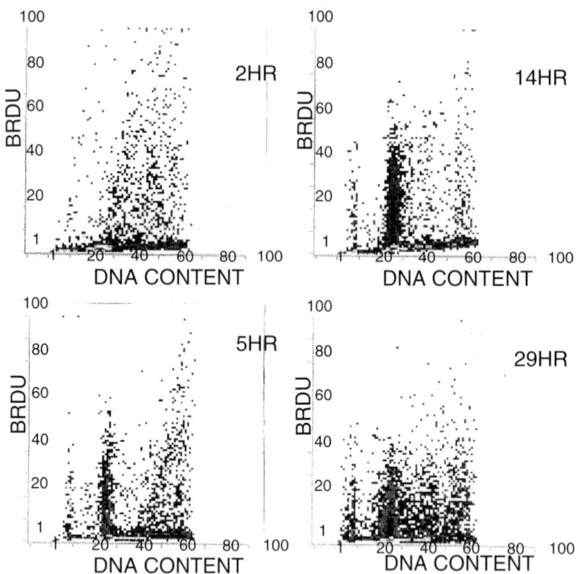

Fig. 1   A study of BrdUrd incorporation after bolus intraperitoneal injection into the human HT29 colorectal tumor grown in a mouse model. The G_0/G_1 phase of the principal tumor cell population is centered on channel 40 on the X axis. BrdUrd fluorescence is represented on the Y axis. By 2 hr, labeled S-phase cells are seen in the G_2/M phase. By 5 hr, many labeled cells have appeared in G_0/G_1 of the daughter cell cycle. By 14 hr, the majority of labeled cells have reached that phase. By 29 hr, many of the daughter cells have reentered the S phase.

D. Correlation of Proliferation Data and Biomarker Expression

The *in vivo* labeling of clinical tumors with a robust S-phase label provides a unique opportunity for correlative studies of biomarker expression in proliferative cells. They provide a framework for the study of proteins that act at specific points in the cycle to initiate, regulate, suppress, or terminate DNA replication, such as the oncoproteins p62-c-Myc and p53 (Rew *et al.*, 1991b, 1996), or proteins associated with apoptosis, such as bc12 (Wilson *et al.*, 1996).

V. Clinical Studies of Cell Production Rates with Thymidine Analogs

Studies of cell production rates *in vivo* using tritiated thymidine and the fraction of labeled mitoses (FLM) long predated the halogenated pyrimidines (Tubiana

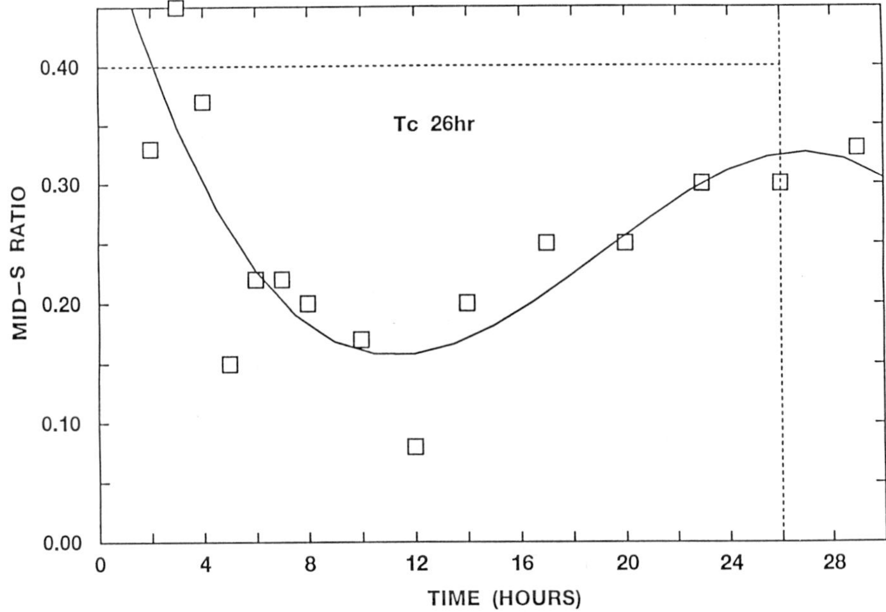

Fig. 2 The method of estimation of the cell cycle duration in the model described in Fig. 1 by plotting the proportion of labeled cells in the mid S phase against time. The cell cycle length in this model is of the order of 26 hr.

and Courdi, 1989). However, the HP/flow cytometry method represented a quantum leap in clinical utility and speed in analysis (Rew and Wilson, 1991). Studies reporting time-dependent parameters derived from the HP/FCM method fall into four general categories: Those descriptive studies that report proliferative data, including heterogeneity studies; those that correlate such data with clinical outcome; those that correlate such data with response to therapy; and those that combine data from immunohistochemical (IHC) measurements of labeling indices with flow cytometric measures of S-phase duration in related samples. These combined studies allow estimates of proliferation rates in selected areas of heterogeneous tissues and tumors, or in tissues such as epithelia where proliferative cells display geographic organization.

In early clinical studies, BrdUrd was shown to be a reliable marker when given intravenously in a subclinical bolus dose of 100–250 mg a few hours before surgical biopsy (Fig. 3). Wilson *et al.* (1988) reported the *in vivo* measurement of the LI, Ts, and Tpot of 26 evaluable human tumors of various lineages obtained by local biopsy. Riccardi *et al.* (1988) reported the cell kinetics of 46 acute leukemias, 27 gastric carcinomas, and 16 gliomas.

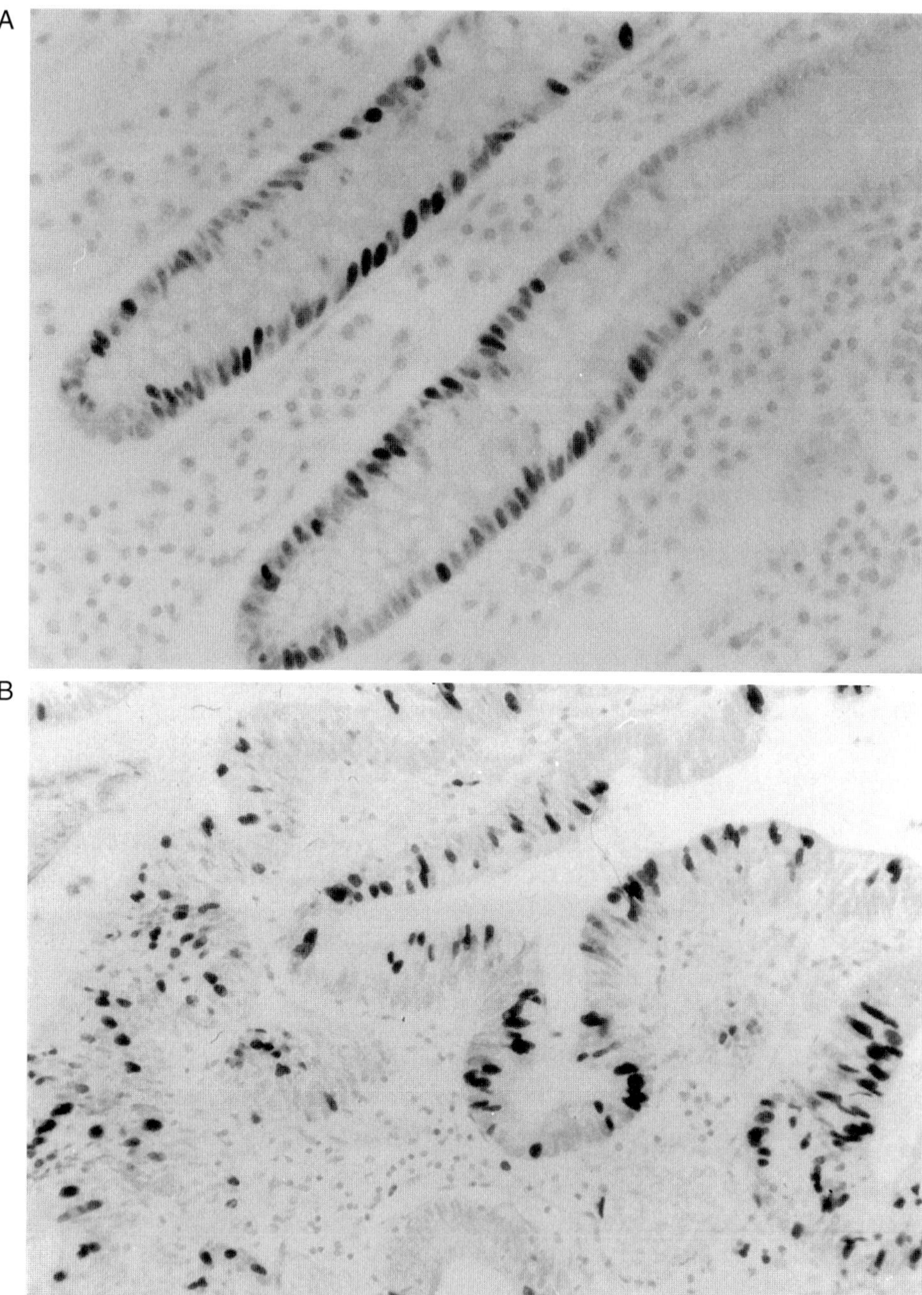

Fig. 3 Series of photomicrographs of colorectal tissues obtained at surgery, where the BrdUrd pulse-labeled proliferating cells are detected immunohistochemically (see text) with peroxidase counterstain. (A) Highly ordered distribution of labeled cells in the base of normal mucosal crypts. (B) Villous adenoma of the rectum. (C) Well-differentiated invasive adenocarcinoma of the colon. Microadenoma in mucosa from a patient with familial polyposis coli. (D) Note the migration of the poliferating compartment towards the luminal face.

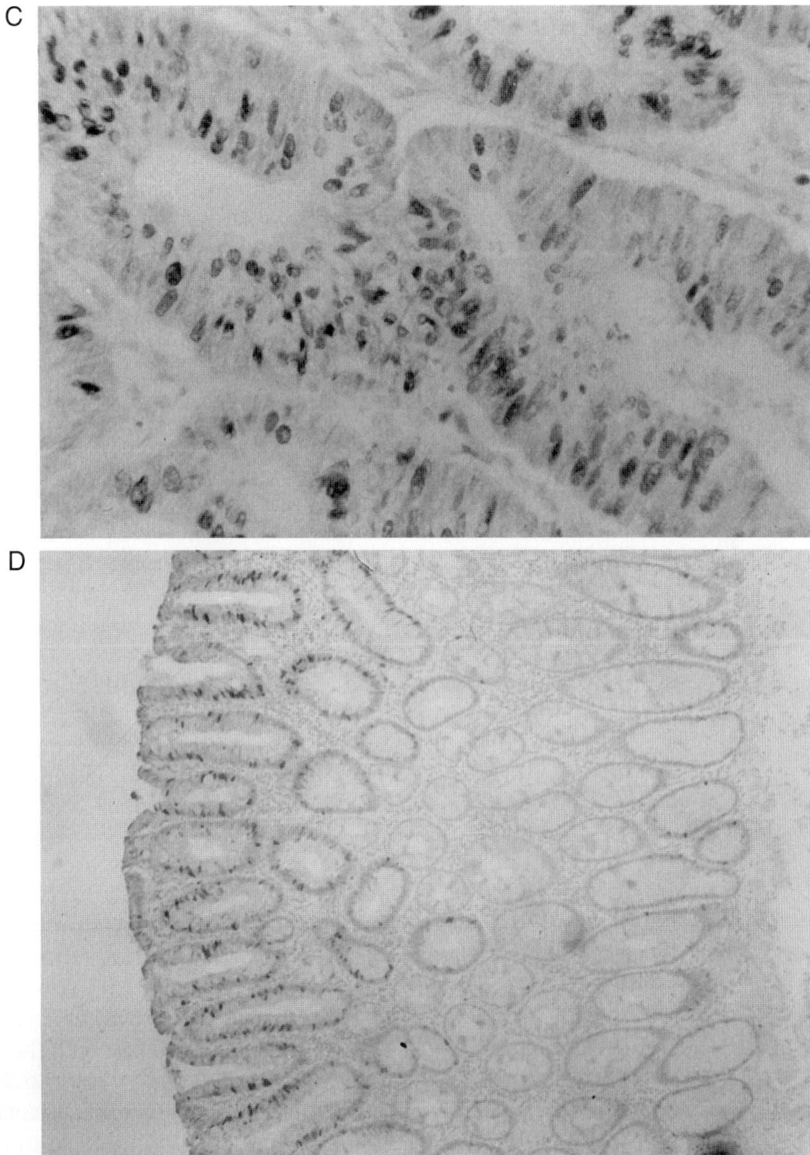

Fig. 3 (*continued*).

A. Descriptive Clinical Studies of Dynamic Tumor Proliferation

1. Squamous Carcinomas

a. Squamous Carcinomas of the Head and Neck

Squamous cell carcinomas of the head and neck (HN-SCC) have been of particular interest in clinical research because of the therapeutic challenge that they pose, their radioresponsiveness, and their accessibility to biopsy and to direct observation of clinical growth rates.

Forster *et al.* (1992) reported cell kinetics in a series of 82 HN-SCC tumors. Cooke *et al.* (1994) reported interim survival data for this series. They found no significant correlation between LI, Ts, and Tpot and tumor stage, node status, or tumor site. Bennett *et al.* (1992) assessed tumor proliferation of HN-SCC by both IHC and flow cytometric (FCM) analysis. Using FCM data alone, 46% of the tumors exhibited a Tpot of less than 5 days. When the Ts from the FCM data was combined with the average IHC LI, 84% of Tpot values were less than 5 days, and with the maximum LI in any one tissue section, 99% were less than 5 days. Jones *et al.* (1994) reported data from 75 patients with HN-SCCs. Nylander *et al.* (1994) studied 31 such tumors by both FCM and IHC following IdUrd labeling. The utility of prognostic correlations were limited by small numbers and incomplete follow-up. Benazzo *et al.* (1995) studied 46 head and neck cancers. BrdUrd LI and TS significantly correlated with histological differentiation grading: Grade III tumors showed higher LI values and shorter TS values than Grade I and Grade II tumors. Similar study results are reported by Kotelnikov *et al.* (1995a,b) and Wilson *et al.* (1995). These data are recorded in Table I. Figure 4A illustrates the typical proliferation patterns of two diploid tumors labeled *in vivo* with BrdUrd, a basal cell carcinoma and a squamous cell carcinoma of the pinna (Rew, 1991).

b. Squamous Esophageal Tumors

The Northwood (England) group has reported the proliferation parameters in the biopsies of 30 esophageal squamous tumors using BrdUrd pulse labeling (Wilson, 1991; Rew *et al.*, 1991a) (Table II). Haustermans *et al.* (1994, 1995, 1997) reported a similar series of 31 patients. IdUrd was injected 6 to 10 hr before surgery, and five biopsies per tumor were taken ($n = 305$). Tumor-stage, pathological node status, and sex significantly influenced the disease free survival (DFS). When DFS was studied as a function of Tpot, no significant difference was found between fast- and slow-proliferating tumors. Haustermans *et al.* (1994) also reported the heterogeneity of proliferation in these tumors. The coefficient of variation (cv) of intratumor Tpot measurements from up to five biopsies from each of 30 tumors ranged from 7.0 to 39%, or up to 5 days. In a sample of 30 biopsies from a single squamous tumor to assess intratumor heterogeneity, the mean Tpot was 4.2 ± 0.9 [standard deviation (SD)] days, with a range of 2.0–6.4 days.

c. Squamous Carcinomas of the Uterine Cervix

Cervical tumors are also significant for their radioresponsiveness and thus have been a focus for several labeling studies. Bolger *et al.* (1993) assessed 120

Table I
Tumors of the Head, Neck, and Lung

Reference	No.	Median LI, %	Median Ts, hr	Median Tpot (range), days	Comments
Head and Neck					
Bennett *et al.* (1992)	123	6.8	9.9	5.7	BrdUrd
Forster *et al.* (1992)	105	7.0 (1.3–21.9)	14.0 (7.0–106)	5.9 (1.3–67.5)	BrdUrd
Cooke *et al.* (1994)					
Jones *et al.* (1994)	75	8.9 (1.6–25.0)	14.8	Check	BrdUrd
Nylander *et al.* (1994)	31 FCM IHC	13.6 (3.6–26.4)	16.1 (3.9–32.4)	4.6 (1.3–12.2)	IdUrd
		9.1 (1.6–35.0)		5.4 (1.1–36.2)	
Wilson *et al.* (1995)	165	5.0 diploid	9.9	1.8 diploid	
		9.3 aneuploid	5.4–21.9	3.2 aneuploid	
Benazzo *et al.* (1995)	46/52	7.9 (2–18)	11.6 (6–28.5)	5.7 (2–30)	BrdUrd
Kotelnikov *et al.* (1995)	12	>20	12.1 (5.1–21.5)	43.2 hr (18.8–84.5 hr)	IdUrd + BrdUrd histochemistry
Bourhis *et al.* (1996)	70	6.3–7.7	8.3–9.3	4.6–5.6	BrdUrd
Corvo *et al.* (1996)	82	8.0 (1.5–28.0)	10.0 (6.0–14.0)	5.0 (2.0–20.0)	BrdUrd
CNS tumors					
Astrocytoma, Shibuya *et al.* (1993)	100		9.2 (6.0–13.7)	1 to 60 days	
Meningioma, Riccardi *et al.* (1988)	22	2.1 (0.9–3.9)	16.7	63.2	BrdUrd
Malignant, Struikmans *et al.* (1997)	71	0.03	4.5	5.4	BrdUrd
Benign, Struikmans *et al.* (1997)	52	0.01	4.7	20.9	BrdUrd
CNS mets, Struikmans *et al.* (1997)	14	0.03	3.9	3.9	BrdUrd

A
80
100
80
60
40
20
1
1   20   40   60   80   100
G1 GATED(P1) RED FL/AREA
P2 GATED(P1) GREEN VS RED
1   200   400   600   800   1000
1200
100
80
60
40
20
1
G1 GATED(P1) RED FL/AREA
P2 GATED(P1) GREEN VS RED
1   200   400   600   800   1000
1   20   40   60   80   100
B
600
100
80
60
40
20
1
G1 GATED(P1) RED FL/AREA
P2 GATED(P1) GREEN VS RED
1   200   400   600   800   1000
1   20   40   60   80   100
600
100
80
60
40
20
1
G1 GATED(P1) RED FL/AREA
P2 GATED(P1) GREEN VS RED
1   200   400   600   800   1000
1   20   40   60   80   100

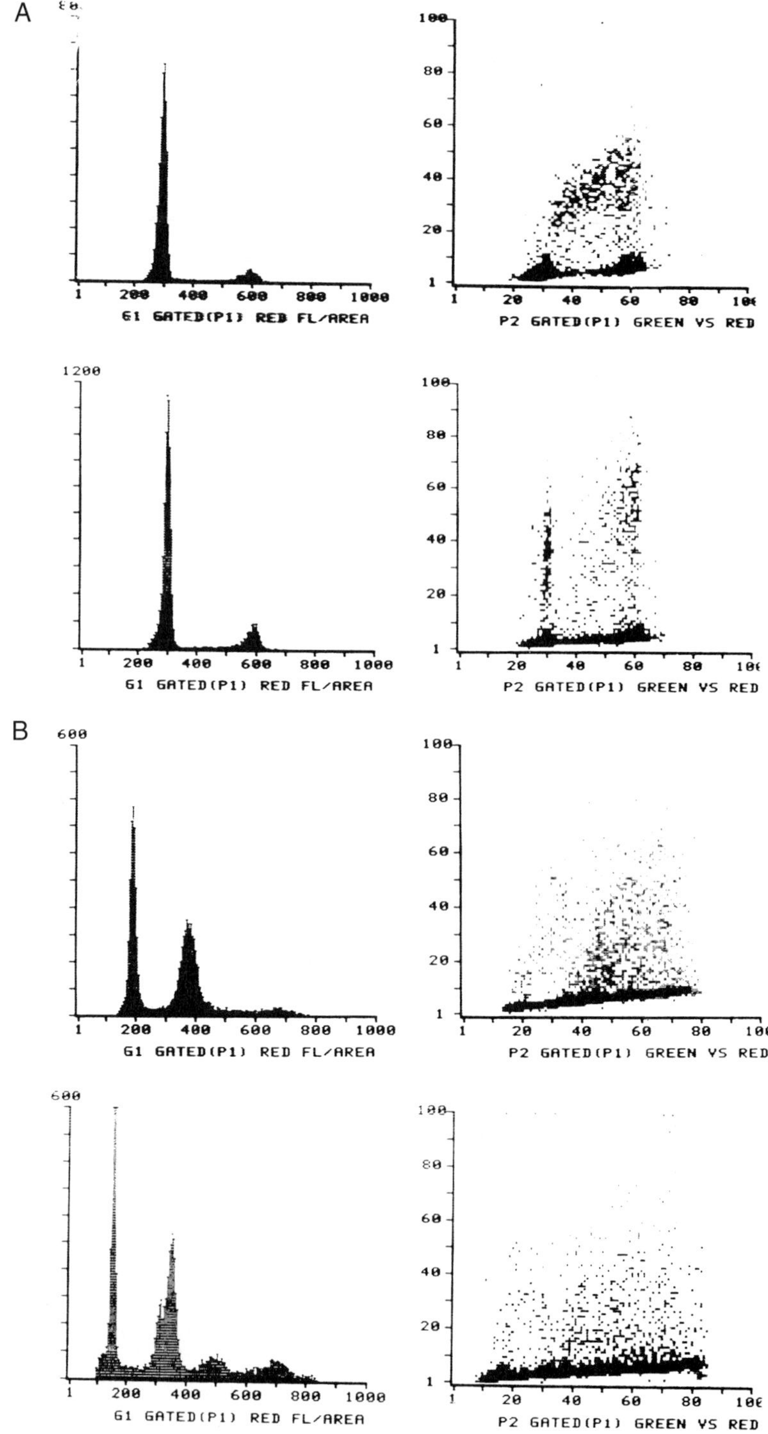

Fig. 4

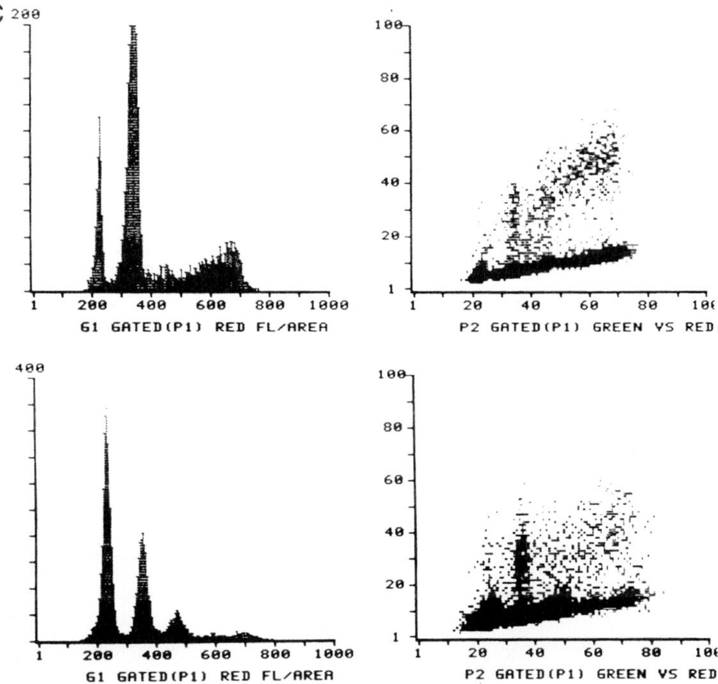

Fig. 4 Paired histograms and dot plots displaying the DNA profiles and corresponding multivariate BrdUrd (*Y* axis) versus DNA (*X* axis) analyses of a series of tumors pulse labeled with 250 mg BrdUrd *in vivo* (see text). All analyses were undertaken on an Orthocytofluorograph cytometer at the GRAY Laboratory, Northwood, England. (A) Two diploid tumors. The upper profile is from a giant basal cell carcinoma of the scalp, and the lower from a squamous cell carcinoma of the pinna. (B) Typical gastric adenocarcinoma (upper profile) and its liver metastasis (lower profile). (C) Two typical aneuploid colorectal tumors.

cervical tumors. In 89% both static and temporal kinetic parameters could be measured. The analysis of multiple biopsies from each tumor revealed marked intratumor heterogeneity (Table III). There was an elevation in the LI, but no difference in Ts, between tumor and non-neoplastic cervical tissue. There was a significant elevation of the LI, in proliferating cells from advanced stage and large size tumors. In a further study, this group reported (Bolger *et al.* (1993) no correlation between these parameters and a response to conventional radiotherapy was found other than an association between a high BrdUrd labeling index and pelvic tumor recurrence. Tsang *et al.* (1995) measured the pretreatment Tpot values in carcinomas of the uterine cervix in 46 patients. For 25 patients where Tpot measurements were performed at two separate laboratories, systematic laboratory differences were detected in the data.

Table II
Gastrointestinal Tract Tumors

Reference	No.	Median or mean LI (ranges), %	Median Ts (range), hr	Mean or median Tpot (range), days	Comments
Upper intestinal tumors					
Esophageal squamous, Haustermans *et al.* (1994)	31	19.0 ± 7.0 (SD)	17.2 ± 8.0	4.4 (2.0–6.4)	IdUrd
Esophageal squamous, Rew (1993)	9	5.3 (1.4–17.4)	9.8 (4.6–17.8)	4.3 (2.7–17.9)	BrdUrd
Esophageal squamous, Wilson (1991)	50	7.8 (0.4–27.5)	12.4 (6.9–28.6)	5.2 (1.6–56.8)	BrdUrd
Gastric adenocarcinoma, Rew (1993)	39	4.9 (0.9–18.5)	10.7 (3.6–31.9)	5.2 (0.8–39.4)	BrdUrd
Gastric adenocarcinoma, Haustermans (1995); Haustermans *et al.,* 1995)	32	16.0 ± 9.0 (SD)	15.8 ± 6.7	5.6 (2.1–9.5)	IdUrd
Gastric adenocarcinoma, Riccardi *et al.* (1988)	22	9.9 (5.7–14.0)	15.2 (13.4–22.7)	9.8 (6.8–13.5)	BrdUrd
Colorectal					
Camplejohn (1982)	19	—	—	192 hr	Stathmokinetics
Bergstrom and Stenling (1990)	?	13 (2.9–27.0)	16 (7.0–35.0)	3.6 (2.5–14.0)	IdUrd
Rectal, Terry *et al.* (1995)	101	21.0 (13.5–27.0)	20.0	3.3 (2.4–5.6)	BrdUrd 6 hr before biopsy
Rew *et al.* (1991a)	100	9.0 (0.7–22.2)	13.1 (4.0–28.6)	3.9 (1.75–21.4)	BrdUrd
Michel *et al.* (1997)	19	5.3–15.4	9.7–16.6	4.5 (1.2–21)	BrdUrd
Wilson *et al.* (1993a,b)	125	12.7 (0.6–33.8)	14.1 (5.2–48.3)	4.5 (0.9–53.8)	IudUrd

Table III
Female Reproductive Tract Tumors

Reference	No.	Median LI (range), %	Median Ts (range), hr	Median Tpot (range), days	Comments
Breast					
Rew *et al.* (1992)	75	4.2 (0.6–15.4)	8.7 (2.7–22.2)	8.2 (1.8–47.5)	BrdUrd
Stanton *et al.* (1996)	84	3.2	12.0	12.5	BrdUrd
Gynecology					
Ovarian, Erba *et al.* (1994)	55	6.1–7.2	14.7	12.5	BrdUrd
Cervix, Bolger *et al.* (1996)	120	9.8 (6.7–14.5) interquartile	12.8 (11.1–14.9) interquartile	4.0 (3.1–6.3) interquartile	BrdUrd
Cervix (squamous), Tsang *et al.* (1995)	39			5.5	BrdUrd
Cervix (adenocarcinoma), Tsang *et al.* (1995)	7			6.6	BrdUrd
Ovarian metastases, Rew (1991)	2	5.1, 5.4	12.5, 13.9	3.5, 6.5	

2. Central Nervous System Tumor

Most studies of HP labeling *in vivo* on brain tumors have measured labeling indices alone (see Dolbeare, 1995a). For example, Shibuya *et al.* (1993) studied 100 brain tumors double labeled with BrdUrd and IdUrd. Struikmans *et al.* (1997) reported on 71 malignant and 52 benign brain tumors and 14 cerebral metastases. Labeling indices were low, and there was little difference between data for benign and malignant tumors. Studies of meningiomas (Riccardi *et al.*, 1988) and astrocytomas (Danova *et al.*, 1991) have also been reported.

3. Adenocarcinomas

a. Gastric Adenocarcinomas

Gastric adenocarcinomas are very heterogeneous in their clinical growth behavior, from the flat, infiltrative to the markedly exophytic lesions. The contribution of cell production rates to their biological aggressiveness is uncertain. Static labeling index measurements of gastric tumors have been studied, particularly by Japanese groups (Miwa *et al.*, 1993; Ohyama *et al.*, 1990). A number of series of time-dependent indices of gastric adenocarcinomas are reported from Europe. We have reported data on 39 such tumors (Rew 1991, 1993). Haustermans *et al.* (1997) reported a series of 32 tumors labeled with IdUrd. Intratumor, intertumor, and interlaboratory heterogeneity of proliferation indices have also been reported by this group. In a sample of 30 biopsies from a single adenocarcinoma, the mean Tpot was 5.2 ± 1.7 days, with a range of 2.1–9.5 days. The cv of intratumor Tpot measurements from up to five biopsies from each of 30 tumors ranged from 16.0 to 51% (Table I). Figure 4b illustrates the proliferation patterns of a typical gastric adenocarcinoma and its liver metastasis following *in vivo* BrdUrd labeling (Rew, 1991).

b. Colorectal Adenocarcinomas

Colorectal proliferation studies are of particular value for the facility to correlate proliferative data and labeling patterns with form and function across the mucosa–adenoma–tumor sequence. They also provide valuable insights into the gross proliferative biology of this common surgical disease.

Cell proliferation studies on these tumors predate the HP assay. Camplejohn (1982) published estimates of colorectal tumor proliferation rates using laborious stathmokinetic measurements of tritiated thymidine labeling. The efficacy of BrdUrd labeling of colorectal tumors *in vivo* was established by Risio *et al.* (1988) and by Khan *et al.* (1988), who published labeling studies of small series.

Three large studies have subsequently provided dynamic data on proliferation rates in primary colonic and rectal cancers using HP labeling *in vivo*. Rew *et al.* (1991a) found no correlation between any kinetic parameters and the Dukes stage or histological classification of 100 colorectal tumors labeled *in vivo* with BrdUrd. A series of tumors labeled with IdUrd (Wilson *et al.*, 1993a,b) yielded similar data. Terry *et al.* (1995) reported the cell kinetics of 101 rectal cancers. A smaller series of 19 tumors was reported by Michel *et al.* (1997).

These series also demonstrated considerable intratumoral heterogeneity of proliferative parameters according to site of biopsy, and between aneuploid and diploid tumors and populations. A cross study of samples between these institutions demonstrated considerable reliability and reproducibility of the measurements, irrespective of institution, choice of HP, or cytometer used. Figure 4c illustrates the proliferation patterns of two typical aneuploid colorectal tumors following *in vivo* BrdUrd labeling from the Northwood series (Rew *et al.*, 1991a).

c. Breast Carcinomas

Breast cancers are an important clinical challenge and a valuable model for proliferation studies. Primary and recurrent lesions are accessible to biopsy, and volume growth is more easily estimated in the absence of exfoliation or necrosis as significant causes of cell loss, as for example, in tumors treated by tamoxifen or other chemotherapy alone.

Two series report the *in vivo* labeling of human breast tumors with BrdUrd. In a series of 69 patients with invasive breast carcinoma (Rew *et al.*, 1992), there were no significant differences in the total LI, Ts, or Tpot when patients were stratified according to lymph node status, tumor size, tumor grade, or menopausal status. Histological counting of labeling indices in this series of breast tumors yielded data comparable to flow cytometric analysis (Ashton-Key *et al.*, 1993). Stanton *et al.* (1996; Stanton, 1996) reported a similar series of 84 cases from Glasgow. LIs were significantly higher in aneuploid tumors and in tumors not expressing estrogen receptors, but they were not correlated with tumor size, nodal status, or expression of c-erbB2 (Table III).

d. Ovarian Carcinomas

Erba *et al.* (1994) reported the kinetic parameters of human ovarian adenocarcinoma *in vivo* using BrdUrd incorporation in 55 untreated patients. LI and TS were not correlated with clinical tumor stage, histological grading, residual tumor size, or DNA ploidy (Table III).

4. Other Tumors

a. Malignant Melanoma

Primary malignant melanoma lesions are difficult to study because of their small size and heterogeneous schirrhous admixture of skin and tumor cells, and because of the limitation of losing key pathological information on depth and planes during experimental biopsy. Laing *et al.* (1992) have reported one large series of 83 primary and metastatic malignant melanomas. Thin, good prognostic primary lesions (<1.5 mm thickness) displayed significantly slower proliferation than did the thicker lesions.

b. Hematological Malignancy

Hematological tumors are of interest for the ease with which multiple analyses can be performed on blood samples. Hematological stem cells are among the most actively proliferating subsets in the body (Raza *et al.*, 1992). Giordano *et al.* (1993) reported the kinetics of acute nonlymphoblastic leukemia (ANLL). Sixty-five patients with untreated ANLL and 15 patients with solid tumors and normal bone marrow (BM) received 250 mg/m^2 BrdUrd. Raza *et al.* (1997) studied bone marrow of 68 patients with myelodysplastic syndromes (MDS) who received sequential infusions of IdUrd and/or BrdUrd (Table IV).

c. Lung Carcinomas

Wilson (1993) reported a series of 38 lung tumors yielding 88 samples (Table IV). Tinnemans *et al.* (1993) studied bronchoscopy specimens of 27 lung cancers and 11 benign biopsies after *in vivo* labeling with 50 mg/m^2 BrdUrd. They obtained cytokinetic data from seven samples of small cell lung cancer (SCLC) and 20 samples of non-small cell lung cancer (NSCLC). No significant differences were observed between the mean values of the cytokinetic parameters of SCLC and NSCLC.

d. Urological and Other Tumors

Other tumor types, including tumors of the urological tract carcinoma of the bladder (Rew *et al.*, 1991c), lymphomas, and sarcomas, have been studied in smaller numbers (Table V).

B. Descriptive Studies of Cell Production Rates in Nonmalignant Epithelium

The *in vivo* labeling of human tumors also provides opportunities for the qualitative and quantitative histological study of patterns of proliferation in histological samples of tumors and adjacent normal tissue. A number of such studies have been reported. They provide valuable descriptive and quantitative information on the proliferation of epithelia in particular (Potten *et al.*, 1992a,b). Labeled cells can be clearly distiguished by conventional immunohistochemical techniques. Figure 3 illustrates this point with a series of photomicrographs of BrdUrd pulse-labeled colorectal tissues obtained from surgical resection specimens (see Rew *et al.*, 1991a). In Fig. 3a, the distribution of labeled cells in the base of normal mucosal crypts is highly ordered. In Fig. 3b, the cells in the epithelium of this villous adenoma of the rectum, a premalignant lesion, have lost their spatial discipline but remain confined to the epithelial plane. In Fig. 3c, a well-differentiated invasive adenocarcinoma of the colon, the proportion of proliferating tumor cells is much higher, associated with a greater degree of spatial indiscipline. Figure 3d shows how HP labeling can highlight disordered proliferation controls in disease states. It illustrates a microadenoma in mucosa from a patient with familial polyposis coli and a concurrent carcinoma. The

Table IV
Other Tumors

Reference	No.	Median LI (range), %	Median Ts (range), hr	Median Tpot (range), days	Comments
Melanoma					
Laing *et al.* (1992)					
Primary tumors	24	4.2 (1.3–13.6)	10.7 (6.3–20.5)	7.2 (3.5–41.3)	BrdUrd
Metastases	61	5.4 (0.6–16.8)	11.6 (6.1–26.2)	7.2 (2.3–139.0)	
Hematology					
Normal bone marrow Giordano *et al.* (1993)	15	11.8 ± 3.1	10.1 ± 2.0	3.5 ± 5.6	BrdUrd
Myelodysplastic syndrome Raza *et al.* (1997)	68	28.4	11.8	40.8 hr cell cycle time	Double labeling
MDS + acute myeloid leukemia Raza *et al.* (1992)	10	27.7 (22–37.5)	15.5 (11.4–24.6)	58.3 hr cell cycle time	BrdUrd
Acute non-lymphocytic leukemia Giordano *et al.* (1993)	54	6.1 ± 2.9	16.2 ± 6.1	18.1 ± 18.9	BrdUrd
ANL leukemia Riccardi *et al.* (1988)	42	6.2 (0.9–11.7)	12.1 (6.9–27.8)	8.5 (2.8–16.7)	BrdUrd
Lung					
Wilson (1993) Tumors	38	8.0 (0.7–28.2)	15.1 (5.5–37.8)	7.3 (1.4–132.0)	
Samples	88				
Normal lung	9	2.7 (1.2–7.8)	11.1 (4.2–41.1)	17.3 (6.5–32.0)	BrdUrd
Diploid tumors	14	5.0 (1.1–11.9)	6.9 (3.6–13.5)	9.6 (1.9–28)	
Aneuploid tumors	13	14.7 (3.8–33.6)	14.0 (5.7–29.4)	6.7 (1.5–20.7)	
Tinnemans *et al.* (1993)					

Table V
Unpublished Data and Rare Tumors[a]

Tumor type	No.	Mean LI (range), %	Mean Ts (range), hr	Mean Tpot (range), days	Comments
Lymphomas	2	5.4 (0.9–13.4)	11.0 (8.3–16.0)	15.8 (2.5–23.2)	
Hodgkin's nodes	3	2.6 (0.9–4.6)	10.4 (9.4–11.0)	17.7 (5.5–39.6)	
Basal cell carcinoma	1	5.4	10.7	6.6	
Sarcoma	2	1.5, 2.7	7.1, 22.0	8.8, 48.8	
Liposarcoma	1	0.6	2.4	13.5	
Osteosarcoma	1	0.8	8.8	36.7	
Urological					
TCC bladder, Rew *et al.* (1991c)	19	2.5 (0.5–10.0)	6.2 (3.5–9.7)	17.1 (3.6–40.0)	BrdUrd
Renal adenocarcinoma, Rew *et al.* (1991c)	2	1.9, 6.6	7.1, 11.9	4.5, 18.0	BrdUrd
Prostate carcinoma, Rew *et al.* (1991c)	1	1.9	9.0	7.7	

[a] G.D. Wilson and D.A. Rew, unpublished data, 1988–1991.

distribution of proliferating cells is completely inverted as compared with normal mucosa.

The rate of cell proliferation in normal tissues is of particular interest in comparison with tumor proliferation rates, where samples are obtained in the course of labeling studies on cancer cases. The ordered architecture of epithelium does not allow direct measurement of labeling indices derived from tissue homogenates using the automated techniques of laser cytometry. The counting of labeled cells must be in a structured fashion, and it must be correlated directly with the tissue architecture (Potten *et al.*, 1992a,b) (Table VI).

1. Skin and Mucosa

Van Erp *et al.* (1996) studied the cell cycle kinetics of normal skin epidermal cells in 14 lymphoma patients with IdUrd using serial biopsies. Esophageal squamous mucosa also has a well-defined proliferative zone at the base of the epidermis. The proliferation rates in labeled cells from normal esophageal mucosa from esophageal tumor resection specimens has been reported (Rew, 1991; Haustermans, 1995). Small numbers have been studied, and there is no information on the relationship between mucosal proliferation rates and squamous malignant change. The complex, convoluted architecture of gastric and duodenal mucosa has also been analyzed. Patel *et al.* (1993) used Ts data from tissue homogenates in conjunction with standard histochemical counting to estimate the cell turnover time of gastric mucosa at between 11.5 and 28.1 days according to anatomical site.

Colorectal mucosa is a structured tissue in which the proliferating cells are linearly ordered in the depths of each columnar mucosal crypt. Potten *et al.* (1992a) conducted a systematic quantitative analysis of the distribution of BrdUrd-labeled cells in human colorectal mucosa, taken from sites throughout the colorectum. A range of proliferative indices were measured, including the crypt labeling index, the peak labeling position, and the detailed distribution of labeled cells, along with the Ts median value of 8.6 hr. The cell cycle time was calculated to be 30 hr. The entire crypt turnover time was thus calculated to be 82 hr. These data were very consistent with earlier calculations made using tritiated thymidine in much smaller series and with animal model studies.

2. Proliferation in Transition Tissues

Colorectal adenomas are of interest for their malignant potential and for the intermediate level of proliferative disorder that they display between mucosa and invasive tumors. In a small study, incidental adenomas analyzed from colorectal tumor resection specimens, including cases of familial polyposis coli, displayed patterns of disorder and values for labeling indices intermediate between mucosal crypts and tumors (Rew, 1993). Ts values were similar to those in mucosal and tumor cells (Table VI).

Table VI
Normal and Nonmalignant Tissues

Reference	No.	Mean LI (range), %	Ts	Tpot	Comments
Skin					
Van Erp *et al.* (1996) Gastrointestinal Epithelium		3.5	9.7 ± 0.6 hr	28.4 hr cell cycle time	
Squamous esophageal mucosa, Rew (1991)	7	3.7 (0.4–8.4)	12.3 (6.5–18.2)	25.3 (4.3–67.6)	FCM data on tissue disaggregates
Squamous esophageal mucosa, Haustermans (1995)	53	5.2 ± 2.1 (SD)	9.6 ± 2.8 (SD)	8.4 ± 3.8 (SD)	IdUrd
Gastric mucosa, Patel *et al.* (1993)	27	2.7 (1.1–4.7)	9.6 (3.8–19.7)	15.6 (3.4–59.7)	FCM data on tissue disaggregates
Colorectal mucosa, Rew (1991) (flow cytometry analysis)	157	2.0 (0.6–8.4)	10.7 (3.1–37.5)	24.8 (2.9–101)	FCM data on tissue disaggregates
Colorectal mucosa, Potten *et al.* (1992a,b) (histometric analysis)	147	Varies with position in crypt	As above	Cell cycle time 30 hr	Same series as Rew, 1991; histochemistry and manual counting
Villous adenoma, Rew *et al.* (1991a)	6	5.3 (2.3–9.1)	8.6 (4.8–12.8)	6.1 (3.6–10.6)	BrdUrd
Metaplastic polyp, Rew *et al.* (1991a) (colorectal mucosa)	10	4.9 (0.6–20.1)	10.6 (5.2–19.6)	6.3 (2.5–53.6)	BrdUrd

C. Cell Production Rates and Therapy

Most clinical research on the correlations between therapy and cell production rates has been conducted during radiotherapy for squamous cell carcinomas of the head and neck. The rate of cell proliferation in individual tumors, as indicated by the local Tpot, may influence the tumor response to fractionated radiotherapy. The presence of rapidly proliferating clones in radiosensitive tumors suggested the design of trials of continuous, hyperfractionated, accelerated radiotherapy (CHART). This is intended to enhance tumor kill by preventing proliferation during the course of treatment. Studies have focused on head and neck cancer in particular (Withers *et al.*, 1988; Dische and Saunders, 1989; Saunders *et al.*, 1991). In a series reported from Northwood, data were complete for 90 patients (Wilson *et al.*, 1995). No kinetic parameter predicted the outcome of patients treated by CHART. When the Tpot was calculated using a combination of histology LI and FCM TS, diploid tumors showed more rapid median proliferation rates (Tpot 1.8 days) than aneuploid tumors (3.2 days).

Bourhis *et al.* (1993) studied the predictive value of pretreatment Tpot and LI in 70 patients with head and neck squamous cell carcinoma treated with conventional radiotherapy. No relationship was found between the Tpot or LI and the tumor stage, nodal status, histological grade, and the site of the primary. The mean Tpot of the tumors that relapsed locally was 5.3 days, compared to 6.1 days for those who did not relapse locally (not significant). The TS, LI, DNA index, and Tpot were not associated with local relapse, nor with disease-free survival (DFS). Corvo *et al.* (1993, 1995) reported a further such series of 82 patients (data shown in Table I).

Interim reports have been published from the EORTC (European Organisation for Research into the Treatment of Cancer) phase III trial comparing conventional fractionation (70–72 Gy in 7–8 weeks, 1.8–2.0 Gy/fraction) to a split-course accelerated treatment (72 Gy in 5 weeks, 1.6 Gy/fraction, three fractions per day with a 12–14 day split after 8 days). An analysis of 60 cases demonstrates that pretreatment Tpot data could not discriminate patients with improved ($>$4 days) or impaired ($<$4 days) survival even though the clinical data have shown benefit for the accelerated schedule (Begg *et al.*, 1990, 1992). A similar finding has emerged from studies from Paris (Bourhis *et al.*, 1996) in which 70 patients with oropharyngeal tumors treated by 70 Gy in 7 weeks were evaluated.

A study from Genoa (Antognoni *et al.*, 1996) of 69 patients with a median follow-up of 47 months treated by conventional fractionation and a boost schedule suggested that a Tpot $>$5 days could predict ($p = 0.04$) improved (68%) local control at 3 years compared to 13% in fast tumors ($<$5 days). This study supports the Tpot as a predictor of outcome. Overall the 3 year local control rate was 54% in patients with a Tpot of $>$5 days and 25% in those with a Tpot $<$5 days ($p = 0.004$). In a report from Sweden (Zackrisson *et al.*, 1997) in which 89 patients were treated by conventional radiation with a median follow-up of 30 months, nodal involvement and Tpot were correlated with local control.

Thus, the ability of Tpot measurements to consistently predict the outcome in conventional radiation treatment remains questionable. Studies of their predictive value in accelerated fractionation schedules have failed to reveal any significance of this measure of proliferation in either concomitant boost, split-course accelerated treatment or in the CHART schedule (Corvo *et al.*, 1996; Horiot *et al.*, 1997).

An analysis by Begg *et al.* (1998) of correlations between kinetic data and the outcome of treatment of 476 patients in 11 centers by at least 6 weeks of conventional radiotherapy demonstrated that the labeling index was the only significant variable in a univariate analysis for no control. Multivariate analysis showed no significant correlations, although comparisons were hampered by intercenter variation in measurements.

The results of these studies have thus not been consistent, and no consensus has yet emerged as to whether Tpot will be accepted as a useful clinical predictive test. There have been several drawbacks to the published studies, in the lack of randomization, small patient numbers, short follow-up, and lack of formal quality control of the Tpot measurements.

D. An Overview of Proliferation Data in Human Tissues and Tumors

The halogenated pyrimidines have proved to be potent tools for measuring tumor cell production rates when analyzed by laser flow or scanning cytometry (Rew *et al.*, 1998). The technique has been evaluated by a number of groups using different equipments and analytical techniques, and results have been found to show high concordance from one center to another. There are nevertheless a number of factors that confound the interpretation of proliferation data derived from human tumor biopsies.

The reliability of analyses is a key factor in the interpretation of proliferation measurements (Wilson, 1993). There is considerable technical, observer, and institutional variability in the data, contributing to the spread of data within and between tumors (Wheeless *et al.*, 1991; Haustermans *et al.*, 1995). Concern has been expressed by one group about the stoichiometry of antibody binding after DNA denaturation (Gilliland *et al.*, 1997; Williamson *et al.*, 1994).

Heterogeneity of architecture and cell content confounds the interpretation of data in solid tumors (Rew, 1996). Site to site variation for proliferation parameters is highly significant in tumors. The labeling index can vary considerably from region to region within a tumor. In any one local area of a tumor it may range from less than 1% to more than 50%, and it may vary according to the technique selected for measurement of the HP LI. For example, histochemical measures of the labeling index derived by manual counting are usually higher than equivalent FCM-derived indices (Bennett *et al.*, 1992; Ashton-Key *et al.*, 1993).

We are also presented with a complex problem in deciding which of the many indices generated by a HP label is most representative of the proliferation of the tumor. For example, the LI(max) and Tpot(max) of the maximally proliferative clones may have the greatest relevance to tumor cell behavior. In the study illustrated in Fig. 5, we have calculated the Tpot by the method of Begg for 60 colorectal tumors from our series using the same Ts values but with labeling index data derived in one of three ways: by flow cytometry, by using average counts from histochemically labeled sections, and by counting the maximally proliferative zones on the tissue sections. The series displayed upward of twofold variation in proliferation rates depending on the LI counting method chosen.

1. The S-Phase Duration

The S-phase duration (Ts) generally shows much less variation than the LI within and between tumors and between tumors and normal tissues. Many studies reveal that Ts in human cells appears to be relatively consistent, with most values falling between 10 and 20 hr across a wide range of tissues and tumors. Given the profound conservation of key biological processes in nature, it seems probable that the time taken to duplicate DNA, the duration of S phase, is a species-specific constant for a given number of chromosomes and volume of DNA. The

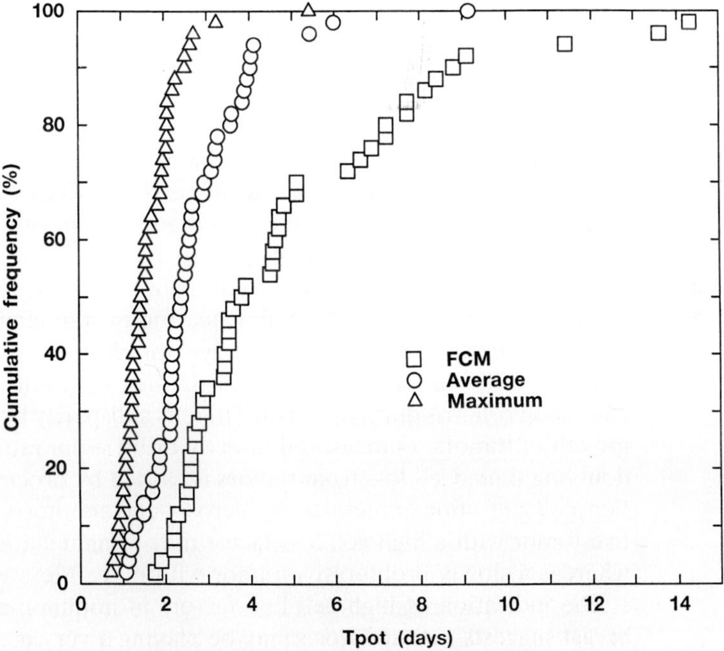

Fig. 5 The problem of heterogeneity of potential doubling time calculations according to the method of estimation of the labeling index in a series of 60 human colorectal tumors: FCM, flow cytometric method; Average, average counts from histochemically labeled sections; Maximum, the counts from maximally proliferative zones on the tissue sections (see text).

variation observed in the Ts measurement may thus be due to experimental artifact. Indeed, DNA duplication is such an evolutionarily conserved and powerfully regulated process that it might be surprising if the duration of DNA synthesis were significantly different between normal cells and viable tumor cells of similar lineages and DNA content.

Measurement of the Ts becomes less reliable at low fractions of HP labeling, because of the method by which the Ts is estimated from the absolute numbers of labeled cells in each phase of the cell cycle. This may give rise to prolonged estimates of the Tpot.

It thus seems likely that the S-phase duration is a species-specific constant in cells with normal DNA content. If DNA duplication proceeds at a constant rate, then one variable in the Ts duration might be the absolute quantity of abnormal DNA in the tumor cells, reflected in the DNA index, or aneuploidy. We found no significant difference in the Ts of cohorts of diploid and aneuploid colorectal tumor samples to support this hypothesis (Rew *et al.*, 1991a), although such differences may be concealed within experimental error.

2. The Potential Doubling Time and the Cell Loss Factor

There are a number of interesting aspects to the Tpot data presented. One feature of the data across the many series of tumors reported is the similarity rather than the differences in calculated cell production rates between classes of tumors of very different histology and behavior. Another is the high rate of cell production in many tumors, with Tpots of the order of 5 days. The observed volume doubling time of tumors in clinical practice is commonly of the order of 100 days or more. This provides strong corroborative evidence that the difference between the calculated and the observed doubling times is due to a high cell loss factor. This might be expected to be due principally to exfoliation and necrosis, as is often observed in tumors in the GI and uropenital tract.

The cell loss factor has a major bearing on the clinical volume growth of a tumor. Tissue and tumor growth is a dynamic disequilibrium between cell production and cell loss. Proliferation measurements do not take into account tumor and tissue cell loss during growth. There is a disparity between the actual volume growth of tumors, as measured directly or by serial radiology, and the potential doubling time. Cell loss from tumors is caused by processes that include exfoliation, cell migration or metastasis, necrosis, and apoptosis. Thus, a highly proliferative tumor with a high cell loss factor may remain static in size or even regress, whereas a slowly proliferative tumor with no cell loss will continue to enlarge.

The indication of high cell loss factors in nonluminal tumors such as of the breast suggests that apoptosis may be playing a very major role in cell loss. This is to be expected where large numbers of abnormal cells are produced in these proliferative neoplasms. These observations indicate that cell loss is a dominant rather than a marginal factor in tumor growth. Apoptosis is most likely to account for a large proportion of cell loss.

3. The Labeling Index as a Surrogate Measure of Dynamic Proliferation

If the Ts is relatively constant from one cell population to another, then the observed experimental variation Tpot is thus largely attributable to site to site and intertumor variation in the LI. If we assume the Ts to be a species constant and take the median value as representative, then we may use the LI as an index of proliferation rate to a reasonable first order approximation. This would simplify the approach to proliferation measurements, as static, *in vitro,* and histochemical measurements of proliferative markers would be considerably simpler to obtain in service laboratory practice than would be the results of dynamic *in vivo* studies.

4. Cell Production Rates in Normal Tissues

The concept of the Tpot is also applicable to normal tissues, where at maturity, cell production is in a steady-state balance with cell loss. Where normal tissue samples such as gastrointestinal mucosa have been studied, cell proliferation rates are similar to those of the tumors. This strengthens the theory that tumor growth may be due not so much to an acceleration of cell production rates as to a reduction in cell loss rates. One proviso to generalizations about normal tissue proliferation rates is that much of the available data on clinical tissue samples has been obtained from surgical tumor resection cases. This raises the question as to whether the proliferation patterns in the normal tissues may have been modified, for example, through exposure to tumor growth factors.

5. Proliferation Data, Biological Aggressiveness, and Clinical Outcome

It is commonly assumed that a higher rate of tumor cell production increases the biological aggressiveness of tumors. There is no absolute measure of "speed" in tumor growth, and it is thus accepted practice to stratify data sets by the median Tpot value in any one tumor series. In those series where survival has been recorded, there has generally been little or no correlation between proliferation parameters and time to death.

Proliferation measures must be regarded as an inadequate measure of tumor growth and of biological aggressiveness. They do not measure cell loss, and they are unable to predict invasive and metastatic potential. There may be differences in proliferative behavior between primary and metastatic lesions, such that assumptions cannot be made about the overall behavior of the tumor mass from measurements on the primary tumor alone. Proliferative measurements are usually made on the primary tumor, and these may have little relevance to the proliferative biology of metastatic or invasive clones.

Proliferative biology may also change with time. Tumor growth fractions and proliferation rates may change with growth and time. A snapshot biopsy measurement taken at any point in the life of the tumor may not be predictive either of the past behavior or of future growth rates of that tumor.

6. Cell Proliferation and Experimental Therapy

Cell proliferation research may help improve adjuvant chemotherapy (Tannock, 1986; Van Putten, 1979) and radiotherapy (Denekamp, 1986; Kummermehr and Trott, 1982; Terry, 1996) in a number of ways. HP labeling is now a standard laboratory technique for measuring normal and perturbed cell proliferation in experimental models. For example, it allows the point of action of drugs in the cell cycle to be inferred from the cell cycle profiles of experimental cell populations. The block to the cell cycle by individual agents causes proliferating cells to "pile up" in specific phases of the cycle (Dolbeare, 1995b). The technique also allows the assessment of the proliferative response of tissues to damage, such as induced by hypoxia, in tumor cell cultures and animal models (Webster *et al.,* 1998).

E. Concluding Comments on Cell Proliferation Studies

In vivo proliferation data provide a sound, evidence-based framework for modeling the growth of tumors and tissues, for studying chemotherapy and radiotherapy in the laboratory, for understanding the importance of cell loss and apoptosis in tumor progression, and for providing new hypotheses for clinical trials of treatment of those many tumors whose behavior defies surgical excision.

Cell production rate measurements are yet to find useful clinical applications, either in confident prognostication or in guiding therapeutic strategy. Nevertheless, they have taught us much about the proliferative biology of human tumors (Rew and Wilson, 2000a,b). They have served to emphasize how cell production is only one side of the equation of tumor growth. The other side, cell loss, continues to be a problem of quantitative analysis. We now recognize the complexity of proliferative behavior within and between tumors and normal tissues, along with the challenge that it poses to the development of better adjuvant therapies. No single or simple proliferation marker yet appears likely to be able to improve on the clinical or histological detection of metastases as an index of clinical prognosis.

VI. Further Applications of Cytometry in Clinical Oncology

A. Laser Scanning Cytometry in Clinical Studies

The inability of flow cytometry to correlate fluorimetric data with cell morphology on a precise, cell by cell basis renders it unsuitable for the study of many aspects of solid tumor biology. It prevents the confident discrimination of cell types in complex populations of cells with very variable morphology, scatter characteristics, lineage, and viability.

Laser scanning cytometry (Kamentsky and Kamentsky, 1991) creates a direct link between cell morphology, laser excitation, and fluorochromatic quantitation. The fixing of the cell and tissue in space and time beneath the familiar microscope objective of a laser scanning cytometer vastly increases the scope of analysis, interpretation, and experimentation when compared with the flow cytometer (Rew *et al.*, 1999). It allows the corroboration of descriptive morphology with qualitative and quantitative measures of light scatter and fluorescence, and their spatial distribution within and around the cell. It allows study of dynamic patterns of transport of fluorochromes, of subcellular concentration of dyes, and of normal and abnormal cell physiology. The visualization capabilities of epifluorescence, bright-field microscopy, and image processing are further enhanced by direct linkage to an image processing system (Woltmann *et al.*, 1998).

Laser scanning cytometry replicates the range of assays on clinical material available in flow cytometry, including DNA ploidy (Martin-Reay *et al.*, 1994; Luther and Kamentsky, 1996; Sasaki *et al.*, 1996; Chapter 31 of this volume), immunophenotyping (Clatch *et al.*, 1998; Clatch and Forman, 1998; Chapter 46 of this volume), and cell proliferation (Rew *et al.*, 1998). It also offers unique capabilities in terms of sample presentation on a microscope slide, accommodating fine needle aspirates, smears, imprints, viable cells in irrigated chambers, cells grown to confluence *in situ* (Yang *et al.*, 1998), and complex samples such as asthmatic sputum (Woltmann *et al.*, 1999) (Fig. 6). It may also be used with limited resolution for subcellular analyses such as fluorescence *in situ* hybridization (Kamentsky *et al.*, 1997) and genotoxicity testing using the mouse micronucleus assay (Rew and Styles, 1998). Cell and tissue culture techniques on a microscope chamber slide also provide for direct quantitative assay of processes such as cell signaling, growth factor response, molecular translocation, and incorporation of viral vectors in contiguous cell samples (Musco *et al.*, 1998). The laser scanning cytometer can also analyze conventional tissue sections to a limited degree, using discriminators such as nuclear staining. Measurement is still constrained by problems with cell overlap and sectioning and by boundary discrimination, but this may improve with better software algorithms and new staining techniques.

B. Cytometric Assays in Cancer Therapy

The LSC offers considerable advantages in the cytometric analysis of complex cell populations from solid tumors, whose variable light scatter characteristics or DNA content do not allow simple classification of cell subtypes. Specific cell types or assay results can be validated by direct inspection, whereas rare event analysis and cell sorting on the slide can be undertaken in the course of normal routines.

Research using laser cytometry has opened up many new vistas on the cancer cell and its behavior. However, none of the applications in solid tumor biology have become a standard part of the clinicopathological assessment. Most cancer treatments carry a significant morbidity, and the selection of prognostic groups also helps to optimize treatments and to avoid unnecessary treatment for some categories of patients. In cancer biology, the greatest immediate value would

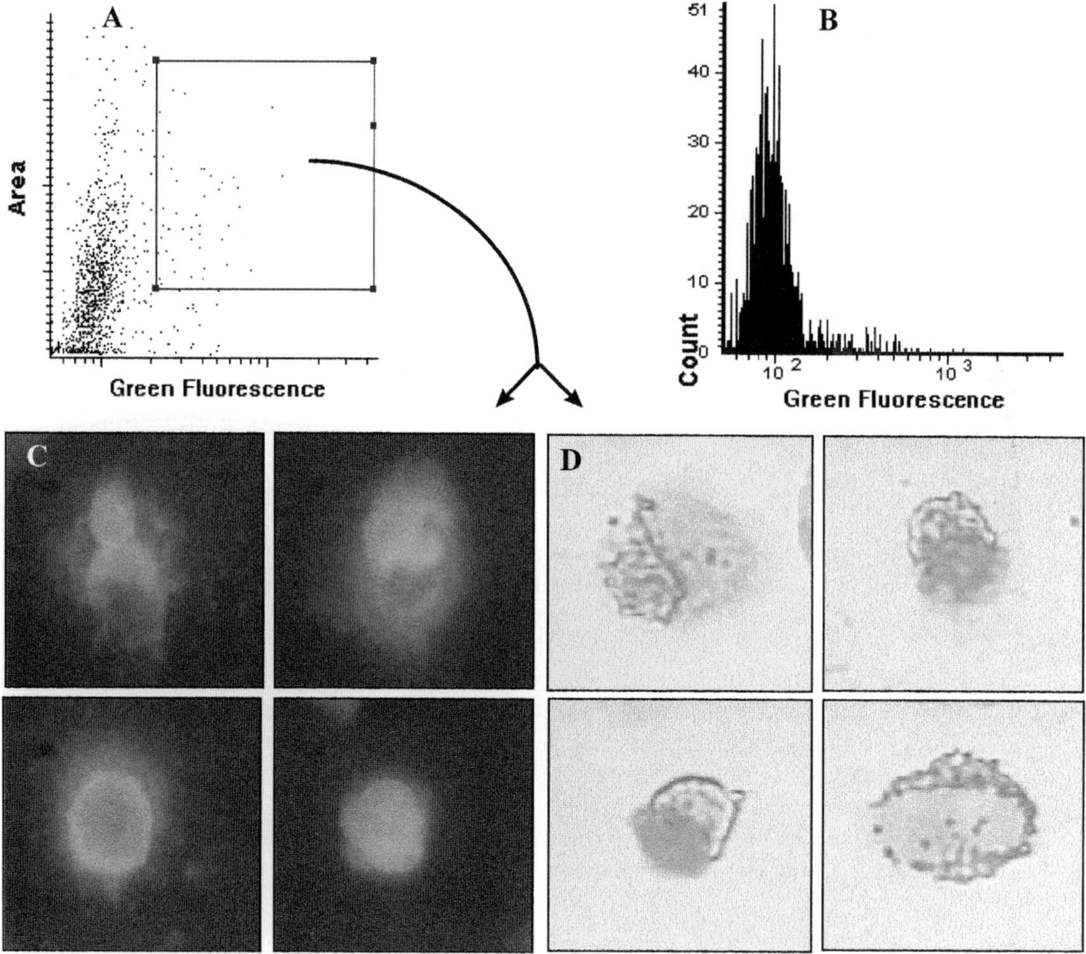

Fig. 6 The capabilities of laser scanning cytometry in clinical sample analysis. A Cytospin of human asthmatic sputum has been stained with monoclonal anti-human major basic protein specific antibody (BMK13) (Bradsure Biologicals, Loughborough, England) and secondary Oregon Green conjugated goat anti-mouse antibody (Molecular Probes, Europe, BV) and analyzed on the Compucyte LSC. Plotted are peak green fluorescence signal against cell size of cells captured with propidium iodide nuclear counterstain (A, B). Examples of relocation and still video image capture of cells within the region of interest under epifluorescence (C) and after chromatic restaining of the same slide under bright-field microscopy with chromotrope 2R (D) (courtesy of Dr. G. Woltmann, Leicester, England) (Woltmann *et al.*, 1999).

derive from techniques that allow us to make better use of existing adjuvant therapies on a case by case basis. Cytometric techniques which thus aided the selection or rejection of specific cancer therapy might identify mechanisms for the circumvention of drug resistance, the specific targeting of individual drugs to tumors, or the optimum use of radiotherapy fractionation and radiosensitizers.

C. Cytometric Assays of Fluorescent Cytotoxic Drugs

The decision to use adjuvant chemotherapy for primary, metastatic, or recurrent tumors, and the selection of particular cytotoxic agents thereafter, is much influenced by empiricism and individual practice in clinical oncology. This may be depriving many patients with cancers of real benefits from individually targeted adjuvant chemotherapy. The absence of simple clinical assays of cytotoxic drug uptake by complex populations of viable, nonviable, drug-sensitive, and drug-resistant tumor cells has been a serious handicap.

The anthracycline/anthraquinone families of cytotoxic agents include Adriamycin, daunorubicin, mitozantrone, epirubicin, and idarubicin. These agents have a number of potent cytotoxic actions, which include intercalation of DNA and interference with the topoisomerase proteins and other DNA repair enzymes. These compounds possess intrinsic fluorescence; they absorb light around 470 nm and emit around 560 nm (Krishan and Ganapathi, 1980; Krishan *et al.*, 1987; Krishan and Sauerteig, 1992). They are expelled from the cell by the p170 glycoprotein pump, which confers multidrug resistance (MDR), and this mechanism may help limit their cytotoxicity (Herweijer *et al.*, 1989; Gheuens *et al.*, 1991; Smith *et al.*, 1992; Carpentier *et al.*, 1992; Tiberghien and Loor, 1996; Homolya *et al.*, 1996; Haugland and Larison, 1996; Landon, 1997). Flow cytometric studies of anthracycline uptake into tumor cells in suspension from hematological tumors, effusions, and ascites have been reported elsewhere in this book. It is also possible to demonstrate fluorochromatic drug uptake into solid tumor disaggregates, as with tumor cells in suspension in blood or effusions, using FCM. However, efforts to assay these drugs are confounded by heterogeneity of cell size, origin, and viability.

These drugs are potent anticancer agents, but their systemic toxicity has restricted their clinical use. Laser cytometry may offer a variety of techniques to circumvent systemic toxicity. The capabilities of the LSC in cell verification by direct visualization suggested a feasible strategy for quantitation of drug uptake and comparative studies of fluorochromatic drug uptake into tumor cells, as validated in our early studies (Reeve, 2000). One way may be to study means by which their target cell toxicity may be promoted at lower systemic doses through the use of MDR blocking agents. Another may be to identify those tumors that fail to concentrate the agents, and to spare these patients from inappropriate therapy. A third way may be to identify from a panel of agents the drug most reliably concentrated in tumor cells from surgical biopsies or fine needle aspiration. Indeed, proof of specific drug uptake would be a considerable advance over the current empiricism inherent in single agent, multiple agent, and sequential adjuvant therapy. Fine needle aspirates and endoscopic biopsies would be well suited to the monitoring of accessible tumors for changes in drug resistance during a course of therapy. Better assays may encourage the earlier use of these agents as prophylaxis against minimal residual disease after surgery over a range of tumor types.

The assay of drug uptake is alone insufficient to indicate cell killing. For example, fluorescent drug may be metabolized, sequestered, or otherwise rendered inert within target cells. Thus, additional assays of drug efficacy must be developed. These might produce evidence of induced cell death (Furuya *et al.,* 1997) or disruption of other vital functions, such as evidenced by the accumulation of cell cycle phase specific markers or mitotic disruption.

More speculatively, cytometric assays may encourage drug designers to develop "detectability," by intrinsic fluorescence or monoclonal immunogenicity, in other anticancer agents. For example, fluorescence techniques have also been used to detect cis-platinum–rhodamin 123 conjugates (Assaraf *et al.,* 1989), and methotrexate (Teicher *et al.,* 1986) in cell models. A more difficult step will be to translate laboratory studies into practical randomized controlled clinical trials, so as to evaluate current unselective, empirical chemotherapy with more selective strategies.

VII. Conclusions

The complex morphology and behavior of human cells, tissues, and tumors challenges the designers of cytometric instrumentation. The range of clinical applications of flow cytometry is well defined, and the inherent constraints of the technology limit its utility in the clinical laboratory. A cytometer must above all provide useful information not otherwise readily obtainable by other means. The new generation of instruments open new horizons in clinical diagnosis and therapeutics, and we may expect further advances and insights in the next few years. Early in the twentieth century, the pathologist James Ewing (1919) hoped that "The twentieth century . . . may, thereby prove to be the era of successful therapeutics and prophylactics." His speculation was not fully vindicated for the solid tumors that preoccupy surgeons and oncologists. Nevertheless, technologies that Ewing could not know, and at which we must still marvel, offer us much excitement and optimism in the new century.

Acknowledgments

I thank Dr. George Wilson for support throughout these studies, and the late Dr. Nic McNally, of the Gray Laboratory, Northwood, England. I thank the many scientists, surgeons, oncologists, and patients who have supported our studies. Elements of our work have been supported in turn by the UK's Cancer Research Campaign, Wessex Cancer Trust, and the NHS Research and Development Executive.

References

Antognoni, P., Bignardi, M., Cazzaniga, L. F., Poli, A. M., Richetti, A., Bossi, A., Rampello, G., Barbera, F., Soatti, C., Bardelli, D., Giordano, M., and Danova, M. (1996). Accelerated radiation therapy for locally advanced squamous cell carcinomas of the oral cavity and oropharynx selected according to tumor cell kinetics—a phase II multicenter study. *Int. J. Radiat. Oncol. Biol. Phys.* **36,** 1137–1145.

Ashton-Key, M., Coddington, R., Rew, D. A., Campbell, I. D., and Taylor, I. (1993). Quantitative indices of proliferation using bromodeoxyuridine: A study of sixty invasive breast carcinomas. *Breast* **2,** 42–47.

Assaraf, Y. G., Seamer, L. C., and Schimke, R. T. (1989). Characterisation by flow cytometry of fluorescein-methotrexate transport in chinese hamster ovary cells. *Cytometry* **10,** 50–55.

Barrett, J. C., Tsutsui, T., and Ts'o P. O. P. (1978). Neoplastic transformation induced by a direct perturbation of DNA. *Nature* **274,** 229–232.

Begg, A. C. (1989). Derivation of cell kinetic parameters from human tumors after labelling with bromodeoxyuridine or iododeoxyuridine. *Br. Inst. Radiol. Rep.* **19,** 113–119.

Begg, A. C., McNally, N. J., Shrieve, D. C., and Karcher, H. (1985). A method to measure the duration of DNA synthesis and the potential doubling time from a single sample. *Cytometry* **6,** 620–626.

Begg, A. C., Moonen, L., Hofland, I., Dessing, M., and Bartelink, H. (1988). Human tumor kinetics using a monoclonal antibody against iododeoxyuridine. Intra-tumor sampling variations. *Radiother. Oncol.* **11,** 337–347.

Begg, A. C., Hofland, I., Moonen, L., Bartelink, H., Schraub, S., Bontempts, P., Le Fir, R., Van den Bogaert, W., Caspers, R., van Glabbeke, M., and Horiot J.-C. (1990). The predictive value of cell kinetic measurements in a European trial of accelerated fractionation in advanced head and neck tumors: An interim report. *Int. J. Radiat. Oncol. Biol. Phys.* **19,** 1449–1453.

Begg, A. C., Hofland, I., Van Glabbeke, M., Bartelink, H., and Horiot, J.-C. (1992). Predictive value of potential doubling time for radiotherapy of head and neck tumor patients: Results from the EORTC cooperative trial 22851. *Semin. Radiat. Oncol.* **2,** 22–25.

Begg, A., Haustermans, K., and Wilson, G. D. (1998). The value of pretreatment cell kinetic parameters as predictors for radiotherapy outcome in head and neck cancer: A multicentre analysis: *Radiother. Oncol.* in press.

Benazzo, M., Mevio, E., Occhini, A., Franchini, G., and Danova, M. (1995). Proliferative characteristics of head and neck tumors. In vivo evaluation by BrdUrd incorporation and flow cytometry Orl. *J. Oto-Rhino-Laryngol.* **57,** 39–43.

Bennett, M. H., Wilson, G. D., Dische, S., Saunders, M. I., Martindale, C. A., Robinson, B. M., O'Halloran, A. E., Leslie, M. D., and Laing, J. H. E. (1992). Tumor proliferation assessed by combined histological and flow cytometric analysis: Implications for therapy in squamous cell carcinoma in the head and neck. *Br. J. Cancer* **65,** 870–878.

Bergstrom, C., and Stenling, R. (1990). Proliferative heterogeneity in colorectal cancer, studied using Iododeoxyuridine. Abstract to European Study Group for cell proliferation. ESPOO XVII, Finland, September.

Bolger, B. S., Cooke, T. G., Symonds, R. P., MacLean, A. B., and Stanton, P. D. (1993). Measurement of cell kinetics in cervical tumors using bromodeoxyuridine. *Br. J. Cancer* **68,** 166–171.

Bolger, B. S., Cooke, T. G., Symonds, R. P., MacLean, A. B., and Stanton, P. D. (1996). Prediction of radio-therapy response of cervical tumors through measurement of proliferation rate. *Br. J. Cancer* **74,** 1223–1226.

Bourhis, J., Wilson, G., Wibault, P., Bosq, J., Chavaudra, N., Janot, F., Luboinski, B., Eschwege, F., and Malaise, E. P. (1993). In vivo measurement of the potential doubling time by flow cytometry in oropharyngeal cancer treated by conventional radiotherapy. *Int. J. Radiat. Oncol. Biol. Phys.* **26,** 793–799.

Bourhis, J., Dendale, R., Hill, C., Bosq, J., Janot, F., Attal, P., Fortin, A., Marandas, P., Schwaab, G., Wibault, P., Malaise, E. P., Bobin, S., Luboinski, B., Eschwege, F., and Wilson, G. (1996). Potential doubling time and clinical outcome in head and neck squamous cell carcinoma treated with 70 GY in 7 weeks. *Int. J. Radiat. Oncol. Biol. Phys.* **35,** 471–476.

Camplejohn, R. S. (1982). Cell Kinetics of colorectal tumors. *Recent Results Cancer Res.* **83,** 21–30.

Carpentier, Y., Gorisse, M. C., and Desoize, B. (1992). Evaluation of a method for the detection of cells with reduced drug retention in solid tumors. *Cytometry* **13,** 630–637.

Christov, K., Chew, K. L., Ljung, B. M., Waldman, F. M., Goodson III, W. H., Smith, H. S., and Mayall, B. H. (1994). Cell proliferation in hyperplastic and in situ carcinoma lesions of the breast estimated by *in vivo* labeling with bromodeoxyuridine. *J Cell. Biochem.* 19 (Suppl.), 165–172.

Clatch, R. J., and Foreman, J. R. (1998). Five color immunophenotyping plus DNA content by laser scanning cytometry. *Cytometry* **34,** 36–38.

Clatch, R. J., Foreman, J. R., and Walloch, J. L. (1998). Simplified immunophenotypic analysis by laser scanning cytometry. *Cytometry* **34,** 3–16.

Cooke, L. D., Cooke, T. G., Forster, G., Jones, A. S., and Stell, P. M. (1994). Prospective evaluation of cell kinetics in head and neck squamous carcinoma: The relationship to tumor factors and survival. *Br. J. Cancer* **69,** 717–720.

Corvo, R., Giaretti, W., Sanguineti, G., Geido, E., Orecchia, R., Barra, S., Margarino, G., Bacigalupo, A., and Vitale, V. (1993). Potential doubling time in head and neck tumors treated by primary radiotherapy: Preliminary evidence for a prognostic significance in local control. *Int. J. Radiat. Oncol. Biol. Phys.* **27,** 1165–1172.

Corvo, R., Giaretti, W., Sanguineti, G., Geido, E., Orecchia, R., Guenzi, M., Margarino, G., Bacigalupo, A., Garaventa, G., Barbieri, M., and Vitale, V. (1995). In vivo cell kinetics in head and neck squamous cell carcinomas predicts local control and helps guide radiotherapy regimen. *J. Clin. Oncol.* **13,** 1843–1850.

Corvo, R., Giaretti, W., Geido, E., Sanguineti, G., Orecchia, R., Scala, M., Garaventa, G., Mora, E., and Vitale, V. (1996). Cell kinetics and tumor regression during radiotherapy in head and neck squamous cell carcinomas. *Int. J. Cancer* **68,** 151–155.

Danova, M., Gaetani, P., Lombardi, D., Giordano, M., Riccardi, A., and Mazzini, G. (1991). Prognostic value of DNA ploidy and proliferative activity in human malignant gliomas. *Med. Sci. Res.* **19,** 613–615.

Darzynkiewicz, Z., Gong, J., Juan, G., Ardelt, B., and Traganos, F. (1996). Cytometry of cyclin proteins. *Cytometry* **25,** 1–13.

Denekamp, J. (1986). Cell kinetics and radiation biology. *Int. J. Radiat. Biol.* **49,** 2, 357–380.

Dische, S., and Saunders, M. I. (1989). Continuous, hyperfractionated, accelerated radiotherapy (CHART). *Br. J. Cancer* **59,** 325–326.

Dolbeare, F. (1995a). Bromodeoxyuridine: A diagnostic tool in biology and medicine, Part I: Historical perspectives, histochemical methods and cell kinetics. [Review] *Histochem. J.* **27,** 339–369.

Dolbeare, F. (1995b). Bromodeoxyuridine: A diagnostic tool in biology and medicine, Part II: Oncology, chemotherapy and carcinogenesis. [Review] *Histochem. J.* **27,** 923–964.

Dolbeare, F. (1996). Bromodeoxyuridine: A diagnostic tool in biology and medicine, Part III. Proliferation in normal, injured and diseased tissue, growth factors, differentiation, DNA replication sites and in situ hybridization. [Review] *Histochem. J.* **28,** 531–575.

Dolbeare, F., Gratzner, H., Pallavicini, M., and Gray, J. W. (1983). Flow cytometric measurement of total DNA content and incorporated bromodeoxyuridine. *Proc. Natl. Acad. Sci. U.S.A.* **80,** 5573–5577.

Dolbeare, F., Beisker, W., Pallavicini, M. G., Vanderlaan, M., and Gray, J. W. (1985). Cyto-chemistry for BrdUrd/DNA analysis. Stoichiometry and sensitivity. *Cytometry* **6,** 521–530.

Erba, E., Giordano, M., Danova, M., Mazzini, G., Ubezio, P., Torri, V., Mangioni, C., Landoni, F., Bolis, P., Tenti, P., *et al.* (1994). Cell kinetics of human ovarian cancer with *in vivo* administration of bromodeoxyuridine. *Ann. Oncol.* **5,** 627–634.

Ewing, J. (1919). "Neoplastic Diseases." Saunders, Philadelphia.

Fine, J. D., and Breathnach, S. M. (1986). Distinctive eruption characterised by linear supravenous papules and erythroderma following bromoxuridine therapy and radiotherapy. *Arch. Dermatol.* **122,** 199–200.

Forster, G., Cooke, T. G., Cooke, L. D., Stanton, P. D., Bowie, G., and Stell, P. M. (1992). Tumor growth rates in squamous carcinoma of the head and neck measured by in vivo bromodeoxyuridine. *Br. J. Cancer* **65,** 698–702.

Frankfurt, O. S., Slokum, H. K., and Rustum, Y. M. (1984). Flow cytometric analysis of DNA aneuploidy in primary and metastatic human solid tumors. *Cytometry* **5,** 71–80.

Frankfurt, O. S., Arbuck, S. G., Chin, J. L., and Greco, W. R. (1986). Prognostic applications of DNA flow cytometry for human solid tumors. *Ann. N.Y. Acad. Sci.* **468,** 276–290.

Furuya, T., Kamada, T., Murakami, T., Kurose, A., and Sasaki, K. (1997). Laser scanning cytometry allows detection of cell death with morphological features of apoptosis in cells stained with PI. *Cytometry* **29**, 173–177.

Gheuens, E. E., van Bockstaele, D. R., van der Keur, M., Tanke, H. J., van Oosterom, A. T., and De Bruijn, E. A. (1991). Flow cytometric double labelling technique for screening of multidrug resistance. *Cytometry* **12**, 636–644.

Gilliland, R., Williamson, K. E., Hamilton, P., Crockard, A., and Spence, R. A. (1997). DNA denaturation sensitivity may invalidate BrdUrd-DNA flow cytometric analysis of the potential doubling time in colorectal tumours. *Br. J. Surg.* **84**, 242–247.

Giordano, M., Danova, M., Mazzini, G., Gobbi, P., and Riccardi, A. (1993). Cell kinetics with *in vivo* bromodeoxyuridine assay, proliferating cell nuclear antigen expression, and flow cytometric analysis. Prognostic significance in acute non-lymphoblastic leukemia. *Cancer* **71**, 2739–2745.

Gonchoroff, N. J., Greipp, P. R., Kyle, R. A., and Katzmann, J. A. (1985). A monoclonal antibody reactive with 5-bromo-2-deoxyuridine that does not require DNA denaturation. *Cytometry* **6**, 506–512.

Goodson, W. H., Ljung, B. M., Moore, D. H., Mayall, B., Waldman, F. M., Chew, K., Benz, C. C., and Smith, H. S. (1993). Tumor labeling indices of primary breast cancers and their regional lymph node metastases. *Cancer* **71**, 3914–3919.

Gratzner, H. G. (1982). Monoclonal antibody to 5-bromo and 5-iodo deoxyuridine: A new reagent for detection of DNA replication. *Science* **218**, 474–475.

Gray, J. W. (1985). Monoclonal antibodies against bromodeoxyuridine. *Cytometry* **6**, 499–662.

Hall, P. A., Levison, D. A., Woods, A. L., Yu, C. C.-W., Kellock, D. B., Watkins, J. A., Barnes, D. M., Gillett, C. E., Camplejohn, R. S., Dover, R., Waseem, N. H., and Lane, D. P. (1990). Proliferating cell nuclear antigen localisation in paraffin sections: An index of cell proliferation with evidence of deregulated expression in some neoplasms. *J. Pathol.* **162**, 285–294.

Haugland, R. P., and Larison, K. D. (eds.) (1996). Probes for live cell function. *In* "Handbook of Fluorescent Probes and Research Chemicals" pp. 377–378. Molecular Probes, Eugene, Oregon.

Haustermans, K. (1995). In vivo growth kinetic measurements in human oesophageal cancer. Ph.D. Thesis, University of Leiden, Holland.

Haustermans, K., Vanuytsel, L., Geboes, K., Lerut, T., Van Thillo, J., Leysen, J., Coosemans, W., and van der Schueren E. (1994). *In vivo* cell kinetic measurements in human oesophageal cancer: What can be learned from multiple biopsies? *Eur. J. Cancer* **30A**, 1787–1791.

Haustermans, K., Hofland, I., Pottie, G., Ramaekers, M., and Begg, A. C. (1995). Can measurements of potential doubling time (Tpot) be compared between laboratories? A quality control study. *Cytometry* **19**, 154–163.

Haustermans, K., Fowler, E., Geboes, K., Lerut, T., and van der Schueren, E. (1997). Do cell kinetics have prognostic and/or predictive value in oesophageal cancer treated by surgery? *Eur. J. Surg. Oncol.* **23**, 293–297.

Hedley, D. W. (1989). Flow cytometry using paraffin-embedded tissue. Five years on. *Cytometry* **10**, 229–241.

Herweijer, H., Van den Engh, G., and Nooter, K. (1989). A rapid and sensitive flow cytometric method for the detection of multi-drug resistant cells. *Cytometry* **10**, 463–468.

Homolya, L., Hoolo, Z., Muller, M., Mechetner, E. B., and Sarkadi, B. (1996). A new method for quantitative assessment of p-glycoprotein related multidrug resistance in tumor cells. *Br. J. Cancer* **73**, 849–855.

Horiot, J. C., Bontemps, P., van den Bogaert, W., Le Fur, R., van den Weijngaert, D., Bolla, M., Bernier, J., Lusinchi, A., Stuschke, M., Lopez-Torrecilla, J., Begg, A. C., Pierart, M., and Collette, L. (1997). Accelerated fractionation (AF) compared to conventional fractionation (CF) improves loco-regional control in the radiotherapy of advanced head and neck cancers: Results of the EORTC 22851 randomized trial. *Radiother. Oncol.* **44**, 111–121.

Hoshino, T., Nagashima, T., Murovic, J., Levin, E., Levin, V., and Rupp, S. (1985). Cell kinetic studies of in situ human brain tumors using bromodeoxyuridine. *Cytometry* **6**, 627–633.

Hoshino, T., Prados, M., Wilson, C. B., Cho, K. G., Lee, K. S., and Davis, R. L. (1989). Prognostic implications of the bromodeoxyuridine labeling index of human gliomas. *J. Neurosurg.* **71,** 335–41.

Howard, A., and Pelc, S. R. (1951). Nuclear incorporation of ^{32}P as demonstrated by autoradiographs. *Exp. Cell Res.* **2,** 178–187.

Jones, A. S., Roland, N. J., Caslin, A. W., Cooke, T. G., Cooke, L. D., and Forster, G. A. (1994). comparison of cellular proliferation markers in squamous cell carcinoma of the head and neck. *J. Laryngol. Otol.* **108,** 859–864.

Kamentsky, L. A., and Kamentsky, L. D. (1991). Microscope based multiparameter laser scanning cytometer which yields data comparable to flow cytometry data. *Cytometry* **12,** 381–387.

Kamentsky, L. A., Kamentsky, L. D., Fletcher, J. A., Kurose, A., and Sasaki, K. (1997). Methods for automatic multiparameter analysis of fluorescence in situ hybridised specimens with a laser scanning cytometer. *Cytometry* **27,** 117–125.

Kaufman, E. R. (1986). Altered CTP synthetase activity confers resistance to 5-bromodeoxyuridine toxicity and mutagenesis. *Mutat. Res.* **161,** 19–27.

Khan, S., Raza, A., Petrelli, N., and Mittleman, A. (1988). *In vivo* determinations of labelling index of metastatic colorectal carcinoma and normal colonic mucosa using intravenous infusions of bromodeoxyuridine. *J. Surg. Oncol.* **39,** 114–118.

Kinsella, T. J., Mitchell, J. B., Russo, A., Aiken, M., Morstyn, G., Hsu, S. M., Rowland, J., and Glatstein, E. (1984). Continuous intravenous infusions of bromodeoxyuridine as a clinical radiosensitizer. *J. Clin. Oncol.* **2,** 1144–1150.

Kotelnikov, V. M., Coon, J. S., Taylor, S., Hutchinson, J., Panje, W., Caldarelli, D. D., LaFollette, S., and Preisler, H. D. (1995a). *In vivo* labeling with halogenated pyrimidines of squamous cell carcinomas and adjacent non-involved mucosa of head and neck region. *Cell Prolif.* **28,** 497–509.

Kotelnikov, V. M., Coon, J. S., Haleem, A., Taylor, S., Hutchinson, J., Panje, W., Caldarelli, D. D., Griem, K., and Preisler, H. D. (1995b). Cell kinetics of head and neck cancers. *Clin. Cancer Res.* **1,** 527–537.

Krishan, A., and Ganapathi, R. (1980). Laser flow studies on the intracellular fluorescence of anthracyclines. *Cancer Res.* **40,** 3895–3900.

Krishan, A., and Sauerteig, A. (1992). Flow cytometric monitoring of cellular resistance to cancer chemotherapy. *In* "Flow Cytometry: Principles and Clinical Application" (K. D. Bauer, R. E. Duque, and T. V. Shankey, eds.), pp. 459–467. Williams & Wilkins, New York.

Krishan, A., Shridar, K. S., Davila, E., Vogel, C., and Sternheim, W. (1987). Patterns of anthracycline retention in human tumor cells. *Cytometry* **8,** 306–314.

Kummermehr, J., and Trott, K. R. (1982). Rate of repopulation in a slow and fast growing mouse tumor. *In* "Progress in Radio-oncology II" (K. H. Karcher *et al*, eds.) pp. 299–308. Raven, New York.

Laing, J. H. E., Rew, D. A., and Wilson, G. D. (1992). Cell kinetics of human solid tumors. *Br. J. Radiol.* 24(Suppl.) 163–167.

Landon, T. M. (ed). (1997). Multidrug resistance assays. *In* "Bioprobes," Issue 25, p. 21. Molecular Probes, Eugene, Oregon.

Langford, L. A., Cooksley, C. S., and DeMonte, F. (1996). Comparison of MIB-1 (Ki-67) antigen and bromodeoxyuridine proliferation indices in meningiomas. *Hum. Pathol.* **27,** 350–4.

Larsson, P., Roos, G., Stenling, R., and Ljungberg, B. (1994). Proliferation of human renal cell carcinoma studied with *in vivo* iododeoxyuridine labeling and immuno-histochemistry. *Scand. J. Urol. Nephrol.* **28,** 135–140.

Lloveras, B., Garin-Chesa, P., Myc, A., and Melamed, M. (1994). *In vitro* bromodeoxyuridine labelling of malignant neoplasms. A comparative study with flow cytometry cell-cycle analysis. *Am. J. Clin. Pathol.* **101,** 703–707.

Luther, E., and Kamentsky, L. A. (1996). Resolution of mitotic cells using laser scanning cytometry. *Cytometry* **23,** 272–278.

Maas, R.A., Bruning, P. F., Breedijk, A. J. and Peterse, J. L. (1996). Iododeoxyuridine labeling of S-phase fraction in fine needle aspirates from breast carcinomas. *J. Clin. Patho.* **49,** 607–609.

Martin-Reay, D. G., Kamentsky, L. A., Weinberg, D. S., Hollister, K. A., and Cibas, E. S. (1994). Evaluation of a new slide based laser scanning cytometer for DNA Analysis of tumors. *Anat. Pathol.* **102,** 432–438.

Meyer, J. S., Koehm, S. L., Hughes, J. M., Higa, E., Wittliff, J. L., Lagos, J. A., and Manes, J. L. (1993). Bromodeoxyuridine labeling for S-phase measurement in breast carcinoma. *Cancer* **71,** 3531–3540.

Michel, P., Hermet, J., Paresy, M., Menard, J. F., Laquerriere, A., Scotte, M., Paillot, B., Peillon, C., and Ducrotte, P. (1997). Comparison between endoscopic and surgical sampling for the measurement of potential doubling time in colorectal cancer. *Cytometry* **29,** 273–278.

Miwa, H., Wada, R., Abe, H., Ohkura, R., Yang, S. W., Watanabe, H., Ogihara, T., Hamada, T., Meyer, J. S., Nauert, J., Koehm, S., and Hughes, J. (1989). Cell kinetics of human tumors by *in vitro* bromodeoxyuridine labeling. *J. Histochem. Cytochem.* **37,** 1449–1454.

Miwa, H., Wada, R., Abe, H., Ohkura, R., Yang, S. W., Watanabe, H., Ogihara, T., Hamada, T., and Sato, N. (1993). Diagnosis of gastric adenoma versus early gastric cancer by bromodeoxyuridine immunohistochemistry from gastric biopsy specimen. *J. Gastroenterol. Hepatol.* **8,** 133–137.

Musco, M. L., Cui, S., Small, D., Nodelman, M., Sugarman, B., and Grace, M. (1998). Comparison of flow and laser scanning cytometry for the intracellular evaluation of adenoviral infectivity and p53 protein expression in gene therapy. *Cytometry* **33,** 290–296.

Nylander, K., Anneroth, G., Gustafsson, H., Roos, G., Stenling, R., and Zackrisson, B. (1994). Cell kinetics of head and neck squamous cell carcinomas. Prognostic implications. *Acta Oncol.* **33,** 23–28.

Ohyama, S., Yonemura, Y., and Miyazaki, I. (1990). Prognostic value of S phase fraction and DNA ploidy studied with in vivo administration of bromodeoxyuridine on human gastric cancers. *Cancer* **65,** 116–121.

Patel, S., Rew, D. A., Taylor, I., Potten, C. S., and Roberts, S. (1993). Studies of proliferation in human gastric mucosa following in vivo bromodeoxyuridine labeling. *Gut* **34,** 893–896.

Popert, R. J., Joyce, A. D., Thomas, D. J., Walmsley, B. H., and Coptcoat, M. J. (1993). Bromodeoxyuridine labeling of transitional cell carcinoma of the bladder—an index of recurrence?. *Br. J. Urol.* **71,** 279–283.

Potten, C. S., Kellett, M., Roberts, S., Rew, D. A., and Wilson, G. D. (1992a). Measurement of *in vivo* proliferation in human colorectal mucosa using bromodeoxyuridine. *Gut* **33,** 71–78.

Potten, C. S., Kellett, M., Roberts, S., and Rew, D. A. (1992b). Proliferation in human gastrointestinal epithelium using bromodeoxyuridine *in vivo:* Data for different sites, proximity to a tumor, and polyposis coli. *Gut* **33,** 524–529.

Quinn, C. M., and Wright, N. A. (1990). Review article: The clinical assessment of proliferation and growth in human tumors: Evaluation of methods and applications as prognostic variables. *J. Pathol.* **160,** 93–102.

Raffel, C., Deen, D. F., and Edwards, M. S. (1988). Bromodeoxyuridine: A comparison of its photosensitizing and radiosensitizing properties. *J. Neurosurg.* **69,** 410–415.

Raza, A., Ucar, K., and Preisler, H. D. (1985). Double labelling and in vitro versus in vivo incorporation of BrdUrd in patients with acute nonlymphocytic leukemia. *Cytometry* **6,** 633–641.

Raza, A., Yousuf, N., Bokhari, S. A., *et al.* (1992). In situ cell cycle kinetics in bone marrow biospies following sequential infusions of IUdR/BrdUrd. *Leuk. Res.* **16,** 299–306.

Raza, A., Alvi, S., Broady-Robinson, L., Showel, M., Cartlidge, J., Mundle, S. D., Gregory, S. A., *et al.* (1997). Cell cycle kinetic studies in 68 patients with myelodysplastic syndromes following intravenous iodo- and/or bromodeoxyuridine. *Exp. Hematol.* **25,** 530–535.

Reeve, L. (2000). The development of tumor specific assays for cellular response to anthracycline drugs using laser cytometry. Ph.D. Thesis, University of Leicester, United Kingdom.

Rew, D. A. (1991). The *in vivo* proliferation kinetics of human solid tumors. M. Chir. Thesis, University of Cambridge, England.

Rew, D. A. (1993). Cell proliferation, tumor growth and clinical outcome: Gains and losses in intestinal cancer. *Ann. R. College Surgeons Engl.* **75,** 397–404.

Rew, D. A. (1994). The significance of aneuploidy. *Br. J. Surg.* **81,** 1416–1422.

Rew, D. A. (1996). Heterogeneity, biodiversity and bioperversity in human solid tumors. *Eur. J. Surg. Oncol.* **22,** 469–473.

Rew, D. A., and Styles, J. A. (1998). Automation of the mouse micronucleus genotoxicity assay by laser scanning cytometry. *Cytometry* ISAC IX Meeting Suppl. 9.

Rew, D. A., and Wilson, G. D. (1991). Advances in cell kinetics: A leading article. *Br. Med. J.* **303,** 532–533.

Rew, D. A., and Wilson, G. D. (2000a). Cell proliferation rates in human tissues and tumors Part I: Methods, techniques and limitations. *Eur. J. Surg. Oncol.* **26**(2), 227–238.

Rew, D. A., and Wilson, G. D. (2000b). Cell proliferation rates in human tissues and tumors Part II: Clinical data. *Eur. J. Surg. Oncol.* **26**(2), 405–417.

Rew, D. A., Wilson, G. D., Taylor, I., and Weaver, P. C. (1991a). Proliferation characteristics of 100 colorectal tumors measured *in vivo. Br. J. Surg.* **78,** 60–66.

Rew, D. A., Taylor, I., Watson, J. V., and Wilson, G. D. (1991b). The c-myc protein product is a marker of DNA synthesis but not of malignancy in intestinal tissues and tumors. *Br. J. Surg.* **78,** 1080–1083.

Rew, D. A., Thomas, D. J., Coptcoat, M. J., and Wilson, G. D. (1991c). *In vivo* measurement of urothelial tumor kinetics. *Br. J. Urol.* **68,** 44–48.

Rew, D. A., Campbell, I., Taylor, I., and Wilson, G. D. (1992). The *in vivo* proliferation kinetics of invasive carcinoma of the breast. *Br. J. Surg.* **79,** 335–339.

Rew, D. A., Karkera, R., Stradling, R., Mullee, M., Julious, S., and Wilson, G. D. (1996). The flow cytometric analysis of total p53 protein content and proliferation indices in colorectal cancer, in relation to clinical outcome. *Eur. J. Surg. Oncol.* **22,** 508–515.

Rew, D. A., Reeve, L., and Wilson, G. D. (1998). A comparison of flow and laser scanning cytometry for the measurement of cell proliferation in human solid tumors. *Cytometry* **33,** 355–361.

Rew, D. A., Woltmann, G., and Wardlaw, A. J. (1999) Laser scanning cytometry. *Lancet* **353,** 255–256.

Riccardi, A., Danova, M., Wilson, G. D., Ucci, G., Dormer, P., Mazzini, G., Brugnatelli, S., Girino, M., McNally, N., and Ascari, E. (1988). Cell kinetics in human malignancies studied with *in vivo* administration of bromodeoxyuridine and flow cytometry. *Cancer Res.* **48,** 6238–6245.

Risio, M., Coverlizza, S., Ferrari, A., Candelaresi, G. L., and Rossini, F. P. (1988). Immunohistochemical study of epithelial cell proliferation in hyperplastic polyps, adenomas, and adenocarcinomas of the large bowel. *Gastroenterology* **94,** 899–906.

Sasaki, K., Ogino, T., and Takahashi, M. (1986). Immunological determination of labeling index on human tumor tissue sections using monoclonal anti-BrdUrd antibody. *Stain Technol.* **61,** 155–161.

Sasaki, K., Murakami, T., and Takahashi, M. (1987). A rapid and simple estimation of cell cycle parameters by continuous labelling with BrdUrd. *Cytometry* **8,** 526–528.

Sasaki, K., Kurose, A., Miura, Y., Sato, T., and Ikeda, E. (1996). DNA ploidy analysis by laser scanning cytometry in colorectal cancers, and comparison with flow cytometry. *Cytometry* **23,** 106–109.

Saunders, M. I., Dische, S., Grosch, E., Fermont, D. C., Ashford, R., Maher, J., and Makepeace, A. R. (1991). Experience with CHART. *Int. J. Radiat. Oncol. Biol. Phys.* **21,** 871–878.

Sawtell, R. M., Rew, D. A., Stradling, R., and Wilson, G. D. (1995). The pan cycle expression of proliferating cell nuclear antigen (PCNA) in human colorectal cancer, and its proliferative correlations. *Commun. Clin. Cytometry* **22,** 190–199.

Shackney, S. E., and Shankey, T. V. (1995). Genetic and phenotypic heterogeneity of human malignancies: Finding order in chaos. [Review]. *Cytometry* **21,** 2–5.

Shackney, S. E., Smith, C. A., Pollice, A. A., *et al.* (1995). Preferred genetic evolutionary sequences in human breast cancer: A case study. *Cytometry* **21,** 6–13.

Shibuya, M., Ito, S., Davis, R. L., Wilson, C., and Hoshino, T. (1993). A new method for analyzing the cell kinetics of human brain tumors by double labeling with bromodeoxy-uridine in situ and with iododeoxyuridine *in vitro. Cancer* **71,** 3109–3113.

Smith, P. J., Sykes, H. R., Fox, M. E., and Furlong, I. J. (1992). Subcellular distribution of mitoxantrone in human and drug resistant murine cells analysed by flow cytometry and confocal microscopy. *Cancer Res.* **52,** 4000–4008.

Speth, P. A., Kinsella, T., Chang, A., Klecker, R., Belanger, K., Smith, R., Rowland, J., Cupp, J., and Collins, J. M. (1989). Iododeoxyuridine incorporation into human haematopoeitic cells, normal liver and hepatic metastases in man as a radiosensitiser and a marker for cell kinetic studies. *Int. J. Radiat. Oncol. Biol. Phys.* **16,** 1247–1250.

Stanton, P. D. (1996). Ph.D. Thesis, University of Glasgow, Scotland.

Stanton, P. D., Cooke, T. G., Forster, G., Smith, D., and Going, J. J. (1996). Cell kinetics *in vivo* of human breast cancer. *Br. J. Surg.* **83,** 98–102.

Steel, G. G. (1977). "Growth Kinetics of Tumors." Oxford Univ. Press (Clarendon), Oxford.

Struikmans, H., Rutgers, D. H., Jansen, G. H., Tulleken, C. A. F., Van der Tweel, I., and Batterman, J. J. (1997). S phase fraction, bromodeoxyuridine labelling index, duration of S phase, potential doubling time and DNA index in benign and malignant brain tumors. *Radiat. Oncol. Invest.* **5,** 170–179.

Tannock, I. (1986). Experimental chemotherapy and concepts related to the cell cycle. *Int. J. Radiat. Biol.* **49,** 335–355.

Teicher, B. A., Holden, S. A., Jacobs, J. L., Abrams, M. J., and Jones, A. G. (1986). Intracellular distribution of a platinum rhodamine 123 complex in cis-platinum sensitive and resistant human squamous carcinoma cell lines. *Biochem. Pharmacol.* **35,** 3365–3369.

Terry, N. H. A. (1996). Predictive assays for radiotherapy: The role of tumor proliferation (Tpot) measurements. *Onkologie* **19,** 322–327.

Terry, N. H., White, R. A., Meistrich, M. I., and Calkins, D. P. (1991). Evaluation of flow cytometric methods for determining population potential doubling times using cultured cells. *Cytometry* **12,** 234–241.

Terry, N. H., Meistrich, M. L., Roubein, L. D., Lynch, P. M., Dubrow, R. A., and Rich, T. A. (1995). Cellular kinetics in rectal cancer. *Br. J. Cancer* **72,** 435–441.

Tiberghien, F., and Loor, F. (1996). Ranking of P-glycoprotein substrates and inhibitors by a calcein AM fluorimetry screening assay. *Anticancer Drugs* **7,** 568–578.

Tinnemans, M. M., Schutte, B., Lenders, M. H., Ten Velde, G. P., Ramaekers, F. C., and Blijham, G. H. (1993). Cytokinetic analysis of lung cancer by *in vivo* bromodeoxyuridine labelling. *Br J. Cancer* **67,** 1217–1222.

Tinnemans, M. M., Lenders, M. H., ten Velde, G. P., Wagenaar, S. S., Blijham, G. H., Ramaekers, F. C., and Schutte, B. (1995). Evaluation of proliferation parameters in *in vivo* BrdUrd labeled lung cancers. *Virchows Arch.* **427,** 295–301.

Tsang, R. W., Fyles, A. W., Kirkbride, P., Levin, W., Manchul, L. A., Milosevic, M. F., Rawlings, G. A., Banerjee, D., Pintilie, M., and Wilson, G. D. (1995). Proliferation measurements with flow cytometry Tpot in cancer of the uterine cervix, correlation between two laboratories and preliminary clinical results. *Int. J. Radiat. Oncol. Biol. Phys.* **32,** 1319–1329.

Tubiana, M., and Courdi, A. (1989). Cell proliferation kinetics in human solid tumors; relation to probability of metastatic dissemination and long term survival. *Radiother. Oncol.* **15,** 1–18.

Vanderlaan, M., and Thomas, C. B. (1985). Characterization of monoclonal antibodies to bromodeoxyuridine. *Cytometry* **6,** 501–505.

Van Erp, P. E., Boezeman, J. B., and Brons, P. P. (1996). Cell cycle kinetics in normal human skin by *in vivo* administration of iododeoxyuridine and application of a differentiation marker— implications for cell cycle kinetics in psoriatic skin. *Anal. Cell. Pathol.* **11,** 43–54.

Van Putten, L. M. (1979). "Cell Kinetics, a Guide for Chemotherapy? Controversies in Cancer" pp. 117–119. Masson Publ., NY.

Waldman, F. M., Dolbeare, F., and Gray, J. (1988). Clinical applications of the Bromodeoxyuridine/ DNA assay. *Cytometry* 5(Suppl. 3), 65–72.

Watson, J. V. (1992). "An Introduction to Flow Cytometry." Cambridge Univ. Press, Cambridge.

Webster, L., Hodgkiss, R. J., and Wilson, G. D. (1998). Cell cycle distribution of hypoxia and progression of hypoxic tumor cells in vivo. *Br. J. Cancer* **77,** 227–234.

Wheeless, L. L., Coon, J. S., Cox, C., *et al.* (1991). Precision of DNA flow cytometry in inter-institutional analyses. *Cytometry* **12,** 405–412.

Williamson, K. E., Gilliland, R., Weir, H., Grimes, J., Hamilton, P., Anderson, N., Crockard, A., and Rowlands, B. (1994). Hydrochloric acid denaturation of colorectal tumour tissue infiltrated with bromodeoxyuridine. *Cytometry* **15,** 162–168.

Wilson, G. D. (1991). Assessment of human tumor proliferation using bromodeoxyuridine—current status. *Acta Oncol.* **30,** 903–910.

Wilson, G. D. (1993). Limitations of the BUdR technique for measurement of tumor proliferation. *In* "Current Topics in Clinical Radiobiology of Tumors: Medical Radiology" (H. P. Beck-Bornholdt, ed.). Springer-Verlag, Berlin.

Wilson, G. D., McNally, N. J., Dunphy, E., Karcher, H., and Pfragner, R. (1985). The labelling index of human and mouse tumors assessed by BrdUrd staining *in vitro* and *in vivo* and flow cytometry. *Cytometry* **6,** 641–647.

Wilson, G. D., McNally, N. J., Dische, S., Saunders, M. L., Des Rochers, C., Lewis, A. A., and Bennett, M. H. (1988). Measurement of cell kinetics in human tumors in vivo using bromodeoxy-uridine incorporation and flow cytometry. *Br. J. Cancer* **58,** 423–431.

Wilson, G. D., Dische, S., and Saunders, M. I. (1995). Studies with bromodeoxyuridine in head and neck cancer and accelerated radiotherapy. *Radiother. Oncol.* **36,** 189–197.

Wilson, G. D., Grover, R., Richman, P. I., Daley, F. M., Saunders, M. I., and Dische, S. (1996). BCL2 expression correlates with favourable outcome in head and neck cancer treated by accelerated radiotherapy. *Anticancer Res.* **16,** 2403–2408.

Wilson, M. S., West, C. M., Wilson, G. D., Roberts, S. A., James, R. D., and Schofield, P. F. (1993a). An assessment of the reliability and reproducibility of measurement of potential doubling times in human colorectal cancers. *Br. J. Cancer* **67,** 754–759.

Wilson, M. S., West, C. M., Wilson, G. D., Roberts, S. A., James R. D., and Schofield, P. F. (1993b). Intratumoral heterogeneity of tumor potential doubling times in colorectal cancers. *Br. J. Cancer* **68,** 501–506.

Withers, H. R., Taylor, J. M. G., and Maciejewski, B. (1988). The hazard of accelerated tumor clonogen repopulation during radiotherapy. *Acta Oncol* **27,** 131–146.

Woltmann, G., Wardlaw, A., and Rew, D. A. (1998). Image analysis enhancement of the laser scanning cytometer. *Cytometry* **33**(3), 362–365.

Woltmann, G., Ward, R. J., Symon, F. A., Rew, D. A., Pavord, I., and Wardlaw, A. J. (1999). Objective quantitative analysis of eosinophils and bronchial epithelial cells in induced sputum by laser scanning cytometry. *Thorax* **54,** 124–130.

Wright, W. E. (1986). Bromodeoxyuridine, probability and cell variants. Towards a molecular under-standing of the decision to differentiate. *BioEssays* **3,** 245–247.

Yang, W.-D., De Bono, D., and Rew, D. A. (1998). Accelerated endothelial cell senescence studied by laser scanning cytometry. Abstract CB18: *Cytometry* ISAC IX Meeting Suppl. 9.

Zackrisson, B., Gustafsson, H., Stenling, R., Flygare, P., and Wilson, G. D. (1997). Predictive value of potential doubling time in head and neck cancer patients treated by conventional radiotherapy. *Int. J. Radiat. Oncol. Biol. Phys.* **38,** 677–683.

CHAPTER 52

Prediction and Precise Diagnosis of Diseases by Data Pattern Analysis in Multiparameter Flow Cytometry: Melanoma, Juvenile Asthma, and Human Immunodeficiency Virus Infection

**Günter Valet,* Hanna Kahle,* Friedrich Otto,[†]
Edeltraut Bräutigam,[‡] and Luc Kestens[§]**

Cell Biochemistry Group
Max-Planck-Institut für Biochemie
D-82152 Martinsried, Germany

[†] Fachklinik Hornheide
Abteilung für Tumorforschung
D-48157 Münster, Germany

[‡] Pathologisches Institut
Klinikum Görlitz
D-02828 Görlitz, Germany

[§] Prince Leopold Institute for Tropical Medicine, Pathology & Immunology
B-2000 Antwerp, Belgium

I. Introduction
II. Material and Methods
    A. Melanoma
    B. Asthma
    C. HIV Infection
    D. Flow Cytometric List Mode Analysis
    E. Data Pattern Classification
III. Results
    A. Melanoma

METHODS IN CELL BIOLOGY, VOL. 64

      B. Asthma
      C. HIV Infection
 IV. Discussion
      References

I. Introduction

The frequent use of multiparameter flow cytometry in the clinical or research laboratory together with the determination of a significant number of humoral biochemical parameters from blood serum, urine, or spinal fluid of patients generates a very substantial amount of information. Such information is only selectively evaluated at present, for example, according to the frequency of lymphocyte subpopulations, cell lineage assignments, abnormal immunophenotype, or cell activation marker description based on flow cytometric histograms or flow cytometric list mode data, as well as according to humoral or clinical parameters. Owing to the lack of suitable tools, the real potential of the multidimensionality of the measured information remains largely inaccessible.

This situation is highly unsatisfactory, considering the theoretical potential of cytometric information, for example, for the prediction of further disease course in individual patients. Cytometrically determined biochemical parameter patterns from directly or indirectly affected cellular systems or organs are of substantial interest for these purposes, because they are typically collected at the very spot of disease action and should therefore represent prime information carriers for disease course predictions, given the generation of diseases from biochemical deviations in cellular systems or organs. Predictions are preferable to statistical disease prognosis estimation. Although statistical disease prognosis is sufficient for therapy development and monitoring, it is of little value for the individual patient as well as for individualized therapy schemes.

The elaboration of general principles for disease course predictions in the clinical environment constitutes an important challenge for optimization of a patient's disease management. There are multiple methodological choices for the determination of structural or functional cell biochemical parameters and also for data evaluation by mathematical result modeling or by algorithmic principles. The task consists therefore of the rational selection of optimal biomolecular parameter patterns and result evaluation strategies for predictive medicine.

The high amount of data from flow cytometric measurements prompted the earlier development of the CLASSIF1 algorithm (Valet *et al.*, 1993) to assure the fast, exhaustive, and unbiased extraction of discriminant biomolecular data pattern from any kind of flow cytometric or other multiparameter data on typical personal computers, that is, out of hundreds or thousands of data columns. CLASSIF1 data classifications require specific and precise measurements according to standardized methods, but no mathematical preconditions or assumptions have to be fulfilled. The resulting classifiers are robust and interlaboratory portable. The intellectual analysis of the selected discriminant data patterns of the classifiers favors scientific hypothesis formation through intuitive result presentation. Thus, large scale information extraction from multiparameter data is available, as an important precondition for individualized disease course predictions in patients.

Data analysis may simultaneously concern flow cytometric, humoral biochemistry, or clinical patient data in order to determine the most discriminant parameter pattern from the totality of the available information. In most instances, only a comparatively small amount of the entire information, that is, typically between 0.5 and 20%, assures the required discrimination.

The use of neural network (Frankel *et al.*, 1996, 1989; Boddy *et al.*, 1994; Molnar *et al.*, 1993; Ravdin *et al.*, 1993), principal component (Leary, 1994), cluster (Verwer and Terstappen, 1993; Demers *et al.*, 1992; Terstappen *et al.*, 1990; Schut *et al.*, 1993), or discriminant and statistical (Davey *et al.*, 1999; Hokanson *et al.*, 1999; Molnar *et al.*, 1993; Rothe *et al.*, 1990) analysis, hierarchical classifiers (Decaestecker *et al.*, 1996), classification and regression trees (CART, Beckman *et al.*, 1995), as well as knowledge based systems (Thews *et al.*, 1996; Diamond *et al.*, 1994) or fuzzy logic (Molnar *et al.*, 1993) describe major other approaches to multiparameter data analysis in cytometry and in the clinical laboratory. Difficulties of handling high parameter numbers, a need for mathematical assumptions, nonintuitive result presentation, problems with missing values, as well as complexity of implementation and operation have so far not led to a widespread application of these methodologies in the clinical or biomedical research environment.

It is the intention of this study to show the potential of the algorithmic CLASSIF1 approach for multiparameter data analysis in various clinical conditions.

II. Material and Methods

A. Melanoma

Clinical parameters, such as tumor thickness (TD, mm), tumor invasion depth into skin layers (LE, Clark level), TK as the mean of TD + LE, tumor ulceration (UL, 1 = no, 2 = yes) as well as flow cytometric DNA ploidy (euploid = 1, aneuploid = 2), and percentage of S-phase cells, that is, a total of six parameters were determined from surgery material and were available for 499 melanoma patients who either had survived 10 years after tumor surgery (A) or died (B) within this time period. The melanomas were localized on different parts of the dermal integument. For flow cytometry, 20–100 mg of tumor tissue was minced with a razor blade, incubated in a 2.1% citric acid, 0.5% Tween 20 solution for 20 min at 22°C, and centrifuged; the pellet was fixed with 70% ethanol, and resuspended in citric acid/Tween 20 solution. The cell nuclei preparation was stained for 30 min in the presence of 1.75 μg/ml DAPI (4′,6-diamidino-2-phenylindole) in a 7.1% Na_2HPO_4 solution (Partec, Münster, Germany) (Otto *et al.*, 1981). The DNA fluorescence of the cell nuclei was measured with a PASII flow cytometer (Partec) using a HBO-100 high pressure mercury arc lamp with a UG1 (Schott, Mainz, Germany) fluorescence excitation filter and a GG435 (Schott) fluorescence emission filter for the determination of DAPI fluorescence. The coefficients of variation (CV) of the DNA distributions were between 1.5 and 3.1%, which is essential for the sensitive detection of DNA aneuploidy. The S-phase fraction was calculated from the DNA distribution according to a rectangular S-phase model.

B. Asthma

Data from 40 juvenile asthma patients with a mean age of 10.00 ± 0.63 years (2.4–16.5 years) and 18 healthy reference children with a mean age of 10.65 ± 0.67 years (4.6–16.4 years) were processed. Available were 49 clinical chemistry parameters per patient as well as 103 lymphocyte frequency and relative fluorescence intensity values obtained by quadrant statistics from two-color whole blood lyse–nonwash direct mouse monoclonal antibody immunofluorescence assays using the following antibody combinations: CD4/8, CD3/19, CD3/HLA-DR, CD3/16+56, CD25/3, CD71/3, CD4/45RA, CD45RO/4, CD62L/4, CD4/29, CD57/8, CD8/11b, CD5/19, CD21/19, CD62L/20, IgG1/IgG2. Furthermore, list mode files from CD45/14, CD4/29, CD4/8, CD56/8, CD3/56, CD25/3, CD3/HLA-DR, CD71/3, CD3/19, CD5/19, and IgG1/IgG2 assays, prepared as outlined, were collected for 10,000 nucleated cells in flow cytometry standard (FCS) 1.0 format. Measured were fluorescein isothiocyanate (FITC) and PE (phycoerythrin) immunofluorescence as well as forward (FSC) and perpendicular (SSC) scatter light with a Becton Dickinson FACScan (Becton Dickinson, Heidelberg, Germany) flow cytometer.

C. HIV Infection

Data of two-parameter peripheral blood lyse–nonwash FITC/PE immunophenotype measurements (CD45/14, CD3/16+56, CD2/19, CD45RA/4, HLA-DR/CD8, CD8/38, CD26/8, and CD26/4) of seronegative or human immunodeficiency virus (HIV)-infected seropositive patients were available as BD-FACScan (Becton Dickinson, Erembodegem, Belgium) list mode files as well as in the form of a 23-parameter dBase3 data base containing percent cell frequency values, manually extracted from two-parameter FITC/PE histograms, gated for lymphocytes by FSC/SSC.

D. Flow Cytometric List Mode Analysis

The list mode files were analyzed with the CLASSIF1 list mode analysis software (Valet *et al.*, 1993). In short, two-parameter FITC/PE histograms of

FSC/SSC-gated lympho-, mono-, and granulocytes were evaluated by quadrant analysis for percent cell frequency, FITC and PE fluorescence intensity, fluorescence ratio, and relative FITC and PE antibody (Ab) surface density (fluorescence/square root of FSC) of the various cell populations using fixed fluorescence thresholds at one-third of the four decade logarithmic scale (channel 85 on a 256 channel scale). The FSC/SSC gates, in contrast, were autoadaptive for the three cell populations such as to always comprise more than 95% of the nucleated cells by nonoverlapping polygons. The calculated parameters of the FSC/SSC, FSC/FITC, and FITC/PE histograms as well as of the four quadrants of the FITC/PE histograms were data based (Valet and Höffkes, 1997) such that 34 data columns per leukocyte population, that is, a total of $3 \times 34 = 102$ data columns were available for lympho-, mono-, and granulocytes per measurement instead of only 12 parameters in case of cell frequency evaluation. To remain comparable with the cell frequency quadrant analysis, only data of single or double fluorescence positive quadrants were further classified; in other words, parameters of the fluorescence double negative cell population were not evaluated. This reduced the available parameters in each two-parameter immunophenotype for lympho-, mono-, and granulocytes to $3 \times 23 = 69$ columns for the subsequent data pattern classification.

E. Data Pattern Classification

The results of the three studies were classified with the CLASSIF1 data pattern analysis algorithm (Valet and Höffkes, 1997; Valet *et al.*, 1993). The data were *a priori* assigned to either the learning set or to the embedded test set, such that patients 1, 5, 10, 15, . . . , of each classification category remained unknown to the algorithm during the learning phase. The algorithm proceeds as follows: Paired percentiles, for example, 10 and 90% percentiles for the value distributions of the learning set reference samples in each data base column are determined. Subsequently, all values of each data base column, that is, the values of the reference as well as of the abnormal samples, are transformed into triple matrix characters by assigning: "−" to values below the lower percentile, "0" to values between the percentiles, and "+" to values above the upper percentile. The resulting triple matrix replica of the numeric data base serves for the subsequent iterative data classification.

The algorithm optimizes during the learning phase the sum of the diagonal values of the confusion matrix, established between, for example, the known disease categories of patients on the ordinate and the computer determined classification of these disease categories on the abscissa. Ideally, that is, when all samples are correctly classified by the algorithm, the values in each of the diagonal boxes of the confusion matrix are 100%, while 0% values are encountered in the nondiagonal boxes. Since this ideal condition is not present at the beginning of the iterative optimization, the algorithm sequentially excludes either single data base columns or paired combinations of columns with any other of the data base columns temporarily from the classification process. Once all permutations have been processed and no further improvement is reached, data base columns that had improved the classification result when excluded either alone or in combination with another column are permanently disregarded for the classification process; thus, only discriminant data base columns survive the selection process. The most frequent triple matrix character for each of the selected data columns of the learning set is entered into the classifier mask for each classification category. The classifier mask of the reference samples contains typically only 0 values because 0 values are the most frequent triple matrix characters in the value distributions of the reference samples from which the percentiles were calculated; for example, for the 10/90% percentile pair, 80% of triple matrix characters are 0, 10% are − and 10% are +.

The best possible classification is determined by successively classifying a data set separately for the different percentile pairs 10/90%, 15/85%, 20/80%, 25/75%, and 30/70% as well as cumulatively such that the most discriminant percentile pair for each data base column is used for the finally learned classifier. The three data sets on melanoma, juvenile asthma, and HIV infection were integrally processed, that is, without exclusion of samples.

As a check for a learned classifier, each patient of the learning set is reclassified according to the highest positional coincidence of the patient classification mask with anyone of the classifier masks at the finally selected classification conditions. Equal coincidence frequency for two classifier masks results in a double classification. Double classifications may represent either a biological transition state or a classification error, for example, in the case of small learning sets. The reclassification of the learning set samples permits a quick visual check of the sample triple matrices for systematic deviations (e.g., with time or on change of reagents) from the classifier masks.

The quality of a learned classifier is judged in a standardized way by the average recognition index (ARI) and by the average multiplicity index (AMI). The ARI is calculated as the sum of the diagonal values of the confusion matrix divided by the number of classification categories. It should be higher than 80% for clinical purposes. The AMI is a measure for the average frequency of assignment of more than one classification category to a sample. The AMI is ideally 1.00 in the absence of multiple classifications, and it is 1.1, 1.2, or 1.33 in case every tenth, fifth, or third example on average is assigned a double classification. AMIs between 1.0 and 1.2 are acceptable in practice. The AMI is calculated as the sum of the classification values in all lines of the confusion matrix divided by the number of classification categories, followed by a further division by 100. All AMIs of the subsequent classification were always 1.00, that is, no multiple classifications occurred.

The classification coincidence factor (CCF) indicates the coincidence of the individual patient/sample classification triple matrix with the best fitting classifier mask. The CCF is used to identify "unknown" sample classification masks that have a lower CCF than the lowest CCF observed during the learning process. Low CCFs may occur through inclusion of wrong samples, systematic errors during parameter measurement, or because of missing values. Although systematically deviating test set samples are definitively rejected, samples with missing values are manually classified according to the best positional coincidence with any one of the classifier masks (e.g., patient 0001 in Fig. 2). No sample in all the processed samples of this study had to be excluded because of systematic deviations.

III. Results

A. Melanoma

The 499-patient data set consisted in sequence of 135 male and 216 female 10-year survivor (A) and of 83 male and 65 female nonsurvivor (B) patients. The first 75 male and 76 female (A) as well as 76 male and 64 female (B) patients were selected as the learning set with the first unknown test set. The learning set of 231 patients [59 male/61 female (A), 60 male/51 female (B)] contained the first unknown test set of 60 patients [16 male/15 female (A), 16 male/13 female (B)] in embedded form as outlined in Section II,E. The remaining 208 patients [60 male/140 female (A) together with 7 male/1 female (B)] served as the second unknown test set for the learned classifier. The total test set contained 268 patients.

The single-parameter sensitivity for correct nonsurvivor prediction was checked prior to data pattern analysis as a reference for data classification improvement by data pattern analysis. Single parameter sensitivity at 90% specificity for the identification of survivors was 53.7, 31.3, 50.9, 38.0, 38.1, and 21.7% for parameters TD, LE, TK, UL, AN, and SP. Individually, the values were too low for clinical predictions.

A first data pattern classification aiming at the discrimination of melanomas according to their location on the dermis was not successful (results not shown). In a second attempt it was investigated whether sex difference for survival existed in the provided data set. Due to identity of the classifier masks for male and female patients at the percentiles 10–90%, 15–85%, 20–80%, 25–75% and 30–70%, no distinction for the available parameters exists between male and female patients with regard to survival. Data from male and female patients can therefore be classified together in the search for melanoma-dependent differences in post tumor surgery survival (Table I). The impossibility of distinction between male and female patients is indirectly reflected by closely similar means and SEMs for each one of the investigated six parameters (Table II). Concerning the distinction between 10-year survivor and nonsurvivor patients, the optimal classifier (Table I) provides predictive values of 80.3% for survivors and 79.8% for nonsurvivors. The selected classification parameters are increased TD, TK, and SP. Only half of the provided parameters are selected for classification, although all six parameters are significantly increased for nonsurvivors (Table II). The triple matrix patterns of the three selected parameters for survivor (A) and nonsurvivor (B) (Fig. 1) show that the classification result remains in many instances correct,

Table I
Melanoma: 10-Year Survivors and Nonsurvivors

Clinical outcome	Number of patients (*n*)	CLASSIF1 prediction (%)[a]		Specificity/ sensitivity
		Survivor	Nonsurvivor	
A. Learning set				
Survivor	120	**81.7**	18.3	81.7
Nonsurvivor	111	21.6	**78.9**	78.9
Negative/positive predictive values		80.3	79.8	ARI 80.0
B. First unknown test set				
Survivor	31	**80.6**	19.4	80.6
Nonsurvivor	29	20.7	**79.3**	79.3
Negative/positive predictive values		80.6	79.3	ARI 80.0
C. Second unknown test set				
Survivor	200	**75.5**	24.5	75.5
Nonsurvivor	8	12.5	**87.5**	87.5
Negative/positive predictive values		99.3	85.4	ARI 81.5

[a] 20–>30% optimized percentile thresholds (data base MELA6.BI4/.BI6).

although not all three parameters coincide with the classifier mask for either survivor or nonsurvivor. In other words, the individual patient classification is robust against a certain degree of noncoincidence with the best fitting classifier mask. The CCF is 0.67, that is, two out of three parameters have to match with the best fitting classifier mask for technically valid classifications.

The embedded first test set is classified with predictive values of 80.6% for the 31 survivors and with 79.3% for the 29 nonsurvivors (Table I). The second test set is classified with predictive values of 99.3% for the 200 survivors and with 85.4% for the 8 nonsurvivors (Table I). The classification of the two unknown test sets shows that the CLASSIF1 classifier provides a robust classification of unknown samples as an important quality criterion for multiparameter data classifiers.

B. Asthma

Data pattern analysis of the clinical chemistry parameters provides a significantly better discriminatory result (Table III) than single parameter discrimination (Table IV). A sensitivity of 70% for the recognition of asthmatic children at 100% specificity for the identification of the healthy reference children is obtained. The respective positive and negative predictive values, that is, the correct prediction of the asthmatic and healthy children from the CLASSIF1 determined optimal parameter pattern (Table III), are 100 and 60% for the learning set. Seven of the 49 clinical chemistry measurements [i.e., thrombocyte (TRCS) and eosinophil (EOS) counts, aspartate aminotransferase (ASAT), thyroid-stimulating hormone (TSH), ferritin, IgE, and β-globulin] were selected by the CLASSIF1 algorithm. The data pattern classification of the clinical chemistry parameters for the test set patients

Table II
Melanoma Parameters[a]

Patients	Number (n)	Diameter (TD, mm)	Infiltr.(LE) (arb.units)	TK = (TD + LE)/2 (mm)	Ulceration (UL)	DNA ploidy (PL)	S-phase (SP, %)
A. Male and female patients (learning set + first unknown test set)							
Male	151	2.54 ± 0.23	3.47 ± 0.06	3.00 ± 0.14	1.33 ± 0.04	1.21 ± 0.03	8.40 ± 0.47
Female	140	3.07 ± 0.35	3.57 ± 0.08	3.35 ± 0.21	1.35 ± 0.04	1.18 ± 0.03	8.21 ± 0.42
B. Survivors and nonsurvivors (learning set)							
Survivor	120	1.52 ± 0.22	3.14 ± 0.07	2.32 ± 0.13	1.14 ± 0.03	1.06 ± 0.02	7.31 ± 0.36
Nonsurvivor	111	4.18 ± 0.32[b]	3.92 ± 0.05[b]	4.09 ± 0.18[b]	1.56 ± 0.04[b]	1.35 ± 0.04[b]	9.39 ± 0.51[b]

[a] Means ± SEM.
[b] $2p < 0.001$, t-test.

MELANOMA: CLASSIF1 TRIPLE MATRIX CLASSIFICATION

A.) RECLASSIFICATION OF LEARNING SET

NR.	CLASSIFIER CATEGOR.	CATEGORY ABBREVIAT.	COIN	CLASSIFIER MASKS
1	SURVIVAL	A	1.00	000
2	DEATH	B	1.00	+++

REC. NR.	DATAB: MELA6.BI4 RECORD LABELS	CLASSIF1-CLASSIFIC.	COIN FACT	SAMPLE CLASSIF.MASKS . = no value
2	47888. A	A	1.00	000
3	49896. A	A	.67	00+
4	51077. A	A	.67	--+
6	58306. A	A	1.00	000
7	63220. A	B	.67	++0
8	63231. A	A	1.00	000
9	67200. A	A	1.00	--0
11	74733. A	B	.67	++0
12	76030. A	A	1.00	---
13	77598. A	B	.67	++-
153	46931. B	B	1.00	+++
154	47071. B	A	1.00	000
155	51680. B	A	1.00	000
157	54535. B	B	.67	++0
158	70983. B	B	1.00	+++
159	77107. B	B	.67	++-
160	77913. B	A	.67	0+0
162	76475. B	B	1.00	+++
163	53494. B	B	1.00	+++
164	71682. B	B	.67	++0

B) CLASSIFICATION OF UNKNOWN TEST SET

REC. NR.	RECORD LABELS	truth	CLASSIF1-CLASSIFIC.	COIN FACT	SAMPLE CLASSIF.MASKS
1	59083. ?	A	A	.67	00+
2	56627. ?	A	A	1.00	0-0
3	74007. ?	A	A	1.00	000
4	79931. ?	A	A	1.00	000
5	59731. ?	A	A	1.00	000
6	54563. ?	A	A	.67	-0+
7	53485. ?	A	B	.67	++0
8	63089. ?	A	A	1.00	--0
9	62437. ?	A	B	1.00	+++
10	75891. ?	A	A	1.00	00-
32	54748. ?	B	B	1.00	+++
33	53400. ?	B	B	.67	++-
34	54520. ?	B	B	.67	++-
35	80046. ?	B	B	1.00	+++
36	74562. ?	B	B	1.00	+++
37	77415. ?	B	B	1.00	+++
38	73113. ?	B	A	.67	0+0
39	71988. ?	B	B	1.00	+++
40	56694. ?	B	B	1.00	+++
41	57738. ?	B	A	.67	00+

pat-ID truth classification   classification
      hidden truth   coincidence factor (CCF)
  during learning phase

Table III
Asthma: Blood Clinical Chemistry

Clinical diagnosis	Number of patients (n)	CLASSIF1 classification (%)[a]		Specificity/ sensitivity
		Healthy	Asthmatic	
A. Learning set				
Healthy children	18	**100.0**	0.0	100.0
Asthmatic children	40	30.0	**70.0**	70.0
Negative/positive predictive values		60.0	100.0	ARI 85.0
B. Unknown test set				
Healthy children	6	**66.7**	33.3	66.7
Asthmatic children	10	30.0	**70.0**	80.0
Negative/positive predictive values		57.1	77.7	ARI 68.5

[a] 25–>30% optimized percentile thresholds (data bases GOERLI1.BI4/.BI6).

as compared to the learning set has a lower discriminatory potential with positive and negative predictive values of 77.7 and 57.1% (Table III).

The means of six [β-globulin, IgG3, complement CH100 titer, T4, and leukocyte (LKCS)] of the clinical chemistry parameters are significantly different between asthmatic and nonasthmatic children ($2p < 0.05$, t-test). Significant mean value differences do not necessarily parallel good discrimination potential. As shown in Table IV, only β-globulin has enough discriminatory potential for data pattern analysis. This is further substantiated by systematic analysis of the discriminatory potential of the statistically most significant single parameters such as β-globulin, by which asthmatic children are detectable with a sensitivity of 40.9% at 90% specificity for the recognition of healthy children, similarly as, for example, IgG3 (41.1%), β-globulin (40.9%), and LKCS (38.9%). Statistically different parameters may, however, also be low discriminating parameters, such as complement CH100 (22.5%) or T4 (19.0%). On the other hand, nonsignificantly different and low discriminating single parameters may prove quite useful in data pattern analysis such as ASAT (10.4%) (Table IV).

Fig. 1 Melanoma classification for the known learning set (A) and the first unknown test set (B) of patients using the classifier of Table I. The classifications for the first 10 patients in each classification category are displayed. Patient group A represents 10-year survivors, while patients in group B did not survive. Patients are classified according to the best positional coincidence of the patient classification mask with one of the two classifier masks. The three selected classification parameters are tumor thickness (TD at position 1 of the classifier masks), the mean value of tumor thickness and tumor infiltration depth (TK at position 2), and percentage of S-phase tumor cells (SP at position 3). Classification is performed down to a CCF of ≥ 0.67, that is, to a positional coincidence of the patient classification mask with the best fitting classifier mask for two of the three classification parameters. The truth positions were left blank (?) for the test set patients (B) to make them invisible for the CLASSIF1 algorithm during the learning process.

Table IV
Asthma: Single Parameter Sensitivity and CLASSIF1 Classificator Masks of Selected Clinical Chemistry Parameters

Classification parameters (selected from 49 parameters)	Single parameter sensitivity (%) at 90% specificity	Classification matrix	
		N	A
1. TRCS	29.2	0	−
2. EOS	30.9	0	+
3. ASAT	10.4	0	−
4. TSH	31.0	0	−
5. Ferritin	23.3	0	−
6. IgE	35.0	0	+
7. β-Globulin	40.9	0	−

The classification (Table V) of manually evaluated flow cytometry histograms for lymphocyte relative cell frequency (%) and relative antigen expression of the 16 two-parameter immunophenotypes indicates ideal classification with 100.0% sensitivity/specificity and positive/negative predictive values for asthmatic and healthy children. This classification is reached not only for the known learning set (Table V), but equally for the unknown test set of patients (Table V); that is, the classification is robust. The discrimination was achieved with data from only four measurements (CD4/45RA, CD8/11b, CD21/19, and CD71/3) (Table VI), whereas the other 12 measurements provided less direct and redundant information that resulted in exclusion during the selection process.

The classification of the list mode files included only partially the same measurements as for the manually evaluated histograms. In particular the most

Table V
Asthma: Flow Cytometry by Cell Frequency and Fluorescence Intensity

Clinical diagnosis	Pat. (*n*)	CLASSIF1 classification (%)[a]		Specificity/ sensitivity
		Healthy	Asthmatic	
A. Learning set				
Healthy children	19	**100.0**	0.0	100.0
Asthmatic children	39	0.0	**100.0**	100.0
Negative/positive predictive values		100.0	100.0	ARI 100.0
B. Unknown test set				
Healthy children	6	**100.0**	0.0	100.0
Asthmatic children	9	0.0	**100.0**	100.0
Negative/positive predictive values		100.0	100.0	ARI 100.0

[a] 10−>15% optimized percentile thresholds (data bases GO5LEARN.BI4/.BI6).

Table VI
Asthma: Selected Cell Frequency and Fluorescence Intensity Parameters[a]

Classification parameters (selected from 103 lymphocyte parameters)	Healthy ($n = 19$)	Asthma ($n = 39$)	Units	Classification matrix	
				N	A
1. CD45RA Ab on CD4[−]/CD45RA[+] lymphocytes	26.94 ± 2.94	103.38 ± 7.68[b]	Arbitrary units	0	+
2. % CD4[−]/CD45RA[+] lymphocytes	11.57 ± 0.92	51.51 ± 1.90[b]	% of lymphocytes	0	+
3. CD4Ab on CD4[+]/CD45RA[−] lymphocytes	115.31 ± 7.88	26.59 ± 1.60[b]	Arbitrary units	0	−
4. % CD4[+]/CD45RA[−] lymphocytes	50.94 ± 2.51	14.17 ± 0.64[b]	% of lymphocytes	0	−
5. % CD5[+]/CD19[+] lymphocytes	7.84 ± 0.86	8.49 ± 1.36[b]	Arbitrary units	0	+
6. CD8 Ab on CD8[+]/CD11b[−] lymphocytes	67.16 ± 7.79	37.62 ± 3.61[b]	% of lymphocytes	0	−
7. CD45RO Ab on CD45RO[+]/CD4[−] lymphocytes	41.36 ± 6.51	26.17 ± 4.35	% of lymphocytes	0	−

[a] Means ± SEM.
[b] $2p < 0.001$, t-test.

discriminant measurements such as CD4/45RA, CD8/11b, and CD21/19 were not available as list mode files. Nevertheless the same ideal 100.0% result for sensitivity/specificity, positive/negative predictive values was obtained by the exhaustive CLASSIF1 list mode evaluation of lympho-, mono-, and granulocyte parameters for the learning set (Table VII) as well as for the unknown test set patients (Table VII). A closer analysis of the selected classification parameters (Table VIII), shows that 7 (CD45/14, CD3/HLA-DR, CD4/29, CD71/3, CD3/

Table VII
Asthma: Exhaustive Flow Cytometric List Mode Analysis

Clinical diagnosis	Pat. (n)	CLASSIF1 classification (%)[a]		Specificity/ sensitivity
		Healthy	Asthmatic	
A. Learning set				
Healthy children	19	**100.0**	0.0	100.0
Asthmatic children	30	0.0	**100.0**	100.0
Negative/positive predictive values		100.0	100.0	ARI 100.0
B. Unknown test set				
Healthy children	5	**100.0**	0.0	100.0
Asthmatic children	9	0.0	**100.0**	100.0
Negative/positive predictive values		100.0	100.0	ARI 100.0

[a] 10–>30% optimized percentile thresholds (data bases KHLEARN.BI4/.BI6).

Table VIII
Asthma: Selected Parameters Exhaustive List Mode Analysis[a]

Classification parameters (selected from 759 lympho-, mono-, granulocyte parameters of 11 two color immunophenotypes)	References ($n = 19$)	Asthma ($n = 30$)	Units	Classification matrix N	A
1. CD45Ab surf. dens.[b] on CD45+ lymphocytes	0.233 ± 0.007	0.170 ± 0.012^c	Arbitrary units	0	−
2. % CD3+/HLA-DR+ lymphocytes	11.26 ± 1.58	3.66 ± 0.31^c	% of lymphocytes	0	−
3. HLA-DR Ab on CD3+/HLA-DR+ lymphocytes	0.965 ± 0.091	1.971 ± 0.254^c	Arbitrary units[d]	0	+
4. CD4 Ab surf. dens. on CD4+ lymphocytes	0.132 ± 0.005	0.095 ± 0.006	Arbitrary units	0	−
5. % CD71+ lymphocytes	8.02 ± 1.24	3.63 ± 0.55^c	% of lymphocytes	0	−
6. CD3 Ab on CD3−/CD56+ granulocytes	0.0894 ± 0.0046	0.0461 ± 0.0032	Arbitrary units[d]	0	−
7. CD56/CD3 Ab ratio on CD3−/CD56+ granulocytes	4.30 ± 0.63	13.67 ± 1.88^c	Arbitrary units	0	+
8. CD4 Ab on CD4−/CD29+ granulocytes	0.1000 ± 0.0054	0.0475 ± 0.0033^c	Arbitrary units[d]	0	−
9. % CD25+ granulocytes	78.28 ± 4.51	11.10 ± 2.93^c	% of granulocytes	0	−
10. CD19/CD5 Ab ratio on CD5−/CD19+ granulocytes	1.66 ± 0.24	4.51 ± 1.76	Arbitrary units	0	+
11. % CD5+/CD19− granulocytes	42.80 ± 6.53	5.97 ± 2.38^c	% of granulocytes	0	−

[a] Means $\pm$ SEM.
[b] Ab surf. dens. = relative antibody surface density (linearized fluorescence/square root of forward light scatter).
[c] $2p < 0.001$, t-test.
[d] Arbitrary unit: 0.001–10 V fluorescence scale, relinearized from the four decade log fluorescence scale of the flow cytometer, ratio calculation from relinearized fluorescence values.

Fig. 2 Juvenile asthma patient classification for the first 10 patients of the learning set (A) and the five and nine patients of the test set (B) using the classifier of Table VII. Eleven classification parameters were automatically selected by the CLASSIF1 algorithm from 759 data columns extracted from the lympho-, mono-, and granulocyte cell populations of 11 FITC/PE immunophenotype list mode files per patient (Table VIII). Parameters from seven immunophenotypes (CD45/14, CD3/HLA-DR, CD71/3, CD3/16+56, CD4/29, CD25/3, and CD5/19) are required for the classification. Classification is performed down to a CCF of ≥ 0.55, that is, to a positional coincidence of the classification mask of the patient with the best fitting classifier mask for 6 of the 11 classification parameters. Patient 0001 of the test set is not classified ($-$) because of a CCF of 0.27 as a consequence of missing values (.) by nonavailable list mode files. Due to coincidence for three positions with the asthma classifier mask (A) and two with the mask of normal children (N), the patient was manually classified as asthma for the test set classification (Table VII). The truth position for all test set patients were left blank (?) during the learning phase as in Fig. 1.

JUVENILE ASTHMA: CLASSIF1 TRIPLE MATRIX CLASSIFICATION

A.) RECLASSIFICATION OF LEARNING SET

NR.	CLASSIFIER CATEGOR.	CATEGORY ABBREVIAT.	COIN	CLASSIFIER MASKS
1	NORMAL	N	1.00	00000000000
2	ASTHMA	A	1.00	--+---+--+-

REC. NR.	DATAB: KHLEARN.BI4 RECORD LABELS	CLASSIF1-CLASSIFIC.	CLAS COIN	SAMPLE CLASSIF.MASKS . = no value
40	0059 N	N	1.00	+00++0-++-+
41	0058 N	N	.64	00+000+0-0-
42	0061 N	N	1.00	0+00000+000
43	0060 N	N	.82	0-000000-00
44	0064 N	N	.73	000+000--0-
45	0066 N	N	1.00	0000+000000
46	0065 N	N	.55	0-00-00-0+-
48	0067 N	N	1.00	00-00+0+0-+
49	0069 N	N	.64	0++0000-++-
50	0070 N	N	.82	0-00-00000+
2	0003 A	A	.55	+-0+-00--+-
3	0004 A	A	.64	+-++00+--+-
5	0008 A	A	.82	0-++--+--+-
6	0009 A	A	1.00	--+---+--+-
7	0010 A	A	1.00	--+---+--+-
9	0014 A	A	.91	--+-+-+--+-
10	0015 A	A	1.00	--+---+--+-
12	0018 A	A	1.00	--+---+--+-
13	0027 A	A	.91	--+--0+--+-
14	0030 A	A	1.00	--+---+--+-

B) CLASSIFICATION OF UNKNOWN TEST SET

REC. NR.	RECORD LABELS		CLASSIF1-CLASSIFIC.	CLAS COIN	SAMPLE CLASSIF.MASKS
47	0068 ?	N	N	.91	0+-+0+-++--
53	0073 ?	N	N	.64	0-+0000-0++
58	0078 ?	N	N	.82	00-+0000-0-
60	0080 ?	N	N	.55	0-+-000-0++
61	0083 ?	N	N	1.00	0+00000+0-+
1	0001 ?	A	-	.27	--+..00....
4	0007 ?	A	A	.82	+-++--+--+-
8	0012 ?	A	A	.82	--+---0-0+-
11	0017 ?	A	A	1.00	--+---+--+-
16	0032 ?	A	A	.64	--+-.-+-...
21	0037 ?	A	A	1.00	--+---+--+-
26	0042 ?	A	A	.91	-0+---+--+-
31	0048 ?	A	A	.91	--+0--+--+-
35	0053 ?	A	A	.82	--+-0-+-0+-

```
      |      |          |          |                        |
  pat-ID   truth        |    classification      classification
              hidden truth         coincidence factor (CCF)
          during learning phase
```

Table IX
Asthma: Exhaustive List Mode Analysis of CD25/3

| | Number of | CLASSIF1 classification (%)[a] | | |
Clinical diagnosis	patients (*n*)	Healthy	Asthmatic	Specificity/ sensitivity
A. Learning set				
Healthy children	20	**100.0**	0.0	100.0
Asthmatic children	31	3.2	**96.8**	96.8
Negative/positive predictive values		95.2	100.0	ARI 98.4
B. Unknown test set				
Healthy children	4	**100.0**	0.0	100.0
Asthmatic children	8	12.5	**87.5**	87.5
Negative/positive predictive values		80.0	100.0	ARI 93.7

[a] 10–>15% optimized percentile thresholds (data bases KQLEARN.BI4/.BI6).

56, CD25/3, and CD5/19) of the 11 measurements were required for classification. The printout of the classification masks (Fig. 2) for the individual patients of the learning and test sets indicates robustness of classification in case of partial nonidentity between the patient classification mask and the best fitting classifier mask. Minimally 6 of the 11 classifier parameters have to match with the selected classifier mask to avoid sample rejection at the observed CCF of 0.55.

The separate classification of each individual two-parameter immunophenotype with simultaneous consideration of lympho-, mono-, and granulocyte data shows that, for example, the analysis of the single CD25/3 immunophenotype alone discriminates already quite well (ARI = 98.4%) between asthmatic and nonasthmatic children in the learning set (Table IX) as well as in the unknown test set patients (Table IX) with a selection of three cell frequency parameters (Table X). Similar results were obtained for CD57/8 (ARI = 95.9%) and CD5/19 (ARI = 95.9%).

Table X
Asthma: Selected Parameters Exhaustive CD25/3 Analysis[a]

Classification parameters (selected from 69 lympho-, mono-, and granulocyte parameters)	Healthy (*n* = 20)	Asthma (*n* = 31)	Classification matrix	
			N	A
1. % CD25[+] granulocytes	78.28 ± 4.51	11.10 ± 2.93[b]	0	–
2. % CD25[-]/CD3[+] granulocytes	0.224 ± 0.068	3.41 ± 1.19[b]	0	+
3. % CD25[+]/CD3[+] granulocytes	9.47 ± 3.16	2.65 ± 1.19[b]	0	–

[a] Means ± SEM (% of granulocytes).
[b] 2*p* < 0.05, *t*-test.

Table XI
HIV Infection: Flow Cytometry by Cell Frequency

| | | CLASSIF1 classification (%)[a] | | |
Clinical diagnosis	Number of patients (n)	Seronegative	Seropositive	Specificity/ sensitivity
A. Learning set				
Seronegative	15	**100.0**	0.0	100.0
Seropositive	55	7.3	92.7	82.6
Negative/positive predictive values		78.9	**100.0**	ARI 96.4
B. Unknown test set				
Seronegative	5	**100.0**	0.0	100.0
Seropositive	14	0.0	**100.0**	100.0
Negative/positive predictive values		100.0	100.0	ARI 100.0

[a] 10–90% percentile thresholds (data base CD26TOT6.BI4/.BI6).

C. HIV Infection

The classification (Table XI) of the 18 parameters from manual analysis of two-color lymphocyte immunophenotype histograms, including the white blood cell and lymphocyte counts (WBC, LYC), provides positive and negative predictive values of 100.0 and 78.9% for HIV seropositive and seronegative patients with similar values for the unknown test set patients (Table XI). The selected three parameters (Table XII) involve CD45RA/4, HLA-DR/CD8, and CD8/38 immunophenotype measurements.

The exhaustive lympho-, mono-, and granulocyte parameter extraction by the CLASSIF1 analysis provided average recognition between 96.1 and 100.0% (ARI) at multiplicity indices between 1.00 and 1.02 (AMI) for the individual evaluation of either the CD2/19, HLA-DR/CD8, CD45RA/4, or the CD8/38 measurement. Evaluation of only the lymphocyte cell population provided ARIs

Table XII
HIV Infection: Selected Lymphocyte Frequency Parameters[a]

Classification parameters (selected from 18 lymphocyte/ leukocyte parameters)	Seronegative ($n = 15$)	Seropositive ($n = 55$)	Units	Classification matrix	
				N	P
1. CD45RA$^+$/CD4$^+$ lymphocytes	23.53 ± 1.94	8.14 ± 0.83[b]	% of lymphocytes	0	–
2. HLA-DR$^+$/CD8$^+$ lymphocytes	7.06 ± 1.04	36.01 ± 1.86[b]	% of lymphocytes	0	+
3. CD8$^+$/CD38$^+$ lymphocytes	13.60 ± 1.04	45.01 ± 2.30[b]	% of lymphocytes	0	+

[a] Means ± SEM.
[b] $2p < 0.001$, t-test.

Table XIII

HIV Infection: Exhaustive Flow Cytometric List Mode Analysis on Lymphocytes (HLA–DR/CD8)

| | | CLASSIF1 classification (%)[a] | | |
Clinical diagnosis	Number of patients (n)	Seronegative	Seropositive	Specificity/ sensitivity
A. Learning set				
Seronegative	15	**100.0**	0.0	100.0
Seropositive	55	0.0	**100.0**	100.0
Negative/positive predictive values		100.0	100.0	ARI 100.0
B. Unknown test set				
Seronegative	5	**100.0**	0.0	100.0
Seropositive	14	0.0	**100.0**	100.0
Negative/positive predictive values		100.0	100.0	ARI 100.0

[a] 15–85% percentile thresholds (data base PRLEARN.BI4/.BI6).

of 100.0% for HLA-DR/CD8 (Table XIII), 99.3% for CD45RA/4, and 97.8% for CD8/38, all at 1.00 multiplicity. HLA-DR/CD8 provided in addition positive and negative predictive values of 100% for the HIV seronegative and the seropositive patients in various disease states (seronegative $n = 15$, seropositive WHO stage 1/2/3/4 $n = 21/9/14/11$) for the learning set (Table XIII) as well as in the test set patients (Table XIII) ($n = 5/5/2/3/4$). Four of the five selected parameters concern antigen expression, relative antigen surface density, and antigen ratios and only one concerns a percent cell frequency parameter (Table XIV). The listing of the triple matrices for the individual patients (Fig. 3) indicates robustness of classification for some degree of positional nonidentity between the patient classi-

Table XIV

HIV Infection: Selected HLA–DR/CD8 Lymphocyte Parameters[a]

Classification parameters (selected from 22 lymphocyte parameters	Seronegative ($n = 15$)	Seropositive ($n = 55$)	Units	Classification matrix	
				N	P
1. CD8 Ab on CD8$^+$ lymphocytes	18.25 ± 1.02	7.91 ± 0.23[b]	Arbitrary units	0	−
2. CD8 rel. Ab surf. dens. on CD8$^+$ lymphocytes	0.761 ± .046	0.328 ± 0.009[b]	Arbitrary units	0	−
3. CD8 Ab on HLA-DR$^-$/CD8$^+$ lymphocytes	18.01 ± 1.04	7.63 ± 0.22[b]	Arbitrary units	0	−
4. CD8/HLA-DR Ab ratio on HLA-DR$^-$/ CD8$^+$ lymphocytes	671.1 ± 47.7	197.3 ± 10.5[b]	Arbitrary units	0	−
5. % HLA-DR$^+$/CD8$^+$ lymphocytes	3.68 ± 0.51	24.12 ± 1.44[b]	% of lymphocytes	0	+

[a] Means ± SEM.
[b] $2p < 0.001$, t-test.

```
HIV INFECTION: CLASSIF1 TRIPLE MATRIX CLASSIFICATION
```

A.) RECLASSIFICATION OF LEARNING SET

NR.	CLASSIFIER CATEGOR.	CATEGORY ABBREVIAT.	COIN	CLASSIFIER MASKS
1	NORMAL	N	1.00	00000
2	SEROPOS	P	1.00	----+

REC. NR.	DATAB: PRLEARN.BI4 RECORD LABELS	CLASSIF1-CLASSIFIC.	CLAS COIN	SAMPLE CLASSIF.MASKS . = no value
55	KEOK02 N	N	1.00	00000
56	KEOK03 N	N	1.00	0000-
57	KEOK04 N	N	.60	000-+
58	KEOK06 N	N	.80	+++++
59	KEOK07 N	N	.80	+++++
60	KEOK08 N	N	1.00	++++0
61	KEOK09 N	N	1.00	00000
62	KEOK11 N	N	1.00	00000
63	KEOK12 N	N	.60	-0-00
1	KE3756 P	P	1.00	----+
2	KE3758 P	P	1.00	----+
3	KE3759 P	P	1.00	----+
4	KE3766 P	P	1.00	----+
5	KE3767 P	P	1.00	----+
6	KE3768 P	P	1.00	----+
7	KE3769 P	P	1.00	----+
8	KE3771 P	P	1.00	----+
9	KE3772 P	P	1.00	----+
10	KE3773 P	P	1.00	----+

B) CLASSIFICATION OF UNKNOWN TEST SET

REC. NR.	RECORD LABELS	CLASSIF1-CLASSIFIC.	CLAS COIN	SAMPLE CLASSIF.MASKS
87	KEOK01 ? N	N	1.00	00000
91	KEOK05 ? N	N	.80	0000+
96	KEOK10 ? N	N	1.00	000+-
101	KEOK15 ? N	N	1.00	00000
106	KEOK20 ? N	N	1.00	00000
17	KE3755 ? P	P	1.00	----+
21	KE3765 ? P	P	1.00	----+
26	KE3770 ? P	P	1.00	----+
31	KE3776 ? P	P	.80	----0
36	KE3782 ? P	P	1.00	----+
41	KE3789 ? P	P	1.00	----+
46	KE3795 ? P	P	1.00	----+
52	KE3993 ? P	P	1.00	----+
57	KE4001 ? P	P	1.00	----+
63	KE4061 ? P	P	1.00	----+

```
        pat-ID    truth |  classification   classification
               hidden truth                 coincidence factor (CCF)
            during learning phase
```

Fig. 3 Classification of the learning (A) and test set (B) of HIV seronegative (N) and seropositive (P) patients using the classifier of Table XIII. Five parameters per patient were selected from 22 lymphocyte data columns of HLA-DR/CD8 immunophenotype list mode analysis (Table XIV). Classification is performed down to a CCF of ≥0.60, that is, to a positional coincidence for three of the five classification parameters of the patient's classification mask with the best fitting classifier mask.

fication mask and the best fitting classifier mask. A minimum of three positional coincidences with the selected five parameters classification matrix is required (CCF = 0.60).

IV. Discussion

The three classification examples from unrelated clinical areas show the potential of data pattern classification for disease course prediction (melanoma) as well as for a precise biomolecular diagnosis (juvenile asthma, HIV infection). Precise diagnosis represents a precondition for the elaboration of predictive classifiers.

The predictive capacity of the melanoma classifier (Table I) is similar to the one for survival prediction in colorectal carcinoma patients (van Driel *et al.*, 1999) that is, lower than for the estimation of sepsis outcome in intensive care medicine (Rothe *et al.*, 1990; Valet *et al.*, 1998) as well as for the preoperative prediction of postcardiotomy syndrome in children with open heart surgery (Tarnok *et al.*, 1997, 1999). This is caused by the comparatively small initial parameter pattern of four clinical and only two flow cytometric parameters (Table II). In spite of the few parameters, the classification of the 231 learning set patients and especially of the 268 unknown test patients provides stable results.

When comparing the information content in the various measurements in juvenile asthma and HIV infected patients, only a small fraction of the available parameters contains the discriminant information. In juvenile asthma, the diagnostic information is encountered in 14.2% (7 of 49) of the clinical chemistry parameters (Table IV), in 6.7% (7 of 103) of the lymphocyte analysis (Table VI) provided by 4 of 12 FITC/PE immunophenotypes, in 1.4% (11 of 759) of the parameters from exhaustive lympho-, mono-, and granulocyte analysis from 7 of 11 FITC/PE immunophenotypes (Table VIII), and in 4.3% (3 of 69) of the parameters from single CD25/3 immunophenotype analysis (Table X). A similar situation was encountered in the analysis of HIV infected patients (16.6%, 22.7%, Tables XII and XIV).

The confinement of discrimination to relatively few biomolecular parameters was similarly encountered in intensive care medicine for the determination of cell function parameters (Valet *et al.*, 1993; Rothe *et al.*, 1990) as well as for immunophenotyping, in particular in lymphomas (Valet and Höffkes, 1997), in the expression of thrombocyte surface antigens for myocardial infarction risk assessment (Valet *et al.*, 1993), in the prediction of the postcardiotomy syndrome in children (Tarnok *et al.*, 1997, 1999), and in the early detection of the overtraining syndrome in competition cyclists (Gabriel *et al.*, 1993, 1998; Valet *et al.*, 1993).

Considering the diversity of these diseases, the restriction of discrimination to a relatively narrow biomolecular parameter pattern seems to represent a more general rule. It comprises the potential for a significantly higher impact of predictive and diagnostic achievements for the individual patient at equal efforts. The advantage of data pattern analysis is that the discriminatory data pattern is provided in a standardized way, accessible to international efforts of consensus formation and optimization as evidence based medicine (EBM) at a cellular level.

The results of the immunophenotype classifications in asthma (Tables VIII and X) reemphasize the earlier observation of lymphoma immunophenotyping (Valet and Höffkes, 1997) showing that a significant amount of discriminatory information is localized on granulocytes or monocytes although the antibody panels are primarily selected for lymphocyte antigens. The reason for this seems to be either reactive adaptation of existing nonlymphocytic cell populations to the disease process or a reactively altered formation of cell populations by the hemopoietic organs.

Concerning the issue of whether the evaluation of percent cell frequency parameters is sufficient or whether the more complex quantitative analysis of antibody binding is required, Tables VI, VIII, X, and XIV clearly show that a substantial number of the discriminatory parameters are antibody intensity and antibody binding ratios. It seems therefore mandatory to routinely evaluate fluorescence intensities, fluorescence ratios, and in the future also coefficients of variations for all cell population parameters in flow cytometric histograms.

Backed by the information provided in this chapter and from earlier results, it seems clear that exhaustive information extraction from clinical multiparameter flow cytometry measurements in combination with discriminant data pattern analysis will constitute an important access route for disease course prediction at the individual patient level. Although the currently presented classification work concerns retrospectively prospective metaanalysis, it can be reasonably assumed that the classifiers will perform equally well in prospective studies. This hope is deduced from the observed robustness of classification of unknown samples in all the various studies performed up to now with the CLASSIF1 algorithm.

References

Beckman, R. J., Salzman, G. C., and Stewart, C. C. (1995). Classification and regression trees for bone marrow immunophenotyping. *Cytometry* **20,** 210–217.

Boddy, L., Morris, C. W., Wilkens, M. F., Tarran, G. A., and Burkill, P. H. (1994). Neural network analysis of flow cytometric data for 40 marine phytoplankton species. *Cytometry* **15,** 283–293.

Davey, H. M., Jones, A., Shaw, A. D., and Kell, D. B. (1999). Variable selection and multivariate methods for the identification of microrganisms by flow cytometry. *Cytometry* **35,** 162–168.

Decaestecker, C., Remmelink, M., Salmon, I., Camby, I., Goldschmidt, D., Patein, M., Van Ham, P., Pasteels, J. L., and Kiss, R. (1996). Methodological aspects of using decision trees to characterise Leiomyomatous tumors. *Cytometry* **24,** 83–92.

Demers, S., Kim, J., Legendre, P., and Legendre, L. (1992). Analyzing multivariate flow cytometric dara in aquatic sciences. *Cytometry* **13,** 291–298.

Diamond, L. W., Nguyen, D. T., Andreeff, M., Maiese, R. L., and Braylan, R. C. (1994). A knowledge-based system for the interpretation of flow cytometric data in leukemia and lymphomas. *Cytometry* **17,** 266–273.

Frankel, D. S., Olsen, R. J., Frankel, S. I., and Chisholm, S. W. (1989). Use of a neural net computer system for analysis of flow cytometric data of phytoplankton populations. *Cytometry* **10,** 540–550.

Frankel, D. S., Frankel, S. L., Binder, B. J., and Vogt, R. F. (1996). Application of neural networks to flow cytometry data analysis and real-time cell classification. *Cytometry* **23,** 290–302.

Gabriel, H., Valet, G., Urhausen, A., and Kindermann, W. (1993). Selbstlernende Klassifizierung durchflußzytometrischer Listendaten von immunphänotypisierten Lymphozyten bei akuter körperlicher Arbeit. *Deutsche Zschr. Sportmedizin* **44,** 461–465.

Gabriel, H., Urhausen, A., Valet, G., Heidelbach, U., and Kindermann, W. (1998). Overtraining and immune system: A prospective longitudinal study in endurance athletes. *Med. Sci. Sports Exerc.* **30,** 1151–1157.

Hokanson, J. A., Rosenblatt, J. I., and Leary, J. F. (1999). Some theoretical and practical considerations for multivariate statistical cell classification useful in autologous stem cell transplantation and tumor cell purging. *Cytometry* **36,** 60–70.

Molnar, B., Szentirmay, Z., Bodo, M., Sugar, J., and Feher, J. (1993). Application of multivariate, fuzzy set and neural network analysis in quantitative cytological examinations. *Anal. Cell. Pathol.* **5,** 161–175.

Leary, J. F. (1994). Strategies for rare cell detection and isolation. *In* "Methods in Cell Biology: Flow Cytometry" (Z. Darzynkiewicz, J. P. Robinson, and H. A. Crissman, eds.), 2nd Ed., Part B, Vol. 42. pp. 331–358. Academic Press, San Diego.

Otto, F. J., Oldiges, H., Göhde, W., and Jain, V. K. (1981). Flow cytometric measurementy of nuclear DNA content variation as a potential in vivo mutagenicity test. *Cytometry* **2,** 188–191.

Ravdin, P. M., Clark, G. M., Hough, J. J., Owens, M. A., and McGuire, W. L. (1993). Neural network analysis of DNA flow cytometry histograms. *Cytometry* **14,** 74–80.

Rothe, G., Kellermann, W., and Valet, G. (1990). Flow cytometric parameters of neutrophil function as early indicators of sepsis or trauma-related pulmonary or cardiovascular failure. *J. Lab. Clin. Med.* **115,** 52–61.

Schut, T. C. B., De Grooth, B. G., and Greve, J. (1993). Cluster analysis of flow cytometric list mode data on a personal computer. *Cytometry* **14,** 649–659.

Tarnok, A., Hambsch, J., Borte, M., Valet, G., and Schneider, P. (1997). Immunological and serological discrimination of children with and without post-surgical capillary leak syndrome. *In* "The Immune Consequences of Trauma, Shock and Sepsis" (E. Faist, ed.), pp. 845–849. Monduzzi Editore, Bologna.

Tarnok, A., Pipek, M., Valet, G., Richter, J., Hambsch, J., and Schneider, P. (1999). Children with post-surgical capillary leak syndrome can be distinguished by antigen expression on neutrophils and monocytes. *In* "Progress in Biomedical Optics, Proceedings Systems and Technologies for Clinical Diagnostics and Drug Discovery II" (G. E. Cohn and J. C. Owicki, eds.), SPIE Vol. 3603, pp. 61–71. Int. Soc. for Optical Engineering, Bellingham, WA.

Terstappen, L. W. M., Mickaels, R. A., Dost, R., and Loken, M. R. (1990). Increased light scattering resolution facilitates multidimensional flow cytometric analysis. *Cytometry* **11,** 506–512.

Thews, O., Thews, A., Huber, C., and Vaupel, P. (1996). Computer-assisted interpretation of flow cytometry data in hematology. *Cytometry* **23,** 140–149.

Valet, G., and Höffkes, H. G. (1997). Automated classification of patients with chronic lymphatic leukemia and immunocytoma from flow cytometric three colour immunophenotypes. *Cytometry (Commun. Clin. Cytometry)* **30,** 275–288.

Valet, G., Valet, M., Tschöpe, D., Gabriel, H., Rothe, G., Kellermann, W., and Kahle, H. (1993). White cell and thrombocyte disorders: Standardized, self-learning flow cytometric list mode data classification with the CLASSIF1 program system. *Ann. N.Y. Acad. Sci.* **677,** 233–251.

Valet, G., Roth, G., and Kellermann, W. (1998). Risk assessment for intensive care patients by automated classification of flow cytometric oxidative burst, serine and cysteine proteinase activity measurements using CLASSIF1 triple matrix analysis. *In* "Cytometric Cellular Analysis" (J. P. Robinson and G. Babcock, eds.), pp. 289–306. Wiley-Liss, New York.

Van Driel, B. E. M., Valet, G. K., Lyon, H., Hansen, U., Song, J. Y., Van Noorden, C. J. F. (1999). Prognostic estimation of survival of colorectal cancer patients with the quantitative histochemical assay of G6PDH activity and the multiparameter classification program CLASSIF1. *Cytometry (Commun. Clin. Cytometry)* **38,** 176–183.

Verwer, B. J. H., and Terstappen, L. W. M. M. (1993). Automatic lineage assignment of acute leukemias by flow cytometry. *Cytometry* **14,** 862–875.

PART XI

Microorganisms and Infectious Diseases

CHAPTER 53

Flow Cytometric Analysis
of Microorganisms

S. A. Sincock* and J. Paul Robinson*,†

*Purdue Cytometry Laboratories
Department of Basic Medical Sciences
School of Veterinary Medicine, and

†Department of Biomedical Engineering
Purdue University,
West Lafayette, Indiana 47907

I. Introduction
    A. Instrument Setup for Microbes
    B. Sample Preparation
II. Experimental Approaches
    A. Detection of Microbes
    B. Identification of Specific Microorganisms
    C. Cell Viability
    D. Identification of Viable Bacteria with Fluorescent *in Situ* Hybridization
    E. Gram Stain
III. Applications in Medical and Food Microbiology
    A. Antimicrobial Agents
    B. Food and Drink
IV. Conclusion
    References

I. Introduction

Conventional techniques (i.e., growth on laboratory media) employed for the detection and enumeration of microbes in clinical and environmental samples require time (24 to 48 hr), and they have a strong bias in that these methods detect only organisms that grow under a selected set of conditions. Problems with the current technology for microbial cell analysis led to development of alternative techniques that include flow cytometry. Flow cytometry allows rapid, multiparameter data acquisition and analysis of individual cells.

Although flow cytometry was rapidly accepted into hospital pathology and immunology laboratories, microbiology laboratories have remained essentially oblivious to the use of this technology. With few exceptions (Dubelaar *et al.*, 1999; Steen, 1980, 1983; Steen and Boye, 1980), flow cytometers were not designed to measure microorganisms, but rather mammalian cells in the range of 5 to 15 μm, the general size of blood cells. In practice the measurement of smaller particles, while possible, often required modifications to the instrument or a greater understanding of and interest in the technological aspects of cytometry than generally possessed by those with expertise in clinical microbiology. In addition, clinical microbiologists generally found the technology expensive and inappropriate for their cells of interest.

Improvements in the sensitivity and specificity of flow cytometric instrumentation have made possible a wide range of techniques to rapidly characterize microbial populations. More importantly, microbiologists have started to recognize the potential of flow cytometry to study the responses of individual cells in environmental and clinical samples and to report their findings. An excellent review describing applications of flow cytometry in the field of microbiology has been published (Davey and Kell, 1996).

This chapter discusses experimental approaches that have been or could be used to study individual microbial cells using flow cytometry and key factors that may impact these studies, including instrument setup, instrument operation, and sample preparation. A brief discussion of flow cytometric applications to the field of medical and food microbiology is also included.

A. Instrument Setup for Microbes

Flow cytometers designed specifically for small particles (i.e., Bio-Rad Bryte HS, Hercules, CA; Skatron, Oslo, Norway) are no longer commercially available, and technical support for existing instruments is limited. An ordinary flow cytometer optimized for mammalian cells can be adapted for microbial cell analysis with a few simple changes in instrument setup and operation. For example, sheath fluid, sample buffer, media used to grow bacteria, and other reagents (i.e., dyes, antibodies) must be filtered (0.2-μm filter or smaller) to remove any particles that could interfere with bacterial measurements. The laboratory water system used to prepare sample buffer and sheath fluid should also be rigorously cleaned and maintained to prevent bacterial contamination.

Daily quality control procedures should include the instrument alignment beads recommended by the manufacturer and latex beads of size similar to that of the microbe of interest (1.0, 1.5, 2.0, 4.0, 6.0 μm). Because small latex beads can give a scatter signal quite different from that of bacteria of similar size, ethanol- or heat-fixed vegetative cells (i.e., *Escherichia coli*) or unfixed spores in water (i.e., *Bacillus subtilis*) should also be used as an internal laboratory standard to check light scatter parameters. Fixed cells or spores can be stored at 4°C for up to 6 months.

Initial light scatter parameters should be established using target microorganisms spiked with latex beads. For example, in Fig. 1 *Bacillus subtilis* cells were spiked with a small number of 1.0-μm beads. Bacteria and beads in the spiked sample were separated using a dual-parameter histogram of log forward scatter (FS) and log side scatter (SS). A region was established for the bacterial population and used as a gating parameter to exclude cell aggregates and debris from further analysis. Sterile, filtered sample buffer was used to set the discriminator or threshold on forward light scatter to eliminate background particles.

In order to reduce the risk associated with analyzing potentially hazardous microorganisms, certain protective measures should be followed and strictly enforced. In particular, the protective doors that shield the instrument sample probe should be kept closed to reduce aerosolization of bacterial particles, bleach should be added to the waste container to kill any harmful organisms, and personal protective gear (i.e., gloves, mask, laboratory coat) should be worn at all times. Laboratory personnel should also avoid contaminating the computer keyboard and mouse with bacteria. Instrument maintenance should include frequent flushing of the system between samples to reduce instrument carryover of bacteria and dyes and rigorous shutdown/cleaning procedures.

B. Sample Preparation

Biological characteristics of bacteria such as size, shape, DNA, RNA, and protein content can change depending on growth conditions and cell source. For example, exponentially growing cells are larger than dormant or starved cells and contain considerably higher levels of nucleic acids. Growing cells have a wide light scatter distribution with a cometlike tail in the direction of increasing

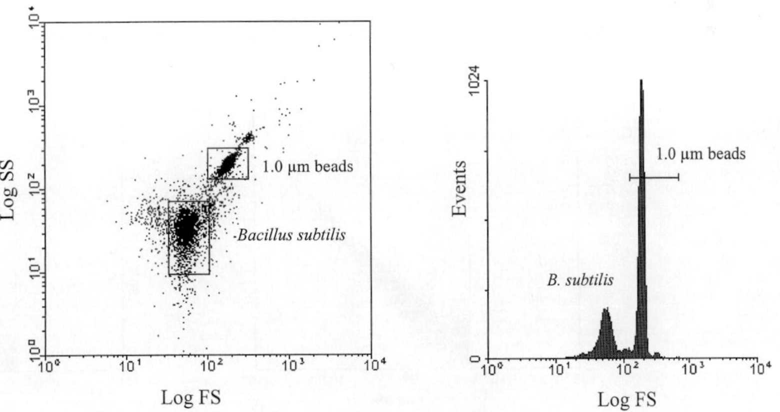

Fig. 1 Light scatter measurements of a mixture of *Bacillus subtilis* cells and latex beads. Fluorescent microspheres (1.0 μm) were added as an internal standard.

scatter (Thomas *et al.*, 1997). Prior to flow cytometric analysis, growing cells should be washed with sterile, filtered buffer to remove debris and reduce cell clumping (Fig. 2). A washing step will also remove medium that may interfere with staining. Bacteria can grow as single cells or in pairs, chains, or clusters. Gentle pipetting or vortexing may be necessary to disrupt the chains or clusters and form a single cell suspension.

Some bacteria have considerable permeability barriers (i.e., cell walls, endospores, capsules, efflux pumps) to fluorescent dyes or DNA probes and may require use of fixatives or EDTA. However, sample preparation methods necessary for efficient penetration of a fluorochrome into target cells may significantly affect light scatter profiles. For example, alcohol fixation can cause considerable cell shrinkage and a reduction in cell size.

II. Experimental Approaches

The basic problem in developing flow cytometric protocols for microbial cell analysis is the assumption that procedures developed and optimized for mammalian cells will work for bacteria. In some cases, ignorance of traditional flow methods is an advantage; however, the fundamentals of microbiology must always be understood. In this chapter, we have outlined a few experimental approaches for using flow cytometry to study microbial cells. These approaches included generic detection of microorganisms, specific identification of target organisms, cell viability determinations, and Gram staining.

A. Detection of Microbes

Nucleic acid dyes can be combined with light scatter measurements to detect bacteria using flow cytometry. A detailed discussion of bacterial DNA appears

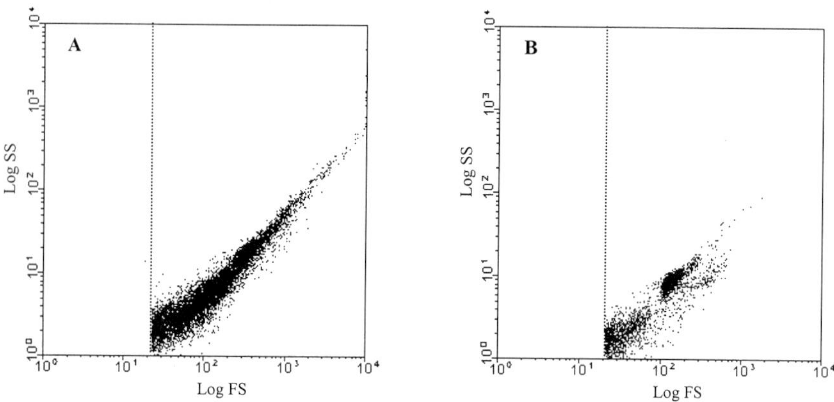

Fig. 2   Changes in light scatter profiles of bacteria due to sample preparation: (A) *Bacillus subtilis* spore slurry in water and (B) washed spores.

in Chapter 54 of this volume. The dye selected for a detection assay should have a high specificity for DNA binding, high extinction coefficient, and high quantum yield. Depending on the available excitation source, 4′,6-diamidino-2-phenylindole (DAPI), Hoechst 33258, propidium iodide, YO-PRO-1, or YOYO-1 could be used for a rapid detection assay. YOYO-1 (Molecular Probes, Inc., Eugene, OR) is a membrane impermeant cyanine dye (excitation 491 nm, emission 509 nm) that is essentially nonfluorescent unless bound to nucleic acids. Dyes that are membrane impermeant will stain only cells that are dead or have compromised membranes. Live cells must be fixed for the dye to pass through the membrane. Rapid fixation with ice-cold 70% ethanol will ensure that the selected dye will enter all cells in the sample and bind to nucleic acids. Alcohol fixation will cause some cell shrinkage and prevent further studies regarding cell viability.

Figure 3 is an example of a rapid detection assay. A "bacteria" region (region F) was created using *E. coli* cells fixed with ice-cold 70% ethanol. Cells were washed briefly with filtered 0.8% NaCl, stained with 0.1 μM solution of YOYO-1 for 5 min in the dark, and then analyzed using flow cytometry (Sincock *et al.*, 1996a). YOYO-1 stained *E. coli* cells were gated on region F; the fluorescence of the gated population was then measured and displayed as a histogram with fluorescence intensity on the *x*-axis and the number of cells on the *y*-axis. The background noise was determined using 0.2 μm-filtered 0.8% NaCl.

Samples containing dust, pollen, fungal spores, or unknown bacteria were tested with this assay. Particles in the test samples that met the light scatter requirements (bacteria region) and stained positive for nucleic acids were classified as bacteria. Although pollen, mold, and fungal spores contain nucleic acids and will stain with YOYO-1, they do not meet the light scatter gating require-

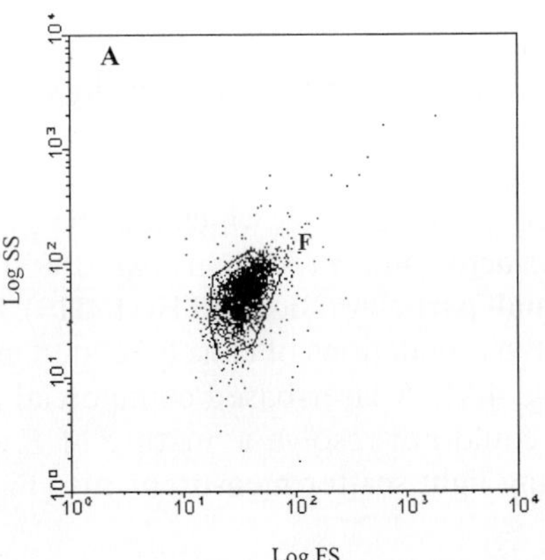
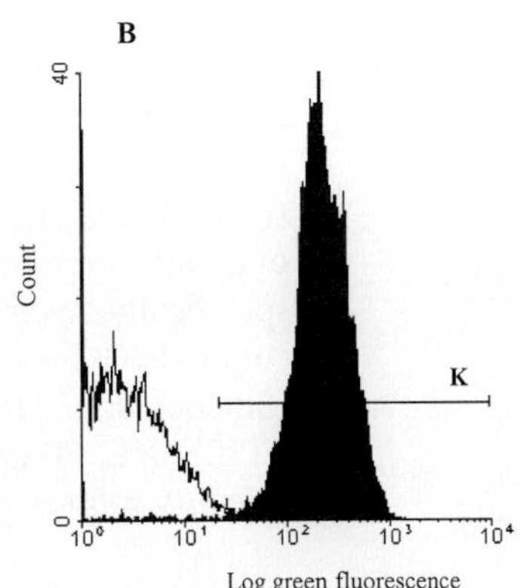

Fig. 3 Detection of *E. coli* cells in environmental samples using YOYO-1 nucleic acid stain. (A) Light scatter measurements of ethanol-fixed *E. coli* cells (region F). (B) Fluorescence histogram overlay of YOYO-1 stained (■) and unstained *E. coli* cells (□).

ments due to their large size (20–100 μm) and therefore can be excluded. YOYO-1 will not stain dust particles that fall within the bacteria region because they do not contain nucleic acids.

This assay can be used for generic detection of bacteria within a heterogeneous sample but cannot be used for specific identification. In mixed populations of bacteria (i.e., *E. coli* combined with *Staphylococcus aureus*), it was not possible to discriminate between bacteria using light scatter or by using differences in relative staining.

B. Identification of Specific Microorganisms

During the 1990s, experimental approaches to identifying microorganisms in liquid samples using flow cytometry have included light scatter profiles, DNA content, immunoassays, neural nets, and rRNA probes, with varying degrees of success. Some of these areas are discussed in detail in other chapters of this volume (Chapters 54, 55); however, a brief introduction to this material follows.

1. Light Scatter Measurements

Light scattering profiles are a function of cellular size, shape, and refractive index of a cell. Morphological features of bacteria that can influence light scatter profiles include shape (rods, cocci, vibrios, spirilla, spirochetes), flagella, pilli, and capsules. Growth conditions, cell source, and responses to stress (i.e., starvation, antimicrobial exposure) can also influence light scatter profiles.

Light scatter profiles are a useful first step in characterizing microorganisms. Identifying specific organisms within mixed populations is difficult. Allman *et al.* (1993) collected dual-parameter contour plots of forward versus side scatter for artificial mixtures of clinically relevant microorganisms using an arc lamp-based cytometer. Mixtures of vegetative cells (i.e., *Salmonella typhimurium, Legionella pneumophila, Staphylococcus aureus*) had overlapping light scatter profiles. However, light scatter profiles could be used to resolve spore-forming bacteria (i.e., *Clostridium perfringens*) from vegetative cells. Spores give a forward light scatter signal that is out of proportion to their size, which may be explained on the basis of a high value for their refractive index (Allman *et al.*, 1993). Using a cytometer specifically designed for small particles (Bio-Rad Bryte HS), light scatter profiles could also be used to resolve populations of closely related gram-positive spores (Sincock *et al.*, 1996b) (Fig. 4A). A laser-based commercial cytometer (Coulter EPICS XL, Hialeah, FL) could not resolve a mixture of *E. coli, S. aureus,* and *Bacillus subtilis* spores using light scatter measurements (Fig. 4B).

2. DNA Content

Using flow cytometry, DNA base composition of individual cells within a bacterial sample can be determined without extraction of DNA. Van Dilla *et al.*

A

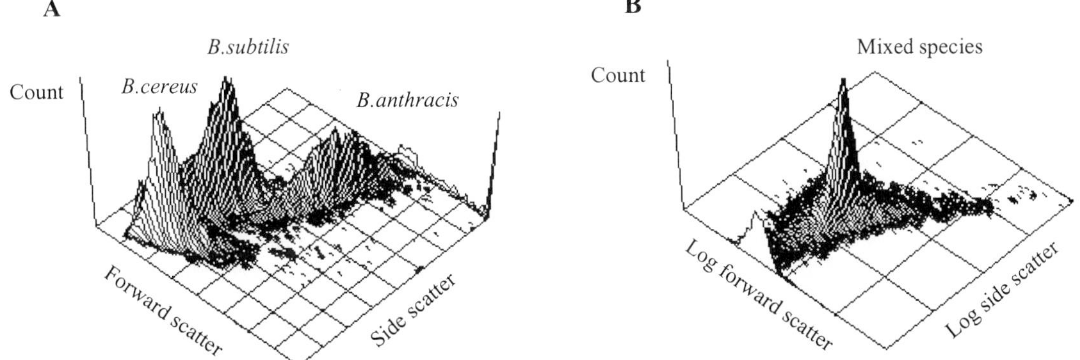

Fig. 4 Isometric plots of forward scatter versus side scatter for artificial mixtures of bacterial species: (A) mixture of three closely related *Bacillus* spores (Bio-Rad Bryte flow cytometer) and (B) mixture of *E. coli* and *S. aureus* and *Bacillus subtilis* spores (Coulter EPICS XL flow cytometer).

(1983) used a combination of DNA-specific fluorochromes to analyze six species of ethanol-fixed bacteria with differing DNA base composition. Using a combination of Hoechst 33258, a fluorochrome that binds preferentially to the regions of DNA rich in AT base pairs, and chromomycin A3, a fluorochrome that binds preferentially to regions of DNA rich in GC base pairs, this group established a direct relationship between the fluorescence dye ratio calculated by flow cytometry and the % Guanine + Cytosine (%[G + C]) content. This method was able to resolve individual species within an artificial mixture of *Staphylococcus aureus*, *Escherichia coli*, and *Pseudomonas aeruginosa* based on differences in DNA content. Each species formed a distinct cluster within the dual-parameter fluorescence histogram. However, further studies using this method suggested that flow cytometric determination of %[G+C] be limited to samples containing only one bacterial species (Sanders *et al.*, 1990).

3. Immunofluorescence Approach

Fluorescently labeled antibodies combined with light scatter measurements can be used for the specific identification of microorganisms. Microbes that have been identified using a flow cytometric immunoassay include pathogenic microorganisms found in food, water, sewage, and aerosols (Table I). An example of a direct flow cytometric immunoassay can be found in Fig. 5. *Escherichia coli* O157:H7 cells at a concentration of 10^6 cells/ml were incubated with a fluorescein isothiocyanate (FITC)-conjugated rabbit anti-*E. coli* O157:H7 polyclonal antibody for 5 min at room temperature and analyzed by flow cytometry. For this assay, the desired population of cells was selected by gating on light scatter signals. A discriminator was set on forward scatter and used to resolve bacteria from noncellular material and electronic noise. The fluorescence of the gated

Table I

Examples of Microbes Identified Using a Flow Cytometric Immunoassay

Microbe	References
Food	
Escherichia coli O157:H7	Seo *et al.* (1998a,b); Tortorello *et al.* (1998)
Listeria monocytogenes	Pinder and McClelland (1994); Donnelly and Baigent (1986)
Salmonella typhimurium	Clarke and Pinder (1998); Pinder and McClelland (1994); McClelland and Pinder (1994b)
Salmonella serotypes	McClelland and Pinder (1994a)
Oral Bacteria	
Streptococcus mutans & *Actinomyces viscosus*	Barnett *et al.* (1984)
Streptococcus pyogenes	Sahar *et al.* (1983)
Dental plaque	Obernesser *et al.* (1990)
Aerosols	
Francisella tularensis	Henningson *et al.* (1998)
Water and sewage	
Legionella pneumophila	Ingram *et al.* (1982)
Nitrosomonas serotypes	Volsch *et al.* (1990)
Salmonella spp.	Desmonts *et al.* (1990)
Fecal bacteria	Apperloo-Renkema *et al.* (1992); van der Waaij *et al.* (1994)
Cryptosporidium parvum	Vesey *et al.* (1993, 1997); Valdez *et al.* (1997); Arrowood *et al.* (1995)
Giardia spp.	Dixon *et al.* (1997); Bruderer *et al.* (1994) Heyworth and Pappo (1989)
Biowarfare Agents	
Bacillus anthracis	Sincock *et al.* (1996b); Phillips and Martin (1983, 1988)
Cell surface polysaccharides or proteins	
Bacteroides fragilis	Lutton *et al.* (1991)
E. coli lipopolysaccharide expression	Nelson *et al.* (1991)
Myxococcus virescens	Martinelli *et al.* (1995)
Pseudomonas aeruginosa outer membrane protein	Hughes *et al.* (1996)
Microsphere-based immunoassays	
Helicobacter pylori	Best *et al.* (1992)
E. coli O157:H7	Seo *et al.* (1998a,b)

population was then measured and displayed as a histogram with fluorescence intensity on the *x*-axis and the number of cells on the *y*-axis (5000 counts). Target *E. coli* O157:H7 cells were identified and enumerated within a few minutes of obtaining the sample. Culturing of the target organism was not necessary for identification in this direct immunoassay.

Low numbers of target organisms can be identified in the presence of large numbers of nontarget organisms or high levels of background particulate material

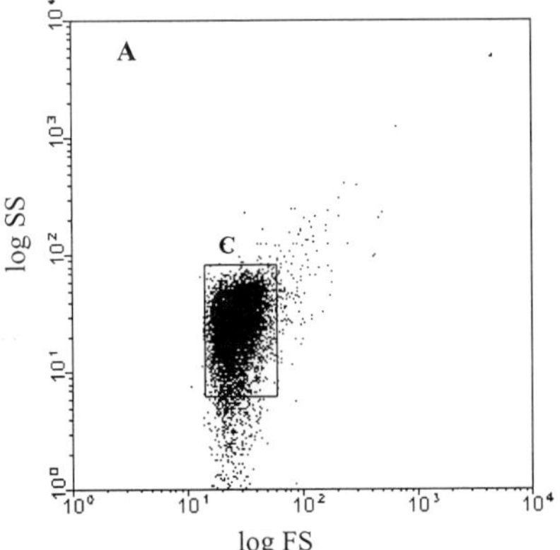
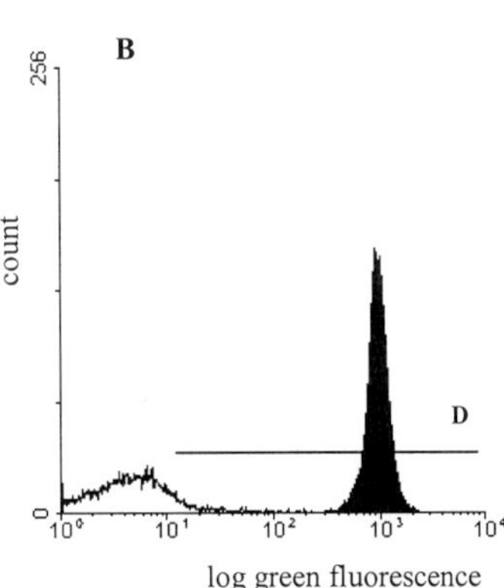

Fig. 5 (A) Dual-parameter histogram of forward versus side scatter for live *E. coli* O157:H7 cells stained with a FITC-labeled anti-*E. coli* O157:H7 polyclonal antbody. (B) Comparison of stained (■) and unstained (□) *E. coli* O157:H7 cells. Region D represents bacteria stained positive with FITC-labeled antibody. Cells were gated on region C (5000 events).

using nonselective media. Unknown samples are incubated in nonselective media for a short period of time, washed with buffer, and then stained with a specific antibody. Flow cytometry can be used to discriminate target organisms from nontarget organisms by means of specific antibody binding. Enrichment media specific to the nutritional requirements of the target organism can also be used. Ideally, only the target organism will grow. Nutritional supplements can also be used to facilitate expression of specific polysaccharides on bacterial cell surfaces that can be used to discriminate between closely related species. For example, viable *Bacillus anthracis* spores were identified after a brief incubation (20 min, 37°C) in media selected to stimulate the outgrowth and expression of specific polysaccharides on the surface of vegetative cells. After exposure to the food source, *Bacillus anthracis* spores were able to sporulate and transition to vegetative cells. Cells were stained with FITC-conjugated monoclonal antibody specific for *B. anthracis* cell wall polysaccharide and analyzed using flow cytometry (Fig. 6) (Sincock *et al.*, 1996b).

Because cell fixation is not necessary for antibody binding, the immunofluorescence approach can be combined with certain stains to identify viable target organisms. For example, the survival ratio of *Francisella tularensis,* the causative agent of tularemia, was determined before and after aerosolization using a specific anti-*F. tularensis* monoclonal antibody to identify the target organisms together with rhodamine 123 to count the number of viable or metabolically active cells (Henningson *et al.*, 1998). In a second example, Red613-conjugated anti-*Salmonella typhimurium* monoclonal antibody combined with Chemchrome, a live cell stain, was used to detect viable *Salmonella typhimurium* cells in the presence of large number of nontarget and dead organisms (Clarke and Pinder, 1998).

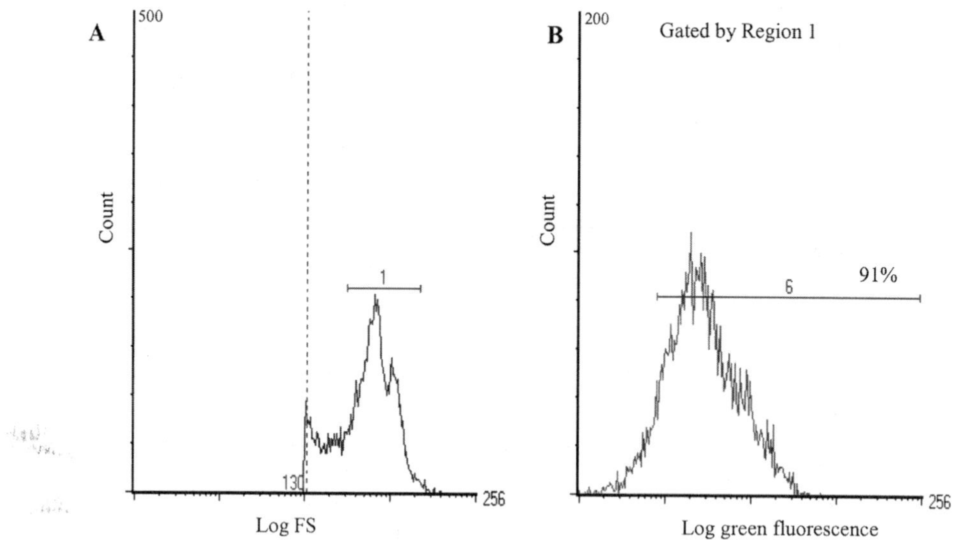

Fig. 6　Flow cytometric analysis of *B. anthracis* spores incubated in polysaccharide media for 30 min at 37°C. A specific FITC-conjugated anti-cell wall polysaccharide MAb stained emerging vegetative *B. anthracis* cells (region 6) (viable cells) but did not stain encapsulated *B. anthracis* cells, dormant *B. anthracis* spores, or other *Bacillus* species. Forward light scatter was used to identify vegetative cells population in the sample (region 1). (See color plates.)

Flow cytometric immunoassays are rapid (usually less than 10 min), sensitive (lower limit of 10^3 cells/ml), specific, require very little sample preparation, and need no cell fixation. Immunoassay-based methods rely on the specificity and sensitivity of the selected antibody to identify the target organism. Unfortunately, very few good antibodies for microbes are commercially available, and in-house antibody development can be time consuming and expensive. High levels of nonbacterial particles, bacterial debris, and antibody aggregates in the test sample can also produce sensitivity problems.

4. Automated Classification and Identification Techniques

Flow cytometry can be used to generate multiparameter data for individual cells. However, the vast quantity of information generated can make data analysis difficult. Artificial neural networks are computing technologies that can be used to discriminate between different cell types based on flow cytometry data (Boddy and Morris, 1993). A computer is "taught" how to recognize data patterns (i.e., staining profiles of different organisms) and to analyze cell populations using examples. Eventually, the neural network can identify specific cell types in real time and adapt to changing conditions (Frankel *et al.*, 1989). Artificial neural networks have been developed for chromosome classification (Errington and

Graham, 1993), leukemia subsets, (Maguire *et al.*, 1994a,b), and phytoplankton populations (Frankel *et al.* 1989). Davey *et al.* (1999) developed an artificial neural network for detection and identification of *Bacillus globigii* spores against a background of other microorganisms (*Escherichia coli, Micrococcus luteus, Saccharomyces cerevisiae*). Data sets were collected for microorganisms stained with six cocktails of fluorescent stains. These stains included Tinopal CBS-X, Nile Red, propidium iodide, FITC, DiSC$_2$(5), Oxonol V, SYTO 17, and TO-PRO-3. Forward scatter, side scatter, and autofluorescence measurements were also included in the data sets. Careful selection of the staining cocktail and data analysis method allowed accurate identification of the target organism (*Bacillus* spores). Trained neural networks may be useful in identifying specific organisms against a high background of particulate matter or discriminating between closely related organisms in real time. Applications may include food analysis, clinical microbiology samples, and identification of biowarfare agents.

C. Cell Viability

Fluorescent dyes have been successfully used as indicators of cell viability in fluorescence microscopy and flow cytometry. Using these dyes, live and dead cells within a heterogeneous sample population can be identified and counted within a few minutes. Traditional methods employed to detect and enumerate bacteria (such as growth on laboratory media) require time (24 to 48 hr) and may underestimate the number of viable bacteria. Therefore, direct methods for the assessment of microbial viability are of increasing importance. Because each technique has its limitations, each investigator must choose the experimental approaches that are best suited for the test organisms and the specific questions being asked.

1. Membrane Integrity

Membrane integrity analysis is based on the capacity of bacterial cells to exclude certain compounds. Stains that are commonly used to determine membrane integrity include ethidium bromide, propidium iodide, and SYTOX Green dead cell stain. These dyes passively enter stressed, injured, or dead cells via damaged membranes and intercalate into DNA and RNA. The fluorescence indicates a loss of viability or membrane integrity. Flow cytometry can be used to quantify the fluorescence associated with dead or injured cells. Because the influx of the dye can be correlated with the extent of the bacterial wall permeability, the number of fluorescent cells counted using flow cytometry is inversely proportional to the number of viable cells. These dye exclusion methods have been successfully used to monitor antibiotic-induced changes in bacterial membrane permeability (Gant *et al.*, 1993). For example, the oral pathogen *Streptococcus mutans* was treated with the antibiotic clindamycin and then stained with

SYTOX Green (Fig. 7). After 2 hr of exposure to the antibiotic, a significant number of cells were dead, as indicated by strong green fluorescence.

Membrane integrity analysis is not suitable for all cell types because some bacteria can rapidly pump out dyes using an efficient efflux pump (Jernaes and Steen, 1994). In this case, damaged or injured cells would not fluoresce and would be counted as viable.

2. Membrane Potential

Membrane potential analysis is based on the selective permeability and active transport of charged molecules through intact membranes. Cells with a membrane potential actively take up lipophilic, cationic dyes or actively exclude lipophilic, anionic dyes. Using flow cytometry, any particle in the approximate size range of bacteria that is found to have a membrane potential can be identified as a viable organism. However, organisms can show considerable variation in dye uptake due to differences in membrane potential (Allman *et al.*, 1993).

Using the lipophilic cation rhodamine 123, which preferentially accumulates within viable cells, several groups have been able to discriminate between live, dead, and dormant cells in culture. Viable and nonviable cells have been enumerated using flow cytometry (Kaprelyants and Kell, 1992, 1993a,b; Kaprelyants *et al.*, 1993). Studies using this dye have determined that dye uptake is variable both between species and among cells from the same culture (Porter *et al.*, 1995). In addition, this dye can be used for gram-negative bacteria only after they have been treated with EDTA (Diaper *et al.*, 1992).

In contrast to rhodamine 123, the lipophilic oxonol dyes are anionic and preferentially accumulate within dead bacteria; they have been used to assess

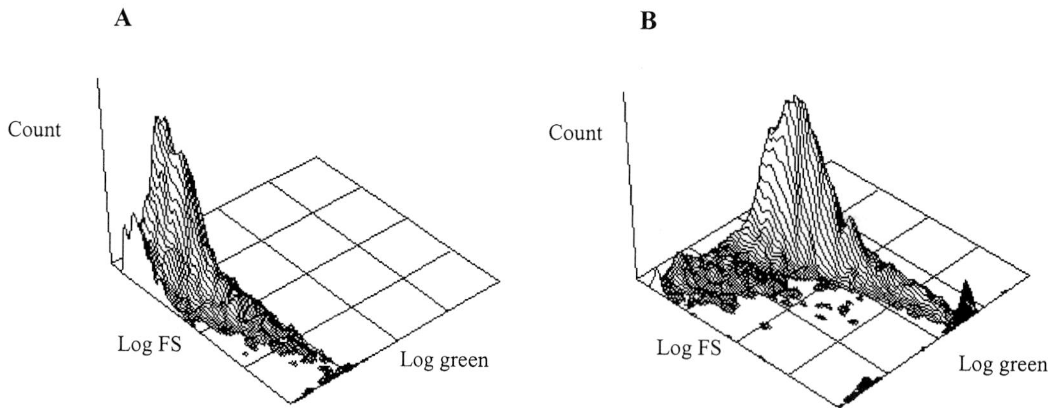

Fig. 7 Isometric plots of log green fluorescence versus log forward scatter for *Streptococcus mutans* cells stained with SYTOX Green after exposure to clindamycin (10× MIC) for (A) 0 hr and (B) 2 hr.

bacterial antibiotic susceptibility (Deere *et al.*, 1995; Mason *et al.*, 1995b) and cell viability (Jepras *et al.*, 1995; Mason *et al.*, 1995a) by flow cytometry. In these studies, either heat or bactericidal antibiotics were used to kill cells prior to oxonol staining, and comparisons were made with untreated cells. Figure 8 is a fluorescence histogram overlay of *E. coli* cells treated with gentamicin at 10 times the mimimum inhibitory concentration (10× MIC) and then stained with bis(1,3-dibutylbarbituric acid) trimethine oxonol [DiBAC$_4$(3)]. Over time, the number of dead cells increased as indicated by an overall shift in green fluorescence.

3. Enzymatic Activity

Flow cytometric detection of intracellular enzymatic activity utilized lipophilic, uncharged, nonfluorescent derivatives such as fluorescein diacetate (FDA) that readily diffuse across cell membranes. Once inside the cell, the derivative is hydrolyzed by nonspecific esterases to release the highly fluorescent parent compound. Because the parent compound is polar and charged, it is retained inside the cells with intact membranes. Dead or dying cells with compromised membranes rapidly leak the dye.

Flow cytometry can be used to detect the number of viable bacteria and to verify the metabolic activity of these cells (Diaper and Edwards, 1994; Diaper *et al.*, 1992). However, FDA does not efficiently penetrate some types of membranes and the fluorescein product tends to leak from cells or can be actively pumped out (Edwards, 1996). Other related fluorescent compounds such as

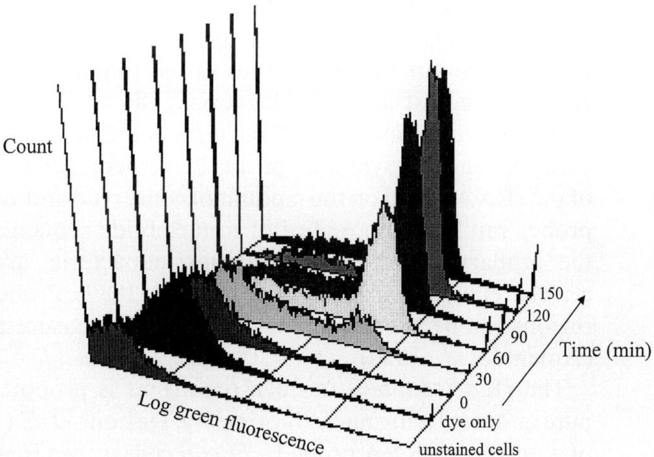

Fig. 8 Fluorescence histogram overlays of *E. coli* cells stained wiith DiBAC$_4$(3) (oxonol) after exposure to 10× MIC gentamicin. Cells were stained with oxonol after 0, 30, 60, 90, 120, or 150 min of drug incubation.

carboxyfluorescein diacetate (CFDA) and sulfofluorescein diacetate (SFDA) exhibit similar problems (Tsuji *et al.*, 1995).

4. Bacterial Respiration

The redox dye 5-cyano-2,3-ditolyl tetrazolium chloride (CTC) was first employed for the direct microscopic enumeration of respiring bacteria in environmental samples. CTC is readily reduced via electron transport activity to insoluble, highly fluorescent, and intracellularly accumulated CTC–formazan through bacterial respiration. Actively respiring bacteria (red fluorescence) can be distinguished from nonrespiring bacteria and abiotic material. More recent studies have used flow cytometry to enumerate respiring bacteria in lakes (del Giorgio *et al.*, 1997), marine systems (López-Amorós *et al.*, 1998), and after exposure to antibiotics (Suller and Lloyd, 1999).

However, several problems associated with using CTC have been identified. The CTC assay may not be sensitive enough to detect low respiration rates of microorganisms, especially in very small bacteria. In addition, not all bacteria are able to reduce tetrazolium salts. It is also thought that CTC may have an inhibitory effect on bacterial metabolism (Ullrich *et al.*, 1996; Yu *et al.*, 1995).

D. Identification of Viable Bacteria with Fluorescent *in Situ* Hybridization

Staining with membrane-integrity or membrane-potential fluorochromes offers limited information on the numbers of viable bacterial cells within a sample and none at all about their identity. Staining with fluorochromes that preferentially bind to specific DNA base pairs offers limited information on species identification but not cell viability. Fluorescent *in situ* hybridization can be used to label specific nucleic acid sequences inside intact, viable cells and identify species of bacteria present in the sample. Probe binding to ribosomal RNA (rRNA) is perhaps the best target for bacterial cells.

rRNA can be found in all bacteria and consists of both highly conserved and variable regions. Synthetic probes have been developed that can target sections of the rRNA based on the amount of conserved and variable regions. Appropriate probes can be composed of oligonucleotide sequences that distinguish between the primary kingdoms (eukaryotes, eubacteria, archaebacteria) and between closely related organisms (DeLong *et al.*, 1989). Probes that target very conserved regions can be used as universal probes to measure total rRNA within a sample (Amann *et al.*, 1990).

The rRNA content of microorganisms is proportional to the growth rate in pure culture. Using microfluorimetry, DeLong *et al.* (1989) quantified the binding of a universal rRNA probe to *E. coli* cells grown in media that support different growth rates. The fluorescence intensity of single cells due to hybridization with the universal probe varies linearly with growth rate and can be used to estimate the growth rate of that particular organism in a natural population. Further studies conducted by Wallner *et al.* (1993) demonstrated that 16S rRNA probe-conferred fluorescence is directly proportional to ribosome content. Because the amount of fluorescence can be correlated with cellular rRNA content, it is possible to obtain information on the physiological state (i.e., growth rate, activity, viability) of specific bacterial cells (Manz *et al.*, 1993; Wallner *et al.*, 1993). Due to the abundance of cellular ribosomes in rapidly growing cells (approximately 10^4 to 10^5 per cell), the binding of fluorescent probes to individual cells can be readily visualized (DeLong *et al.*, 1989).

After appropriate selection, rRNA-targeted oligonucleotides can be sequenced, labeled with an appropriate fluorochrome, and used as probes in hybridization experiments. After hybridization, the fluorescence conferred by rRNA-targeted oligonucleotide probes can be analyzed by flow cytometry (Rice *et al.*, 1997; Thomas *et al.*, 1997; Simon *et al.*, 1995; Lange *et al.*, 1997; Wallner *et al.*, 1993, 1995; Amann *et al.*, 1990) or confocal microscopy (Amann *et al.*, 1996).

E. Gram Stain

Gram staining is the most commonly used procedure in clinical microbiology laboratories. Specimens are smeared on glass slides, heat fixed, Gram stained, and examined microscopically. Based on the outcome of the Gram reaction, bacteria are divided into two taxonomic groups. Cells stained purple-blue are gram-positive; cells stained red are gram-negative. This technique is relatively simple, albeit messy. However, some organisms can show gram variability (i.e., *Acinetobacter* species), particularly anaerobes.

Sizemore *et al.* (1990) reported on the use of a fluorescently labeled lectin as an alternative Gram staining technique. Lectin isolated from *Triticum vulgaris*, or wheat germ agglutinin (WGA), will bind specifically to *N*-acetylglucosamine in the outer peptidoglycan layer of gram-positive bacteria. Gram-negative bacteria have an outer membrane covering the peptidoglycan layer that prevents lectin binding. Using this method, heat-fixed bacterial smears were covered with a small aliquot of FITC-conjugated WGA (100 μg/ml), washed briefly with phosphate buffer, and observed using fluorescence microscopy. Unlike the Gram staining method, culture age did not affect lectin binding, suggesting that this technique can be used directly on samples without culturing and may offer an alternative method to classify fastidious, slowly growing, or viable but nonculturable organisms. In theory, flow cytometry could be used to extend this technique.

Flow cytometry has been used to determine the Gram stain of unfixed cells using $DiIC_1(5)$ (Shapiro, 1995) or rhodamine 123 (Allman *et al.*, 1993). More recently, Mason *et al.* (1998) developed a two-color flow assay for mixed populations of bacteria in suspension. Bacterial strains isolated from clinical specimens were cultured overnight, washed, and then stained with a combination of fluorescent nucleic acid-binding dyes hexidium iodide (excitation 488 nm, emission 605 nm) and SYTO 13 (excitation 488 nm, emission 509 nm). Hexidium iodide (HI) preferentially penetrates gram-positive bacteria, whereas SYTO 13 enters both gram-positive and gram-negative bacteria. When used in combination, these dyes allow differential labeling of unfixed gram-positive bacteria (HI and SYTO 13, red-orange fluorescence) and gram-negative bacteria (SYTO 13 only, green fluorescence) in suspension (Mason *et al.*, 1998). Using this method, artificial mixtures of *E. coli* and *S. aureus* cells analyzed using flow cytometry were clearly separated using fluorescence. Total time needed for this assay was 15 min.

III. Applications in Medical and Food Microbiology

To date, the most frequent application of flow cytometry to the study of microorganisms is the field of environmental microbiology, where rapid assessment of bacterial viability in natural samples is important. The rapid methods first described in these studies have been adapted for use in medical and food microbiology. In these areas, flow cytometry can significantly shorten the analysis time required for detection and identification of bacteria compared with conventional detection procedures and provide additional information on responses of individual cells.

A. Antimicrobial Agents

Flow cytometry permits rapid analysis of individual bacterial, fungal, or protozoan responses to antimicrobial agents. Antimicrobial agents such as antibiotics, disinfectants, and antiseptics are used to reduce the number of microorganisms to a level that is insufficient to transmit infection. Antibiotics are products of the metabolism of a microorganism that are inhibitory to other microorganisms. Disinfectants are chemical or physical agents used to kill pathogenic microorganisms on nonliving objects (i.e., sink, table); antiseptics are chemicals used to kill microbes on a living object (i.e., skin, mouth). Flow cytometry can be used to investigate physiological and morphological changes that can occur after drug exposure, even if little is known about a particular antimicrobial agent.

1. Exposure to Antibiotics

Clinical microbiology laboratories devote a great deal of resources to antibiotic susceptibility testing. Routine analysis is limited to growth inhibition assays using fast growing, nonfastidious bacteria. Flow cytometry can supply valuable additional information on the response of individual cells to antibiotic exposure within a short period of time and provide an indication of population dynamics within the heterogeneous test sample. For example, gentamicin was added to early exponential phase *E. coli* cells in broth and incubation was allowed to continue for 5 hr. Untreated *E. coli* cells were used as controls. At timed intervals, aliquots of treated and untreated cells were removed, stained, and analyzed. Membrane perturbation was assessed using the membrane potential-sensitive dye

DiBAC$_4$(3) and the membrane integrity dye propidium iodide. Dual-parameter histograms of log forward scatter versus log fluorescence suggest that membrane potential of the treated cells collapsed after 5 hr; however, a subpopulation of treated cells maintained membrane integrity (Fig. 9).

Table II is a brief summary of work using flow cytometry to investigate the effect of antibiotic and antifungal agents on target organisms. Procedures for antibiotic susceptibility testing using flow cytometry are described in detail in Chapter 55 of this volume.

2. Exposure to Disinfectants or Antiseptics

Traditional assessment of disinfectant efficacy involves the incubation of microbes in liquid or on solid media for 24 to 48 hr. Most bacteria will not grow in the presence of low concentrations of disinfectants. To avoid this inhibitory effect, disinfectant compounds must be inactivated or neutralized before treated cells are incubated in media or plated. In addition, some cells will experience a

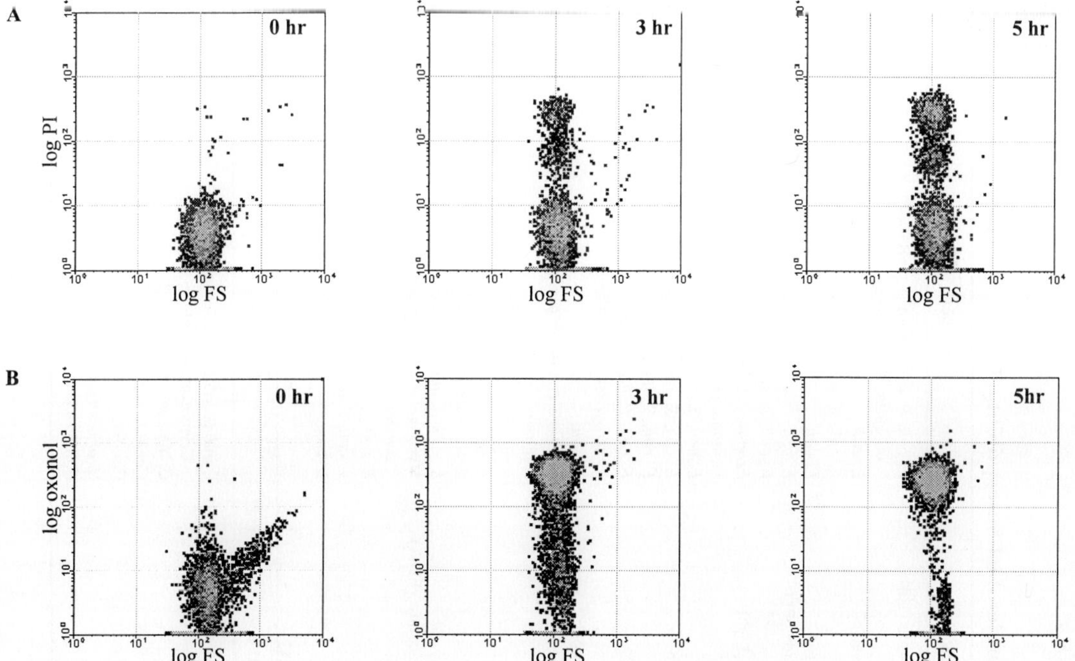

Fig. 9 Dual-parameter density dot plots of log forward scatter versus log fluorescence for *E. coli* cells stained with (A) propidium iodide (PI) or (B) DiBAC$_4$(3) (oxonol), after exposure to gentamicin for 0, 3, and 5 hr.

Table II
Rapid Antimicrobial Susceptibility Testing Using Flow Cytometry

Bacteria: Dye	Species	Antibiotic	References
Ethidium bromide (EtBr)	*E. coli, P. aeruginosa, S. aureus, P. mirabilis, S. pyogenes*	Amikacin	Cohen and Sahar (1989)
EtBr + mithramycin	*E. coli*	Ceftazidime, ciprofloxacin, gentamicin	Walberg *et al.* (1997a)
	E. coli, K. pneumoniae	Ampicillin	Walberg *et al.* (1997b)
Acridine orange	*E. coli*	Gentamicin	Mason and Lloyd (1997)
Propidium iodide (PI)	*E. coli*	Gentamicin, mecillinam, cefotaxime, ampicillin, ciprofloxacin	Gant *et al.* (1993)
	E. coli, P. aeruginosa	Ampicillin, ceftriaxone, ciprofloxacin, rifampin, imipenem	Gottfredsson *et al.* (1998)
Live/Dead BacLight	*Propionibacterium acnes*	Lymecycline, minocycline	Arrese *et al.* (1998)
	L. monocytogenes	Bacteriocin	Swarts *et al.* (1998)
	E. coli, S. aureus, P. aeruginosa	Ceftazidime, ampicillin, vancomycin	Suller and Lloyd (1999)
SYTOX Green	*E. coli, S. aureus, B. cereus*	Ampicillin, amoxicillin, penicillin G, vancomycin	Roth *et al.* (1997)
	E. coli, S. aureus, P. aeruginosa	Ceftazidime, ampicillin, vancomycin	Suller and Lloyd (1999)
Rhodamine 123	*E. coli, P. fluorescens, E. aerogenes, A. globiformis*	Valinomycin	Porter *et al.* (1995)
DiBAC$_4$(3) (oxonol)	*E. coli*	Azithromycin, cefuroxime, ciprofloxacin	Jepras *et al.* (1997)
	E. coli, S aureus, P. aeruginosa	Gramicidin S	Jepras *et al.* (1995)
	S. aureus	Methicillin	Suller and Lloyd (1998); Suller *et al.* (1997)
	E. coli, S. aureus	Ampicillin, gentamicin, ciprofloxacin	Mason *et al.* (1994)
	Aeromonas salmonicida	Gentamicin	Deere *et al.* (1995)
	E. coli, S. aureus, P. aeruginosa	Ceftazidime, ampicillin, vancomycin	Suller and Lloyd (1999)
Fluorescein diacetate (FDA)	*Mycobacterium tuberculosis*	Ethambutol, isoniazid, rifampin	Kirk *et al.* (1998)
CTC	*S. aureus*	Methicillin	Suller and Lloyd (1998)
	E. coli	Ciprofloxacin	Mason *et al.* (1995b)
	E. coli, S. aureus, P. aeruginosa	Ceftazidime, ampicillin, vancomycin	Suller and Lloyd (1999)
FITC	*E. coli*	Amoxycillin, mecillinam, chloramphenicol, ciprofloxacin trimethoprim	Durodie *et al.* (1995)

Yeast: Dye	Species	Antifungal Agent	References
Propidium iodide	*Candida albicans, S. cerevisiae, Cryptococcus neoformans*	Amphotericin B, fluconazole, cilofungin	Green *et al.* (1994)
	C. albicans, C. krusei, C. parapsilosis	Amphotericin B, fluconazole	Ramani *et al.* (1997)
Ethidium bromide	*Candida* spp.	Amphotericin B	O'Gorman and Hopfer (1991)
DiOC$_5$(3) (oxonol)	*Candida* spp., *T. glabrata*	Amphotericin B	Peyron *et al.* (1997)
	C. albicans, C. tropicalis	Amphotericin B	Ordóñez and Wehman (1995)
FUN-1	*C. albicans*	Amphotericin B, flucytosine, fluconazole, ketoconazole	Wenisch *et al.* (1997)

lag of regrowth, similar to the postantibiotic effect, after exposure to disinfectants. For example, chlorhexidine delays regrowth after exposure for more than 2 hr. Flow cytometry combined with fluorescent probes allows the activity of disinfectant compounds on target organisms to be ascertained within a few minutes and provides information on the heterogeneity of the sample population.

Sheppard *et al.* (1997) used oxonol and propidium iodide to monitor chlorhexidine-induced membrane damage in stationary and log phase *E. coli* cells. Their results indicated that membrane potential (oxonol) of cells collapsed prior to loss of membrane integrity (propidium iodide). Increased light scattering properties of organisms exposed to higher chlorhexidine concentrations suggest that there are also major changes to internal cellular structure. Comas and Vives-Rego (1997) used rhodamine 123, bis-oxonol, propidium iodide, SYTO-13, and SYTO-17 to assess the effect of formaldehyde and surfactants [i.e., sodium dodecyl sulfate (SDS), benzalkonium chloride] on *E. coli*.

Paul *et al.* (1996) used oxonol to determine the effectiveness of oral antiseptics found in mouthwash and toothpaste to kill bacteria such as *Streptococcus mutans,* *Streptococcus sanguis,* and *Streptococcus oralis* that cause tooth decay and gum disease. Membrane potential damage after 30 sec of exposure to triclosan, chlorohexidine, or cetylpyridinium chloride at $5\times$ MIC was assessed using flow cytometry and compared to plate-count data. Flow cytometry provided information within minutes on the immediate effect of oral antiseptics on target bacteria; plate-count data required 24 to 48 hr.

In our laboratory, we have developed a rapid flow cytometric assay to evaluate alternative disinfectant processes. Outbreaks of cryptosporidiosis have been attributed to the inability of chlorine to inactivate the oocyst form of *Cryptosporidium parvum*. Gamma (γ) irradiation may be a viable alternative to conventional chlorine-based wastewater disinfection processes.

Purified *Cryptosporidium parvum* oocysts were exposed in batch reactors to γ-irradiation from a ^{60}Co source. Exposures to γ-irradiation ranged from 50 to 800 krad. Untreated oocysts, heat-killed (70°C for 30 min) control oocysts, and irradiated oocysts were stained with SYTOX Green dead cell stain (10 μM final concentration), incubated at 37°C for 1 hr, and counted using flow cytometry (Fig. 10). Differences in light scattering properties were used to differentiate oocysts from sporozoites, ghosts (oocyst shells), and debris. After exposure to γ-irradiation, the oocysts were morphologically intact, but the process damaged the oocyst wall and allowed SYTOX Green, a membrane integrity stain, to enter and bind to nucleic acids. Nonviable oocysts with damaged but intact walls fluoresced bright green; viable oocysts and ghosts did not stain. Flow cytometry was used to count the number of damaged or inactivated oocysts after disinfectant exposure (Sincock *et al.,* 1998).

B. Food and Drink

Flow cytometry has been used to detect and identify pathogenic microorganisms in food samples and to monitor food and drink products for spoilage microor-

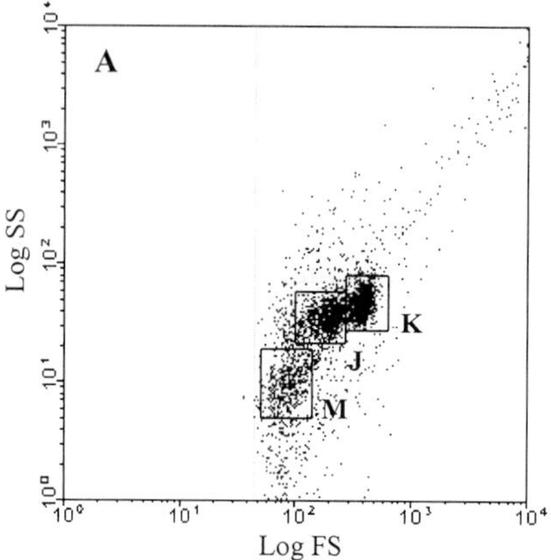

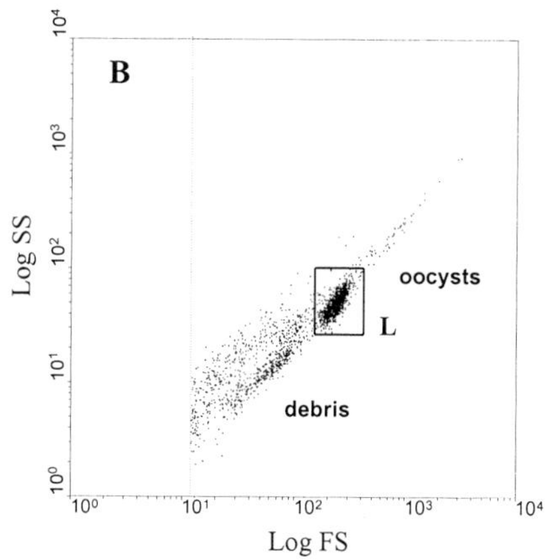

Fig. 10 Flow cytometric analysis of *Cryptosporidium parvum* oocysts exposed to γ-irradiation from a ^{60}Co source: (A) 200-krad dose: (B) 800-krad dose. Intact cysts (region L, K) can be differentiated from sporozoites (region J), ghosts (region M), and debris using light scatter measurements. High doses of γ-irradiation prevented excystation of sporozoites from oocysts.

ganisms. Food pathogens that can be detected and identified using flow cytometry include *Listeria monocytogenes* in raw milk (Donnelly and Baigent, 1986), *Salmonella typhimurium* in eggs and milk (Pinder and McClelland, 1994; McClelland and Pinder, 1994b), and *Escherichia coli* O157:H7 in ground beef, apple juice, and milk (Seo *et al.*, 1998a,b; Tortorello *et al.*, 1998). In general, flow cytometry requires specific monoclonal antibodies to detect and identify food pathogens.

Spoilage microorganisms are not necessarily harmful but can interfere with the quality of a food or drink product and can cause delays in product releases at great economic cost to the manufacturer. To guarantee that food or drink products conform to specifications, flow cytometry has been used to detect spoilage caused by yeast in soft drinks (Pettipher, 1991), yogurt, and fruit juice (Mulard, 1995), and to monitor the viability of yeast used for beer (Jespersen *et al.*, 1993; Jespersen and Jakobsen, 1994), wine (Bruetschy *et al.*, 1994), and cider (Willetts *et al.*, 1997; Lloyd *et al.*, 1996).

In order to identify sources of food contamination and spoilage, a large number of samples need to be tested. Flow cytometry allows the rapid and semiautomated analysis of heterogeneous food samples; however, extensive sample preparation is needed to isolate target organisms from high background levels of nonpathogenic microflora and particulate matter found in food samples. Sample preparation may include homogenization of solid food using a stomacher, filtering of large food particles, serial dilutions, or special reagent addition (i.e., clearing solution to remove micelles in milk and egg samples). After the cells have been isolated from the food sample, enrichment media can be used to increase the number of target organisms and to allow recovery of stressed or injured cells. Increasing the number of target organisms in food samples is extremely important because the infective dose for some foodborne illnesses can be as low as 10 cells.

IV. Conclusion

It is clear that there are a tremendous number of excellent uses of flow cytometry in the field of microbiology; however, there are some valid problems in implementing this technology. Listed below are what are considered to be the main advantages and disadvantages in the application of flow cytometry to microbial systems.

1. Advantages of using flow cytometry to analyze microbes

a. Technology is clinically proven in areas such as leukemia/lymphomas, HIV monitoring, platelet studies, and functional studies. Most hospitals and research centers have already purchased the instrument.

b. Assays can be performed rapidly, usually taking less than 1 min with little to no sample preparation.

c. Sensitivity is high, as low as 10^3 cells/ml reported.

d. Direct detection and identification can be achieved without elaborate or time-consuming culturing of microbes.

e. Cost per test, after initial investment in instrument, is low.

f. Automation/walk-away capability is available in most instruments as well as report generation for clinicians.

g. Flow cytometry is user friendly once the initial protocol is developed.

h. Instrument maintenance (daily/monthly quality control) is minimal.

2. Disadvantages to using flow cytometry to analyze microbes

a. Most microbiologists are not comfortable with using nontraditional, high technology procedures to run routine tests.

b. Initial cost of instrument is high.

c. Few microbial reagents or kits are commercially available for use with flow cytometry. Reagents (i.e., antibodies, DNA probes, control cells) must be developed in-house.

d. Protocols for microbes need to be developed in-house. Because most instruments were designed for mammalian cells, instrument setup and operation must also be modified.

e. Few, if any, instruments are designed specifically for small particles.

f. Little support is available from instrument manufacturers. Service technicians and technical support personnel are not familiar with procedures or methods utilized in microbial flow cytometry.

Clearly, the application of flow cytometry to the field of microbiology involves many unresolved problems; however, the continuing development of detailed protocols, appropriately designed instruments, and fluorescent probes will enable flow cytometry to solidify its position as the technology of preference.

References

Allman, R., Manchee, R., and Lloyd, D. (1993). Flow cytometric analysis of heterogeneous bacterial populations. *In* "Flow Cytometry in Microbiology" (D. Lloyd, ed.), pp. 27–48. Springer-Verlag, New York.

Amann, R. I., Binder, B. J., Olson, R. J., Chisholm, S. W., Devereux, R., and Stahl, D. A. (1990). Combination of 16S rRNA-targeted oligonucleotide probes with flow cytometry for analyzing mixed microbial populations. *Appl. Environ. Microbiol.* **56,** 1919–1925.

Amann, R., Snaidr, J., Wagner, M., Ludwig, W., and Schleifer, K.-H. (1996). In situ visualization of high genetic diversity in a natural microbial community. *J. Bacteriol.* **178,** 3496–3500.

Apperloo-Renkema, H. Z., Wilkinson, M. H. F., and van der Waaij, D. (1992). Circulating antibodies against faecal bacteria assessed by immunomorphometry: Combining quantitative immunofluorescence and image analysis. *Epidemiol. Infect.* **109,** 497–506.

Arrese, J. E., Goffin, V., Avila-Camacho, M., Greimers, R., and Piérard, G. E. (1998). A pilot study on bacterial viability in acne. Assessment using dual flow cytometry on microbials present in follicular casts and comedones. *Int. J. Dermatol.* **37,** 461–464.

Arrowood, M. J., Hurd, M. R., and Mead, J. R. (1995). A new method for evaluating experimental cryptosporidial parasite loads using immunofluorescent flow cytometry. *J. Parasitol.* **81,** 404–409.

Barnett, J. M., Cuchens, M. A., and Buchanan, W. (1984). Automated immunofluorescent speciation of oral bacteria using flow cytometry. *J. Dent. Res.* **63,** 1040–1042.

Best, L. M., Veldhuyzen van Zanten, S. J. O., Bezanson, G. S., Haldane, D. J. M., and Malatjalian, D. A. (1992). Serological detection of *Helicobacter pylori* by a flow microsphere immunofluorescence assay. *J. Clin. Microbiol.* **30,** 2311–2317.

Boddy, L., and Morris, C. W. (1993). Neural network analysis of flow cytometry data. *In* "Flow Cytometry in Microbiology" (D. Lloyd, ed.), pp. 159–170. Springer-Verlag, New York.

Bruderer, T., Niederer, E., and Köhler, P. (1994). Separation of a cysteine-rich surface antigen-expressing variant from a cloned *Giardia* isolate by fluorescence-activated cell sorting. *Parasitol. Res.* **80,** 303–306.

Bruetschy, A., Laurent, M., and Jacquet, R. (1994). Use of flow cytometry in oenology to analyse yeasts. *Lett. Appl. Microbiol.* **18,** 343–345.

Clarke, R. G., and Pinder, A. C. (1998). Improved detection of bacteria by flow cytometry using a combination of antibody and viability markers. *J. Appl. Microbiol.* **84,** 577–584.

Cohen, C. Y., and Sahar, E. (1989). Rapid flow cytometric bacterial detection and determination of susceptibility to amikacin in body fluids and exudates. *J. Clin. Microbiol.* **27,** 1250–1256.

Comas, J., and Vives-Rego, J. (1997). Assessment of the effects of gramicidin, formaldehyde, and surfactants on *Escherichia coli* by flow cytometry using nucleic acid and membrane potential dyes. *Cytometry* **29,** 58–64.

Davey, H. M., and Kell, D. B. (1996). Flow cytometry and cell sorting of heterogeneous microbial populations: The importance of single-cell analyses. *Microbiol. Rev.* **60,** 641–696.

Davey, H. M., Jones, A., Shaw, A. D., and Kell, D. B. (1999). Variable selection and multivariate methods for the identification of microorganisms by flow cytometry. *Cytometry* **35,** 162–168.

Deere, D., Porter, J., Edwards, C., and Pickup, R. (1995). Evaluation of the suitability of *bis*-(1,3-dibutylbarbituric acid)trimethine oxonol, (diBAC$_4$(3)($^-$)), for the flow cytometric assessment of bacterial viability. *FEMS Microbiol. Lett.* **130,** 165–169.

del Giorgio, P. A., Prairie, Y. T., and Bird, D. F. (1997). Coupling between rates of bacterial production and the abundance of metabolically active bacteria in lakes, enumerated using CTC reduction and flow cytometry. *Microb. Ecol.* **34,** 144–154.

DeLong, E. F., Wickham, G. S., and Pace, N. R. (1989). Phylogenetic stains: Ribosomal RNA-based probes for identification of single cells. *Science* **243**, 1360–1362.

Desmonts, C., Minet, J., Colwell, R., and Cormier, M. (1990). Fluorescent-antibody method useful for detecting viable but nonculturable *Salmonella* spp. in chlorinated wastewater. *Appl. Environ. Microbiol.* **56**, 1448–1452.

Diaper, J. P., and Edwards, C. (1994). The use of fluorogenic esters to detect viable bacteria by flow cytometry. *J. Appl. Bacteriol.* **77**, 221–228.

Diaper, J. P., Tither, K., and Edwards, C. (1992). Rapid assessment of bacterial viability by flow cytometry. *Appl. Microbiol. Biotechnol.* **38**, 268–272.

Dixon, B. R., Parenteau, M., Martineau, C., and Fournier, J. (1997). A comparison of conventional microscopy, immunofluorescence microscopy and flow cytometry in the detection of *Giardia lamblia* cysts in beaver fecal samples. *J. Immunol. Methods* **202**, 27–33.

Donnelly, C. W., and Baigent, G. J. (1986). Method for flow cytometric detection of *Listeria monocytogenes* in milk. *Appl. Environ. Microbiol.* **52**, 689–695.

Dubelaar, G. B. J., Gerritzen, P. I., Beeker, A. E. R., Jonker, R. R., and Tangen, K. (1999). Design and first results of CytoBuoy: A wireless flow cytometer for *in situ* analysis of marine and fresh waters. *Cytometry* **37**, 247–254.

Durodie, J., Coleman, K., Simpson, I. N., Loughborough, S. H., and Winstanley, D. W. (1995). Rapid detection of antimicrobial activity using flow cytometry. *Cytometry* **21**, 374–377.

Edwards, C. (1996). Assessment of viability of bacteria by flow cytometry. *In* "Flow Cytometry Applications in Cell Culture" (M. Al-Rubeai and A. N. Emery, eds.), pp. 291–310. Dekker, New York.

Errington, P. A., and Graham, J. (1993). Application of artificial neural networks to chromosome classification. *Cytometry* **14**, 627–639.

Frankel, D. S., Olson, R. J., Frankel, S. L., and Chisholm, S. W. (1989). Use of a neural net computer system for analysis of flow cytometric data of phytoplankton populations. *Cytometry* **10**, 540–550.

Gant, V. A., Warnes, G., Phillips, I., and Savidge, G. F. (1993). The application of flow cytometry to the study of bacterial responses to antibiotics. *J. Med. Microbiol.* **39**, 147–154.

Gottfredsson, M., Erlendsdottir, H., Sigfusson, A., and Gudmunsson, S. (1998). Characteristics and dynamics of bacterial populations during postantibiotic effect determined by flow cytometry. *Antimicrob. Agents Chemother.* **42**, 1005–1011.

Green, L., Petersen, B., Steimel, L., Haeber, P., and Current, W. (1994). Rapid determination of antifungal activity by flow cytometry. *J. Clin. Microbiol.* **32**, 1088–1091.

Henningson, E. W., Krocova, Z., Sandström, G., and Forsman, M. (1998). Flow cytometric assessment of the survival ratio of *Francisella tularensis* in aerobiological samples. *FEMS Microbiol. Ecol.* **25**, 241–249.

Heyworth, M. F., and Pappo, J. (1989). Use of two-colour flow cytometry to assess killing of *Giardia muris* trophozoites by antibody and complement. *Parasitology* **99**, 199–203.

Hughes, E. E., Matthews-Greer, J. M., and Gilleland, H. E., Jr. (1996). Analysis by flow cytometry of surface-exposed epitopes of outer membrane protein F of *Pseudomonas aeruginosa. Can. J. Microbiol.* **42**, 859–862.

Ingram, M., Cleary, T. J., Price, B. J., Price, R. L., and Castro, A. (1982). Rapid detection of *Legionella pneumophila* by flow cytometry. *Cytometry* **3**, 134–137.

Jepras, R. I., Carter, J., Pearson, S. C., Paul, F. E., and Wilkinson, M. J. (1995). Development of a robust flow cytometric assay for determining numbers of viable bacteria. *Appl. Environ. Microbiol.* **61**, 2696–2701.

Jepras, R. I., Paul, F. E., Pearson, S. C., and Wilkinson, M. J. (1997). Rapid assessment of antibiotic effects on *Escherichia coli* by *bis*-(1,3-dibutylbarbituric acid) trimethine oxonol and flow cytometry. *Antimicrob. Agents Chemother.* **41**, 2001–2005.

Jernaes, M. W., and Steen, H. B. (1994). Staining of *Escherichia coli* for flow cytometry: Influx and efflux of ethidium bromide. *Cytometry* **17**, 302–309.

Jespersen, L., and Jakobsen, M. (1994). Use of flow cytometry for rapid estimation of intracellular events in brewing yeasts. *J. Inst. Brew.* **100**, 399–403.

Jespersen, L., Lassen, S., and Jakobsen, M. (1993). Flow cytometric detection of wild yeast in lager breweries. *Int. J. Food Microbiol.* **17,** 321–328.

Kaprelyants, A. S., and Kell, D. B. (1992). Rapid assessment of bacterial viability and vitality by rhodamine 123 and flow cytometry. *J. Appl. Bacteriol.* **72,** 410–422.

Kaprelyants, A. S., and Kell, D. B. (1993a). The use of 5-cyano-2,3-ditolyl tetrazolium chloride and flow cytometry for the visualisation of respiratory activity in individual cells of *Micrococcus luteus. J. Microbiol. Methods* **17,** 115–122.

Kaprelyants, A. S., and Kell, D. B. (1993b). Dormancy in stationary-phase cultures of *Micrococcus luteus:* Flow cytometric analysis of starvation and resuscitation. *Appl. Environ. Microbiol.* **59,** 3187–3196.

Kaprelyants, A. S., Gottschal, J. C., and Kell, D. B. (1993). Dormancy in non-sporulating bacteria. *FEMS Microbiol. Rev.* **104,** 271–286.

Kirk, S. M., Schell, R. F., Moore, A. V., Callister, S. M., and Mazurek, G. H. (1998). Flow cytometric testing of susceptibilities of *Mycobacterium tuberculosis* isolates to ethambutol, isoniazid, and rifampin in 24 hours. *J. Clin. Microbiol.* **36,** 1568–1573.

Lange, J. L., Thorne, P. S., and Lynch, N. (1997). Application of flow cytometry and fluorescent in situ hybridization for assessment of exposures to airborne bacteria. *Appl. Environ. Microbiol.* **63,** 1557–1563.

Lloyd, D., Moran, C. A., Suller, M. T. E., and Dinsdale, M. G. (1996). Flow cytometric monitoring of rhodamine 123 and a cyanine dye uptake by yeast during cider fermentation. *J. Inst. Brew.* **102,** 251–259.

López-Amorós, R., Comas, J., García, M. T., and Vives-Rego, J. (1998). Use of the 5-cyano-2,3-ditolyl tetrazolium chloride reduction test to assess respiring marine bacteria and grazing effects by flow cytometry during linear alkylbenzene sulfonate degradation. *FEMS Microbiol. Ecol.* **27,** 33–42.

Lutton, D. A., Patrick, S., Crockard, A. D., Stewart, L. D., Larkin, M. J., Dermott, E., and McNeill, T. A. (1991). Flow cytometric analysis of within-strain variation in polysaccharide expression by *Bacteroides fragilis* by use of murine monoclonal antibodies, *J. Med. Microbiol.* **35,** 229–237.

McClelland, R. G., and Pinder, A. C. (1994a). Detection of low levels of specific *Salmonella* species by fluorescent antibodies and flow cytometry. *J. Appl. Bacteriol.* **77,** 440–447.

McClelland, R. G., and Pinder, A. C. (1994b). Detection of *Salmonella typhimurium* in dairy products with flow cytometry and monoclonal antibodies. *Appl. Environ. Microbiol.* **60,** 4255–4262.

Maguire, D., King, G. B., Kelley, S., and Robinson, J. P. (1994a). Neural network classification of acute leukemias using flow cytometry analysis, data. *Proc. 13th Southern Biomed. Eng. Conf.,* 645–648.

Maguire, D. J., King, G. B., and Robinson, J. P. (1994b). A comparison of hard and soft boundaries of intensity regions for the statistical classification of acute leukemias. *Cytometry* 7(Suppl.), 48 (Abstract).

Manz, W., Szewzyk, U., Ericsson, P., Amann, R., Schleifer, K.-H., and Stenström, T.-A. (1993). In situ identification of bacteria in drinking water and adjoining biofilms by hybridization with 16S and 23S rRNA-directed fluorescent oligonucleotide probes. *Appl. Environ. Microbiol.* **59,** 2293–2298.

Martinelli, F., Pizzi, R., Cabibbo, E., Licenziati, S., Dima, F., Canaris, A. D., Crea, G., Ravizzola, G., Caruso, A., and Turano, A. (1995). Monoclonal antibodies against antigens exposed on the surface of vegetative forms and spores of *Myxococcus virescens. Microbiologica* **18,** 399–407.

Mason, D. J., and Lloyd, D. (1997). Acridine orange as an indicator of bacterial susceptibility to gentamicin. *FEMS Microbiol. Lett.* **153,** 199–204.

Mason, D. J., Allman, R., Stark, J. M., and Lloyd, D. (1994). Rapid estimation of bacterial antibiotic susceptibility with flow cytometry. *J. Microsc.* **176,** 8–16.

Mason, D. J., López-Amorós, R., Allman, R., Stark, J. M., and Lloyd, D. (1995a). The ability of membrane potential dyes and calcafluor white to distinguish between viable and non-viable bacteria. *J. Appl. Bacteriol.* **78,** 309–315.

Mason, D. J., Power, G. M., Talsania, H., Phillips, I., and Gant, V. A. (1995b). Antibacterial action of ciprofloxacin. *Antimicrob. Agents Chemother.* **39,** 2752–2758.

Mason, D. J., Shanmuganathan, S., Mortimer, F. C., and Gant, V. A. (1998). A fluorescent gram stain for flow cytometry and epifluorescence microscopy. *Appl. Environ. Microbiol.* **64,** 2681–2685.

Mulard, Y. (1995). Flow cytometry: Real time microbiology testing. *Food Technol. Eur.* **2,** 72–76.

Nelson, D., Bathgate, A. J., and Poxton, I. R. (1991). Monoclonal antibodies as probes for detecting lipopolysaccharide expression on *Escherichia coli* from different growth conditions. *J. Gen. Microbiol.* **137,** 2741–2751.

Obernesser, M. S., Socransky, S. S., and Stashenko, P. (1990). Limit of resolution of flow cytometry for the detection of selected bacterial species. *J. Dent. Res.* **69,** 1592–1598.

O'Gorman, M. R. G., and Hopfer, R. L. (1991). Amphotericin B susceptibility testing of *Candida* species by flow cytometry. *Cytometry* **12,** 743–747.

Ordóñez, J. V., and Wehman, N. M. (1995). Amphotericin B susceptibility of *Candida* species assessed by rapid flow cytometric membrane potential assay. *Cytometry* **22,** 154–157.

Paul, F., Jepras, R., Hynes, D., Smith, A., and Marken, B. (1996). Activity of common oral antiseptics against bacteria assessed using the oxonol DiBAC$_4$(3). *Cytometry* 8(Suppl.), 117 (Abstract).

Pettipher, G. L. (1991). Preliminary evaluation of flow cytometry for the detection of yeasts in soft drinks. *Lett. Appl. Microbiol.* **12,** 109–112.

Peyron, F., Favel, A., Guiraud-Dauriac, H., el Mzibri, M., Chastin, C., Duménil, G., and Regli, P. (1997). Evaluation of a flow cytofluorometric method for rapid determination of amphotericin B susceptibility of yeast isolates. *Antimicrob. Agents Chemother.* **41,** 1537–1540.

Phillips, A. P., and Martin, K. L. (1983). Immunofluorescence analysis of Bacillus spores and vegetative cells by flow cytometry. *Cytometry* **4,** 123–131.

Phillips, A. P., and Martin, K. L. (1988). Limitations of flow cytometry for the specific detection of bacteria in mixed populations. *J. Immunol. Methods* **106,** 109–117.

Pinder, A. C., and McClelland, R. G. (1994). Rapid assay for pathogenic salmonella organisms by immunofluorescence flow cytometry. *J. Microsc.* **176,** 17–22.

Porter, J., Pickup, R., and Edwards, C. (1995). Membrane hyperpolarisation by valinomycin and its limitations for bacterial viability assessment using rhodamine 123 and flow cytometry. *FEMS Microbiol. Lett.* **132,** 259–262.

Ramani, R., Ramani, A., and Wong, S. J. (1997). Rapid flow cytometric susceptibility testing of *Candida albicans. J. Clin. Microbiol.* **35,** 2320–2324.

Rice, J., Sleigh, M. A., Burkill, P. H., Tarran, G. A., O'Connor, C. D., and Zubkov, M. V. (1997). Flow cytometric analysis of characteristics of hybridization of species-specific fluorescent oligonucleotide probes to rRNA of marine nanoflagellates. *Appl. Environ. Microbiol.* **63,** 938–944.

Roth, B. L., Poot, M., Yue, S. T., and Millard, P. J. (1997). Bacterial viability and antibiotic susceptibility testing with SYTOX Green nucleic acid stain. *Appl. Environ. Microbiol.* **63,** 2421–2431.

Sahar, E., Lamed, R., and Ofek, I. (1983). Rapid identification of *Streptococcus pyogenes* by flow cytometry. *Eur. J. Clin. Microbiol.* **2,** 192–195.

Sanders, C. A., Yajko, D. M., Hyun, W., Langlois, R. G., Nassos, P. S., Fulwyler, M. J., and Hadley, W. K. (1990). Determination of guanine-plus-cytosine content of bacterial DNA by dual-laser flow cytometry. *J. Gen. Microbiol.* **136,** 359–365.

Seo, K. H., Brackett, R. E., and Frank, J. F. (1998a). Rapid detection of *Escherichia coli* O157:H7 using immunomagnetic flow cytometry in ground beef, apple juice, and milk. *Int. J. Food Microbiol.* **44,** 115–123.

Seo, K. H., Brackett, R. E., Frank, J. F., and Hilliard, S. (1998b). Immunomagnetic separation and flow cytometry for rapid detection of *Escherichia coli* O157:H7. *J. Food Protect.* **61,** 812–816.

Shapiro, H. M. (1995). "Practical Flow Cytometry," 3rd ed. Wiley-Liss, New York.

Sheppard, F. C., Mason, D. J., Bloomfield, S. F., and Gant, V. A. (1997). Flow cytometric analysis of chlorhexidine action. *FEMS Microbiol. Lett.* **154,** 283–288.

Simon, N., LeBot, N., Marie, D., Partensky, F., and Vaulot, D. (1995). Fluorescent in situ hybridization with rRNA-targeted oligonucleotide probes to identify small phytoplankton by flow cytometry. *Appl. Environ. Microbiol.* **61,** 2506–2513.

Sincock, S. A., Anderson, P. E, and Stopa, P. J. (1996a). New fluorescent stains for flow cytometric detection of bacteria. *Cytometry* 8(Suppl.), 128 (Abstract).

Sincock, S. A., Anderson, P. E., Stopa, P. J., and Ezzell, J. (1996b). Rapid detection of pathogenic bacteria using flow cytometry. *Cytometry* 8(Suppl.), 117 (Abstract).

Sincock, S. A., Thompson, J. E., Blatchley III., E. R., Ragheb, K. E., and Robinson, J. P. (1998). Flow cytometry can detect inactivation of *Cryptosporidium parvum* oocysts by γ irradiation. *Cytometry* 9(Suppl.), 46–47 (Abstract).

Sizemore, R. K., Caldwell, J. J., and Kendrick, A. S. (1990). Alternate gram staining technique using a fluorescent lectin. *Appl. Environ. Microbiol.* **56,** 2245–2247.

Steen, H. B. (1980). Further developments of a microscope-based flow cytometer: Light scatter detection and excitation intensity compensation. *Cytometry* **1,** 26–31.

Steen, H. B. (1983). A microscope-based flow cytophotometer. *Histochem. J.* **15,** 147–160.

Steen, H. B., and Boye, E. (1980). Bacterial growth studied by flow cytometry. *Cytometry* **1,** 32–36.

Suller, M. T. E., and Lloyd, D. (1998). Flow cytometric assessmant of the postantibiotic effect of methicillin on *Staphylococcus aureus. Antimicrob. Agents Chemother.* **42,** 1195–1199.

Suller, M. T. E., and Lloyd, D. (1999). Fluorescence monitoring of antibiotic-induced bacterial damage using flow cytometry. *Cytometry* **35,** 235–241.

Suller, M. T. E., Stark, J. M., and Lloyd, D. (1997). A flow cytometric study of antibiotic-induced damage and evaluation as a rapid antibiotic susceptibility test for methicillin-resistant *Staphylococcus aureus. J. Antimicrob. Chemother.* **40,** 77–83.

Swarts, A. J., Hastings, J. W., Roberts, R. F., and von Holy, A. (1998). Flow cytometry demonstrates bacteriocin-induced injury to *Listeria monocytogenes. Curr. Microbiol.* **36,** 266–270.

Thomas, J.-C., Desrosiers, M., St.-Pierre, Y., Lirette, P., Bisaillon, J.-G., Beaudet, R., and Villemur, R. (1997). Quantitative flow cytometric detection of specific microorganisms in soil samples using rRNA targeted fluorescent probes and ethidium bromide. *Cytometry* **27,** 224–232.

Tortorello, M. L., Stewart, D. S., and Raybourne, R. B. (1998). Quantitative analysis and isolation of *Escherichia coli* O157:H7 in a food matrix using flow cytometry and cell sorting. *FEMS Immunol. Med. Microbiol.* **19,** 267–274.

Tsuji, T., Kawasaki, Y., Takeshima, S., Sekiya, T., and Tanaka, S. (1995). A new fluorescence staining assay for visualizing living microorganisms in soil. *Appl. Environ. Microbiol.* **61,** 3415–3421.

Ullrich, S., Karrasch, B., Hoppe, H.-G., Jeskulke, K., and Mehrens, M. (1996). Toxic effects on bacterial metabolism of the redox dye 5-cyano-2,3-ditolyl tetrazolium chloride. *Appl. Environ. Microbiol.* **62,** 4587–4593.

Valdez, L. M., Dang, H., Okhuysen, P. C., and Chappell, C. L. (1997). Flow cytometric detection of *Cryptosporidium* oocysts in human stool samples. *J. Clin. Microbiol.* **35,** 2013–2017.

van der Waaij, L. A., Mesander, G., Limburg, P. C., and van der Waaij, D. (1994). Direct flow cytometry of anaerobic bacteria in human feces. *Cytometry* **16,** 270–279.

Van Dilla, M. A., Langlois, R. G., Pinkel, D., Yajko, D., and Hadley, W. K. (1983). Bacterial characterization by flow cytometry. *Science* **220,** 620–622.

Vesey, G., Slade, J. S., Byrne, M., Shepherd, K., Dennis, P. J., and Fricker, C. R. (1993). Routine monitoring of *Cryptosporidium* oocysts in water using flow cytometry. *J. Appl. Bacteriol.* **75,** 87–90.

Vesey, G., Deere, D., Weir, C. J., Ashbolt, N., Williams, K. L., and Veal, D. A. (1997). A simple method for evaluating *Cryptosporidium*-specific antibodies used in monitoring environmental water samples. *Lett. Appl. Microbiol.* **25,** 316–320.

Volsch, A., Nader, W. F., Geiss, H. K., Nebe, G., and Birr, C. (1990). Detection and analysis of two serotypes of ammonia-oxidizing bacteria in sewage plants by flow cytometry. *Appl. Environ. Microbiol.* **56,** 2430–2435.

Walberg, M., Gaustad P., and Steen, H. B. (1997a). Rapid assessment of ceftazidime, ciprofloxacin, and gentamycin susceptibility in exponentially-growing *E. coli* cells by means of flow cytometry. *Cytometry* **27,** 169–178.

Walberg, M., Gaustad, P., and Steen, H. B. (1997b). Rapid discrimination of bacterial species with different ampicillin susceptibility levels by means of flow cytometry. *Cytometry* **29,** 267–272.

Wallner, G., Amann, R., and Beisker, W. (1993). Optimizing fluorescent in situ hybridization with rRNA-targeted oligonucleotide probes for flow cytometric identification of microorganisms. *Cytometry* **14,** 136–143.

Wallner, G., Erhart, R., and Amann, R. (1995). Flow cytometric analysis of activated sludge with rRNA-targeted probes. *Appl. Environ. Microbiol.* **61,** 1859–1866.

Wenisch, C., Linnau, K. F., Parschalk, B., Zedtwitz-Liebenstein, K., and Georgopoulos, A. (1997). Rapid susceptibility testing of fungi by flow cytometry using vital staining. *J. Clin. Microbiol.* **35,** 5–10.

Willetts, J. C., Seward, R., Dinsdale, M. G., Suller, M. T. E., Hill, B., and Lloyd, D. (1997). Vitality of cider yeast grown micro-aerobically with added ethanol, butan-1-ol or iso-butanol. *J. Inst. Brew.* **103,** 79–84.

Yu, W., Dodds, W. K., Banks, M. K., Skalsky, J., and Strauss, E. A. (1995). Optimal staining and sample storage time for direct microscopic enumeration of total and active bacteria in soil with two fluorescent dyes. *Appl. Environ. Microbiol.* **61,** 3367–3372.

CHAPTER 54

Staining and Measurement of DNA in Bacteria

Harald B. Steen

Department of Biophysics
Institute for Cancer Research
0310 Oslo, Norway

I. Introduction
II. Basic Considerations
    A. Size and Shape
    B. DNA Staining
    C. DNA Structure
    D. Cell Wall
    E. Active Efflux
    F. Dependence on Growth Conditions
III. Experimental Methods
    A. Staining of Fixed Bacteria
    B. Vital Staining
    C. Spores
    D. Sheath Fluid
    E. Flow Cytometers
IV. Standards and Controls
    References

I. Introduction

Since its introduction some 30 years ago, flow cytometry has been a major tool in studies of eukaryotic and particularly mammalian cells, and it has found numerous applications in measurement of the composition and physiological status of cells. Similar measurements of bacteria are just in their infancy. This is in spite of the fact that the cell cycle of bacteria and the associated physiology present a number of essential problems which are difficult to answer without a

METHODS IN CELL BIOLOGY, VOL. 64

method that facilitates precise measurements of large numbers of individual cells. The most important parameter in this respect is the cellular DNA content.

Flow cytometry is also an obvious approach to detection and counting of bacteria in various applications, such as clinical bacteriology, food analysis, and environmental monitoring including analysis of air, water, and sewage. The most direct and general way to distinguish biological cells from other types of microscopical particles that are often present in such samples is to stain with a DNA specific, fluorescent dye.

To simplify sample preparation and to avoid loss of cells, aggregation, and denaturation, it is preferable in many such applications to eliminate fixation and stain the cells vitally. Vital staining of bacteria presents other problems, but also other opportunities, compared to those encountered with mammalian cells. One the one hand, one has to cope with problems related to the impermeability of the cell wall and the active excretion of dye. On the other hand, the structure of the bacterial cell wall provides new opportunities for staining that are not available with mammalian cells.

The cell cycle of bacteria is basically different from that of mammalian cells in that more than one replication cycle can be in progress in the same chromosome at the same time. This makes DNA histograms of bacteria look very different from what we are used to seeing for mammalian cells and complicates cell cycle analysis significantly. Furthermore, the cell cycle of bacteria varies much more with growth conditions and the physiological state of the cell than does that of mammalian cells. For example, the DNA histogram of bacteria growing very slowly may look very much like what we are used to seeing for mammalian cells, whereas for cells growing under optimal conditions it may look like one broad, featureless peak with an average DNA content several times that of the slowly growing cells. This fact complicates cell cycle studies, but more importantly, it represents an opportunity to learn more about the biochemistry and physiology of bacteria.

It is now possible to carry out DNA measurements of bacteria with sufficient sensitivity and precision to allow assessment of essential cell cycle parameters (Boye *et al.,* 1983; Skarstad *et al.,* 1983, 1985; Steen, 1990a) and to study the mechanisms of the replication of the chromosome in more detail than was previously possible (Skarstad *et al.,* 1986; Skarstad and Boye, 1988; Løbner-Olesen, *et al.,* 1989; Boye and Løbner-Olesen, 1990). DNA staining of both fixed and vital cells has also facilitated accurate cell counting and rapid assessment of the effects of various antibiotics (see Chapter 55 in this volume).

II. Basic Considerations

From an experimental, flow cytometric point of view, bacteria differ from mammalian cells in several important respects.

A. Size and Shape

Bacteria are typically some three orders of magnitude smaller than mammalian cells with regard to the volume of cells as well as their content of DNA and various proteins. For example, the DNA content of the *Escherichia coli* chromosome is about 1400 times less than that of diploid human cells. The majority of the flow cytometers currently available were designed to measure primarily mammalian cells. Hence, flow cytometry of bacteria puts great demands on instrument sensitivity for both light scattering and fluorescence. In particular, the light scattering sensitivity of many instruments is not sufficient to measure bacteria reliably.

Many bacterial species, such as *E. coli,* are rods rather than spheres and may therefore create orientation artifacts, which is to say that the fluorescence and light scattering signals may depend on the orientation of the cell relative to the direction of the excitation light (Pinkel and Stovel, 1985). This may be a problem especially in laser-based instruments with near parallel excitation light.

In contrast to mammalian cells, which, with few exceptions, are spherical in suspension and have a nucleus that is roughly concentric with the outer cell wall, with a size roughly constant relative to that of the cell, bacteria may vary greatly with respect to the intracellular distribution of their DNA, depending on growth conditions, exposure to drugs, and other factors. Thus, although under certain conditions the DNA appears to be evenly distributed in all of the cytoplasm, it may be concentrated into a minor portion of the cell volume in other cases. Again, this may cause artifacts in some instruments. The structural difference that may be reflected in the different distribution of DNA may also affect staining, as discussed subsequently.

B. DNA Staining

Complicating the matter further, bacteria in some situations may have a relatively much higher RNA content than typical mammalian cells, notably when they grow under optimal conditions. This means that dyes with some affinity for RNA, such as ethidium bromide and propidium iodide (PI), are not suitable, except if RNA has been removed, for example, by treating the cells with RNase. Working with *E. coli,* however, we have not been able to obtain consistent results subsequent to RNase treatment of bacteria. The reason may be that the cell wall, even after fixation, is not sufficiently permeable to the enzyme. Consequently, dyes with higher DNA specificity have to be used in order to obtain DNA histograms of adequate quality, namely, dyes such as 4',6-diamidino-2-phenylindole (DAPI), the bisbenzimide dyes Hoechst 33342 and Hoechst 33258, chromomycin A, mithramycin, and 7-aminoactinomycin D (7-AMD). DAPI and the Hoechst dyes require ultraviolet (UV) excitation (i.e., wavelengths around 360 nm), which means that the flow cytometer light source must be either an argon or krypton ion laser running at about 350 nm or a mercury high pressure arc lamp. These lamps have a strong emission line at about 366 nm that coincides with the peak absorption of these dyes. In laser instruments mithramycin, and its analog chromomycin A, can be excited only marginally by means of the 458 nm line of a tunable, high power argon laser, but more efficiently with the 413 nm line of the krypton laser. Furthermore, owing to a relatively low fluorescence quantum yield, these dyes are not very bright. The fluorescence yield of 7-AMD is too low to facilitate measurements on bacteria.

The use of UV-excitable DNA stains may present problems owing to spectral overlap in cases where, in addition to DNA, one wants to measure a fluorescence conjugated antibody. With mammalian cells this is typically done by combining a fluorescently conjugated antibody with a DNA specific dye that has its fluorescence at a higher wavelength than the antibody in order to reduce the problem of spectral overlap, for example, a fluorescein isothiocyanate (FITC)-conjugated antibody with PI staining of the DNA (subsequent to treatment with RNase and/or removal of the cytoplasm). Because with bacteria we have to use DNA specific dyes excited in the UV or deep blue, this option is not available, and we may often meet the situation that the weak antibody fluorescence may "drown" in the long-wavelength tail of the fluorescence spectrum of the very much stronger DNA-associated fluorescence. In some cases the only solution to this problem is a dual laser instrument that allows excitation and measurement of the two chromophores in separate foci.

C. DNA Structure

Bacteria differ from eukaryotic cells in that the chromosome does not contain histones and other proteins that can affect dye binding and thereby possibly destroy the stoichiometry of the staining. Presumably bacterial DNA is more loosely packed so that "chromatin structure" should not be expected to affect the staining. Nevertheless, we have seen many cases where DNA-associated dye fluorescence appears to depend on the physiological status of the cells. One explanation may be that the binding of dye depends on the degree of supercoiling of the DNA. The interaction of bacterial DNA with proteins and the importance of the DNA conformation in general are still very much an open question, a question that may be elucidated by means of flow cytometry.

D. Cell Wall

The cell wall of bacteria is quite different from that of mammalian cells. In addition to the cytoplasmic membrane, the main structure of which is a double lipid layer similar to that of mammalian cells, bacteria exhibit a complex cell wall consisting primarily of peptidoglycans, lipoproteins, and lipopolysaccharides. The permeability of this envelope is significantly different from that of the plasma membrane. For example, the cytoplasmic membrane may contain porins admitting molecules like ethidium bromide. On the other hand, the cell wall may be impermeable to dyes that enter mammalian cells, such as Hoechst 33342. Hence, the knowledge one may have on staining of mammalian cells, and especially that of vital cells, is not necessarily applicable to bacteria.

E. Active Efflux

Vital staining of bacteria is further complicated by the fact that some bacteria have the ability to excrete some dyes very efficiently (Lewis, 1994). For example, some DNA-binding dyes that readily permeate the cell wall may be pumped out so rapidly that hardly any staining occurs. This excretion is an active pumping that depends on metabolic energy. Being one of the mechanisms of resistance against antibiotics, it is a phenomenon of considerable interest, a phenomenon that lends itself to study by flow cytometry.

F. Dependence on Growth Conditions

Although bacteria are easy to grow and process for measurement, we have found that the results are generally much more sensitive to growth conditions than are corresponding data for mammalian cells. Strict reproducibility of culture conditions are essential to obtain consistent data. For example, the cell cycle distribution is significantly affected by cell density, even at densities where growth is still exponential (Steen, 1990a). Although this susceptibility to growth conditions may be an experimental problem, it also presents an inroad to studies of the physiology of bacteria.

III. Experimental Methods

A. Staining of Fixed Bacteria

Ethanol fixation has been found to permeabilize bacteria to all of the DNA specific dyes dealt with in this chapter, and to preserve the cells for long periods of time. We have not seen any exception to this among a wide variety of bacterial species, both gram negative and gram positive, provided they are not spores. In order to avoid aggregation, fixation is carried out by squirting aliquots of a fluid cell culture, or cell suspension, directly into stirred, ice-cold ethanol to yield a final ethanol concentration of 70%.

For assessment of DNA we are routinely using mithramycin. The fluorescence quantum yield of mithramycin is relatively low, that is, about 0.05% (Langlois and Jensen, 1979), thus putting great demands on instrument sensitivity. We therefore use it in combination with ethidium bromide in order to obtain a higher fluorescence yield. Ethidium bromide, which also binds to RNA and consequently does not produce clean DNA histograms when excited directly, has negligible absorption at the excitation wavelength used for mithramycin (mainly the strong 436 nm line of the Hg arc lamp), and it is therefore excited almost exclusively by resonance energy transfer from adjacent (closer than 5 nm) mithramycin molecules (Langlois and Jensen, 1979). Hence, fluorescence from RNA-bound ethidium bromide appears to be negligible with this staining and excitation wavelength. Because of the higher fluorescence quantum yield of ethidium bromide, excitations transferred to this dye produce more fluorescence than mithra-mycin alone. Thus, the DNA specificity of mithramycin is combined with the higher fluorescence yield of ethidium bromide. The net result is an increase in fluorescence intensity by a factor of about 2. It is interesting that this increase is significantly smaller than the value reported for mammalian cells, that is, 3.4 (Zante *et al.,* 1976; Langlois and Jensen, 1979). It has been reported that DAPI and Hoechst 33258 give roughly the same results as mithramycin/ethidium bro-mide in terms of resolution [coefficient of variation (cv) values] (Bernander *et al.,* 1998).

With all of these dyes the cells are in a 10 mM Tris buffer with 10 mM $MgCl_2$ at pH 7.4 subsequent to washing the cells in the same buffer. Suitable dye concentrations are as follows: mithramycin, 90 μg/ml; ethidium bromide, 20 μg/ml; DAPI, 1 μg/ml; and Hoechst 33258, 1 μg/ml.

Fixed cells are stable in 70% ethanol at 4°C for many weeks. In the staining solution, however, results may sometimes be affected when stored on ice for more than a few hours, possibly owing to the presence of traces of DNase. Stock solutions of the dyes in the Tris buffer may be stable for many months when stored in the dark at 4°C. The final staining solution should be stored in the dark at 4°C and used within 1 week of being prepared.

B. Vital Staining

Vital *E. coli* and many other species of bacteria, both gram negative and gram positive, do not stain under physiological conditions with commonly used DNA-binding dyes such as mithramycin, ethidium bromide, DAPI, and Hoechst 33258. The reason for this appears to be a combination of low permeability and the ability of many species to actively excrete dye molecules that enter the cell (Jernaes and Steen, 1994). When vital *E. coli,* harvested in the exponential phase, were incubated with dye at 0°C, they stained readily with ethidium bromide, with the degree of staining depending on the buffer in which the staining was performed. Thus, the cells appear to be permeable to this dye, at least at low temperature. However, when the temperature is raised to about 20°C, the staining drops almost to zero within a few seconds (Table I). The reduction in fluorescence

Table I
Median Fluorescence Intensity of Vital *E. coli*[a]

Conditions	PBS	Tris	EDTA
Room temperature	2 ± 1	5 ± 1	16 ± 3
Ice	597 ± 38	1004 ± 16	988 ± 0
Room temperature with metabolic inhibitor	66 ± 7	83 ± 4	959 ± 61

[a] Cells were stained with 20 μg/ml ethidium bromide in the buffers noted. The fluorescence median for a similar sample fixed in 70% ethanol was 898 ± 0. Error limits represent range of two measurements.

intensity observed on raising the temperature was a factor of a few hundred. The interpretation is that when cells are brought to room temperature, their metabolism increases so as to activate the efflux pump that brings the dye out of the cells. This interpretation was confirmed by the observation that when the metabolism of the cells were paralyzed by metabolic inhibitors, sodium azide (4 g/liter) and 2-deoxy-D-glucose (5 mM), the staining, in the presence of EDTA (10 mM), was about the same as at 0°C and was similar also to that obtained for cells permeabilized by 70% ethanol by the method previously described (Fig. 1). On the other hand, in phosphate-buffered saline (PBS) and Tris (100 mM) the metabolic inhibitors increased the staining by more than a factor 10, but still only to a level more than one order of magnitude below complete staining (Walberg *et al.*, 1997, 1998). Hence, it appears that even in the presence of the metabolic inhibitors the cells were able to excrete dye, although at a greatly reduced rate. The effect of EDTA seems to be to increase permeability to a

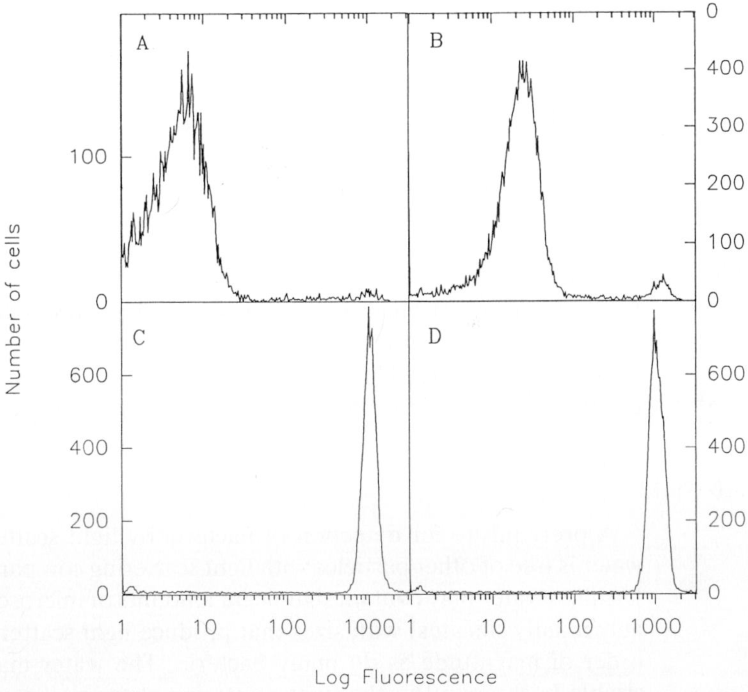

Fig. 1 Fluorescence histograms of vital *E. coli* harvested during exponential growth and stained with 20 μg/ml ethidium bromide in the following solutions: (A, C) phosphate-buffered saline (PBS), pH = 7.3, (B) PBS with 10 mM EDTA, (D) PBS with 10 mM EDTA, 4 g/liter sodium azide and 5 mM 2-deoxy-D-glucose. All samples were made isomolar with NaCl. For (C) the cells were measured within a few seconds after the sample was removed from ice, whereas the other samples were prepared and measured at room temperature.

level where a fully effective efflux pump is required to keep the dye out, whereas when the pump is inhibited it is not able to keep up with dye influx in this buffer. In the other buffers the permeability is so low that even a strongly inhibited pump is able to keep the intracellular dye concentration at a low level, although noticeably above zero.

The practical consequence of these results is that vital staining of *E. coli* can be carried out by the following procedure. Add to a cell suspension an equal volume of 40 μg/ml ethidium bromide in 20 mM EDTA, 8 g/liter sodium azide, and 10 mM 2-deoxy-D-glucose, and measure in the flow cytometer at room temperature within 2 hr. The cells are washed once before staining if required to remove excess amounts of debris.

It must be emphasized that this procedure cannot be applied to all kinds of bacteria, nor necessarily to other dyes. The "rules" that govern vital staining of bacteria appear to be quite complex. We have found great variation between different gram-positive species, as well as significant differences between different DNA specific dyes (Walberg *et al.*, 1998). For example, ethidium bromide stains both *Bacillus cereus* and *Bacillus aureus* to the level of ethanol fixed cells or above in PBS at room temperature, but not *Enterococcus faecalis*. Hoechst 33258 stains only *Bacillus cereus,* and only in the presence of a metabolic inhibitor carbonyl cyanide m-chlorophenylhydrazone (CCCP). And so on. Interestingly, the vital staining in some cases exceeds that of fixed cells by as much as a factor of 2, supposedly reflecting differences in DNA conformation. This was found also for nonintercalating dyes such as Hoechst 33342.

C. Spores

Some species of bacteria can sporulate. The spores appear to be quite dry, which supposedly is the reason why we have not been able to stain their DNA. Neither have we found literature reports to show that others have been more successful.

D. Sheath Fluid

A prerequisite for detection of bacteria by light scattering is that the sheath water is free of other particles with light scattering comparable to that of bacteria. Distilled water may contain significant amounts of microscopic particles (presumably mainly silicates) with sizes that produce light scattering signals of the same order of magnitude as do many bacteria. The water may therefore produce a sizable background in the light scattering channel. To avoid this artifact, both sheath water and reagent water should be filtered, preferably with a somewhat finer filter than the standard 0.22 μm pore size, which transmits significant numbers of particles larger than 0.3 μm. Because these particles may be silicates with a much higher refractive index than biological cells, they may have light scattering at the level of bacteria.

E. Flow Cytometers

Owing to the small size and low DNA content of bacteria, the sensitivity of the flow cytometer with regard to light scattering and fluorescence is critical. We have obtained data on fluorescence as well as low- and large-angle light scattering with adequate quality using an arc lamp-based instrument (Steen, 1990b). (A commercial model of this instrument, Bryte HS, has been available from Bio-Rad, Hercules, CA.) The optical configuration of the arc lamp-based instrument is essentially similar to that of the epifluorescence microscope (Steen and Lindmo, 1979). In addition it comprises a dark-field configuration that facilitates measurement of light scattering at small and large scattering angles in separate detectors (Steen and Lindmo, 1985; Steen, 1986). The instrument employs a 100 W high pressure mercury arc lamp as the excitation light source. For measurement of cells stained with mithramycin/ethidium bromide, a filter combination is used (B1 filter block) that contains a dichroic excitation filter with a transmission band between 390 and 440 nm, a dichroic beam splitter having a characteristic wavelength of 460 nm, a long-pass emission filter with transmission from 470 nm, and a short-pass emission filter with transmission below 720 nm. The excitation filter of the B1 block transmits primarily the 405 and 436 nm mercury emission lines. In particular, the 436 nm line is quite intense and coincides with the absorption peak of DNA-bound mithramycin. The 720 nm short-pass filter is used to eliminate red and infrared background light, which would otherwise increase the shot noise on the signal and thereby reduce the effective sensitivity. For the measurement of vital cells stained with ethidium bromide we use a G1 filter block that has an excitation filter, transmitting the strong Hg line at 546 nm, a long-pass emission filter with transmission above 570 nm, plus the previously mentioned short-pass 720 nm filter.

Instrument sensitivity is the major concern in measurement of bacteria. It is a function of the background of shot noise on the signal as much as of the signal itself. This noise is primarily due to the stochastic nature of the emission of light and of the photoelectrons that the light induces in the light detectors. Thus, the signal to noise ratio, S/N, is the essential parameter in this regard (Steen, 1992). In order to reduce noise the flow cytometer should be used with the smallest permissible excitation and emission slits in order to eliminate light from sources other than the cells. (The emission slit is equivalent to the "pinhole" in other types of flow cytometers.) To increase the signal the flow velocity may be reduced by lowering the sheath pressure. (In the Bryte instrument a pressure around 0.5 kg/cm^2 is suitable.) This can be done without affecting measuring precision because the cell density of samples from bacteria cultures are typically high, that is, 10^8–10^9 cells/ml, and the sample flow rate can therefore be kept correspondingly low, that is, 0.1–1 μl/min. An instrument with volumetric sample injection, that is, injection of the sample by means of a syringe pump, is preferable to achieve a stable sample flow at this level. Volumetric sample injection also facilitates direct determination of the cell density of the sample.

Simultaneous light scattering detection to gate the fluorescence measurement is essential in order to distinguish degraded cells and debris from intact cells. With samples having a high content of debris, light scattering can be a prerequisite for adequate data quality. Measurement of light scattering is not only to gate out irrelevant particles, it may also be used to determine cellular size and dry weight. Furthermore, dual parameter low-angle versus large-angle light scattering measurement can be used to distinguish dead cells from vital ones (Steen, 1986). Thus, as with mammalian cells, dead cells have a higher ratio of large-angle to low-angle scattering. Cells treated with β-lactam antibiotics, for example, penicillin, may be an exception to this (see Chapter 55 of this volume).

The performance of the arc lamp-based Bryte instrument was compared to that of a FACStar Plus (Becton Dickinson, San Jose, CA) (Bernander *et al.,* 1998). Whereas the fluorescence sensitivity appeared to be roughly the same for the two instruments, the laser-based instrument could just detect, but not really measure, the light scattering.

IV. Standards and Controls

When bacteria grow sufficiently slowly, or when they are treated with drugs such as rifampicin, they yield DNA histograms with fairly sharp peaks. In some applications it may be of interest to know how much of the variation represented by the width of these peaks represents the precision of measurement and how much is due to variation of the sample. The latter, of course, is the combined result of the biological variation of the sample and the variation in staining. In measurements of bacteria, which typically challenges the sensitivity of the instrument, the variation related to the flow cytometer, cv_p, is limited primarily by photon noise according to Eq. (1):

$$cv_p = [(f + b)/a\Phi_e i_x f^2]^{1/2} \tag{1}$$

where f is the number of fluorescent molecules in the cell that yields the signal, b is a measure of the background fluorescence (i.e., light due to Raman scattering of the sheath fluid, fluorescence from free dye in the sample and from optical components, and scattered excitation light leaking through filters), Φ_e is the photoelectron quantum yield of the light detector, i_x is the excitation light intensity, and a is a constant (Steen, 1992). The variation measured for a given sample will be the result of the variation associated with the instrument, cv_p, and the variation within the sample, cv_s, according to the laws of statistics:

$$cv_m^2 = cv_p^2 + cv_s^2 \tag{2}$$

Combining Eqs. (1) and (2) we have

$$cv_m^2 = [(f + b)/a\Phi_e i_x f^2] + cv_s^2 \tag{3}$$

From Eq. (3) it is seen that by plotting cv_m^2 versus $1/i_x$ one obtains a straight line that extrapolates to the value of cv_s^2. Hence, cv_s can be determined by measurement of cv_m with two or more values of i_x. In some instruments i_x may be varied simply by the laser power. If the laser runs at a fixed power, or if it is an arc lamp-based instrument, i_x can be varied by means of gray filters in the excitation light path. The relative values of i_x can be taken directly from the position of the corresponding histogram peak(s).

Essentially the same equations apply to light scattering detection, except that f is replaced by a number which is a function of the parameters that determine the light scattering characteristics of the cell, such as size and refractive index.

Monodisperse fluorescent particles (available from several companies) should always be run after the instrument has been turned on in order to check its performance, and the instrument should be adjusted for optimum performance if required. The particles may also be used as a standard of signal intensity for both fluorescence and light scattering.

DNA histograms of bacteria in many cases have no sharp peaks that can be used to check the preparation of the sample as well as the instrument function. One or more standard biological samples are therefore most helpful. For this purpose we use *E. coli* cells treated with the antibiotic rifampicin. This treatment produces cells giving DNA histograms with two major peaks, representing cells with either two and four chromosomes or four and eight chromosomes, depending on the growth rate of the bacterial culture when the drug was applied (Fig. 2). Such histograms may thus be used both to calibrate the fluorescence scale versus DNA content and to check the resolution of the measurement. (Rifampicin inhibits initiation of chromosome replication, while replication forks already started when the drug is given are allowed to run to completion.) A most convenient control sample may be prepared from *E. coli* lacking a functional Dam methyltransferase. Such mutant cells lack the ability to coordinate multiple initiations and therefore contain all integral numbers of chromosomes (one, two, three, four, etc., and not just two and four) (Boye *et al.,* 1988; Boye and Løbner-Olesen, 1990).

When examining bacteria for the contents of replication origins by treatment with rifampicin, it is important that cell division is stopped as rapidly as possible, so that the number of origins per cell at the time of rifampicin addition can be measured. If cell division occurs after replication initiation has been stopped, the number of origins per cell will be too low. For *E. coli* this can be achieved by using the cell division inhibitors furazlocillin or cephalexin. At low concentrations, these drugs inhibit septum formation within 1 to 2 min (Boye and Løbner-Olesen, 1991)

The procedure for preparing this standard sample is as follows. An *E. coli* wild-type strain or a *dam* mutant is grown in LB mcdium with 0.2% glucosc at 37°C to an optical density (OD, 600 nm) of 0.1. At this point, when the cells are in exponential growth, rifampicin (15 mg/ml in ethanol) is added to the culture to a final concentration of 150–300 μg/ml. If required, cell division is stopped by the addition of furazlocillin (4 μg/ml) or cephalexin (10 μg/ml). After a further 3 hr of culture, cells are harvested and fixed according to the above procedure.

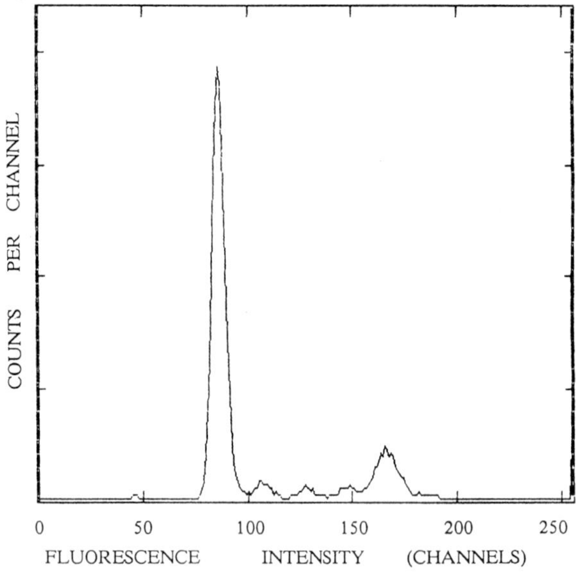

Fig. 2 Fluorescence histogram of *E. coli* K-12 cells, strain CM735, harvested from a culture grown for 3 hr in the presence of the antibiotic rifampicin, which was given while the cells were in rapid exponential growth. The cells were fixed and stained with a combination of mithramycin and ethidium bromide according to the present procedure. The two prominent peaks represent cells with four and eight chromosomes, respectively. Peaks due to cells with two, five, six, and seven chromosomes are also evident. The histogram can thus be used to calibrate the fluorescence axis with regard to cellular DNA content. The cv of the main histogram peak is 3.5%.

Provided the instrument has sufficient sensitivity and is properly tuned, the cv of the fluorescence peaks should be of the order of 5% or lower.

References

Bernander, R., Stokke, T., and Boye, E. (1998). Flow cytometry of bacterial cell: Comparison between different flow cytometers and different DNA stains. *Cytometry* **31,** 29–36.

Boye, E., and Løbner-Olesen, A. (1990). The role of Dam methyltransferase in control of DNA replication in *Escherichia coli. Cell* **62,** 981–989.

Boye, E., and Løbner-Olesen, A. (1991). Bacterial growth studied by flow cytometry. *Res. Microbiol.* **142,** 131–135.

Boye, E., Steen, H. B., and Skarstad, K. (1983). Flow cytometry of bacteria: A promising tool in experimental and clinical microbiology. *J. Gen. Microbiol.* **129,** 973–980.

Boye, E., Løbner-Olesen, A., and Skarstad, K. (1988). Timing of chromosomal replication in *Escherichia coli. Biochim. Biophys. Acta* **951,** 359–364.

Langlois, R. G., and Jensen, R. H. (1979). Interactions between pairs of DNA-specific fluorescent stains bound to mammalian cells. *J. Histochem. Cytochem.* **27,** 72–79.

Løbner-Olesen, A., Skarstad, K., Hansen, F. G., von Mayenburg, K., and Boye E. (1989). The DnaA protein determines the initiation mass of *Escherichia coli* K-12. *Cell* **57,** 881–889.

Jernaes, M., and Steen, H. B. (1994). Rapid staining of *E. coli* cells for flow cytometry: Influx and efflux of ethidium bromide. *Cytometry* **17,** 302–309.

Lewis, K. (1994). Multidrug resistance pumps in bacteria: Variations on a theme. *Trends Biochem. Sci.* **19,** 119–123.

Pinkel, D., and Stovel, R. (1985). Flow chambers and sample handling. *In* "Flow Cytometry: Instrumentation and Data Handling" (M. A. Van Dilla, P. N. Dean, O. D. Laerum, and M. R. Melamed, eds.), pp. 77–128. Academic Press, London.

Skarstad, K., and Boye, E. (1988). Perturbed chromosomal replication in recA mutants of *Escherichia coli. J. Bacteriol.* **170,** 2549–2554.

Skarstad, K., Steen, H. B., and Boye, E. (1983). Cell cycle parameters of slowly growing *E. coli* B/r studied by flow cytometry. *J. Bacteriol.* **154,** 656–663.

Skarstad, K., Steen, H. B., and Boye, E. (1985). *Escherichia coli* DNA distributions measured by flow cytometry and compared to theoretical computer simulations. *J. Bacteriol.* **163,** 661–668.

Skarstad, K., Boye, E., and Steen, H. B. (1986). Timing of initiation and replication in individual *Escherichia coli* cells. *EMBO J.* **5,** 1711–1717.

Steen, H. B., and Lindmo, T. (1979). Flow cytometry: A high resolution instrument for everyone. *Science* **204,** 403–404.

Steen, H. B., and Lindmo, T. (1985). Differential light scattering detection in an arc lamp-based epi-illumination flow cytometer. *Cytometry* **6,** 281–285.

Steen, H. B. (1986). Simultaneous separate detection of low angle and large angle light scattering in an arc lamp-based flow cytometer. *Cytometry* **7,** 445–449.

Steen, H. B. (1990a). Flow cytometric studies of microorganisms. *In* "Flow Cytometry and Sorting" (M. R. Melamed, T. Lindmo, and M. L. Mendelsohn, eds.), pp. 605–622. Alan R. Liss, New York.

Steen, H. B. (1990b). Characteristics of flow cytometers. *In* "Flow Cytometry and Sorting" (M. R. Melamed, T. Lindmo, and M. L. Mendelsohn, eds.), pp. 11–25. Alan R. Liss, New York.

Steen, H. B. (1992). Noise, sensitivity, and resolution of flow cytometers. *Cytometry* **13,** 822–830.

Walberg, M., Gaustad, P., and Steen, H. B. (1997). Rapid assessment of ceftazidime, ciprofloxacin, and gentamicin susceptibility in exponentially-growing *E. coli* cells by means of flow cytometry. *Cytometry* **27,** 169–178.

Walberg, M., Gaustad, P., and Steen, H. B. (1998). Rapid preparation procedure for staining of exponentially growing *P. vulgaris* cells with ethidium bromide: A flow cytometry-based study of probe uptake under various conditions. *J. Microbiol. Methods* **34,** 49–58.

Zante, J., Schumann, J., Barlogie, B., Goehde, W., and Buchner, T. (1976). New preparations and staining procedures for specific and rapid analysis of DNA distributions. *In* "Pulsecytophotometry. Second International Symposium" (W. Goehde, J. Schumann, and T. Buchner, eds.), pp. 97–106. European Press, Gent, Belgium.

CHAPTER 55

Flow Cytometric Monitoring of Bacterial Susceptibility to Antibiotics

Mette Walberg* **and Harald B. Steen**[†]

* Institute of Medical Microbiology
National Hospital, University of Oslo
0027 Oslo, Norway

† Department of Biophysics
Institute for Cancer Research
0310 Oslo, Norway

I. Introduction
II. Fluorescent Dyes
III. Uptake of Fluorescent Dyes
    A. Viable Cells
    B. Energy-Dependent Drug Efflux
    C. Fixed versus Viable Cells
IV. Effects of Antibiotics
V. Assessment of Drug Effects
    A. Rapidly Growing Bacteria
    B. Mycobacteria
VI. Applications of Flow Cytometry to Medical Microbiology
References

I. Introduction

In the 1980s many people still believed that the era of bacterial diseases had passed because bacterial infections could be successfully controlled. However, despite the increasing number of antibacterial drugs, there is now growing concern that infectious diseases may get out of control. Emergence of new pathogens and increasing resistance toward antibacterial drugs have increased the need for rapid diagnostics.

The essential steps of the diagnostic procedure are detection and counting, identification, and susceptibility testing. A microbiological diagnosis is not fin-

METHODS IN CELL BIOLOGY, VOL. 64

ished until these crucial steps have been carried out. The majority of assays used today depend on formation of visible colonies on agar. This implies 24 hr of incubation for the step of detection and counting, and another 24 hr or more for identification and susceptibility testing. Thus, the minimum time required from sampling to a full diagnosis of a bacterial infection is at least 48 hr. For slowly growing bacteria, such as mycobacteria, the corresponding intervals are drastically increased, that is, from days to weeks. The result of this practice is often suboptimal treatment of the patient, even use either of drugs that have no effect on the actual pathogen or, even worse, drugs that favor growth conditions of the pathogen through suppression of the normal flora of the patient.

Although new techniques based on nucleic acid hybridization and direct detection of the infectious organism through binding of specific antibodies have contributed increasingly within medical microbiology, agar incubation remains the standard. On this background flow cytometry may have a significant potential within medical microbiology, not least within monitoring of bacterial drug susceptibility testing. This potential is demonstrated by the increasing number of articles published within this field (Allman *et al.*, 1992; Amann *et al.*, 1990; Bernander *et al.*, 1998; Bownds *et al.*, 1996; Boye *et al.*, 1983; Clarke and Pinder, 1998; Cohen and Sahar, 1989; Comas and Vives-Rego, 1997, 1998; Dennis *et al.*, 1983, 1985; Diaper and Edwards, 1994; Diaper *et al.*, 1992; Durodie *et al.*, 1995; Gant *et al.*, 1993; Guindulain *et al.*, 1997; Humphreys *et al.*, 1994; Ianelli *et al.*, 1998; Jepras *et al.*, 1995, 1997; Jernaes and Steen, 1994; Kirk *et al.*, 1997, 1998; Mansour *et al.*, 1985; Martinez *et al.*, 1982; Mason *et al.*, 1994; Mason and Lloyd, 1997; Miller and Quarles, 1990; Monfort and Baleux, 1996; Mortimer *et al.*, 2000; Nebe-von Caron and Badley, 1995; Nebe-von Caron *et al.*, 1998; Norden *et al.*, 1995; Ordonez and Wehman, 1993, 1995; Phillips and Martin, 1988; Porter *et al.*, 1995a,b; Ramani *et al.*, 1997; Robertson and Button, 1989; Skarstad *et al.*, 1986; Steen *et al.*, 1982; Suller and Lloyd, 1998; Suller *et al.*, 1997; Trousselier *et al.*, 1993; van Dilla *et al.*, 1983; Walberg *et al.*, 1996, 1997a,b, 1998a; Wickens *et al.*, 2000).

II. Fluorescent Dyes

The main applications of flow cytometry to bacteria, using fluorescent dyes, have been measurements of the cellular content of nucleic acids and protein, membrane potential, and enzyme activities, all of which have been used to monitor bacterial drug susceptibility.

One obvious target for fluorescent probes in bacteria is the nucleic acids. In determination of drug susceptibility DNA appears as the better target since the cellular RNA content may vary quite significantly with growth conditions. Ethidium bromide (EB) as well as propidium iodide (PI) form complexes with double stranded DNA and RNA by intercalating between base pairs (LePeck and Paoletti, 1967). In contrast, mithramycin (Mi), which binds to G–C-rich sequences of DNA, is highly DNA specific (Ward *et al.*, 1965). The fluorescence yield of Mi is relatively low, but it may be increased by combining it with ethidium bromide. Both EB, PI, and the Mi/EB combination have been used in flow cytometry (FCM)-based studies to assess drug effects in bacteria (Boye *et al.*, 1983; Cohen and Sahar, 1989; Gant *et al.*, 1993; Martinez *et al.*, 1982; Steen *et al.*, 1982; Walberg *et al.*, 1996, 1997a,b). The DNA specific Hoechst dyes (H 33342 and H 33258) and 4′,6-diamidino-2-phenylindole (DAPI) are likely to perform in much the same way as Mi/EB (Bernander *et al.*, 1998), although they have still not been used in bacterial drug susceptibility studies.

Fluorescein isothiocyanate (FITC) was applied by Durodie *et al.* (1995) in a drug susceptibility assay based on the measurement of total cellular protein content.

Membrane potential dependent dyes, or so-called viability markers, such as oxonols, accumulate inside dead cells owing to their lipophilic anionic structure. In contrast to oxonols, rhodamine 123 and carbocyanines are accumulated cytosolically by cells with membrane potential, that is, viable cells, thanks to their lipophilic cationic nature. For assessment of bacterial susceptibility toward antibiotics, the oxonol DiBAC$_4$(3) has already been used quite extensively (Jepras *et al.*, 1997; Mason *et al.*, 1994; Suller *et al.*, 1997; Suller and Lloyd, 1998). According to Jepras *et al.* (1995), DiBAC$_4$(3) fluorescence correlates well with plating efficiency. A carbocyanine dye has also been used in one susceptibility assay (Ordonez and Wehman, 1993). Also 5-cyano-2,3-di-4-tolyltetrazolium chloride (CTC) and fluorescein diacetate (FDA), which are cleaved intracellularly by the metabolically active cell to yield the corresponding fluorochrome, have been used; namely, FDA was used in mycobacterial susceptibility studies (Bownds *et al.*, 1996; Kirk *et al.*, 1998; Norden *et al.*, 1995), and CTC was used in analyses of staphylococcal drug response (Suller and Lloyd, 1998).

In addition to bacterial drug susceptibility assays, several studies addressing bacterial viability in general have been published. These articles have been based mainly on measurement of rhodamine 123 and cyanine associated fluorescence, sometimes in combination with the oxonol DiBAC$_4$(3). CTC and FDA have also been used to monitor bacterial metabolic status (Comas and Vives-Rego, 1997; Diaper *et al.*, 1992; Diaper and Edwards, 1994; Kapreylants and Kell, 1992, 1993a,b; Lopez-Amoros *et al.*, 1995, 1997; Monfort and Baleux, 1996; Nebe-von Caron and Badley, 1995; Nebe-von Caron *et al.*, 1998; Porter *et al.*, 1995a,b; Sheppard *et al.*, 1997).

III. Uptake of Fluorescent Dyes

A. Viable Cells

The bacterial cell wall is a complex barrier that facilitates permeation of essential molecules such as nutrients and waste products, while the intake of a wide range of other compounds, among them most fluorescent dyes, is strongly restricted. Intracellular binding of dye is the net result of influx and efflux through the bacterial wall. Influx is believed to be a passive process driven by the concentration gradient. The efflux is an energy-dependent, active process that can be inhibited by metabolic inhibitors, such as carbonyl cyanide *m*-chlorophenylhydrazone (CCCP). In gram-negative bacteria the outer membrane (OM) constitutes the main permeation barrier (Fig. 1). The OM is impermeable to macromolecules and allows only limited diffusion of hydrophobic substances owing to its lipopolysaccharide (LPS)-covered surface. Small hydrophilic compounds may diffuse through the OM via porin channels, which are believed to be water filled, but the narrowness of the porins restricts their diffusion (Nikaido and Vaara, 1985). Polymyxin and Tris are cations that may permeabilize the OM (Hancock, 1984; Nikaido and Vaara, 1985; Storm *et al.*, 1977). The OM disorganization and the resulting permeabilizing effect of chelators such as EDTA are also well known (Leive, 1965). In contrast to other members of the family Enterobacteriaceae, *Proteus* species are not affected by EDTA and other commonly used OM permeabilizers, apparently as a result of the high content of phosphate-linked 4-aminoarabinose in their LPS (Sidorczyk *et al.*, 1983).

Gram-positive cells have no OM. The corresponding barrier in these bacteria is represented by the thicker peptidoglycan (PG) layer (Fig. 1). The structure of the PG layer is the product of the cooperative interaction of a number of proteins including the transpeptidases or so-called penicillin binding proteins (PBPs). β-Lactam antibiotics bind to these PBPs, thereby inhibiting the building of the PG layer and increasing cellular permeability (Boneca *et al.*, 1997; Severin

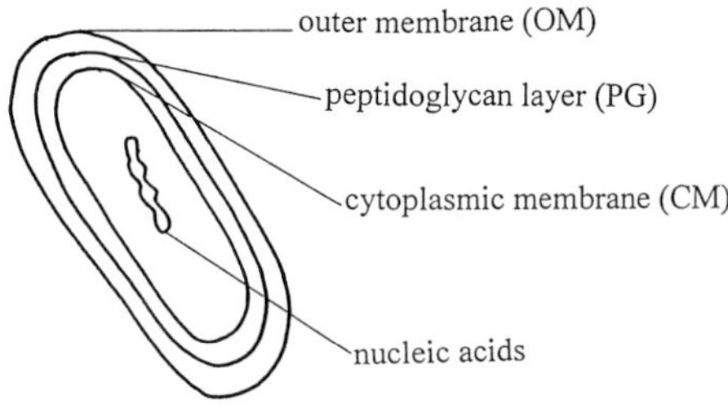

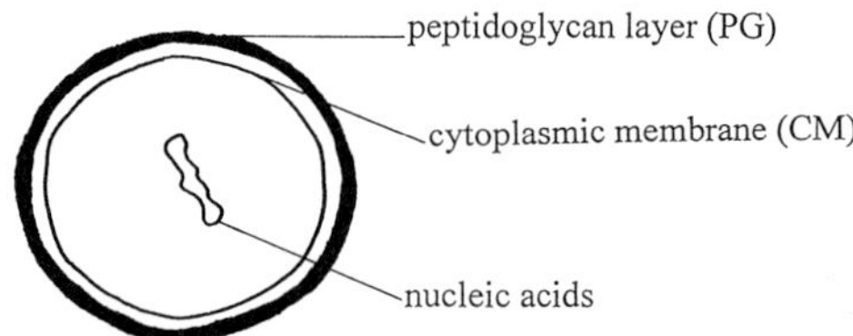

Fig. 1 Schematic drawing of the bacterial cell wall layers.

et al., 1997). Also, electroporation is known to permeabilize the PG layer; this procedure is commonly used to increase uptake of foreign DNA in bacteria (Belliveau and Trevors, 1989; Cruz-Rodz and Gilmore, 1990; Dunny *et al.,* 1991), but it has not been used in bacterial drug susceptibility assays. Substances such as EDTA and Tris are known not to affect PG layer permeability.

B. Energy–Dependent Drug Efflux

In addition to low permeability, active efflux may contribute to reduce the net uptake of noxious agents. The multidrug resistance pumps (MDRs) expel a wide range of molecules from microorganisms, thus playing an important role in conferring resistance toward toxic compounds, such as antibiotics, metal ions, detergents, as well as some fluorescent dyes. During the last 15 years ethidium resistance mediated by efflux has been discovered in several bacterial species, including *Escherichia coli* (Lewis, 1994; Jernaes and Steen, 1994), *Staphylococcus aureus* (Ng *et al.,* 1994), *Bacillus subtilis* (Ahmed *et al.,* 1995), and *Neisseria gonorrhoeae* (Hagmann *et al.,* 1995). Through the last few years increasing interest has been directed toward efflux, clearly because it emerges as an important mechanism in antibiotic drug resistance. Among the antibiotics that are subject to efflux in bacteria are tetracycline, fluoroquinolones, and β-lactam antibiotics (Poole *et al.,* 1993). It is obvious that efflux may be a severe problem in bacterial assays that employ vital staining (Fig. 1). However, it has been shown that dyes, which are substrates for such highly efficient efflux, may still be used to monitor drug effects in some bacteria, provided metabolic inhibitors, such as CCCP or sodium azide, are used (Walberg *et al.,* 1997a).

C. Fixed versus Viable Cells

Ethanol fixation appears to permeabilize the cell wall completely for smaller molecules, such as most fluorescent dyes (Fig. 1) (Allman *et al.,* 1992; Amann *et al.,* 1990; Boye *et al.,* 1983; Jernaes and Steen, 1994; Mansour *et al.,* 1985; Martinez *et al.,* 1982; Phillips and Martin, 1988; Robertson and Button, 1989; Skarstad *et al.,* 1986; van Dilla *et al.,* 1983; Walberg *et al.,* 1996, 1997a,b), and it abolishes energy-dependent efflux (Jernaes and Steen, 1994; Midgley, 1986, 1987). Thus, the diffusion of fluorescent dyes into fixed cells levels off only at dye equilibrium across the cell wall, and the binding of dye is determined solely by the concentration of the dye and its binding constant. However, ethanol fixation has several disadvantages: (1) It abolishes vital functions such as membrane potential and efflux of molecules. (2) It may cause aggregation of cells and thereby falsely low counts and distorted fluorescence and light scattering histograms. If the tendency to aggregate differs between subpopulations, fixation may cause errors in the relative magnitude of these populations. (3) Fixation, which requires centrifugation and washing, may cause further cell loss and is time-consuming. The latter point is an important consideration in designing protocols for routine assays.

An alternative that avoids most of these problems is heat treatment. Cohen and Sahar (1989) showed that both gram-negative and gram-positive bacteria could be stained after 10 min at 70°C. However, this procedure causes DNA denaturation and may thus interfere with staining. Cold shock (0°C) in combination with metabolic inhibitors was found to permeabilize gram-negative cells, and it does not cause DNA denaturation (Jernaes and Steen, 1994; Walberg *et al.*, 1998a). However, this procedure did not work for gram-positive cells. For these cells ethanol fixation may still be required (Walberg *et al.*, 1999). Some bacteria may be stained with nucleic acid targeting dyes without prior fixation. Thus, in 1993 Gant *et al.* showed that nucleic acid associated fluorescence may facilitate measurements of antibiotic effects in unfixed *E. coli* cells. However, some of their results were affected by active efflux of their fluorescent dye, PI.

IV. Effects of Antibiotics

Antibacterial drugs inhibit bacteria via different modes of action, and they may be classified as bactericidal or bacteriostatic. The ultimate effect of any bactericidal and bacteriostatic antibiotic substance on susceptible cells is proliferation arrest, in some cases followed by cell death and ensuing disintegration (bactericidal antibiotics). β-Lactams, for example, penicillin, and fluoroquinolones, for example, ciprofloxacin, are examples of bactericidal antibiotics. The β-lactam antibiotics inhibit PG layer synthesis via binding to various PBPs, thereby giving rise to swelling or filament formation in susceptible cells (gram-positive cocci and Enterobacteriaceae, respectively). DNA synthesis, on the other hand, is not affected directly. Hence, the DNA content will continue to increase in the presence of β-lactams. Fluoroquinolones are DNA gyrase inhibitors that stop DNA replication. The fluoroquinolones, such as ciprofloxacin, cause filament formation in susceptible *E. coli* cells via induction of the cellular SOS response that is part of the bacterial DNA repair mechanisms. Among the antibiotics that may be classified as bacteriostatic are protein synthesis inhibitors such as chloramphenicol, tetracycline, and the macrolides. The bactericidal action of aminoglycosides, which inhibit protein synthesis, is explained by their additional permeabilizing effect on the OM.

Rifampicin and isoniazide are used to treat mycobacterial infections. They exert their action through inhibition of RNA polymerase and cell wall synthesis, respectively. Vancomycin, which is classified as a bactericidal drug, affects the bacterial cell via several targets (inhibition of wall and protein synthesis and cytoplasmic membrane permeabilization).

V. Assessment of Drug Effects

A. Rapidly Growing Bacteria

The most obvious effect of antibiotics is arrest of proliferation, as reflected by a constant or declining cell number. Precise determination of cell number is best performed with an instrument with volumetric sample injection that delivers a calibrated sample flow, thus allowing direct calculation of the cell density.

The extensive morphological changes obtained by β-lactams are a result of their binding to the PBPs in the PG layer. Drugs that bind to the PBP 3 receptor in *E. coli* (e.g., ampicillin, ceftazidime) lead to the formation of filaments, whereas PBP 2 receptor binding by, for example, mecillinam and carbapenems, inhibits the formation of the cylindrical part of the rod-shaped bacteria, thus giving rise to formation of spheroids. Quinolones inhibit DNA replication through binding to DNA gyrase. The filamentation that occurs as a result of quinolone incubation is not a result of wall targeting, but rather an effect of the cellular SOS response induction (Friedman *et al.*, 1995). In contrast to the lethal filamentation caused by β-lactams, the filamentation caused by quinolones is part of a cellular defense response toward the DNA damaging effect of these drugs, a response that allows cellular repair mechanisms to operate to avoid division of cells with defective DNA. Despite the attack of quinolone on DNA gyrase, the content of nucleic acids tends to increase in such cells. This increase is a result of the DNA repair mechanisms that are induced as part of the SOS response (Friedman *et al.*, 1995).

As the cell walls of cells treated with β-lactams become thinner and more fragile, further incubation eventually causes disintegration. Since cell division, but not DNA synthesis, is inhibited, the DNA content continues to grow up to that point. Hence, adding these drugs to exponentially growing bacterial cultures will cause (a) stagnation of the increase of cell density, (b) an increase in light scattering, (c) an increase in DNA content, (d) an increase in the ratio of low-angle and large-angle light scattering, reflecting a reduced refractive index [this reduction of (average) refractive index is due to the fact that the cell wall, which represents a significant fraction of the cell dry weight, becomes increasingly thinner due to the inhibition of PG synthesis, so that the ratio of dry weight to cell volume decreases], (e) disintegration of cells, as detected as an increased number of small particles [i.e., increasing number of counts in the lower (left) channels of the light scattering histograms, with little or no DNA-associated fluorescence], and (f) a strongly reduced membrane potential. Figure 2 shows the effects on light scattering, DNA content, cell number, and optical density in *E. coli* following exposure to β-lactam antibiotics (ampicillin, mecillinam, ceftazidime) and the quinolone ciprofloxacin.

Significantly, all of these parameters, except DNA content, can be assessed without staining of the cells. Hence, they can be measured directly on vital cells without fixation. This should allow additional assessment of membrane potential, for example, by oxonol-associated fluorescence, and enzyme activities, as measured by CTC and FDA associated fluorescence (see earlier). As noted already, assays that do not require fixation should also have a big advantage in clinical routine applications. Similar to the drugs that cause filamentation, the effect of mecillinam may be detected as increase in light scattering and fluorescence intensities, although the effect is not as great as that resulting from filamentation (Fig. 2) (Walberg *et al.*, 1996).

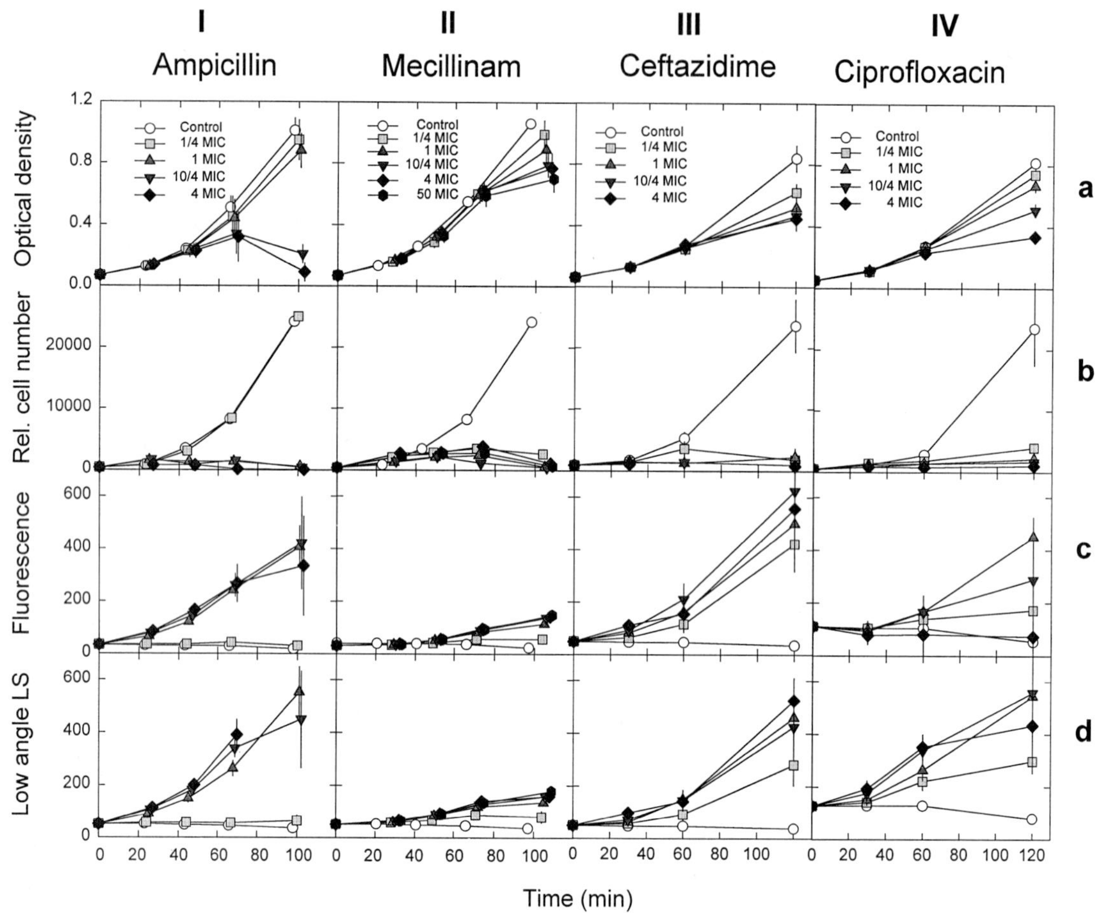

Fig. 2   Optical density ($OD_{600\ nm}$) (a), relative cell number (b), fluorescence (c), and low-angle light scattering (d) of exponentially growing *E. coli* cells as a function of time with ampicillin, mecillinam, ceftazidime, and ciprofloxacin. Fluorescence is a measure of cellular DNA content (cells were stained with Mi/EB), while light scattering reflects cellular size. The data in (b–d) are the mean of the median values obtained from flow cytometric histograms, whereas the data in (a) are the mean values of measurements of the cell culture $OD_{600\ nm}$. The increase in light scattering with time reflects the formation of filaments (ampicillin, ceftazidime, ciprofloxacin) and spheroids (mecillinam) (d) of cells unable to divide (b). The corresponding increase in fluorescence reflects cellular DNA synthesis that continues despite antibiotic exposure (c). The OD reflects the product of cell size and number (a). Note that the light scattering measurements (d) can be carried out on unfixed and unstained cells, which implies that the cells can be prepared for measurement in just a few minutes. The cells were analyzed by means of an Argus flow cytometer, a mercury lamp-based instrument equipped with volumetric sample injection, which is necessary for direct determination of cell number (b).

The time required for the effects of antibiotics to become statistically significant depends on the cell doubling time. Typically, however, they become evident within once and twice the doubling time, that is, typically within 30 min for rapidly growing bacteria (Fig. 2). The exception is the formation of debris, signifying cell disintegration, which may take an hour or more before it commences. For example, the *E. coli* filament formation that follows incubation with 1 MIC (minimum inhibitory concentration) ampicillin or ciprofloxacin may be detected within 30 min by an increase in light scattering and DNA-associated fluorescence, while disintegrating cells become evident after about 2 hr when the cells begin to disintegrate (Fig. 2) (Walberg *et al.*, 1996, 1997a). Several authors have shown that various β-lactam drugs as well as other types of antibiotics lead to an increase of DNA content in cells within 30 min of incubation. Martinez *et al.* (1982) treated *E. coli* with moxalactam, cefamandole, and cefazoline in clinical concentrations, that is, about the MIC value, and stained the cells with the DNA specific dye combination of Mi/EB. Steen *et al.* (1982) and Boye *et al.* (1983) described effects both on light scattering and DNA-associated fluorescence resulting from treatment of *E. coli* cells with penicillin, chloramphenicol, erythromycin, streptomycin, and doxycycline. However, their results were based on drug concentrations far exceeding the clinical range.

Cohen and Sahar (1989) screened the amikacin susceptibility of 46 clinical isolates from serous cavities, wound exudate, bile, and bronchial lavage. These authors based their results on cell count only, using EB staining to distinguish cells from debris. Drug effects were apparent within 1 hr after drug addition. The cells were stained after a permeabilizing procedure that was based on heat shock.

Ordonez and Wehman (1993) showed in a clinical assay that in *S. aureus* effects of oxacillin as well as penicillin could be detected sensitively by flow cytometry. Their results were based on measurement of membrane potential-associated fluorescence arising from a cyanine (3′,3′-dipentyloxacarbocyanine), which is a dye accumulating over an energized membrane. The dose-dependent reduction in fluorescence of drug-treated cells was monitored 90 min after drug addition.

The clinical potential of flow cytometry in *E. coli* drug responses was also highlighted by Gant *et al.* (1993) in a paper that included a wide range of drugs. Light scattering, both low and large angle, as well as PI-associated fluorescence were the basis for their analyses, that were performed after 3 and 6 hr of drug incubation. The effect of ampicillin and cefotaxime on light scattering was significant, that is, a significant increase of low intensity events occurred with prolonged incubation, demonstrating the formation of debris following β-lactam incubation. Significant effects on light scattering were also observed after mecillinam and ciprofloxacin treatment, whereas the effects of gentamicin on light scattering were not obvious. The Gant *et al.* (1993) paper is based on analysis of nonfixed cells at room temperature, and although PI efflux probably interfered with some of their results, significant effects of the β-lactams and ciprofloxacin were described, whereas effects of gentamicin were hardly detected. The latter drug, which is an aminoglycoside that inhibits ribosome function, is an example that not all drugs give effects that are readily detectable by flow cytometry, except inhibition of proliferation (Walberg *et al.,* 1997a).

The membrane potential sensitive dye DiBAC$_4$(3) has been used in several bacterial drug susceptibility studies. Mason *et al.* (1994) showed that DiBAC$_4$(3) fluorescence could facilitate detection of effects of ampicillin, ciprofloxacin, and gentamicin on *E. coli,* as well as the effect of gentamicin on *S. aureus* after 30 min of incubation. The drugs were used in clinically relevant concentrations. Increasing light scattering and fluorescence were taken as a measure of drug effect. The usefulness of light scattering measurements, which may be performed directly on unstained samples, was pointed out by these authors. Similar effects on DiBAC$_4$(3) fluorescence were demonstrated by Jepras *et al.* (1997) for cefuroxim, ciprofloxacin, and azithromycin. In contrast to Mason *et al.,* however, these authors applied drug concentrations five times the MIC value. DiBAC$_4$(3) has also been applied in assays addressing methicillin responses in *S. aureus* cells (Suller *et al.,* 1997; Suller and Lloyd, 1998). According to these authors, antibiotic susceptibility patterns for penicillin, methicillin, and vancomycin may be determined in 2–4 hr from an overnight culture. In a study focusing on postantibiotic effects in *S. aureus,* they emphasized the potential of flow cytometry in studies of nonhomogeneous cellular responses. Their results, which show active subpopulations 4 hr after the removal of methicillin, suggest that the postantibiotic effect is more complex than what is traditionally assumed (Suller and Lloyd, 1998).

In a study testing sub-MIC drug concentrations on *E. coli,* Durodie *et al.* (1995) observed effects of amoxillin and mecillinam, ciprofloxacin, chloramphenicol, and trimethoprim as early as 1 hr after drug addition. Their analyses were based on measurement of the FITC fluorescence/forward scattering ratio. Sub-MIC effects may be due to a temporal inhibition.

B. Mycobacteria

For slowly growing bacteria such as mycobacteria, the timesaving aspect of flow cytometry is obvious, since this technique is not dependent on proliferation provided the number of cells is sufficient to obtain statistically reliable data. Current diagnostic assays employ radioactively labeled carbon liquid media as indicators of mycobacterial growth. This technique requires days or even weeks for results to be available since bacterial multiplication is necessary. So far, three papers have been published on flow cytometric assessment of mycobacterial drug susceptibility. In all cases the drug effect was observed by a decrease in FDA associated fluorescence. The results demonstrate that drug effects may be detected independent of mycobacterial multiplication. They have all led to the same conclusion: susceptibility testing of these slowly growing microorganisms may be provided within 24 hr with assays that are simple to perform. All the common antituberculosis drugs have been tested in clinical concentrations: rifampicin, isoniazide, ethambutol, and streptomycin in assays including *Mycobacte-*

rium tuberculosis (Kirk *et al.*, 1998; Norden *et al.*, 1995) as well as nontuberculous mycobacteria (Bownds *et al.*, 1996). The latter were tested against ciprofloxacin, clarithromycin, erythromycin, kanamycin, tobramycin, and rifampicin.

VI. Applications of Flow Cytometry to Medical Microbiology

Flow cytometry may be used at different stages of the diagnostic procedure. First, its application to isolates that have been cultured after overnight incubation on agar plates, for example, isolates from urine and wound exudate samples, seems obvious. In these cases, one would deal with pure cultures. Second, the technique could be applied directly to clinical samples such as cerebrospinal fluid, ascites, pericardial and pleural exudates, and synovial fluid, and to positive BACTEC blood culture flasks. In such cases antibiotics susceptibility testing is vital, whereas bacterial identification is less important, at least in the short term. Both approaches will make it possible to significantly reduce the time required for susceptibility results to be available. In the case of samples analyzed without prior culture, susceptibility results could be available almost 2 days earlier than those obtained by conventional diagnostics, while 24 hr could be gained for cultured isolates. In both cases, the ability of the flow cytometer to reveal heterogeneous responses could be used for rapid detection of drug-resistant subpopulations, that is, within few hours after drug addition, or even less for rapidly growing cells like *E. coli* (Walberg *et al.*, 1997b). The timesaving potential of flow cytometry in susceptibility monitoring also makes this method an interesting tool for screening of new antibiotics.

References

Ahmed, M., Lyass, L., Markham, P. N., Taylor, S. S., Vazquez-Laslop, N., and Neyfakh, A. A. (1995). Two highly similar multidrug transporters of *Bacillus subtilis* whose expression is differentially regulated. *J. Bacteriol.* **177,** 3904–3910.

Allman, R., Hann, A. C., Manchee, R., and Lloyd, D. (1992). Characterization of bacteria by multiparameter flow cytometry. *J. Appl. Bacteriol.* **73,** 438–444.

Amann, R. I., Binder, B. J., Olson, R. J., Chisholm, S. W., Deveraux, R., and Stahl, D. A. (1990). Combination of 16S rRNA-targeted oligonucleotide probes with flow cytometry for analyzing mixed microbial populations. *Appl. Environm. Microbiol.* **56,** 1919–1925.

Belliveau, B. H., and Trevors, J. T. (1989). Transformation of *Bacillus cereus* vegetative cells by electroporation. *Appl. Environ. Microbiol.* **55,** 1649–1652.

Bernander, R., Stokke, T., and Boye, E. (1998). Flow cytometry of bacterial cells: Comparison between different flow cytometers and different DNA stains. *Cytometry* **31,** 29–36.

Boneca, I. G., Xu, N., Gage, D. A., de Jonge, B. L., and Tomasz, A. (1997). Structural characterization of an abnormally cross-linked muropeptide dimer that is accumulated in the peptidoglycan layer of methicillin and cefotaxime-resistant mutants of *Staphylococcus aureus. J. Biol. Chem.* **272,** 29053–29059.

Bownds, S. E., Kurzynski, T. A., Norden, M. A., Dufek, J. L., and Schell, R. F. (1996). Rapid susceptibility testing for nontuberculous mycobacteria using flow cytometry. *J. Clin. Microbiol.* **34,** 1386–1390.

Boye, E., Steen, H. B., and Skarstad, K. (1983). Flow cytometry of bacteria: A promising tool in experimental and clinical microbiology. *J. Gen. Microbiol.* **129,** 973–980.

Clarke, R. G., and Pinder, A. C. (1998). Improved detection of bacteria by flow cytometry using a combination of antibody and viability markers. *J. Appl. Microbiol.* **84,** 577–584.

Cohen, C. Y., and Sahar, E. (1989). Rapid flow cytometric detection and determination of susceptibility to amikacin in body fluids and wound exudates. *J. Clin. Microbiol.* **27,** 1250–1256.

Comas, J., and Vives-Rego, J. (1997). Assessment of the effects of gramicidin, formaldehyde, and surfactants on *E. coli* by flow cytometry using nucleic acid and membrane potential dyes. *Cytometry* **29,** 58–64.

Comas, J., and Vives-Rego, J. (1998). Enumeration, viability, and heterogeneity in *S. aureus* cultures by flow cytometry. *J. Microbiol. Methods* **32,** 45–53.

Cruz-Rodz, A. M., and Gilmore, M. S. (1990). High efficiency introduction of plasmid DNA into glycine treated *Enterococcus faecalis* by electroporation. *Mol. Gen. Genet.* **224,** 152–154.

Dennis, K., Srienc, F., and Bailey, J. E. (1983). Flow cytometric analysis of plasmid heterogeneity in *E. coli* populations. *Biotechnol. Bioeng.* **25,** 2485–2490.

Dennis, K., Srienc, F., and Bailey, J. E. (1985). Ampicillin effects on five recombinant *E. coli* strains: Implications for pressure design. *Biotechnol. Bioeng.* **27,** 1490–1494.

Diaper, J. P., and Edwards, C. (1994). The use of fluorogenic esters to detect viable bacteria by flow cytometry. *J. Appl. Bacteriol.* **77,** 221–228.

Diaper, J. P., Tither, K., and Edwards, C. (1992). Rapid assessment of bacterial viability by flow cytometry. *Appl. Microbiol. Biotechnol.* **38,** 268–272.

Dunny, G. M., Lee, L. M., and LeBlanc, D. J. (1991). Improved electroporation and cloning vector system for gram-positive bacteria. *Appl. Environ. Microbiol.* **57,** 1194–1201.

Durodie, J., Coleman, K., Simpson, I. N., Loughborough, S. H., and Winstanley, D. W. (1995). Rapid detection of antimicrobial activity using flow cytometry. *Cytometry* **21,** 374–377.

Friedberg, E. C., Walker, G. C., and Siede, W. (1995). "DNA Repair and Mutagenesis." ASM Press, Washington, DC.

Gant, V. A., Warnes, G., Phillips, I., and Savidge, F. (1993). The application of flow cytometry to the study of bacterial response to antibiotics. *J. Med. Microbiol.* **39,** 147–154.

Guindulain, T., Comas, J., and Vives-Rego, J. (1997). Use of nucleic acid dyes Syto-13, TOTO-1, and YOYO-1 in the study of *E. coli* and marine prokaryotic populations by flow cytometry. *Appl. Environm. Microbiol.* **63,** 4608–4611.

Hagman, K. E., Pan, W., Spratt, B. G., Balthazar, J. T., Judd, R. C., and Schafer, W. M. (1995). Resistance of *N. gonorhoeae* to antimicrobial hydrophobic agents is modulated by the mtrCDE efflux system. *Microbiology* **141,** 611–622.

Hancock, R. E. W (1984). Alteration in outer membrane permeability. *Annu. Rev. Microbiol.* **38,** 237–264.

Humphreys, M. J., Allman, R., and Lloyd, D. (1994). Determination of the viability of *Trichomonas vaginalis* using flow cytometry. *Cytometry* **15,** 343–348.

Ianelli, D., D'Apice, L., Fenizia, D., Serpe, L., Cottone, E., Viscardi, M., and Capparelli, R. (1998). Simultaneous identification of antibodies to *Brucella abortus* and *Staphylococcus aureus* in milk samples by flow cytometry. *J. Clin. Microbiol.* **36,** 802–806.

Jepras, R. I., Carter, J., Pearson, S. C., Paul, F. E., and Wilkinson, M. J. (1995). Development of a robust flow cytometric assay for determining numbers of viable bacteria. *Appl. Environ. Microbiol.* **61,** 2696–2710.

Jepras, R. I., Paul, F. E., Pearson, S. C., and Wilkinson, M. J. (1997). Rapid assessment of antibiotic effects on *E. coli* by bis(1,3-dibutylbarbutyric acid)trimethine oxonol and flow cytometry. *Antimicrob. Agents Chemother.* **41,** 2001–2005.

Jernaes, M. W., and Steen, H. B. (1994). Staining of *E. coli* for flow cytometry: Influx and efflux of ethidium bromide. *Cytometry* **17,** 302–309.

Kapreylants, A., and Kell, D. B. (1992). Rapid assessment of bacterial viability and vitality by Rhodamine 123 and flow cytometry. *J. Appl. Bacteriol.* **72,** 410–422.

Kapreylants, A., and Kell, D. B. (1993a). Dormancy in stationary-phase cultures of *M. luteus:* Flow cytometric analysis of starvation and resuscitation. *Appl. Environ. Microbiol.* **59,** 3187–3196.

Kapreylants, A., and Kell, D. B. (1993b). The use of 5-cyano-2,3-ditolyl tetrazolium chloride and flow cytometry for the visualization of respiratory activity in individual cells of *M. luteus*. *J. Microbiol. Methods* **17**, 115–122.

Kirk, S. M., Callister, S. M., Lim, L. C. M., and Schell, R. F. (1997). Rapid susceptibility testing of *Candida albicans* by flow cytometry. *J. Clin. Microbiol.* **35**, 358–363.

Kirk, S. M., Schell, R. F., Moore, A. V., Callister, S. M., and Mazurek, G. H. (1998). Flow cytometric testing of *Mycobacterium tuberculosis* isolates to ethambutol, isoniazid, and rifampicin in 24 hours. *J. Clin Microbiol.* **36**, 1568–1573.

Leive, L. (1965). Release of lipopolysaccharide by EDTA treatment of *E. coli*. *Biochem. Biophys. Res. Commun.* **21**, 290–296.

LePeck, J. B., and Poaletti, C. (1967). A fluorescent complex between ethidium bromide and nucleic acids. *J. Mol. Biol.* **27**, 87–106.

Lewis, K. (1994). Multidrug resistance pumps in bacteria: Variation on a theme. *Trends Biochem. Sci.* **19**, 119–123.

Lopez-Amoros, R., Mason, D. J., and Lloyd, D. (1995). Use of two oxonols and a fluorescent tetrazolium dye to monitor starvation of *E. coli* in seawater by flow cytometry. *J. Microbiol. Methods* **22**, 165–176.

Lopez-Amoros, R., Castel, S., Comas-Riu, J., and Vives-Rego, J. (1997). Assessment of *E. coli* and *Salmonella* viability and starvation by confocal laser microscopy and flow cytometry using rhodamine 123, DiBAC$_4$(3), propidium iodide, and CTC. *Cytometry* **29**, 289–305.

Mansour, J. D., Robson, J. A., Arndt, C. W., and Schulte, T. H. (1985). Detection of *E. coli* in blood using flow cytometry. *Cytometry* **6**, 186–190.

Martinez, O. M., Gratzner, H. G., Malinin, T. I., and Ingram, M. (1982). The effect of some betalactam antibiotics on *E. coli* studied by flow cytometry. *Cytometry* **3**, 129–133.

Mason, D. J., and Lloyd, D. (1997). Acridine orange as indicator of bacterial susceptibility to gentamicin. *FEMS Microbiol. Lett.* **153**, 199–204.

Mason, D. J., Allman, R., Stark, J. M., and Lloyd, D. (1994). Rapid estimation of bacterial antibiotic susceptibility with flow cytometry. *J. Microsc.* **176**, 8–16.

Midgley, M. (1986). The phosphonium ion efflux system of *E. coli*: Relationship to the ethidium efflux system and energetics study. *J. Gen. Microbiol.* **132**, 3187–3193.

Midgley, M. (1987). An efflux system for cationic dyes and related compounds in *E. coli*. *Microbiol. Sci.* **4**, 125–128.

Miller, J. S., and Quarles, J. M. (1990). Flow cytometric identification of microorganisms by dual staining with FITC and PI. *Cytometry* **11**, 667–675.

Monfort, P., and Baleux, B. (1996). Cell cycle characteristics and changes in membrane potential during growth of *E. coli* as determined by a cyanine fluorescent dye and flow cytometry. *J. Microbiol. Methods* **25**, 79–86.

Mortimer, F. C., Mason, D. J., and Gant, V. A. (2000). Flow cytometric monitoring of antibiotic-induced injury in *E. coli* using cell-impermeant fluorescent probes. *Antimicrob. Agents Chemother.* **44**, 676–681.

Nebe-von Caron, G., and Badley, R. A. (1995). Viability assessment of bacteria in mixed populations using flow cytometry. *J. Microsc.* **179**, 55–66.

Nebe-von Caron, G., Stephens, P., and Badley, R. A. (1998). Assessment of bacterial viability status by flow cytometry and single cell sorting. *J. Appl. Microbiol.* **84**, 988–998.

Ng, E. Y., Trucksis, M., and Hooper, D. C. (1994). Quinolone resistance mediated by *norA*: Physiologic characterization and relationship to *flqB*, a quinolone resistance locus on the *S. aureus* chromosome. *Antimicrob. Agents Chemother.* **38**, 1345–1355.

Nikaido, H., and Vaara, M. (1985). Molecular basis of bacterial outer membrane permeability. *Microbiol. Rev.* **49**, 1–32.

Norden, M. A., Kurzynski, T. A., Bownds, S. E., Callister, S. M., and Schell, R. F. (1995). Rapid susceptibility testing of *Mycobacterium tuberculosis* (H37Ra) by flow cytometry. *J. Clin. Microbiol.* **33**, 1231–1237.

Ordonez, J. V., and Wehman, N. M. (1993). Rapid flow cytometric antibiotic susceptibility assay for *S. aureus*. *Cytometry* **14**, 811–818.

Ordonez, J. V., and Wehman, N. M. (1995). Amphotericin B susceptibility of *Candida* species assessed by rapid flow cytometric membrane potential assay. *Cytometry* **22,** 154–157.

Phillips, A. P., and Martin, K. L. (1988). Limitation of flow cytometry for the specific detection of bacteria in mixed populations. *J. Immunol. Methods* **106,** 109–117.

Poole, K., Krebes, K., McNally, C., and Neshat, S. (1993). Multiple antibiotic resistance of *Ps. aeruginosa:* Evidence for involvement of an efflux operon. *J. Bacteriol.* **175,** 7363–7372.

Porter, J., Edwards, C., and Pickup, R. W. (1995a). Rapid assessment of physiological status in *E. coli* using fluorescent probes. *J. Appl. Bacteriol.* **75,** 399–408.

Porter, J., Pickup, R., and Edwards, C. (1995b). Membrane hyperpolarization by valinomycin and its limitations for bacterial viability assessment using rhodamine 123 and flow cytometry. *FEMS Microbiol. Lett.* **132,** 259–262.

Ramani, R., Ramani, A., and Wong, S. J. (1997). Rapid flow cytometric susceptibility testing of *Candida albicans. J. Clin. Microbiol.* **35,** 2320–2324.

Robertson, B. R., and Button, D. K. (1989). Characterization aquatic bacteria according to population cell size, and apparent DNA content by flow cytometry. *Cytometry* **10,** 70–76.

Severin, A., Severina, E., and Tomasz, A. (1997). Abnormal physiological properties and altered cell wall composition in *Streptococcus pneumoniae* grown in the presence of clavulanic acid. *Antimicrob. Agents Chemother.* **41,** 504–510.

Sheppard, F. C., Mason, D. J., Bloomfield, S. F., and Gant, V. A. (1997). Flow cytometric analysis of chlorhexidine action. *FEMS Microbiol. Lett.* **154,** 283–288.

Sidorczyk, Z., Zahringer, U., and Rietschel, E. T. (1983). Chemical structure of the lipid A component of the lipopolysaccharide from a *Proteus mirabilis* Re-mutant. *Eur. J. Biochem.* **137,** 15–22.

Skarstad, K., Boye, E., and Steen, H. B. (1986). Timing of initiation of chromosome replication in individual *E. coli* cells. *EMBO J.* **5,** 1711–1717.

Steen, H. B., Boye, E., Skarstad, K., Bloom, B., Godal, T., and Mustafa, S. (1982). Application of flow cytometry on bacteria: Cell cycle kinetics, drug effects, and quantitation of antibody binding. *Cytometry* **2,** 249–257.

Storm, D. R., Rosenthal, K. S., and Swanson, P. E. (1997). Polymyxin and related peptide antibiotics. *Annu. Rev. Biochem.* **46,** 723–763.

Suller, M. T. E, and Lloyd, D. (1998). Flow cytometric assessment of the postantibiotic effects of methicillin on *S. aureus. Antimicrob. Agents Chemother.* **42,** 1195–1199.

Suller, M. T. E., Stark, J. M., and Lloyd, D. (1997). A flow cytometric study of antibiotic-induced damage and evaluation as a rapid antibiotic susceptibility test for methicillin-resistant *S. aureus. J. Antimicrob. Chemother.* **40,** 77–83.

Trousselier, M., Courties, C., and Vaquer, A. (1993). Recent application of flow cytometry in aquatic microbial ecology. *Biol. Cell* **78,** 111–121.

van Dilla, M. A., Langlois, R. G., Pinkel, D., Yajko, D., and Hadley, W. K. (1983). Bacterial charaterization by flow cytometry. *Science* **220,** 620–622.

Walberg, M. Gaustad, P., and Steen, H. B. (1996). Rapid flow cytometric assessment of mecillinam and ampicillin bacterial susceptibility. *J. Antimicrob. Chemother.* **37,** 1063–1075.

Walberg, M., Gaustad, P., and Steen, H. B. (1997a). Rapid assessment of ceftazidime, ciprofloxacin, and gentamicin susceptibility in exponentially growing *E. coli* cells by means of flow cytometry. *Cytometry* **27,** 169–178.

Walberg, M., Gaustad, P., and Steen, H. B. (1997b). Rapid discrimination of bacterial species with different ampicillin susceptibility levels by means of flow cytometry. *Cytometry* **29,** 267–272.

Walberg, M., Gaustad, P., and Steen, H. B. (1998a). Rapid preparation procedure for staining of exponentially growing *P. vulgaris* cells with ethidium bromide: A flow cytometry-based study of probe uptake under various conditions. *J. Microbiol. Methods* **34,** 49–58.

Walberg, M., Gaustad, P., and Steen, H. B. (1999). Uptake kinetics of nucleic acid targeting dyes in *S. aureus, E. faecalis,* and *B. cereus:* A flow cytometric study. *J. Microbiol. Methods* **35,** 167–176.

Ward, D. C., Reich, E., and Golberg, I. H. (1965). Base specificity in the interaction of polynucleotides with antibiotic drugs. *Science* **149,** 1259.

Wickens, H. J., Pinney, R. J., Mason, D. J., and Gant, V. A. (2000). Flow cytometric investigation of filamentation, membrane patency, and membrane potential in *E. coli* following ciprofloxacin exposure. *Antimicrob. Agents Chemother.* **44,** 682–687.

CHAPTER 56

Flow Cytometry for Evaluation and Investigation of Human Immunodeficiency Virus Infection

Thomas W. Mc Closkey

Department of Pediatrics
Division of Allergy and Immunology
North Shore University Hospital
New York University School of Medicine
Manhasset, New York 11030, USA

I. Introduction
II. Application of Flow Cytometry to Monitor HIV Infection
    A. T Lymphocyte Immunophenotyping: General Procedure
    B. Data Management and Verification
    C. Multicolor Fluorescence Immunophenotyping
III. Application of Flow Cytometry to Investigate HIV Disease
    A. Immunophenotypic Research Applications
    B. Additional Research Applications
IV. Conclusions
    References

I. Introduction

Our understanding of human immunodeficiency virus (HIV) disease continues to increase in regard to both prognostic indicators of disease progression (deWolf *et al.*, 1997) and host/virus interactions (Pantaleo *et al.*, 1998). Advances in available therapies have seen the introduction of powerful antiretroviral drugs (Flexner, 1998) that are able to dramatically reduce the level of virus in adults (Hammer *et al.*, 1997) and in children (Mueller *et al.*, 1998). However, a latent virus reservoir persists despite apparent successful treatment (Chun *et al.*, 1997); thus, science continues in its pursuit of knowledge about HIV. The technology

METHODS IN CELL BIOLOGY, VOL. 64

of flow cytometry has played a major role in the battle against HIV, in both patient monitoring and in research that has elucidated disease pathogenesis.

II. Application of Flow Cytometry to Monitor HIV Infection

The major use of flow cytometry to evaluate HIV infected persons is for the immunophenotypic analysis of lymphocytes. Whole blood samples are labeled with fluorescent dye conjugated monoclonal antibodies, and the proportion of T cells that express CD4 or CD8 is determined. Changes in these populations can indicate progression of disease, which may influence treatment decisions.

A. T Lymphocyte Immunophenotyping: General Procedure

1. Guidelines

Published guidelines are available for laboratories performing lymphocyte phenotyping. The Centers for Disease Control (www.cdc.gov) has published revised guidelines that specifically address obtaining CD4 T cell immunophenotyping results from samples acquired from persons with HIV infection (Centers for Disease Control and Prevention, 1997). This document provides comprehensive instructions on specimen preparation, sample analysis, and data reporting and serves as a guide from obtaining the blood sample to providing results to the physician. The latest revision reflects the growth of flow technology in the 1990s; for example, new reagents allow three- and four-color analyses, and it has become possible to obtain absolute counts directly on the cytometer. The National Institute of Allergy and Infectious Disease (www.niaid.nih.gov) has also published guidelines for flow cytometric immunophenotyping (Calvelli *et al.*, 1993), and a three-color supplement to this document (Nicholson *et al.*, 1996a) again reflected the rapidly changing methodology in this field.

An individual laboratory may implement minor modifications to the suggested procedures; however, these changes should only be instituted following careful consideration. Small-scale experiments can be performed to determine optimized conditions for the entire procedure that will provide evidence on which modifications can be based. In studies that examined laboratories performing lymphocyte immunophenotyping, lack of adherence to guidelines may lead to variability in sample processing (Harwell *et al.*, 1995) and in reporting results (Peddecord *et al.*, 1993).

2. Antigens to Distinguish Major Lymphocyte Subpopulations

A basic phenotyping panel applicable to HIV infected persons includes CD3/CD4 for quantitation of CD4 T cells and CD3/CD8 for CD8 T cells. For pediatric samples, the measurement of CD19 for B cells is included. For quality control purposes, enumeration of other antigens is required. For example, by also labeling for natural killer (NK) cells (with CD16 and/or CD56), a lymphosum (T + B + NK) may be attained, and by examining the combination CD45/CD14, information regarding the purity of the lymphocyte gate becomes available. A basic six-tube immunophenotyping panel for the analysis of HIV patients has been described consisting of isotype control, CD45/14, CD3/4, CD3/8, CD3/16,56, and CD19 (Schenker *et al.*, 1993). This panel allows detection of all lymphocyte subsets and provides important quality control information.

3. Sample Handling: Biosafety

"Universal precautions"should be used for handling all blood samples, which include appropriate use of gloves, masks, and laboratory coats; avoidance of sharp implements; and careful sample handling. Specific guidelines for appropriate sample handling are available [Centers for Disease Control (www.cdc.gov), 1987; Occupational Safety and Health Administration (www.osha.gov), 1991]. Strict adherence to the correct procedures combined with the relatively low risk of infection to exposed health care workers (Henderson *et al.*, 1986; McCray, 1986; Cardo *et al.*, 1997) make accidental infection with HIV unlikely, and recommendations for chemoprophylaxis after accidental exposure exist (Centers for Disease Control, 1996). Latex gloves provide superior protection to those made from vinyl (Kotilainen *et al.*, 1989; Klein *et al.*, 1990). There are a number of disinfecting agents that may be used for the work area and the cytometer (Spire *et al.*, 1984; Martin *et al.*, 1985; Sattar and Springthorpe, 1991), with household bleach among the most effective. In addition, lysing and fixing reagents can reduce the infectious activity of specimens from HIV infected persons (Lifson *et al.*, 1986; Cory *et al.*, 1990; Nicholson *et al.*, 1993a; Nicholson and Browning, 1994). When processing unfixed HIV specimens on the cytometer for cell sorting, extra precautions are necessary (Schmid *et al.*, 1997).

4. Specimen Preparation

The following is a generalized protocol for immunophenotyping peripheral blood samples for monitoring HIV infection. A blood sample is acquired by venipuncture (or heelstick for an infant) into a tube containing anticoagulant (ethylenediaminetetraacetate, heparin, or acid citrate dextrose). Samples should always be processed as soon as possible; if whole blood storage is required, room temperature provides the most stable conditions (Thornthwaite *et al.*, 1984), whereas storage at 4°C results in significant alterations (Ekong *et al.*, 1992). Acid citrate dextrose or heparin outperform ethylenediaminetetraacetate for whole blood immunophenotypic analysis more than 24 hr after collection (Nicholson and Green, 1993). Whole blood preparatory methods should be used, as Ficoll separation of mononuclear cells increases the obtained percentages of CD4 and CD19 lymphocytes while decreasing the percentage of CD8 cells (Renzi and

Ginns, 1987; Romeu *et al.*, 1992), and samples from HIV infected individuals are more prone to these changes (Tamul *et al.*, 1994).

Optimal amounts of fluorochrome conjugated monoclonal antibodies (Table I) are added, with care taken to place the reagent at the bottom of the tube. Determining the optimal concentration for a particular monoclonal antibody involves labeling cells with different amounts of reagent around the manufacturer's suggested amount. The amount required for good resolution of positive cells may be less than that suggested (Edwards and Shopp, 1989), which will reduce overall costs and may also decrease background binding. Each new lot of reagent should be tested in such a manner, as lot to lot variation could result from differences in protein concentration or fluorochrome to protein ratio. Following addition of antibody, whole blood is added, again verifying that it is placed at the bottom of the tube. Tubes are mixed well and then incubated for 10 min in the dark at room temperature. A lysing solution is added to eliminate red blood cells, and samples are fixed in 1% paraformaldehyde. Many different commercial lysing and fixing reagents are currently available (Carter *et al.*, 1992; Ericson *et al.*, 1994); each laboratory should decide which one works best in their hands. Fixation should not alter the light scatter or fluorescence properties of the cells (Lanier and Warner, 1981); it eliminates the need for immediate analysis and reduces the infectious potential of specimens from HIV infected persons.

5. Flow Cytometric Analysis

The instrument should be aligned with fluorescent beads. On a machine with fixed alignment, this step consists of checking the alignment; for cytometers with

Table I
Monoclonal Antibody Combinations for Evaluation of HIV Infection

Fluorescence channels[a]	Antibody combinations
Two-color fluorescence	CD45/14
	CD3/4
	CD3/8
	CD3/19
	CD3/16,56
Three-color fluorescence (light scatter gate)	CD45/14/3
	CD4/8/3
	CD3/16,56/19
Three-color fluorescence (CD45 gate)	CD3/4/45
	CD3/8/45
	CD3/19/45
	CD3/16,56/45
Four-color fluorescence	CD45/14/3/19
	CD3/16,56/4/8

[a] Fluorochrome combinations include FITC/PE/ECD or perCP/PE-Cy5 or APC.

adjustable alignment, it should be optimized. In any case, results should always be compared to those from previous days to detect variation or drift that may be occurring over time. As a second step, the flow cytometer should be calibrated by setting a known fluorescence quantity at the same point on the histogram. Resolution of the instrument should also be checked, ensuring its ability to differentiate positive from negative cells. All of these procedures are an attempt to eliminate day to day variation in machine parameters by eliminating potential sources of change.

Care must be taken in drawing the light scatter lymphocyte gate, which should be drawn by back gating from a fluorescence histogram of CD45/CD14 (Fig. 1). The cells that are negative for CD14 and express high levels of CD45 are defined as lymphocytes (Loken *et al.*, 1990). This gate is assigned to a histogram of forward scatter versus side scatter (log side scatter may be used), and then a gate is drawn around the lymphocytes. Thus, the lymphocyte population is defined by its surface antigen expression, not by its light scattering characteristics. The major advantage of this technique is that the ability to resolve the lymphocytes from monocytes and debris is greater by fluorescence than by light scatter, allowing a region to be created that maximizes inclusion of lymphocyte events and minimizes inclusion of nonlymphocyte events. As a second step in this procedure, this gate should now be checked for the percentage of events that are lymphocytes [CD45^{bright}/CD14^{negative}]. Once this gate is drawn, it will be carried over for the remainder of the samples for a given individual. The importance of correctly drawing the lymphocyte gate cannot be overemphasized; in fact, laboratories may decide to include CD45 in each tube for gating purposes (Nicholson *et al.*, 1996b). Fluorescence results are entirely dependent on this gate, as its placement affects the obtained percentage of T (Kromer and Grossmuller, 1994) and B (Storek *et al.*, 1992) cells. Generally, 2500 gated lymphocytes are counted on the cytometer; however, counting 500 or 1000 events may give comparable accuracy (Lape-Nixon and Prince, 1996).

Compensation must be set to correct for spectral overlap. This is generally accomplished by running single-color tubes and setting the appropriate subtraction values. A useful procedure is to adjust the compensation until the fluorescence intensity of the single-color positive events approximates that of the negative events. Analysis regions should be set to best separate clusters; these are generally set based on isotype control antibodies, however, the operator should move the regions if necessary. Particularly with multiple color immunophenotyping, and especially when positive clusters are clearly separated from negative cells, isotype controls become less useful. With the current availability of storage media, data should be saved as list mode files. This allows reanalysis of patient samples if required and access to prior data should a question arise at a later time. Media for storage include zip drives, jazz drives, recordable CDs, tape backup systems, optical drives, digital video disks, or downloading onto a main server; ideally, two different methods should be used.

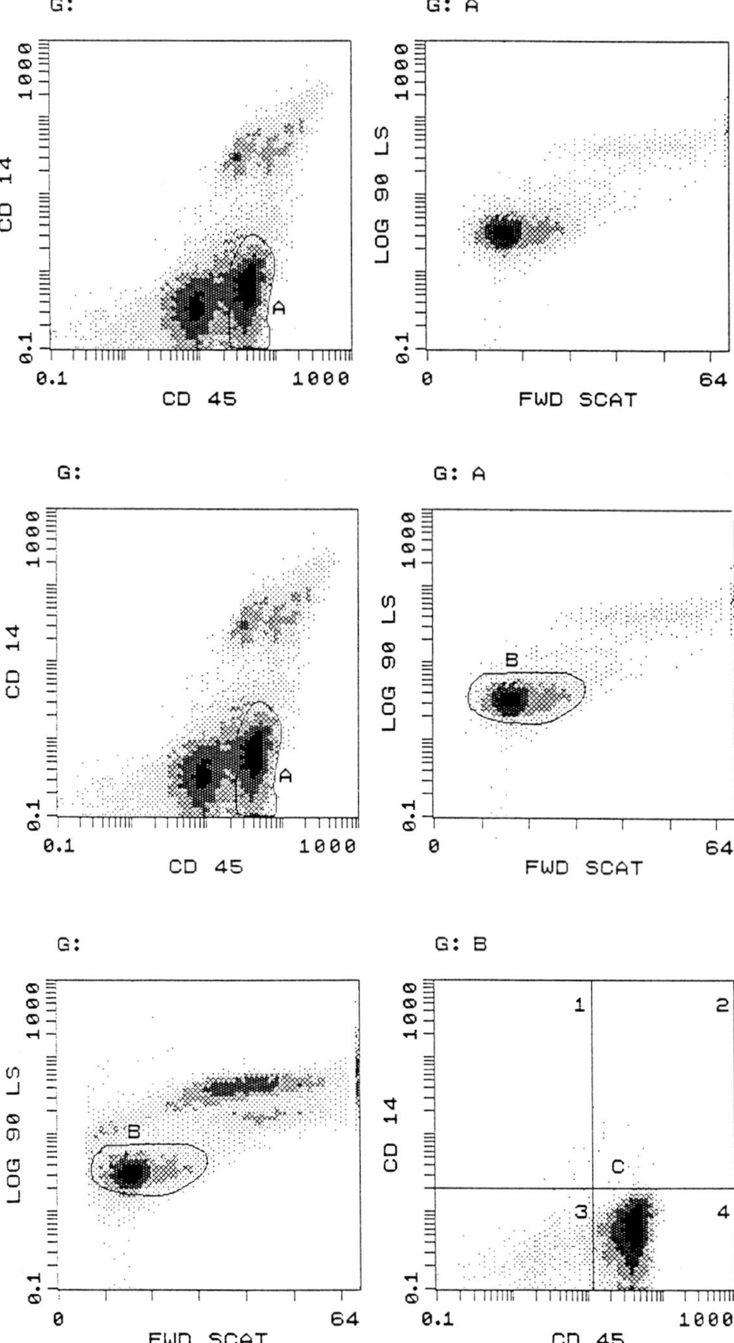

6. Rescuing Problem Samples

For laboratories that do not include CD45 as a gating reagent in every tube, it is possible to use the following technique to rescue problem samples that do not meet predefined acceptance criteria. For example, if the lymphocyte purity of the gate is below the set limit, the sample should not be analyzed. When possible, leftover blood from patient samples should be stored overnight at room temperature. If immunophenotyping samples are analyzed the following morning, unacceptable samples can be detected within 24 hr of blood collection. The whole blood sample is then reprocessed utilizing CD45 as a gating reagent. This allows the lymphocyte population to be defined by both light scatter and CD45 fluorescence (Nicholson *et al.*, 1996b), and it allows them to be separated from, for example, unlysed red blood cells present in the tube. The anti-CD45 monoclonal antibody can be combined with CD3/4 and CD3/8 to obtain accurate percentages of CD4 T cells and CD8 T cells even in a problem sample (Fig. 2). In fact, CD45 labeling of samples may be performed after paraformaldehyde fixation (Pizzolo and Melamed, 1996) such that the same tube can be rerun on the cytometer if another fluorescence channel is available.

7. Absolute Counts

The major use of flow cytometry to evaluate individuals infected with HIV is in the calculation of CD4 cell counts. The CD4 absolute count is calculated by determining the total white blood cell count, the percentage of white cells that are lymphocytes, and the percentage of lymphocytes that express CD4 by a combination of hematology cell counter and flow cytometry instrumentation. Changes in CD4 absolute number in the peripheral blood has proved to be a useful marker of disease status as well as response to treatment, and it remains the standard on which patient care decisions are based. The identification of inflection points in T cell counts of HIV infected persons (Gange *et al.*, 1998) have been noted in relationship to disease progression. Variability in absolute lymphocyte counts obtained from automated cell counters can be significant (Simson and Groner, 1995), which is then passed on to the CD4 count. Reagents

Fig. 1 Histograms demonstrating procedure for optimizing the light scatter lymphocyte gate. (Top) The monoclonal antibody combination of CD45/CD14 allows a gate to be drawn on lymphocytes that are CD45[bright] CD14[negative] (gate A) which is then assigned to a histogram of forward scatter versus 90° scatter. (Middle) Now a gate (gate B) can be drawn around the lymphocyte population. Lymphocyte recovery can be determined from this step. (Bottom) Finally the light scatter lymphocyte gate (gate B) is assigned back to a fluorescence histogram. The purity of the gate can be measured by determining the percentage of events within the gate that are lymphocytes (region C4). Sample was labeled with CD45 FITC and CD14 PE and analyzed on a Coulter Epics Elite cytometer.

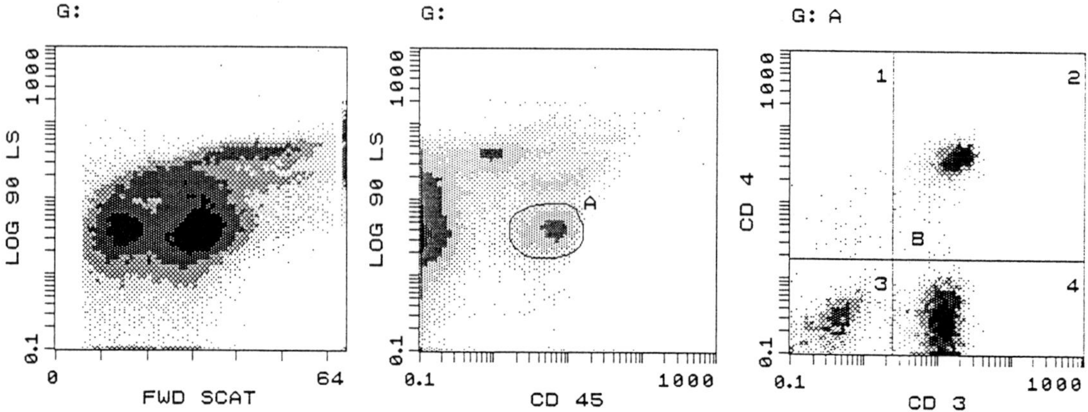

Fig. 2 A common sample preparation problem, especially with pediatric samples, is incomplete red cell lysis. In such a sample, it is impossible to utilize a light scatter gate to analyze the lymphocyte population. The inclusion of an anti CD45 monoclonal antibody allows the identification of a well-resolved lymphocyte population (gate A) and the subsequent determination of the CD3 and CD4 positive population. Sample was labeled with CD3 FITC/CD4 PE/CD45 perCP and analyzed on a Coulter Epics Elite cytometer.

are available to obtain absolute counts from the cytometer (Nicholson *et al.,* 1997), and alternative technologies for obtaining CD4 counts without the need for complex instrumentation are available for developing countries [World Health Organization (www.who.int) Global Program on AIDS, 1994].

8. Special Considerations for HIV Infected Children

Perinatal transmission of HIV superimposes a viral infection onto a developing immune system; therefore, the study of this infection in children takes on added complexity. In comparison to adults, infants exhibit differences in absolute numbers, percentages, and functions of leukocytes; thus, the impact of perinatal HIV infection on a newborn requires a different view than when the virus is acquired by an adult (European Collaborative Study, 1994; Barnhart *et al.,* 1996). A classification system for HIV infected children has been devised (Centers for Disease Control, 1994), immune defects and immunological characteristics have been described (Pahwa, 1990; Chirmule *et al.,* 1995), and guidelines for medical management of pediatric HIV infection have been established (Working Group on Antiretroviral Therapy and Medical Management of Infants, Children, and Adolescents with HIV Infection, 1998).

B. Data Management and Verification

1. Data Base

Patient data should be entered and stored in a data base. Data should be entered by patient identification number, and access should be limited via password protection or some other manner to protect patient confidentiality. The records in the data base, ideally on a dedicated computer, should be backed up regularly in the event of a hard drive failure. The input of biostatisticians is valuable in developing the format of the data base such that later analyses of patients or patient groups may be accomplished. Each laboratory should design its own system, but general information to be entered includes patient identification number, date of visit, age, treatment, and disease stage. The data base should also provide means to generate a report form, with the results from the current specimen and control values for the measured phenotypes. Another useful feature is the ability to output chronological results for a patient, which may be used to examine response to therapy. The data base can be programmed to calculate absolute values for subsets examined. In addition, quality control can be performed automatically by the data base. A lymphosum (T + B + NK) and a T sum (CD4 + CD8 = CD3) can be generated and a specimen tagged if out of range. The correction factor for the percentage of lymphocytes in the gate (determined by CD45/14 combination) can be applied to all data, and the report form can contain the corrected values. A tube to tube consistency check for the same antigen can be performed, and the computer can alert for data entry errors, for example, if data obtained from quadrant analysis does not sum to 100.

2. Reference Ranges

The percentages and subset distribution of lymphocyte subpopulations are relatively stable in healthy adults. To establish normal values for adult immuno-phenotyping (Ohta *et al.*, 1986; Reichert *et al.*, 1991), a laboratory can generate ranges with a group of volunteers. The same antibody panel used for patient samples for enumeration of T, B, and NK cells should be applied. For the clinical immunophenotyping of pediatric samples, however, determining normal values is more difficult, as the immune system of the child is maturing, with lymphocyte percentages and absolute numbers changing over the first few years of life. Thus, to evaluate lymphocyte immunophenotyping results from a child, the age-matched normal range values must be known. Although an effort should be made by the laboratory to establish their own pediatric reference values, ranges for the major lymphocyte subsets in children can be gathered from the existing litererature (Denny *et al.*, 1992; Kotylo *et al.*, 1993; Aldhous *et al.*, 1994; Comans-Bitter *et al.*, 1997), including studies that present immunophenotyping data in newborns (Mc Closkey *et al.*, 1997; O'Gorman *et al.*, 1998).

3. Quality Control

Various aspects are important (Table II) in verifying that the obtained flow cytometric results are correct (Lewis and Rickman, 1992; Muirhead, 1993). Laser alignment can be verified with beads, and instrument fluorescence can be standardized so that relative fluorescence intensities can be compared. In addition, the ability of the cytometer to resolve dimly fluorescent populations should be

Table II
Important Components of the Lymphocyte
Immunophenotyping Procedure

Step	Important considerations
Review guidelines	Consider sources of variation
	Review procedures, consider improvements
Blood collection	Specimen integrity
	Storage temperature
	Time until processing
Specimen processing	Antibody concentration
	Red cell lysis
	Fixation
Running the sample	Alignment and standardization
	Compensation
	Resolution
	Variations in machine operation
Analyzing and reporting the data	Lymphocyte gate (purity and recovery)
	Analysis region placement
	Check lymphosum and T sum
	Check tube to tube variation
	Correct for percentage lymphocytes
Participate in proficiency programs	Determine actual laboratory performance
	Improve; achieve consistency

verified as a test of performance. All these results should be tabulated and monitored, which can serve to reveal changes in laser output, optical alignment, fluorescence intensity, or compensation settings. The linearity of the machine amplifiers should be verified (Bagwell *et al.,* 1989), and details regarding detector calibration for accurate detection of fluorescence signals are available (Durand, 1994). The goal is to minimize the likelihood of the instrument being the source of variation.

A major source of variation is in sample preparation. Reagents are available to verify the lysing and antibody labeling procedures, or, alternatively, preparing a specimen from a healthy control will allow evaluation of correct cell preparatory procedures. An isotype control tube or the negative cells in the test sample can serve as a guide for analysis region placement. A new lot of antibody should be titered next to the old or on standardized cells such as frozen lymphocytes. Personnel should receive training in both cell preparation and cytometer operation. Although knowledge of cell–reagent interactions, optics, and computers is helpful, competence in this field requires a strong background in biology, biochemistry, and physics.

4. Proficiency Testing Programs

Various testing programs exist in which laboratories may participate to evaluate and improve their flow cytometric immunophenotyping assays. These programs involve a shipment of samples to the laboratory where they are processed, and then results are sent in and compiled. Reports are provided to the laboratory so that their values can be compared with those obtained by other institutions for the same sample. These programs are available through Fast Systems (Kagan *et al.*, 1993), the College of American Pathologists (www.cap.org) (Homburger *et al.*, 1993), and the Centers for Disease Control (www.cdc.gov) (Gerber, 1993). Participation in these testing programs may help to reveal laboratories having difficulty with the assay (Paxton *et al.*, 1989), for example, the inability to detect an abnormal specimen. Regular processing of proficiency samples improves the overall performance by a laboratory in immunophenotyping.

5. Potential Problems

Common problems with immunophenotyping involve sample preparation. Incomplete lysis of red blood cells can contaminate the lymphocyte gate and preclude accurate data analysis, whereas other causes of poor gate purity may result from debris or granulocyte contamination possibly due to medications or biological factors (Calvelli *et al.*, 1993). Regarding instrument setup, poor alignment or incorrect compensation or analysis region placement can yield erroneous values. Data entry errors should also be considered.

In addition to these typical problems, there is also variation that derives from the choice of reagent. Commercially available lysing reagents from different companies may result in differences in light scatter, amounts of debris, and percentages of lymphocytes (Bossuyt *et al.*, 1997; Macey *et al.*, 1997; Campo, 1997). In addition, antibodies conjugated to the fluorochrome fluorescein may yield lower values for the percentage of CD4 lymphocytes when compared to phycoerythrin (Gelman *et al.*, 1993). Artifactual labeling of lymphocytes with monoclonal antibodies can occur. For example, patients whose therapy has included treatment with murine monoclonal antibodies exhibit high levels of nonspecific binding (Wilks *et al.*, 1990), and factors present in the serum of certain individuals may induce artifactual antibody labeling (Ekong *et al.*, 1993; Nicholson *et al.*, 1994). Washing the cells may overcome this problem when it is encountered.

Because the results obtained with fluorescence data are limited to events within the light scatter gate, the presence of lymphocytes outside the gate can alter the percentage phenotype values. Escapees are cellular aggregates that are lost from analysis because they fall outside the conventional lymphocyte light scatter gate. These cell clusters can be induced by binding of certain antibodies, and thus the percentage obtained for that reagent will actually be less than the true percentage. Escapee events may be lymphocytes complexed to myeloid cells (Prince *et al.*, 1994); they can be reduced by the addition of paraformaldehyde and are prevalent in fluorescein-labeled populations. The formation of escapees has also been linked with an interaction between antibody and CD32 (Gratama *et al.*, 1997), and blocking this interaction can decrease their incidence. Other factors shown to influence this phenomenon were the specificity, subclass, and amount of monoclonal antibody.

Other potential cell/fluorochrome interactions need to be considered as well. The tandem conjugate of phycoerythrin–cyanine 5 can bind to CD64 (van Vugt *et al.,* 1996) on monocytes and activated granulocytes. The fluorochrome R phycoerythrin has been reported to bind specifically to the murine immunoglobulin receptor (Takizawa *et al.,* 1993), though no binding to human cells was observed. Care should especially be exercised when working with new antibody reagents to ensure that they are suited for the assay system, as demonstrated by the observation that antibodies against human Fas ligand designed for immunohistochemical analysis were unsuitable for use in flow cytometry (Smith *et al.,* 1998). Even well-characterized clones have the potential to yield surprises; the BB1 monoclonal antibody directed against CD80 was later shown to also bind CD74 (Freeman *et al.,* 1998). Contamination with another antibody or fluorochrome is possible, as is loss of binding due to degradation. Unusual phenotyping results require knowledge of all these potential sources of variation; it is worthwhile to consider the possibility that a certain reagent may not perform as expected.

C. Multicolor Fluorescence Immunophenotyping

1. Reagents and Strategies

The introduction of new fluorochromes and new instrumentation continues to expand the possibilities of multicolor fluorescence flow cytometry. Multiple indirect labeling steps have become unnecessary due to the current availability of directly labeled monoclonal antibodies against most human leukocyte antigens. Adding more fluorochromes to the analysis allows the gathering of more information, uses less sample volume which results in a lower number of tubes, and requires less preparation and analysis time per patient. In short, multicolor phenotyping provides increased information at reduced cost. Two-color analysis most commonly uses the fluorochromes fluorescein and phycoerythrin. Reagents are now available to analyze whole blood specimens from HIV infected persons by three- or four-color methods with combinations of energy-coupled dye, peridinin chlorophyll protein, phycoerythrin–cyanine 5, and allophycocyanin which provide third and fourth options (Yeh *et al.,* 1987; Afar *et al.,* 1991; Lansdorp *et al.,* 1991; Gruber *et al.,* 1993). Typical four-color clinical cytometry (Fig. 3) may combine fluorescein and phycoerythrin with energy-coupled dye and phycoerythrin–cyanine 5 or peridinin chlorophyll protein and allophycocyanin. A comparison of two- to three-fluorochrome or three- to four-fluorochrome analysis should be performed for laboratories switching to a new method by preparing the same samples with both methods to ensure consistent results.

A single tube three-color assay for determination of CD4 has been described (Nicholson *et al.,* 1993b). This assay utilizes whole blood and labels cells with CD45/CD3/CD4. So one approach to three-color cytometry is to add CD45 as

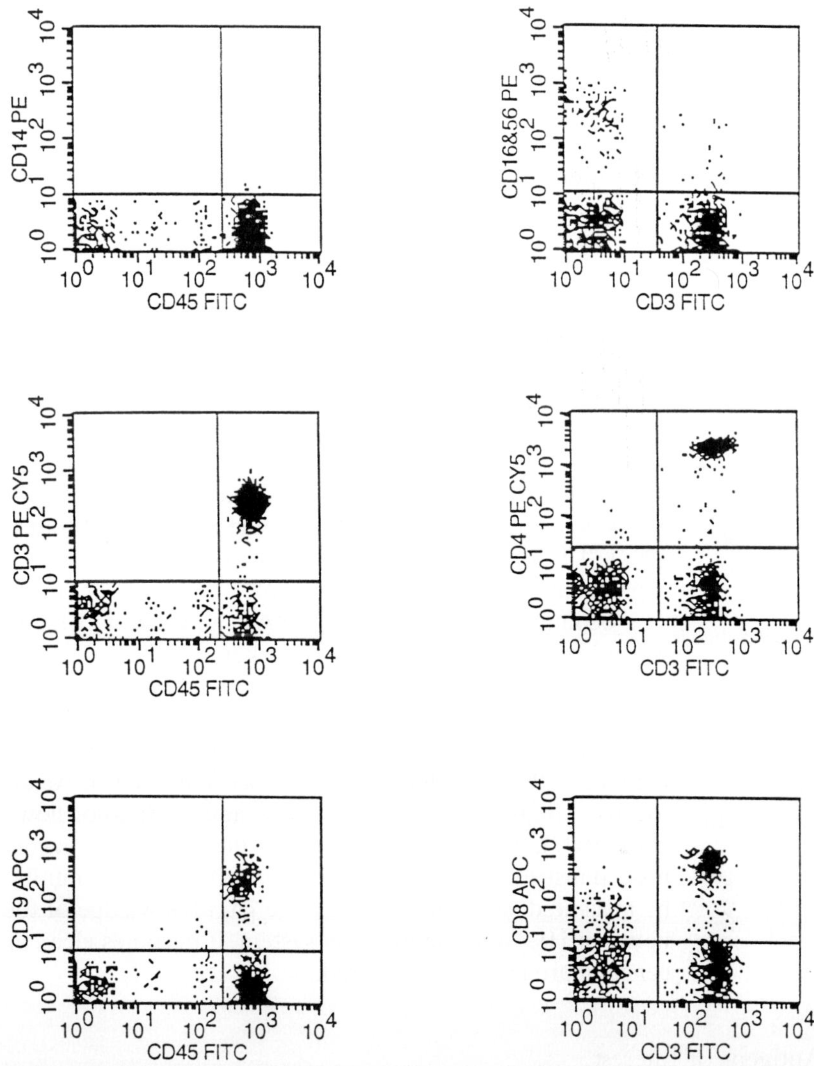

Fig. 3 Histograms demonstrating four color immunophenotyping panel for determining major lymphocyte subpopulations. Tube one (left) used the combination CD45 FITC/CD14 PE/CD 3 PE-Cy5/CD19 APC; tube two (right) used the combination CD3 FITC/CD16,56PE/CD4 PE-Cy5/CD8 APC. These two tubes yield the following information: total T, B, and NK cells, CD4 and CD8 T cells, as well as quality control information of lymphocyte purity, lymphocyte recovery, lymphosum, T sum, and tube to tube consistency check. Samples were analyzed on a Becton Dickinson FACSCalibur cytometer.

a gating reagent to every tube. This is a highly accurate method, as the lymphocyte population can be defined by forward scatter, 90° scatter, and CD45 fluorescence, which provides separation from red blood cells and debris. However, this choice omits one fluorescence channel from analysis. A different approach is to use one marker to define a subset; for example, a T gating protocol (Mandy *et al.*, 1992) focuses on CD3 expressing cells. This design allows the detection of two antigens of interest on a specific lymphocyte subset as when various two-color combinations were evaluated on CD8 cells (Prince and Jensen, 1991) of HIV infected persons.

Multicolor flow cytometry is particularly suited to answering complex biological questions such as those that arise during investigation of the human immune system. Four-color flow cytometry has been used to identify natural killer cell subpopulations (Srour *et al.*, 1990) and subpopulations within the stem cell lineage (Bender *et al.*, 1991). Currently available cytometers for clinical analysis have four-color capability, so panels designed for evaluation of HIV disease can pinpoint even more specific subsets. The number of described antigens on leukocytes (www.ncbi.nlm.nih.gov/prow) has increased tremendously (Kishimoto *et al.*, 1997). The ongoing development of new fluorochrome reagents continues to expand the possibilities for multicolor fluorescence. Fluorochromes such as 7-aminocoumarin (Khalfan *et al.*, 1986) and Cascade Blue (Whitaker *et al.*, 1991) will allow addition of another marker for those instruments with ultraviolet capability. The introduction of the tandem conjugates phycoerythrin–cyanine 7 and allophycocyanin–cyanine 7 has provided new reagents for immunofluorescence that can be measured in the red portion of the spectrum (Beavis and Pennline, 1996; Roederer *et al.*, 1996a). Three-laser excitation has been utilized for five- (Beavis and Pennline, 1994) and eight-color flow cytometry (Roederer *et al.*, 1997), demonstrating that new fluorochrome reagents combined with improved instrumentation will allow us to increase the number of antigens that can be measured simultaneously. As we gain knowledge of the complex interaction between HIV and the immune system, it becomes clear that multicolor fluorescence approaches are required.

2. Antigens of Interest

The ability to measure multiple fluorochromes simultaneously allows the possibility to determine expression of additional lymphocyte antigens that undergo alterations during HIV infection. Many of these changes result from the state of chronic immune activation that occurs due to the ongoing viral infection. A loss of CD8 naive T cells was demonstrated in adults with HIV (Roederer *et al.*, 1995) by applying multicolor immunophenotyping using the markers CD11a, CD45RA, and CD62L. A reduction in the percentage of CD4 and CD8 T lymphocytes that express the costimulatory molecule CD28 occurs during HIV infection (Caruso *et al.*, 1994), and expansion of the CD8 T cell population has been attributed to lymphocytes negative for CD28 (Roos *et al.*, 1994). Increased percentages of CD8 T lymphocytes that express the activation antigens HLA-DR and CD38 have been noted during HIV infection (Kestens *et al.*, 1992), and elevated levels of CD8$^+$ CD38$^+$ T lymphocytes are an important indicator of HIV disease progression (Giorgi *et al.*, 1993); in fact, the percentage of CD8 lymphocytes expressing CD38 is different in long-term survivors as compared to progressors (Barker *et al.*, 1998). Other antigens of interest include CD25, which can differentiate lymphocytes with productive infection versus those with latent infection (Borvak *et al.*, 1995), and the apoptosis-related antigen CD95, which increases during HIV infection in adults (Mc Closkey *et al.*, 1995). Expression of CD69 occurs at an early time after activation, and impaired expression of CD69 has been noted in HIV infected patients (Perfetto *et al.*, 1997), suggesting this marker may be a useful measurement to assess immune function. The CD4 lymphocyte subset that expressed CD57 has been reported to be expanded during HIV infection (Legac *et al.*, 1992), and the expression of CD30 on CD8 cells may be a marker of lymphocytes producing interleukin 4 (IL-4) and IL-5 (Manetti *et al.*, 1994).

Expression of certain antigens changes with age, so evaluation of pediatric HIV infection requires knowledge of age-related variables for any parameter being measured. The following observations confirm that changes in many lymphocyte subsets also occur during pediatric HIV infection. Using the antigens CD11a, CD45RA, and CD62L to distinguish naive from memory lymphocytes, Rabin *et al.* (1995) found that HIV infected children exhibited a marked decrease in naive CD8 T cells. Alterations in CD45RA, CD62L, CD38, CD57, and HLA-DR have been shown in HIV infected children (Plaeger-Marshall *et al.*, 1993), as have changes in percentages of cells coexpressing CD38 and HLA-DR (Plaeger-Marshall *et al.*, 1994). Loss of CD28 expression on CD8 T lymphocytes occurs during pediatric HIV disease (Niehues *et al.*, 1998), and expression of CD95 increases early following perinatal transmission (Mc Closkey *et al.*, 1998a). Another feature to consider in HIV infected children is the larger thymic output, which may lead to a greater ability to renew the lymphocyte pool; for example, it has been suggested that in HIV infected children the CD8 cells expressing CD38 represent activated lymphocytes (DeMartino *et al.*, 1998) while the CD4 cells expressing CD38 are immature cells that have recently entered the peripheral circulation.

Besides the utility of measuring expression of certain lymphocyte antigens to assess disease status, alterations in particular subsets may be useful indicators of response to therapy. Treatment of HIV infection with highly active antiretroviral therapy has resulted in an increased interest in understanding the immune reconstitution that takes place, such as the renewal of CD4 T cells (Fleury *et al.*, 1998). Flow cytometry has contributed to our understanding of the alterations in memory and naive cells during therapy (Autran *et al.*, 1997; Gray *et al.*, 1998), and decreased virus load has been correlated with reduction in numbers of CD8$^+$ CD38$^+$ HLA-DR$^+$ T lymphocytes. T lymphocyte subsets defined by expression of CD45RA, CD45RO, HLA-DR, CD25, and CD28 have been quantitated

(Arno *et al.*, 1998) to determine the degree of immune restoration following a prolonged period of undetectable levels of virus.

III. Application of Flow Cytometry to Investigate HIV Disease

The tests that are performed routinely in clinical laboratories today were, at one time, experimental assays conducted in a research setting. A decade ago clinical laboratories were not processing specimens for immunophenotyping by four-color flow cytometry. The progression from basic research to clinical research to clinical test is a natural one that ensures progress in the area of patient care. For example, the measurement of CD8 T lymphocytes that express CD38 can add prognostic value to the determination of CD4 count by helping to predict disease progression (Giorgi *et al.*, 1993). In this regard, what follows are areas of scientific investigation that, though experimental today, may enter the clinical laboratory in the future. Even if a certain assay proves not to provide useful information to the treating physician, these methods have enhanced our understanding of disease pathogenesis; perhaps they will enable improved therapeutic options in the future.

A. Immunophenotypic Research Applications

1. Antigen Quantitation

An area of increasing interest is the quantitation of the number of molecules of a particular antigen expressed on the surface of the cell. Although determining the level of expression requires careful instrument setup and controls, the technology and reagents are available (Lenkei *et al.*, 1998). The goal of this assay is to detect biologically relevant changes; in fact, an analysis of the level of expression of leukocyte surface antigens revealed alterations in HIV infected individuals (Roederer *et al.*, 1996b), leading to the possibility that these changes might serve as surrogate markers for reduced functional capacity. Furthermore, the process of performing quantitative flow cytometric determinations has the added benefit of focusing attention on proper instrument setup (Nicholson and Stetler-Stevenson, 1998). The observation that increased relative fluorescence intensity of CD38 on CD8 T cells serves as a marker of poor prognosis during HIV infection (Liu *et al.*, 1996) provides evidence for the utility of quantitating antigen expression.

2. Intracellular Cytokine Labeling

A flow cytometric assay can measure levels of intracellular cytokines in specific subpopulations by labeling permeabilized cells with monoclonal antibodies following stimulation (Jason and Larned, 1997; Maino and Picker, 1998). This assay has been used to quantitate cytokine production in both CD4 and CD8 cells from HIV infected persons (Meyaard *et al.*, 1996) and to determine antigen specific T cell frequencies during HIV infection (Waldrop *et al.*, 1997).

3. Chemokine Receptors

Monoclonal antibodies are available against chemokine receptors, which act as cellular coreceptors for HIV and play an important role in disease pathogenesis. Determining levels of chemokine receptors and how they differ in cell populations (Bleul *et al.*, 1997) and between individuals has become an area of great interest, both as applied to the study of disease pathogenesis and in terms of potential for therapy (Moore, 1997).

4. T Cell Receptor Vβ Analysis

The differential response to HIV of different populations of T lymphocytes defined by expression of Vβ antigens and changes in these subsets during infection (Gorochov *et al.*, 1998) is currently under investigation. T cell receptor (TCR) Vβ distributions are also being examined as an indicator of immune reconstitution following therapy.

5. HIV p24 Measurement

An assay is available to detect HIV p24 antigen using flow cytometry. Although early results measuring p24 antigen appeared promising, the low specificity of this assay has detracted from its utility (Kux *et al.*, 1996; Cameron *et al.*, 1998) to the extent that it is not useful for detecting infected cells.

B. Additional Research Applications

1. Cell Death

The observation that HIV can induce apoptosis has generated interest in its role in disease pathogenesis (Oyaizu and Pahwa, 1995). Multiple flow cytometric methods are applicable for the quantitative investigation of HIV-induced apoptosis (Mc Closkey *et al.*, 1998b), including the possibility for simultaneous immunophenotypic analysis (Mc Closkey *et al.*, 1998c).

2. Lymphocyte Proliferation

Understanding the kinetics of replication of the lymphocyte pool is important during the course of HIV infection. Expression of the proliferation marker Ki-67 in T lymphocytes (Sachsenberg *et al.*, 1998) revealed that the growth fraction was elevated in HIV infected individuals, and bromodeoxyuridine (BrdU) labeling was used to reveal a rapid turnover of T lymphocytes in SIV infected macaques (Mohri *et al.*, 1998).

3. Flow Cytometry–Polymerase Chain Reaction

The flow cytometric analysis of polymerase chain reaction (PCR) samples has been utilized to detect HIV at the DNA and RNA levels (Patterson *et al.*, 1993; Yang *et al.*, 1995). Furthermore, this procedure can be combined with measurement of other parameters, for example, the detection of HIV DNA simultaneously with the enumeration of CD4 positive cells (Patterson *et al.*, 1995).

4. Monitoring Gene Therapy

With the advent of gene therapy approaches to treating human disease, including HIV infection (Su *et al.*, 1997), flow cytometry may be used to monitor outcome in individual patients. The ability to detect and monitor the genetically altered cells will be critical in assessing the efficacy of this technique.

IV. Conclusions

Flow cytometry continues as a technology for monitoring the course of HIV infection by determining percentages of major lymphocyte subsets. New research has indicated the importance of measuring additional antigens, and as clinical instruments attain multicolor fluorescence capabilities, antibody panels can be designed that can detect narrowly defined lymphocyte subpopulations. Flow cytometry has also played a major role in the investigation of HIV infection. The ability to measure multiple parameters at once lends itself well to the complex spectrum of events that occur during this syndrome; as new fluorescent reagents are introduced, applications of flow cytometry to the study of HIV continue to expand. As progress continues toward a better understanding of this disease at both the cellular and molecular levels, perhaps the future will witness better days for those affected by this epidemic.

References

Afar, B., Merrill, J., and Clark, E. A. (1991). Detection of lymphocyte subsets using three color single laser flow cytometry and the fluorescent dye peridinin chlorophyll *a* protein. *J. Clin. Immunol.* **11**, 254–261.

Aldhous, M. C., Raab, G. M., Doherty, K. V., Mok, J. Y. Q., Bird, A. G., and Froebel, K. S. (1994). Age related ranges of memory, activation, and cytotoxic markers on CD4 and CD8 cells in children. *J. Clin. Immunol.* **14**, 289–298.

Arno, A., Ruiz, L., Juan, M., Zayat, M. D. K., Puig, T., Balague, M., Romeu, J., Pujol, R., O'Brien, W. A., and Clotet, B. (1998). Impact on the immune system of undetectable plasma HIV RNA for more than 2 years. *AIDS* **12,** 697–704.

Autran, B., Carcelain, G., Li, T. S., Blanc, C., Mathez, D., Tubiana, R., Katlama, C., Debre, P., and Leibowitch, J. (1997). Positive effects of combined antiretroviral therapy on CD4 T cell homeostasis and function in advanced HIV disease. *Science* **277,** 112–116.

Bagwell, C. B., Baker, D., Whetstone, S., Munson, M., Hitchcox, S., Ault, K. A., and Lovett, E. J. (1989). A simple and rapid method for determining the linearity of a flow cytometer amplification system. *Cytometry* **10,** 689–694.

Barker, E., Mackewicz, C. E., Reyes-Teran, G., Sato, A., Stranford, S. A., Fujimura, S. H., Christopherson, C., Chang, S. Y., and Levy, J. A. (1998). Virological and immunological features of long term HIV infected individuals who have remained asymptomatic compared with those who have progressed to AIDS. *Blood* **92,** 3105–3114.

Barnhart, H. X., Caldwell, M. B., Thomas, P., Mascola, L., Ortiz, I., Hsu, H. W., Schulte, J., Parrott, R., Maldonado, Y., and Byers, R. (1996). Natural history of HIV disease in perinatally infected children: Analysis from the pediatric spectrum of disease project. *Pediatrics* **97,** 710–716.

Beavis, A. J., and Pennline, K. J. (1994). Simultaneous measurement of five cell surface antigens by five color immunofluorescence. *Cytometry* **15,** 371–376.

Beavis, A. J., and Pennline, K. J. (1996). Allo 7: A new fluorescent tandem dye for use in flow cytometry. *Cytometry* **24,** 390–394.

Bender, J. G., Unverzagt, K. L., Walker, D. E., Lee, W., vanEpps, D. E., Smith, D. H., Stewart, C. C., and To, L. B. (1991). Identification and comparison of CD34 positive cells and their subpopulations from normal peripheral blood and bone marrow using multicolor flow cytometry. *Blood* **77,** 2591–2596.

Bleul, C. C., Wu, L., Hoxie, J. A., Springer, T. A., and Mackay, C. R. (1997). The HIV co-receptors CXCR4 and CCR5 are differentially expressed and regulated on human T lymphocytes, *Proc. Natl. Acad. Sci. U.S.A.* **94,** 1925–1930.

Borvak, J., Chou, C. S., Bell, K., VanDyke, G., Zola, H., Ramilo, O., and Vitetta, E. S. (1995). Expression of CD25 defines peripheral blood mononuclear cells with productive versus latent HIV infection. **155,** 3196–3204.

Bossuyt, X., Marti, G. E., and Fleisher, T. A. (1997). Comparative analysis of whole blood lysis methods for flow cytometry. *Cytometry* **30,** 124–133.

Calvelli, T., Denny, T. N., Paxton, H., Gelman, R., and Kagan, J. (1993). Guideline for flow cytometric immunophenotyping: A report from the National Institute of Allergy and Infectious Diseases, Division of AIDS. *Cytometry* **14,** 702–715.

Cameron, P. U., Hunter, S. D., Jolley, D., Sonza, S., Mijch, A., and Crowe, S. M. (1998). Specificty of binding of HIV anti p24 antibodies to CD4 lymphocytes from HIV infected subjects. *Cytometry* **33,** 83–88.

Campo, J. A. (1997). No wash step required. *Cytometry* **30,** 324–325.

Cardo, D. M., Culver, D. H., Ciesielski, C. A., Srivastava, P. U., Marcus, R., Abiteboul, D., Heptonstall, J., Ippolito, G., Lot, F., McKibben, P. S., and Bell, D. M. (1997). A case control study of HIV seroconversion in health care workers after percutaneous exposure. *N. Engl. J. Med.* **337,** 1485–1490.

Carter, P. H., Resto-Ruiz, S., Washington, G. C., Ethridge, S., Palini, A., Vogt, R., Waxdal, M., Fleisher, T., Noguchi, P. D., and Marti, G. E. (1992). Flow cytometric analysis of whole blood lysis, three anticoagulants, and five cell preparations. *Cytometry* **13,** 68–74.

Caruso, A., Cantalamessa, A., Licenziati, S., Peroni, L., Prati, E., Martinelli, F., Canaris, A. D., Folghera, S., Gorla, R., Balsari, A., Cattaneo, R., and Turano, A. (1994). Expression of CD28 on CD4 and CD8 lymphocytes during HIV infection. *Scand. J. Immunol.* **40,** 485–490.

Centers for Disease Control (1987). Recommendations for prevention of HIV transmission in health care settings. *Morbid. Mortal. Wkly. Rep.* **36,** 1–18.

Centers for Disease Control (1994). 1994 revised classification system for HIV infection in children less than 13 years of age. *Morbid. Mortal. Wkly. Rep.* **43,** 1–10.

Centers for Disease Control (1996). Update: Provisional public health service recommendations for chemoprophylaxis after occupational exposure to HIV. *J. Am. Med. Assoc.* **276,** 90–92.

Centers for Disease Control and Prevention (1997). 1997 revised guidelines for performing CD4 T cell determinations in persons with HIV. *Morbid. Mortal. Wkly. Rep.* **46,** 1–29.

Chirmule, N., Lesser, M., Gupta, A., Ravipati, M., Kohn, N., and Pahwa, S. (1995). Immunological characteristics of HIV infected children: relationship to age, CD4 counts, disease progression, and survival. *AIDS. Res. Hum. Retro.* **11,** 1209–1219.

Chun, T. W., Stuyver, L., Mizell, S. B., Ehler, L. A., Mican, J. A. M., Baseler, M., Lloyd, A. L., Nowak, M. A., and Fauci, A. S. (1997). Presence of an inducible HIV latent reservoir during highly active antiretroviral therapy. *Proc. Natl. Acad. Sci. U.S.A.* **94,** 13193–13197.

Comans-Bitter, W. M., deGroot, R., van den Beernd, R., Neijens, H. J., Hop, W. C. J., Groeneveld, K., Hooijkaas, H., and van Dongen, J. J. M. (1997). Immunophenotyping of blood lymphocytes in childhood: Reference values for lymphocyte subpopulations. *J Pediatr.* **130,** 388–393.

Cory, J. M., Rapp, F., and Ohlsson-Wilhelm, B. M. (1990). Effects of cellular fixatives on HIV production. *Cytometry* **11,** 647–651.

De Martino, M., Rossi, M. E., Azzari, C., Gelli, M. G., Galli, L., and Vierucci, A. (1998). Different meaning of CD38 molecule expression on CD4 and CD8 cells of children perinatally infected with HIV surviving longer than five years. *Pediatr. Res.* **43,** 752–758.

Denny, T., Yogev, R., Gelman, R., Skuza, C., Oleske, J., Chadwick, E., Cheng, S. C., and Connor, E. (1992). Lymphocyte subsets in healthy children during the first five years of life. *JAMA* **267,** 1484–1488.

deWolf, F., Spijkerman, I., Schellekens, P. T., Langendam, M., Kuiken, C., Bakker, M., Roos, M., Coutinho, R., Miedema, F., and Goudsmit, J. (1997). AIDS prognosis based on HIV RNA, CD4 T cell count, and function: Markers with reciprocal predictive value over time after seroconversion. *AIDS* **11,** 1799–1806.

Durand, R. E. (1994). Calibration of flow cytometer detector systems. *In* "Methods in Cell Biology" (Z. Darzynkiewicz, J. P. Robinson, and H. A. Crissman, eds.), Vol. 42, pp. 597–604. Academic Press, San Diego.

Edwards, B. S., and Shopp, G. M. (1989). Efficient use of monoclonal antibodies for immunofluorescence. *Cytometry* **10,** 94–97.

Ekong, T., Hill, A. M., Gompels, M., Brown, A., and Pinching, A. J. (1992). The effect of the temperature and duration of sample storage on the measurement of lymphocyte subpopulations from HIV positive and control subjects. *J. Immunol. Methods* **151,** 217–225.

Ekong, T., Gompels, M., Clark, C., Parkin, J., and Pinching, A. (1993). Double staining artifact observed in certain individuals during dual color immunophenotyping of lymphocytes by flow cytometry. *Cytometry* **14,** 679–684.

Ericson, J. G., Trevino, A. V., Toedter, G. P., Mathes, L. E., Newbound, G. C., and Lairmore, M. D. (1994). Effects of whole blood lysis and fixation on the infectivity of HTLV-1. *Cytometry* **18,** 49–54.

European Collaborative Study (1994). Natural history of vertically acquired HIV infection. *Pediatrics* **94,** 815–819.

Fleury, S., DeBoer, R. J., Rizzardi, G. P., Wolthers, K. C., Otto, S. A., Welbon, C. C., Graziosi, C., Knabenhans, C., Soudeyns, H., Bart, P. A., Gallant, S., Corpataux, J. M., Gillet, M., Meylan, P., Schnyder, P., Meuwly, J. Y., Spreen, W., Glauser, M. P., Miedema, F., and Pantaleo, G. (1998). Limited CD4 T cell renewal in early HIV infection: Effect of highly active antiretroviral therapy. *Nat. Med.* **4,** 794–801.

Flexner, C. (1998). HIV protease inhibitors. *N Engl. J. Med.* **338,** 1281–1292.

Freeman, G. J., Cardoso, A. A., Boussiotis, V. A., Anumanthan, A., Groves, R. W., Kupper, T. S., Clark, E. A., and Nadler, L. M. (1998). The BB1 monoclonal antibody recognizes both cell surface CD74 as well as B7-1, resolving the question regarding a third CD28/ctla-4 counter-receptor. *J. Immunol.* **161,** 2708–2715.

Gange, S. J., Munoz, A., Chmiel, J. S., Donnenberg, A. D., Kirstein, L. M., Detels, R., and Margolick, J. B. (1998). Identification of inflections in T cell counts among HIV infected individuals and relationship with progression to clinical AIDS. *Proc. Natl. Acad. Sci. U.S.A.* **95,** 10848–10853.

Gelman, R., Cheng, S. C., Kidd, P., Waxdal, M., and Kagan, J. (1993). Assessment of the effects of instrumentation, monoclonal antibody, and fluorochrome on flow cytometric immunophenotyping: A report based on 2 years of the NIAID DAIDS flow cytometry quality assessment program. *Clin. Immunol. Immunopathol.* **66,** 150–162.

Gerber, A. R. (1993). Centers for Disease Control programs in flow cytometric immunophenotyping. *Ann. NY. Acad. Sci.* **677,** 40–42.

Giorgi, J. V., Liu, Z., Hultin, L. E., Cumberland, W. G., Hennessey, K., and Detels, R. (1993). Elevated levels of CD38 CD8 T cells in HIV infection add to the prognostic value of low CD4 T cell levels: Results of 6 years of followup. *J. Acq. Immune Defic. Synd.* **6,** 904–912.

Gorochov, G., Neumann, A. U., Kereveur, A., Parizot, C., Li, T., Katlama, C., Karmochkine, M., Raguin, G., Autran, B., and Debre, P. (1998). Perturbation of CD4 and CD8 T cell repertoires during progression to AIDS and regulation of the CD4 repertoire during antiviral therapy. *Nat. Med.* **4,** 215–221.

Gratama, J. W., van der Linden, R., van der Holt, B., Bolhuis, R. L. H., and van de Winkel, J. G. J. (1997). Analysis of factors contributing to the formation of mononuclear cell aggregates ("escapees") in flow cytometric immunophenotyping. *Cytometry* **29,** 250–260.

Gray, C. M., Schapiro, J. M., Winters, M. A., and Merigan, T. C. (1998). Changes in CD4 and CD8 T cell subsets in response to highly active antiretroviral therapy in HIV infected patients with prior protease inhibitor experience. *AIDS Res. Hum. Retro.* **14,** 561–569.

Gruber, R., Reiter, C., and Riethmuller, G. (1993). Triple immunofluorescence flow cytometry, using whole blood, of CD4 and CD8 lymphocytes expressing CD45RO and CD45RA. *J. Immunol. Methods* **163,** 173–179.

Hammer, S. M., Squires, K. E., Hughes, M. D., Grimes, J. M., Demeter, L. M., Currier, J. S., Eron, J. J., Feinberg, J. E., Balfour, H. H., Deyton, L. R., Chodakewitz, J. A., and Fischl, M. A. (1997). A controlled trial of two nucleoside analogs plus indinavir in persons with HIV infection and CD4 counts of 200 per cubic millimeter or less. *N. Engl. J. Med.* **337,** 725–733.

Harwell, T. S., Ferbas, J., and Logar, A. J. (1995). Are there differences between laboratories that use or fail to use the CDC's guidelines to measure CD4 and CD8 T cells?. *Cytometry* **21,** 256–257.

Henderson, D. K., Saah, A. J., Zak, B. J., Kaslow, R. A., Lane, H. C., Folks, T., Blackwelder, W. C., Schmitt, J., LaCamera, D. J., Masur, H., and Fauci, A. S. (1986). Risk of nosocomial infection with human T cell lymphotropic virus type III/lymphadenopathy associated virus in a large cohort of intensively exposed health care workers. *Ann. Intern. Med.* **104,** 644–647.

Homburger, H. A., Rosenstock, W., Paxton, H., Paton, M. L., and Landay, A. L. (1993). Assessment of interlaboratory variability of immunophenotyping: results of the college of American pathologists flow cytometry survey. *Ann. N.Y. Acad. Sci.* **677,** 43–49.

Jason, J., and Larned, J. (1997). Single cell cytokine profiles in normal humans: comparison of flow cytometric reagents and stimulation protocols. *J. Immunol. Methods* **207,** 13–22.

Kagan, J., Gelman, R., Waxdal, M., and Kidd, P. (1993). NIAID division of AIDS flow cytometry quality assessment program. *Ann. N.Y. Acad. Sci.* **677,** 50–52.

Kestens, L., Vanham, G., Gigase, P., Young, G., Hannet, I., Vanlangendonck, F., Hulstaert, F., and Bach, B. A. (1992). Expression of activation antigens, HLADR and CD38, on CD8 lymphocytes during HIV infection. *AIDS* **6,** 793–797.

Khalfan, H., Abuknesha, R., Rand-Weaver, M., Price, R. G., and Robinson, D. (1986). Aminomethyl coumarin acetic acid: A new fluorescent labeling agent for proteins. *Histochem J.* **18,** 497–499.

Kishimoto, T., Goyert, S., Kikutani, H., Mason, D., Miyasaka, M., Moretta, L., Ohno, T., Okumura, K., Shaw, S., Springer, T. A., Sugamura, K., Shgawara, H., von dem Borne, A. E. G. K., and Zola, H. (1997). Update of CD antigens 1996. *J. Immunol.* **158,** 3035–3036.

Klein, R. C., Party, E., and Gershey, E. L. (1990). Virus penetration of examination gloves. *Biotechniques* **9,** 196–199.

Kotilainen, H. R., Brinker, J. P., Avato, J. L., and Gantz, N. M. (1989). Latex and vinyl examination gloves: Quality control procedures and implications for health care workers. *Arch. Intern. Med.* **149,** 2749–2753.

Kotylo, P. K., Fineberg, N. S., Freeman, K. S., Redmond, N. L., and Charland, C. (1993). Reference ranges for lymphocyte subsets in pediatric patients. *Am. J. Clin. Pathol.* **100**, 111–115.

Kromer, E., and Grossmuller, F. (1994). Light scatter based lymphocyte gate: Helpful tool or source of error? *Cytometry* **15**, 87–89.

Kux, A., Bertram, S., Hufert, F. T., Schmitz, H., and von Laer, D. (1996). Antibodies to p24 antigen do not specifically detect HIV infected lymphocytes in AIDS patients. *J. Immunol. Methods* **191**, 179–186.

Lansdorp, P. M., Smith, C., Safford, M., Terstappen, L. W. M. M., and Thomas, T. E. (1991). Single laser three color immunofluorescence staining procedures based on energy transfer between phycoerythrin and cyanine 5. *Cytometry* **12**, 723–730.

Lanier, L. L., and Warner, N. L. (1981). Paraformaldehyde fixation of hematopoietic cells for quantitative flow cytometry analysis. *J. Immunol. Methods* **47**, 25–30.

Lape-Nixon, M. L., and Prince, H. E. (1996). How many gated lymphocytes are needed for accurate assessment of T subset percentages by flow cytometry? *Cytometry* **26**, 223–226.

Legac, E., Autran, B., Merle-Beral, H., Katlama, C., and Debre, P. (1992). CD4$^+$ CD7$^-$ CD57$^+$ T cells: A new T subset expanded during HIV infection. *Blood* **79**, 1746–1753.

Lenkei, R., Mandy, F., Marti, G., and Vogt, R. (1998). Quantitative fluorescence cytometry: An emerging consensus. *Cytometry* **33**, 93–280.

Lewis, D. E., and Rickman, W. R. (1992). Methodology and quality control for flow cytometry. In "Manual of Clinical Laboratory Immunology" (N. R. Rose, E. C. deMacario, J. L. Fahey, H. Friedman, and G. M. Penn, eds.), pp. 164–173, American Society for Microbiology, Washington, D.C.

Lifson, J. D., Sasaki, D. T., and Engleman, E. G. (1986). Utility of formaldehyde fixation for flow cytometry and inactivation of the AIDS associated retrovirus. *J. Immunol. Methods* **86**, 143–149.

Liu, Z., Hultin, L., Cumberland, W. G., Hultin, P., Schmid, I., Matud, J. L., Detels, R., and Giorgi, J. V. (1996). Elevated relative fluorescence intensity of CD38 antigen expression on CD8 T cells is a marker of poor prognosis in HIV infection: Results of 6 years of followup. *Cytometry* **26**, 1–7.

Loken, M. R., Brosnan, J. M., Bach, B. A., and Ault, K. A. (1990). Establishing optimal lymphocyte gates for immunophenotyping by flow cytometry. *Cytometry* **11**, 453–459.

Mc Closkey, T. W., Oyaizu, N., Kaplan, M., and Pahwa, S. (1995). Expression of the Fas antigen in patients infected with HIV. *Cytometry* **22**, 111–114.

Mc Closkey, T. W., Cavaliere, T., Bakshi, S., Harper, R., Fagin, J., Kohn, N., and Pahwa, S. (1997). Immunophenotyping of T lymphocytes by three color flow cytometry in healthy newborns, children, and adults. *Clin. Immunol. Immunopathol.* **84**, 46–55.

Mc Closkey, T. W., Oyaizu, N., Bakshi, S., Kowalski, R., Kohn, N., and Pahwa, S. (1998a). CD95 expression and apoptosis during pediatric HIV infection: Early upregulation of CD95 expression. *Clin. Immunol. Immunopathol.* **87**, 33–41.

Mc Closkey, T. W., Chavan, S., Tamma, S. M. L., and Pahwa, S. (1998b). Comparison of seven quantitative assays to assess lymphocyte cell death during HIV infection: Measurement of induced apoptosis in anti Fas treated Jurkat cells and spontaneous apoptosis in PBMC from children infected with HIV. *AIDS Res. Hum. Retro.* **14**, 1413–1422.

Mc Closkey, T. W., Bakshi, S., Than, S., Arman, P., and Pahwa, S. (1998c). Immunophenotypic analysis of peripheral blood mononuclear cells undergoing *in vitro* apoptosis after isolation from HIV infected children. *Blood* **92**, 1–9.

McCray, E. (1986). Occupational risk of the acquired immunodeficiency syndrome among health care workers. *N. Engl. J. Med.* **314**, 1127–1132.

Macey, M. G., McCarthy, D. A., Davies, C., and Newland, A. C. (1997). The Qprep system: Effects on the apparent expression of leukocyte cell surface antigens. *Cytometry* **30**, 67–71.

Maino, V. C., and Picker, L. J. (1998). Identification of functional subsets by flow cytometry: Intracellular detection of cytokine expression. *Cytometry* **34**, 207–215.

Mandy, F. F., Bergeron, M., Recktenwald, D., and Izaguirre, C. A. (1992). A simultaneous three color T cell subset analysis with single laser flow cytometers using T cell gating protocol: Comparison with conventional two color method. *J. Immunol. Methods* **156**, 151–162.

Manetti, R., Annunziato, F., Biagiotti, R., Giudizi, M. G., Piccinni, M. P., Giannarini, L., Sampognaro, S., Parronchi, P., Vinante, F., Pizzolo, G., Maggi, E., and Romagnani, S. (1994). CD30 expression by CD8 T cells producing type 2 helper cytokines: Evidence for large numbers of CD8$^+$ CD30$^+$ T cell clones in HIV infection. *J. Exp. Med.* **180**, 2407–2411.

Martin, L. S., McDougal, J. S., and Loskoski, S. L. (1985). Disinfection and inactivation of the human T lymphotropic virus III/lymphadenopathy associated virus. *J. Infect. Dis.* **152**, 400–403.

Meyaard, L., Hovenkamp, E., Keet, I. P. M., Hooibrink, B., de Jong, I. H., Otto, S. A., and Miedema, F. (1996). Single cell analysis of IL-4 and IFN-γ production by T cells from HIV infected individuals: Decreased IFN-γ in the presence of preserved IL-4 production. *J. Immunol.* **157**, 2712–2718.

Mohri, H., Bonhoeffer, S., Monard, S., Perelson, A. S., and Ho, D. D. (1998). Rapid turnover of T lymphocytes in SIV infected rhesus macaques. *Science* **279**, 1223–1227.

Moore, J. P. (1997). Coreceptors: Implications for HIV pathogenesis and therapy. *Science* **276**, 51–52.

Mueller, B. U., Nelson, R. P., Sleasman, J., Zuckerman, J., Heath-Chiozzi, M., Steinberg, S. M., Balis, F. M., Brouwers, P., Hsu, A., Saulis, R., Sei, S., Wood, L. V., Zeichner, S., Katz, T. T. K., Higham, C., Aker, D., Edgerly, M., Jarosinski, P., Serchuck, L., Whitcup, S. M., Pizzuti, D., and Pizzo, P. A. (1998). A phase I/II study of the protease inhibitor ritonavir in children with HIV infection. *Pediatrics* **101**, 335–343.

Muirhead, K. A. (1993). Establishment of quality control procedures in clinical flow cytometry. *Ann. N.Y. Acad. Sci.* **677**, 1–20.

Nicholson, J. K. A., and Browning, S. (1994). Ability of Optilyse lysing/fixing reagents to inactivate HIV infected H9 cells in whole blood. *J. Immunol. Methods* **168**, 283–284.

Nicholson, J. K. A., and Green, T. A. (1993). Selection of anticoagulants for lymphocyte immunophenotyping: Effect of specimen age on results. *J. Immunol. Methods* **165**, 31–35.

Nicholson, J. K. A., and Stetler-Stevenson, M. (1998). Quantitative fluorescence: To count or not to count. Is that the question? *Cytometry* **34**, 203–204.

Nicholson, J. K. A., Browning, S. W., Orloff, S. L., and McDougal, J. S. (1993a). Inactivation of HIV infected H9 cells in whole blood preparations by lysing/fixing reagents used in flow cytometry. *J. Immunol. Methods* **160**, 215–218.

Nicholson, J. K. A., Jones, B. M., and Hubbard, M. (1993b). CD4 T lymphocyte determinations on whole blood specimens using a single tube three color assay. *Cytometry* **14**, 685–689.

Nicholson, J. K. A., Rao, P. E., Calvelli, T., Stetler-Stevenson, M., Browning, S. W., Yeung, L., and Marti, G. (1994). Artifactual staining of monoclonal antibodies in two color combinations is due to an immunoglobulin in the serum and plasma. *Cytometry* **18**, 140–146.

Nicholson, J., Kidd, P., Mandy, F., Livnat, D., and Kagan, J. (1996a). Three color supplement to the NIAID DAIDS guideline for flow cytometric immunophenotyping. *Cytometry* **26**, 227–230.

Nicholson, J. K. A., Hubbard, M., and Jones, B. M. (1996b). Use of CD45 fluorescence and side scatter characteristics for gating lymphocytes when using whole blood lysis procedure and flow cytometry. *Cytometry* **26**, 16–21.

Nicholson, J. K., Stein, D., Mui, T., Mack, R., Hubbard, M., and Denny, T. (1997). Evaluation of a method for counting absolute numbers of cells with a flow cytometer. *Clin. Diagn. Lab. Immunol.* **4**, 309–313.

Niehues, T., Ndagijimana, J., Horneff, G., and Wahn, V. (1998). CD28 expression in pediatric HIV infection. *Pediatr. Res.* **44**, 265–268.

O'Gorman, M. R. G., Millard, D. D., Lowder, J. N., and Yogev, R. (1998). Lymphocyte subpopulations in healthy 1–3 day old infants. *Cytometry* **34**, 235–241.

Ohta, Y., Fujiwara, K., Nishi, T., and Oka, H. (1986). Normal values of peripheral blood lymphocyte populations and T cell subsets at a fixed time of day: A flow cytometric analysis with monoclonal antibodies in 210 healthy adults. *Clin. Exp. Immunol.* **64**, 146–149.

Occupational Safety and Health Administration (OSHA) (1991). Occupational exposure to blood-borne pathogens. *Fed. Register* **56**, 64174–64182.

Oyaizu, N., and Pahwa, S. (1995). Role of apoptosis in HIV disease pathogenesis. *J. Clin. Immunol.* **15**, 217–231.

Pahwa, S. (1990). Immune defects in pediatric AIDS, their pathogenesis, and role of immunotherapy. *Crit. Care Med.* **18,** s138–s143.

Pantaleo, G., Cohen, O. J., Schacker, T., Vaccarezza, M., Graziosi, C., Rizzardi, G. P., Kahn, J., Fox, C. H., Schnittman, S. M., Schwartz, D. H., Corey, L., and Fauci, A. S. (1998). Evolutionary pattern of HIV replication and distribution in lymph nodes following primary infection: Implications for antiviral therapy. *Nat. Med.* **4,** 341–345.

Patterson, B. K., Till, M., Otto, P., Goolsby, C., Furtado, M. R., McBride, L. J., and Wolinsky, S. M. (1993). Detection of HIV DNA and messenger RNA in individual cells by PCR driven in situ hybridization and flow cytometry. *Science* **260,** 976–979.

Patterson, B. K., Goolsby, C., Hodara, V., Lohman, K. L., and Wolinsky, S. M. (1995). Detection of CD4 T cells harboring HIV DNA by flow cytometry using simultaneous immunophenotyping and PCR driven in situ hybridization: Evidence of epitope masking of the CD4 cell surface molecule in vivo. *J. Virol.* **69,** 4316–4322.

Paxton, H., Kidd, P., Landay, A., Giorgi, J., Flomenberg, N., Walker, E., Valentine, F., Fahey, J., and Gelman, R. (1989). Results of the flow cytometry ACTG quality control program: Analysis and findings. *Clin. Immunol. Immunopathol.* **52,** 68–84.

Peddecord, K. M., Benenson, A. S., Hofherr, L. K., Francis, D. P., Garfein, R. S., Cross, G. D., and Schalla, W. O. (1993). Variability of reporting and lack of adherence to consensus guidelines in human T lymphocyte immunophenotyping reports: Results of a case series. *J. Acq. Immune Defic. Synd.* **6,** 823–830.

Perfetto, S. P., Hickey, T. E., Blair, P. J., Maino, V. C., Wagner, K. F., Zhou, S., Mayers, D. L., Louis, D. S., June, C. H., and Siegel, J. N. (1997). Measurement of CD69 induction in the assessment of immune function in asymptomatic HIV infected individuals. *Cytometry* **30,** 1–9.

Pizzolo, J. G., and Melamed, M. R. (1996). Third color CD45 staining of paraformaldehyde fixed peripheral blood lymphocytes. *Cytometry* **23,** 67–71.

Plaeger-Marshall, S., Hultin, P., Bertolli, J., O'Rourke, S., Kobayashi, R., Kobayashi, A. L., Giorgi, J. V., Bryson, Y., and Stiehm, E. R. (1993). Activation and differentiation antigens on T cells of healthy, at risk, and HIV infected children. *J Acq. Immune Defic. Synd.* **6,** 984–993.

Plaeger-Marshall, S., Isacescu, V., O'Rourke, S., Bertolli, J., Bryson, Y. J., and Stiehm, E. R. (1994). T cell activation in pediatric AIDS pathogenesis: Three color immunophenotyping. *Clin. Immunol. Immunopathol.* **71,** 19–26.

Prince, H. E., and Jensen, E. R. (1991). Three color cytofluorometric analysis of CD8 cell subsets in HIV infection. *J. Acq. Immune Defic. Synd.* **4,** 1227–1232.

Prince, H. E., York, J., and Kuttner, D. K. (1994). Reduction of escapee formation in flow cytometric analysis of lymphocyte subsets. *J. Immunol. Methods* **177,** 165–173.

Rabin, R. L., Roederer, M., Maldonado, Y., Petru, A., Herzenberg, L. A., and Herzenberg, L. A. (1995). Altered representation of naive and memory CD8 T cell subsets in HIV infected children. *J. Clin. Invest.* **95,** 2054–2060.

Reichert, T., deBruyere, M., Deneys, V., Totterman, T., Lydyard, P., Yuksel, F., Chapel, H., Jewell, D., van Hove, L., Linden, J., and Buchner, L. (1991). Lymphocyte subset reference ranges in adult caucasians. *Clin. Immunol. Immunopathol* **60,** 190–208.

Renzi, P., and Ginns, L. C. (1987). Analysis of T cell subsets in normal adults: comparison of whole blood lysis technique to ficoll-hypaque separation by flow cytometry. *J. Immunol. Methods* **98,** 53–56.

Roederer, M., Kantor, A. B., Parks, D. R., and Herzenberg, L. A. (1996a). Cy7 PE and Cy7 APC: Bright new probes for immunofluorescence. *Cytometry.* **24,** 191–197.

Roederer, M., Herzenberg, L. A., and Herzenberg, L. A., (1996ba). Changes in antigen densities on leukocyte subsets correlate with progression of HIV disease. *Inter. Immunol.* **8,** 1–11.

Roederer, M., DeRosa, S., Gerstein, R., Anderson, M., Bigos, M., Stovel, R., Nozaki, T., Parks, D., Herzenberg, L., and Herzenberg, L. (1997). 8 color, 10 parameter flow cytometry to elucidate complex leukocyte heterogeneity. *Cytometry* **29,** 328–339.

Romeu, M. A., Mestre, M., Gonzalez, L., Valls, A., Verdaguer, J., Corominas, M., Bas, J., Massip, E., and Buendia, E. (1992). Lymphocyte immunophenotyping by flow cytometry in normal adults:

Comparison of fresh whole blood lysis technique, ficoll-paque separation, and cryopreservation. J. Immunol. Methods **154,** 7–10.

Roos, M. T. L., deLeeuw, A. S. M., Claessen, F. A. P., Huisman, H. G., Kootstra, N. A., Meyaard, L., Schellekens, P. T. A., Schuitemaker, H., and Miedema, F. (1994). Viro immunological studies in acute HIV infection. *AIDS.* **8,** 1533–1538.

Sachsenberg, N., Perelson, A. S., Yerly, S., Schockmel, G. A., Leduc, D., Hirschel, B., and Perrin, L. (1998). Turnover of CD4 and CD8 T lymphocytes in HIV infection as measured by Ki-67 antigen. *J. Exp. Med.* **187,** 1295–1303.

Sattar, S. A., and Springthorpe, V. S. (1991). Survival and disinfectant inactivation of HIV: A critical review. *Rev. Infect. Dis.* **13,** 430–447.

Schenker, E. L., Hultin, L. E., Bauer, K. D., Ferbas, J., Margolick, J. B., and Giorgi, J. V., (1993). Evaluation of a dual color flow cytometry immunophenotyping panel in a multicenter quality assurance program. *Cytometry* **14,** 307–317.

Schmid, I., Nicholson, J. K. A., Giorgi, J. V., Janossy, G., Kunkl, A., Lopez, P. A., Perfetto, S., Seamer, L. C., and Dean, P. N. (1997). Biosafety guidelines for sorting unfixed cells. *Cytometry* **28,** 99–117.

Simson, E., and Groner, W. (1995). Variability in absolute lymphocyte counts obtained by automated cell counters. *Cytometry* **22,** 26–34.

Smith, D., Sieg, S., and Kaplan, D. (1998). Technical note: Aberrant detection of cell surface Fas ligand with anti peptide antibodies. *J. Immunol.* **160,** 4159–4160.

Spire, B., Montagnier, L., Barre-Sinoussi, F., and Chermann, J. C. (1984). Inactivation of lymphadenopathy associated virus by chemical disinfectants. *Lancet.* **2,** 899–901.

Srour, E. F., Leemhuis, T., Jenski, L., Redmond, R., and Jansen, J. (1990). Cytolytic activity of human natural killer cell subpopulations isolated by four color immunofluorescence flow cytometric cell sorting. *Cytometry* **11,** 442–446.

Storek, J., Ferrara, S., and Isacescu, V. (1992). Blood B cell subpopulations: The effect of narrow versus wide forward scatter x side scatter gating. *J. Immunol. Methods* **156,** 129–133.

Su, L., Lee, R., Bonyhadi, M., Matsuzaki, H., Forestell, S., Excaich, S., Bohnlein, E., and Kaneshima, H. (1997). Hematopoietic stem cell based gene therapy for AIDS: Efficient transduction and expression of RevM10 *in vivo* and *in vitro. Blood* **89,** 2283–2290.

Takizawa, F., Kinet, J. P., and Adamczewski, M. (1993). Binding of phycoerythrin and its conjugates to murine low affinity receptors for IgG. *J. Immunol. Methods* **162,** 269–272.

Tamul, K. R., O'Gorman, M. R. G., Donovan, M., Schmitz, J. L., and Folds, J. D. (1994). Comparison of a lysed whole blood method to purified cell preparations for lymphocyte immunophenotyping: Differences between healthy controls and HIV positive specimens. *J. Immunol. Methods* **167,** 237–243.

Thornthwaite, J. T., Rosenthal, P. K., Vazquez, D. A., and Seckinger, D. (1984). The effects of anticoagulant and temperature on the measurements of helper and suppressor cells. *Diag. Immunol.* **2,** 167–174.

vanVugt, M. J., van den Herik-Oudijk, I. E., and van de Winkel, J. G. J. (1996). Binding of PECy5 conjugates to the human high affinity receptor for IgG CD64. *Blood* **88,** 2358–2361.

Waldrop, S. L., Pitcher, C. J., Peterson, D. M., Maino, V. C., and Picker, L. J. (1997). Determination of antigen specific memory effector CD4 T cell frequencies by flow cytometry: Evidence for a novel, antigen specific homeostatic mechanism in HIV associated immunodeficiency. *J. Clin. Invest.* **99,** 1739–1750.

Whitaker, J. E., Haugland, R. P., Moore, P. L., Hewitt, P. C., Reese, M., and Haugland, R. P. (1991). Cascade blue derivatives: Water soluble, reactive, blue emission dyes evaluated as fluorescent labels and tracers. *Anal. Biochem.* **198,** 119–130.

Wilks, D., Byrom, N., Walker, L., Habeshaw, J., and Dalgleish, A. (1990). Characteristic immunophenotyping artifact seen in patients with anti mouse immunoglobulin antibodies. *Cytometry* **11,** 318–319.

Working Group on Antiretroviral Therapy and Medical Management of Infants, Children, and Adolescents with HIV Infection (1998). Antiretroviral therapy and medical management of pediatric HIV infection. *Pediatrics* **102,** 1005–1062.

World Health Organization Global Program on AIDS (1994). Report of a WHO workshop on flow cytometry and alternative methodologies for CD4 lymphocyte determinations: applications for developing countries. *AIDS* **8,** WHO1–WHO4.

Yang, G., Olson, J. C., Pu, R., and Vyas, G. N., (1995). Flow cytometric detection of HIV proviral DNA by the PCR incorporating digoxigenin or fluorescein labeled dUTP. *Cytometry* **21,** 197–202.

Yeh, S. W., Ong, L. J., Clark, J. H., and Glazer, A. N. (1987). Fluorescence properties of allophycocyanin and a crosslinked allophycocyanin trimer. *Cytometry* **8,** 91–95.

INDEX

A

Acute lymphoblastic leukemia
  B-lineage, 392–393
  IgH gene rearrangements, 406
  molecular analysis, 412
  T-cell, 392
Acute myeloid leukemia
  FAB classification, 345, 348–351
  false negativity and false positivity, 352
  intermediate cell populations in, 353–354
  MRD detection in, 393–395
  showing features of CD45 expression, 354
Adenocarcinomas, clinical studies of proliferation, 462–463
Adriamycin, *see* Doxorubicin
Agar, incubation on, for colony formation, 553–554
Algorithm
  built into LSC software, 337
  for multiparameter data analysis, 488–492
  quantification, IFI, 88–91
  segmentation, IFI, 80–82
  TFL-TELO, 77
Alignment
  fixed, checking for, 570–571
  instrument for flow karyotyping, 15–16
Alkaline comet assay
  developed by Olive, 256–257
  measurement of DNA damage, 253–254
Alkaline comet method, 239
Alkaline lysis solutions, in QDFM, 36, 41–42
ALL, *see* Acute lymphoblastic leukemia
Allophycocyanin, emission distribution, 296–297
7-Amino-actinomycin, DNA-specific, 423–425
3-Aminopropyltriethoxy silane, slide preparation, 38
Amiprophos-methyl solutions, 8
AML, *see* Acute myeloid leukemia
Anthracyclines
  fluorescence quenching, 199
  transport and retention monitoring, 195
Anthraquinones, cytotoxicity, 177

Antibiotics
  effects on target organisms, 526–527
  inhibition of bacteria, 558
  time parameters for effects of, 561
Antibodies
  in chromosome translocation study, 99
  conjugated, brightness, 298, 300
  exhaustion, fluorescence saturation due to, 150
  monoclonal, bacterium-specific, 519–520
  in QDFM, 35
  reliable, for immunophenotyping, 314
  single antibody histogram analysis, 289–290
Anticancer agents, cytotoxic
  disruption of DNA integrity, 174–175
  DNA topoisomerase inhibitors, 175–177
  drug–DNA interactions, 183–188
  flow cytometric analysis, quality control, 178–182
  inhibition of
    DNA metabolism, 175
    mitotic spindle function, 175
Antigen quantitation, 582
Antigens
  alterations during HIV infection, 580–582
  major lymphocyte subpopulations distinguished by, 568–569
  multiple, simultaneous assessment, 314
  proliferation-associated, 364
    LSCM analysis, 430–434
  suitable for monitoring AML, 394–395
  surface, blast cell populations, 351–352
Antimicrobial agents, flow cytometric analyses, 526–529
Antitumor agents, delivery
  clinical implications, 224–225
  physiological considerations, 223–224
  slow drug penetration, 224
Aphidicolin, synchronization of cells by, 276–277
Apoptosis, LSCM analysis, 425–430
Artificial neural networks, in detection of bacteria, 520–521

Asthma
  data pattern classification, 494, 497–503
  immunophenotype classifications, 507
  multiparameter flow cytometric analysis,
    490
Average multiplicity index, 492

B

Bacteria
  antibiotic effects, 558
  cell cycle, 540
  cell viability, 521–524
  detection, 514–516
    by artificial neural networks, 520–521
  DNA content, 516–517
  DNA staining, 541–542
  dye efflux, 543
  enzymatic activity, 523–524
  fixed, staining, 543–544
  flow cytometer setup for, 512–513
  flow cytometric measurement, 547–548
  flow cytometry, fluorescent dyes for,
    554–555
  Gram stain, 525–526
  immunofluorescence, 517–520
  light scatter measurements, 516
  membrane potential, 522–523
  rapidly growing, drug effects, 558–562
  sample preparation, 513–514
  size and shape, 541
  spores and sheath fluid, 546
  viable, FISH identification, 524–525
  vital cells, fluorescent dye uptake, 555–558
  vital staining, 544–546
Beads
  brightness, 298, 300
  fluorescence, IFI value, 88–90
  for instrument performance verification,
    294–295
Biomarkers
  clinical, cytometric quantitation, 449
  expression, correlation with proliferation
    data, 452
Biosafety, blood sample handling, 569
Biotin, labeled DNA: visualization, 65–66
Bivariate analysis, cyclins, 433
Bivariate profiles, hemoglobin
  effect of TGF-β, 145–147
  flow data, 143–144
  isolation of fetal nucleated red cells, 144–145
  problems and limitations, 149–151
  study of sickle-cell erythropoiesis, 148–149

Blast cells
  and monocytes: intermediate cell population
    between, 353–354
  populations, surface antigens, 351–352
  studied within bone marrow, 344
Blastomeres
  abnormal, 111
  preparation for
    chromosome translocation study, 102–103
    PRINS, 58–59
Blood cultures, maternal, isolation of fetal
    nucleated red cells, 144–145
Bone marrow
  light microscopic examination, 347–348
  for MRD detection in leukemias, 386
  sample processing, 345
BrdUrd, *see* 5′-Bromodeoxyuridine
Breast cancer
  BrdUrd labeling, 463
  endocrine therapy, 376–377
  proliferation markers and prognosis,
    368–370
5′-Bromodeoxyuridine
  in early clinical studies, 453
  incorporated
    immunostaining of, 276
    measurement by flow cytometry, 136–137
  labeling of
    cells, 131
    RNA, 134–135
  subcellular distribution, 132–133, 135
  substituted RNA, 130–131
  as tumor radiosensitizer, 450

C

Calibration
  fluorescence axes, problems with,
    149–150
  results of telomere length measurement,
    85–86
Camptothecin
  apoptosis induction, 429–430
  cytotoxicity, 177
Carbocyanine dyes, symmetrical and
    asymmetric, 119–120
Carboxyfluorescein diacetate succinimidyl
    ester, *see* CFDA-SE
Cardiolipin dye, for mitochondria, 120,
    124–125
CD10, expression patterns, 390
CD11b, coexpression with CD15, 156
CD15, coexpression with CD11b, 156

CD34
  expression on immature HP, 154
  lost by developing HP, 158
CD45
  B and T cells positive for, 333
  expression, AMLs showing original features
    of, 354
  as gating reagent, 573
  intensity, combined with linear side scatter,
    344
CD45/SSC gating, 345, 348–356
CD61, and cell differentiation, 159
Cell biopsy, and fixation, 111
Cell counts
  CD4 absolute, 573–574
  for erythropoiesis study, 142–143
Cell culture
  integrity: quality control, 178–179
  for MRD detection, 397
Cell cycle
  bacteria, 540
  effect of TGF-β, 146
  related expression of DNA topoisomerases,
    180
  and RNA synthesis, 134–135
  synchronization, plant chromosomes, 19, 21
Cell death
  HIV-induced, 583
  nuclear protein-induced, 270
Cell heterogeneity, detection by comet assay,
  237
Cell lineages, bone marrow, identification,
  348
Cell lines, for studies of drug efflux and MDR,
  196–197
Cell loss factor, and potential doubling time,
  472
Cell lysis, optimization of conditions, 261
Cell populations
  fluorescence histograms and plots, 291–293
  intermediate
    between blast cells and monocytes,
      353–354
    between lymphocytes and blast cells, 353
Cell preparation
  analysis of RNA synthesis, 131–132
  mitochondria fluorescent detection, 122
  study of erythropoiesis, 141–142
Cell production rates
  clinical studies with thymidine analogs,
    452–474
  in normal tissues, 473

studies in nonmalignant epithelium,
  464–468
and therapy, 469–470
Cell shape, and gene expression, as mechanism
  for contact effect, 221–222
Cell size, forward scatter reflecting, 289–290
Cell sorting
  methodology, 215
  from multicell culture systems, 214–215
Cell survival, in complex systems, predication
  by DNA damage, 244–246
Cell viability, bacteria
  enzymatic activity, 523–524
  membrane integrity, 521–522
  membrane potential, 522–523
Cell wall, bacteria, 542
CFDA-SE
  bright and dim cells, 160, 167
  diminished fluorescence, 157
  halving, 159–160
  in HP maturation studies, 155–156
Charge-coupled device, in halo–comet assay,
  265
Chemokine receptors, 583
Chemotherapeutic agents
  activity patterns in spheroids, 222–223
  contact effect, 218–219
Chemotherapy
  correlation with TLI, 373, 375–376
  neoadjuvant, 375–376
Children, perinatal transmission of HIV,
  574
Chromatin packaging, as mechanism for
  contact effect, 220–221
Chromosomes
  border, overlaid with telomere border,
    87–88
  human, application of PRINS, 56
  painting, combined with PRINS, 61–64
  plant
    flow cytogenetics problems, 4–5
    mitotic, 5
    preparation of suspensions, 14, 19, 21–22
    sorted fraction purity, 17–18, 20–21,
      25–26
    sorting, 17, 20, 24–26
    staining, 14–15
  prematurely condensed, 398
  protective cap provided by telomeres, 70
  segmentation, 82, 84–85, 92–93
Chromosome translocation
  detection in interphase nuclei, 98–99

PRINS–painting of, 63–64
probe preparation
    applications, 112
    buffers and other solutions, 101–102
    critical aspects, 111–112
    discussion of techniques, 109–111
    instruments, 102
    materials, 99–101
    protocols, 102–108
CLASSIF1 algorithm
    asthma, 494, 499
    data classifications, 488–492
    HIV infection, 503
Classification coincidence factor, 492
    asthma, 502
    melanoma, 494
Classifier masks, 492–494
Clinical oncology, future of cytometry in,
    474–478
Clones, selection from library, 111–112
Clone-specific markers, in PCR, 404–406
Clone-specific PCR, 408
Clone-specific primers, in limiting dilution,
    410–411
CMXRosamine, probe for mitochondria, 119,
    125–126
Colorectal adenocarcinomas, BrdUrd labeling,
    462–463
Colorectal cancer, proliferation markers and
    prognosis, 370–372
Comet assay
    alkaline method (by Olive), 256–257
    alkaline method (by Singh), 253–254
    application to clinical samples, 247
    development, 236–237
    neutral method (by Olive), 257–258
    neutral method (by Singh), 254–256
    types of DNA damage detected by,
        242–246
Comets
    antibody-stained, 237
    image analysis, 240–242
    preparation steps, 238–240
Common acute lymphoblastic leukemia
    antigen, see CD10
Communication, intracellular, as mechanism
    for contact effect, 220
Compensation matrix, manual application, 307
Compensation standard, files for, 304–305
Competetive PCR assay, 409
Confocal microscopy, DNA–drug interactions,
    184–185

Contact effect
    chemotherapeutic agents, 218–219
    ionizing radiation, 217–218
    proposed mechanisms for, 219–222
Contouring, critical step in LSCM, 318–319
Contour plots, for cell populations, 292–293
Correlated list mode data
    acquisition and analysis, 290–293
    multiparameter analysis produced by,
        311
Cross-linking, DNA, 182
Culture system
    3-D: cellular and environmental
        heterogeneity, 212–213
    multicell spheroids, 213–215
        chemotherapy in, 222–225
Cyclin-dependent kinase inhibitors, 378
Cyclins, staining for, 432–433
Cyclophosphamide, plus adriamycin and 5-
    fluorouracil, 374
Cytogenetics
    FISH, 398, 400–401
    for MRD detection, 397–401
Cytokines
    differentiation driven by, HP subsets during,
        157–167
    intracellular, labeling, 582–583
Cytospin, specimen preparations, 317
Cytospin chamber, in LSCM of human tumors,
    424, 428–429
Cytotoxicity
    anthraquinones, 177
    Hoechst 33342, 215, 217

D

DAPI
    staining of plant chromosomes, 14–15
    stock solution, 10
Data management, HIV patients
    data base, 574–575
    problems with immunophenotyping,
        577–578
    proficiency testing programs, 576–577
    quality control, 575–576
    reference ranges, 575
Data pattern classification
    asthma, 497–502
    CLASSIF1 algorithm for, 491–493
    for diagnosis, 506
    melanoma, 493–494
Daunomycin, cellular retention, 203, 205
Deterioration, chronic, in optical system, 295

Diagnosis
  acute leukemias, 355–356
  with data pattern classification, 506
DiBAC$_4$(3), membrane potential sensitive,
  562
Dideoxynucleotides, blocking with, 61, 66
Difference filter, chromosomes, 84–85
Digital image analysis, for QDFM, 46
Digoxigenin, labeled DNA: visualization, 65
Disinfectants, exposure to, 527, 529
DNA
  bacteria
    staining, 541–542
    structure, 542
  complementary, mapping, 51
  fragmentation
    calculation, 179
    and cell lysis assay, 178
  high-molecular-weight, purification, 38–40
  integrity, drug-induced disruption, 174–175
  labeled, visualization, 65–66
  labeling, primed *in situ*, 17–18
  libraries, chromosome-specific, 27–28
  loops, relaxed, 265–266
  metabolism, inhibitors, 175
  percentage in comet tail, 241
  prepared from YAC clones, 42–43,
    104–107
  repeat sequences, detection with
    oligonucleotide primers, 59–60
  supercoiling, 252–254
DNA amplification, *in vitro*, generation of
  probes by, 43–44
DNA content
  analytic methods for, 131–134
  bacteria, 516–517, 558–559, 561
  in human tumors, LSC applications,
    423–425
  nuclear quantification in S phase, 363–364
DNA damage
  electrophoretic detection, 236–237
  measurement on cell-by-cell basis, 253–266
  prediction of cell survival in complex
    systems, 244–246
  radiation-induced, 251–252
  types detected by comet assay, 242–244
DNA–drug interactions
  confocal microscopy, 184–185
  flow cytometric analysis of
    drug uptake, 184
      Hoechst 33342–DNA binding, 185–188
  ligand characteristics, 183

  spectrofluorimetry, 183–184
DNA fibers
  genomic, gene mapping by hybridization
    onto, 48, 50
  preparation for QDFM, 40–41
DNA fragments
  cloned, probes generated from, 41–43
  and ethanol fixation, 426
  mapped by FISH, 34
DNA histograms
  bacteria, 540
    standards and controls, 548–550
  in bivariate profiles of hemoglobin, 144
  and gating, 142
DNA index, 425
  in monitoring MRD, 402
DNA repair, chromatin structure effect,
  220–221
DNA replication
  alteration after heat shock, 276–278
  patterns, 281–282
DNA strand breaks, labeling, 427–428
DNA topoisomerase I, inhibitors, 177, 429–430
DNA topoisomerase II
  drug-induced trapping, 181–182
  inhibitors, 177
DNA topoisomerases, cell cycle related
  expression, 180
DoOC$_6$(3), mitochondrial dye, 119
Dot plots
  for cell populations, 293
  two-parameter, 198, 200
Doxorubicin
  cellular retention, 203, 205
  fluorescence, 219
  penetration problem, 224
  plus 5-fluorouracil and cyclophosphamide,
    374
  plus vincristine, 373–374
Drug binding, DNA-targeting, 187–188
Drug dose, intensity, as determinant of
  treatment efficacy, 375
Drug efflux, energy-dependent, in bacteria, 557
Drug efflux blockers
  combinations, 197
  effects on drug retention, 195–196
  fluorochrome retention with or without,
    201–206
Drugs
  β-lactams, inhibition of peptidoglycan layer,
    558–559
  fluorescent cytotoxic, cytometric assays,
    477–478

Dye combinations, protocols for mitochondria
analysis, 125–126
Dye efflux, bacteria, 543, 545–546

E

Embryos, whole, preparation for PRINS,
58–59
Endocrine therapy, proliferating tumors,
376–377
Enzymatic activity, bacteria, flow cytometry,
523–524
Enzymes, target, availability: quality control,
180–182
Epifluorescence microscopy, 329
images, 332–333
Epithelium, nonmalignant, cell production
rates, 464–468
Equipment, for plant flow cytogenetics, 12
Erythrocytes, lysis, and immunofluorescence
staining, 345–346
Erythroid cultures, for study of erythropoiesis,
141
Erythropoiesis
erythroid cultures for study of, 141
fetal, 140
sickle-cell, 148–149
Esophageal tumors, squamous, 456
Estrogen receptors, LSCM analysis, 435–436
Etoposide
effect on topoisomerase II, 177
spheroids repeatedly treated with, 226–227

F

FAB classification, AML, 345, 348–351
False positivity, AML, 352
Filamentation, bacterial, drug-induced, 559
Filters
difference: chromosome analysis, 84–85
optical, LP and BP, 297
Fingerprinting, and PCR products, 407–408
FISH, *see* Fluorescence *in situ* hybridization
FITC, *see* Fluorescein isothiocyanate
Fixation
cell biopsy and, 111
ethanol, bacterial cells, 557
protocols for mitochondria analysis,
125
Flow cytogenetics, plant
chromosome sorting, 26–28
problems with, 4–5

Flow cytometers
bacterial sample preparation, 513–514
instrument performance, verification,
293–295
instrument setup, 142
for microbes, 512–513
measurement of bacteria, 547–548
Flow cytometry, *see also* Multiparameter data
analysis
analysis of
drug uptake, 184
Hoechst 33342–DNA binding, 185–188
responses to antimicrobial agents,
526–529
RNA synthesis, 129–137
application to
medical microbiology, 563
microbial systems, 531–532
bacteria
basic considerations, 541–543
fluorescent dyes for, 554–555
rapidly growing, 558–562
standards and controls, 548–550
BrdUrd/DNA distribution, 132
compared with
immunohistochemistry, 315, 340
LSCM, 326–327
detection of
microbes, 514–516
MRD, 401–402
nuclear hsp70, 278
determination of Gram stain, 525–526
fluorescent detection of mitochondria, 118
fluorescent halo assay based on, 262–263
HIV disease investigation, 582–584
HIV infection monitoring, 568–582
identification of
food and drink contaminants, 529–531
specific microorganisms, 516–521
viable bacteria, with FISH, 524–525
identification of MDR, 194–196
controls, standards, and instruments,
199–201
critical aspects, 198–199
list mode analysis, 490–491
measurement of
hemoglobin profiles, 141, 150–151
nuclear matrix protein content, 272–273
nuclear protein content, 271–272
S-phase fraction, 448–449
membrane integrity analysis, 521–522
membrane potential analysis, 522–524

monitoring of HP maturation, 155–156
  assessment of HP subsets, 157–167
  critical aspects, 156–157
MRD analysis by immunophenotyping, 389,
  391–393
multicolor fluorescence, 578–582
mycobacteria, 562–563
nuclear matrix stability assay after heat
  shock, 274–276
and sorting
  application to plant chromosomes, 15–18
  chromosome analysis, 22–24
  theoretical flow karyotypes, 22
S-phase cell fraction, 363–364, 369–372
Flow karyotyping
  and chromosome sorting, 19–21
  detection of MRD, 401–402
  instrument alignment, 15–16
  univariate and bivariate analysis, 16, 22–24
Fluorescein isothiocyanate
  plus phycoerythrin: histograms, 490–491
  scattergrams, 324–326
Fluorescence axes, calibration, problems with,
  149–150
Fluorescence beads, IFI value, 88–90
Fluorescence *in situ* hybridization
  diagnostic screening, 400
  DNA fragment mapping, 34
  identification of viable bacteria, 524–525
  interphase nuclei, 98
  Ki-67 expression determined by, 434
  in LSCM, measurement of nucleoli, 438–441
  monitoring MRD, 400–401
  PGD based on, 99
  protocols, 108
  for QDFM, 45–46
  relocalization for, 327, 329, 331–333,
    335–336
  and slide pretreatment, 112
  target DNA, 398, 400
Fluorescence intensity
  flow cytometric analysis of asthma, 498–499
  integrated, *see* Integrated fluorescence
    intensity
Fluorescence microscopy
  BrdUrd subcellular distribution, 132–133
  chromosome identification, 17
  imaging system for telomeres, 73–77
Fluorescence quenching
  anthracyclines, 199
  and DNA-targeting drug binding, 187–188
Fluorescence resonance energy transfer,
  189–190

Fluorescent detection, mitochondria, 118
Fluorescent dyes
  in flow cytometry of bacteria, 554–555
  uptake, vital cells, 555–558
Fluorescent halo assay, measurement of DNA
  damage, 258, 261–263
Fluorochromes
  cellular retention, 201–206
  in detection of MRD, 391
  DNA-specific, for LSCM, 423–425
  multiple, simultaneous measurement,
    580–582
  overlapping fluorescence spectra, 295–298
  phycoerythrin/Texas Red, 339
  study of drug retention and efflux, 197
5-Fluorouracil, plus adriamycin and
  cyclophosphamide, 374
Food and drink, contaminants, flow cytometric
  analysis, 529–531
Formaldehyde, fixative, 8
Forward scatter
  acquisition by correlated list mode data,
    290–291
  vs. side scatter display, 309
Fractionation, and cell/culture integrity,
  178–179
Francisella tularensis, survival ratio, 519

G

Gastric adenocarcinomas, potential doubling
  time, 462
Gating
  CD45 as reagent, 573
  CD45 scattergram, 325–326
  CD45/SSC, 345, 348–356
  color, 311
  and DNA histograms, 142
  for immunophenotyping, 321–322
  light scatter lymphocyte, 571
  multiparameter data analytic approach, 308
Gene expression, and cell shape, as mechanism
  for contact effect, 221–222
Gene mapping, by hybridization onto genomic
  DNA fibers, 48, 50
Gene rearrangements, and PCR, 404–407
Gene therapy, HIV infection, 584
GFP, *see* Green fluorescent protein
Gram stain, bacteria, 525–526
Granulocytic/monocytic pathway, 160, 167
Green fluorescent protein, subcellularly
  targeted, 121

H

Hairy root cultures, 26

Halo–comet method, measurement of DNA
  damage, 263–266

Halogenated pyrimidines
  in cell proliferation research, 449–452
  in vivo labeling of brain tumors, 462

Heat shock
  altered DNA replication patterns after,
    276–278
  hsp70 localization after, 283
  nuclear matrix stability assay after,
    274–276
  oxidative potential of intracellular milieu
    after, 280–281

Heat treatment, for bacterial staining,
  558

Helium neon laser, LSC with, 337

Hematological malignancy, BrdUrd infusions,
  464

Hemoglobin
  accumulation pattern, 140
  bivariate profiles
    effect of TGF-β, 145–147
    flow data, 143–144
    isolation of fetal nucleated red cells,
      144–145
    problems and limitations, 149–151
    study of sickle-cell erythropoiesis,
      148–149

Hemopoietic progenitors
  circulating, 155
  committed, 154
  maturation, flow cytometric monitoring,
    155–156
  progenitor subsets, assessment, 157–167

Heterochromatin regions, DNA replication in,
  282

Hexidium iodide, gram-positive bacterial stain,
  525–526

High-power fields, and mitotic index, 362

Histograms
  bivariate, 301–303
  CD45/SSC, 347
  DNA
    in bivariate profiles of hemoglobin,
      144
    and gating, 142
  FITC/PE, 490–491
  fluorescence, 291–292
  single antibody histogram analysis, 289–290
  tail moment, 246

HIV infection
  CLASSIF1 analysis, 503–506
  data management and verification, 574–578
  immunophenotyping
    multicolor fluorescence, 578–582
    T cell, 568–574
  multiparameter flow cytometric analysis, 490
  research applications
    cell death, 583
    immunophenotypic, 582–583
    monitoring gene therapy, 584

HLA-DR, HP stained with, 158–160

HLA-DR/CD8, HIV immunophenotype,
  503–504

Hoaglund's nutrient solution, 7–8

Hoechst 33342
  cytotoxicity, 215, 217
  DNA binding, flow cytometric analysis,
    185–188

HP, *see* Hemopoietic progenitors

hsp70
  localization after heat shock, 283
  nuclear
    flow cytometric detection, 278
    immunofluorescence localization, 279–280

Hybridization master mix, 101

Hyperthermia, effects on cellular organelles,
  270

I

IFI, *see* Integrated fluorescence intensity

Illumination
  source, for fluorescence microscopy, 74
  stability, 75–76

Image-based cytometry, for fluorescent halo
  assay, 258, 261–262

Images, comet, analysis, 240–242

Immunofluorescence
  identification of microorganisms, 517–520
  in situ, localization of hsp70, 279–280

Immunohistochemistry
  compared with flow cytometry, 315, 340
  single antigen assay, 314

Immunophenotyping
  detection of MRD, 388–397
  leukemia and lymphoma, 313–316
  by LSCM, 316–327
  multicolor fluorescence, 578–582
  T lymphocytes, for HIV, 568–574

Instrumentation
  for chromosome translocation study, 102
  comet assay (Olive methods), 257–258

flow cytometer
  improvements, 512
    sensitivity for bacteria, 547–548
    verification, 293–295
  flow cytometry-based fluorescent halo assay,
    262–263
  image-based cytometry, 261
  immunophenotyping of malignancy, 314
  nuclear matrix stability assay, 275–276
  for PRINS, 57–58
  for QDFM, 37–38
Integrated fluorescence intensity, telomeres, 73
  quantification algorithm, 88–91
  segmentation algorithm, 80–82
  validation of method, 92
Iododeoxyuridine, 449–450
Ionizing radiation
  contact effect, 217–218
  inhibition of DNA loop rewinding, 261–262
  producing strand breaks, 245
Isolation
  fetal nucleated red cells: bivariate profiles,
    144–145
  nuclear matrix protein, 273
  nuclear protein, 271–272

J

JC-1, mitochondrial dye, 119–120

K

Ki-67
  analysis in
    fresh cytological specimens, 431–432
    histological tissue sections, 432
  antibody, observed in nucleus, 433–434
  proliferation marker, 364
Kinetics, tumor cell, 361
  relation to tumor biology, 365–366

L

Labeling index
  in clinical studies of tumor proliferation, 456,
    459, 462–463
  in vivo halogenated pyrimidine, 450–451
  proliferation in human tissues and tumors,
    470–471
  quantification of biomarker expression, 449
  as surrogate measure of dynamic
    proliferation, 473

[3]H-thymidine, *see* ^{3}H-Thymidine labeling
  index
β-Lactams
  effect on DNA content, 561
  inhibition of peptidoglycan layer,
    558–559
Laser scanning cytometer
  analysis of apoptosis, 425–430
  cell attributes measured by, 422–423
  estrogen receptor analysis, 435–436
  FISH protocol, measurement of nucleoli,
    438–441
  in MRD detection, 389–390
  proliferation-associated antigen analysis,
    430–434
  transcription factor analysis, 437–438
Laser scanning cytometry
  assays
    in cancer therapy, 475–476
    on fluorescent cytotoxic drugs, 477–478
  cellular DNA content analysis, 423–425
  in clinical studies, 474–475
  compared with flow cytometry, 326–327
  description, 316
  immunophenotyping by, 316–327
LB01 lysis buffer, 10
Leukemia
  acute
    detection of MRD, 388
    follow-up and prognosis, 356
  acute lymphoblastic, *see* Acute
    lymphoblastic leukemia
  acute myeloid, *see* Acute myeloid leukemia
  chronic lymphocytic, detection of MRD,
    395–396
  immunophenotyping, 313–316
    results, 325–326
  PCR analysis, 412–413
Leukocyte common antigen, *see* CD45
Leukocytes
  clinical chemistry parameter in asthma, 497
  three scan pass system, 338–339
Light microscopy
  bone marrow examination, 347–348
  cell lineages, identification, 348
  relocalization for, 327, 329, 331–333,
    335–336
Light scatter
  as contouring parameter in LSCM, 319
  flow cytometric analysis of bacteria,
    512–513
  lymphocyte gate, 571, 577

plus nucleic acid dyes: detection of microbes,
514–516
profiles, microorganisms, 516
Limiting dilution method, PCR, 409–411
List mode analysis
exhaustive, asthma, 499–500, 502
flow cytometry, 490–491
Lung resistance related protein, 194
Lymphadenopathy, immunophenotyping
results, 324–325
Lymphocytes
B, pattern of antigenicity, 331
B and T
CD45 positive, 333
scattergrams, 322
and blast cells: intermediate cell population
between, 353
B-lineage ALL, 392–393
metaphase spreads, 102
subpopulations, reference ranges, 575
T, immunophenotyping, 568–574
Lymphoma, immunophenotyping, 313–316
Lymphoproliferative disorders, detection of
MRD, 395–396
Lymphoreticular cells, LSC analysis,
319–320

M

Magnesium sulfate, stock solution, 11
Mapping, *see also* Quantitative DNA fiber
mapping
cDNA, 51
cloned probes onto DNA molecules, 50–51
Masking, incorporated BrdUrd, 133
MDR, *see* Multiple drug resistance
Melanoma
data pattern classification, 493–494
malignant, proliferation, 463
multiparameter flow cytometric analysis,
489–490
Membrane integrity, bacteria, 521–522
Membrane potential analysis, bacteria,
522–523
Metaphase
chromosomes, preparation for PRINS, 58, 66
lymphocyte, spreads, 102
Metaphase accumulation, plant chromosomes,
19, 21
Microbial systems, application of flow
cytometry, 531–532
Microbiology, medical, flow cytometric
applications, 563

Minimal residual disease
cell culture, 397
cytogenetics, 397–401
detection
requirements for, 386
tissue sources for, 387
flow cytometry, 401–402
immunophenotyping, 388–397
PCR, 402–414
technical problems, 414–415
Mithramycin
staining of plant chromosomes, 15
stock solution, 10
Mitochondria, fluorescent detection, 118
cardiolipin dye, 120, 124–125
cell preparation for, 122
critical aspects, 125–126
dyes for, 121–122
mitochondrial oxidative turnover,
123–124
mitochondrial protein level, 124
normalized mitochondrial membrane
potential, 122–123, 126
reduced dyes, 120–121
subcellularly targeted GFP, 121
symmetrical and asymmetric carbocyanine
dyes, 119–120
xanthylium dyes, 119
Mitochondrial membrane potential,
normalized, 122–123, 126
Mitotic index, 362
Mitotic spindle function, inhibitors, 175
MitoTracker Green, mitochondrial dye, 120
Molecular analysis, monitoring MRD by,
412–414
Monoblastic cells, component of bone marrow
infiltration, 349, 351
MRD, *see* Minimal residual disease
Mucosa, proliferation rates, 467
Multicolor detection, PRINS, 60–61
Multifraction studies, spheroids for,
226–227
Multiparameter data analysis
algorithms, 488–489
asthma, 490, 494–502
cell gating approach, 308
HIV infection, 490
immunophenotyping study, 346–347
melanoma, 489–490, 493–494
numbers of analysis parameters, 309, 311
Multipass system, immunophenotyping,
336–340

Multiple drug resistance
  drugs, 197
    identification methods, 194–196
    as phenotypic characteristic, 193
Multiple myeloma, detection of MRD, 396
Mycobacteria, diagnostic assays, 562–563

N

NADH, mitochondrial, 122
NF-κB, LSCM analysis, 437–438
Noise, in fluorescence microscopy, 76–77
Non-Hodgkin's lymphoma, 386
  immunophenotyping, 396
  PCR analysis, 413
Non-small cell lung cancer
  BrdUrd labeling, 464
  proliferation markers and prognosis, 372
  proliferative activity, 365
Nonyl acridine orange, mitochondrial dye, 120
Nuclear matrix protein, content, flow cytometric analysis, 272–273
Nuclear matrix stability assay, after heat shock, 274–276
Nuclear protein, content
  flow cytometric analysis, 271–272
  temperature effect, 281
Nucleic acid dyes, in flow cytometric detection of bacteria, 514–516
Nucleoli, measurement with LSCM FISH, 438–441

O

Oligonucleotide primers
  for PCR, 106–107, 405
  in PRINS, 55, 59–60
Operational amplifier, 297–298
Optical filtration
  for four-color compensation, 303–307
  PECY5 conjugated antibodies: problems, 300–301
  problem of overlapping fluorochrome spectra, 295–298
  for two- or three-color compensation, 301–303
Outer membrane, bacteria, 556
Ovarian carcinomas, BrdUrd incorporation, 463
Oxidative turnover, mitochondrial, 123–124

P

p24, HIV: measurement, 583
Painting, chromosome, combined with PRINS, 61–64
Particles
  in bacteria test samples, 515–516
  cellular, loss during nuclear isolation, 272
  monodisperse fluorescent, 549
PCR, see Polymerase chain reaction
PECY5
  CD64 binding, 578
  compensation requirements, 305–307
  plus CD8, 303
  properly compensated instrument for, 300–301
Peptidoglycan layer
  gram-positive bacteria, 556–557
  inhibition by b-lactams, 558–559
PFGE, see Pulsed field gel electrophoresis
PGD, see Preimplantation genetic diagnosis
p53 gene, LSCM analysis, 437–438
P-glycoprotein, methotrexate as substrate, 194
Phenotype, asynchronous, detection in B-lineage ALL, 393
Phospholipid, mitochondrial membrane, 124–125
Photobleaching, fluorescent images, 75
Photomultiplier tubes
  fluorescence 1 and 2, 297–298
  in LSC, 337–338
  for spectral shift analysis, 186
Phycoerythrin
  plus FITC: histograms, 490–491
  plus Texas Red: fluorochromes, 339
  scattergrams, 324–326
Plants
  flow cytogenetics
    chromosome sorting, 26–28
    problems with, 4–5
  seeds
    germination media, 13
    with reconstructed karyotypes, 6
Plasmids, IFI value, 90
Ploidy
  measurements, as prognostic markers, 448
  in surgical oncology, 447–448
Plugs
  containing YAC, 39
  preparation by PFGE, 105
Polymerase chain reaction
  analysis of MRD, 402–414
    technical considerations, 406–412

clone-specific markers, 404–406
degenerate oligonucleotide-primed,
106–107
estimation of chromosome purity, 18
in HIV detection, 584
inter-Alu, 106
premix, 11–12
real-time quantitative, 411–412
tumor-specific markers, 403–404
Potential doubling time
adenocarcinomas, 462–463
and cell loss factor, 472
pretreatment, 469–470
squamous carcinomas, 456, 459
Preimplantation genetic diagnosis, 98–99
selection of probes for, 103
YAC probes, 112
Primed *in situ* labeling
applications, 56
chemical and instruments for, 56–58
critical aspects, 66–67
protocols, 58–66
reaction mix, 11
PRINS, *see* Primed *in situ* labeling
Probes
cloned, mapping onto DNA molecules,
50–51
generation
from cloned DNA fragments, 41–43
by *in vitro* DNA amplification, 43–44
labeling via random priming, 44–45,
107–108
preparation for chromosome translocation
study, 99–112
unlabeled, as primer in PRINS, 55–56
Prognosis
acute leukemia, 356
proliferation markers and, 367–372
Prognostic markers
ploidy measurements as, 448
S-phase fraction measurements as, 448–449
Prognostic studies, in surgical oncology,
447–448
Proliferation
activity of tumor cell subpopulations,
360–361
cytometric studies, 448–449
data, and biological aggressiveness, 473
human tissues and tumors, overview of data
on, 470–474
lymphocyte, in HIV infection, 583–584
research with halogenated pyrimidines,
449–452

in surgical oncology, 447–448
TGF-β effect, 145–147
in transition tissues, 467
Proliferation-associated antigens, 364
LSCM analysis, 430–434
Proliferation markers
first generation translational studies,
377–378
human solid tumors, 361
Ki-67, 364
and prognosis, 367–372
and response to systemic treatments,
372–377
Prooxidant measurement
reagents and staining procedure, 280–281
special considerations, 281
Prooxidant production, heating protocol-
dependent, 283, 285
Propidium iodide, as contouring parameter in
LSCM, 318
Protein
complex with DNA, precipitation,
181–182
mitochondrial, 124
nucleolar, stained by silver methods, 439
Pulsed field gel electrophoresis
for preparation of DNA from YAC, 105
for QDFM, 39–40
Purity, sorted fraction of chromosomes, 17–18,
20–21, 25–26

Q

QDFM, *see* Quantitative DNA fiber mapping
Quality control
data management for HIV evaluation,
575–576
flow cytometric analysis of drug–DNA
interactions, 178–182
Quantitative DNA fiber mapping
applications, 34
measuring clone overlap, 51
chemicals and buffers for, 35–37
critical aspects, 46–47
digital image analysis, 46
FISH, 45–46
generation of probes, 41–44
instruments for, 37–38
preparation of DNA fibers, 40–41
probe labeling via random priming, 44–45
protocols, 38–40

R

Radiation, ionizing
  contact effect, 217–218
  inhibition of DNA loop rewinding, 261–262
  producing strand breaks, 245
Random priming, probe labeling via, 44–45,
    107–108
Reagents
  choices, for immunophenotyping, 577
  gating, CD45 as, 573
  labeling of DNA strand breaks, 427–428
  multicolor fluorescence immunophenotyping,
    578, 580
  for plant flow cytogenetics, 6–12
Recurrence, local, proliferating tumors, 369
Red cells, fetal nucleated, isolation, 144–145
Relocalization, for light microscopy and FISH,
    327, 329, 331–333, 335–336
Repeated-PRINS, stronger PRINS signals
    obtained by, 64
Resistance
  drug
    genetic or kinetic basis, 227
    in large spheroids, 223
    kinetic, in multifraction exposures, 225–227
Resolution
  in fluorescence microscopy, 77
  increased due to changes in fluorescence
    signals, 189
Retention
  anthracycline, 195
  fluorochromes, 201–206
Retrospective studies, ploidy and proliferation
    in surgical oncology, 447–448
Rhodamine 123, probe for mitochondria, 119
RNA
  BrdUrd-substituted, 130–131
  labeling with BrdUrd, 134–135
  ribosomal, content of microorganisms,
    524–525
  synthesis, flow cytometric analysis, 129–137
RNase protection assay, 408–409
Root tip cells, accumulation in metaphase,
    13–14

S

Sample preparation
  bacteria standard, 549–550
  for flow cytometry
    bacteria, 513–514
    food and drink, 530–531

for telomere length measurement, 76, 79–80
  whole blood, 569–570
Scattergrams
  cell location, 322
  FITC *vs.* phycoerythrin, 324–326
  immunofluorescence, 319
Seeds
  germination media, 13
  with reconstructed karyotypes, 6
Segmentation
  chromosomes, 82, 84–85, 92–93
  telomere, and IFI measurements, 80–82
Sheath fluid, bacteria, 546
Sheath fluid SF 50, 10–11
Sickle-cell erythropoiesis, 148–149
Signals, from repeated-PRINS, 64, 67
Single-cell gel electrophoresis
  in alkaline conditions, 254
  development, 236–237
Slide blocking solution, in QDFM, 37
Slides
  chamber, for immunophenotyping, 320–321
  equilibrated, 239
  prepared with 3-aminopropyltriethoxy silane,
    38
  pretreatment
    and FISH, 112
    for QDFM, 46–47
  three-layer agarose, 255
Solid tumors
  growth, 225
  malignancies
    dual-color staining, 396–397
    PCR analysis, 413–414
  resistance, 227
Sorting
  cell, from multicell culture systems, 214–215
  plant chromosomes
    after univariate analysis, 17
    and flow karyotyping, 19–21
Southern analysis, telomere length, 71–72, 91
Specimens
  fresh cytological, Ki-67 analysis in, 431–432
  handling
    biosafety, 569
    for immunophenotyping study, 352
  hematologic, immunophenotyping, 319–321,
    336–340
  for LSCM, 317–318
  small, for immunophenotyping, 314
Spectral compensation
  four-color, 303–307

and optical filtration, 295–307
set to correct spectral overlap, 571
two- or three-color, 301–303
Spectral shift analysis, Hoechst 33342–DNA
  binding, 185–187
Spectrofluorimetry, DNA–drug interactions,
  183–184
Sperm sample, preparation for PRINS, 58
S phase
  duration, 471–472
  incorporation of DNA precursors, 363
  quantification of nDNA content,
    363–364
S-phase fraction, measurements, as prognostic
  markers, 448–449
Spheroids
  alteration of damage induction in,
    221–222
  cells resistant to chemotherapeutic agents,
    218–219
  culture system, 213–214
  drug activity patterns in, 222–223
  multidimensional complexity, 212
  for multifraction studies, 226–227
  slow drug penetration in, 224
Spores, bacterial, 546
Squamous cell carcinomas, clinical studies of
  proliferation, 456–461
Stability, illumination, for fluorescent
  microscopy, 75–76
Staining
  for analysis of RNA synthesis, 132
  costaining with PECY5-CD8, 303
  for cyclins, 432–433
  DNA, bacteria, 541–542
  for evaluating drug retention and efflux,
    197–198
  fixed bacteria, 543–544
  immunofluorescence, and erythrocyte lysis,
    345–346
  nuclear isolation and, 271–272
  nuclear matrix isolation and, 273
  protocols for mitochondria analysis, 125
  vital *E. coli,* 544–546
Stock solutions, for plant flow cytogenetics,
  6–12
Surgical oncology, ploidy and proliferation in,
  447–448
Surrogate
  appropriate cellular, 295
  measure of dynamic proliferation: labeling
    index as, 473

Synchronization, cell cycle, plant
  chromosomes, 19, 21
SYTO 13, Gram stain, 525–526

T

Tail moment threshold, 244–245
Taq polymerase, for PRINS, 56–57
Target enzymes, availability: quality control,
  180–182
Target organisms, flow cytometric
  identification, 518–520
T-cell receptor
  as clone-specific markers, 404–405
  Vb antigen analysis, 583
Telomere length measurement
  biological studies, 93
  calibration of results, 85–86
  chromosome segmentation for, 82–85
  fluorescence labeling of sample, 72–73
  fluorescence microscopy imaging system,
    73–77
  human cells, Southern analysis, 91
  presentation of results to user, 86–88
  sample preparation for, 76, 79–80
  system description, 77–79
  TFL-TELO algorithm, 77
Telomeres
  integrated fluorescence intensity, 73
    algorithm, 88–90
    validation of method, 92
  quantitative FISH and image analysis, 72
  segmentation, 80–82
  Southern analysis, 71–72
  structure–function, 70–71
Terminal restriction fragments, 71
Testing programs, for flow cytometric
  immunophenotyping assays, 576–577
TFL-TELO algorithm, 77
TGF-β, *see* Transforming growth factor β
Therapy
  cancer, cytometric assays in, 475–476
  cell production rates and, 469–470
  experimental, and cell proliferation, 474
Thymidine analogs, in clinical studies of cell
  production rates, 452–474
3H-Thymidine labeling index, 363, 365–366
  breast cancer, 368–369
  correlation with chemotherapy, 373, 375–376
Time
  in flow cytometric diagnostic procedure,
    553–554
  role in tumor proliferation cytometry, 451

for statistical significance of antibiotic
effects, 561
Tirapazamine, induced DNA damage, 243
Tissue sections, Ki-67 analysis in, 432
Tissue sources, for detection of MRD, 387
TLI, *see* ^{3}H-Thymidine labeling index
Transcription factors, LSCM analysis,
437–438
Transcription foci, nuclear detection,
136–137
Transforming growth factor b, effect on
proliferation and hemoglobin profiles,
145–147
Transition tissues, proliferation in, 467
Translational studies
first generation, with proliferation markers,
377–378
human tumors, role of proliferation indices,
366–367
proliferation markers
and prognosis, 367–372
and response to systemic treatments,
372–377
Translocation, t(9;22), 335
Tumor cells
difference from those in tissue culture,
211–212
subpopulations, proliferative conditions, 360
Tumors
endocrine therapy, 376–377
estrogen receptor-stained, 436
heterogeneity, 446–447
human
classifications, 446

and experimental tumors, 360–361
proliferation
adenocarcinomas, 462–463
overview of data on, 470–474
squamous cell carcinomas, 456–461
time-dependent parameters, 451
solid, *see* Solid tumors
stain concentration profile, 217
translational research, 366–377
Tumor-specific markers, in PCR, 403–404

U

User-interaction features, telomere length
measurement, 86–88
Uterine cervix, squamous carcinomas, 459

V

Vincristine, plus adriamycin, 373–374
Visualization, labeled DNA, for PRINS, 65–66

X

Xanthylium dyes, probes for mitochondria, 119

Y

YAC, *see* Yeast artificial chromosomes
Yeast artificial chromosomes
in chromosome translocation study, 100
clones
DNA prepared from, 42–43, 104–107
spanning break point, 109–111
plugs containing, 39

VOLUMES IN SERIES

Founding Series Editor
DAVID M. PRESCOTT

Volume 1 (1964)
Methods in Cell Physiology
Edited by David M. Prescott

Volume 2 (1966)
Methods in Cell Physiology
Edited by David M. Prescott

Volume 3 (1968)
Methods in Cell Physiology
Edited by David M. Prescott

Volume 4 (1970)
Methods in Cell Physiology
Edited by David M. Prescott

Volume 5 (1972)
Methods in Cell Physiology
Edited by David M. Prescott

Volume 6 (1973)
Methods in Cell Physiology
Edited by David M. Prescott

Volume 7 (1973)
Methods in Cell Biology
Edited by David M. Prescott

Volume 8 (1974)
Methods in Cell Biology
Edited by David M. Prescott

Volume 9 (1975)
Methods in Cell Biology
Edited by David M. Prescott

Volume 10 (1975)
Methods in Cell Biology
Edited by David M. Prescott

Volume 11 (1975)
Yeast Cells
Edited by David M. Prescott

Volume 12 (1975)
Yeast Cells
Edited by David M. Prescott

Volume 13 (1976)
Methods in Cell Biology
Edited by David M. Prescott

Volume 14 (1976)
Methods in Cell Biology
Edited by David M. Prescott

Volume 15 (1977)
Methods in Cell Biology
Edited by David M. Prescott

Volume 16 (1977)
Chromatin and Chromosomal Protein Research I
Edited by Gary Stein, Janet Stein, and Lewis J. Kleinsmith

Volume 17 (1978)
Chromatin and Chromosomal Protein Research II
Edited by Gary Stein, Janet Stein, and Lewis J. Kleinsmith

Volume 18 (1978)
Chromatin and Chromosomal Protein Research III
Edited by Gary Stein, Janet Stein, and Lewis J. Kleinsmith

Volume 19 (1978)
Chromatin and Chromosomal Protein Research IV
Edited by Gary Stein, Janet Stein, and Lewis J. Kleinsmith

Volume 20 (1978)
Methods in Cell Biology
Edited by David M. Prescott

Advisory Board Chairman
KEITH R. PORTER

Volume 21A (1980)
Normal Human Tissue and Cell Culture, Part A: Respiratory, Cardiovascular, and Integumentary Systems
Edited by Curtis C. Harris, Benjamin F. Trump, and Gary D. Stoner

Volume 21B (1980)
Normal Human Tissue and Cell Culture, Part B: Endocrine, Urogenital, and Gastrointestinal Systems
Edited by Curtis C. Harris, Benjamin F. Trump, and Gary D. Stoner

Volume 22 (1981)
Three-Dimensional Ultrastructure in Biology
Edited by James N. Turner

Volume 23 (1981)
Basic Mechanisms of Cellular Secretion
Edited by Arthur R. Hand and Constance Oliver

Volume 24 (1982)
The Cytoskeleton, Part A: Cytoskeletal Proteins, Isolation and Characterization
Edited by Leslie Wilson

Volume 25 (1982)
The Cytoskeleton, Part B: Biological Systems and *in Vitro* Models
Edited by Leslie Wilson

Volume 26 (1982)
Prenatal Diagnosis: Cell Biological Approaches
Edited by Samuel A. Latt and Gretchen J. Darlington

Series Editor
LESLIE WILSON

Volume 27 (1986)
Echinoderm Gametes and Embryos
Edited by Thomas E. Schroeder

Volume 28 (1987)
***Dictyostelium discoideum*: Molecular Approaches to Cell Biology**
Edited by James A. Spudich

Volume 29 (1989)
Fluorescence Microscopy of Living Cells in Culture, Part A: Fluorescent Analogs, Labeling Cells, and Basic Microscopy
Edited by Yu-Li Wang and D. Lansing Taylor

Volume 30 (1989)
Fluorescence Microscopy of Living Cells in Culture, Part B: Quantitative Fluorescence Microscopy—Imaging and Spectroscopy
Edited by D. Lansing Taylor and Yu-Li Wang

Volume 31 (1989)
Vesicular Transport, Part A
Edited by Alan M. Tartakoff

Volume 32 (1989)
Vesicular Transport, Part B
Edited by Alan M. Tartakoff

Volume 33 (1990)
Flow Cytometry
Edited by Zbigniew Darzynkiewicz and Harry A. Crissman

Volume 34 (1991)
Vectorial Transport of Proteins into and across Membranes
Edited by Alan M. Tartakoff

Selected from Volumes 31, 32, and 34 (1991)
Laboratory Methods for Vesicular and Vectorial Transport
Edited by Alan M. Tartakoff

Volume 35 (1991)
Functional Organization of the Nucleus: A Laboratory Guide
Edited by Barbara A. Hamkalo and Sarah C. R. Elgin

Volume 36 (1991)
***Xenopus laevis:* Practical Uses in Cell and Molecular Biology**
Edited by Brian K. Kay and H. Benjamin Peng

Series Editors
LESLIE WILSON AND PAUL MATSUDAIRA

Volume 37 (1993)
Antibodies in Cell Biology
Edited by David J. Asai

Volume 38 (1993)
Cell Biological Applications of Confocal Microscopy
Edited by Brian Matsumoto

Volume 39 (1993)
Motility Assays for Motor Proteins
Edited by Jonathan M. Scholey

Volume 40 (1994)
A Practical Guide to the Study of Calcium in Living Cells
Edited by Richard Nuccitelli

Volume 41 (1994)
Flow Cytometry, Second Edition, Part A
*Edited by Zbigniew Darzynkiewicz, J. Paul Robinson,
    and Harry A. Crissman*

Volume 42 (1994)
Flow Cytometry, Second Edition, Part B
*Edited by Zbigniew Darzynkiewicz, J. Paul Robinson,
    and Harry A. Crissman*

Volume 43 (1994)
Protein Expression in Animal Cells
Edited by Michael G. Roth

Volume 44 (1994)
***Drosophila melanogaster:* Practical Uses in Cell and Molecular Biology**
Edited by Lawrence S. B. Goldstein and Eric A. Fyrberg

Volume 45 (1994)
Microbes as Tools for Cell Biology
Edited by David G. Russell

Volume 46 (1995)
Cell Death
Edited by Lawrence M. Schwartz and Barbara A. Osborne

Volume 47 (1995)
Cilia and Flagella
Edited by William Dentler and George Witman

Volume 48 (1995)
***Caenorhabditis elegans:* Modern Biological Analysis of an Organism**
Edited by Henry F. Epstein and Diane C. Shakes

Volume 49 (1995)
Methods in Plant Cell Biology, Part A
Edited by David W. Galbraith, Hans J. Bohnert, and Don P. Bourque

Volume 50 (1995)
Methods in Plant Cell Biology, Part B
Edited by David W. Galbraith, Don P. Bourque, and Hans J. Bohnert

Volume 51 (1996)
Methods in Avian Embryology
Edited by Marianne Bronner-Fraser

Volume 52 (1997)
Methods in Muscle Biology
Edited by Charles P. Emerson, Jr. and H. Lee Sweeney

Volume 53 (1997)
Nuclear Structure and Function
Edited by Miguel Berrios

Volume 54 (1997)
Cumulative Index

Volume 55 (1997)
Laser Tweezers in Cell Biology
Edited by Michael P. Sheez

Volume 56 (1998)
Video Microscopy
Edited by Greenfield Sluder and David E. Wolf

Volume 57 (1998)
Animal Cell Culture Methods
Edited by Jennie P. Mather and David Barnes

Volume 58 (1998)
Green Fluorescent Protein
Edited by Kevin F. Sullivan and Steve A. Kay

Volume 59 (1998)
The Zebrafish: Biology
Edited by H. William Detrich III, Monte Westerfield, and Leonard I. Zon

Volume 60 (1998)
The Zebrafish: Genetics and Genomics
Edited by H. William Detrich III, Monte Westerfield, and Leonard I. Zon

Volume 61 (1998)
Mitosis and Meiosis
Edited by Conly L. Rieder

Volume 62 (1999)
Tetrahymena Thermophila
Edited by David J. Asai and James D. Forney

Volume 63 (2000)
Cytometry, Third Edition, Part A
Edited by Zbigniew Darzynkiewicz, J. Paul Robinson, and Harry Crissman

Volume 64 (2000)
Cytometry, Third Edition, Part B
Edited by Zbigniew Darzynkiewicz, J. Paul Robinson, and Harry Crissman

ISBN 0-12-544167-3

90018

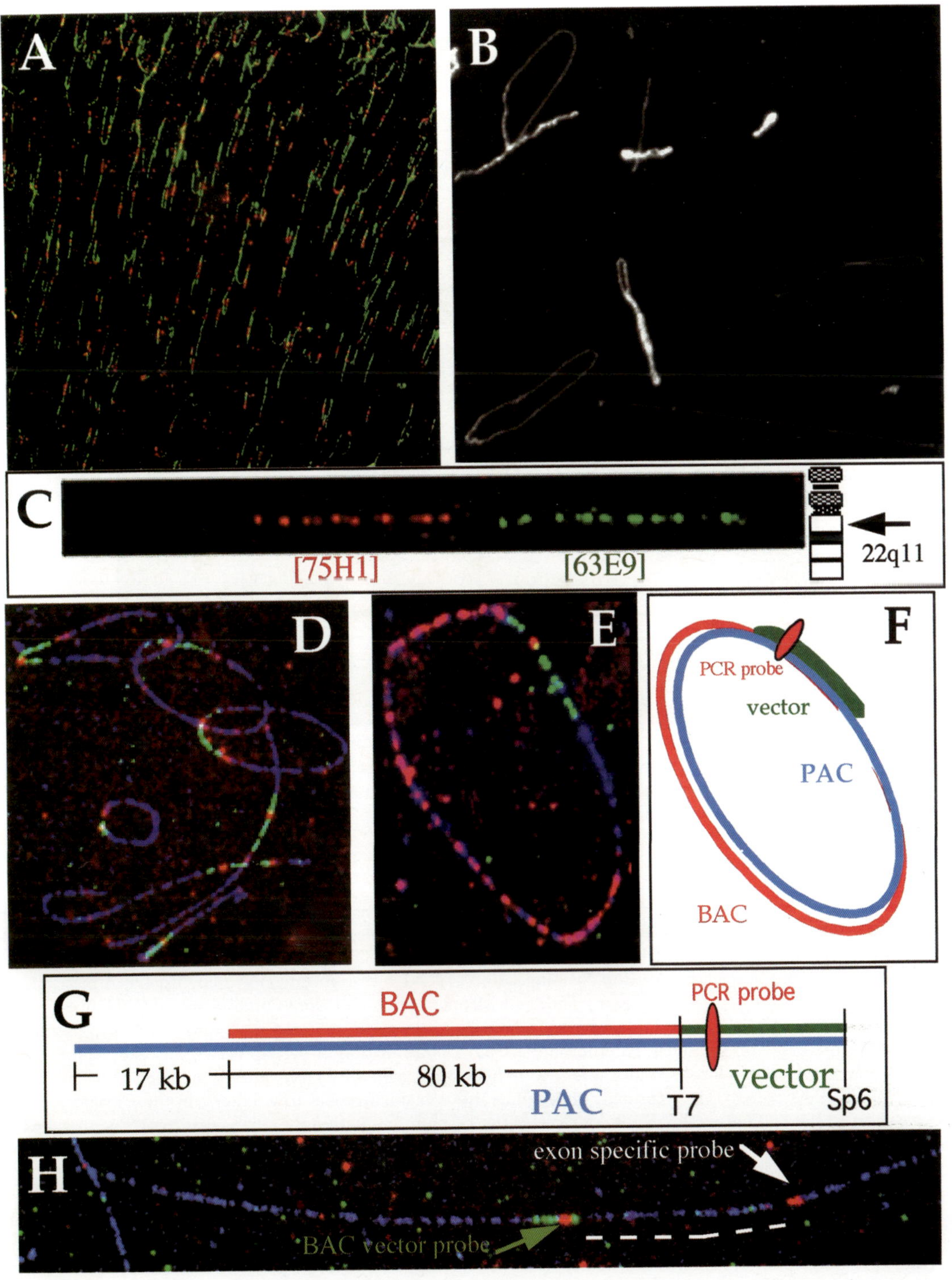

Chapter 31, Fig. 1 *Caption found on next page.*

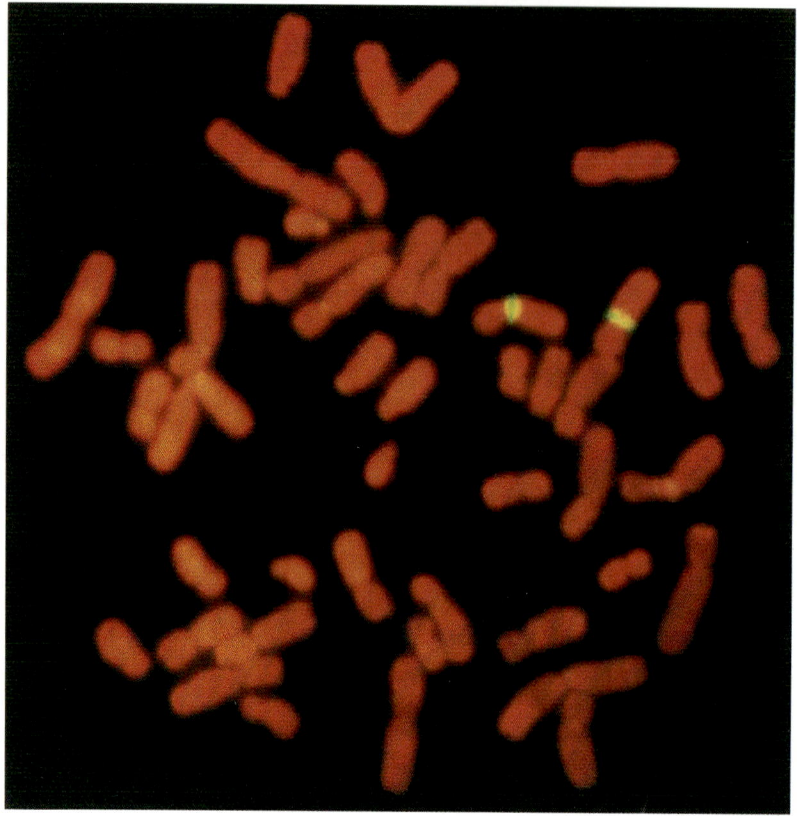

Chapter 32, Fig. 1 A PRINS reaction performed with a primer (5′-CCATTCCATTCCAYTCGGGTT-3′) detecting satellite III DNA on chromosome 9. Fluorescein-12-dUTP was used as the labeled nucleotide. The annealing and chain elogation were performed at 60°C for 5 min.

Chapter 31, Fig. 1 Quantitative DNA fiber mapping. (A) λDNA molecules bound to an APS–derivatized glass slide were hybridized with red or green labeled restriction fragments. (B) Circular DNA molecules were excised from a PFGE gel and purified by agarase digestion. Staining with YOYO-1 reveals mostly closed circular DNA molecules in the presence of linear molecules of different length. (C) Mapping of cosmid clones onto genomic DNA fibers reveals the distance between linked clones. The arrow and idiogram to the right indicated the approximate map position of the clones on the long arm of human chromosome 22. (D) Determination of overlap between two P1 clones (clone A and B). Circular DNA molecules from clone A were hybridized with a P1 DNA probed prepared from the partially overlapping clone B (red). The hybridization mixture also contained probes to counterstain the P1 vector part in green and an ~1300 bp PCR probe that binds close to the T7 promoter within the vector part (red). The stretched P1 molecules are counterstained by hybridization of a P1 probe (clone A) detected in blue. Overlap between the clones appears as a small red domain adjacent to the P1 vector part. (E–G) Measurement of overlap between a PAC and a BAC clone. The BAC DNA was labeled with digoxigenin and detected in red after hybridization to circular PAC molecules (counterstained in blue). Probes for the vector part contain P1 vector DNA (green) and a PCR-made probe (red) shown schematically in (F). The overlap between the clones extends for about 80 kb adjacent to the T7 promoter end of the vector part. (H) Cloned cDNA fragments can be mapped into larger genomic intervals by QDFM. Here, DNA fibers prepared from a BAC clone were hybridized with an ~2-kb insert of a plasmid specific for exon 2 of the human band 4.1 gene. The dashed line indicates the distance of the exon from the T7 end of the vector part.

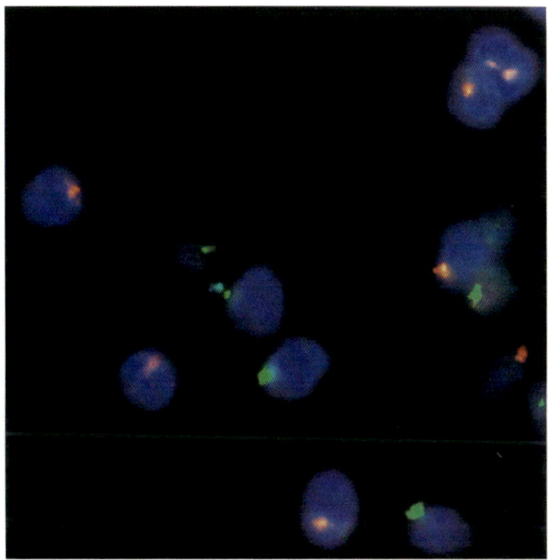

Chapter 32, Fig. 2 Multicolor-PRINS on sperm cells. Sperm cells were denatured in 3 *M* NaOH for 4 min at room temperature. The first PRINS reaction was performed at 60°C for 30 min using an X chromosome specific primer (5′-CCTGAAAGCCTTTTCGCTTTATCTTCACAGAAAGA-3′) with digoxigenin-11-dUTP as the labeled nucleotide. The second PRINS reaction was performed at 60°C for 30 min using a Y chromosome-specific primer (5′-TTTTTTTTTCATTTAAAAAAATATGGATCTTGGCTCAGGCC-3′) with biotin-16 dUTP as the labeled nucleotide. The dideoxy reaction was omitted. The X signal was visualized in red with anti-digoxigenin-rhodamine and the Y signal was visualized in green with fluorescein-conjugated avidin.

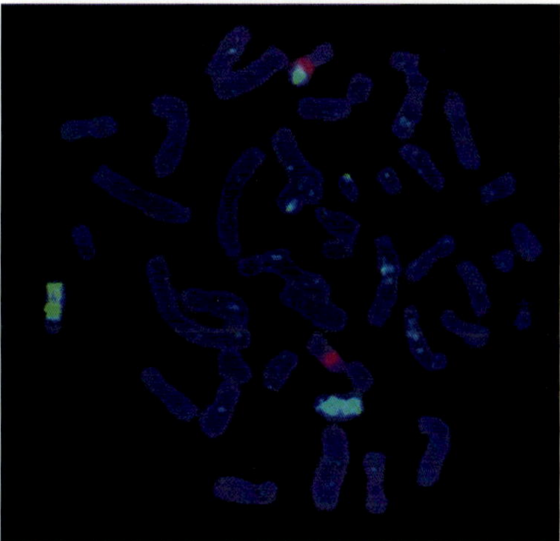

Chapter 32, Fig. 3 PRINS–painting was used for clarification of complex structural rearrangements in bone marrow cells from a patient with acute lymphoblastic leukemia (ALL). PRINS was performed with an oligonucleotide representing the α-satellite DNA of chromosome 17 (5′-ACGTGAACTTTGAAAGGAAAGTTCAACTCGGGGAT-3′) at 60°C for 30 min with digoxigenin-11-dUTP as the labeled nucleotide. After washing and dehydration a biotin-labeled chromosome 13 library was hybridized to the slide overnight at 37°C. The painting library was visualized with fluorescein-conjugated avidin, and the PRINS reaction was visualized with anti-digoxigenin-rhodamine, Fab fragments. Chromosomes were counterstained with DAPI. The result shows a derivative chromosome with a chromosome 17 centromere stained red and a part of the chromosome stained green by the chromosome 13 library, showing that the chromosome is a der(17)t(13;17).

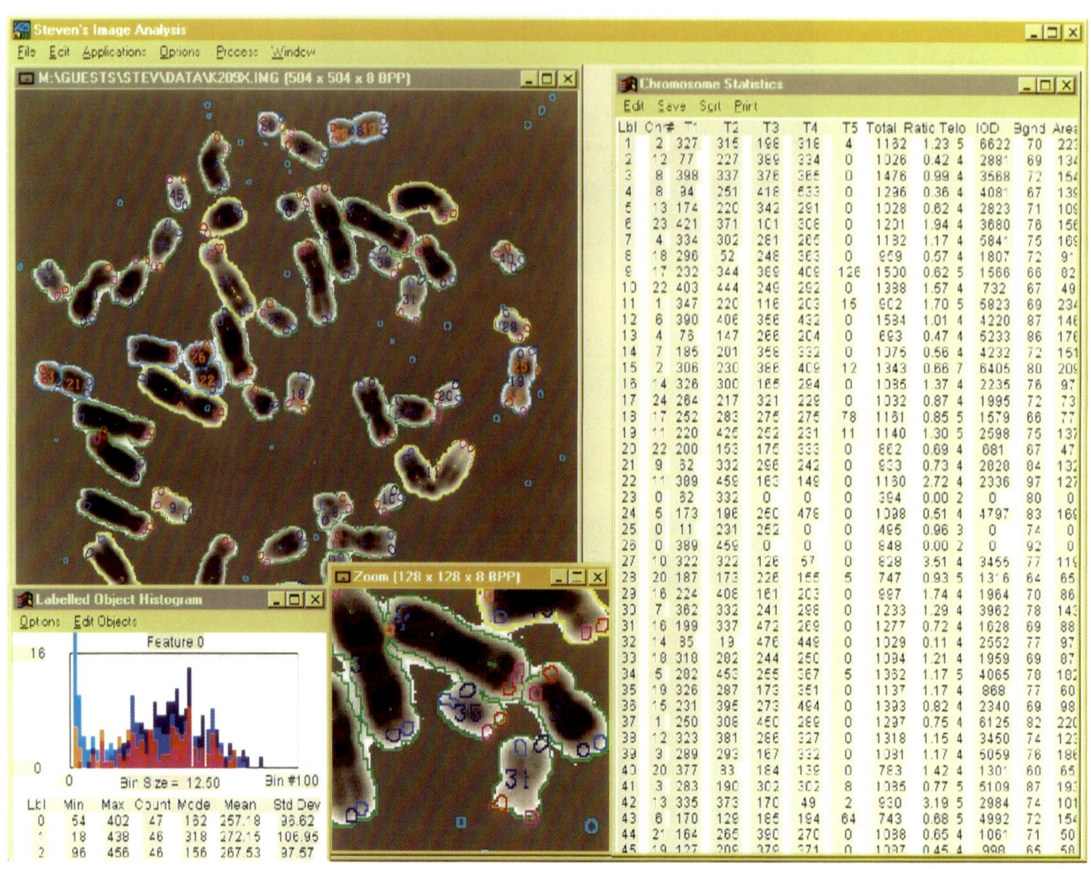

Chapter 33, Fig. 4   Sample output of TFL-TELO program. The segmented telomere boundaries are superimposed on the segmented chromosome image. The IFI values of each chromosome are listed in the table. A 2×-magnified window and the different boundary colors help with the verification and editing of the results generated by the program.

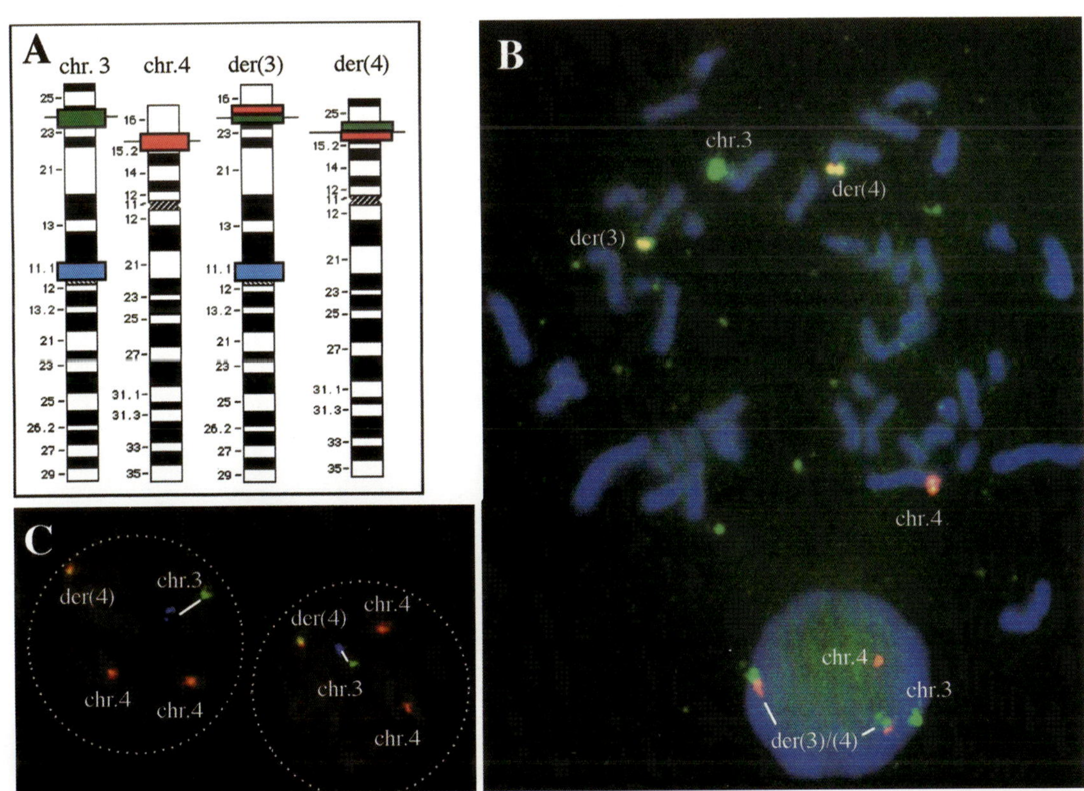

Chapter 34, Fig. 2 Hybridization results. (A) Idiogram showing the probe scheme used; besides the two break point spanning probes for chromosomes 3 (green) and 4 (red), a chromosome 3-specific centromeric probe (blue) is included.

DNA Replication Morphology

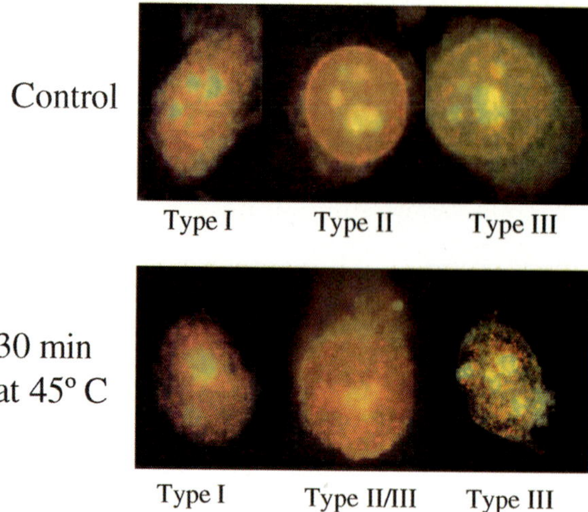

Chapter 44, Fig. 2 DNA replication patterns observed in control and heat-shocked HeLa cells. Type I replication pattern replicates euchromatin, leaving heterochromatic regions (nucleoli) bright green. Types II and III (and unresolved II and III following heat shock) are involved in heterochromatin replication, resulting in the yellow nucleolar region (red plus green).

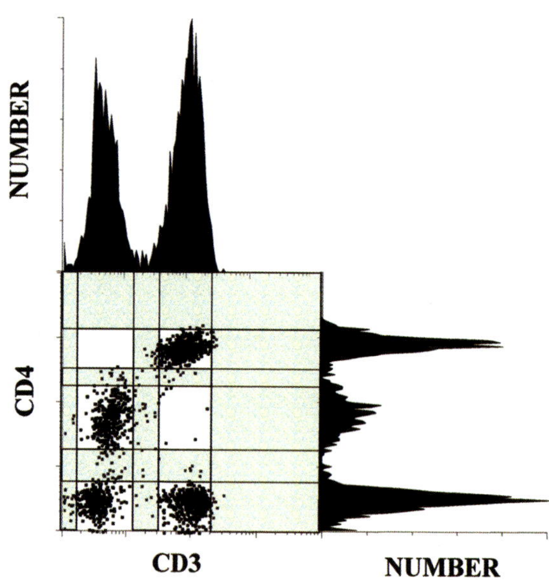

Chapter 45, Fig. 2 Bivariate dot plot to provide correlation of parameters. When the univariate histograms displayed in Fig. 1 for CD3 and CD4 are arranged and parallel markers describing the position of each Gaussian peak are drawn across the bivariate space, the x and y coordinates of each event produce clusters of dots that form a population of cells. In this example, four distinct populations, CD3−CD4+, CD3+CD4+, CD3+CD4−, and CD3−CD4− are clearly resolved. Their number can then be computed to provide quantitative information on the frequency of each subset.

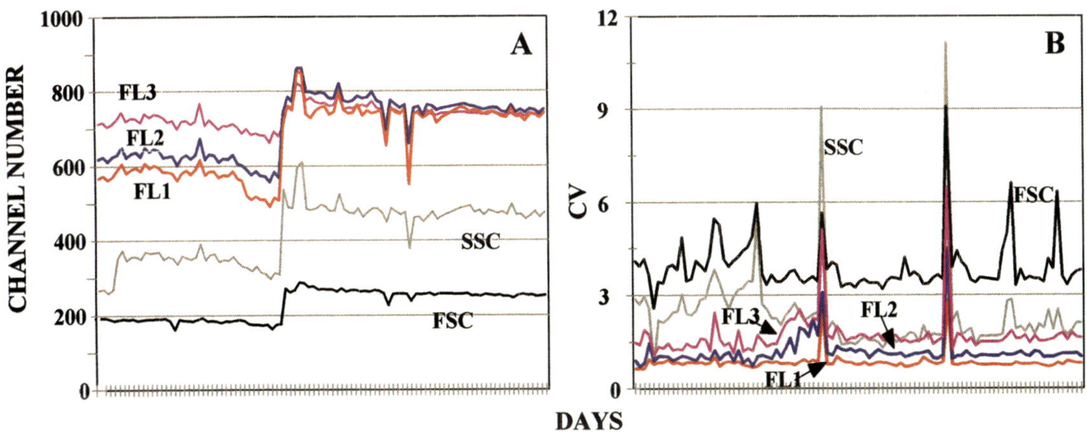

Chapter 45, Fig. 3 Performance verification of a FACScan. Microspheres (DNA check) were used daily to measure the median channel for each parameter (A). The threshold channel number was allowed to vary ±10% for acceptable instrument performance per median; channel value depended on the lot of beads as shown by their abrupt change when the lot changed, indicating that the lot-to-lot reproducibility was far worse than the instrument's reproducibility. The coefficient of variation (CV) was established at less than 3% for all parameters except FSC and SSC, which were 6% (B). Instrument failure shown as spikes required maintenance to reestablish performance prior to use. At no time were instrument settings changed to provide compliance; instead, the reason for noncompliance was sought and repaired.

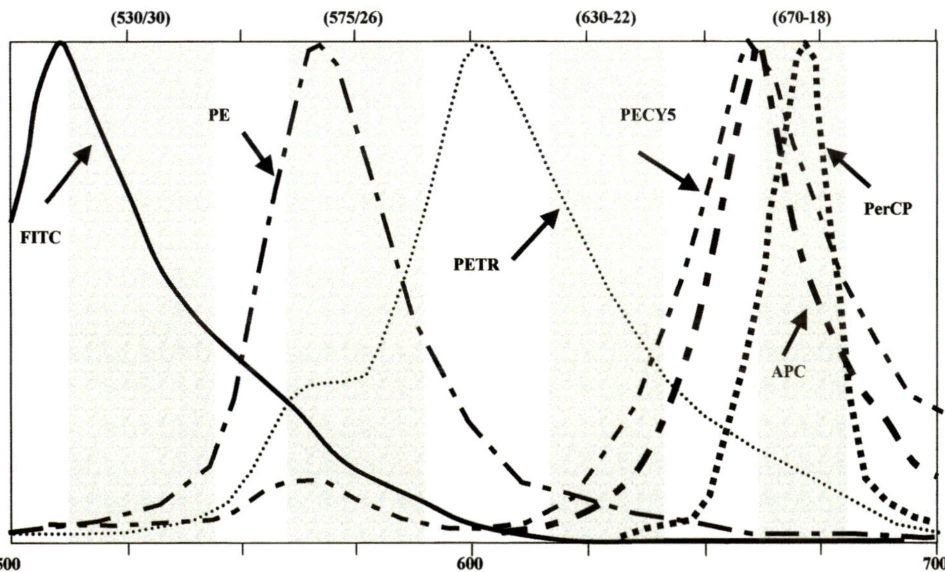

Chapter 45, Fig. 4 Emission spectra from commonly used fluorochromes. For each fluorochrome, the relative photon frequency (intensity) as a function of wavelength (energy) is shown. The commonly used band-pass filter system is also shown for each fluorochrome. For the tandem complexes, the emission spectrum is composed of a PE component and the acceptor component (TR or CY5). The height of the PE component, identical to the PE emission spectrum, is greater for Texas Red than for CY5. Note also there is virtually no spectral separation possible between CY5, PerCP, and APC. The band-pass filter range usually is shown by the strippled bars.

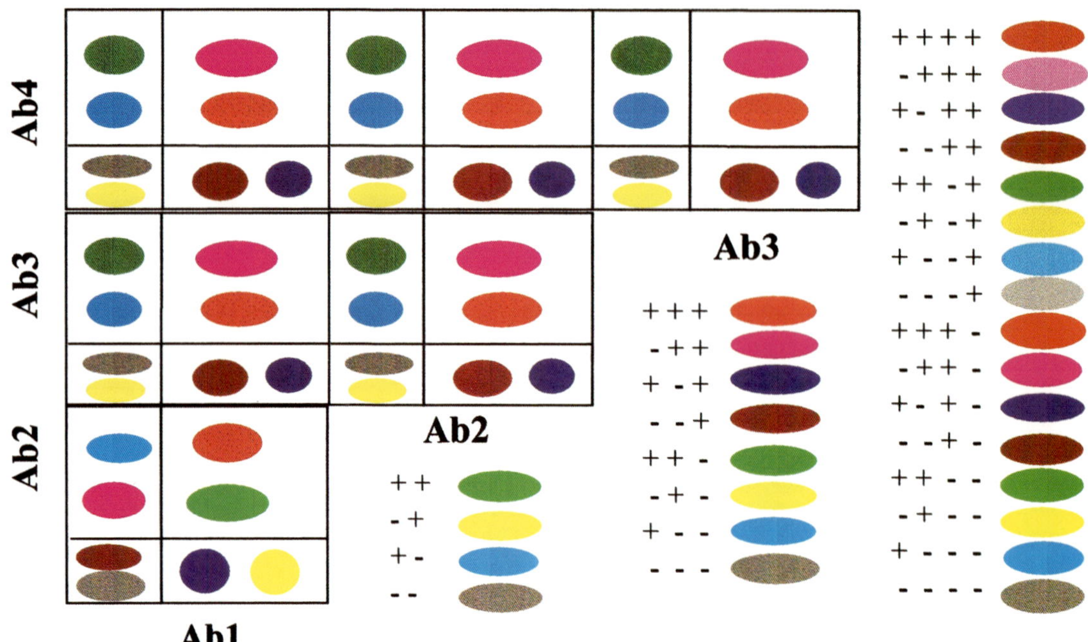

Chapter 45, Fig. 14 Bivariate plots for binary combinations of multiparameter data. For two-parameter fluorescence measurements, there are four possible binary combinations, and each is assigned a different color. These same colors along with four additional colors are assigned to the eight binary combinations for three-parameter fluorescence measurements. When four colors are used, there are eight additional binary combinations for a total of 16, so we repeat the eight colors again. This process is continued for five parameters, which produces 32 binary combinations, etc.

Chapter 45, Fig. 15 Multiparameter data analysis. To illustrate the strategy for evaluating multiparameter data analysis, human blood was stained with the antibody combination FITC-CD3, PE-CD56, PECY5-CD19, and APC-CD45. A bivariate plot of SSC versus each fluorescence parameter is displayed in the top row (A). A region (R1-R4) is drawn on the desired population of positive cells or to all events above a nagative control, as shown. Next the FSC versus SSC for each selected cluster is displayed by gating them on each of the regions. Another region, R5, can be drawn to further define the desired population (A, bottom row). Finally, the bivariate plots of each antibody combination are displayed (B). For four-color data, there are six bivariate combinations. The frequency and color of each cluster of events defined by a Boolean expression are shown in Table III. In this example, the gate R4 and R5 that defines CD45$^+$ mononuclear cells was used. Because there are no CD45$^-$ events defined by the gate used, these values have not been included in Table III.

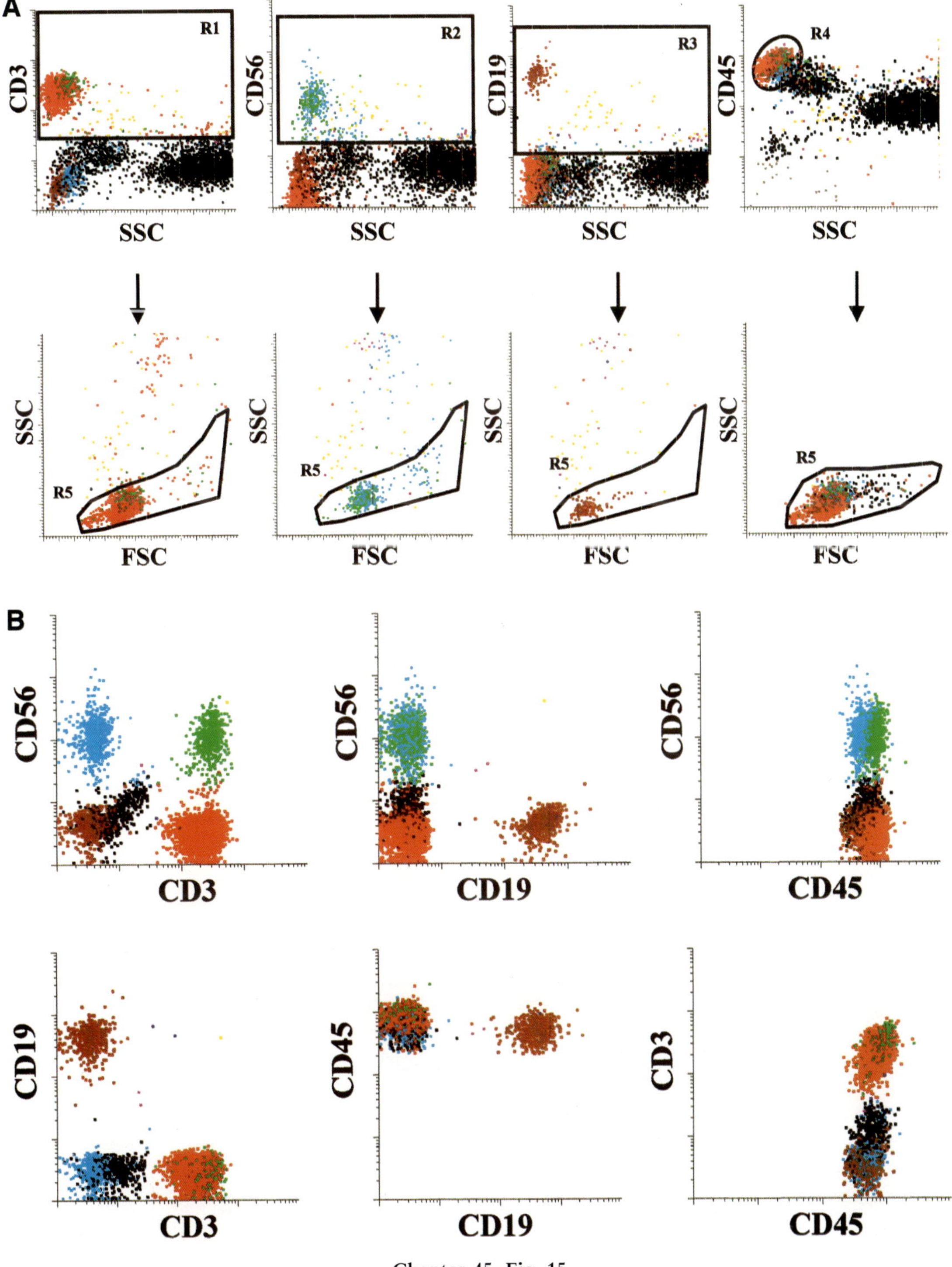

Chapter 45, Fig. 15

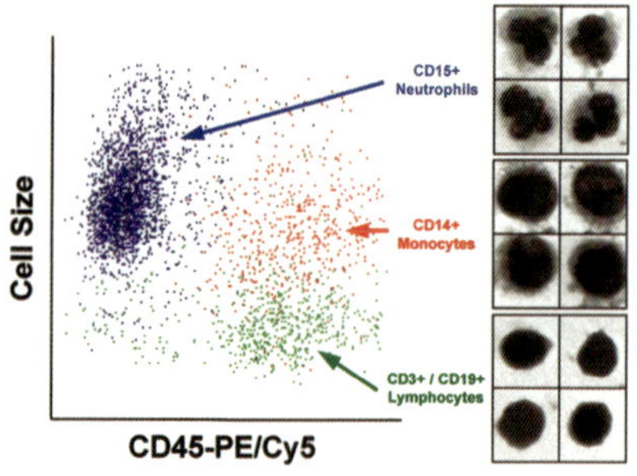

Chapter 46, Fig. 3 Initial gating based on cell size and CD45 positivity.

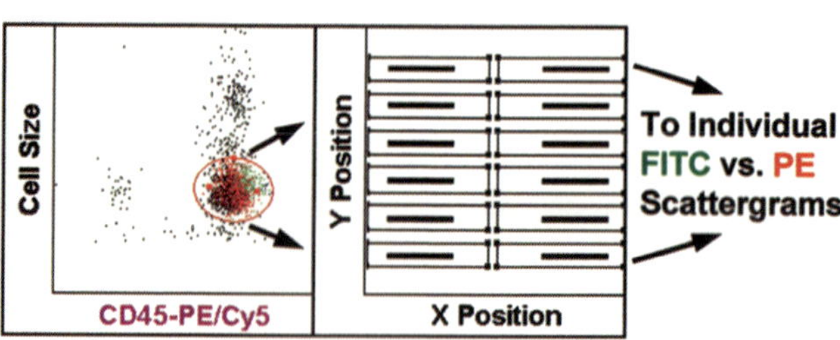

Chapter 46, Fig. 4 Subsequent gating based on cell position.

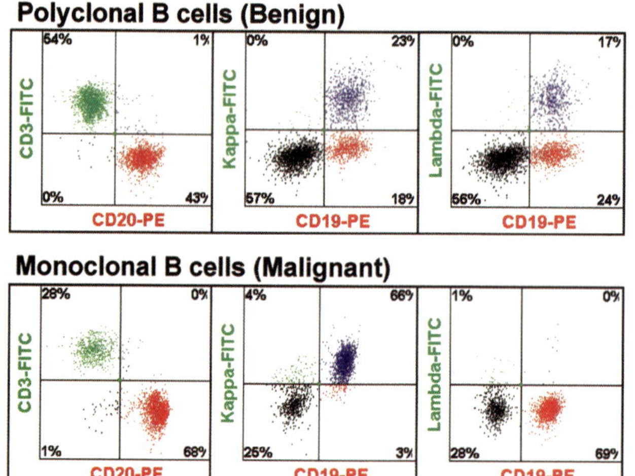

Chapter 46, Fig. 5 Example results. (Top) Benign lymph node with polyclonal B cells. (Bottom) Malignant lymph node with monocelonal B cells.

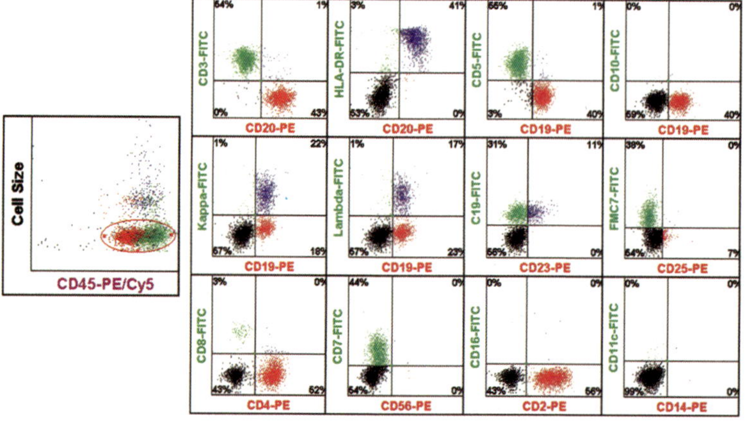

Chapter 46, Fig. 6 Example clinical case 1: Fine needle aspiration biopsy of a benign lymph node.

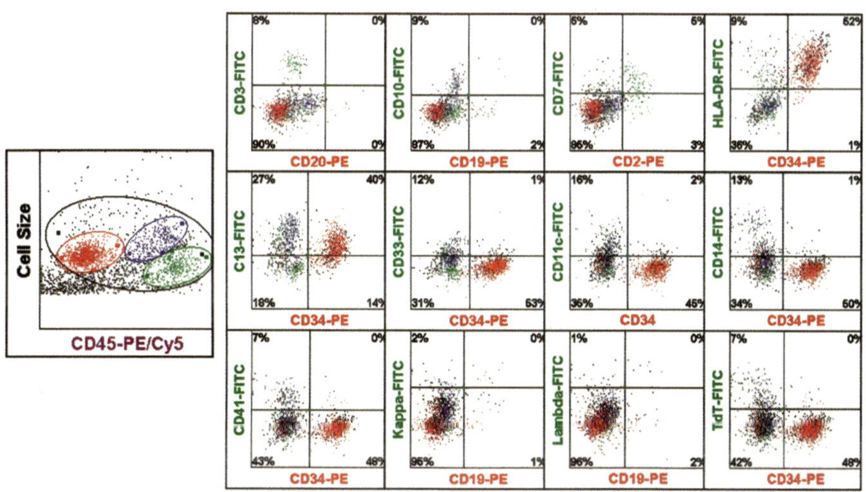

Chapter 46, Fig. 7   Example clinical case 2: Peripheral blood with acute myelogenous leukemia.

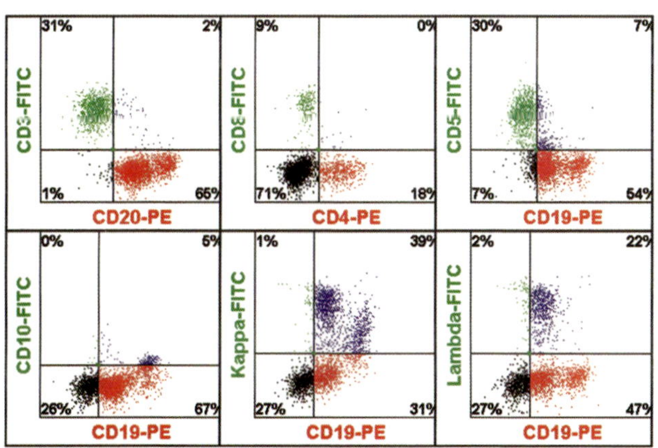

Chapter 46, Fig. 9   Example clinical case 3: Fine needle aspiration biopsy of a parotid mass with Burkitt's lymphoma.

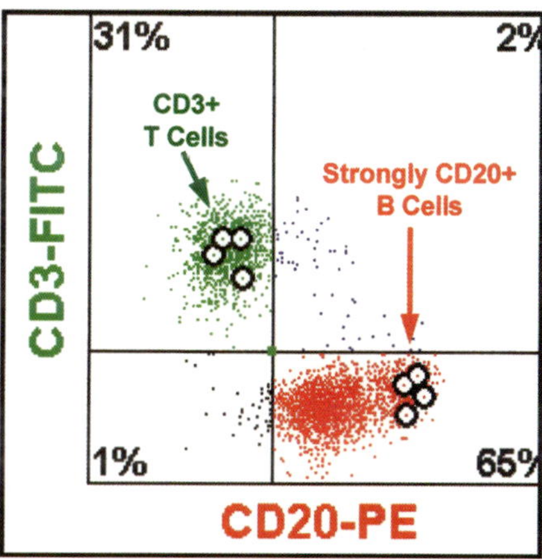

Chapter 46, Fig. 10   Example clinical case 3: Selection of cells for relocalization.

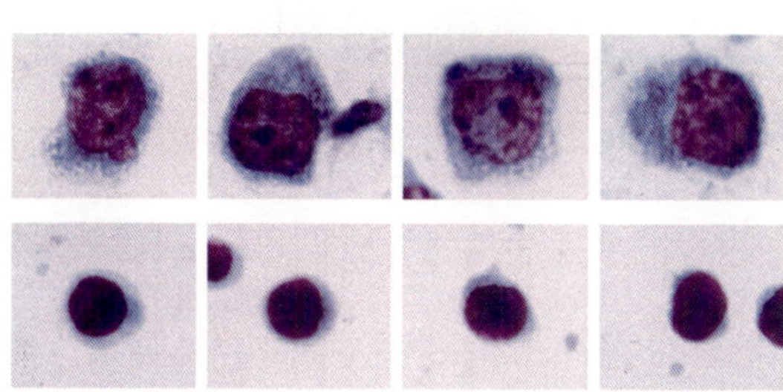

Chapter 46, Fig.11   Example clinical case 3: Relocalization for light microscopy.

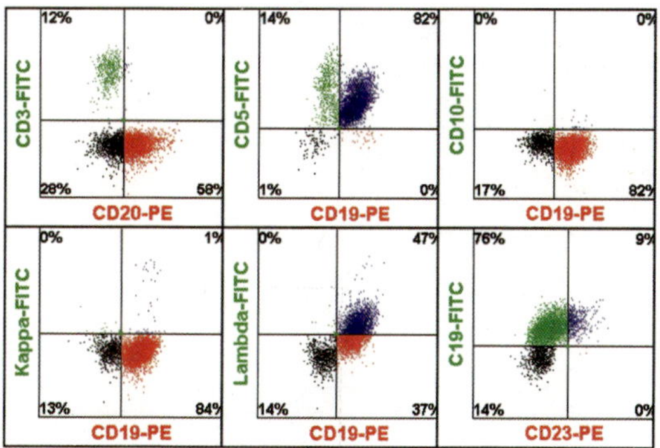

Chapter 46, Fig. 12   Example clinical case 4: Peripheral blood with chronic lymphocytic leukemia.

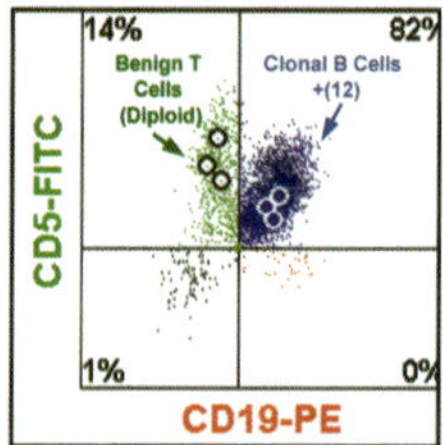

Chapter 46, Fig. 13 Example clinical case 4: Selection of cells for relocalization.

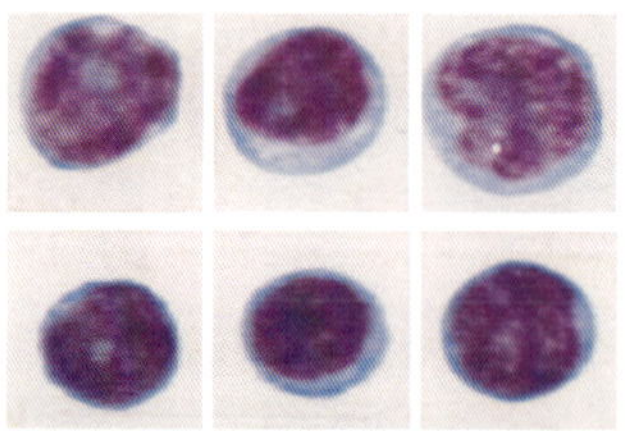

Chapter 46, Fig. 14 Example clinical case 4: Relocalization for light microscopy.

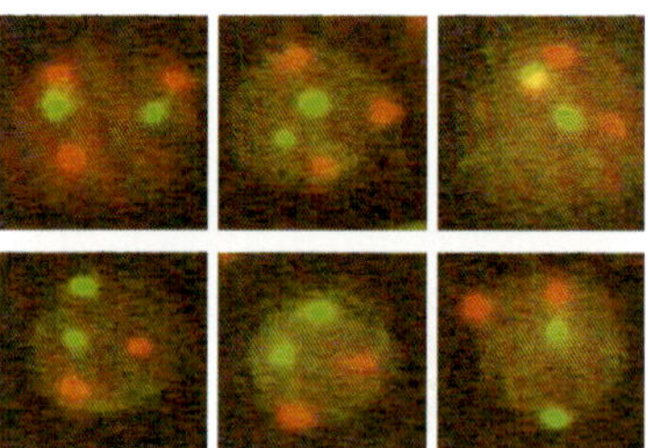

Chapter 46, Fig. 15 Example clinical case 4: Relocalization for epifluorescence microscopy, FISH.

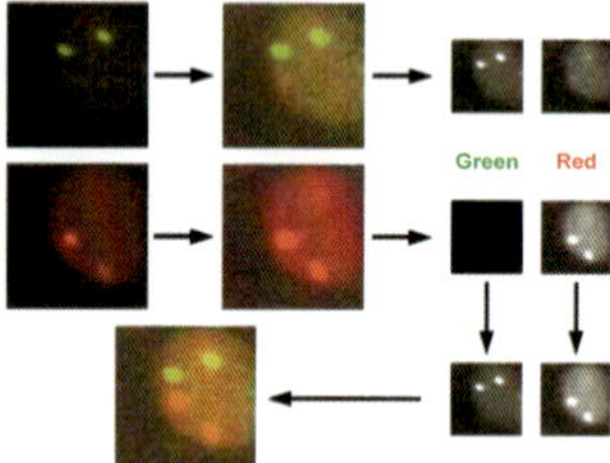

Chapter 46, Fig. 16 Enhancement of epifluorescence video images of FISH probe spots.

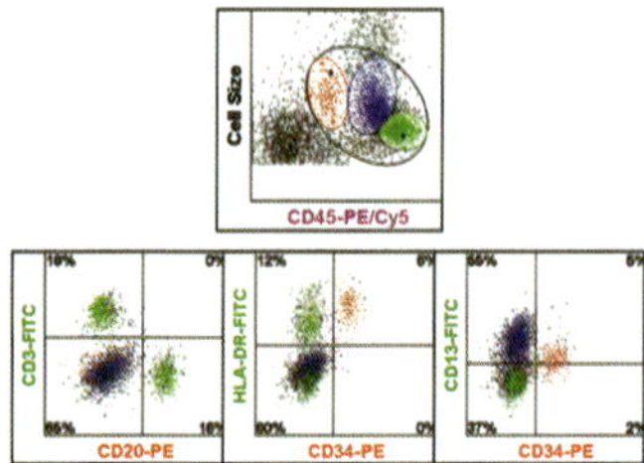

Chapter 46, Fig. 17 Example clinical case 5: Bone marrow aspirate with chronic myelogenous leukemia in blast crisis.

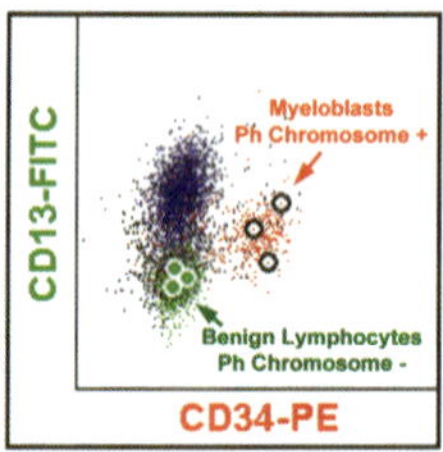

Chapter 46, Fig. 18 Example clinical case 5: Selection of cells for relocation.

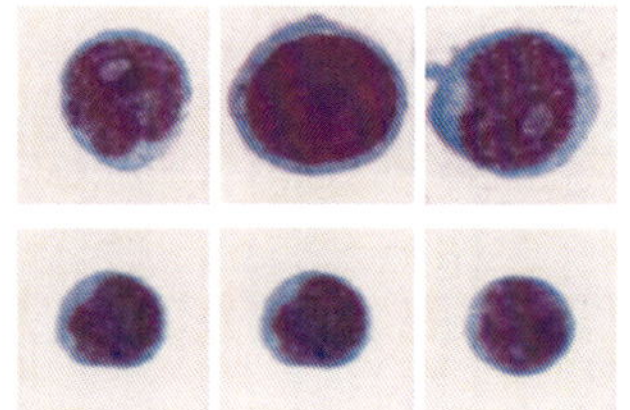

Chapter 46, Fig. 19 Example clinical case 5: Relocalization for light microscopy.

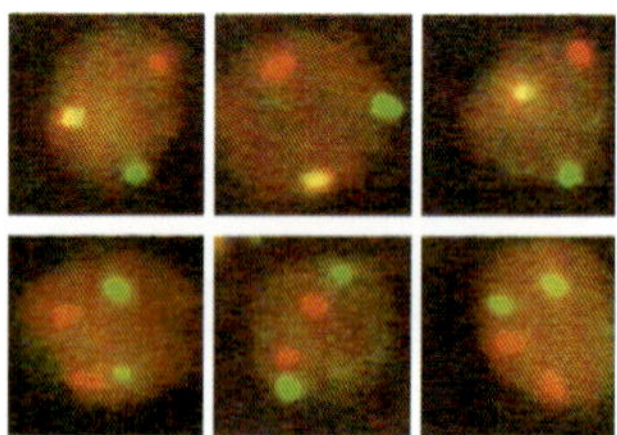

Chapter 46, Fig. 20 Example clinical case 5: Relocalization for epifluorescence microsopy, FISH.

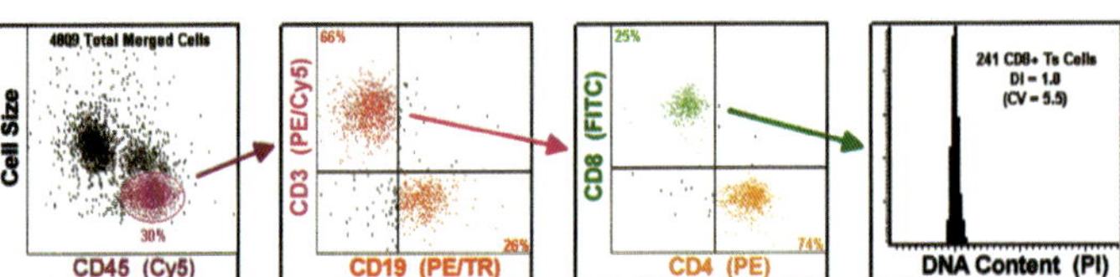

Chapter 46, Fig. 22 Five-color immunophenotyping plus DNA content analysis of peripheral blood lymphocytes.

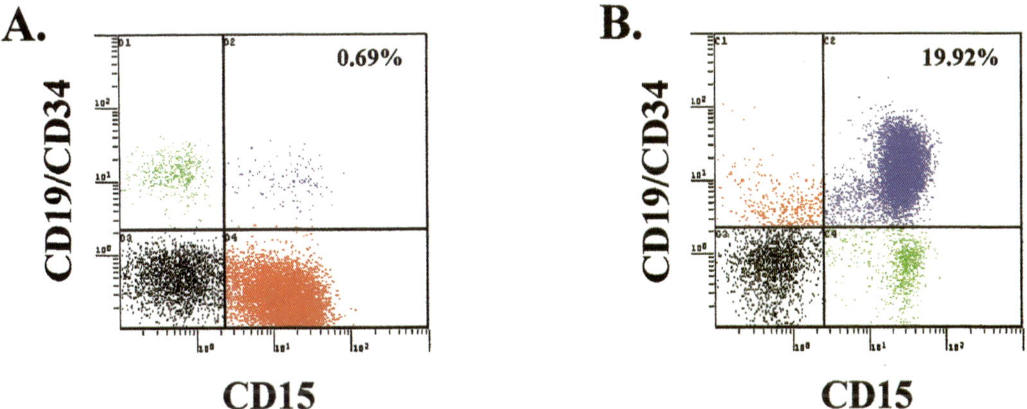

Chapter 49, Fig. 1 CD19 PE/CD34 PerCP versus CD15 FITC bivariate distributions of bone marrow cells of B precursor ALL at time of complete clinical remission (A) and relapse (B). Percentages indicate resileukemic cells.

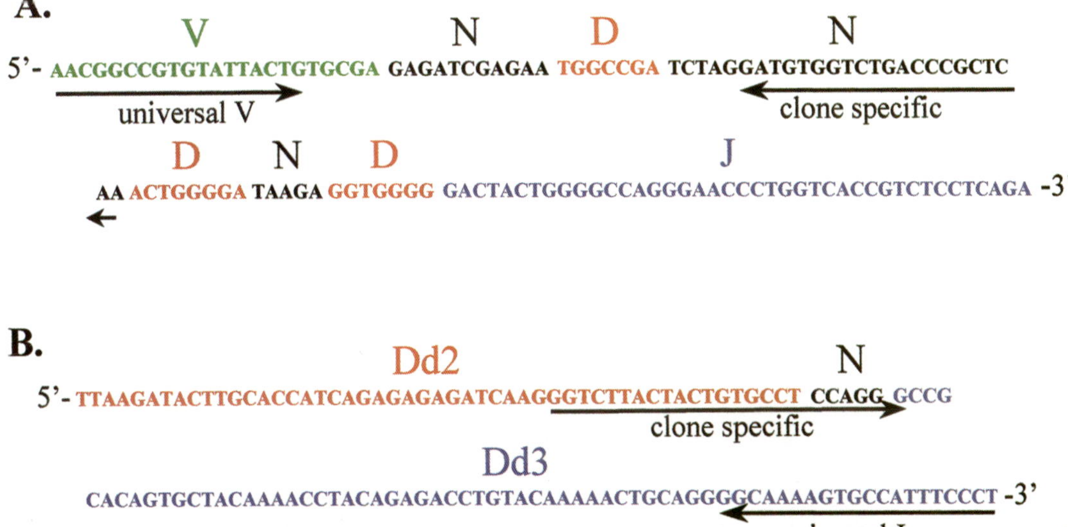

Chapter 49, Fig. 3 Sequences of IgH (A) and Vδ2–Dδ3 (B) gene rearrangements from a pediatric pre-B ALL patient at diagnosis. Primers for clone-specific PCR are indicated.

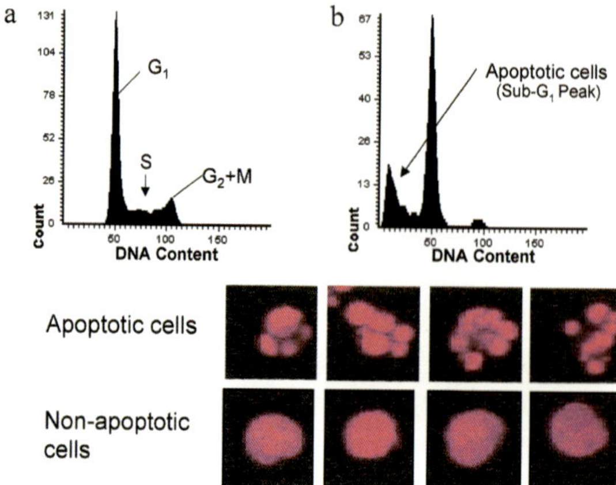

Chapter 50, Fig. 2   DNA frequency histograms of HL-60 human leukemic cells, untreated (a) and treated with camptothecin to induce apoptosis (b). Note the subdiploid peak in treated cells reflecting apoptosis. Also shown are LSC computer images of apoptotic and nonapoptotic cells. Propidium iodide staining, 20% objective.

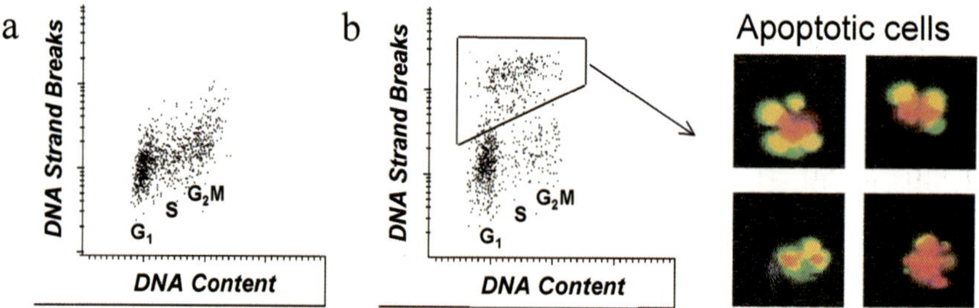

Chapter 50, Fig. 3   DNA strand labeling by TdT assays (TUNEL method) in HL-60 cells treated with camptothecin to induce apoptosis (b). In contrast to control (nontreated) cells (a), apoptotic cells have high values of green fluorescence (b, box), reflecting the incorporation of FITC-labeled nucleotide. When relocated on a LSC computer screen, the positively labeled cells displayed characteristic features of apoptosis (right).

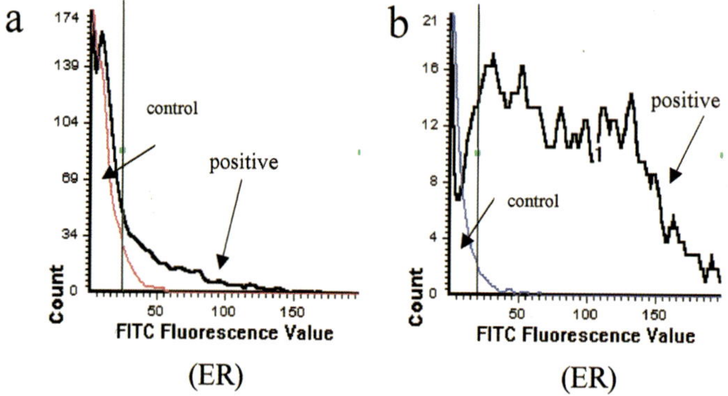

Chapter 50, Fig. 6   Histograms of weakly positive (a) and strongly positive (b) breast carcinoma stained for estrogen receptor in routine paraffin-embedded tissue sections. The threshold between negative and positive populations is set individually for each tumor based on the fluorescence of the negative controls.

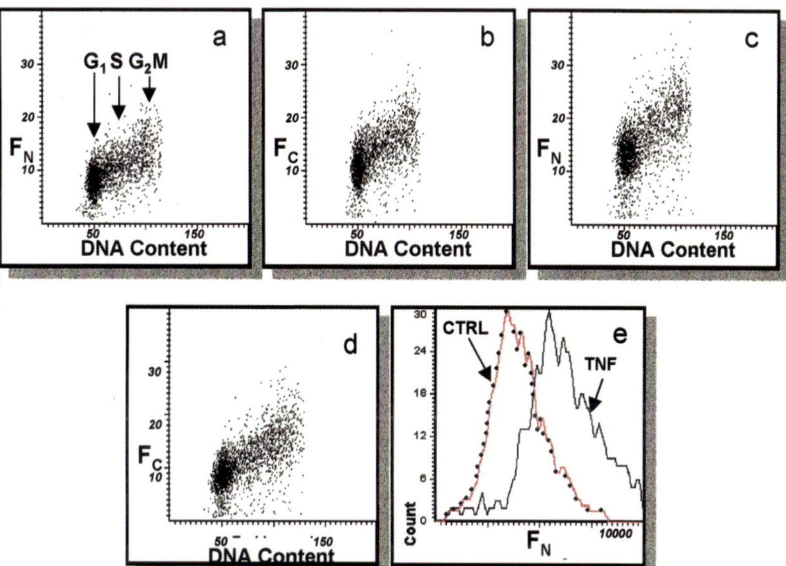

Chapter 50, Fig. 7 Analysis of the changes in expression of nuclear factor κB (NF–κB) in U-937 cells treated with 10 ng/ml of tumor necrosis factor f (TNF–β) for 1 hr. (a, b) Scattergrams representing NF–κB located in the nucleus (F_N) and cytoplasm (F_C) versus DNA content prior to the treatment. (c, d) Scattergrams for F_N and F_C, respectively, after treatment. (e) Change in F_N after treatment (ctrl is control; TNF, tumor necrosis factor). From Deptala *et al.*, (1998), with permission.

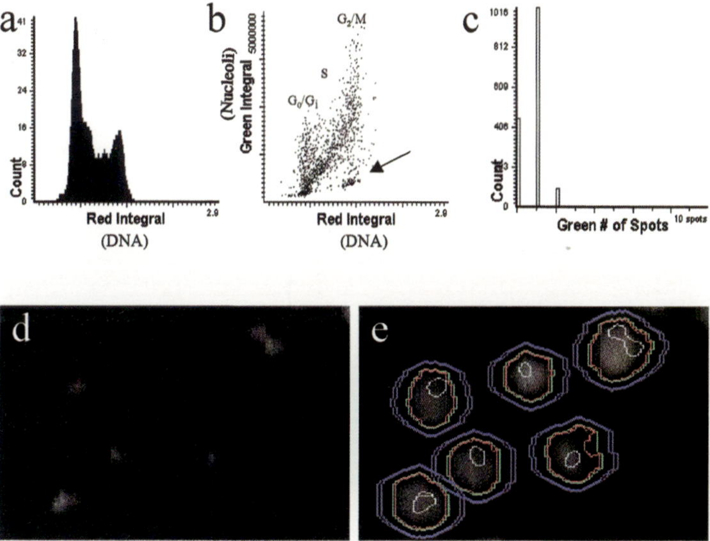

Chapter 50, Fig. 8 Human lymphocytes from peripheral blood, stimulated by phytohemagglutinin for 48 hr, stained with anti-nucleolar monoclonal antibody, and analyzed on LSC using the FISH protocol. (a) DNA histogram (PI staining). (b) Values of green ("nucleolar") versus red (DNA) fluorescence (arrow, mitotic cells), (c) Data showing the number of spots (number of nucleoli) per nucleus. Actual LSC images are also shown [(d), green fluorescence (nucleoli); (e), red fluorescence with LSC contouring of the nucleoli].

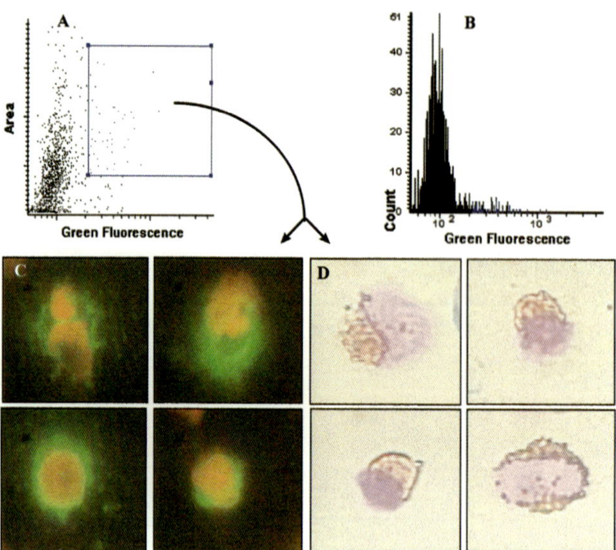

Chapter 51, Fig. 6 The capabilities of laser scanning cytometry in clinical sample analysis. A Cytospin of human asthmatic sputum has been stained with monoclonal anti–human major basic protein specific antibody (BMK13) and secondary Oregon Green conjugated goat anti-mouse antibody and analyzed on the Compucyte LSC. Plotted are peak green fluorescence signal against cell size of cells captured with propidium iodide nuclear counterstain (A, B). Examples of relocation and still video image capture of cells within the region of interest under epifluorescence with the IB excitation cube, wide band (C) and after chromatic restaining of the same slide under bright-field microscopy with chromotrope 2R (D) (with thanks to Dr. G. Woltmann, Leicester, England).

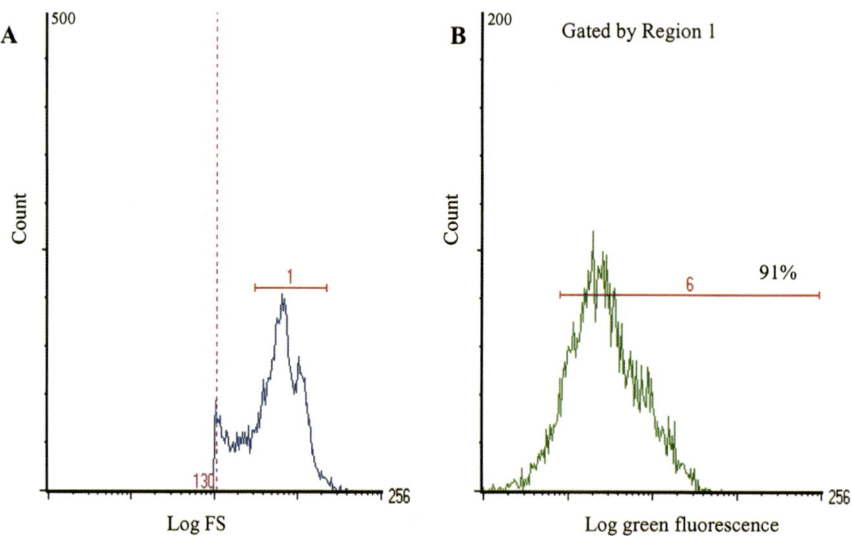

Chapter 53, Fig. 6 Flow cytometric analysis of *B. anthracis* spores incubated in polysaccharide media for 30 min at 37°C. A specific FITC-conjugated anti-cell wall polysaccharide MAb stained emerging vegetative *B. anthracis* cells (region 6) (viable cells) but did not stain encapsulated *B. anthracis* cells, dormant *B. anthracis* spores, or other *Bacillus* species. Forward light scatter was used to identify vegetative cell population in the sample (region 1).